Acoustics, Aeroacoustics and Vibrations

To Maria, Sandra & Yvette (F. A.)

To Béatrice, Romane & Jeanne (P.-O. M.)

Acoustics, Aeroacoustics and Vibrations

Fabien Anselmet
Pierre-Olivier Mattei

iSTE

WILEY

First published 2016 in Great Britain and the United States by ISTE Ltd and John Wiley & Sons, Inc.

ISTE Ltd
27-37 St George's Road
London SW19 4EU
UK

www.iste.co.uk

John Wiley & Sons, Inc.
111 River Street
Hoboken, NJ 07030
USA

www.wiley.com

Library of Congress Control Number: 2015958611

British Library Cataloguing-in-Publication Data
A CIP record for this book is available from the British Library
ISBN 978-1-84821-861-1

Contents

Preface

This book is an introduction to the theories of vibroacoustics and aeroacoustics. It cannot therefore be seen as exhaustive. It only presents one of the many ways of presenting the bases of acoustics, the science of vibrations of fluid and solid continua. The scope of applications is extremely large and varied, from musical instruments to ultrasound echography including the active control of vibrations or the transmission of vibrations between different propagation media. The basic equations of acoustics are very simple (in appearance at least), small in number (less than half a dozen) and the underlying hypotheses elementary. The bulk of the difficulties lie in solving these equations.

Whereas numerous books exist in acoustics and in fluid mechanics, a much smaller number exists about the field that is the subject of this book and, to our knowledge, none about this form in which each chapter, which although may be used independently, is based on knowledge that was developed in several others. We hope that the curious or the simply interested readers will find food to satisfy their appetite before indulging by reading more specialized books whose references can be found in the bibliography presented at the end of this publication and that will enable them to deepen their knowledge.

In this book, we did not use the duality system of theory and exercises in a systematical manner. Thus, the reader is strongly encouraged to carry out the demonstrations again and to finish those left aside. Since we have chosen to demonstrate the simplest results only, this method allows for the refining of the understanding of simple ideas without cluttering the text with too cumbersome results.

Fabien ANSELMET
Pierre-Olivier MATTEI
November 2015

1

A Bit of History

In the beginning, there was music... So could begin a history of acoustics, a very old science that has played a primordial role in the process of scientific development. As a matter of fact, it appears that music was one of the first elaborate means of communication. The oldest known instruments date back to 35,000 years ago (lower Paleolithic era called aurignacian) and were simple flutes carved in bones. We had to wait until Greek antiquity to understand the fact that sound is produced by a deformation movement of the body that creates it. However, the process of transmission of sound from the instrument to the ear was unknown to these people. The "Theory of Sound", published in 1895 by Lord Rayleigh [STR 45] and which inspired this short introductory chapter, established what is known as classical acoustics.

Conventionally, acoustics is split into three parts: the production of sound, its propagation and reception. For each of them, it can be useful to present a brief historical primer, focusing particularly on the period which witnessed the birth of the basic concepts, between the 16th and 19th Century.

Finally, we will conclude this chapter by mentioning aeroacoustics, a very recent science that has gained momentum only with the pioneering works of Sir James Lighthill in the 1950s.

1.1. The production of sound

It is commonly accepted that, as early as the 6th Century BC, Pythagoras was the first Greek to study the origin of musical sounds. He showed that the highest pitches are produced by the shortest strings and that a string half as long as another emits a pitch an octave above. The method of plucking the strings was gradually developed, although without its relationship to the concept of frequency being established. It does not appear that this approximation had been made before Galileo (1564–1642). In his treatise, he discusses the influence of the length, tension and density of the string. He further observes that those sounds whose frequencies are integers multiple of the lowest frequency combine pleasantly to the ear.

In 1636, the Franciscan friar Mersenne (1588–1648) carried out, in Paris, the first serious publication about the vibration of strings. He was the first to measure the frequency of a musical sound.

The pioneer of experimentation on the connection between the frequency of the sound and the way the string is plucked was Sauveur (1653–1716), who incidentally suggested the term "acoustics" for the science of sounds. Sauveur, as well as Wallis (1616–1705) at the same time in England, observed that vibrating strings showed motionless points which he named "nodes", while others were animated by a maximal amplitude motion called today "antinodes". The English mathematician Taylor (1685–1731) built the first strictly dynamic solution to the problem of the vibration of strings. His calculations were in adequate agreement with Galileo and Mersenne's experiments. Although he had focused on a specific problem, Taylor has paved the way for Bernoulli (1700–1782), d'Alembert (1717–1783) and Euler's (1707–1783) more elaborate mathematical techniques (partial differential equations).

The phenomenon of the succession of nodes and antinodes on a string characterizes multiple frequencies of the simple vibration frequency of the string. The latter produces a fundamental sound, while the former were baptized harmonics. Experimentally, Sauveur noted that a plucked string emitted a sound with a complex structure in which many harmonics were present. The theoretical explanation of this phenomenon was given by Bernoulli in 1755. The resulting vibration is the algebraic sum of the partial vibrations, an idea that will be later called superposition principle. This drove Fourier, in 1822, to build his famous decomposition theorem whose scope of application is much wider than that of acoustics.

Lagrange (1736–1813) gave an elegant analytical explanation in 1759 to the problem of the strings. He focused on the sounds produced by pipe organs and woodwinds, facing the problem of boundary conditions.

The extension of the methods previously described requires knowledge of the relation of the behavior of the body that relates its deformation to the stress that is imposed on it. This issue had been experimentally addressed on solid bodies, between 1660 and 1676 by Hooke who derived on this occasion the concept of elasticity. This law, more comprehensive than that introduced by Hooke, forms the current basis of the concept of linear elasticity, either in the static (strength of materials), dynamic (viscoelasticity) or quasi-static (vibration) fields. This last field is the basis for the study of noise emissions. Hooke's law was used for theoretical means for the first time in 1744 by Euler, and then in 1751 by Bernoulli in cantilevers or supported beam vibration problems. They based their studies on the deformation energy which later led Rayleigh (1842–1919) to the well-known fourth-order spatial differential equation. This equation, which governs the vibration behavior of the beams, is known as Euler's equation.

The vibration of elastic plates had been studied by Chladni (1756–1824). The results, published in 1787, show the existence of nodal lines that are the two-dimensional equivalent of the vibration nodes of strings. The experiment consists of sprinkling very fine sand on a plate. When it is subjected to vibrations, for example using a violin bow, the sand gathers at the vibration nodes of the plate, the nodal lines. A few examples of one of the plates proposed by Chladni in his book [CHL 09] are presented in Figure 1.1. Following these works, Napoleon in 1802 granted the Institut de France 3,000 F with the subject of the prize being to "present the mathematical theory of vibrations of elastic surfaces, and compare them to the experiment" [CHL 09]. The winner, Sophie Germain (1776–1831), who was the first French self-taught female mathematician, gave an exact fourth-order equation in 1816.

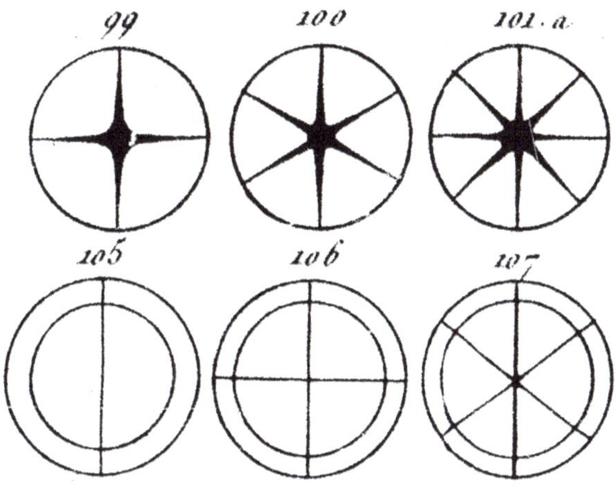

Figure 1.1. *A few examples of Chladni's figures (engraving from [CHL 09])*

Currently, only the structures with a simple form can possibly be solved analytically. This mainly concerns plates (circular, elliptical and rectangular) and shells (cylindrical and spherical), which is sufficient in most industrial applications. All these models are originating from the three-dimensional elasticity equations in which a simplification is introduced assuming that a dimension is small compared with the others, usually the thickness. The complexity of the obtained models suggests that it is unrealistic to expect that geometrical structures of complicated form be addressed in a simple way. It is necessary to resort to approximate numerical methods such as finite elements, and there again, models do not guarantee entire satisfaction. Needless to say that nonlinear elasticity, anisotropic or coupling (interaction) problems between the vibrating structure and the fluid that surrounds it are still the focus of research.

1.2. The propagation of sound

As has already been stressed, the Greeks were aware of the importance of the air in the propagation of sound, without, however, understanding by which mechanism fluids intervened. Since, during observations, the air remains motionless during the "transportation" of acoustic waves, some philosophers did not admit this point of view. For example, Gassendi from France (1592–1655) thought that sound spreads using beams of fine particles capable of impressing the ear. During this time, experiments related to the propagation of sound in a rarefied atmosphere did not invalidate the ideas of Gassendi. These latter were only challenged by Boyle (1627–1691) who carried out a significant experiment that allowed the observation of the decay of the sound intensity transmitted with the intensification of the vacuum. He concluded that the air conveys sound, without being the only material to exhibit this property. The question was to know how fast the air conveys the sound. By using firearms, Gassendi came up with a speed of 480 m/s. Better experiments driven by Mersenne gave 450 m/s. While Aristotle argued that treble pitches were more rapidly transmitted than low ones, Gassendi achieved an important observation by highlighting the fact that the speed of sound is independent of its tonal pitch. In 1656, the Italians Borelli (1608–1679) and Viviani (1622–1703), without specifying the air temperature nor its dampness, found a value of 350 m/s. In 1740, the Italian Branconi showed that the speed of sound increases with the temperature. The first significant measurement in open air was without any doubt conducted in 1822 by the members of the Bureau des Longitudes in Paris by means of a cannon. The observers were divided into two groups situated at two stations distant from each other by 18,700 m. Each group determined the time interval between the perceptions of light and sound of the cannon fired by the other group and then fired a cannon shot in turn so that the first group could perform a similar measurement. Measurements invariably gave 55 s at 15 degree C (therefore, a speed of 340 m/s). Brought down to zero degrees Celsius, the results yielded 332 m/s. During the 18th and the first half of 19th Century, a very large number of experiments confirmed this value. The accepted value is 331.36 ± 0.08 m/s with air at rest at 0 C and 1013.25 HPa.

In 1808, the physicist Biot (1774–1862) made the first experiments on the speed of sound inside a solid medium by using a 951-m long cast-iron pipe which had to be used to provide drinking water in the city of Paris. The observer measured a difference of 2.5 s in the travel times taken by sound through the metal and air. If c_0 and a refer to the velocities of the waves in the air and metal, Biot immediately obtained

$951/c_0 - 951/a = 2.5$ or $a = 10.5c_0$ that is a wave velocity in cast iron used to make the pipe 10 times greater than that of the air.

In 1827, Colladon and Sturm studied the speed of sound in Lake Geneva by measuring the time difference between the image of an inflammation produced by the shock of a hammer on a bell and the sound perceived by an observer located on the shore as indicated in the engraving reproduced in Figure 1.2. They found 1,435 m/s at 8°C.

A first theorization test was proposed by Newton in 1687. Circa 1760, Lagrange proposed another model which yielded once again Newton's results. However, the numerical value that they obtained is significantly different from that obtained experimentally. No significant progress was made before 1816, date at which Laplace criticized the isotherm hypothesis of his predecessors. He replaced it by an adiabaticity hypothesis that seemed to him more appropriate. The two velocities, obtained by Laplace and Newton, are in the ratio of specific heats. They showed the existence of a specific heat at constant volume and constant pressure whose values were known only with a very relative accuracy. Laplace used the numerical value of 1.5 for their ratio γ obtained by experimenters Laroche and Bernard. He obtained a value of the speed of sound of 346 m/s at 6 C. He estimated this value as almost identical to the experimental value of 337 m/s. A few years later, γ was again measured and the value still accepted nowadays of $\gamma = 1.41$ was obtained. This corresponds to a speed of sound in perfect agreement with the experimental values. It should be noted that if Lagrange had had full confidence in his theory, he could have indirectly measured γ very precisely.

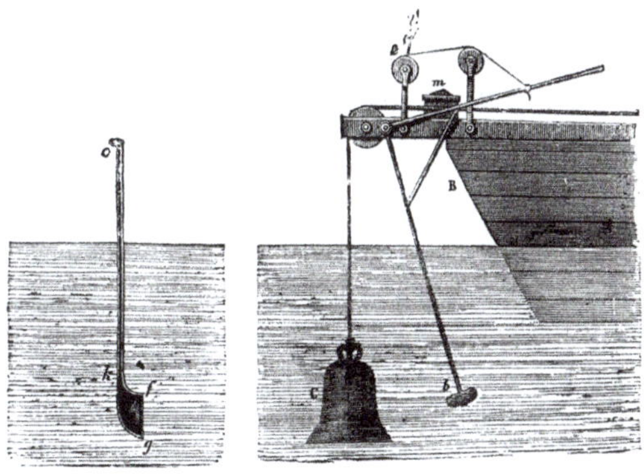

Figure 1.2. *Device setup for measuring sound in Lake Geneva (engraving from [DRI 73])*

Tests for solving the equation of propagation of sound followed d'Alembert's works on the vibration of bodies. In particular, at the end of the 18th Century, the problem of the propagation of sound in pipes started to be well known. Kundt (1839–1894) developed a method that enabled the measurement of the "speed" of stationary waves. Poisson derived in 1820 a theory of propagation in pipes for almost all cases. This work was finished by Helmholtz (1821–1894) in 1860. The particular case of the abrupt change in cross-section was studied by Poisson, as well as the reflection and transmission of sound under normal incidence at the boundary of two different fluids. The case of oblique incidence was treated by Green (1793–1841) in 1838. This had the purpose of bringing forward the similarities and differences between sound and light.

Nowadays, the interest is toward problems of propagation in limited (shallow-water propagation and cavity resonance) inhomogeneous media as well as in random media (turbulence).

1.3. The reception of sound

During the historical development of acoustics, the primary receiver of the studied sound was the human ear. However, no satisfactory or complete theory of hearing has been elaborated, leaving this psychoacoustics problem open, as are also most aspects that relate to the brain and its ability to interpret its environment.

After establishing the evidence of the relationship between the frequency and the plucking of strings, an important task was the determining of the frequency limits of human hearing. Savart (1791–1841) placed these boundaries between 8 and 24,000 Hz. Later Seebeck (1770–1831), using tuning forks, Biot, Koënig and Helmholtz obtained values between 16 and 32 Hz for low frequencies. Nowadays, an audible range between 20 and 20,000 Hz is commonly accepted.

In 1843, Ohm showed that the ear is capable of achieving a spectral analysis of complex sounds. This created a renewed interest in physiologic acoustics. Helmholtz, in 1862, proposed a theoretical model of the mechanism of the ear. It was during this work, called "Theory of Resonance", that he invented the resonator that now bears his name and which is presented in Figure 1.3. He developed the theory of summation and difference of pitches and laid the foundations for all the future research in this area [HEL 68].

From the middle of the 19th Century, acoustics has undergone considerable development and here it is impossible to continue our history of acoustics which would otherwise take us in too many different directions. The interested readers can refer to the historical book (and still actually relevant in many aspects) written by Strutt and Rayleigh [STR 45], as well as to the summary book edited by

Rossing [ROS 07] which contains a more complete historical introduction than this in addition to a list of historical references.

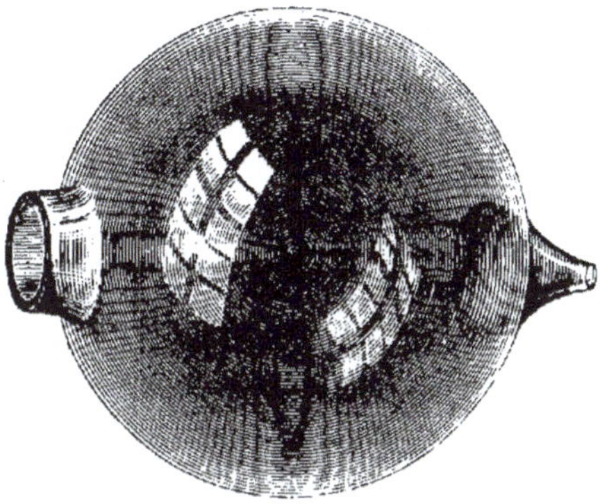

Figure 1.3. *Helmholtz resonator (engraving from [HEL 68])*

1.4. Aeroacoustics

Aeroacoustics is a discipline that involves varied aerodynamic and acoustic phenomena, which are in addition tightly coupled. Its real development as a special branch of physics is fairly recent, since it dates back to the founding works published by Lighthill in 1952. In general, two stages can be distinguished in the sound radiation by flows: the generation of noise in the areas of turbulence where nonlinear effects are very important, and the linear propagation of acoustic waves in a medium at rest until far-field. In order to identify the sources of noise, it is, therefore, necessary to have a good knowledge of turbulent flows; this explains why the progress of aeroacoustics was directly related to that of fluid mechanics, both at the experimental and the numerical level. However, in order to take into account the physical characteristics of acoustic fluctuations, a number of original techniques had to be also developed.

It is, therefore, in 1952 that Lighthill suggested an analogy from which most of the aeroacoustic theories have been developed. By recombining the fluid mechanics equations to show the noise produced by a flow as the solution of an equation of propagation in a medium at rest, Lighthill has enabled the foundation, on rigorous mathematical bases, of aeroacoustics, which then was stumbling a lot: in particular, in this framework, the experimental observation of the variation of noise radiated by a subsonic engine jet as a power eight function of the speed, previously of mysterious origin, is explained in a very simple way. In the context of the Lighthill analogy, the generation of noise is effectively identified by the term on the right of the Lighthill equation, by means of acoustic sources terms constructed from the velocity field and, in particular, from fourth-order correlations of turbulent velocity fluctuations which constitute, in the majority of cases, the primary source term by identifying the source to a quadripole.

The reduction, or even the control (passive or active), of the noise of aerodynamic origin constitutes today a major industrial challenge. Thus, for example, in the nuclear sector, the acoustic radiation produced in the pipes by fast flows is likely to severely damage the structures due to resonant effects being induced. Nonetheless, it is of course in the fields of aviation and land transportation that noises are more directly felt, such that gradually more stringent standards should be respected, and that the concept of acoustic comfort of users must be taken into account from the beginning by the manufacturers. It is, therefore, desirable to be able to accurately predict the sound field generated by turbulent flows if the main purpose is to address it efficiently.

2

Elements of Continuum Mechanics

The subjects covered in this book all base themselves on the approximation of continua, be it to characterize the properties of solid entities under deformation, or those of fluids with sound waves moving though them. This continuum approximation is very simple in the physical concepts it employs, that are nothing more than the principles of conservation of mass, momentum and energy, but transposing them mathematically to use them to solve problems can be slightly arduous and highly complex as it requires tensor calculus. Thus, compared to the relative simplicity of particle mechanics, the formal complexity of the mechanics of deformable bodies is due to it taking into account their deformations and the strain resulting within these bodies (or vice-versa). This chapter therefore presents, in an intentionally succinct fashion, the main elements that will be necessary to move on to the problems in the chapters further on. It also refers to specialist books that will, if necessary, allow the reader to go deeper into continuum mechanics. We will begin this chapter with a general presentation of continuum mechanics and then of the laws of conservation of mass, momentum and energy. We will then end by describing the behaviours of common media that are the elastic solid media, thermoelastic, viscoelastic as well as fluid media

2.1. Mechanics of deformable media

2.1.1. *Continuum*

A medium is called a continuum in a domain $\Omega \in \mathbb{R}^3$ if its measurable physical properties vary in a continuous manner in Ω. If ρ is the density of the material, the mass of the material that occupies Ω is given by:

$$m = \int_\Omega \rho dV, \; \rho : \text{density}. \tag{2.1}$$

We define the notion of particle by the smallest set of Ω for which the density keeps a constant value. A continuum can in some cases present discontinuities of the first kind such as shock waves. More formally, it is desirable to assume that the physical quantities be integrable within Lebesgue's sense (see, for example, relation [3.4]) which allows zero-measure discontinuities to be addressed (such as cracks in a solid medium and shock waves in a fluid).

2.1.2. *Kinematics of deformable media*

In order to describe a continuum, it is important to have some way of describing it. Two different visions exist, the Lagrange and Euler kinematics. In Lagrange's kinematics, which addresses the motion from a local perspective, the main focus is on the evolution (transport and deformation) over time of a small area of fluid. In Euler's kinematics, on the contrary, it is sought to describe the motion of the continuum in its entirety[1].

2.1.2.1. *Lagrange's kinematics*

This kinematics is based on the concept of fluid particle. Suppose that at an instant $t = t_0$, the position of the particles that constitute the continuum is known. If each of them can be followed over time, the motion of the continuum may be known in its entirety. Given a_j, the position of the particle at instant t_0, the Lagrange kinematics is equivalent to providing a relation of the form $x_i = g_i(a_j, t)$. The independent parameters a_j are named Lagrange variables. The curve traced in the referential relatively to time is called trajectory. At any time, the following properties are verified:

1) the application which from the initial point allows shifting to the current point is one-to-one;

2) the partial derivatives $x_{i,j} = \partial x_i / \partial x_j$ exist and are continuous;

3) $du_i = \sum_{j=1}^{3} (\partial x_i / \partial a_j) da_j = (\partial x_i / \partial a_j) da_j$ with Einstein's summation convention on the repeated indices. This determines a Cramer system. If its determinant is non-zero: $da_i = (\partial a_i / \partial x_j) du_j$.

Considering the particle which is located in P at time t, $d\vec{P}$ is its displacement during dt. The vector $\vec{v} = d\vec{P}/dt$, of components $v_i = dx_i/dt$, is named velocity of the continuum medium. Since this derivative is calculated by following a particle during its motion, this derivative is named material derivative. In this case, the terms a_j are constants and it yields $v_i = dx_i/dt = \partial g_i(a_j, t)/\partial t$. Similarly, the particle

1 A witticism says that Euler describes phenomena as they should be, while Lagrange describes them as they are.

acceleration is defined as $\vec{\gamma}$:

$$\vec{\gamma} = \frac{d\vec{v}}{dt} = \frac{d^2\vec{P}}{dt^2}, \gamma_i = \frac{\partial v_i}{\partial t} = \frac{\partial^2 g_i\left(a_j, t\right)}{\partial t^2}.$$

2.1.2.2. *Euler's kinematics*

We will now set aside the idea of following each particle motion to the benefit of looking for the description of the motion of the whole continuum medium. For example, we will replace the notion of the particles trajectory to rather focus on the information given by a velocity field in the set Ω. It is upon this idea that the Euler kinematics relies. Consider a velocity field $\vec{u}\left(x_i, t\right)$ where x_i and t are the Euler variables. The x_i are no longer the coordinates of a particle but independent space variables.

The problem that arises is the representation into Euler variables of the motion in Lagrange variables. Let by hypothesis $x_i = x_i\left(a_j, t\right)$ and $v_i = v_i\left(a_j, t\right)$. Thus, with $x_i = x_i\left(a_j, t\right)$, we have $a_j = a_j\left(x_l, t\right)$ and therefore $v_i = v_i\left(a_j\left(x_l, t\right), t\right) = v_i\left(x_l, t\right)$, where the x_l are the independent space variables. The inverse problem which is solved by integration is much more demanding.

The trajectories are defined by the three relations $dx_i/dt = v_i\left(x_i, t\right) = 0$ which determine a set of curves with three parameters. A streamline is the curve tangent to the velocity vector at each of the considered points of the space. Consider a differential motion $d\vec{P}$ on the streamline, $d\vec{P} = \lambda\vec{v}_P, \lambda \in \mathbb{R}$. In addition, v_1, v_2, v_3 are the components of $\vec{v}$ and dx_1, dx_2, dx_3 those of $d\vec{P}$ (for example, in $\mathbb{R}^3$). The streamlines equation is given by $dx_1/v_1 = dx_2/v_2 = dx_3/v_3 = \lambda$, for an arbitrary λ. A streamtube is the set of streamlines with regard to a closed surface. In the case of a permanent or a stationary flow, streamlines and pathlines are indistinct.

2.1.2.3. *Kinematics of a surface*

We will follow the motion of a surface in Eulerian coordinates. $g\left(x_i, t\right) = 0$ is the equation of the surface. We name H the norm of $\overrightarrow{\mathrm{grad}}g$ and $\vec{n}$ the normal external to the surface. Between t and $t + \tau$, the function g remains zero. If $\vec{v}$ is the velocity of the surface, we have: $\partial g/\partial t + v_i \partial g/\partial x_i = 0 \Rightarrow \partial g/\partial t + H\vec{v}\cdot\vec{n} = 0 \Rightarrow \vec{v} = -1/H\partial g/\partial t\vec{n}$.

2.1.2.4. *Material derivatives*

The material derivative is the derivative with respect to time when following the motion of the particle. The variation of a scalar function $f(M,t)$ is expressed in the form $df(M,t) = (\partial f(M,t)/\partial t)dt + (\partial f(M,t)/\partial x_i)dx_i$, or:

$$\begin{aligned}\frac{df(M,t)}{dt} &= \frac{\partial f(M,t)}{\partial t} + \frac{\partial f(M,t)}{\partial x_i}\frac{dx_i}{dt} \\ &= \frac{\partial f(M,t)}{\partial t} + \left(\vec{v}(M,t)\cdot\overrightarrow{\mathrm{grad}}\right) f(M,t).\end{aligned} \quad [2.2]$$

This last formula applies to vector functions. It then yields:

$$\frac{d\vec{v}(M,t)}{dt} = \frac{\partial \vec{v}(M,t)}{\partial t} + \left(\vec{v}(M,t)\cdot\overrightarrow{\text{grad}}\right)\vec{v}(M,t), \qquad [2.3]$$

Note that since $\overrightarrow{\text{grad}}\left(\vec{v}\cdot\vec{v}\right) = 2\left(\vec{v}\cdot\overrightarrow{\text{grad}}\right)\vec{v} + 2\vec{v}\wedge\overrightarrow{\text{curl}}\vec{v}$, taking into account the anti-symmetry of the vector product $\left(\vec{v}\cdot\overrightarrow{\text{grad}}\right)\vec{v} = \overrightarrow{\text{grad}}\left(\vec{v}\cdot\vec{v}\right)/2 + \overrightarrow{\text{curl}}\vec{v}\wedge\vec{v}$, the following formulation is thus obtained for the material derivative:

$$\frac{d\vec{v}(M,t)}{dt} = \frac{\partial \vec{v}(M,t)}{\partial t} + \overrightarrow{\text{grad}}\frac{\vec{v}(M,t)\cdot\vec{v}(M,t))}{2} + \overrightarrow{\text{curl}}\vec{v}(M,t)\wedge\vec{v}(M,t). \qquad [2.4]$$

2.1.3. *Deformation tensor (or Green's tensor)*

Consider a small volume element of length dx_i placed in A, of Eulerian coordinates x_i (see Figure 2.1)

This small volume element is transported and deformed during the motion of the continuum (fluid or solid). During the time interval δt, A moves to $A+\delta A$, δA has for coordinates $\delta x_i = \partial x_i/\partial t \delta t$. Since the elementary volume underwent a deformation in addition to the translation δA, B is not found in $B+\delta A$ but in B'. To estimate the deformation of this volume, it is necessary to calculate the variation of the length $\widehat{AB}$.

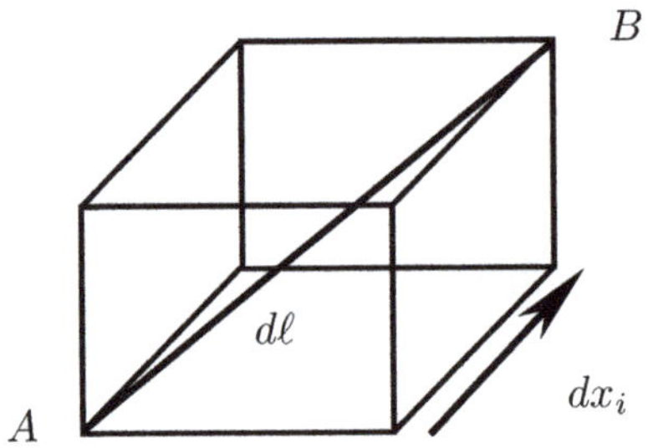

Figure 2.1. *Elementary volume*

We will characterize the vector $B' - (B+\delta A)$. Before deformation, the distance $\widehat{AB}$ was $d\ell = \sqrt{dx_i^2}$. After deformation, it is $d\ell' = \sqrt{dx_i'^2}$. Consider u_i, the displacement vector. In Lagrange coordinates, $u_i = x_i - a_i$. Since $dx_i' = dx_i + du_i$, then $d\ell'^2 = \left(dx_i + du_i\right)^2 = dx_i^2 + du_i^2 + 2dx_i du_i$. Considering $du_i = \partial u_i/\partial x_j dx_j$, it appears by replacing:

$$d\ell'^2 = d\ell^2 + 2\frac{\partial u_i}{\partial x_j}dx_j dx_i + \frac{\partial u_i}{\partial x_j}\frac{\partial u_i}{\partial x_k}dx_j dx_k, \qquad [2.5]$$

since $2(\partial u_i/\partial x_j)dx_j dx_i = (\partial u_i/\partial x_j + \partial u_j/\partial x_i)\, dx_i dx_j$ and $(\partial u_i/\partial x_j)\times(\partial u_i/\partial x_k)dx_j dx_k = (\partial u_j/\partial x_i) \times (\partial u_j/\partial x_k)dx_i dx_k$. Thus, the length after deformation dl' given by relation [2.5] is written $dl'^2 = dl^2 + 2d_{ik}dx_i dx_k$, where we have defined d_{ik} by:

$$d_{ik} = \frac{1}{2}\left(\frac{\partial u_i}{\partial x_k} + \frac{\partial u_k}{\partial x_i} + \frac{\partial u_j}{\partial x_i}\frac{\partial u_j}{\partial x_k}\right). \qquad [2.6]$$

The tensor $\vec{\vec{D}}$ of components d_{ij} is named the deformation tensor or Green's tensor. In the case of small deformations, the quadratic term is ignored: $(\partial u_j/\partial x_i) \times (\partial u_j/\partial x_k) \ll \partial u_i/\partial x_k$ and $(\partial u_j/\partial x_i) \times (\partial u_j/\partial x_k) \ll \partial u_k/\partial x_i$. Then, it gives:

$$d_{ik} = \frac{1}{2}\left(\frac{\partial u_i}{\partial x_k} + \frac{\partial u_k}{\partial x_i}\right). \qquad [2.7]$$

A synthetic notation is often used:

$$\vec{\vec{D}} = \frac{1}{2}\left(\overrightarrow{\text{grad}}\vec{u} + {}^t\overrightarrow{\text{grad}}\vec{u}\right), \qquad [2.8]$$

where the gradient vector $\overrightarrow{\text{grad}}\vec{u}$ is a tensor of components $\partial u_i/\partial x_j$. This notation is advantageous since it allows the deformation tensor to be easily written in any coordinate system, while everything that has been covered so far was done in Cartesian coordinates.

2.2. Conservation laws

Here, we introduce the three fundamental equations of mechanics, the law of conservation of mass, the fundamental principle of dynamics or law of conservation of momentum and the equation of conservation of energy. As we will see later, the principle of conservation of energy is indirectly introduced in the form of a condition at infinity (Sommerfeld's condition or principle of limit absorption in harmonic regime and outgoing wave condition in temporal regime).

2.2.1. *Conservation of mass*

Let Ω, an arbitrary fixed area with boundary $\partial\Omega$. Ω is named control volume. At instant t, the control volume has a mass m given by relation [2.1] $m(\Omega, t) = \int_\Omega \rho(\Omega, t)dV$. Over time, with the exception of the sources of mass, there is no change in the mass of the medium $d/dt \int_\Omega \rho(\Omega, t)dV = 0$, where d/dt is the material time derivative. The lemma on the derivatives of the integrals of physical volumes can be used to give outside the surfaces of discontinuity $d/dt \int_\Omega \phi dV = \int_\Omega (\partial\phi\partial t + \text{div}(\phi\vec{v})) dV$, where $\vec{v}$ is the velocity of the volume Ω. If $\vec{n}$ is the normal external to Ω,$d/dt \int_\Omega \rho(\Omega,t)dV = \int_\Omega (\partial\rho/\partial t + \text{div}(\rho\vec{v})) dV = 0$ is obtained, where further, since the volume control is arbitrary:

$$\frac{\partial\rho}{\partial t} + \text{div}(\rho\vec{v}) = 0. \qquad [2.9]$$

By introducing the material derivative in the expression above and taking into account the relation $\text{div}(\rho\vec{v}) = \rho\text{div}\vec{v} + \vec{v}\cdot\overrightarrow{\text{grad}}\rho$, the equation of conservation of mass yields:

$$\frac{d\rho}{dt} + \rho\text{div}\vec{v} = 0. \qquad [2.10]$$

2.2.2. *Conservation of momentum*

We are going to apply the fundamental principle of dynamics. It is necessary to make an evaluation of the surface and volume forces acting on the various volume elements. Considering again our control volume Ω, with boundaries $\partial\Omega$, the momentum $\vec{J} = \rho\vec{v}$ contained in Ω is given by the integral $\vec{J}(\Omega,t) = \int_\Omega \rho\vec{v}(\Omega,t)dV$. The variation of this quantity over time is balanced by the constraints applied on the boundaries of the domain $\partial\Omega$ and by the volume forces applied to the volume $d/dt \int_\Omega \rho(\Omega,t)\vec{v}(\Omega,t)dV = \int_{\partial\Omega} \vec{\vec{\sigma}}\vec{n}dS + \int_\Omega \vec{f}_v dV$. $\vec{\vec{\sigma}}$ is the stress tensor which relates the stresses within a body[2] to the deformations it undergoes. We obtain $\int_\Omega (\partial(\rho\vec{v})\partial t + \text{div}(\rho\vec{v}\otimes\vec{v})) dV = \int_{\partial\Omega} \vec{\vec{\sigma}}\vec{n}dS + \int_\Omega \vec{f}_v dV$, where $\otimes$ is the tensor product. If we use Ostrogradsky's theorem for the boundary integral:

$$\int_\Omega \left(\frac{\partial\rho\vec{v}}{\partial t} + \text{div}(\rho\vec{v}\otimes\vec{v}) - \text{div}\vec{\vec{\sigma}}\right) dV = \int_\Omega \vec{f}_v dV.$$

Since the control volume is arbitrary stop, the equality only takes place by equating the integrands $\partial(\rho\vec{v})\partial t + \text{div}(\rho\vec{v}\otimes\vec{v}) - \text{div}\vec{\vec{\sigma}} = \vec{f}_v$. In scalar form, this relation is written for all indices i (with Einstein's convention):

$$\frac{\partial\rho v_i}{\partial t} + \frac{\partial(\rho v_i v_j - \sigma_{ij})}{\partial x_j} = f_{vi}. \qquad [2.11]$$

2 This is true only for a body that has a proper shape, in a viscous fluid, the stress is related to the deformation velocity; this is the point that differentiates solid mechanics from fluid mechanics.

In order to write this law in matrix form, it should be noted that div $(\rho\vec{v}\otimes\vec{v}) = \rho\vec{v}\text{div}\vec{v} + \vec{v}\overrightarrow{\text{grad}}(\rho\vec{v})$, $d(\rho\vec{v})\,dt + \rho\vec{v}\text{div}\vec{v} - \text{div}\vec{\vec{\sigma}} = \vec{f}_v$. If the equation of conservation of mass [2.10] is introduced, we obtain the equation of conservation of momentum:

$$\rho\frac{d\vec{v}}{dt} - \text{div}\vec{\vec{\sigma}} = \rho\vec{f}_v \text{ or even } \rho\left(\frac{\partial\vec{v}}{\partial t} + \vec{v}\overrightarrow{\text{grad}}\vec{v}\right) - \text{div}\vec{\vec{\sigma}} = \rho\vec{f}_v. \qquad [2.12]$$

2.2.3. *Conservation of energy*

We are going to apply the first law of thermodynamics which states that the variation of internal energy e specific to a control volume Ω over time depends only on the variations of energy corresponding to the work exchanged with the external continuum and the amount of energy being considered in the form of heat. We will call $\vec{q}$ the heat flux[3] and ρq_e the heat sources and $\vec{n}$ the normal external to the control volume Ω. The first law of thermodynamics is written as [FIL 99] $d/dt\int_\Omega \rho\left(e + 1/2\vec{v}^2\right)dV = \int_{\partial\Omega}\left(\vec{\vec{\sigma}}\cdot\vec{v} - \vec{q}\right)\cdot\vec{n}dS + \int_\Omega\left(\vec{f}_v\cdot\vec{v} + \rho q_e\right)\cdot\vec{n}dV$. With the relation $d/dt\int_\Omega \phi dV = \int_\Omega \partial\phi/\partial t + \phi\text{div}\vec{v}dV$, Ostrogradsky's theorem and by noting that the relation is true for an arbitrary control volume Ω, it gives $\partial\left(\rho\left(e + 1/2\vec{v}^2\right)\right)/\partial t + \rho\left(e + 1/2\vec{v}^2\right)\text{div}\vec{v} - \text{div}\left(\vec{\vec{\sigma}}\cdot\vec{v} - \vec{q}\right) - \left(\rho\vec{f}_v\cdot\vec{v} + \rho q_e\right) = 0$. Considering the equations of conservation of mass [2.10] and momentum (or impulse) [2.12], the equation of conservation of energy is obtained:

$$\rho\left(\frac{\partial e}{\partial t} + \vec{v}\cdot\overrightarrow{\text{grad}}e\right) + \text{div}\vec{q} = \vec{\vec{\sigma}} : \vec{\vec{D}} + \rho q_e, \qquad [2.13]$$

where the Green tensor $\vec{\vec{D}}$ has been defined by relation [2.8]. The notation $\vec{\vec{\sigma}} : \vec{\vec{D}}$ characterizes the product twice contracted of tensors $\vec{\vec{\sigma}}$ and $\vec{\vec{D}}$; as these tensors are of order 2, the contracted product corresponds to a scalar.

2.3. Constitutive laws

We have briefly seen in previous sections how a continuum medium is deformed when it is subjected to a displacement or a set of forces. In order to complete the description of this continuum, it should be characterized more specifically by seeking how it will react to this deformation. For more clarity, we will consider two extremely simple examples. The first example is that of a metal bar. When it is deformed, bent, for example, it will "resist" the deformation and return to its initial

---

3 For example, given by Fourier's law $\vec{q} = -k_\theta\overrightarrow{\text{grad}}(T)$, with $k_\theta > 0$ the coefficient of thermal conductivity.

position if the applied force is not too strong. The second example is that of a volume of liquid. It is general experience to know that liquids have no proper shape[4] and that the concept of deformation has little meaning. More specifically, according to Boussinesq's definition, "a fluid is a homogeneous and an isotropic medium that can restore its isotropy after all the possible deformations and keep it during these deformations as long as these are carried out sufficiently slowly". These simple examples remind us that continuum media can show radically opposite behaviors. Of course, there are continua that have intermediate behaviors, or even completely different ones. For example, invar, which is a magnetostrictive alloy of iron and nickel that has practically invariable dimensions (hence its name) when it is subjected to temperature variations. It is widely used in the construction of liquid gas tanks (at -160^oC). Any other metal that would suffer such thermal contraction would certainly cause the rupture of the tanks. Some polymers with a solid consistency can behave like fluids, while certain fluids present a constraint threshold below which they behave like solids. There are heterogeneous fluid continua such as smokes or mists, alloys with form memory[5], not to mention metamaterials which possess amazing properties (with negative refractive light index, or with a negative Poisson ratio).

Practically, the only means of description of the constitutive laws of continuum media originate from experience. The objective is to establish a relationship that relates deformation (or the velocity of deformation for liquids) to the stress.

2.3.1. *Elasticity*

We will subject a test tube of steel of initial length l_0 to tractions $\vec{F}$ applied at both ends in parallel to its axis and we are going to plot the elongation curve $l - l_0$.

Three zones, very schematically represented in Figure 2.2, can be distinguished. Region I is the elastic region where the elongation remains proportional to the applied force. This constitutes the definition of linear elasticity and is called Hooke's law. The mathematical writing of this law is $\sigma_{ij} = C_{ijkl}d_{kl}$ where σ_{ij} are the components of the stress tensor (or Kirchhoff's tensor) , C_{ijkl} are the components of the fourth-order tensor that characterizes the material and d_{kl} are the components of the Green deformation tensor. Region II is the plastic region in which the body shows a residual deformation. In this region, despite a force that increases only very lightly, the material significantly elongates. Region III or necking region is the region in

4 It is often said that a fluid is a body likely to flow easily and it does not have any form; but, in practice, this definition is insufficient.

5 These alloys present an equilibrium state which can be recovered by heating after a mechanical deformation or sometimes two states of equilibrium and it is possible to shift from one to the other by altering the temperature.

which the body breaks. The behavior of the material is no longer uniform. These regions have characteristics which vary considerably from one body to another and are shown here only as mnemonics. Their study largely falls beyond the scope of this book.

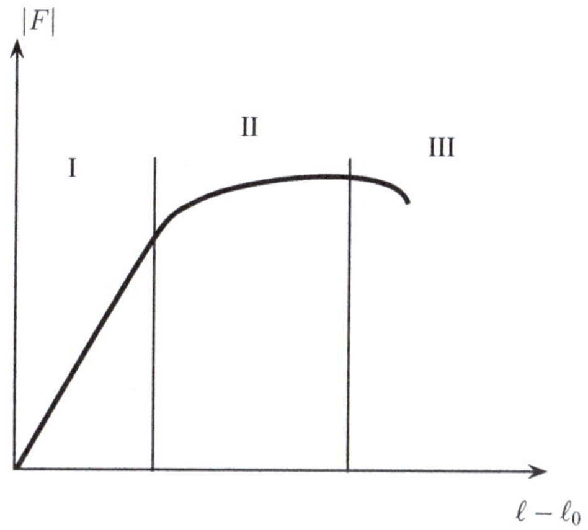

Figure 2.2. *Typical traction curve*

Note that the 81 coefficients C_{ijkl} are reduced in the case of homogeneous (the material consists of a single component) and isotropic bodies (the properties of the medium are independent of the direction) to two independent coefficients. This pair of coefficients takes several expressions: (λ, μ): Lamé's coefficients, (E, ν): the elasticity modulus or Young's modulus (E) and Poisson's ratio (ν) or even (K, μ): the compression modulus K and the shear modulus or Lamé's modulus μ. These coefficients are related by the relations:

$$\mathrm{K} = \lambda + \frac{2}{3}\mu = \frac{E}{3(1-2\nu)}, \lambda = \frac{\mathrm{E}\nu}{(1-2\nu)(1+\nu)}, \mu = \frac{\mathrm{E}}{2(1+\nu)}. \qquad [2.14]$$

$$\mathrm{E} = \frac{\mu(3\lambda+2\mu)}{\lambda+\mu} = \frac{9\mathrm{K}\mu}{3\mathrm{K}+\mu}, \nu = \frac{\lambda}{2(\lambda+\mu)} = \frac{1}{2}\frac{3\mathrm{K}-2\mu}{3\mathrm{K}+\mu}. \qquad [2.15]$$

Note that since K and μ are positive, we easily derive that if $\mathrm{K} = 0$, $\nu = -1$, and that if $\mu = 0$, $\nu = 0.5$ or $-1 < \nu < 0.5$. Young's modulus (or elasticity modulus), which represents the ability of the continuum to withstand pressure, and Lamé's coefficients are expressed in Pascals. Poisson's ratio (or lateral contraction coefficient) is the ratio of the lateral contraction to the elongation. As we have just seen, ν is comprised between -1 and 1/2; but in practice, no natural material (with the exception of pyrite [ZEN 48]) presents a Poisson ratio $\nu < 0$ at the macroscopic level (it is then referred to as auxetic material) that is to say whose elongation is accompanied by a transversal dilatation; however, it is commonly known how to manufacture wire mesh-based materials that exhibit a negative Poisson ratio. In practice, $0 < \nu < 1/2$. For most metals, $\nu \approx 1/3$. It should also be noted that all the coefficients K, E, λ and μ have a pressure dimension. Hooke's law is written in the case of a homogeneous and isotropic body as:

$$\sigma_{ij} = \lambda d_{ll}\delta_{ij} + 2\mu d_{ij}, \text{and in matrix form } \vec{\vec{\sigma}} = \lambda \text{Tr}\vec{\vec{D}} + 2\mu\vec{\vec{D}}, \qquad [2.16]$$

where δ_{ij} is the Kronecker's symbol that equals 1 if $i = j$ and 0 otherwise. The term Tr$\vec{\vec{D}}$ is the trace of the tensor $\vec{\vec{D}}$ which is the sum of the terms of the diagonal of the tensor. Hooke's law is easily expressed relatively to Young's modulus and to Poisson's ratio as $\sigma_{ij} = E(1+\nu)\left(d_{ij} + \nu/(1-2\nu)d_{ll}\delta_{ij}\right)$. The inverse formula, known as Kirchhoff–St Venant's law. Expresses the deformation based on the stress $d_{ij} = \left((1+\nu)\sigma_{ij} - \nu\sigma_{ll}\delta_{ij}\right)/E$.

2.3.1.1. *Stress-deformation tensor*

The stress tensor components, given by Hooke's law, are written in expanded form:

$$\begin{array}{l} \sigma_{11} = \frac{E}{(1+\nu)(1-2\nu)}\left((1-\nu)d_{11} + \nu(d_{22}+d_{33})\right) \\ \sigma_{22} = \frac{E}{(1+\nu)(1-2\nu)}\left((1-\nu)d_{22} + \nu(d_{11}+d_{33})\right) \\ \sigma_{33} = \frac{E}{(1+\nu)(1-2\nu)}\left((1-\nu)d_{33} + \nu(d_{11}+d_{22})\right) \\ \sigma_{12} = \frac{E}{(1+\nu)}d_{12}, \sigma_{13} = \frac{E}{(1+\nu)}d_{13}, \sigma_{23} = \frac{E}{(1+\nu)}d_{23} \end{array} \qquad [2.17]$$

2.3.1.2. *Infinitesimal strain tensor*

In order to ease the writings, we use the following synthetic notation: $u_{i,j} = \partial u_i/\partial x_j$. In all cases, the displacement field of the solid is defined by $\vec{u} = (u_1, u_2, u_3)$.

CARTESIAN COORDINATES.– in the Cartesian coordinate system (O, x_1, x_2, x_3), we have:

$$\begin{array}{l} d_{11} = u_{1,1}, d_{22} = u_{2,2}, d_{33} = u_{3,3} \\ d_{12} = \frac{1}{2}\left(u_{1,2} + u_{2,1}\right), d_{13} = \frac{1}{2}\left(u_{1,3} + u_{3,1}\right), d_{23} = \frac{1}{2}\left(u_{2,3} + u_{3,2}\right) \end{array} \qquad [2.18]$$

CYLINDRICAL COORDINATES.– in the cylindrical coordinate system (O, x_1, x_2, x_3) (compared to the usual notations (O, r, θ, z), $x_1 \equiv r$, $x_2 \equiv \theta$ and $x_3 \equiv z$) we have:

$$\begin{array}{l} d_{11} = u_{1,1}, d_{22} = \frac{u_{2,2}}{x_1} + \frac{u_1}{x_1}, d_{33} = u_{3,3} \\ d_{12} = \frac{1}{2}\left(\frac{u_{1,2}}{x_1} + u_{2,1} - \frac{u_2}{x_1}\right), d_{13} = \frac{1}{2}\left(u_{1,3} + u_{3,1}\right), d_{23} = \frac{1}{2}\left(u_{2,3} + \frac{u_{3,2}}{x_1}\right) \end{array} \qquad [2.19]$$

SPHERICAL COORDINATES.– in the spherical coordinate system (O, x_1, x_2, x_3) (compared to the usual notations (O, r, θ, ϕ), $x_1 \equiv r$, $x_2 \equiv \theta$ and $x_3 \equiv \phi$) we have:

$$\begin{aligned} &d_{11} = u_{1,1}, d_{22} = \frac{u_{2,2}}{x_1} + \frac{u_1}{x_1}, d_{33} = \frac{u_{3,3}}{x_1 \sin x_2} + \frac{u_2}{x_1 \tan x_2} + \frac{u_1}{x_1} \\ &d_{12} = \frac{1}{2}\left(\frac{u_{1,2}}{x_1} + u_{2,1} - \frac{u_2}{x_1}\right), d_{13} = \frac{1}{2}\left(\frac{u_{1,3}}{x_1 \sin x_2} + u_{3,1} - \frac{u_3}{x_1}\right), \\ &d_{23} = \frac{1}{2}\left(\frac{u_{2,3}}{x_1 \sin x_2} + \frac{u_{3,2}}{x_1} - \frac{u_3}{x_1 \tan x_2}\right) \end{aligned} \qquad [2.20]$$

2.3.2. *Thermoelasticity and effects of temperature variations*

Although it is a common experience to observe that a heated body expands, a solid that is deformed suffers a change of temperature. Reversible elastic processes are coupled to the irreversible processes produced by the thermal conduction of the warmer areas toward the colder areas. As a result of the second law of thermodynamics, by creation of entropy, a part of the mechanical energy is converted into heat. In a number of cases, this energy loss although small is not negligible. When a material is stressed in a reversible adiabatic process, the heterogeneity of the strain field creates a temperature distribution. Local gradients and the thermal conductivity of the material then initiate a flux, described by Fourier's law, which allows the solid to regain its equilibrium state. Coupled thermoelastic equations, using the isotherm constants of the material, are derived from the free energy as a function of the temperature and the infinitesimal strain tensor around a reference state.

The classical elasticity and heat equations when they are decoupled have very damped and dispersive elastic waves and heat waves as solution. The coupled equations have for solutions dispersive quasi-elastic and damped waves and quasi-thermal waves. Helmholtz theorem of the decomposition of the displacement field in the sum of an irrotational term with a term without divergence makes it possible to show that transverse waves are decoupled from thermal phenomena. Therefore, only bending and compression stresses cause a coupling between the mechanical and thermal fields.[6]

The concept of usual relaxation time in viscoelasticity appears naturally in the thermoelastic theory and justifies the *a posteriori* usage of an equivalent rheological model. This thermoelastic relaxation time defines a crucial region in which the conduction process manages to balance the inequalities of temperature (isothermal or relaxed regime) or not (adiabatic or non-relaxed regime) during a vibration period. In the adiabatic regime, the thermal and the mechanical fields are in phase while in an isothermal regime, they are out of phase by 90°. In the transition area, these fields are out of phase by 45° and the damping is maximum.

6 This justifies the use of torsion clocks to measure the viscoelasticity of metals.

Even though the process of dissipation is similar, the nature of the coupling depends a lot on the studied elastic wave. The thermoelastic coupling introduced by longitudinal waves and bending waves is qualitatively different. The distance between cold and warm regions of a structure greatly differs depending on the considered wave [LIF 99]. It is connected as a first approximation to the cross-sectional dimension for bending waves[7], the characteristic distance is proportional to a half wavelength in the case of a compressional wave. A significant consequence in the latter case is that the high-frequency thermal relaxation time is very short and that the process is isothermal (or relaxed), whereas at low frequencies the process is adiabatic: it is exactly the reverse for bending waves. A certain number of distances and relaxation times related to thermoelastic coupling characterize the propagation of a wave in an infinite plate. The discreet relaxation spectrum resulting therefrom is altered in the presence of boundary conditions. In fact, most of the models are based on a simplified geometric description. The Zener model [ZEN 48], commonly used in order to describe the thermal field, retains only the first transverse mode, to which a single and unique distance is associated, the transverse distance. As a result, only the heat transfers relatively to the thickness are taken into account; transfers in the plane of the plate such as the influence of the boundary conditions on the edges of the plate are ignored.

Suppose that a body undergoes a temperature variation[8] $T - T_0$, where T_0 is the temperature at rest and that this temperature variation is assumed to be sufficiently low to only induce low-intensity distortions (by thermal expansion). The relative variation in volume during the deformation is given by $d_{ll} = \alpha(T - T_0)$ where α is the coefficient of thermal expansion. The relations which express the deformation based on the stress in the presence of thermal deformation generalize Hooke's law and are called Duhamel–Neumann's law [PAR 84]:

$$d_{ij} = \frac{1}{E}\left((1+\nu)\sigma_{ij} - \frac{\nu}{1-2\nu}\sigma_{ll}\delta_{ij}\right) + \alpha(T - T_0)\delta_{ij}. \qquad [2.21]$$

And therefore:

$$\sigma_{ij} = -3K\alpha(T - T_0)\delta_{ij} + \frac{E}{1+\nu}\left(d_{ij} + \frac{\nu}{1-2\nu}d_{ll}\delta_{ij}\right). \qquad [2.22]$$

These given equations introduce an additional variable, the temperature T. Either it is a quantity of the problem (it is the case for external heat input), or it is an unknown of the problem as in the case of heat produced by the deformation. In the latter case, an additional equation is necessary to calculate it. To this end, we use the law of conservation of energy [2.13]. The internal energy for thermoelastic solids is given

7 Zener uses the characteristic distance h/π in his model, where h is the thickness.

8 This temperature variation can be caused by the movement itself or by external causes.

by $e = e(s, \vec{\vec{D}})$ [FIL 99], the first differential of the state equation is $de = Tds + (\vec{\vec{\sigma}}/\rho) : d\vec{\vec{D}}$, where was defined the temperature $T = (\partial e/\partial s)_{\vec{\vec{D}}}$ and the stress tensor $\sigma_{ij} = \rho(\partial e/\partial d_{ij})_{s,d_{kl_{kl \neq ij}}}$. The equation of energy conservation [2.13] written on the entropy s becomes:

$$\rho T \frac{ds}{dt} = -\text{div}\vec{q} + \rho q_e. \quad [2.23]$$

If in this equation, Fourier's law is introduced that characterizes the thermal conduction $\vec{q} = -k_\theta \overrightarrow{\text{grad}} T$ as well as the expression of the entropy [LAN 89b] $\rho s = \rho s_0 + \rho c_v(T - T_0)/T_0 + 3\alpha\sigma_{ll}$, where s_0 is the entropy at rest and c_v is the specific heat per unit volume at constant strain, the linearized heat conduction equation is obtained:

$$\rho c_v \frac{dT}{dt} - k_\theta \Delta T + \alpha T_0 \frac{d\sigma_{ll}}{dt} = \rho q_e. \quad [2.24]$$

The third term of the left member of equation [2.24] represents the thermomechanical coupling. The Duhamel–Neumann law coupled with the thermal conduction equation allows thermoelastic losses to be characterized in structures. To solve these coupled equations, it is assumed that fluctuations around the equilibrium quantities are very small and dT/dt is replaced by $\partial T/\partial t$ and $d\sigma_{ll}/dt$ by $\partial\sigma_{ll}/\partial t$ and, in the case of thin structures, it will be considered that the thermal field is one-dimensional and leads to a transverse gradient (Zener hypothesis for the bending of thin structures [ZEN 48]).

2.3.3. *Viscoelasticity*

Viscoelasticity characterizes the ability of certain bodies to store some of the energy and dissipate another part. These two combined behaviors put any material between the two ideals that are the perfectly elastic material and the Newtonian viscous fluid. Viscoelasticity phenomena are various (frequency dependence, creep and dissipation) and are extensively addressed in many reference books [CHR 82, PER 60]. In mechanics, the viscoelastic equations comprise a time dependency other than inertial and model damping when the local dissipation[9] can be characterized by a constitutive law introducing a temporal dependency between stress and strain. As we have mentioned in the previous section, since thermoelastic dissipation emerges from a local process in the structure plane, thermoelastic damping can be represented by viscoelastic operators in the case of thin structures.

9 The term "local" means that the dissipation in a point depends exclusively on the dissipative processes involved in the immediate vicinity of this point.

The approximation obtained by the Zener model is such that the homogeneous damping model includes thickness as a geometric parameter; it is, therefore, not intrinsic and does only model damping during bending at low frequency . This model reflects, however, how extrapolations are easily permitted by the viscoelastic models.

Linear viscoelasticity is modeled after two formalisms that allow us to account for the viscoelastic effects (memory effect, frequency dependence, phase shift between dual quantities and dissipation). The first is based on the introduction in the constitutive equations of temporal differential operators $\sum_i \alpha_i \partial^i / \partial t^i$ in combination with conventional operators such as the rigidity operator. The second is based on integral convolution operators whose kernel is called delay function or relaxation. This last formulation consists of describing the dissipation by the introduction in the partial differential equations of motion[10] $\mathcal{O}u(x,t) - \ddot{u}(x,t) = 0$ of a convolution kernel $\mathcal{G}(t,\tau)$ which leads to an integro-differential equation $\mathcal{O}u(x,t) - \ddot{u}(x,t) + \int_0^t \mathcal{G}(t,\tau)\dot{u}(x,\tau)d\tau = 0$; the particular case of the viscosity $\eta\dot{u}(x,t)$ being taken into account by this representation if it is assumed that the dissipation kernel takes the form $\mathcal{G}(t,\tau) = \eta\delta_\tau(t)$, where $\delta_\tau(t)$ is the delayed Dirac distribution at time t of time τ. These formalisms characterize differently a behavior which can be equivalent, and then symbolized by a same rheological model [CHR 82]. These operators are used with both integer and fractional partial derivative (or integral) models.

We assume in the remainder of this section that the motion that we have to consider is one-dimensional. For an elastic medium with $\sigma = \sigma_{11}$, $\epsilon = d_{11}$ and $D = E(1-\nu)/\left[(1+\nu)(1-2\nu)\right]$ such a motion is characterized by Hooke's law which is written as $\sigma = D\epsilon$.

2.3.3.1. *Partial differential operator*

The partial differential operator coupled to the rigidity operator leads to general constitutive equations, acceptable from a thermodynamic perspective if $J \geq I$, such as:

$$\sum_i^I \alpha_i \frac{\partial^i \sigma(t)}{\partial t^i} = D \sum_j^J \beta_j \frac{\partial^j \epsilon(t)}{\partial t^j}. \quad [2.25]$$

Through a time Fourier transform, we obtain $\hat{\sigma}\sum_i^I \alpha_i(-\imath\omega)^i = D\hat{\epsilon}\sum_j^J \beta_j(-\imath\omega)^j$, which is written in compact form $\hat{\sigma} = \hat{D}(\omega)\hat{\epsilon}$. In the homogeneous and isotropic case,

10 The general form of the equations of motion $\mathcal{O}u(x,t) - \ddot{u}(x,t) = 0$ reflects the balance between the potential elastic or thermoelastic energy, which is described by an operator $\mathcal{O}u(x,t)$ with space partial derivatives, and the kinetic energy that is described by an operator $\ddot{u}(x,t)$ with time partial derivatives.

assuming the Poisson ratio real, the complex rigidity modulus $\hat{E}(\omega) = \hat{D}(\omega)(1+\nu)(1-2\nu)/(1-\nu)$ can be expressed in the form of a Padé series:

$$\hat{E}(\omega) = \frac{(1+\nu)(1-2\nu)}{1-\nu} \frac{\sum_j^J \beta_j(-\imath\omega)^j}{\sum_i^I \alpha_i(-\imath\omega)^i}. \qquad [2.26]$$

If this writing is very convenient, it has the disadvantage of lacking rigor because it implies that Young's modulus E depends on time and therefore that the equations by inverse Fourier transform should strictly involve convolution products (which will be detailed in the following section). Describing the viscosity by a partial differential operator allows a damping spectrum to show up consisting of a series of Debye peaks; each peak is associated with a relaxation mechanism and with a characteristic time constant.

2.3.3.1.1. Elementary models

Among the simplest rheological models, there is the Kelvin–Voigt model which corresponds to a spring of stiffness D and a damper of damping coefficient c mounted in parallel: $\sigma = D(\epsilon + \tau\dot{\epsilon})$ where $\tau = c/D$ is the characteristic time of the model, and Maxwell's model which corresponds to a spring and a damper mounted in series: $\sigma + \tau\dot{\sigma} = \tau D\dot{\epsilon}$.

We will briefly study the Kelvin–Voigt model. Supposing that the deformation is periodic of the form $\epsilon(t) = \hat{\epsilon}\cos\omega t$, we thus have $\sigma(t) = D(\hat{\epsilon}\cos\omega t - \tau\omega\hat{\epsilon}\sin\omega t)$, or furthermore:

$$\sigma(t) = D\hat{\epsilon}\sqrt{1+(\omega\tau)^2}\cos(\omega t + \arctan\omega\tau). \qquad [2.27]$$

The deformation and stress are not in phase, which involves the transformation of the mechanical energy into heat. The writing of Young's modulus in complex form is easy to obtain. Consider $\sigma(t) = \Re\left[\hat{D}\hat{\epsilon}\exp(-\imath\omega t)\right]$. Define $\hat{D} = D' - \imath D'' = D'(1-\imath\eta)$, $\eta = D''/D'$. It is easy to obtain $\sigma(t) = D'\hat{\epsilon}\cos(\omega t) - D''\hat{\epsilon}\sin(\omega t)$. Equating the two expressions of stress yields $D(\hat{\epsilon}\cos\omega t - \tau\omega\hat{\epsilon}\sin\omega t) = D'\hat{\epsilon}\cos(\omega t) - D''\hat{\epsilon}\sin(\omega t)$, by identifying $\hat{D}$ and D, it can be seen that $\eta = \omega\tau$. This quantity is named loss factor or damping factor. It can be observed in the expression $\sigma(t) = D\hat{\epsilon}\sqrt{1-\omega^2\tau^2}\cos(\omega t + \arctan\omega\tau)$, the term $\arctan\eta = \delta$ corresponding to the phase angle δ[11].

EXERCISE.– Establish the relation 2.27.

SOLUTION.– *We have* $\sigma(t) = D\hat{\epsilon}(\cos\omega t - \omega\tau\sin\omega t)$. *Define* $\tan\delta = \omega\tau$. *It is easy to obtain* $\sigma(t) = D\hat{\epsilon}(\cos\delta\cos\omega t - \sin\delta\sin\omega t)/\cos\delta)$, *or further*

11 The angle δ is often designated as the loss angle.

$\sigma(t) = D\hat{\epsilon}\cos(\omega t + \delta) / \cos\delta$. *However, if* $\tan\delta = \phi$, *then* $\cos\delta = \left(1+\phi^2\right)^{-1/2}$. *Finally,* $\sigma(t) = D\hat{\epsilon}\sqrt{1+(\omega\tau)^2}\cos(\omega t + \delta)$ *is obtained. This last relation justifies the designation of loss angle for* δ.

Starting again from equation [2.27] in which we set $\epsilon(t) = \hat{\epsilon}\cos\omega t$, it then yields:

$$\sigma^2(t) - 2D'\sigma(t)\epsilon(t) + D'^2\epsilon^2(t)(1+\eta^2) - \eta^2 D'^2\hat{\epsilon}^2 = 0. \quad [2.28]$$

The classical ellipse in stress-strain representation is obtained, whose area is given by $S = \pi\eta$ [CRE 05]. An illustration is given in Figure 2.3 for $\eta = 0.5$. The dissipated energy E_d over a period T is given by $E_d = \oint \sigma d\epsilon = D'\hat{\epsilon}^2 S$, which is proportional to the area of the ellipse and therefore to the damping factor η. Thus, the energy dissipated as heat during a time interval t is given by $E_d(t) = E_d t/T = D'\hat{\epsilon}^2/2\eta\omega t = E_{R_0}\eta\omega t$, where E_{R_0} is the initial mechanical energy of the system. It is then shown that after cutting off the excitation, the energy decreases exponentially as $E_{R_0}(1 - exp(-\omega\eta t))$ with a time constant $\omega\eta$.

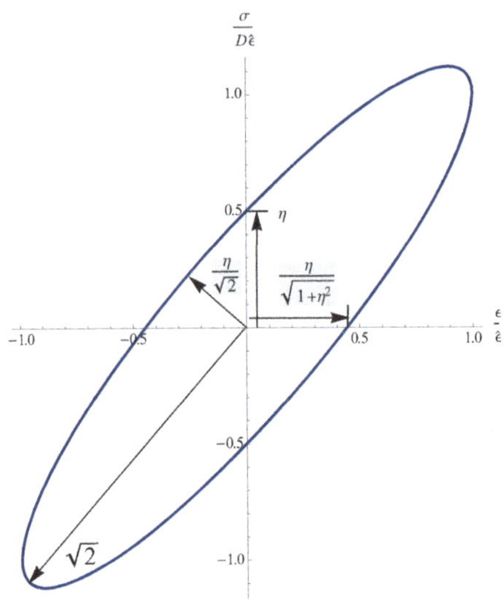

Figure 2.3. *Theoretical stress-strain curve of a dissipative material under cyclic stress ($\eta = 0.5$)*

Two examples of the stress-strain curves of two different speakers under cyclic stress [MAR 11] are presented in Figure 2.4. The ellipse deformations demonstrate that these two speakers both show a low nonlinearity in their behavior. The speaker that corresponds to the right curve is a standard loudspeaker with elastomer half-roll suspension (the non-symmetrical suspension geometry explaining the asymmetry of the curve); it presents a low stiffness and high damping $\eta \approx 0.2$. The speaker which corresponds to the left curve is a low-range corrugated paper suspension speaker; it presents high stiffness and low damping $\eta \approx 0.07$.

Another commonly used simple model is Zener's model which corresponds to Maxwell's model mounted in parallel of a spring:

$$\sigma + \tau_\epsilon \dot{\sigma} = D(\epsilon + \tau_\sigma \dot{\epsilon}). \qquad [2.29]$$

We will briefly study it. In this model, Zener [ZEN 48] has introduced two time constants: τ_ϵ, which is the relaxation time of the effort under constant strain[12] and τ_σ, which is the relaxation time of the deformation under constant effort. If we apply a strain ϵ_0 at $t = 0$, equation [2.29] has then $\sigma(t) = D\epsilon_0 + (\sigma_0 - D\epsilon_0)\exp(-t/\tau_\epsilon)$ for solution, where σ_0 is the original stress. D is named the relaxed elastic modulus D_R. Conversely, if we apply an effort σ_0 at $t = 0$, it is shown that equation [2.29] has then for solution $\epsilon(t) = D_R^{-1}\sigma_0(\epsilon_0 - D_R^{-1}\sigma_0)\exp(-t/\tau_\sigma)$. Now, by integrating equation [2.29] over an interval, $\int_{\delta t} \sigma dt + \tau_\epsilon d\sigma \delta t = D_R \int_{\delta t} \epsilon dt + D_R \tau_\sigma d\epsilon \delta t$ is obtained. If δt tends toward zero, the integrated terms cancel out and the instantaneous elastic modulus $D_I = D_R \tau_\sigma / \tau_\epsilon$ such as $d\sigma = D_I d\epsilon$.

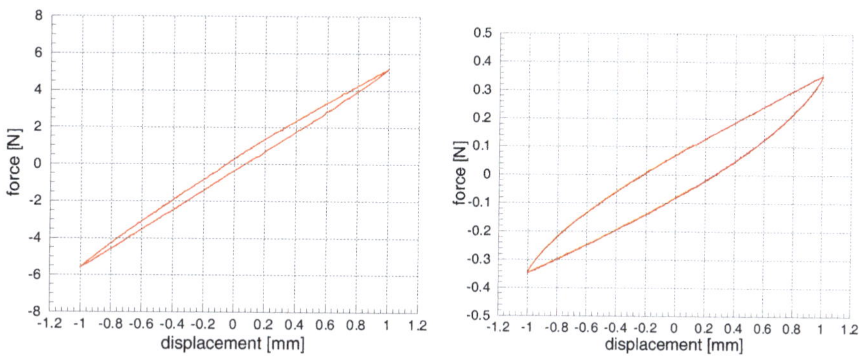

Figure 2.4. *Examples of stress-strain curves for two speakers under cyclic stress for a velocity of 0.5 mm/s*

As previously, we will introduce the complex elastic modulus $\hat{D}$ and the loss angle. To this end, we search for the solutions in the form $\sigma(t) = \hat{\sigma}\exp(\imath\omega t)$ and $\epsilon(t) = \hat{\epsilon}\exp(\imath\omega t)$. It then gives $\hat{\sigma} = \hat{D}\hat{\epsilon}$ where $\hat{D} = (1 + \imath\omega\tau_\sigma) / (1 + \imath\omega\tau_\epsilon)$. We

12 As a matter of fact, if we impose ϵ and $\dot{\epsilon}$ equal to zero, it can be easily shown that the solution to $\sigma + \tau_\epsilon \dot{\sigma} = 0$ is given by $\sigma(t) = \sigma_0 \exp(-t/\tau_\epsilon)$.

still have $\tan\delta = \Im\left(\hat{D}\right)/\Re\left(\hat{D}\right)$, therefore $\tan\delta = \omega\left(\tau_\sigma - \tau_\epsilon\right)/\left(1+\omega^2\tau_\sigma\tau_\epsilon\right)$, or furthermore with $\overline{\tau} = \sqrt{\tau_\sigma\tau_\epsilon}$ and $\overline{D} = \sqrt{D_R D_I}$,

$$\tan\delta = \frac{D_I - D_R}{\overline{D}}\frac{\omega\overline{\tau}}{1+\omega^2\overline{\tau}^2} = \Delta E\frac{\omega\overline{\tau}}{1+\omega^2\overline{\tau}^2}, \qquad [2.30]$$

for a real ν. The loss angle presents a maximum for $\omega\overline{\tau} = 1$ and is then equal to $\tan\delta_m = \Delta E/2$. The usual value of Young's modulus is the relaxed modulus which is valid in low frequency (below the pulse $1/\overline{\tau}$). At high frequencies (beyond the pulse $1/\overline{\tau}$), the instantaneous modulus should be used.

Zener has characterized this loss angle for thin metal structures. He has demonstrated that nearly 99% of the relaxation is brought by the first transverse thermal mode and that the relaxation time is then given by $\overline{\tau} = (h/\pi)^2/\mathcal{D}$, where $\mathcal{D}$ is the thermal diffusivity with $\mathcal{D} = k_\theta/(\rho c_v)$ m^2/s, where h is the thickness of the structure, c_v is the specific heat per volume unit at constant strain and k_θ is the thermal conductivity. Equation [2.30] shows that the maximum value of the loss angle depends on the material by means of $\Delta E = E\alpha^2 T_0/(\rho c_v)$, where α is the thermal expansion coefficient, and T_0 is the ambient temperature[13]. For the usual metals, $0.001 < \Delta E < 0.01$. For example, in the case of an aluminum plate $c_v = 900$ J/kg/K, $k_\theta = 210$ W/m/K, $\rho = 2700$ kg/m^3 and $\alpha = 23.10^{-6}$ K^{-1}. Let a value $\Delta E \approx 0.004$, it thus yields $\tan\delta_m \approx 0.002$. The effect is significant because the commonly accepted value of the damping factor of the aluminum is 20 times lower, $\eta \approx 0.0001$ [LES 88]. For a 1.5 mm thick plate $\mathcal{D} = 8.64.10^{-5}$ m^2/s and therefore a characteristic time $\overline{\tau} \approx 2.6$ ms. The frequency in the vicinity of which damping is maximal is $f = 1/(2\pi\overline{\tau}) \approx 60$ Hz.

In the case of steel, the thermoelastic damping is about twice lower than for aluminum ($\tan\delta_m \approx 0.0012$). In addition, its thermal diffusivity, and thus the frequency at which it is maximal, is 4 times lower than for aluminum. In practice, for the dimensions that usually come across, it is fairly insignificant for acoustic phenomena (about 8 Hz for a 2 mm thick plate).

2.3.3.2. *Convolution operator*

The convolution operator reflects a memory effect. At a given moment, this effect characterizes the dependence of the response to previous modifications whose respective importance is weighted by a relaxation kernel (or a function)[14]. In the

13 The Zener model is suitable to describe in a simplified manner the thermoelastic effects as if they were viscoelastic effects and achieves this with an excellent accuracy, including for two-dimensional structures such as plates [ZOG 06].

14 It is Boltzmann's superposition principle which postulates that two stresses successively applied to a viscoelastic body act independently and that the overall strain results from the sum of the two separate strains.

general three-dimensional isothermal case, the dependence of the current value of the stress tensor $\sigma_{ij}(t)$ on the complete history of the components of the strain tensor $d_{ij}(t)$ is expressed by a convolution integral [SAL 09, ZOG 06]:

$$\sigma_{ij}(t) = C_{ijkl}(0)d_{kl}(t) + \int_0^{\infty} d_{kl}(t-s)\frac{dC_{ijkl}(s)}{ds}ds, \qquad [2.31]$$

where $d_{ij}(t)$ is continuous relatively to time and $C_{ijkl}(t)$ is causal. The kernel $C_{ijkl}(t)$ is comparable to the usual stiffness constants tensor. It can be reduced to two independent scalar functions in the case of isotropic materials (comparable to the Lamé parameters). In practice, it is even further reduced yet retaining a single and unique relaxation function, for example assuming that Poisson's ratio is real.

To simplify the presentation, we will assume again that the motion that we have to consider is one-dimensional, the convolution integral is written in this case [CRE 05]:

$$\sigma(t) = D_1\epsilon(t) - \int_0^{\infty} \epsilon(t-s)\phi(s)ds, \qquad [2.32]$$

where $\phi(s)$ is the relaxation function whose analytically known term is being assumed. For example, for Boltzmann's model, it gives: $\phi(s) = D_2\exp(-s/\tau)/\tau$, where D_2 is a constant and τ is the relaxation time. Consider here again $\epsilon(t) = \hat{\epsilon}\cos\omega t$. Carrying forward this expression into [2.31] thus yields:

$$\sigma(t) = \left(D_1 - \frac{D_2}{1+(\omega\tau)^2}\right)\hat{\epsilon}\cos\omega t - D_2\frac{\omega\tau}{1+(\omega\tau)^2}\hat{\epsilon}\sin\omega t. \qquad [2.33]$$

Relation [2.33] shows that these relaxation processes also result into a gap between the strain and stress and thus to energy dissipation.

EXERCISE.– Calculate the phase shift between stress and strain in Boltzmann's model.

SOLUTION.–*The relation [2.33] is written:*

$$\sigma(t) = \frac{D_1\left(1+(\omega\tau)^2\right) - D_2}{1+(\omega\tau)^2}\hat{\epsilon}\left(\cos\omega t - \frac{D_2\omega\tau}{-D_2 + D_1\left(1+(\omega\tau)^2\right)}\sin\omega t\right)$$

Defining $\tan\delta = D_2\omega\tau/\left([-D_2 + D_1\left(1+(\omega\tau)^2\right)\right)$*, it yields:*

$$\sigma(t) = \frac{D_2\hat{\epsilon}}{1+(\omega\tau)^2}\sqrt{\left(1 + D_1/D_2\left(1+(\omega\tau)^2\right)\right)^2 + (\omega\tau)^2}\cos(\omega t + \delta).$$

There is thus a phase shift between constraint and deformation characterized by the loss angle δ.

2.3.4. *Fluid medium*

The constitutive law associates the stress tensor to the strain rate tensor which is by definition the time derivative of the strain tensor. This is due to the fact that in a fluid, shear effects are transmitted by viscosity which creates a force linked to the strain rate. Given $\vec{\vec{T}}$, the viscous stress tensor which is written for a Newtonian fluid in a form similar to Hooke's law:

$$\vec{\vec{T}} = \lambda \text{Tr} \frac{d\vec{\vec{D}}}{dt} + 2\mu \frac{d\vec{\vec{D}}}{dt}. \qquad [2.34]$$

Here, λ and μ are the viscosity coefficients. The coefficient λ is related to the normal stress, while μ is related to the shear stress of fluid layers. K, the volume viscosity coefficient, is defined for Newtonian fluids by $\text{K} = \lambda + \frac{2}{3}\mu$. In the case of monoatomic gases, $\lambda + 2/3\mu = 0$. For the air, it is often chosen to follow $\lambda = -2/3\mu$. Nevertheless, it should be noted that this choice implies that the volume viscosity coefficient K is zero and that we are faced with the presence of an isentropic motion (without dissipation), while in acoustics the motion is adiabatic (without heat transfer). The viscosity coefficients depend on the temperature and μ follows Sutherland's law [PIE 81]:

$$\frac{\mu(T)}{\mu_0} = \left(\frac{T}{T_0}\right)^{3/2} \frac{T_0 + S}{T + S} \qquad [2.35]$$

with for the air $\mu_0 = 1.71\ 10^{-5}$ Pa.s, $T_0 = 273.15$ K, $S = 110.4$ K and $\lambda + 2/3\mu = 7.82 \exp(-16.8\, T^{-1/3})\mu$. In practice, as a first approximation, it should be sufficient to define $\lambda + 2/3\mu = 0$.

The constitutive law that relates the stress to the strain rate is written in the case of an adiabatic motion:

$$\vec{\vec{\sigma}} = \vec{\vec{T}} - p\vec{\vec{I}}, \qquad [2.36]$$

where $\vec{\vec{I}}$ is the identity matrix and p is the pressure. In the case of a perfect fluid for which the viscosity effects are negligible, the relation above is reduced to $\vec{\vec{\sigma}} = -p\vec{\vec{I}}$ and shows that stress is reduced to a pressure. In the case of an incompressible viscous fluid, the normal stresses, proportional to the cubic expansion rate, are zero and the viscous stress tensor takes the following form : $\vec{\vec{T}} = 2\mu \frac{d\vec{\vec{D}}}{dt}$.

2.4. Hamilton principle

We will use this principle to establish the propagation equations of mechanical waves. It is, therefore, beneficial to remember [GER 93, CRE 05]. Let a mechanical system be described by a displacement field $\vec{u}$. We assume this system as continuous and conservative. Among the trajectories accepted by the system, subjected to the restrictive conditions $\delta\vec{u}\,(t_1) = \delta\vec{u}\,(t_2) = 0$, at the ends of the considered time interval $[t_1, t_2]$ $(t_1, t_2 = \pm\infty$ is permitted), the actual trajectory of the system is the one that makes the Lagrangian action stationary (in fact, minimal). We have:

$$\delta \int_{t_1}^{t_2} L(\vec{u})dt = 0, \qquad [2.37]$$

where $L(\vec{u}) = T - P$, T is the kinetic energy and P is the potential energy. Usually, the Lagrangian action L is known as a Lagrangian action density and can be expressed in the form of an integral over the volume of the domain occupied by the system. This principle, which characterizes the balance between kinetic and potential energies, is equivalent to the principle of least action and is the generalization of the virtual work principle (hence, the virtual elementary variation δ) to a continuous system.

The kinetic energy takes the following form:

$$T(\vec{u}) = \frac{1}{2}\int_{\Omega} \rho\dot{u}_i\dot{u}_i dV, \qquad [2.38]$$

where Ω is the volume occupied by the solid and $\dot{u} = \partial u/\partial t$. The potential energy density, when gravity is neglected, is an elastic potential energy density (the solid tends to return to its initial state when the applied force stops) and is written:

$$P(\vec{u}) = \frac{1}{2}\int_{\Omega} \sigma_{ij}d_{ij}dV. \qquad [2.39]$$

2.5. Characteristics of materials

Table 2.1 presents the mechanical characteristics of a few common materials.

	Density ρ (kg/m^3)	Poisson's ratio ν	Young's modulus E (Gpa)	Expansion coefficient α (K^{-1})	Thermal conductivity k_θ (W/m/K)	Bending loss factor η	Compression wave velocity c_L (m/s)
Water	1,000	-	2	<0 between 0 and 4 ^oC 2.10^{-4} at 20 ^oC 5.10^{-4} at 60 ^oC	0.6	-	1,500
Air	1.3	-	0.00014	1/273	0.025	-	340
Steel	7,800	0.27–0.31	210	12.10^{-6}	25–52	0.0002	5,200
Aluminum	2,700	0.33	70	23.10^{-6}	238	0.0001	5,100
Concrete	2,300	0.2	10–50	11.10^{-6}	1–2	0.01–0.02	2,100–4,600
Wood (// fibers) ($\perp$ fibers)	500	0.2	9–16 0.6 - 1	$3 - 6.10^{-6}$ $20 - 50.10^{-6}$	0.09-0.36	0.005–0.05	4,200–5,600 1,100–1,400
Rubber	980	0.49	0.001	15.10^{-4}	0.16–0.4	0.04–0.1	320
Carbon	1,900	0.3–0.45	240–390	3.10^{-7}	129	0.005	1,300–2,100
Copper	8,900	0.33	125	17.10^{-6}	380	0.002	3,700
Epoxy	1,500	0.2–0.38	5–10	$60 - 90.10^{-6}$	0.25	0.02	1,800–2,600
Magnesium	1,700	0.35	45	26.10^{-6}	1.56	0.0001	5,100
Plaster	700–900	0.1	2.5	10.10^{-6}	0.5–0.7	0.01	1,900–2,800
Plexiglas	1,200	0.35–0.4	3–6	$80 - 90.10^{-6}$	0.08	0.02–0.04	1,600–2,200
Polyester	1,300	0.35–0.37	1–5	$70 - 100.10^{-6}$	0.15–0.4	0.1	900–2,000
Lead	11,500	0.44	17	27.10^{-6}	35	0.02	1,250
Sand	1,500	0.2–0.45	0.03	9.10^{-6}	0.2(sec)-3	0.06–0.12	-
Titanium	4,500	0.35	110	12.10^{-6}	15–20	0.0001	4,900
Glass	2,500	0.2 - 0.3	70	$4 - 9.10^{-6}$	0.8–1.4	0.001	5,300
Zinc	7,100	0.21	120	30.10^{-6}	112	0.0003	4,100

Table 2.1. *Mechanical characteristics of a few common materials*

3

Small Mathematics Travel Kit

This chapter is dedicated to a mathematics complement. Its sometimes light-hearted style and prevailing non-formalism deserve the wrath of mathematicians. The theorems are practically never demonstrated and the readers will more often find a list of the properties useful to physicists rather than a real rigorous construction. However, its content allows more difficult vibration dynamics problems to be addressed. The readers are required to refer to more complete manuals for a proper use of the techniques presented here.

On the other hand, the presentation of this chapter can suggest that the techniques presented here are mere recipes taken from a larger publication. In addition, the belief that by resorting to more complete works to find the right techniques to employ is enough most certainly leads to great frustration. Mathematics (even when applied to physics) is not a distributor of off-the-shelf formulas, On the other hand, the techniques presented here are often sufficient to deal with simplified problems, to grasp their essence and to refine the understanding of the phenomenon being considered, at the risk of having to apply a large-scale numerical arsenal afterward (always hard to implement) for more accurate results, in principle.

We will begin with an introduction to the theory of measure, from the book by Roddier [ROD 78], which is a generalization of the concept of function. This allows us to introduce the concept of a function (or measure) continuous almost everywhere. Then, of course, the Lebesgue integral theory emerges. We then introduce the concept of distribution and give some of the properties of these "strange beings" which are a generalization of the concept of measure. With the definition of the convolution product, we introduce the notion of convolution as well as the basic kernel operators. These kernels, and their corollaries, Green's kernels, play a major role in all the problems of wave propagation. This formalism allows us to introduce the Green representation that allows a very difficult boundary value problem to be solved to be transformed into a simpler integral problem on the boundary of the domain. All these notions, initially quite different from each other, in fact form a unified whole with each part depending heavily on the others and is thereof thus inseparable. Rather than formulating a list of properties, we have voluntarily set aside the "big chunk" that the Fourier transform constitutes. This integral transform plays a primordial role in the computations of the Green kernel. It is very often the only way to explicitly compute these kernels. In practice, since we use this transform throughout the document, we have decided to present its principal properties (at least those being used in acoustics) in the body of this chapter. It is, therefore, essential to refer to more complete manuals in order to at least acquire the basic concepts of this formidable tool that the Fourier transform represents. We will end with an introduction to modal methods[1].

3.1. Measure theory and Lebesgue integration

In the physical sciences, and more generally in the observation sciences, any knowledge is related to the introduction of quantification and measure. In the sections that follow, we will consider an introduction to the mathematical modeling of this notion.

3.1.1. *Boolean algebra*

DEFINITION 3.1.– let Ω be any set, and F a collection of subsets of Ω. F forms a Boolean algebra if:

$$(i)\exists P \in F(F \neq \emptyset), (ii)\forall P \in F, \complement_{\Omega} P \in F, (iii)\forall P_i, P_j \in F, P_i \cup P_j \in F$$

We derived that $\emptyset, \Omega \in F$. Moreover, we can see that $\forall P_i, P_j \in F, P_i \cap P_j \in F$. If the property (iii) remains true for any countable set of P_i, F is called σ-algebra. A set is countable if it is possible to establish a bijection between these points and the set of natural integers $\mathbb{N}$. The set $\mathbb{Q}$ of rational numbers is countable. The set of real numbers, $\mathbb{R}$, is itself not countable. In a σ-algebra, we have $\forall P_i (i \in \mathbb{N}) \in F, \bigcap_{i=1}^{n} P_i \in F$.

The set most often encountered in applications is $\Omega = \mathbb{R}$, the set of real numbers. Consider the Boolean algebra that contains all intervals of the type $]-\infty, a]$, $-\infty < a \leq +\infty$. This algebra generates a σ-algebra known as Borel algebra. Any open or closed interval, especially any element of $\mathbb{R}$, belongs to the Borel algebra. This σ-algebra is generalized to the space $\mathbb{R}^n$ considering the intervals $]-\infty, a_i], i \in \mathbb{N}$, $-\infty < a_i \leq +\infty$.

1 These methods are nothing other than series expansions of functions and remain one of the most used tools for solving vibration dynamics problems for bounded media.

3.1.2. *Measure on a σ-algebra*

DEFINITION 3.2.– Consider Ω, a set having a σ-algebra B, of subsets P. We call measure μ on B any application that assigns each subset $P \in B$ a real or complex scalar denoted by $\mu(P)$ such that:

$$(i)\exists P, \mu(P) < \infty, (ii)\forall P_i, i \in \mathbb{N}, P_i \cap P_j = \emptyset, i \neq j \Rightarrow \mu\left(\bigcup_{i=1}^{\infty}\right) = \sum_{i=1}^{\infty} \mu(P_i).$$

In particular, $\mu(\emptyset) = 0$. A subset E of Ω is measurable if it belongs to the σ-algebra B.

EXAMPLE 3.1.–

– The notion (in $\mathbb{R}$) of the length of a segment results in assuming the quantity $\mu(]a, b]) = b - a$ as the measure of an interval $]a, b]$ on $\mathbb{R}$. This suffices to define a measure on the Borel algebra known as the Lebesgue measure. This measure is the only measure invariant by translation. Given the properties of the measure, a countable reunion of elements $x_i \in \mathbb{R}$, $i \in \mathbb{N}$, (of measure zero) forms a set of measure zero.

– Let $\phi(x)$, a locally integrable function of $x \in \mathbb{R}$, the measure of an interval $]a, b]$ can be defined by the Riemann integral $\mu(]a, b]) = \int_a^b \phi(x)dx$ which is another measure on the Borel algebra.

– The Dirac measure in a is the measure defined by $\mu(P) = 1$ if $a \in P, \mu(P) = 0$ if $a \notin P$.

Generally speaking, a set E is measurable if its characteristic function χ_E is itself measurable[2]. The measure of E is denoted by $\mu(E)$ and $\mu(E) = \int \chi_E \geq 0$. If χ_E is integrable, it is said that E is also integrable.

3.1.3. *Convergence and integration of measurable functions*

DEFINITION 3.3.–

– Measurable function. Given two sets Ω and Ω' having, respectively, the σ-algebras B and B'. The application (or function) $f : \Omega \mapsto \Omega'$ is measurable if:

$$\forall P' \in B', f^{-1}(P') \in B.$$

In practice, the functions encountered in physics are measurable. As a matter of fact, it is not known how to explain a function that is not measurable.

2 It should be remembered that the characteristic function, or indicator function, χ_E of a set E characterizes its membership or not to a subset E of Ω. $\chi_E : \Omega \to [0, 1], x \mapsto \begin{cases} 1 \text{ if } x \in E \\ 0 \text{ if } x \notin E \end{cases}.$

– Uniform convergence. A function $f_k(x)$ converges uniformly to $f(x)$ if:

$$\forall\epsilon > 0, \exists N(\epsilon), k \geq N, |f_k(x) - f(x)| \leq \epsilon. \qquad [3.1]$$

– Simple convergence:

$$\forall\epsilon > 0, \exists N(\epsilon, x), k \geq N, |f_k(x) - f(x)| \leq \epsilon. \qquad [3.2]$$

– Pointwise almost everywhere convergence. A function $f_k(x)$ converges pointwise almost everywhere to $f(x)$ if $f_k(x)$ converges pointwise to $f(x)$ at any point x external to a zero measure set.

– Convergence in measure.

$$\forall\epsilon, \eta > 0, \exists N(\epsilon, \eta), k \geq N, \mu\left\{|f_k(x) - f(x)| \geq \epsilon\right\} \leq \eta. \qquad [3.3]$$

It is possible to find the sequences of functions that converge in measure without converging pointwise everywhere. For example: $f_k(x) = 1$ if $x \in [p/n, (p+1)/n], =$ 0 elsewhere, where $n(n-1) \leq 2k \leq n(n+1), p = k - n(n-1)/2$. This sequence converges in measure to zero but fails to converge at any point of $[0, 1]$.

INTEGRATION.– The concept of measure allows the concept of integration to be generalized to all measurable functions. Let $\phi(x)$, a measurable constant piecewise function (stepwise), the set E_i of values of x for which $\phi(x) = \phi_i$ has measure $\mu(E_i)$. The integral, in Lebesgue's sense, is then:

$$\int \phi d\mu = \sum_i \phi_i \mu(E_i). \qquad [3.4]$$

THEOREM 3.1.– Any positive real function is the limit of an increasing sequence of step functions.

The integral of a measurable positive real function is thus defined by $\int f d\mu = \sup \int \phi d\mu$, where $0 \leq \phi \leq f$. Unlike the Riemann integral, which imposes a slicing along the x-axis (and is the limit of a summation of elementary "rectangles" whose base is given by the segment thus defined on the abscissa and the summit is the average ordinate over this segment), the Lebesgue integral performs a slicing of the y-axis. This definition is both more subtle and cleverer than the Riemann integral. As a matter of fact, the set of values of the function should first be calculated and then the intervals in which these values are obtained should be measured. The Riemann integral is more "risky" because it must be implicitly assumed that the function to be integrated is not too irregular. Nonetheless, when the Riemann integral exists, the Lebesgue integral also exists and they are equal. The reciprocal is false as shown in the sample Dirichlet function $D(x)$ which is zero for $x \in \mathbb{Q}$ and which is 1 for $x \notin \mathbb{Q}$. For this function, all the points are discontinuity points. $D(x)$ takes the simple following form $D(x) = \lim_{m\to\infty}[\lim_{n\to\infty}\cos(m!\pi x)^{2n}]$, this function is not integrable in Riemann's sense, is integrable in Lebesgue's meaning, is well defined and equals 0. In effect, $\int D(x)dx = 1 \times \mu\,(x \in \mathbb{Q}) + 0 \times \mu\,(x \notin \mathbb{Q})$, or $\mu\,(x \in \mathbb{Q})$ is the measure of the set of rational numbers. This set is measurable and is formed by a union of countable points (of zero measure). Thus, $\mu\,(x \in \mathbb{Q}) = 0$. Furthermore, $\mu\,(x \notin \mathbb{Q})$ is the measure of a finite measure set over any bounded interval, therefore, $\mu\,(x \notin \mathbb{Q}) < \infty$. Thus, $\int D(x)dx = 0$.

Let $f(x)$ be any function, f is summable if $(i) f(x)$ is measurable $(ii) \int |f|d\mu < \infty$.

PROPERTIES 3.1.–

– The set of the summable functions is a vector space $\mathcal{L}_1$. The integral is a linear form (functional) on $\mathcal{L}_1$.

– For f to be integrable, it is necessary and sufficient that $|f|$ be itself integrable.

– If $g \geq 0$ is integrable and if f is measurable and such that $|f| \leq g$, then f is summable and $0 \leq \left|\int f d\mu\right| \leq \int |f|\, d\mu \leq \int g d\mu$, the Lebesgue integral is a positive linear form.

– Given $f \geq 0$, a measurable function. For $\int f d\mu = 0$, it is necessary and sufficient that f be zero almost everywhere.

– If f and g are integrable and equal almost everywhere, then $\int f d\mu = \int g d\mu$.

– LEBESGUE'S THEOREM: if a sequence of integrable functions f_k converges almost everywhere to f for $k \to \infty$ and there exists $g \geq 0$ integrable $|f_k| \leq g \forall k$, then f is integrable and $\lim_{k\to\infty} \int f_k d\mu = \int \lim_{k\to\infty} f_k d\mu = \int f d\mu$.

– FUBINI–LEBESGUE'S THEOREM: let $f(x,y)$, a measurable function with two variables x and y. Its integral with respect to the Lebesgue measure on $\mathbb{R}^2$ is:

$$\int\int f(x,y)dxdy = \int dx \int f(x,y)dy = \int dy \int f(x,y)dx. \qquad [3.5]$$

3.1.4. *Functional space – functional*

DEFINITION 3.4.–

– A functional space is a set of functions which has a vector space structure. The most common examples are:

- L_α, space classes of functions f measurable such as $\int |f|^\alpha < \infty, \alpha \geq 1$. A special case very important in physics is the L_2 space, space of square modulus integrable functions,

- $\mathcal{C}^m, m \geq 0$, space of continuous mth-derivative functions,

- $\mathcal{D}_0$, set of bounded measurable functions.

– A functional space is topological if a meaning has been given to the expression $\Phi_n(x) \to \Phi(x)$, that is to say if a choice of notion of convergence has been made.

– A functional T of a functional space E is an application of E in $\mathbb{C}$. The number $T(\Phi)$ associated with $\Phi \in E$ by the functional T is denoted by $\langle T, \Phi \rangle$. Two classical examples are $E = \mathcal{C}^m, \langle T, \phi \rangle = \phi^{(m)}(a), a \in \mathbb{R}$ and $E = L_\alpha, \langle T, \phi \rangle = \int |\phi(x)|^\alpha dx$.

– A functional is zero if $\forall \Phi \in E$, $\langle T, \phi \rangle = 0$. When a functional is linear, common usage refers to it as "linear". A functional T, defined on a topological space E, is continuous if for every sequence $\Phi_n \to \Phi$, within the meaning of the chosen convergence, the numerical sequence $\langle T, \phi_n \rangle \to \langle T, \phi \rangle$.

– The dual $E^\star$ of a functional space E is the set of the linear forms defined on E. $E^\star$ is a vector space if $\forall \Phi \in E$, $\lambda_1 \in \mathbb{C}$ and $\lambda_2 \in \mathbb{C}$ such that $\langle \lambda_1 T_1 + \lambda_2 T_2, \phi \rangle = \langle \lambda_1 T_1, \phi \rangle + \langle \lambda_2 T_2, \phi \rangle$. If E is a topological space, the topological dual E' is the set of the continuous linear functional defined on E. $E' \subseteq E^\star$, a linear functional, is not always continuous. If E_1 is dense in E_2, then $E_1 \subset E_2 \Rightarrow E_2^\star \subset E_1^\star$, in particular, any element of $E_2^\star$ belongs to $E_1^\star$. E_1 is dense in E_2 if any function of E_2 is the limit of a sequence of functions of E_1 within the meaning of the topology of E_2. More "practically", it should be retained that the "smaller" E is, the "larger" its dual is and *vice versa*.

3.1.5. *Measure as linear functional*

We have seen earlier examples of measures on the Borel $sigma$-algebra of $\mathbb{R}$. A measurable function may be integrable for a measure but not for another. The simplest example is the function e^x that is integrable and with a sum equal to 1 for the Dirac measure without being integrable for the Lebesgue measure. However, in the case of a bounded support function, the situation is entirely different. We have the following theorem:

THEOREM 3.2.– Any continuous function with bounded support is integrable for any measure on the Borel σ-algebra.

Consider $\Phi(x)$, $x \in \mathbb{R}^n$, a bounded support function, then $\forall \mu \int \Phi d\mu < \infty$. The support of a function is the smallest closed set $K_f \subset \mathbb{R}^n$ of values x outside of which $\Phi(x) = 0$. In $\mathbb{R}^n$, a closed set is a compact.

If we consider the space $\mathcal{D}_0$, any measure μ appears as a functional: $\mu(\Phi) = \int \Phi d\mu = \langle \mu, \Phi \rangle$. It is obvious that these functionals are linear forms. In addition, the measures are continuous functionals in the following sense: given $\phi_n \in \mathcal{D}_0$, $n \in \mathbb{N}$, if $K_{\phi_n} = K, \forall n$ converges uniformly toward $\phi \in \mathcal{D}_0$ and $\int \phi_n d\mu \to \int \phi d\mu$. Conversely:

THEOREM 3.3 (Riesz).– any linear form continuous on $\mathcal{D}_0$ defines a measure.

The measures form a vector space $\mathcal{D}'_0$, topological dual of $\mathcal{D}_0$.

3.2. Distributions

3.2.1. *The space $\mathcal{D}$ of test functions*

By definition, $\mathcal{D}$ is the space of the indefinitely differentiable functions with bounded support. For example:

– $\phi_0(x) = 0$ if $|x| \geq 1, \phi_0(x) = \mathrm{e}^{-1/(1-x^2)}$ if $|x| < 1$;

– $\phi_{ab}(x) = 0$ if $|x| \notin]a, b[, \phi_{ab}(x) = \mathrm{e}^{-1/2(1/(x-b)-1/(x-a))}$ if $|x| \in]a, b[$.

The space $\mathcal{D}$ is not empty but is nevertheless very "small". In addition, it is a topological space: if $\Phi_n \in \mathcal{D}$, $n \in \mathbb{N}$, if $K_{\Phi_n} = K, \forall n$ then $\Phi_n^{(m)}$ uniformly converges to $\Phi^{(m)} \in \mathcal{D}, \forall m$. All the derivatives of Φ_n converge uniformly to the corresponding derivative of Φ. The convergence is "very" strong.

The generalization to $\mathbb{R}^n$ of this space is obvious, it is advisable to replace $|x|$ by the Euclidean norm $\|x\|^2$ in the definition of the test functions. $\mathcal{D}$ is contained in all the known spaces of remarkable functions.

3.2.2. *Distributions definition*

DEFINITION 3.5.–

– A distribution is a continuous linear form on $\mathcal{D}$. Note T, this distribution. $T \in \mathcal{D}'$. Theoretically, there exist non-continuous linear functionals on $\mathcal{D}$ but in practice it is impossible to explain one of them. $\mathcal{D}'$ is practically confounded with $\mathcal{D}^\star$. For $\mathcal{D}$, it is not justified to distinguish between dual and topological dual.

– Two distributions are equal if $T_1 - T_2 = 0$. Given $\forall \Phi \in \mathcal{D}, \langle T_1 - T_2, \Phi\rangle = 0 \Rightarrow \forall \Phi \in \mathcal{D}, \langle T_1, \Phi\rangle = \langle T_2, \Phi\rangle$.

EXAMPLES OF DISTRIBUTIONS.–

– To any locally integrable function $f(x)$, we can associate a distribution T_f defined by:

$$\forall \Phi(x) \in \mathcal{D}, \langle T_f, \Phi\rangle = \int f(x)\Phi(x)dx. \qquad [3.6]$$

The integral, considered within Lebesgue's sense, exists because $|f(x)\Phi(x)| < |f(x)| \sup(\Phi(x))$ and $|f(x)|$ is integrable. The distribution T_f is actually associated with the class of functions equal almost everywhere (ae) to f because $f \overset{\text{ae}}{=} g \Rightarrow T_f = T_g$. By misuse of language, we say that the locally integrable functions are distributions and we can write:

$$\langle f(x), \Phi(x) \rangle = \int f(x)\Phi(x)dx, \tag{3.7}$$

the confusion comes from the fact that a function, which is an application of a set of *numbers* in a set of *numbers*, is mistaken for a functional that is an application of a set of *functions* in a set of *numbers*.

– Singular distributions. The most common example is the Dirac distribution δ: $\forall \Phi \in \mathcal{D}, \langle \delta, \Phi \rangle = \Phi(0)$. More generally, we can define the Dirac distribution in a, δ_a by: $\forall \Phi \in \mathcal{D}, \langle \delta_a, \Phi \rangle = \Phi(a)$. A -bad- notation often used in physics is $\delta_a = \delta(x-a)$. This notation suggests that δ is a function and can raise -stupid- questions such as "what is the value of $\delta(x-a)$ in $x = b \neq a$". Similarly, it is often written $\int \delta(x)\Phi(x)dx = \phi(0)$ instead of $\langle \delta, \Phi \rangle = \Phi(0)$.

– The pseudo-functions and the distribution $\text{Vp}(1/x)$. The rule of the first example does apply to summable functions only. It does not indicate that $1/x$ defines a distribution since $1/x$ is not originally integrable. The concept of pseudo-function allows for further development. Consider $u(x)$, an integrable function on every compact contained in the complementary of the origin and $\Phi(x)$, a function of $\mathcal{D}$. Given:

$$J(\epsilon) = \int_{-\infty}^{-\epsilon} u(x)\Phi(x)dx + \int_{+\epsilon}^{+\infty} u(x)\Phi(x)dx, \tag{3.8}$$

for some functions $u(x)$, we can write $J(\epsilon) = \sum_p A_p/\epsilon^p + B\ln\epsilon + F(\epsilon)$, $F(\epsilon)$ is the finite part of the integral $J(\epsilon)$. The variable p takes a finite number of real values > 1, A_p and B are constants and $\lim_{\epsilon\to 0} F(\epsilon) = F < \infty$. In this case, u is the linear functional that associates F with Φ. $\text{Fp}(u)$ is defined by:

$$\forall \Phi \in \mathcal{D}, \langle \text{Fp}(u), \Phi \rangle = F. \tag{3.9}$$

EXAMPLE 3.2.– Study of $\text{Fp}(1/x)$ and $\text{Vp}(1/x)$. We have:

$$\forall \Phi \in \mathcal{D}, \int_{|x|>\epsilon} \frac{\Phi(x)}{x}dx = -\left(\Phi(\epsilon) - \Phi(-\epsilon)\right)\ln\epsilon - \int_{-\infty}^{-\epsilon} \Phi'(x)\ln x dx$$
$$+ \int_{+\epsilon}^{+\infty} \Phi'(x)\ln x dx,$$

nevertheless $\Phi(\epsilon) < \epsilon \max|\phi|$ and $\lim_{\epsilon\to 0} \epsilon \ln \epsilon = 0$, F is therefore the whole second member with $A_p = B = 0$ and:

$$F = -\int_{-\infty}^{+\infty} \Phi'(x) \ln x dx = \lim_{\epsilon\to 0}\left[\int_{-\infty}^{-\epsilon} \frac{\Phi(x)}{x} dx + \int_{+\epsilon}^{+\infty} \frac{\Phi(x)}{x} dx\right],$$

we recognize here a Cauchy principal value integral, thus:

$$\forall \Phi \in \mathcal{D}, \langle \mathrm{Vp}(1/x), \Phi\rangle = \oint \frac{\Phi(x)}{x} dx.$$

Consequently, $\mathrm{Vp}(1/x)$ and $\mathrm{Fp}(1/x)$ are two symbols that designate the same distribution.

EXAMPLE 3.3.– Study of $\mathrm{Fp}(x^{-s})$. The distribution $\mathrm{Fp}(x_+^{-s})$ where $s \in \mathbb{C}$ is defined by Boutet de Montvel [BOU 01]:

$$\forall \Phi \in \mathcal{D}, \mathrm{Fp} \int_0^{\infty} \frac{\Phi(x)}{x^s} dx = \mathrm{Fp}\langle x_+^{-s}, \Phi\rangle.$$

Writing the order N Taylor expansion of $\Phi(x)$: $\Phi(x) = \sum_{k<N} \Phi^{(k)}(0) x^k/k! + \Phi_N$, where Φ_N is the remainder of order N. For N large enough: $-\Re s + N > 0$, it gives:

$$\forall \Phi \in \mathcal{D}, \mathrm{Fp} \int_{\epsilon}^{1} \frac{\Phi(x)}{x^s} dx = F(\epsilon) + I_N(\epsilon) + \mathcal{O}(\epsilon^{-s+N}),$$

where $F(\epsilon)$ is a constant and:

$$I_N(\epsilon) = -\sum_{k<N, -s+k+1\neq 0} \frac{\Phi^{(k)}(0)}{k!} \frac{\epsilon^{-s+k+1}}{-s+k+1} - \sum_{k<N, -s+k+1=0} \frac{\Phi^{(k)}(0)}{k!} \ln(\epsilon).$$

The finite part in Hadamard's sense is defined by $\mathrm{Fp}(x_+^{-s}) = \lim_{\epsilon\to 0}\left[\int_{\epsilon}^{1} \Phi(x)/x^s dx - I_N(\epsilon)\right]$. Note that the logarithmic term is present only if s is an integer, which is usually the case.

3.2.3. *Operations on distributions*

LINEAR COMBINATION.– $\forall \lambda_1, \lambda_2 \in \mathbb{C}$, $\forall T_1$ and T_2, two distributions, the resulting distribution is such that:

$$\forall \Phi \in \mathcal{D}, \langle \lambda_1 T_1 + \lambda_2 T_2, \Phi\rangle = \lambda_1\langle T_1, \Phi\rangle + \lambda_2\langle T_2, \Phi\rangle. \quad [3.10]$$

TRANSLATION OF A DISTRIBUTION.– if $f(x)$ is a locally integrable function, by a simple change of variable, we verify that: $\forall\Phi \in \mathcal{D}, \int_{-\infty}^{\infty} f(x-a)\Phi(x)dx = \int_{-\infty}^{\infty} f(x)\Phi(x+a)dx$ that is $\langle f(x-a), \Phi(x)\rangle = \langle f(x), \Phi(x+a)\rangle$ and thus $\langle \tau_a f(x), \Phi(x)\rangle = \langle f(x), \Phi(x+a)\rangle$. By definition, the translated $\tau_a T$ of any distribution T is given by:

$$\forall\Phi \in \mathcal{D}, \langle \tau_a f(x), \Phi(x)\rangle = \langle f(x), \Phi(x+a)\rangle. \quad [3.11]$$

For the Dirac distribution, this leads to:

$$\begin{aligned}\forall\Phi \in \mathcal{D}, \langle \tau_a \delta, \Phi(x)\rangle &= \langle \delta(x), \Phi(x+a)\rangle = \Phi(0+a) = \Phi(a) \\ &\Rightarrow \langle \tau_a \delta, \Phi(x)\rangle = \langle \delta_a, \Phi\rangle\end{aligned}$$

that is $\tau_a \delta = \delta_a$. Similarly, the distribution T, periodic and of period a is defined by $\tau_a T = T$, that is:

$$\forall\Phi \in \mathcal{D}, \langle \tau_a T - T, \Phi(x)\rangle = 0.$$

DISTRIBUTION PARITY.– T is pair if $T(x) = T(-x)$ that is:

$$\begin{aligned}\forall\Phi \in \mathcal{D}, \langle T(x), \Phi(x)\rangle &= \langle T(-x), \Phi(x)\rangle = \langle T(x), \Phi(x)\rangle \\ &\Rightarrow \langle T(x), \Phi(x) - \Phi(-x)\rangle = 0.\end{aligned}$$

It is thus obvious that δ is pair. Similarly, T is odd if $T(x) = -T(-x)$, that is $\forall\Phi \in \mathcal{D}, \langle T(x), \Phi(x) + \Phi(-x)\rangle = 0$.

CHANGE OF SCALE.– by analogy with the locally integrable functions, $T(ax)$ is defined by:

$$\forall\Phi \in \mathcal{D}, \langle T(ax), \Phi(x)\rangle = \frac{1}{|a|}\langle T(x), \Phi\left(\frac{x}{a}\right)\rangle. \quad [3.12]$$

PRODUCT OF TWO DISTRIBUTIONS.– in contrast with the direct product of two functions, there is no natural definition of the product of two distributions. This is easily verified by the fact that two functions can be locally integrable without their product being itself integrable (for example, $1/\sqrt{x}$). The only case in which the product of distributions exists is the following: if S and T are two distributions associated with locally integrable functions f and g, under the condition they exist:

$$\forall\Phi \in \mathcal{D}, \langle S \times T, \Phi\rangle = \int fg\Phi = \langle f, g\Phi\rangle = \langle g, f\Phi\rangle. \quad [3.13]$$

This shows that for the product $S \times T$ to exist, it is necessary and sufficient that one of the functions be indefinitely differentiable. We have the following result: for a distribution T on $\mathbb{R}$ to verify $x \times T = 0$, it is necessary and sufficient that $T = C\delta$ where C is a complex constant. Similarly, if $\alpha(x)$ is an indefinitely differentiable function such that $\alpha(x) = 0$ has only simple roots a_i, i.e. $\alpha(a_i) = 0, \alpha'(a_i) \neq 0$, all solutions of the equation $\alpha(x) \times T(x) = 0$ are of the form $T(x) = \sum_i C_i \delta(x - a_i), C_i \in \mathbb{C}$. It should be noted that this last distribution, which is known under the name of "Dirac comb", is of great importance in signal processing. It is especially essential to demonstrate the theorem for the reconstruction of a Shannon sampled signal. However, if T is "not too irregular", for example if T is an "order m" distribution, that is it can be extended in a continuous linear form on $\mathcal{D}^m$, the space of the m-times continuously differentiable functions with bounded support, we can calculate $\alpha \times T$ if α is an m-times continuously differentiable function in the usual sense.

DISTRIBUTIONS DERIVATION.– let f, a locally integrable function. Let T_f, the distribution associated with this function. We have:

$$\begin{aligned}\forall \Phi \in \mathcal{D}, \langle (T_f)', \Phi \rangle = \langle T_{f'}, \Phi \rangle &= \int_{-\infty}^{\infty} f'(x)\Phi(x)dx \\ &= [f(x)\Phi(x)]_{-\infty}^{+\infty} - \int_{-\infty}^{\infty} f(x)\Phi'(x)dx,\end{aligned}$$

however, Φ is a bounded support function, therefore, $[f(x)\Phi(x)]_{-\infty}^{+\infty} = 0$, so:

$$\forall \Phi \in \mathcal{D}, \langle (T_f)', \Phi \rangle = -\langle T_f, \Phi' \rangle. \qquad [3.14]$$

By definition, $\Phi' \in \mathcal{D}$, the second member of the above equation still exists. The following essential result – which is the reason for the existence of this theory – should be retained: any distribution is differentiable and even indefinitely differentiable. It is easily shown that:

$$\forall \Phi \in \mathcal{D}, \langle T^{(m)}, \Phi \rangle = (-1)^{(m)} \langle T, \Phi^{(m)} \rangle. \qquad [3.15]$$

Now, consider the distribution T_f, associated with a locally integrable function which presents a discontinuity point of the first kind in a:

$$\lim_{x \to a^-} f(x) \neq \lim_{x \to a^+} f(x), \; \lim_{x \to a^-} f'(x) = \lim_{x \to a^+} f'(x) \qquad [3.16]$$

We have:

$$\begin{aligned}\forall \Phi \in \mathcal{D}, (\langle T_f\rangle', \Phi\rangle &= -\langle T_f, \Phi'\rangle \\ &= -\int_{-\infty}^{\infty} f(x)\Phi'(x)dx \\ &= -\int_{-\infty}^{a} f(x)\Phi'(x)dx - \int_{a}^{+\infty} f(x)\Phi'(x)dx \\ &= f(a^+)\Phi(a^+) - f(a^-)\Phi(a^-) + \int_{-\infty}^{+\infty} f'(x)\Phi(x)dx.\end{aligned}$$

Defining $f(a^+) - f(a^-) = \sigma_f(a)$, the jump of the function at point a. We thus obtain:

$$\begin{aligned}\forall \Phi \in \mathcal{D}, \langle (T_f)', \Phi\rangle &= \sigma_f(a) + \langle T_{f'}, \Phi\rangle = \sigma_f(a)\langle \delta_a, \Phi\rangle + \langle T_{f'}, \Phi\rangle \\ &= \langle \sigma_f(a)\delta_a + T_{f'}, \Phi\rangle.\end{aligned} \quad [3.17]$$

Given finally:

$$(T_f)' = \sigma_f(a)\delta_a + T_{f'}. \quad [3.18]$$

The commonly accepted notation dictates that, if the function has several discontinuity points a_i, we should define:

$$f' = \{f'\} + \sum_i \sigma_f(a_i)\delta_{a_i} \quad [3.19]$$

where $\{f'\}$ is the "careless derivative of f outside of its discontinuity points". Similarly, if the a_i are discontinuity points of the first kind for f and f':

$$f'' = \{f''\} + \sum_i \sigma_{f'}(a_i)\delta_{a_i} + \sigma_f(a_i)\delta'_{a_i}. \quad [3.20]$$

EXAMPLE 3.4.–

– Consider H, the Heaviside step:

$$H(x) = 1, \text{ if } x > 0, H(x) = 0, \text{ if } x < 0, \quad [3.21]$$

this function is undefined at the origin (sometimes $H(0) = 1/2$ and even $H(0) = 1$ are imposed). Thus,

$$\forall \Phi \in \mathcal{D}, \langle H(x), \Phi(x)\rangle = \int_0^{\infty} \Phi(x)dx.$$

Therefore:

$$\langle H'(x), \Phi(x)\rangle = -\langle H(x), \Phi'(x)\rangle = -\int_0^{\infty} \Phi'(x)dx = -[\Phi(x)]_0^{\infty} = \Phi(0)$$
$$= \langle \delta(x), \Phi(x)\rangle. \quad [3.22]$$

Thus, $H'(x) = \delta(x)$.

– Let $f(x) = |x|$. We obtain:

$$f'(x) = \text{sgn}(x) \text{ and } f''(x) = \{\text{sgn}(x)\}' + 2\delta(x) = 2\delta(x) \quad [3.23]$$

– Expression of the product $f\delta'$. We have $\forall \Phi \in \mathcal{D}$, $\langle f\delta', \Phi(x)\rangle = \langle \delta', f\Phi(x)\rangle$ that is $\langle f\delta', \Phi(x)\rangle = -\langle \delta, (f\Phi(x))'\rangle$ and $\langle f\delta', \Phi(x)\rangle = -\langle \delta, f'\Phi(x) + f\Phi'(x)\rangle = -f'(0)\langle \delta, \Phi(x)\rangle + f(0)\langle \delta', \Phi\rangle$. Let

$$f\delta' = -f'(0)\delta + f(0)\delta'. \quad [3.24]$$

In particular, $x\delta' = -\delta$, $x\delta^{(m)} = -m\delta^{(m-1)}$.

DERIVATIVE OF A PRODUCT OF DISTRIBUTIONS.– under the condition that the product $T \times S$ exists, we have $(T \times S)' = T \times S' + T' \times S$.

3.2.4. *N-dimensional generalization*

We will consider in this section the space $\mathbb{R}^n$ where the variable x is written: $x = (x_1, x_2, \cdots, x_n)$. The space $\mathcal{D}$ is the space of the test functions in which all partial derivatives of all orders exist. First, it is necessary to retain the generalization of the Lebesgue measure in $\mathbb{R}$ for which we replace the concept of open interval by that of open ball. If in $\mathbb{R}$, the point is a set of measure zero, in $\mathbb{R}^2$, the straight line (more generally, any curve segment) is a set of measure zero. In $\mathbb{R}^n$, any hyperplane (of size $n-1$) is a set of measure zero.

DEFINITION 3.6.– Let $p = (p_1, p_2, \cdots, p_n)$, a set of n integers such that $p_1 + p_2 + \cdots + p_n \leq n$, we define D^p, the derivation:

$$D^p = \left(\frac{\partial}{\partial x_1}\right)^{p_1} \left(\frac{\partial}{\partial x_2}\right)^{p_2} \cdots \left(\frac{\partial}{\partial x_n}\right)^{p_n}. \quad [3.25]$$

We have by definition $\forall \Phi \in \mathcal{D}$, $\langle D^p T, \Phi\rangle = (-1)^{|p|}\langle T, D^p\Phi\rangle$, where $|p| = \sum_i p_i$ is the order of the derivation. Consider $\mathcal{O}$, a differential operator with coefficients A_p, constant or not, such that $\mathcal{O} = \sum_p A_p D^p$. $\mathcal{O}^t$, the adjoint operator of $\mathcal{O}$ (sometimes named transposed) is defined by:

$$\mathcal{O}^t = \sum_p (-1)^p D^p A_p, \tag{3.26}$$

if all coefficients A_p are constant, then $\mathcal{O}^t = \sum_p (-1)^p A_p D^p$. We have $\forall \Phi \in \mathcal{D}, \langle \mathcal{O}T, \Phi\rangle = \langle T, \mathcal{O}^t \Phi\rangle$.

It should be noted that if in $\mathbb{R}$, $1/r$ is not a locally integrable function at the origin, it is in $\mathbb{R}^2$. More generally, in $\mathbb{R}^n$, if $|r|^2 = \sum_i x_i^2$, $1/r^{(n-1)}$ is an integrable function at the origin.

Given a surface S, of normal $\vec{n} = \overrightarrow{(n_1, n_2, \cdots, n_n)}$, and θ_i, the angle of the normal with the axis x_i. We get:

$$\forall \Phi \in \mathcal{D}, \langle \frac{\partial T_f}{\partial x_i}, \Phi\rangle = \langle \sigma_f \cos\theta_i \delta_S, \Phi\rangle + \langle T \frac{\partial f}{\partial x_i}, \Phi\rangle, \tag{3.27}$$

where σ_f is the jump of f when crossing S in the direction of the normal (value of f after S minus value of f before S for a normal oriented outward) and δ_S is the Dirac distribution carried by the surface S. In synthetic notation, this is written:

$$\frac{\partial f}{\partial x_i} = \left\{\frac{\partial f}{\partial x_i}\right\} + n_i \sigma_f \delta_S. \tag{3.28}$$

For example, in $\mathbb{R}^3$, we have:

$$\vec{\nabla} f = \left\{\vec{\nabla} f\right\} + \vec{n}\sigma_f \delta_S \tag{3.29}$$

$$\vec{\nabla} \cdot \vec{f} = \left\{\vec{\nabla} \cdot \vec{f}\right\} + \vec{n} \cdot \sigma_{\vec{f}} \delta_S \tag{3.30}$$

$$\vec{\nabla} \wedge \vec{f} = \left\{\vec{\nabla} \wedge \vec{f}\right\} + \vec{n} \wedge \sigma_{\vec{f}} \delta_S \tag{3.31}$$

$$\Delta f = \{\Delta f\} + \sigma_{\partial_{\vec{n}} f} \delta_S + \vec{\nabla} \cdot (\vec{n} \sigma_f \delta_S), \tag{3.32}$$

where $\partial_{\vec{n}} = \frac{\partial}{\partial \vec{n}} = \vec{n} \cdot \vec{\nabla}$ is the normal derivative.

THEOREM 3.4.– We can show that in $\mathbb{R}^n$, $n \neq 2$, $\Delta\left(\frac{1}{r^{n-2}}\right) = -(n-2) S_n \delta$, where S_n is the surface of the sphere of radius 1 in $\mathbb{R}^n$, given by $S_n = 2\pi^{n/2}/\Gamma(n/2)$. The function $\Gamma(z)$ is defined by $\Gamma(z) = \int_0^{+\infty} \mathrm{e}^{-t} t^{z-1} dt$, $z \in \mathbb{C}$. This function presents the following remarkable properties $\Gamma(z+1) = \Gamma(z)$, $\Gamma(1/2) = \sqrt{\pi}$, $\Gamma(n+1) = n!$,

$\Gamma(z)\Gamma(1-z) = \pi/\sin(\pi z)$. It is considered to be the extension to the complex plane of the factorial. If $n = 2$, we have:

$$\Delta\left(\ln\frac{1}{r}\right) = -2\pi\delta. \quad [3.33]$$

Consider the following integral along a particular direction x_α, $I = \int_{x_\alpha} \phi(\vec{x})\delta\left(f(\vec{x})\right) dx_\alpha$. From what we know of the properties of the Dirac distribution, it is obvious that only the points for which $f(\vec{x}) = 0$ will contribute to the integral I. To calculate I, we will start with a change of variable from x_α to $f(\vec{x})$. We will have, the variables x_i are obviously constant for $i \neq \alpha$, $dx_\alpha = df(\vec{x})/\left(\partial f(\vec{x})/\partial x_\alpha\right)$. I thus becomes $I = \int_{f(\vec{x})\gtreqless 0}^{f(\vec{x})\lesseqgtr 0} \phi(\vec{x})\delta\left(f(\vec{x})\right) df(\vec{x})/(\partial f(\vec{x})/\partial x_\alpha)$, where the inequalities that appear in the bounds of the integral must be chosen according to the sign of $(\partial f(\vec{x})/\partial x_\alpha)$. It is then obvious that it yields:

$$I = \int_{f(\vec{x})<0}^{f(\vec{x})>0} \phi(\vec{x})\delta\left(f(\vec{x})\right) d\frac{f(\vec{x})}{\left|\frac{\partial f(\vec{x})}{\partial x_\alpha}\right|} = \sum\left\{\frac{\phi(\vec{x})}{|\partial f(\vec{x})/\partial x_\alpha|}\right\}_{\vec{x}=\vec{x}_\alpha^\star}, \quad [3.34]$$

where the sum is extended to all zeros of $f(\vec{x})$ denoted by $\vec{x}_\alpha^\star$ (i.e. such that $f(\vec{x}_\alpha^\star) = 0$).

APPLICATION TO VECTOR ANALYSIS.– let $\mathbb{R}^n = \mathbb{R}^3$ and a surface $S \subset \mathbb{R}^3$, $\vec{n}$ is the exterior normal to S. Let $\vec{A}\,(x_1, x_2, x_3)$ a differentiable vector function and define $\vec{f}\,(x_1, x_2, x_3)$ which is equal to $\vec{A}$ inside S and which is zero outside. We get:

$$\vec{\nabla}\cdot\vec{f} = \left\{\vec{\nabla}\cdot\vec{f}\right\} - \vec{n}\cdot\vec{A}\delta_S = \sum_i \frac{\partial f_i}{\partial x_i}. \quad [3.35]$$

Consider $\Phi \in \mathcal{D}$, of unit measure on a domain of $\mathbb{R}^3$ that contains S, we get:

$$\begin{aligned}\langle\vec{\nabla}\cdot\vec{f}, \Phi\rangle &= \int\!\!\int\!\!\int_V \vec{\nabla}\cdot\vec{f}\Phi dV - \int\!\!\int_S \vec{n}\cdot\vec{A}\Phi dS \\ &= \int\!\!\int\!\!\int_V \vec{\nabla}\cdot\vec{f} dV - \int\!\!\int_S \vec{n}\cdot\vec{A}\, dS,\end{aligned} \quad [3.36]$$

where V is the volume bounded by S. However, by definition:

$$\begin{aligned}\langle\vec{\nabla}\cdot\vec{f}, \Phi\rangle &= \langle\sum_i \frac{\partial f_i}{\partial x_i}, \Phi\rangle = -\sum_i \langle f_i, \frac{\partial \Phi}{\partial x_i}\rangle = -\sum_i \int\!\!\int\!\!\int f_i \frac{\partial \Phi}{\partial x_i} dV \\ &= -\sum_i \int\!\!\int\!\!\int_V f_i \frac{\partial \Phi}{\partial x_i} dV, \text{ since } f = 0 \text{ outside of } S.\end{aligned}$$

On the other hand, in V, Φ has been chosen with unit measure. Each of its partial derivative is zero. The last integral is thus zero. If we compare the two expressions of the duality product $\langle \vec{\nabla} \cdot \vec{f}, \Phi \rangle$, we simply derive that:

$$\iiint_V \vec{\nabla} \cdot \vec{f} dV = \iint_S \vec{n} \cdot \vec{A} dS. \qquad [3.37]$$

We recognize there the Ostrogradski theorem. Similarly, we can show that:

$$\iiint_V \vec{\nabla}(f) dV = \iint_S \vec{n} f dS \qquad [3.38]$$

$$\iiint_V \vec{\nabla} \wedge \vec{f} dV = \iint_S \vec{n} \wedge \vec{A} dS. \qquad [3.39]$$

We will now demonstrate Green's formula. Still consider a function defined in S and zero outside this surface. We assume that f is twice differentiable inside S. We have, according to [3.32]:

$$\Delta f = \{\Delta f\} + \sigma_{\partial_{\vec{n}} f} \delta_S + \vec{\nabla} \cdot (\vec{n} \sigma_f \delta_S),$$

Take the duality product of this equality with the test function Φ. The second expression of the second member of equation [3.32] becomes:

$$\langle \sigma_{\partial_{\vec{n}} f} \delta_S, \Phi \rangle = -\langle \frac{\partial f}{\partial \vec{n}} \delta_S, \Phi \rangle, \qquad [3.40]$$

because f and its partial derivatives are zero outside the surface. It should be remembered that the jump of a function when crossing a discontinuity surface is defined by "value after minus value before the surface". Similarly, the third expression of the second member of equation [3.32] is written:

$$\langle \vec{\nabla} \cdot (\vec{n} \sigma_f \delta_S), \Phi \rangle = -\langle \sigma_f \delta_S, \vec{\nabla} \cdot (\vec{n} \Phi) \rangle = +\langle f \delta_S, \frac{\partial \Phi}{\partial \vec{n}} \rangle \qquad [3.41]$$

Let $\langle \Delta f, \Phi \rangle = \langle \{\Delta f\}, \Phi \rangle - \langle \frac{\partial f}{\partial \vec{n}}, \Phi \rangle + \langle f \delta_S, \frac{\partial \Phi}{\partial \vec{n}} \rangle = \langle f, \Delta \Phi \rangle$, $\vec{n}$ is the normal external to the surface S. If we express the duality product within the meaning of the functions (the duality product is then an integral scalar product), we easily obtain Green's formula:

$$\iiint_V \Delta f \Phi dV - \iint_S \frac{\partial f}{\partial \vec{n}} \Phi dS + \iint_S f \frac{\partial \Phi}{\partial \vec{n}} dS = \iiint_V f \Delta \Phi dV. \qquad [3.42]$$

This expression is usually written in the following form:

$$\int\int\int_V \Delta f\Phi - f\Delta\Phi dV = \int\int_S \frac{\partial f}{\partial \vec{n}}\Phi - f\frac{\partial \Phi}{\partial \vec{n}}dS. \qquad [3.43]$$

If the normal is selected inside the surface S, the sign of the second member should be changed. This relation is very important in acoustics (and in all physical sciences). We will see how it allows a differential problem defined in a domain to be transformed into an integral problem defined on the boundary of the domain (which allows us to obtain a dimension of the space to be discretized).

3.2.5. *Distributions tensor product*

While the generalization to the distributions of the ordinary product of two functions $f(x)g(x)$ cannot be totally achieved, the direct product $f(x)g(y)$ is very easily generalized. If $f(x)$ and $g(y)$ are two locally integrable functions, their product $f(x)g(y)$ is itself also integrable and we have:

$$\begin{aligned}\forall\Phi(x,y) \in \mathcal{D}, \langle f(x)g(y), \Phi(x,y)\rangle &= \int\int f(x)g(y)\Phi(x,y)dxdy \\ &= \int f(x)dx \int g(y)\Phi(x,y)dy \\ &= \int g(y)dy \int f(x)\Phi(x,y)dx.\end{aligned}$$

The last two relations are obtained by applying the Fubini–Lebesgue theorem [3.5]. From there:

$$\begin{aligned}\forall\Phi(x,y) \in \mathcal{D}, \langle f(x)g(y), \Phi(x,y)\rangle &= \langle f(x), \langle g(y), \Phi(x,y)\rangle\rangle \\ &= \langle g(y), \langle f(x), \Phi(x,y)\rangle\rangle. \qquad [3.44]\end{aligned}$$

The tensor product of two distributions, denoted by $T(x) \otimes S(y)$, is defined by:

$$\begin{aligned}\forall\Phi(x,y) \in \mathcal{D}, \langle T(x) \otimes S(y), \Phi(x,y)\rangle &= \langle T(x), \langle S(y), \Phi(x,y)\rangle\rangle \\ &= \langle S(y), \langle T(x), \Phi(x,y)\rangle\rangle. \qquad [3.45]\end{aligned}$$

THEOREM 3.5.– Let $\theta(y) = \langle T(x), \Phi(x,y)\rangle$. Then, if $\Phi(x,y) \in \mathcal{D}$, $\theta(y) \in \mathcal{D}$ and $\theta(y)$ have the same properties as Φ compared to y and we have:

$$\frac{d\theta(y)}{dy} = \langle T(x), \frac{\partial\Phi(x,y)}{\partial y}\rangle. \qquad [3.46]$$

PROPERTIES 3.2.–

– $T(x) \otimes S(y) = S(y) \otimes T(x)$.

– The support for the tensor product is the product of supports of S and T.

– This product is generalized to any number of distributions.

– $\forall \Phi(x,y) \in \mathcal{D}$, then $\delta(x) \otimes \delta(y) \otimes \delta(z) = \delta(x,y,z)$. Indeed:

$$\begin{aligned} \langle \delta(x) \otimes \delta(y) \otimes \delta(z), \Phi(x,y,z) \rangle &= \langle \delta(x) \otimes \delta(y), \langle \delta(z), \Phi(x,y,z) \rangle \rangle \\ &= \langle \delta(x) \otimes \delta(y), \Phi(x,y,0) \rangle \\ &= \Phi(0,0,0) = \langle \delta(x,y,z), \Phi(x,y,z) \rangle . \end{aligned} \quad [3.47]$$

3.3. Convolution

3.3.1. *Definition and first properties*

Under the condition that it exists, the convolution product $h(x)$ of two functions $f(x)$ and $g(x)$ is defined by the following integral:

$$h(x) = \int_{-\infty}^{+\infty} f(t)g(x-t)dt = \int_{-\infty}^{+\infty} f(x-t)g(t)dt. \quad [3.48]$$

$h(x)$ is often denoted by $h(x) = f(x) \star g(x)$. This notation – abusive as it should – is sometimes dangerous for the user because one can be coerced into writing $h(a) = f(a) \star g(a)$ where a is a fixed point, which of course is an aberration. The generalization of this to n-dimensional convolution product is trivial.

Similarly, the convolution product of two distributions, denoted by $S \star T$, is defined by:

$$\forall \Phi \in \mathcal{D}, \langle S \star T, \Phi \rangle = \langle S(x) \otimes T(y), \Phi(x+y) \rangle . \quad [3.49]$$

This convolution product does not always have a meaning, however, we have the existence theorem: if A is the support of S and B that of T, let E be the support of $\Phi(x+y)$, the convolution product $S \star T$ has a meaning if $E \cap \times B$ is bounded. In particular, for the product $S \star T$ to exist, it suffices that one of the distributions S or T has bounded support. The bounded support distributions on the right (respectively, on the left) form a vector space denoted by $\mathcal{D}'_-$ (respectively, $\mathcal{D}'_+$) in which the convolution always has a meaning.

Let now n be a number of distributions, if any of the following properties is verified:

1) $(n-1)$ distributions are of bounded support;

2) all distributions belong to $\mathcal{D}'_-$ ($\mathcal{D}'_+$);

3) all convolution products exist in pairs;

then their convolution product exists and is associative.

PROPERTIES 3.3.–

– The convolution product is commutative: $f \star g = g \star f$.

– The convolution product is distributive with respect to addition: $f \star (g_1 + g_2) = f \star g_1 + f \star g_2$.

– If one of the functions is the identity, then:

$$1 \star f = \int_{-\infty}^{+\infty} f(t)dt \qquad [3.50]$$

$$1 \star f \star g = \int_{-\infty}^{+\infty} f(t) \star g(t)dt = \int_{-\infty}^{+\infty} f(t)dt \int_{-\infty}^{+\infty} g(t)dt. \qquad [3.51]$$

– $\delta \star f = f$. The Dirac distribution is the neutral element of the convolution. Moreover, if T is any distribution, $\delta \star T = T$.

– $\delta_a \star T(x) = \delta(x-a) \star T(x) = T(x-a)$. The convolution by a Dirac distribution in a amounts to perform a translation.

– $\delta^{(m)} \star T(x) = T^{(m)}(x)$.

– Consider $R = S \star T$. From $R' = \delta' \star R$, it is easy to see that $R' = \delta' \star S \star T = S' \star T = S \star T'$. The derivation of a convolution product amounts to derive any one distribution of this convolution product.

THEOREM 3.6.– Let T be a distribution and f be an indefinitely differentiable function:

– If $T \star f$ exists, $h(x) = T \star f = \langle T(t), f(x-t) \rangle$.

– If $f \in \mathcal{D}$, $T \star f$ has still a meaning and $h(x)$ is an indefinitely differentiable function $h^{(m)}(x) = \langle T(t), f^{(m)}(x-t) \rangle$. It is then said that the convolution regulates.

– The convolution by a distribution T is a continuous operation in $\mathcal{D}'$ within the meaning of the hypocontinuity: the sequence S_n converges in $\mathcal{D}'$ to S, if the supports $K_{S_n} \subset K$ are such that all the convolution products $S_n \star T$ have a meaning, then the sequence $S_n \star T$ converges in $\mathcal{D}$ to $S \star T$.

– If D is a linear differential operator with constant coefficients, then $D(S \star T) = D(S) \star T = S \star D(T)$.

3.3.2. *Convolution algebra and Green's function*

Consider a linear differential operator with constant coefficients D such as:

$$D = a_0 + a_1 \frac{d}{dx} + \cdots + a_n \frac{d^n}{dx^n}, \qquad [3.52]$$

we can write the equation $DX = B$ in the form:

$$\begin{aligned} DX &= a_0 X + a_1 \frac{dX}{dx} + \cdots + a_n \frac{d^n X}{dx^n} = B \\ &= \left[a_0\delta + a_1\delta' + \cdots + a_n\delta^{(n)}\right] \star X = B. \end{aligned}$$

If we note $D^\star = a_0\delta + a_1\delta' + \cdots + a_n\delta^{(n)}$, then $DX = B$ is written as $D^\star \star X = B$. More generally, any convolution operator $\mathcal{O}$ is given by:

$$\mathcal{O}^\star \star X = B, \qquad [3.53]$$

$\mathcal{O}^\star$ can be a differential operator, an integral operator or an operator that "mixes" differential operator and integral operator. It should be noted that this type of mixed operator may appear when dealing with problems of vibrating structures in contact with fluids.

To solve this equation, it is important to look for $\mathcal{O}^{\star^{-1}}$, inverse convolution of $\mathcal{O}^\star$, such that $\mathcal{O}^\star \star \mathcal{O}^{\star^{-1}} = \delta$. If we calculate the convolution inverse (the Fourier transform is one of the means of calculation), then X is given by $X = \mathcal{O}^{\star^{-1}} \star B$. For this equation to exist, it is necessary and sufficient that the convolution product $\mathcal{O}^{\star^{-1}} \star B$, thus that $\mathcal{O}^\star$, $\mathcal{O}^{\star^{-1}}$ and B belong to the same convolution algebra, for example $\mathcal{E}'$, set of the limited support distributions or $\mathcal{D}'_+$ ($\mathcal{D}'_-$). A very important property of this inverse convolution is that when it exists, it is unique.

CONVOLUTION OPERATORS IN PHYSICS.– When a physical system can be described by an operator that maps any excitation signal S a unique response R, we have $\mathcal{O}S = R$. The study of the system in question leads to the study of the operator $\mathcal{O}$. A system is linear if $\forall\lambda_1, \lambda_2$ constants, $\mathcal{O}(\lambda_1 S_1 + \lambda_2 S_2) = \lambda_1 R_1 + \lambda_2 R_2$. This system is continuous if for $S_k \overset{\mathcal{D}'}{\to} S$, $\mathcal{O}S_k \overset{\mathcal{D}'}{\to} \mathcal{O}S$. Virtually, all linear physical systems are continuous. On the other hand, this system is invariant by translation if $\mathcal{O}S(x) = R(x) \Rightarrow \mathcal{O}S(x-a) = R(x-a)$. A physical system that exhibits these properties is called a linear filter. We have the following theorem: for a linear system to be described by a convolution operator, it is necessary and sufficient that it be a linear filter.

GREEN'S FUNCTION.– When the properties of the physical system are no longer derived from experiment but from theoretical analysis, the relationship between the response and excitation is usually of the form:

$$\mathcal{O}R = S,$$

R and S are generally functions and $\mathcal{O}$ is a differential operator. If $\mathcal{O}$ is linear with constant coefficients, the system is a linear filter. The differential equation takes the form $\mathcal{O}\delta \star R = S$. The impulse response of the system is a convolution inverse of $\mathcal{O}\delta$ and is known as Green's function, denoted by G or Γ. More specifically, Green's function is a special case of the notion of elementary solution.

By definition, an elementary solution is the solution of a non-homogeneous differential equation whose second member is a Dirac function (or distribution). Naming such a solution γ:

$$\mathcal{O}\gamma(M) = \delta(M), \qquad [3.54]$$

where $\mathcal{O}$ is a differential operator. Physically, γ represents the response of the medium to a point source with unity amplitude. The importance of this function is that when it is known, the response of the medium is known for almost all excitation forces. As a matter of fact, to solve $\mathcal{O}p(M) = F(M)$, due to the properties of the Dirac distribution with regard to the convolution product $\star$, it is obvious that the solution is given by $p(M) = F(M) \star \gamma(M)$, assuming that this convolution product exists. This latter exists if at least one of the two functions is a bounded support function[3]. We can now verify that $F(M) \star \gamma(M)$ is the solution of the problem raised. By hypothesis, $\mathcal{O}p(M) = F(M)$, if $p(M) = F(M) \star \gamma(M)$, $\mathcal{O}\left(F(M) \star \gamma(M)\right) = F(M)$ and therefore $[\mathcal{O}\gamma(M)] \star F(M) = F(M)$, however $\mathcal{O}\gamma(M) = \delta(M)$. Thus, $\delta(M) \star F(M) = F(M)$. We can see that the solution to the problem is very simple to calculate. All the difficulty lies in the calculation of this elementary solution which is possible only in a number of extremely limited cases (mainly linear operators with constant coefficients). Note that the presence of boundaries further adds to the difficulty.

An elementary kernel is solution of:

$$\mathcal{O}_M\gamma(M, S) = \delta_S(M). \qquad [3.55]$$

Unlike elementary solutions, which are single variable functions, an elementary kernel is a function of two variables. The index that appears in the operator $\mathcal{O}_M$ indicates that the operator acts on the variable M. Physically, it represents the

3 If $F(M)$ represents a physical source, it is reasonable to assume that it is finite-dimensional.

response of the medium measured in M and to a source placed in S. Two elementary kernels (or two solutions) differ from each other only by a solution of the associated homogeneous problem (or general solution). It should be recalled that the general solution $u_g(M)$ of the differential operator $\mathcal{O}$ is the solution of $\mathcal{O}u_g(M) = 0$. In order to choose between several elementary kernels, a condition for the conservation of energy should be imposed in addition. This allows the Green kernel $G(M, S)$ to be defined as an elementary kernel which satisfies a condition of conservation of energy. Note that this notion of elementary kernel extends to integro-differential operators. Many authors mistake the concepts of Green's kernel and Green's function. This language misnomer – unfortunate – is due to the fact that when the Green kernel can be calculated, it typically takes the form $G(M, S) = G(|M - S|)$, where $|M - S|$ is the distance between M and S and does no longer appear as a function of one variable (the distance between the two points being considered).

3.4. Modal methods

Modal analysis and the spectral operator theory play a central role in numerous engineering problems. It is not possible here to develop a complete theory of modal methods, which can be found in more specialized books [FAH 07, GER 93, LES 88]. The concepts of eigenmodes and resonance modes have a particular significance in all acoustics problems of bounded media damped in stationary or transitory regime. With a very simple example, we will, after having defined these two modes categories, present their implementation as well as the advantages and disadvantages of the resulting decompositions.

3.4.1. *Eigenmodes of a conservative system*

The concept of eigenmode can be seen in two ways. In the first approach[4], the eigenmode is seen as the generalization to linear differential operators with constant coefficients of the notions of eigenvectors and eigenvalues of matrix analysis. In the second approach, the eigenmode is seen as an extension of the concept of general solution of a differential equation to the case of homogeneous boundary values problems (i.e. homogeneous differential equation and homogeneous boundary conditions)[5].

4 We are here talking about the spectral theory of self-adjoint operators in Hilbert spaces [ARV 00].

5 For example, the general solution $y_g(x)$ of the homogeneous differential Helmholtz equation $y''(x) + k^2 y(x) = 0$, where $k = \omega/c$ where $k = \omega/c$ is the wavenumber, ω is the pulse and c is the wave velocity, is given by a linear combination of complex exponentials $y_g(x) = A\exp(-\imath kx) + B\exp(+\imath kx)$. The Helmholtz equation is obtained by the time Fourier transform of d'Alembert's wave propagation equation $\partial^2 Y(x,t)/\partial x^2 +$

Direct and inverse Fourier transforms are given by:

$$f(\omega) = \int_{-\infty}^{+\infty} F(t)e^{\imath\omega t}dt, F(t) = \frac{1}{2\pi}\int_{-\infty}^{+\infty} f(\omega)e^{-\imath\omega t}d\omega \quad [3.56]$$

Note that here we apply Morse and Ingard's convention [MOR 68] for the inverse Fourier transform, which states that the integration path is not exactly the real axis but a parallel line thereto slightly above it, which amounts to considering that all the real poles of $f(\omega)$ are limits of complex poles with a negative imaginary part.

Next, we will look at the following homogeneous differential problem:[6]

$$y''(x) + k^2y(x) = 0, \forall x \in]0, L[, y(x) = 0, x = 0, x = L \quad [3.57]$$

The set of non-zero solutions $[y_n(k_n, x), k_n]$ of this problem, the eigenmodes, allows the expression of the general solution of the problem at the boundaries as a linear combination of elementary functions $y_g(x) = \sum_n A_n y_n(k_n, x)$. We will calculate these modes. Given $y_g(x) = A\exp(-\imath kx) + B\exp(+\imath kx)$, to calculate A and B we apply the boundary conditions. The first gives us $y_g(0) = A + B = 0$ whence $A = -B$. Let still $y_g(x) = 2\imath BA\sin(kx)$. The second gives $y_g(L) = 2\imath BA\sin(kL) = 0$. Since the interest is about non-zero solutions, $B \neq 0$ and we must have $\sin(kL) = 0$ that is $k_n = n\pi/L, n \in \boldsymbol{N}^*$. The eigenmodes are then $y_n(x) = A_n\sin(k_nx) = A_n\sin(n\pi/Lx)$. It is shown that these modes are orthogonal to the scalar product$\langle f, g\rangle = \int_0^L f(x)g^*(x)dx$:

$$\langle y_m, y_n^*\rangle = \int_0^L y_m(x)y_n(x)dx = A_mA_n\delta_{mn} \quad [3.58]$$

Since the eigenmodes are defined up to a multiplicative constant (since they are solutions of a homogeneous problem), the value of each constant A_m can be determined by imposing the unit norm on the modes, that is $\langle y_m, y_m^*\rangle = 1$, from which it is easy to obtain $A_m = \sqrt{2/L}$. The general solution is written as $y_g(x) = \sum_m C_m y_m(x)$ and *a priori* contains an infinite number of terms. The difference between the general solution of the single differential equation (in which there are simply two waves – one in forward direction and the other in backward direction) and that of the differential equation with boundary conditions is due to the fact that these involve a total reflection of waves at each end and therefore that each

$1/c^2\partial^2Y(x,t)/\partial t^2 = 0$; we have defined $y(x) = \mathcal{F}[Y(x,t)]$, and for convenience the dependency with respect to ω of $y(x)$ was omitted.

6 In acoustics, this corresponds to the problem of beam compression in steady-state vibration, of a string fixed at both ends or that of the vibrations of a column of air contained in a circular pipe of small diameter when compared to its length and open at both ends.

two-way wave is reflected an infinite number of times at each end. Eigenmodes form a complete orthonormal system of the space in which the solutions of the problem at the boundaries of the conservative system exist[7] and thus allow the calculation of all the solutions using series expansion of eigenmodes.

This notion of eigenmodes, which is strictly comparable with that of eigenvectors and eigenvalues in linear algebra[8], is an extraordinarily powerful tool for solving the problems of dynamics of elastic structures (or of proportional damping) and bounded acoustic media without dissipation. For example, if we have to solve the following boundary value problem:

$$y''(x) + k^2 y(x) = f(x), \forall x \in]0, L[, y(x) = 0, x = 0, x = L, \qquad [3.59]$$

where $f(x)$ is a source term. The solution is looked for in the form $y(x) = \sum_m g_m y_m(x)$ that is introduced in the previous system:

$$\left(\sum_m g_m y_m(x)\right)''(x) + k^2 \sum_m g_m y_m(x) = f(x), \forall x \in]0, L[,$$
$$\sum_m g_m y_m(0) = 0, \sum_m g_m y_m(L) = 0. \qquad [3.60]$$

The two boundary conditions are already satisfied because of the very definition of the eigenmodes. The first equation is transformed as follows $\sum_m g_m y_m(x)'' + k^2 \sum_m g_m y_m(x) = f(x)$, that is $\sum_m g_m \left(-k_m^2 + k^2\right) y_m(x) = f(x)$ because $y_m(x)''(x) + k_m^2 y_m(x) = 0$. Each member of the last equation above is multiplied by $y_n(x)$ and the scalar product is performed. It gives $\sum_m g_m \left(-k_m^2 + k^2\right) \langle y_m, y_n \rangle = \langle f, y_n \rangle$. Therefore, according to the definition of the eigenmodes[9] $\langle y_m, y_n \rangle = \delta_{mn}$. It is then obvious that $g_n \left(-k_n^2 + k^2\right) = \langle f, y_n \rangle$

7 As a warning, if the system is not conservative, the difficulties will start. According to the loss of energy at stake, the damping of the modes over time can be non-homogeneous and the concept of harmonic regime becomes much less adapted to describe the concept of mode.

8 If A is a square matrix and I is the identity matrix, the eigenvector u_n and the eigenvalue λ_n are the non-zero solutions of $Au_n = \lambda_n I u_n$.

9 This is where the great efficiency of these modal methods can be seen, which transform a differential problem extremely hard into a much simpler scalar problem to solve and valid for a very large number of second members. Obviously, the effort of calculating the modes in advance is something that has to be endured, but when seeking to solve the problem by a finite element-based method, eigenmodes solvers are implemented in almost all codes of finite elements commercially available.

and if $k \neq k_n$ (i.e. if the excitation frequency is not a natural frequency of the system) it yields:

$$y(x) = c^2 \sum_m \frac{\langle f, y_n \rangle}{\omega^2 - \omega_n^2} y_m(x). \qquad [3.61]$$

3.4.2. *Eigenmodes of a non-conservative system*

Any mechanical system that vibrates can reasonably, if its resonance frequencies are sufficiently spaced, be approached around each of its resonance frequencies by a system with one degree of freedom of the viscous mass/spring/damper kind. This system with one degree of freedom presents a damped resonance which is a periodic oscillation with exponential decay. The free motion $u(t)$ of this system is given by the solution of the differential equation $m\ddot{u}(t) + 2c\dot{u}(t) + ru(t) = 0$ where r is the stiffness of the spring, m is the mass of the oscillator and where $2c$ is the viscous damping coefficient. It is trivial to demonstrate[10] that the movement is of the form $u(t) = \exp(-\imath\omega t)$ with $\omega = \dot{\omega} + \imath\hat{\omega} = -\imath c/m \pm \sqrt{-c^2/m^2 + r/m}$. The oscillation pulse $\dot{\omega}$ is given, if the damping is low enough compared with the mass, by $\dot{\omega} = \sqrt{r/m}$. The exponential decay rate is given by the ratio $\hat{\omega} = -c/m$. It should be noted that a measure of this decay thus allows the damping of this system to be fully qualified.

When addressing dissipative problems, a number of problems appear. For example, it is easily understandable that for a non-conservative system, therefore damped, the eigenmodes can/must evolve over time and that the scalar product involves not only space, but also time. In addition, the very definition of the eigenmodes is questionable. We will introduce a new concept, the modes of resonances or generalized eigenmodes. Consider the following problem: let $\mathcal{O}$ be a linear differential operator that characterizes the mechanical behavior of the medium (the Laplacian in the case of fluids or any other operator in spatial partial derivatives). For any given boundary conditions (to simplify the notation, these are included in the definition of the operator) in harmonic regime, the eigenmodes of the problem $\mathcal{O}y + \omega^2 y = 0$ are defined as the discrete set of the pairs $((y_n, \omega_n)$, non-zero solutions of the homogeneous problem $\mathcal{O}y_n = -\omega_n^2 y_n$. If the considered system presents damping, in harmonic regime, this is usually taken into account by the introduction of a complex function that depends on the frequency (such as the surface impedance, the acoustic radiation and the viscous damping), and the operator which describes the behavior of the medium becomes dependent on the frequency $\mathcal{O}(\omega)$, it is then easily derived that the definition of the modes of the problem

---

10 The characteristic polynomial of the equation obtained by defining $u(t) = \exp(\alpha t)$ is of course given by $m\alpha^2 + 2c\alpha + r = 0$, whose roots are $\alpha_{1,2} = -c/m \pm \sqrt{c^2/m^2 - r/m}$.

$\mathcal{O}(\omega)y + \omega^2 y = 0$ becomes, because of the frequency dependence of $\mathcal{O}(\omega)$, significantly less natural.

3.4.2.1. *Eigenmodes-resonance modes*

Non-zero solutions of $\mathcal{O}(\omega)\tilde{y}_n + \tilde{\omega}_n^2\tilde{y}_n = 0$ are retained as definition of the eigenmodes $(\tilde{y}_n, \tilde{\omega}_n)$. It can be shown that the eigenmodes depend on the frequency[11] since we have $\tilde{y}_n = \tilde{y}_n(\omega)$ and $\tilde{\omega}_n = \tilde{\omega}_n(\omega)$. Eigenmodes are complex and are orthogonal with respect to the scalar product $\langle f, g\rangle$ defined in equation [3.58]. The difficulty is that these modes no longer correspond to physical behavior; in particular, these modes, depending on the frequency, are not those that can be observed during transients (such as a music note). Modes best suited to the description of transients should be introduced: resonance modes $(\hat{y}_n, \hat{\omega}_n)$ which are complex non-zero solutions[12] of $\mathcal{O}(\hat{\omega}_n)\hat{y}_n + \hat{\omega}_n^2\hat{y}_n = 0$. Resonance modes are also called complex eigenmodes or generalized eigenmodes. The eigenmodes and the resonance modes are related by the following relations:

$$\hat{y}_m = \tilde{y}_m(\hat{\omega}_m), \hat{\omega}_m^2 = \tilde{\omega}_m^2(\hat{\omega}_m). \qquad [3.62]$$

Generally, the calculation of the eigenmodes presents no insurmountable difficulty. The previous relations show that the crux of the problem lies in the calculation of the resonance pulsations (the zeros of the function $f(\omega) = \omega^2 - \tilde{\omega}_m^2(\omega)$). Once they are known, it suffices to calculate the eigenmode of the resonance frequency mode to obtain the corresponding resonance mode. It can be shown that resonance pulsations $\hat{\omega}_m$ come up in pairs with the same negative imaginary part (for a time dependence such as $\exp(-\imath\omega t)$, this ensures the decay of the modes over time) and opposite real parts[13]. The resonance modes are not orthogonal with respect to the scalar product $\langle f, g\rangle$; this is normal since these modes represent damping modes and their deformation evolves over time, which a purely scalar product space is not able to properly describe. In fact, an orthogonality relation may be established between resonance modes (it will be carried out further, equation [3.77]) which depends on resonance pulsations[14]. Even if this relation is not usable to build expansions in series of resonance modes, it is essential for

11 And thereby, eigenmodes must be recalculated for each frequency; note, however, that in numerous practical cases of slightly damped modes, the frequency dependence of the eigenmodes can be ignored.

12 The calculation of resonance modes is a lot (really a lot) more difficult than that of eigenmodes as the dependence of the operator $\mathcal{O}$ with regard to the frequency is in general nonlinear.

13 It is referred to as anti-Hermitian symmetry: the resonance pulsations $\hat{\omega}_m^{\pm}$ are such that $\hat{\omega}_m^{-} = -\overline{\hat{\omega}_m^{+}}$, where $\overline{z}$ is the conjugate of z.

14 In some way, it allows the introduction of the time dependence of the modes in the orthogonality relation.

establishing a norm of these modes (see equation [3.78]). However, under certain conditions (for example, when the dependence with respect to the frequency of the differential operator that governs the behavior of the medium is of the viscous kind), it may be established that these modes be orthogonal with regard to a particular scalar product.

3.4.2.2. *Series expansion of resonance modes*

The difficulty in defining a scalar product is very troublesome because in fact, it is not possible to build series expansions of resonance modes as easily as for eigenmodes. Nevertheless, through a procedure based on a double temporal Fourier transform, this difficulty can be overcome. The idea is to start from the series expansion of eigenmodes in harmonic regime, to go back to time regime through a Fourier transform which involves the resonance modes that are independent of the time and then to return to the frequency domain by a new Fourier transform of the series expansion of the resonance modes to achieve a series expansion of resonance modes in harmonic regime. If the procedure may seem slightly cumbersome, it is relatively easy to implement in practice. To solidify the concepts, we will calculate the solution $y(\omega, x)$ of the problem at the boundaries according to $\mathcal{O}(\omega)y(\omega, x) + \omega^2 y(\omega, x) = f(\omega, x)$. The objective is to find $y(\omega, x)$ in the form of an expansion series of the eigenmodes $\tilde{y}_m(\omega, x)$, with norm[15] $N_m(\omega)$ in the form $y(\omega, x) = \sum_m g_m \tilde{y}_m(\omega, x)$, which is carried forward in the above equation. It yields:

$$\sum_m g_m \left(\mathcal{O}(\omega)\tilde{y}_m(\omega, x) + \omega^2 \tilde{y}_m(\omega, x)\right) = f(\omega, x). \qquad [3.63]$$

The dot product of each member of this equation is carried out with $\tilde{y}_n(\omega, x)$; the orthogonality relation $\langle \left(\mathcal{O}(\omega) + \omega^2\right) \tilde{y}_m(\omega), \tilde{y}_n(\omega)\rangle = N_m(\omega) \left(\tilde{\omega}_m^2(\omega) - \omega^2\right) \delta_{mn}$, where $N_m(\omega)$ plays the role of a norm, gives:

$$y(\omega, x) = \sum_m \frac{\langle f(\omega), \tilde{y}_m(\omega)\rangle}{\omega^2 - \tilde{\omega}_m^2(\omega)} \frac{\tilde{y}_m(\omega, x)}{N_m(\omega)}. \qquad [3.64]$$

Through inverse Fourier transform:

$$Y(x, t) = \sum_m \frac{1}{2\pi} \int_{-\infty}^{+\infty} \frac{\langle f(\omega), \tilde{y}_m(\omega)\rangle}{\omega^2 - \tilde{\omega}_m^2(\omega)} \frac{\tilde{y}_m(\omega, x)}{N_m(\omega)} \mathrm{e}^{-\imath\omega t} d\omega \qquad [3.65]$$

The integrand is a meromorphic function and the explicit calculation of this integral can be done with the residue theorem on one or the other contour $C^{\pm}$

15 This norm generally depends on the pulse and we will see later with an example how to calculate it.

consisting of the real straight line limited by $\pm R$ and the half-circle located in the upper or lower half-plane (see Figure 3.1). To simplify the calculations a bit, it can be assumed that the source term is independent of the frequency[16] $f(\omega, x) = f(x)$ which corresponds to the calculation of the impulse response of the system.

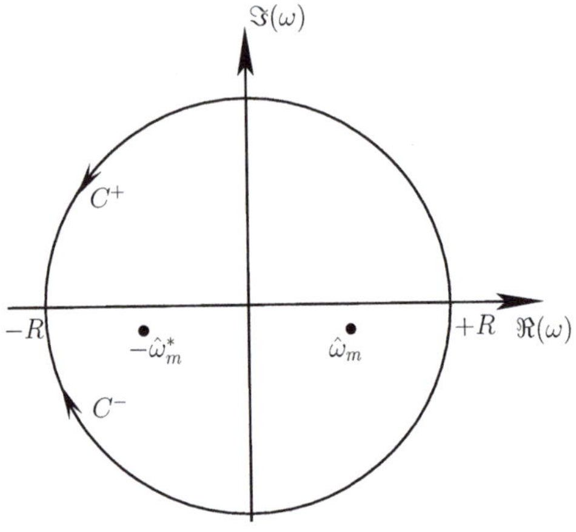

Figure 3.1. *Integration contour*

A priori, it is permissible to integrate on one or the other of the contours $C^{\pm}$, or even on any contour that contains the real axis. It then yields:

$$\int_{C^\pm} \frac{\langle f, \tilde{y}_m(\omega)\rangle}{\omega^2 - \tilde{\omega}_m^2(\omega)} \frac{\tilde{y}_m(\omega, x)}{N_m(\omega)} \mathrm{e}^{-\imath\omega t} = \pm 2\imath\pi \sum_{res(C^\pm)} \frac{\langle f, \tilde{y}_m(\omega)\rangle}{\omega^2 - \tilde{\omega}_m^2(\omega)} \frac{\tilde{y}_m(\omega, x)}{N_m(\omega)} \mathrm{e}^{-\imath\omega t} \quad [3.66]$$

where $\sum_{res(C^\pm)}$ represents the sum extended to all residues of the integrand in the contour being considered. We get a + sign when integrating over C^+ and a $-$ sign when integrating over C^- because we are turning in the inverse direction of the trigonometric sense. Nonetheless, to ensure the method is beneficial for our problem, the contour that allows, through Jordan's lemma, the cancelation of the integral over the semi-circle when switching to the limit $R \to \infty$ should be chosen. The choice of the contour is very simple. In order to keep a finite value in the exponential and to be able to apply Jordan's lemma (zero integrand at infinity on $C^{\pm}$), it is necessary to integrate on C^+ for $t < 0$ and on C^- for $t > 0$. We will now examine the zeros of the denominator of the integrand. For each mode m, these roots are the zeros of the function $\omega^2 - \tilde{\omega}_m^2(\omega)$, that is to say the resonance pulses $\hat{\omega}_m$. These roots come up by pairs with the same negative imaginary part, the contour C^+ does, therefore, not contain any pole and thus, for $t < 0$, we get $Y(x, t) = 0$ that satisfies correctly the

16 For example, for a source such as $\mathcal{F}[\delta(t)] = 1$.

causality principle. For $t > 0$, we integrate over C^- which contains the resonance pulsations, and we get:

$$Y(x,t) = -H(t)\imath \sum_m \left[\frac{\langle f, \tilde{y}_m(\hat{\omega}_m)\rangle}{2\hat{\omega}_m\left(1-\tilde{\omega}'_m(\hat{\omega}_m)\right)} \frac{\tilde{y}_m(\hat{\omega}_m, x)}{N_m(\hat{\omega}_m)} \mathrm{e}^{-\imath\hat{\omega}_m t} \right.$$
$$\left. + \frac{\langle f, \tilde{y}_m(-\hat{\omega}^*_m)\rangle}{-2\hat{\omega}^*_m\left(1+\tilde{\omega}'_m(-\hat{\omega}^*_m)\right)} \frac{\tilde{y}_m(-\hat{\omega}^*_m, x)}{N_m(-\hat{\omega}^*_m)} \mathrm{e}^{\imath\hat{\omega}^*_m t} \right], \qquad [3.67]$$

where $\tilde{\omega}'_m(\omega)$ is the pulse with regard to the derivative of $\tilde{\omega}_m(\omega)$ and $H(t)$ is the Heaviside step. However, by definition, an eigenmode calculated at the resonance frequency is a resonance mode $\hat{y}_m(x) = \tilde{y}_m(\hat{\omega}_m)(x)$ and $-\hat{y}^*_m(x) = \tilde{y}_m(-\hat{\omega}^*_m)(x)$. A resonance mode decomposition in series is obtained in temporal regime, with $\tilde{\omega}'_m(-\hat{\omega}^*_m) = -\tilde{\omega}'^*_m(\hat{\omega}_m)$, which can be demonstrated as being real[17]:

$$Y(x,t) = -H(t)\imath \sum_m \left[\frac{\langle f, \hat{y}_m\rangle}{2\hat{\omega}_m\left(1-\tilde{\omega}'_m(\hat{\omega}_m)\right)} \frac{\hat{y}_m(x)}{N_m(\hat{\omega}_m)} \mathrm{e}^{-\imath\hat{\omega}_m t} \right.$$
$$\left. - \frac{\langle f, \hat{y}^*_m(x)\rangle}{2\hat{\omega}^*_m\left(1-\tilde{\omega}'^*_m(\hat{\omega}_m)\right)} \frac{\hat{y}^*_m(x)}{N_m(-\hat{\omega}^*_m)} \mathrm{e}^{\imath\hat{\omega}^*_m t} \right]. \qquad [3.68]$$

We will write $\hat{\omega}_m = \Omega_m - \imath\alpha_m$ and $-\hat{\omega}^*_m = -\Omega_m - \imath\alpha_m$. The time series [3.68] is then written:

$$Y(x,t) = -H(t)\imath \sum_m \left[\frac{\langle f, \hat{y}_m\rangle}{2\hat{\omega}_m\left(1-\tilde{\omega}'_m(\hat{\omega}_m)\right)} \frac{\hat{y}_m(x)}{N_m(\hat{\omega}_m)} \mathrm{e}^{-\imath\Omega_m t} \right.$$
$$\left. - \frac{\langle f, \hat{y}^*_m(x)\rangle}{2\hat{\omega}^*_m\left(1-\tilde{\omega}'^*_m(\hat{\omega}_m)\right)} \frac{\hat{y}^*_m(x)}{N_m(-\hat{\omega}^*_m)} \mathrm{e}^{\imath\Omega_m t} \right] \mathrm{e}^{-\alpha_m t}, \qquad [3.69]$$

which shows that a behavior is obtained as the superposition of a series of oscillations (terms $\mathrm{e}^{\pm\imath\Omega_m t}$) that are damped (term $\mathrm{e}^{-\alpha_m t}$). In order to obtain a harmonic series decomposition of resonance modes (decomposition that cannot be obtained directly since there is no scalar product in harmonic regime available), it suffices to just take

17 It can be seen that the localization of the resonances in the complex plane is primordial. They must all have negative imaginary parts (for the time dependence that we have chosen in this chapter) which ensures causality. In addition, the Hermitian symmetry of resonances on the imaginary axis, and the fact that all invoked functions satisfy $F(-\omega^*) = -\left(F(\omega)\right)^*$, allows a real solution to be obtained because if $z = a + \imath b$, then $z - z^* = 2\imath b$.

the time Fourier transform of relation [3.69]. This is trivial since only the exponentials depend explicitly on time:

$$\int_{-\infty}^{+\infty} H(t)\mathrm{e}^{-\imath\Omega_m t}\mathrm{e}^{-\alpha_m t}\mathrm{e}^{\imath\omega t}dt = \int_0^{+\infty} \mathrm{e}^{(-\imath(\Omega_m-\omega)+\alpha_m)t}dt$$
$$= \frac{1}{\imath\,(\Omega_m-\omega)+\alpha_m}.$$

We then get the harmonic series decomposition of resonance modes:

$$y(x,\omega) = -\imath\sum_m \left[\frac{\langle f,\hat{y}_m\rangle}{2\hat{\omega}_m\left(1-\tilde{\omega}'_m(\hat{\omega}_m)\right)}\frac{\hat{y}_m(x)}{N_m(\hat{\omega}_m)\left(\imath\,(\Omega_m-\omega)+\alpha_m\right)}\right.$$
$$\left.+\frac{\langle f,\hat{y}^*_m(x)\rangle}{2\hat{\omega}^*_m\left(1-\tilde{\omega}'_m(\hat{\omega}^*_m)\right)}\frac{\hat{y}^*_m(x)}{N_m(-\hat{\omega}^*_m)\left(\imath\,(\Omega_m+\omega)-\alpha_m\right)}\right]. \quad [3.70]$$

3.4.2.3. *Damped beam*

In order to consolidate these ideas, consider the case of a beam of length L in viscosity damped compression fixed at both ends. It is assumed that the beam is damped by viscosity that is most simply modeled in the form of a constant damping η whose dimension is given in kg/(s.m^3). The differential operator that governs this damping is given by $\eta\partial U(x,t)/\partial t = \eta\dot{U}(x,t)$. The equation of motion of the beam, subject to homogeneous initial conditions, is then given, if E is Young's modulus of the beam and ρ is its density, by:

$$-E\frac{\partial^2 U(x,t)}{\partial x^2} + \rho\ddot{U}(x,t) + \eta\dot{U}(x,t) = -F(x,t),$$
$$U(0,t) = 0, U(l,t) = 0, \forall t, U(x,0) = 0, \dot{U}(x,0) = 0, \forall x.$$

The equations in a harmonic regime are obtained by replacing $U(x,t)$ with $u(x,\omega)\exp(-\imath\omega t)$ in the previous system. After derivation with respect to time and simplification by $\exp(-\imath\omega t)$, we obtain:

$$Eu''(x,\omega) + \rho\omega^2 u(x,\omega) + \imath\omega\eta u(x,\omega) = f(x,\omega), u(0,\omega) = 0, u(L,\omega) = 0.$$

3.4.2.4. *Eigenmodes and resonance modes*

Define $\breve{\rho}(\omega) = \rho(1+\epsilon\imath/\omega)$, with $\epsilon = \eta/\rho$. The eigenmodes $\tilde{u}_m(x,\omega)$ and the eigenpulses $\tilde{\omega}_m(\omega)$ are the non-zero solutions of:

$$E\tilde{u}''_m(x,\omega) + \breve{\rho}(\omega)\tilde{\omega}^2_m(\omega)\tilde{u}_m(x,\omega) = 0, \tilde{u}_m(0,\omega) = 0, \tilde{u}_m(l,\omega) = 0.$$

It is easy to show that it yields:

$$\tilde{u}_m(x,\omega) = \sqrt{\frac{2}{L}}\sin\frac{m\pi}{L}x, \tilde{\omega}_m(\omega) = \frac{m\pi}{L}\sqrt{\frac{E}{\rho}}\frac{1}{\sqrt{1+\epsilon\frac{\imath}{\omega}}}. \qquad [3.71]$$

The resonance modes $\hat{u}_m(x)$ and the resonance pulsations $\hat{\omega}_m$ are the solutions of:

$$E\hat{U}_m''(x) + \rho\hat{\omega}_m^2\left(1+\epsilon\frac{\imath}{\hat{\omega}_m}\right)\hat{u}_m(x) = 0, \hat{u}_m(0) = 0, \hat{u}_m(l) = 0.$$

We know that we obtain $\hat{u}_m(x) = \tilde{u}_m(x,\hat{\omega}_m)$ and $\hat{\omega}_m^2 = \tilde{\omega}_m^2(\hat{\omega}_m)$. Since the eigenmodes are not dependent on the frequency, they are coinside with the resonance modes. The resonance pulsations are the solutions of $\hat{\omega}_m^2 = \tilde{\omega}_m^2(\hat{\omega}_m)$, that is $\hat{\omega}_m^2 = (m\pi/L)^2(E/\rho 1)/(1+\epsilon\imath/\hat{\omega}_m)$. This leads to solve the equation $\rho/E\hat{\omega}_m^2 + \imath\epsilon\rho/E\hat{\omega}_m - (m\pi/L)^2 = 0$. This quadratic equation has two roots in $\hat{\omega}_m$:

$$\hat{\omega}_m = -\frac{\imath\epsilon}{2} \pm \sqrt{\frac{E}{\rho}\left(\frac{m\pi}{l}\right)^2 - \left(\frac{\epsilon}{2}\right)^2} \qquad [3.72]$$

The anti-Hermitian symmetry of the roots with respect to the imaginary axis is recovered and ensures the causality of the system and a real response. The two effects of dissipation can be noted: the addition of an imaginary part which characterizes the damping and the decrease in the real part that characterizes either an added mass effect or a decrease in stiffness.

3.4.2.4.1. Norm and scalar product

If, at first glance, the scalar product $\langle f, g\rangle$ seems to be the natural scalar product and adapted to the problem, we will see that this is not, in general, the case for resonance modes.

First, we will start again with the eigenvalue equation $Eu''(x,\omega) + \rho\omega^2(1+\epsilon\imath/\omega)u(x,\omega) = 0$, having *ad hoc* boundary conditions that we multiply member-wise by a function $v(x)$ that satisfies the boundary conditions and that we integrate over the length of the beam $E\int_0^L u''(x)v(x)dx + \rho\omega^2(1+\epsilon\imath/\omega)\times \int_0^L u(x)v(x)dx = 0$. The first integral is integrated by parts. The fully integrated terms cancel out due to the boundary conditions and we obtain $-E\int_0^L u'(x)v'(x)dx + \rho\omega^2(1+\epsilon\imath/\omega)\int_0^L u(x)v(x)dx = 0$. This equation (which is none other than the weak formulation of the homogeneous problem) is also written in the following general form:

$$a(u,v^*) + \rho\omega^2\left(\langle u,v^*\rangle + \epsilon\Gamma_\omega(u,v^*)\right) = 0 \qquad [3.73]$$

where $a(u, v^*) = -E \int_0^L u'(x)v'(x)dx$ is the potential elastic energy of the beam, the kinetic energy of the beam is $\rho\omega^2\langle u, v^*\rangle = \omega^2\rho \int_0^L u(x)v(x)dx$ and the energy of the beam lost by viscosity is given by $\omega^2\rho\epsilon\Gamma_\omega(u, v^*) = \rho\omega^2\epsilon\imath/\omega \int_0^L u(x)v(x)dx$. Eigenvalue equation [3.73] satisfies, for any $v(x)$,

$$a(\tilde{u}_m, v^*) + \tilde{\omega}_m^2(\omega)\left(\rho\langle\tilde{u}_m, v^*\rangle + \rho\epsilon\Gamma_\omega(\tilde{u}_m, v^*)\right) = 0. \quad [3.74]$$

In particular, if we take for $v(x)$ an eigenmode $\tilde{u}_n(x)$, we get:

$$a(\tilde{u}_m, \tilde{u}_n^*) + \tilde{\omega}_m^2(\omega)\left(\rho\langle\tilde{u}_m, \tilde{u}_n^*\rangle + \rho\epsilon\Gamma_\omega(\tilde{u}_m, \tilde{u}_n^*)\right) = 0.$$

From there, it is trivial to see that there are two relations:

$$a(\tilde{u}_m, \tilde{u}_n^*) + \tilde{\omega}_m^2(\omega)\left(\rho\langle\tilde{u}_m, \tilde{u}_n^*\rangle + \rho\epsilon\Gamma_\omega(\tilde{u}_m, \tilde{u}_n^*)\right) = 0$$
$$a(\tilde{u}_n, \tilde{u}_m^*) + \tilde{\omega}_n^2(\omega)\left(\rho\langle\tilde{u}_n, \tilde{u}_m^*\rangle + \rho\epsilon\Gamma_\omega(\tilde{u}_n, \tilde{u}_m^*)\right) = 0$$

and by difference, knowing that there are symmetries $a(\tilde{u}_m, \tilde{u}_n^*) = a(\tilde{u}_n, \tilde{u}_m^*)$, $\langle\tilde{u}_m, \tilde{u}_n^*\rangle) = \langle\tilde{u}_n, \tilde{u}_m^*\rangle$ and $\Gamma_\omega(\tilde{u}_m, \tilde{u}_n^*) = \Gamma_\omega(\tilde{u}_n, \tilde{u}_m^*)$, we get, if $m \neq n$ (if $m = n$ we get $0 = 0$):

$$\left(\tilde{\omega}_m^2(\omega) - \tilde{\omega}_n^2(\omega)\right)\left(\rho\langle\tilde{u}_m, \tilde{u}_n^*\rangle + \rho\epsilon\Gamma_\omega(\tilde{u}_m, \tilde{u}_n^*)\right) = 0.$$

When modes are not degenerate, that is if $\tilde{\omega}_m^2(\omega) \neq \tilde{\omega}_n^2(\omega)$, the orthonormality relation of the eigenmodes is obtained:

$$\rho\langle\tilde{u}_m, \tilde{u}_n^*\rangle + \rho\epsilon\Gamma_\omega(\tilde{u}_m, \tilde{u}_n^*) = N_m(\omega)\delta_{mn}, \quad [3.75]$$

where $N_m(\omega)$ plays the role of a norm that depends on frequency. It can be observed that in the general case this relationship depends on the frequency by means of the term $\Gamma_\omega(u, v^*)$ and therefore the norm of the modes depends in general on frequency. In the case that matters to us here, the dissipation term has a trivial expression and the value of the eigenmodes is defined to be unity, the norm is written: $N_m(\omega) = \rho(1 + \imath\epsilon/\omega)\langle\tilde{u}_m, \tilde{u}_m^*\rangle$.

For the resonance modes, the same cannot be said. The equation satisfied by the resonance modes is, for any $v(x)$ which satisfies the boundary conditions, given by:

$$a(\hat{u}_m, v^*) + \hat{\omega}_m^2\left(\rho\langle\hat{u}_m, v^*\rangle + \rho\epsilon\Gamma_{\hat{\omega}_m}(\hat{u}_m, v^*)\right) = 0. \quad [3.76]$$

From there, it is trivial to see that there are two relations:

$$a(\hat{u}_m, \hat{u}_n^*) + \hat{\omega}_m^2\left(\rho\langle\hat{u}_m, \hat{u}_n^*\rangle + \rho\epsilon\Gamma_{\hat{\omega}_m}(\hat{u}_m, \hat{u}_n^*)\right) = 0$$
$$a(\hat{u}_n, \hat{u}_m^*) + \hat{\omega}_n^2\left(\rho\langle\hat{u}_n, \hat{u}_m^*\rangle + \rho\epsilon\Gamma_{\hat{\omega}_n}(\hat{u}_n, \hat{u}_m^*)\right) = 0$$

whose difference is taken. We obtain, as for eigenmodes, the symmetries $a(\hat{u}_m, \hat{u}_n^*) = a(\hat{u}_n, \hat{u}_m^*)$ and $\langle \hat{u}_m, \hat{u}_n^* \rangle = \langle \hat{u}_n, \hat{u}_m^* \rangle$. On the other hand, unlike the eigenmodes, we generally have $\Gamma_{\hat{\omega}_m}(\hat{u}_m, \hat{u}_n^*) \neq \Gamma_{\hat{\omega}_n}(\hat{u}_n, \hat{u}_m^*)$. If $m \neq n$ and if the modes are non-degenerate, the orthogonality relationship of the resonance modes is obtained:

$$(\hat{\omega}_m + \hat{\omega}_n)\,\rho\langle \hat{u}_m, \hat{u}_n^* \rangle + \rho\epsilon \frac{\hat{\omega}_m^2 \Gamma_{\hat{\omega}_m}(\hat{u}_m, \hat{u}_n^*) - \hat{\omega}_n^2 \Gamma_{\hat{\omega}_n}(\hat{u}_n, \hat{u}_m^*)}{\hat{\omega}_m - \hat{\omega}_n} = 0, m \neq n. \quad [3.77]$$

In practice, this orthogonality relationship is not very useful because it depends on the resonance pulses. Nevertheless, it is useful to calculate the norm of the resonance modes. Starting from this relation and defining $\delta\hat{\omega} = \hat{\omega}_m - \hat{\omega}_n$ [ADH 00], we obtain:

$$(\delta\hat{\omega} + 2\hat{\omega}_n)\,\rho\langle \hat{u}_m, \hat{u}_n^* \rangle + \rho\epsilon \frac{(\delta\hat{\omega} + \hat{\omega}_n)^2\, \Gamma_{\delta\hat{\omega}+\hat{\omega}_n}(\hat{u}_m, \hat{u}_n^*) - \hat{\omega}_n^2 \Gamma_{\hat{\omega}_n}(\hat{u}_n, \hat{u}_m^*)}{\delta\hat{\omega}} = 0.$$

For $\hat{\omega}_m \to \hat{\omega}_n$, we have, by passing to the limit $\delta\hat{\omega} \to 0$, and since the second term can be recognized as a derivative with respect to the pulsations; therefore it yields:

$$2\rho\hat{\omega}_n \langle \hat{u}_n, \hat{u}_n^* \rangle + \rho\epsilon \left[\frac{\partial \left(\omega^2 \Gamma_\omega(\hat{u}_n, \hat{u}_n^*)\right)}{\partial\omega} \right]_{\omega=\hat{\omega}_n} = \theta_n$$

$$\Rightarrow 2\rho\hat{\omega}_n \langle \hat{u}_n, \hat{u}_n^* \rangle + \rho\epsilon \left(2\hat{\omega}_n \Gamma_{\hat{\omega}_n}(\hat{u}_n, \hat{u}_n^*) + \hat{\omega}_n^2 \Gamma'_{\hat{\omega}_n}(\hat{u}_n, \hat{u}_n^*)\right) = \theta_n \quad [3.78]$$

where θ_n is a constant to be chosen. The natural choice is to take $\theta_n = 2\rho\hat{\omega}_n$ to obtain the usual normalization condition $\langle \hat{u}_n, \hat{u}_n^* \rangle = 1$ when damping is zero. Naturally, other choices are possible. In the case of the beam damped by viscosity, we have $\Gamma_\omega(u, v^*) = \imath/\omega \langle u, v^* \rangle$ thus, the orthogonality relation of the resonances [3.77] is reduced to the usual relation $\rho(\Re(\hat{\omega}_n) + \Re(\hat{\omega}_m))\langle \hat{u}_m, \hat{u}_n^* \rangle = 0$, if $m \neq n$ and the norm relation [3.78] is written $2\rho\Re(\hat{\omega}_n)\langle \hat{u}_n, \hat{u}_n^* \rangle = \theta_n$, which shows that the natural choice here is $\theta_n = 2\rho\Re(\hat{\omega}_n)$.

whose difference is taken. We obtain, as for eigenmodes, the symmetries [illegible] and [illegible]. On the other hand, unlike the eigenmodes, we [illegible]

[illegible]

4

Fluid Acoustics

This chapter is devoted to the study of acoustic wave propagation in infinite fluid media. We begin by describing the establishment of propagation equations and then we move on to find solutions to these equations. This should allow us to clearly highlight the various types of waves which can propagate in fluid media. After presenting the classic results for homogeneous, isotropic and other fluids, we present, at the end of this chapter, some effects related to inhomogeneity, anisotropy and motions of the medium. For the mathematical problem that models sound propagation to be posed correctly, when the propagation domain presents obstacles, we must add boundary conditions to the wave propagation equation. If the domain is infinite, in order to satisfy the conservation of energy principle, we impose "outgoing wave"-type conditions. This implies that no energy comes from infinity. Especially when we calculate Green's functions, among all the elementary solutions, we will choose those that satisfy this condition. For example, in the two-dimensional case, for a time dependence such as the one we have chosen, the basic solution is a Hankel's function of order zero of the first kind; if we had chosen a time dependence of $\mathrm{e}^{+\imath\omega t}$, we would have to choose a Hankel's function of the second kind.

In order for us to model propagation in a real medium, we must distinguish two cases. The first case is at the interface of two propagating media, and the second case is, of course, at the interface of a propagating medium and a non-propagating medium. In fact, the distinction is slightly more subtle, since in the first case we focus only on propagating media of the same type (two fluids), while in most cases, the physical problem is the propagation between two propagating media of different types, typically a fluid and an elastic solid. The complexity of these phenomena of vibrational coupling between the waves with different behaviors induces a significant comprehension difficulty. For example, in acoustics of buildings, one of the major problems is the acoustic insulation of rooms. The sources of noise are generally air sources (road noise, music and talking). Sound propagates into the room and seemingly "crosses" the walls, spreading out to neighboring rooms. In fact, the wall (or the floor) begins to vibrate because of the acoustic wave and acts as a loudspeaker membrane, making the air in the room vibrate. It is essential to consider the whole problem in order to understand the physics of the phenomenon and to deal with the nuisance efficiently. With respect to what is of interest to us, the two cases mentioned at the beginning of the paragraph allow us to simply process a significant number of phenomena and to identify essential ideas.

4.1. Acoustics equations

In order for us to establish acoustics equations, we must return to the conservation equations (mass, impulse and energy) within which we introduce a small perturbation. If the amplitude of the perturbation is small, it is possible to linearize the previous equations around a state of equilibrium. It is there that we derive our definition of acoustics:

"Acoustics is the motion which results from the first order of the conservation equations".

4.1.1. *Conservation equations*

We examine three conservation equations: for mass, impulse and energy. As previously discussed, if ρq_m is the mass source, the mass conservation equation can be written as $\partial\rho/\partial t + \text{div}\,(\rho\vec{v}) = \rho q_m$. With $\text{div}\,(\rho\vec{v}) = \rho\text{div}\,(\vec{v}) + \vec{v}\cdot\overrightarrow{\text{grad}}\rho$ and the material derivative $d/dt = \partial/\partial t + \vec{v}\cdot\overrightarrow{\text{grad}}$, we obtain $\partial\rho/\partial t + \vec{v}\cdot\overrightarrow{\text{grad}}(\rho) + \rho\text{div}\,(\vec{v}) = \rho q_m$. Similarly, if $\rho\vec{f_v}$ are the volume forces, the impulse conservation equation (or Newton's law) can be given by $\rho\left(\partial\vec{v}/\partial t + \vec{v}\cdot\overrightarrow{\text{grad}}\,(\vec{v})\right) + \overrightarrow{\text{grad}}(p) - \overrightarrow{\text{grad}}\,(\vec{\tau}) = \rho\vec{f_v}$, where $\vec{\tau}$ is the viscous stress tensor whose components are $\tau_{ij} = \mu(\partial u_i/\partial x_j + \partial u_j/\partial x_i) - 2/3\mu\partial u_k/\partial x_k\delta_{ij}$. μ is the dynamic viscosity ($\mu = \rho\nu$, ν: kinematic viscosity). Finally, if ρq_e represents the sources of heat (for example, due to combustion) and if Φ stands for the dissipation function by the viscous stresses ($\Phi = \tau_{ij}\partial v_i/\partial x_j \geq 0$), the energy conservation equation (with the entropy s) can be given by $\rho T\left(\partial s/\partial t + \vec{v}\cdot\overrightarrow{\text{grad}}(s)\right) = \text{div}(\vec{q}) + \rho q_e + \Phi$, where $\vec{q}$ is the heat flux given by Fourier's law $\vec{q} = -k_\theta\overrightarrow{\text{grad}}(T)$, with k_θ as the thermal conductivity coefficient.

To these three equations we then add an equation of state of two independent thermodynamic variables ($p = p(\rho, s)$) to close the system $dp = (\partial p/\partial\rho)_s\,d\rho + (\partial p/\partial s)_\rho\,ds$. For an ideal gas, we have $dp = c_0^2 d\rho + p/C_v ds$, where C_v is the isochoric (at a constant volume) heat capacity. If the fluid is an ideal gas, the adiabaticity hypothesis leads to the barotropic state law $p = p_0 (\rho/\rho_0)^\gamma$. Since $\rho = \rho_0 + \rho'$ and $p = p_0 + p'$, the barotropic state law can be written as $1 + p'/p_0 = (1 + \rho'/\rho_0)^\gamma$. As the ratio ρ'/ρ_0 is hypothetically small, we have, by limited development, $p'/p_0 = \gamma\rho'/\rho_0$, but for an ideal gas Mariotte's law gives $p_0 = \mathrm{R}\rho_0 T_0$ from where it is easy to deduce that $c_0^2 = \gamma \mathrm{R} T_0$. In standard conditions (normal atmospheric pressure and 0°C), $\gamma = 1.41$, $\mathrm{R} = 286.9$, $T_0 = 273.15$ K $c = 331\text{m/s} \sim 20\sqrt{T}$. It should be noted that in the general case, we have $c_0^2 = (\partial p/\partial \rho)_s$.

4.1.2. *Establishment of general equations*

For an ideal fluid (not viscous) and non-conducting ($k_\theta = 0$), let us introduce at a point in space a small amplitude fluctuation of mass $dq_m = q'_m$ and heat $dq_e = q'_e$. This source will create movements with a small amplitude ($d\rho = \rho'$, $d\vec{v} = \vec{v}'$, $dp = p'$ and $dT = T'$) around a time-independent incompressible state of equilibrium (such that $\operatorname{div}(\vec{v}_0) = 0$), and we obtain $\rho = \rho_0 + \rho'$, $\vec{v} = \vec{v_0} + \vec{v}'$, $p = p_0 + p'$, $T = T_0 + T'$. We thus have

$$\frac{\partial(\rho_0 + \rho')}{\partial t} + (\vec{v_0} + \vec{v}') \cdot \overrightarrow{\operatorname{grad}}(\rho_0 + \rho') + (\rho_0 + \rho') \operatorname{div}(\vec{v_0} + \vec{v}') = (\rho_0 + \rho') q'_m$$

$$(\rho_0 + \rho')\left(\frac{\partial(\vec{v_0} + \vec{v}')}{\partial t} + (\vec{v_0} + \vec{v}') \cdot \overrightarrow{\operatorname{grad}}(\vec{v_0} + \vec{v}')\right) + \overrightarrow{\operatorname{grad}}(p_0 + p') = (\rho_0 + \rho') \vec{f_v}$$

$$\frac{\partial(s_0 + s')}{\partial t} + (\vec{v_0} + \vec{v}') \cdot \overrightarrow{\operatorname{grad}}(s_0 + s') = (\rho_0 + \rho') q'_e$$

$$p' = c_0^2 \rho' + \frac{p_0 + p'}{C_v} s'$$

As we are seeking to obtain equations of the first order, it is appropriate to develop various products and regroup terms of the same order. Let us take, for example, $(\vec{v_0} + \vec{v}') \cdot \overrightarrow{\operatorname{grad}}(\rho_0 + \rho') = \vec{v_0} \cdot \overrightarrow{\operatorname{grad}}\rho_0 + \vec{v}' \cdot \overrightarrow{\operatorname{grad}}\rho_0 + \vec{v_0} \cdot \overrightarrow{\operatorname{grad}}\rho' + \vec{v}' \cdot \overrightarrow{\operatorname{grad}}\rho'$, if we keep only the terms of the orders one and zero, we should eliminate $\vec{v}' \cdot \overrightarrow{\operatorname{grad}}\rho'$. Finally, we obtain two systems:

At order zero, for an ideal gas we obtain:

$$\underbrace{\frac{\partial \rho_0}{\partial t}}_{=0} + \vec{v}_0 \cdot \overrightarrow{\operatorname{grad}}(\rho_0) + \rho_0 \underbrace{\operatorname{div}(\vec{v}_0)}_{=0} = 0$$

$$\rho_0 \left(\underbrace{\frac{\partial \vec{v}_0}{\partial t}}_{=0} + \vec{v}_0 \cdot \overrightarrow{\operatorname{grad}}(\vec{v}_0) \right) + \overrightarrow{\operatorname{grad}}(p_0) = \rho_0 \vec{f}_v$$

$$\underbrace{\frac{\partial s_0}{\partial t}}_{=0} + \vec{v}_0 \cdot \overrightarrow{\operatorname{grad}}(s_0) = 0$$

At order one, we obtain:

$$\frac{\partial \rho'}{\partial t} + \rho_0 \text{div}\,(\vec{v}') + \vec{v}_0 \cdot \overrightarrow{\text{grad}}(\rho') + \rho' \underbrace{\text{div}\,(\vec{v}_0)}_{=0} + \vec{v}' \cdot \overrightarrow{\text{grad}}(\rho_0) = \rho_0 q_m' \qquad [4.1]$$

$$\rho_0 \left(\frac{\partial \vec{v}'}{\partial t} + \vec{v}_0 \cdot \overrightarrow{\text{grad}}\,(\vec{v}') + \vec{v}' \cdot \overrightarrow{\text{grad}}\,(\vec{v}_0) \right) + \rho' \vec{v}_0 \cdot \overrightarrow{\text{grad}}\,(\vec{v}_0) + \overrightarrow{\text{grad}}(p') = \rho' \vec{f}_v \qquad [4.2]$$

$$\rho_0 T_0 \left(\frac{\partial s'}{\partial t} + \vec{v}_0 \cdot \overrightarrow{\text{grad}}(s') + \vec{v}' \cdot \overrightarrow{\text{grad}}(s_0) \right) = \rho_0 q_e' \qquad [4.3]$$

$$p' - c_0^2 \rho' - \frac{p_0}{C_v} s' = 0 \qquad [4.4]$$

For any fluid, we must modify the equation of state [4.4]. To do this, we return to the state relationship $dp = (\partial p/\partial \rho)_s\, d\rho + (\partial p/\partial s)_\rho\, ds$ and we describe the conservation of entropy as $ds/dt = 0$:

$$\frac{dp}{dt} = \left(\frac{\partial p}{\partial \rho}\right)_s \frac{d\rho}{dt} + \left(\frac{\partial p}{\partial s}\right)_\rho \frac{ds}{dt} = c^2 \frac{d\rho}{dt},$$

$$\text{or} \quad \frac{\partial p}{\partial t} + \vec{v} \cdot \overrightarrow{\text{grad}}(p) = c^2 \left(\frac{\partial \rho}{\partial t} + \vec{v} \cdot \overrightarrow{\text{grad}}(\rho) \right),$$

which at order one gives (with $c = c_0 + c'$):

$$\frac{\partial p'}{\partial t} + \vec{v}_0 \cdot \overrightarrow{\text{grad}}(p') + \vec{v}' \cdot \overrightarrow{\text{grad}}(p_0) = c_0^2 \left(\frac{\partial \rho'}{\partial t} + \vec{v}_0 \cdot \overrightarrow{\text{grad}}(\rho') + \vec{v}' \cdot \overrightarrow{\text{grad}}(\rho_0) \right) + c'^2 \vec{v}_0 \cdot \overrightarrow{\text{grad}}(\rho_0) = 0.$$

In the following, so as to not lead to confusion, we will remove the "prime".

4.1.3. *Establishment of the wave equation*

Consider a homogeneous ideal gas at rest, where ρ_0 =cte, p_0 =cte, $\vec{v}_0 = 0$. In addition, for an ideal fluid, the viscosity does not intervene and $\tau_{ij} = 0$. We define $\rho \vec{f}_v = \vec{F}_v$ as the volume forces created by the density fluctuations which are exerted on the media. The equilibrium equations of order one [4.1], [4.2], [4.3] and [4.4] can be written as $\partial \rho/\partial t + \rho_0 \text{div}\,(\vec{v}) = \rho_0 q_m$, $\rho_0 \partial \vec{v}/\partial t + \overrightarrow{\text{grad}}(p) = \vec{F}_v$, $\rho_0 T_0 \partial s/\partial t = \rho_0 q_e$ and $\rho = 1/c_0^2 p - p_0/\left(c_0^2 C_v\right) s$. We introduce the state equation in the mass conservation equation which is derived with respect to time and from the result we remove the divergence of the impulse conservation equation. We obtain:

$$\frac{1}{c_0^2}\frac{\partial^2 p}{\partial t^2}+\rho_0 \mathrm{div}\left(\frac{\partial \vec{v}}{\partial t}\right)=\frac{\partial \rho_0 q_m}{\partial t}-\frac{p_0}{c_0^2 C_v}\frac{\partial^2 s}{\partial t^2} \text{ or } \rho_0 \mathrm{div}\frac{\partial \vec{v}}{\partial t}+\underbrace{\mathrm{div}(\overrightarrow{\mathrm{grad}}(p))}_{\Delta p}=\mathrm{div}(\vec{F}_v).$$

We introduce the entropy equation $\rho_0 T_0 \partial s/\partial t = \rho_0 q_e$ in the second term of the first equation of the previous system and by taking the difference with the second term, we finally obtain:

$$\underbrace{\frac{1}{c_0^2}\frac{\partial^2 p}{\partial t^2}}_{\text{kinetic energy}} - \underbrace{\Delta p}_{\text{potential energy}} = \underbrace{\frac{\partial \rho_0 q_m}{\partial t}+\mathrm{div}(\vec{F}_v)-\frac{\gamma-1}{c_0^2}\frac{\partial \rho_0 q_e}{\partial t}}_{\text{injected energy/dissipated energy}},$$

where $\vec{\nabla}^2 = \Delta = \partial^2/\partial x_1^2 + \partial^2/\partial x_2^2 + \partial^2/\partial x_3^2$ is the Laplacian. This differential equation governs the propagation of waves and is called d'Alembert's equation. The left side of this equation characterizes the dynamic exchange between the kinetic energy of the medium $\partial^2 p/\partial t^2$ and the elastic potential energy (due to the compressibility) of the medium $c_0^2 \Delta p$. The terms of the second member characterize energies injected and dissipated in the medium; in general the volume effects encountered in acoustics are related to gravity and are disregarded with respect to sources of mass flux and heat $\partial \rho_0 q_{m,e}/\partial t$, which is what we will do afterwards.

4.1.4. *Velocity potential*

The Helmholtz–Hodge theorem states that a vector field can be expressed as the sum of potential and rotational fields. Bearing this in mind, consider the velocity field $\vec{v} = \overrightarrow{\mathrm{grad}}\ \Phi + \overrightarrow{\mathrm{curl}}\vec{\Psi}$. It is possible to show that outside the physical sources of a rotational field, unlike in conventional fluid mechanics where there can be spontaneous creation of rotational fields in the wall vicinity, fluid acoustics is irrational. Thus, $\vec{\Psi} = \vec{0}$. The different terms can be written as $\vec{v} = \vec{\nabla}\Phi$, $p = -\rho_0 \partial\Phi/\partial t$, $\rho = -\rho_0/c_0^2 \partial\Phi/\partial t$ and $T = -\alpha_{p_0} T_0/c_{p_0} \partial\Phi/\partial t$, where Φ, the velocity potential, is the solution to the wave equation: $\vec{\nabla}^2\Phi - 1/c_0^2 \partial^2\Phi/\partial t^2 = 0$.

4.2. Propagation and general solutions

4.2.1. *One-dimensional motion*

In the case of a one-dimensional motion, everything depends on the space variable x and the homogeneous wave equation becomes: $\partial^2 p/\partial x^2 - 1/c_0^2 \partial^2 p/\partial t^2 = 0$. To solve this differential equation, let us introduce as new variables $\zeta = x - c_0 t$ and $\eta = x + c_0 t$. Thus, the differential operator can be written as a function of the variables η and ζ, as $d = \partial/\partial\eta d\eta + \partial/\partial\zeta d\zeta$, from where it simply becomes:

$$\frac{\partial}{\partial x} = \frac{\partial}{\partial \eta}\frac{\partial \eta}{\partial x} + \frac{\partial}{\partial \zeta}\frac{\partial \eta}{\partial x} = \frac{\partial}{\partial \eta} + \frac{\partial}{\partial \zeta} \text{ and } \frac{\partial}{\partial t} = \frac{\partial}{\partial \eta}\frac{\partial \eta}{\partial t} + \frac{\partial}{\partial \zeta}\frac{\partial \eta}{\partial t} = c_0\left(\frac{\partial}{\partial \eta} - \frac{\partial}{\partial \zeta}\right)$$

and

$$\frac{\partial^2}{\partial x^2} = \frac{\partial^2}{\partial \eta^2} + 2\frac{\partial^2}{\partial \zeta \partial \eta} + \frac{\partial^2}{\partial \zeta^2} \text{ and } \frac{\partial^2}{\partial t^2} = c_0^2\left(\frac{\partial^2}{\partial \eta^2} - 2\frac{\partial^2}{\partial \zeta \partial \eta} + \frac{\partial^2}{\partial \zeta^2}\right).$$

After introducing these expressions and simplifying, the wave equation can be written as: $\partial^2 p/(\partial\zeta\partial\eta) = 0$. And, sequentially, $\partial p/\partial\zeta = F(\eta)$ and $p = F(\eta) + G(\zeta) + ctte$, where F and G are sufficiently differentiable functions. Finally, the solution of the wave equation is given by $p(x,t) = F(x + c_0 t) + G(x - c_0 t)$. We must check that there is indeed propagation. To keep the solution at a constant value, it is enough to keep $x + c_0 t$ and $x - c_0 t$ constant. Let us examine a small displacement dx for a duration of dt. From $x - c_0 t = ctte$ and $x + c_0 t = ctte$, it is easy to obtain $dx - c_0 dt = 0$ and $dx + c_0 dt = 0$. As well as $c_1 = dx/dt$ and $c_2 = -dx/dt$. If $dx > 0$, the displacement takes place at a velocity of $c_1 = c_0$ and if $dx < 0$, it takes place at a velocity of $c_2 = c_0$. An acoustic wave whose amplitude remains constant moves toward $x > 0$ and toward $x < 0$ with the same velocity c_0.

4.2.2. *Three-dimensional motion*

In the case of a three-dimensional motion in a homogeneous and isotropic medium, the waves are symmetrically spherical. The acoustic pressure depends only on the distance from the origin. Let us thus use a spherical coordinate system (O, r, θ, ϕ). The wave equation can be written as:

$$\frac{1}{r^2 \sin\theta}\left[\sin\theta\frac{\partial}{\partial r}\left(r^2\frac{\partial p}{\partial r}\right) + \frac{\partial}{\partial \theta}\left(\sin\theta\frac{\partial p}{\partial \theta}\right) + \frac{1}{\sin\theta}\frac{\partial^2 p}{\partial \phi^2}\right] - \frac{1}{c_0^2}\frac{\partial^2 p}{\partial t^2} = 0. \quad [4.5]$$

If p depends only on the radius r, the previous equation can be written as $1/r^2\partial\left(r^2\partial p/\partial r\right)/\partial r - 1/c_0^2\partial^2 p/\partial t^2 = 0$. This equation can be written more simply as $\partial^2(rp)/\partial r^2 - 1/c_0^2\partial^2(rp)/\partial t^2 = 0$. The desired solution is obviously $p(r,t) = \left[F(r + c_0 t) + G(r - c_0 t)\right]/r$.

The case of the plane wave

We can also consider a similar type of solution for a single dimension: the plane wave. In this case the wave propagates in a direction parallel to a vector denoted by

$\vec{n}$. The plane wave is invariant in the directions orthogonal to $\vec{n}$. The solution is thus given by:

$$p\left(\vec{x}\cdot\vec{n},t\right)=F\left(\vec{x}\cdot\vec{n}-c_0t\right)+G\left(\vec{x}\cdot\vec{n}+c_0t\right)$$

where $\vec{x}\cdot\vec{n}$ is the scalar product of the vectors $\vec{n}$ and $\vec{x}$. Similarly, the velocity potential can be written as:

$$\Phi\left(\vec{x}\cdot\vec{n},t\right)=f\left(\vec{x}\cdot\vec{n}+c_0t\right)+g\left(\vec{x}\cdot\vec{n}-c_0t\right).$$

Based on the velocity potential, let us calculate the acoustic pressure and the vibration velocity. We have:

$$p\left(\vec{x}\cdot\vec{n},t\right)=-\rho\frac{\partial\Phi}{\partial t}=-\rho\left(-c_0f'\left(\vec{x}\cdot\vec{n}-c_0t\right)+c_0g'\left(\vec{x}\cdot\vec{n}+c_0t\right)\right)$$
$$\vec{v}\left(\vec{x}\cdot\vec{n},t\right)=\vec{\nabla}\Phi=\vec{n}\left(f'\left(\vec{x}\cdot\vec{n}-c_0t\right)+g'\left(\vec{x}\cdot\vec{n}+c_0t\right)\right).$$

The vibration velocity is collinear to the direction of propagation. The plane wave is polarized longitudinally. The particle motion is in the direction of the propagation. It is customary in this case to speak of a compression wave. On the other hand, in the case of a progressive wave, it is obvious that we have $\rho_0c_0\vec{v}=p\vec{n}$. The vibration velocity and the acoustic pressure are in phase. In the case of a regressive wave, the velocity and the pressure are out of phase.

4.3. Permanent regime: Helmholtz equation

Let us resume with d'Alembert's equation $\Delta p-1/c_0^2\partial^2p/\partial t^2=s$, in which s is the acoustic source. Let us suppose that this source has a periodic time dependence. The most general form can be given by $s\left(\vec{x},t\right)=s\left(\vec{x}\right)\exp\left(-\imath\omega t\right)$. It is clear that the acoustic pressure can be written as $p\left(\vec{x},t\right)=p\left(\vec{x}\right)\exp\left(-\imath\omega t\right)$. This acoustic pressure is complex. We must remember that this is just a mathematical way of simplifying the calculations. The "real" acoustic pressure which is measured by a microphone or captured by the ear is the real part of the pressure given by $\mathcal{P}=\Re\left(p\left(\vec{x}\right)\exp\left(-\imath\omega t\right)\right)$. D'Alembert's equation thus becomes:

$$\Delta p\left(\vec{x}\right)\mathrm{e}^{-\imath\omega t}-\frac{\imath^2\omega^2}{c_0^2}p\left(\vec{x}\right)\mathrm{e}^{-\imath\omega t}=s\left(\vec{x}\right)\mathrm{e}^{-\imath\omega t}\Rightarrow\Delta p\left(\vec{x}\right)+\frac{\omega^2}{c_0^2}p\left(\vec{x}\right)=s\left(\vec{x}\right).$$

To simplify, let us take $s\left(\vec{x}\right)=0$, and define $k=\omega/c_0$, the wave number. Thus, the homogeneous equation of wave propagation in a harmonic regime can be written as $\Delta p\left(\vec{x}\right)+k^2p\left(\vec{x}\right)=0$ and is called the Helmholtz equation. We must note that it is possible to choose a time dependence in the form of $\exp\left(+\imath\omega t\right)$ and to arrive at the same equations. We can simply specify that in this case, the conservation equation of energy will take a slightly different form.

If the propagation domain is unbounded, the Helmholtz equation does not have a unique solution. We have to add to it a condition of energy conservation which represents the outgoing wave condition in a harmonic regime. The latter may be given by the Sommerfeld conditions which characterize the fact that there is no energy that comes from infinity:

$$\lim_{r\to\infty} p(r) = \mathcal{O}(r^{-(n-1)/2}) \qquad [4.6]$$

$$\lim_{r\to\infty}\left(\frac{\partial p(r)}{\partial r} - \imath k p(r)\right) = o(r^{-(n-1)/2}) \qquad [4.7]$$

where r is the distance from M to the origin and n is the size of the space[1]. If the temporal dependence were of the type $\exp(+\imath\omega t)$, the second condition should be written as $\lim_{r\to\infty}(\partial p(r)/\partial r + \imath k p(r)) = o(r^{-(n-1)/2})$.

Another way to satisfy the principle is to stop considering the medium as non-dissipating (which is convenient when facilitating the resolutions of many problems) but to limit it to a weakly absorbing medium. In an environment like this we can consider that the wave number k is in reality a complex $k_\epsilon = k + \imath\epsilon$ with a weakly positive imaginary part. In fact, a plane wave which propagates towards $x > 0$ is such that $\exp(\imath k_\epsilon x) = \exp(\imath k x)\exp(-\epsilon x)$, and it becomes zero at infinity. We should then find the solution for p_ϵ, its bounded solution:

$$(\Delta + k_\epsilon^2)p_\epsilon(M) = f(M).$$

The solution which has a physical sense is the limit of p_ϵ for $\epsilon \to 0$. This principle is the limit-absorption principle. It is strictly equivalent to the Sommerfeld conditions [EGO 92]. These two principles lead to the same solution that satisfies the conservation of energy.

4.3.1. *General solutions*

4.3.1.1. *One-dimensional motion*

Let us assume that the propagation of an acoustic wave depends only on the variable x. The Helmholtz equation thus becomes: $p''(x) + k^2 p(x) = 0$. To solve this linear differential equation with constant coefficients, it is natural to use the

1 Let us recall that the Landau notation $f(x) = \mathcal{O}(x^\alpha)$ for $x \to \infty$ can be seen as $f(x)$ behaving as x^α for $x \to \infty$ or $\lim_{x\to\infty} f(x)/x^\alpha = $ ctte. $f(x) = o(x^\alpha)$ can be seen as $f(x)$ decreasing faster than x^α for $x \to \infty$ or $\lim_{x\to\infty} f(x)/x^\alpha = 0$.

Laplace method which consists of writing the solution in the form of an exponential $p = A\exp(\alpha x)$. The Helmholtz equation thus leads to $A(\alpha^2\exp(\alpha x) + k^2\exp(\alpha x)) = 0$. It is obvious that $A \neq 0$ and $\alpha = \pm\imath k$. The solution can thus be given by $p(x,t) = (A\exp(-\imath kx) + B\exp(\imath kx))\exp(-\imath\omega t)$. The amplitude of the wave remains constant. This solution can also be written as: $p(x,t) = A\exp(-\imath\omega(t + x/c_0)) + B\exp(-\imath\omega(t - x/c_0))$ which demonstrates the presence of a wave which moves toward $x > 0$ (the term $\exp(-\imath\omega(t - x/c_0))$ has a constant phase as the time increases) and another one toward $x < 0$.

4.3.1.2. *Two-dimensional motion*

Let us now suppose that the wave propagation occurs in a plane medium. The fluid is always assumed to be homogeneous and isotropic. The acoustic pressure only depends on the distance from the origin. It is therefore natural to use the polar coordinate system (O, r, ϕ) as indicated in Figure 4.1. The Laplacian in this case can be written as $\Delta = \partial^2/\partial r^2 + 1/r\partial/\partial r + 1/r^2\partial^2/\partial\phi^2$. The Helmholtz equation can be written as $\partial^2 p(r,\phi)/\partial r^2 + 1/r\partial p(r,\phi)/\partial r + 1/r^2\partial^2 p(r,\phi)/\partial\phi^2 + k^2 p(r,\phi) = 0$.

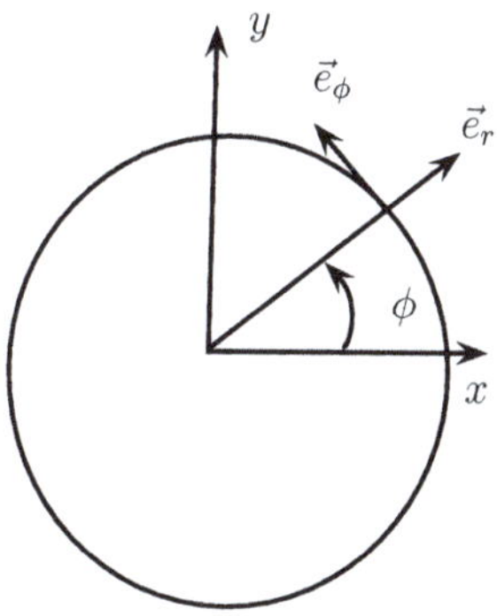

Figure 4.1. *Geometry in polar coordinates*

Let us search for the solution $p(r,\phi)$ in the form of the Laplace product $p(r,\phi) = R(r)\Phi(\phi)$. As k_p is a constant independent of r and ϕ, we obtain:

$$\frac{\partial^2 R(r)\Phi(\phi)}{\partial r^2} + \frac{\partial R(r)\Phi(\phi)}{r\partial r} + \frac{\partial^2 R(r)\Phi(\phi)}{r^2\partial\phi^2} + k_p^2 R(r)\Phi(\phi) = 0$$

$$\Rightarrow \frac{R''(r)}{R(r)} + \frac{R'(r)}{rR(r)} + \frac{\Phi''(\phi)}{r^2\Phi(\phi)} + k_p^2 = 0, r^2\left\{\frac{R''(r)}{R(r)} + \frac{R'(r)}{rR(r)} + k_p^2\right\} + \frac{\Phi''(\phi)}{\Phi(\phi)} = 0.$$

From here, we obtain two ordinary differential equations with respect to the angular variable ϕ: $\Phi''(\phi) + \alpha^2\Phi(\phi) = 0$ and the radial variable r:

$R''(r) + 1/rR'(r) + \left(k_p^2 - \alpha^2/r^2\right) R(r) = 0$. The solution of the first equation is a linear combination of complex exponentials. In addition, it is evident that the solution is 2π-periodic in ϕ, therefore it is necessary to use $\alpha = n$, with n being an integer. We can recognize in the second equation a Bessel equation of the order n [WAT 44]. This solution is a Bessel function with an integer as an index. As the solution to the problem is regular, we must choose the regular solution: J_n. Finally, we obtain $\Phi(\phi) = A_{1n}\mathrm{e}^{\imath n\phi} + A_{2n}\mathrm{e}^{-\imath n\phi}$ and $R(r) = B_n \mathrm{J}_n(k_p r)$. If A_n is a constant, the general solution to the Helmholtz equation in polar coordinates can be given by:

$$p(r, \phi) = \sum_{n=-\infty}^{+\infty} A_n \mathrm{J}_n(kr)\mathrm{e}^{\imath n\phi}. \qquad [4.8]$$

Let us recall the asymptotic behavior of Bessel functions for $z \to \infty$. We have $\mathrm{J}_n(z) \sim 1/\sqrt{z}\cos(z - n\pi/2 - \pi/4)$ which shows that "far" from the origin the behavior is pseudoperiodical. The acoustic wave while oscillating presents an amplitude decreasing as the inverse of the square root of the distance from the origin. This decrease is called geometric dispersion. If the domain in question does not contain the origin, we must add to the general solution [4.8], composed of a simple progressive wave $\mathrm{J}_n(kr)$, the regressive wave $\mathrm{Y}_n(kr)$, which is a Bessel function of the second kind, singular at the origin.

4.3.1.3. *Three-dimensional motion*

Similarly to d'Alembert's equation in a homogeneous and isotropic environment, the Helmholtz equation can be written for a symmetrically spherical wave $[rp(r)]'' + k^2 rp(r) = 0$, thus the solution can be given by $p(r,t) = \left(A\exp\left(-\imath\omega\left(t + r/c_0\right)\right) + B\exp\left(-\imath\omega\left(t - r/c_0\right)\right)\right)/r$. In a three-dimensional environment, the geometric dispersion leads to an amplitude decreasing with the inverse of the distance from the origin. When the wave is not simply spherical, we calculate the solution in spherical variables (O, r, θ, ϕ) in the form of the Laplace product $p(r, \theta, \phi) = R(r)\Theta(\theta)\Phi(\phi)$ [MOR 53, vol. 2, p.1462]. Without going into detail, from it, we can obtain the following system:

$$\begin{aligned} \Phi''(\phi) + n^2\Phi(\phi) &= 0 \\ \frac{1}{\sin(\theta)}\frac{d}{d\theta}\left(\sin\theta\frac{d\Theta(\theta)}{d\theta}\right) + \left(m(m+1) - \frac{n^2}{\sin^2(\theta)}\right)\Theta(\theta) &= 0 \\ \frac{1}{r^2}\frac{d}{dr}\left(r^2\frac{dR(r)}{dr}\right) + \left(k^2 - \frac{m(m+1)}{r^2}\right)R(r) &= 0, \end{aligned}$$

where ϕ varies from 0 to 2π and θ from 0 to π. It is easy to see that $\Phi(\phi) = \exp(\imath n\phi)$, $n \in \mathbb{Z}$. We can show that $\Theta(\theta) = P_m^n(\cos\theta)$ where $m, n = 0, 1, 2, \cdots$ and $n \leq m$ which are the Legendre functions such that

$P_m^n(x) = (1-x^2)^{n/2} d^n P_m(x)/dx^n$, where $P_m(x)$ are the Legendre polynomials such that $P_0(x) = 1$, $P_1(x) = x$, $P_2(x) = (3x^2-1)/2$ and that satisfy the recursion relation $(m+1)P_m(x) = (2m+1)xP_m(x) - mP_{m-1}(x)$. We define $Y_{enm} = \cos(n\phi)P_m^n(\cos\theta)$ and $Y_{onm} = \sin(m\phi)P_m^n(\cos\theta)$ as the spherical harmonics. Finally, we show that $R(r) = \mathrm{j}_m(kr)$ which are the spherical Bessel functions of the first kind such that $\mathrm{j}_m(kr) = \sqrt{\pi/2}\mathrm{J}_{m+1/2}(kr)/\sqrt{kr}$. The spherical Bessel functions for a large r present the asymptotic behavior $\mathrm{j}_m(kr) \approx 1/(kr)\sin(kr - m\pi/2)$ which shows a decrease in $1/r$. When the field domain does not contain the origin, to these outgoing waves, we must add the incoming waves $\mathrm{y}_m(kr)$ which are the spherical Bessel functions of the second kind, singular at the origin. The general solution is a linear combination over all the indices between n and m of the solutions $\exp(\imath n\phi)$, $P_m^n(\cos\theta))$ and $\mathrm{j}_m(kr)$.

The case of the plane wave

As we have seen previously, we can introduce the notion of a plane wave in a three-dimensional environment. In the two-dimensional case, the modifications to the following are trivial. It is obvious that we have the velocity potential $\Phi(\vec{x},t) = A\exp(-\imath\omega(t+\vec{n}\cdot\vec{x}/c_0)) + B\exp(-\imath\omega(t-\vec{n}\cdot\vec{x}/c_0))$. It is natural to take $\vec{k} = k\vec{n}$ as the wave vector and to obtain:

$$\Phi(\vec{x},t) = A\mathrm{e}^{-\imath(\omega t - \vec{k}\cdot\vec{x})} + B\mathrm{e}^{-\imath(\omega t + \vec{k}\cdot\vec{x})}.$$

4.3.1.4. *Acoustic intensity*

Let us imagine a surface with a unit area perpendicular to the direction of wave propagation. The flux of sound energy that passes through this surface, the intensity of the wave, is given by the time average (denoted by an overbar) of the product between the pressure p and the acoustic velocity $\vec{v}$:

$$\vec{I} = \overline{p\vec{v}}.$$

When quantities are represented by complex values, care must be taken since the product of real values is not the real value of the product. In complex notation, we must modify the definition of the intensity as $\vec{I} = \overline{\Re(p)\Re(\vec{v})}$. If $p = p_0\exp(-\imath(\omega t - \varphi))$ and $\vec{v} = \vec{v}_0\exp(-\imath(\omega t - \varphi))$, we thus have (see, for example, section 9 of [BRI 46]):

$$\vec{I} = \overline{\Re(p)\Re(\vec{v})} = \frac{1}{2}\Re(p\vec{v}^\star). \quad [4.9]$$

In the case of the plane wave, such as $v(x,t) = p(x,t)/(\rho_0 c_0)$, we have $p\vec{v}^\star = |p^2|/(\rho_0 c_0)$ where, if P_0 is the modulus of the pressure, we have $|p^2| = P_0^2$. The intensity of the plane wave will be $I = P_0^2(2\rho_0 c_0)$. Over the course of its path, the wave travels the distance of c_0 meters over a time period of 1 s. The energy density

w per unit volume will thus become $W = I/c_0 = P_0^2/\left(2\rho_0 c_0^2\right)$. In the case of a spherical wave in a homogeneous and isotropic medium, the pressure is given by $p(r,t) = P_0 \exp(-\imath kr)/r$ and the radial velocity by $v_r(r,t) = p(r,t)/(\rho_0 c_0)(1+1/(\imath kr))$. In the far-field ($kr \gg 1$), the ratio $p/v - r$ approaches the characteristic impedance of the medium $\rho_0 c_0$. In this region, the relations for intensity and energy density in a plane wave apply. By multiplying the first by the surface of a sphere with the radius r, we obtain the energy carried by a wave every second:

$$W_0 = 4\pi r^2 I = 2\pi r^2 P_0^2/(\rho_0 c_0). \quad [4.10]$$

4.3.2. *Green's kernels*

It is obvious that in practice, we will have to solve problems for which not only the acoustic source term but also the presence of boundaries will play an important role. To do this, we often use a representation of a pressure field which describes the direct fields and those diffracted by the boundaries based on Green's representation of pressure. This representation, which we will introduce in the following chapter, is based on the products of convolution between the sources (either physical or describing boundaries) and the corresponding Green kernels. Physically, a kernel represents the acoustic field measured at point M of the field and created by a source placed at another point S. Let us calculate the Green kernel in $\mathbb{R}$. We are now searching for the solution of the following non-homogeneous equation:

$$G''(x,x_0) + k^2 G(x,x_0) = \delta_{x_0}(x).$$

The solution of this equation is obtained by a direct and inverse Fourier transform of this equation. We know that a Green kernel is an elementary kernel that satisfies the condition of conservation of energy. This last condition is satisfied by applying the limit-absorption principle. In a broad outline, this method consists of looking for a solution to $u_\epsilon(x,x_0)$ such that $G''_\epsilon(x,x_0) + (k+\imath\epsilon)^2 G_\epsilon(x,x_0) = \delta_{x_0}(x)$, where ϵ is a positive real which represents the loss of energy in the fluid. The solution G is obtained by passing to the limit $\epsilon \to 0$. Let us examine:

$$\hat{G}_\epsilon(\zeta) = \int_{-\infty}^{+\infty} G_\epsilon(x) \mathrm{e}^{-2\imath\pi x\zeta} dx \text{ and } G_\epsilon(x) = \int_{-\infty}^{+\infty} \hat{G}_\epsilon(\zeta) \mathrm{e}^{2\imath\pi x\zeta} d\zeta$$

which are the direct and inverse Fourier transforms. By using a Fourier transform, the Helmholtz equation becomes $\left(-4\pi^2\zeta^2 + (k+\imath\epsilon)^2\right) \hat{G}_\epsilon(\zeta,x_0) = \exp(-2\imath\pi x_0)$. It is obvious that the solution $\hat{G}_\epsilon(\zeta)$ can be given by $\hat{G}_\epsilon(\zeta, x_0) = \exp(-2\imath\pi x_0\zeta)/\left(-4\pi^2\zeta^2 + (k+\imath\epsilon)^2\right)$. We thus have:

$$G_\epsilon(x, x_0) = \int_{-\infty}^{+\infty} \left\{ \frac{e^{2\imath\pi(x-x_0)\zeta}}{\left(-4\pi^2\zeta^2 + (k+\imath\epsilon)^2\right)} \right\} d\zeta.$$

Let us first assume that $x > x_0$. To keep the exponential a finite value, we must integrate into the half-plane $\Im \geq 0$. The calculation of this integral is carried out by the method of residues on the contour indicated in Figure 4.2. Because of the limit-absorption principle, the pole $k/2\pi$ has a non-zero positive imaginary part which makes it possible to very simply apply the residue theorem (absence of poles on the real axis).

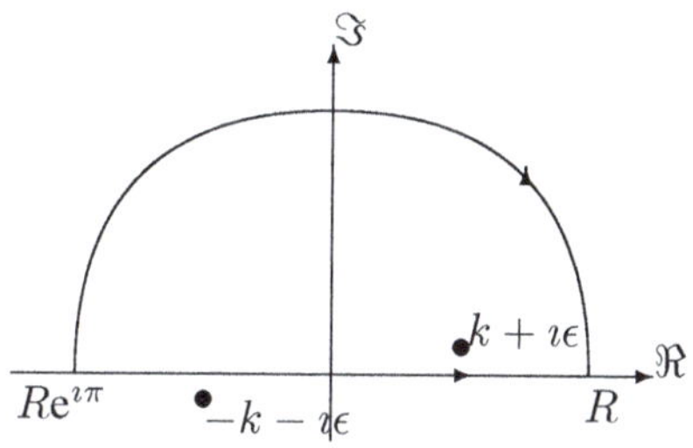

Figure 4.2. *Integration contour*

We obtain, after passing to the limit, $G(x, x_0) = 1/(2\imath k)\exp(\imath k(x - x_0))$, $x > x_0$. Similarly, if $x < 0$, to always keep the exponential a finite value, we must integrate into the half-plane $\Im < 0$. Thus, we obtain $G(x, x_0) = 1/(2\imath k)$ $\exp(-\imath k(x - x_0)), x < x_0$. Finally, when we regroup the results we obtain:

$$G(x, x_0) = \frac{1}{2\imath k} e^{\imath k_l |x - x_0|}. \quad [4.11]$$

We see that Green's kernel, as the general solution, consists of a propagating wave. For the Helmholtz equation in $\mathbb{R}^2$ and $\mathbb{R}^3$, these kernels take the following form:

in $\mathbb{R}^2$: $G(S, M) = -\frac{\imath}{4} \mathrm{H}_0^{(1)}(kr(S, M))$.

in $\mathbb{R}^3$: $G(S, M) = -\frac{e^{\imath kr(S,M)}}{4\pi r(S, M)}$.

$r(S,M)$ is the distance (Euclidean) between S and M. We must note that $G(M,S) = G(S,M)$; this relation is called the reciprocity principle which stipulates in acoustics that, if we were to invert the source and measure, the results would remain unchanged. It is important to note that the expressions of these kernels depend on the chosen time dependence. If we had chosen a dependency of the $\exp(+\imath\omega t)$ type, $I\!R^3$ Green's kernel would have taken the following expression: $G(S,M) = -\exp(-\imath kr(S,M))/(4\pi r(S,M))$ and $G(S,M) = -\imath/4\mathrm{H}_0^{(2)}(kr(S,M))$ in $I\!R^2$. When this does not cause confusion in the following we will note that $\mathrm{H}_0^{(1)}(z) = \mathrm{H}_0(z)$. Let us suppose that for temporal dependence we were to choose $\exp(-\imath\omega t)$. In $I\!R^3$, each of Green's kernels $G(S,M) = -\exp(\pm\imath kr(S,M))/\ (4\pi r(S,M))$ is a basic kernel. Thus we have:

$$p(M,t) = -\frac{\mathrm{e}^{\pm\imath kr(S,M)-\imath\omega t}}{4\pi r(S,M)} = -\frac{\mathrm{e}^{\imath\frac{\omega}{c}(\pm r(S,M)-ct)}}{4\pi r(S,M)}.$$

The condition of conservation of energy indicates that there is no wave that can "arrive" from infinity. Only progressive waves can exist. That is:

$$p(M,t) = -\frac{\mathrm{e}^{\imath\frac{\omega}{c}(+r(S,M)-ct)}}{4\pi r(S,M)}.$$

The choice of the elementary solution is thus evident.

4.3.3. *Wave group, phase velocity and group velocity*

A homogeneous medium without dispersion is characterized by a constant wave propagation velocity c[2]. Let us now study a dispersive medium for which the celerity of the waves depends on the frequency. In such a medium, any signal deforms during its propagation. We must then consider how a wave propagates. For this, the wave is decomposed into sinusoidal waves, solutions of the Helmholtz equation. Let us consider the simple example of a signal composed of two sinusoidal plane waves whose frequencies are very similar and that propagate along the x axis. For simplicity, we will only consider the real part of the sinusoidal signal. Let us consider an acoustic field formed by the superposition of the waves $\Psi_1(x,t)$ and $\Psi_2(x,t)$, solutions of $\left(\Delta + k_i^2\right)\Psi_i(x) = 0, i = 1,2$, such that $\Psi_1(x,t) = A\cos(\omega_1 t - k_1 x)$ and $\Psi_2(x,t) = A\cos(\omega_2 t - k_2 x)$. For simplicity, we have chosen two waves with the same amplitude. The pulsations are such that $\omega_1 = \omega + \epsilon$ and $\omega_2 = \omega - \epsilon$. Similarly, the wave numbers are $k_1 = k + \kappa$ and $k_2 = k - \kappa$. By superposition, the resulting wave $\Psi(x,t) = \Psi_1(x,t) + \Psi_2(x,t)$ will present beats with a total amplitude that ranges from 0 to $2A$ according to the reciprocal phases of the two

2 In a heterogeneous medium the celerity depends on the position of the considered point.

components Ψ_1 and Ψ_2. We have $\Psi(x,t) = 2A\cos(\epsilon t - \kappa x)\cos(\omega t - kx)$. The wave can be described as a plane wave with an average pulsation ω and a constant wave number of k but a variable amplitude of $A'(x,t) = 2A\cos(\epsilon t - \kappa x)$. Another way to look at this phenomenon is to consider that the wave is modulated in amplitude. ω is the pulsation of the carrier wave which propagates at a velocity of $c_0 = \omega/k$. ϵ is the pulsation of the modulation which propagates with a velocity of $v_g = \epsilon/\kappa = d\omega/dk = 1/(dk/d\omega)$. In fact, A' includes a phase factor $\epsilon t - \kappa x = \epsilon\,(t - x/v_g)$, which clearly features the velocity v_g.

The rapid oscillations, with a period of $2\pi/\omega$, propagate with a phase velocity of c_0. The slow modulations with the period $2\pi/\epsilon$ glide across the wave system with a different velocity, the group velocity. Of course, when the celerity of the medium does not depend on the frequency, which is the case for the infinite fluid media studied so far, the wave number can be given by the relationship $c_0 = \omega/k$ and the group velocity $v_g = d\omega/dk = c_0$.

However, when the medium is dispersive, the group velocity and the phase velocity are different. This phenomenon can be seen on the waves visible at the surface of the water. In this case, the phase velocity very strongly depends on the wavelength. Let g be the acceleration of gravity and $\gamma = 7310^{-3}$N/m the surface tension of water at 20°C. In deep water, the phase velocity is given by the equation $c_0^2 = g\lambda/(2\pi) + 2\pi\gamma/(\rho_0\lambda)$. This velocity has a minimum for the wavelength λ_0 given by $2c_0 dc_0 = \left[\left(g/(2\pi) - 2\pi\gamma/(\rho_0\lambda^2)\right)d\lambda\right]_{\lambda_0} = 0$, and $\lambda_0 = 2\pi\sqrt{\gamma/(\rho g)} \approx 1.7$ cm. For long waves, for example, gravity waves, the phase velocity is proportional to $\sqrt{\lambda}$. For capillary waves, the velocity varies as $1/\sqrt{\lambda}$. In both cases, the group velocity is given by:

$$v_g = \frac{d\omega}{dk} = \frac{d\frac{c_0}{\lambda}}{d\frac{1}{\lambda}} = c_0 - \lambda\frac{dc_0}{d\lambda}.$$

For gravity waves, we thus have: $c_0 \sim \sqrt{g\lambda/(2\pi)} \Rightarrow v_g = 1/2c_g, \lambda \gg \lambda_0$ and for capillary waves: $c_0 \sim \sqrt{2\pi\gamma/\rho_0\lambda} \Rightarrow v_g = 3/2c_g, \lambda \ll \lambda_0$. In the first case, the waves propagate twice as fast as the groups (or beats). We will see the waves forming at the back of the group, then crossing over creating a large amplitude in the middle and then disappearing in the front. This phenomenon is observed at sea. The waves arrive on the beach in groups of three or four, the highest are usually found in the middle. In fact, at sea, this phenomenon is more complicated since the wave trains have different propagation directions. Conversely, the capillary waves show the opposite example: the beats propagate themselves faster than the capillary waves. These are formed at the front of the wave group, they are passed by the latter and then disappear at the back.

4.4. Discontinuity equations

In all cases at the interface between two media, we can write the continuity of the conservation equations of mass and impulse. For example, let us consider the interface between two non-viscous fluids. If Σ is the surface discontinuity, simply writing the conservation equations linearized in the sense of distributions gives:

$$\frac{d\rho}{dt} + \rho\vec{\nabla}\cdot(\vec{v}) = \frac{d\rho}{dt} + \left\{\rho\vec{\nabla}\cdot(\vec{v})\right\} + \rho\|\vec{v}\cdot\vec{n}\|_\Sigma$$

$$\rho\frac{d\vec{v}}{dt} + \vec{\nabla}\cdot\left(p\vec{\vec{I}}\right) = \rho\frac{d\rho\vec{v}}{dt} + \left\{\vec{\nabla}\cdot\left(p\vec{\vec{I}}\right)\right\} + \|p\vec{\vec{I}}\cdot\vec{n}\|_\Sigma$$

where $\|f\|_\Sigma$ is the jump of the quantity f at the crossing of Σ, $\vec{n}$ is the outward normal at the surface Σ and, if O is a differential operator, $\{Of\}$ represents the value of Of outside the discontinuity surface Σ. From there, it is obvious to deduce the continuity conditions $\|\vec{v}\cdot\vec{n}\|_\Sigma = 0$ and $\|p\vec{\vec{I}}\cdot\vec{n}\|_\Sigma = 0$. The first equation expresses the continuity of the normal vibration velocities, and the second equation expresses the continuity of acoustic pressures. If the media are plain, the last equation can be written as $\|\vec{\vec{\sigma}}\cdot\vec{n}\|_\Sigma = 0$. We must note that in the case of an interface between a non-viscous fluid medium and a solid medium, the conservation equation of mass does not change, so there is always continuity in the normal vibration velocity. However, the conservation equation of impulse changes (the constitutive law is linked to the solid studied), and there is thus the conservation of the normal component of the stress tensor.

4.4.1. *Interface between two propagating media*

To gain a better understanding of these ideas, let us consider the significant example of a plane diopter attacked by a plane wave in a two-dimensional case in a harmonic regime. Let us consider the plane (O, x, y). A plane wave propagates in the medium I, characterized by a density of ρ_1 and a wave celerity of c_1. It reaches at $y = 0$ a medium II -ρ_2, c_2- located in the half-space $y < 0$.

The incoming wave arrives at $y = 0$ with an angle with respect to the normal to the interface θ_i. A reflected wave is then emitted to the medium I at an angle θ_r. A wave is transmitted in the medium II with an angle θ_t. The different acoustic waves can be described by the velocity potential $\Phi(x, y)$. In the medium I, the total field is $\Phi^I = \Phi_i + \Phi_r$. In the medium II, the field is $\Phi^{II} = \Phi_t$. The boundary conditions are given by the conservation of normal vibration velocity and by the continuity of the acoustic pressure. These conditions can be described for the velocity potential ($\vec{v} = \vec{\nabla}\Phi, p = -\rho\dot{\Phi} = \imath\rho\omega\Phi$) by:

$$\begin{array}{ll}\text{Continuity of normal velocities} & \vec{\nabla}\Phi^I \cdot \vec{n} = \vec{\nabla}\Phi^{II} \cdot \vec{n} \\ \text{Continuity of pressures} & \rho_1\Phi^I = \rho_2\Phi^{II}.\end{array}$$

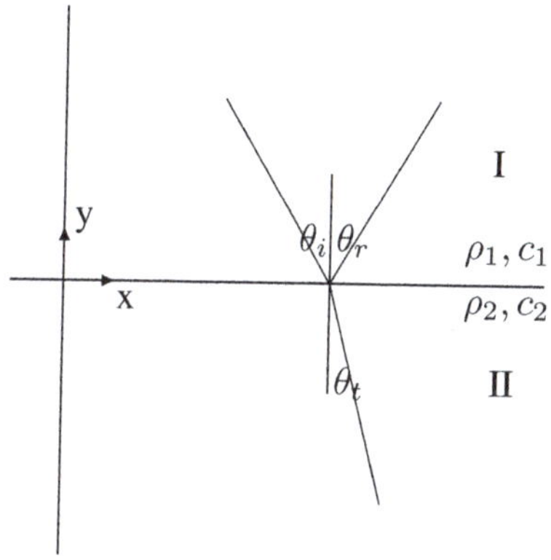

Figure 4.3. *Plane diopter attacked by a plane wave*

Let us show that the continuity conditions lead to Descartes relations which govern reflection and refraction and allow us to calculate the reflection and transmission coefficients. The velocity potential for each wave is given by:

$$\begin{aligned}\Phi^i &= \mathrm{e}^{\imath\vec{k}_1\cdot\vec{n}^i} = \mathrm{e}^{-\imath k_1(x\sin\theta_i - y\cos\theta_i)} \\ \Phi^r &= R_\Phi \mathrm{e}^{\imath\vec{k}_1\cdot\vec{n}^r} = R_\Phi \mathrm{e}^{-\imath k_1(x\sin\theta_r + y\cos\theta_r)} \\ \Phi^t &= T_\Phi \mathrm{e}^{\imath\vec{k}_2\cdot\vec{n}^t} = T_\Phi \mathrm{e}^{-\imath k_2(x\sin\theta_t - y\cos\theta_t)}\end{aligned}$$

where R_Φ is the reflection coefficient and T_Φ is the transmission coefficient. Let us choose the normal directed along the y axis. The first boundary condition is written for the velocity potential:

$$\vec{\nabla}\Phi^I \cdot \vec{n} = \vec{\nabla}\Phi^{II} \cdot \vec{n} \Rightarrow \left(\frac{\partial \Phi^I}{\partial y}\right)_{y=0} = \left(\frac{\partial \Phi^{II}}{\partial y}\right)_{y=0},$$

the second relation is obvious.

The field in the half-space $y > 0$ is, of course, the sum of the incoming and reflected fields $\Phi^I = \Phi^i + \Phi^r$. The two boundary conditions result in:

$$\begin{aligned}-k_1\cos\theta_i \mathrm{e}^{\imath k_1 x\sin\theta_i} + R_\Phi k_1\cos\theta_r \mathrm{e}^{\imath k_1 x\sin\theta_r} &= -T_\Phi k_2 \cos\theta_t \mathrm{e}^{\imath k_2 x\sin\theta_t}\ \forall x \\ \text{and}\quad \rho_1 \mathrm{e}^{\imath k_1 x\sin\theta_i} + R_\Phi \rho_1 \mathrm{e}^{\imath k_1 x\sin\theta_r} &= T_\Phi \rho_2 \mathrm{e}^{\imath k_2 x\sin\theta_t}\ \forall x.\end{aligned}$$

As these relations must be true for all x, it is obvious that we have $k_1 \sin\theta_i = k_1 \sin\theta_r = k_2 \sin\theta_t$. Thus, $\theta_i = \theta_r$ and $c_2 \sin\theta_i = c_1 \sin\theta_t$. These relations are called the Snell–Descartes laws. With these results, the condition of continuity of normal velocities can be written as $k_1 \cos\theta_i(1 - R_\Phi) = k_2 \cos\theta_t T_\Phi$. The equality of acoustic pressures becomes $\rho_1(1 + R_\Phi) = \rho_2 T_\Phi$. From these relationships, it is easy to obtain $T_\Phi = 2\rho_1/\rho_2 z_2/(z_1 + z_2)$ and $R_\Phi = (z_2 - z_1)/(z_1 + z_2)$ where $z_1 = \rho_1 c_1/\cos\theta_i$ and $z_2 = \rho_2 c_2/\cos\theta_t$ are specific impedances.

Similarly, we can define the reflection and transmission coefficients for pressure waves. Acoustic pressure, as we know, is very simply connected to potential velocities by: $p = -\rho\partial\Phi/\partial t$. It is obvious that $T_p = \rho_2/\rho_1 T_\Phi = 2z_2/(z_1 + z_2)$ and $R_p = R_\Phi = (z_2 - z_1)/(z_1 + z_2)$. If $c_2 < c_1$, there is always a wave being transmitted because $\sin\theta_t < \sin\theta_i$. Otherwise, there must be an incidence angle, called the critical angle, beyond which there is no transmission ($\theta_t = \pi/2$). This is total reflection.

4.4.2. *Interface between a propagating and a non-propagating medium*

When the propagating media have very different physical characteristics, such as air and water or air and steel, a good approximation of acoustic behavior of the medium in which the wave propagates can be obtained by neglecting the transmission phenomenon. For example, an acoustic wave propagates in the marine environment and arrives at the interface (the surface of the sea) at normal incidence. The total pressure at the surface of the water is the sum of the acoustic pressure and the atmospheric pressure. However, hypothetically, the acoustic pressure is very low (small perturbation). The total pressure is thus basically equal to atmospheric pressure. At the surface, we can thus write that the acoustic pressure is zero. Conversely, the interface between an acoustic wave that propagates in air and arrives at the sea surface can be considered as an infinitely rigid surface.

The three classic boundary conditions are the following:

1) *Dirichlet condition*: it can be written as $p = 0$ at the boundary and models an infinitely soft surface (the interface water $\rightarrow$ air).

2) *Neumann condition*: it can be written as $\vec{\nabla} p \cdot \vec{n} = 0$ where $\vec{n}$ is a normal unit vector to the surface considered. It is used to model the reflection of a wave on an infinitely rigid surface (the interface air $\rightarrow$ concrete or air $\rightarrow$ water). In practice, if $\vec{n} = (n_1, n_2, n_3)$,

$$\vec{\nabla} p \cdot \vec{n} = n_1 \frac{\partial p}{\partial x_1} + n_2 \frac{\partial p}{\partial x_2} + n_3 \frac{\partial p}{\partial x_3}.$$

3) *Impedance condition*: the previous conditions allow us to treat the phenomena of wave reflection by surfaces with no loss of energy. One of the means available to the

acoustician for describing surfaces that absorb energy is the condition of impedance. In acoustics, the impedance is defined as the ratio between the pressure and the acoustic velocity $Z = p/v$. This notion of acoustic impedance is similar to its electric counterpart. If two media have close impedance values, any wave that propagates a medium can "pass" into the second medium without significant loss of energy (the transmission coefficient is not zero). For example, let us consider a plane wave which propagates along the Ox axis. The fluid can be characterized by a density of ρ and a wave celerity of c. The progressive plane wave is given by $p(x) = \exp(\imath kx)$. Let us calculate the expression of the vibration velocity $v(x)$. The conservation equation of impulse $\rho_0 \partial v/\partial t + \partial p/\partial x = 0$, in a harmonic regime, gives $-\rho_0 \imath\omega v + \partial p/\partial x = 0$ or $v(x) = \imath k/(\imath\omega\rho_0)\exp(\imath kx)$. Thus, the impedance can be given by $Z = \rho_0 c_0$. The impedance can also be regarded as a characteristic of the balance between kinetic energy $e_c = 1/2\rho_0 v^2$ and potential energy $e_p = 1/2\chi_s p^2$, where $\chi_s = \rho_0/c_0^2$ [FIL 99] is the isentropic compressibility (or adiabatic). If $e_c = e_p$, we can deduce that $p^2/v^2 = \rho_0/\chi_s = (\rho_0 c_0)^2$, the result of which is the characteristic impedance of an infinite medium $Z = p/v = \rho_0 c_0$.

On the surface of the domain, we also define the normal impedance $Z_n = p/v_n$ where v_n is the normal vibration velocity. Then, we can define the reduced normal impedance ζ_n which allows us to compare the normal impedance Z_n to the impedance of the propagation medium, $\zeta_n = Z_n/(\rho_0 c_0)$. It is this reduced normal impedance that allows us to write the impedance condition at the boundary of a domain $\vec{\nabla}(p) \cdot \vec{n} + (\imath k/\zeta_n)p = 0$. This last condition is sometimes called the Robin condition. It allows us to regroup the Neumann condition $|\zeta_n| = \infty$ and the Dirichlet $|\zeta_n| = 0$ and the absorption at the wall, if ζ_n is complex. The imaginary part characterizes the absorption of energy at the boundary.

This condition of impedance is only an approximate way of characterizing the acoustic behavior of a material according to a non-rigorous law. However, experience shows that this model is often sufficient. The validity of this approximation depends on the material and the frequency band considered. We must note that ζ_n usually depends on the frequency.

4.5. Impedance: measurement and model

4.5.1. *Kundt's tube*

In this section, we describe the experimental setup for measuring the impedance of an absorbent material by the standing wave tube method or Kundt's tube.

A sample of the material is placed at one end of the tube $(x = L)$. A loudspeaker is placed at the other end of the tube $(x = 0)$. The loudspeaker (the membrane of which vibrates in the piston) emits a monofrequential signal. If the section of the tube is sufficiently small compared to the length of the acoustic wave (therefore at low frequency), only the waves that propagate along the axis of the tube will be considered. The waves emitted by the acoustic source are partly reflected by the sample surface. There establishes a regime of standing waves in the tube.

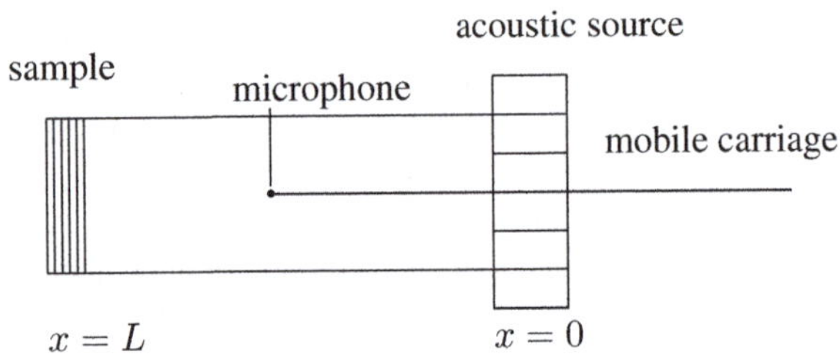

Figure 4.4. *Experimental setup*

Mathematically, the physical phenomenon is well modeled by a one-dimensional problem comprising the following wave propagation equation in the tube $p''(x) + k^2 p(x) = 0,\ x \in]0, L[$. The condition of impedance on the sample $\partial_{\vec{n}} p(x) + \frac{\imath k}{\zeta_n} p(x) = 0$ at $x = L$ is the condition of continuity of normal velocity on the loudspeaker $\partial_{\vec{n}} p(x) - \imath \omega \rho v = 0$ at $x = 0$, where it is assumed that the velocity $v = v e^{-\imath \omega t}$ is constant over the entire surface of the loudspeaker. $\zeta_n \in \mathbb{C}$ is the reduced normal impedance to be determined and $k = \omega / c$. The pressure in the tube is, of course, composed of the sum of the incoming pressure $\exp(+\imath k x)$ and the reflected one $\exp(-\imath k x)$, and we have $p(x) = A \exp(+\imath k x) + B \exp(-\imath k x)$. The constants A and B are calculated by imposing the boundary conditions. Let us choose, for example, for $\vec{n}$ the normal to the interior of the tube. Thus:

$$\frac{\partial p(x)}{\partial \vec{n}} = \frac{dp(x)}{dx} \text{ at } x = 0 \text{ and } \frac{\partial p(x)}{\partial \vec{n}} = -\frac{dp(x)}{dx} \text{ at } x = L$$

with this convention, let us write the boundary conditions $-p'(L) + \imath k / \zeta_n p(L) = 0$ and $p'(0) - \imath \omega \rho v = 0$. The acoustic pressure in the tube is thus given by:

$$p(x) = \frac{\rho c v}{\Delta} (1 + \zeta_n) \left[e^{\imath k (x - L)} + \frac{\zeta_n - 1}{\zeta_n + 1} e^{-\imath k (x - L)} \right]$$

where $\Delta = (1 + \zeta_n) \exp(-\imath k L) + (1 - \zeta_n) \exp(+\imath k L)$ is never zero if ζ_n has a real part and an imaginary part both of which are non-zero (which reflects dissipation in the sample). Let us have $\mathcal{R} = (\zeta_n - 1) / (\zeta_n + 1) = R \exp(\imath \gamma)$. $\mathcal{R} = \mathcal{R}(f)$ is the complex reflection coefficient. It depends on the frequency as well as its modulus $R = R(f) < 1$. A part of the energy of the incoming wave is absorbed by the material and is defined by the absorption coefficient of the material $\alpha(f) = 1 - |R(f)|^2$ which is between 0 (no absorption) and 1 (total absorption).

Due to the steady state present in the tube, the acoustic pressure passes through a succession of minima and maxima, of which the values and positions depend on $\mathcal{R}$ and thus ζ_n. Let us calculate R and γ, as these two values are sufficient to characterize the impedance ζ_n. To do this, let us express $|p(x)|^2$:

$$|p(x)|^2 = |\frac{v\rho c\,(1+\zeta_n)}{\Delta}|^2 \left[1 + R^2 + 2R\cos(2k(L-x)+\gamma)\right],$$

where $|p(x)|^2$ is maximum for $\cos(2k(L-x)+\gamma) = 1$ and the minimum for $\cos(2k(L-x)+\gamma) = -1$. Thus:

$$\frac{|p(x)|^2_{\text{max}}}{|p(x)|^2_{\text{min}}} = \left(\frac{1+R}{1-R}\right)^2 \Rightarrow R = \frac{|p(x)|_{\text{max}} - |p(x)|_{\text{min}}}{|p(x)|_{\text{max}} + |p(x)|_{\text{min}}}.$$

γ is obtained by stating that $|p(x)|^2$ is a minimum for $2k(L-x)+\gamma = (2n+1)\pi, n \in \mathbb{Z}$. The minimum closest to the sample corresponds to $n = 0$. Let us recall that λ is the signal wavelength $\lambda = c/f = 2\pi/k$. The distance between two minima corresponds to a half wave-length wave. Thus, since $L - x$ is the distance between the first minimum and the sample, we have $L - x = -\gamma/(2k) + \lambda/4$. Finally, $\gamma = 2k\,(L - x + \lambda/4)$. We see that $\mathcal{R}$ is completely determined by the measure of two pressures, $|p(x)|^2_{\text{max}}$ and $|p(x)|^2_{\text{min}}$, given by a microphone and the distance $L - x$.

4.5.2. *Delany–Bazley model*

Most of the absorbent materials consist of open-cell porous materials (such as foams and fibers). When the sound wave encounters the material, it penetrates it by passing through the pores. The path traveled by the wave through the pores increases considerably due to successive reflections on the fibers and the main mechanism of wave damping is related to the dissipation by viscosity within the fluid. The Delany–Bazley model [DEL 70] is one of the most used models for describing the impedances of fibrous materials. This model is based on an empirical adjustment of the characteristic impedance and on the wave number based on experimental data depending on the frequency and the parameter σ of the material. σ is the specific flow resistance which is expressed as N.m^{-4}.s and characterizes the density of the porous material; for example, we have $\sigma \approx 20000$ N.m^{-4}.s for rock wool. The characteristic impedance Z_c and the wave number k are given by:

$$Z_c = \rho_0 c_0 \left[1 + a\left(\frac{f}{\sigma}\right)^b + \imath c\left(\frac{f}{\sigma}\right)^d\right], k = \frac{\omega}{c_0}\left[\imath p\left(\frac{f}{\sigma}\right)^q + 1 + r\left(\frac{f}{\sigma}\right)^s\right]$$

with $a = 0.0497$, $b = -0.754$, $c = 0.0758$, $d = -0.732$, $p = 0.169$, $q = -0.595$, $r = 0.0858$ and $s = -0.70$. This model is valid for $0.01 < f/\sigma < 1$ for many materials. The material thickness is taken into consideration through the specific flow resistance, an example of the Delany–Bazley and the Miki models is given in Figure 4.5 for rock wool with a density of 60 kg/m^3.

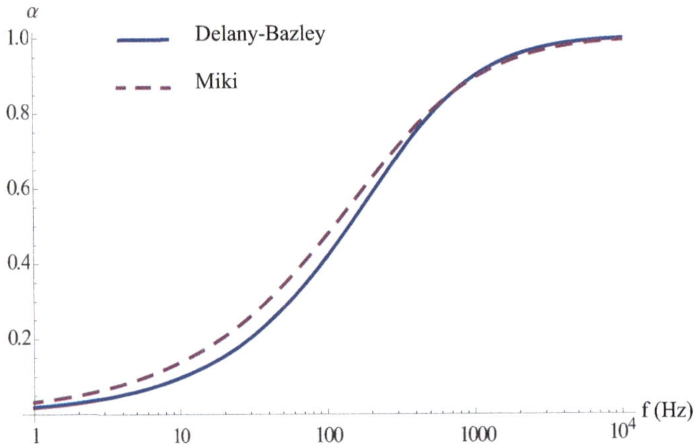

Figure 4.5. *Example of absorption coefficients $\alpha(f) = 1 - |\mathcal{R}(f)|^2$ for the Delany–Bazley and Miki models for rock wool.*

The parameters of this model have been modified by Miki [MIK 90] using the Kramers–Krönig relations[3] in order to make them causal. It suggests $a = 0.0699$, $b = -0.632$, $c = 0.107$, $d = b$, $p = 0.160$, $q = -0.618$, $r = 0.109$ and $q = s$.

Certain materials, such as foams, have a different dissipating behavior because the viscous dissipation within the fluid adds to the motion initiation in the foam backbone and an additional attenuation by vibration dissipation in the backbone material. The dynamics of such systems are a lot more complicated than those of simple fibers since there is a strong interaction between the two components of the material coupled to the effects due to the geometry of the foam (backbone resonance). In fact, they cannot be described by models as simple as the Delany–Bazley–Miki models.

3 Let us recall that the causality of $Z(\omega)$ includes a relationship between its real and imaginary parts [BER 01]. Let us have $Z(\omega) = R(\omega) + \imath X(\omega)$, then the acoustic resistance $R(\omega)$ and reactance $X(\omega)$ are bound by the Plemelj relations $R(\omega) = 1/\pi \int_{-\infty}^{+\infty} \frac{X(\omega')}{\omega'-\omega} d\omega'$ and $X(\omega) = -1/\pi \int_{-\infty}^{+\infty} \frac{R(\omega')}{\omega'-\omega} d\omega'$ or in a similar manner by the Kramers-Krönig relations $R(\omega) = 2/\pi \int_0^{+\infty} \omega' \frac{X(\omega')}{\omega'^2-\omega^2} d\omega'$ and $X(\omega) = -2/\pi \int_0^{+\infty} \omega' \frac{R(\omega')}{\omega'^2-\omega^2} d\omega'$.

4.6. Homogeneous anisotropic medium

In general, it is impossible to write an equation that relates only to pressure. But in an isotropic medium at rest (in the ocean or atmosphere), we have $c = c(M) \neq c_0$, $\rho_0 = \rho_0(M)$, $s_0 = s_0(M)$ and $v_0 = 0$. In the absence of sources, equations [4.1], [4.2] and [4.5] become:

$$\frac{\partial \rho}{\partial t} + \text{div}\,(\rho_0 \vec{v}) = 0 \qquad [4.12]$$

$$\rho_0 \frac{\partial \vec{v}}{\partial t} + \overrightarrow{\text{grad}}(p) = 0 \qquad [4.13]$$

$$\frac{\partial p}{\partial t} + \vec{v} \cdot \overrightarrow{\text{grad}}(p_0) - c_0^2 \left(\frac{\partial \rho}{\partial t} + \vec{v} \cdot \overrightarrow{\text{grad}}(\rho_0) \right) = 0. \qquad [4.14]$$

The conservation equation of impulse at the order zero [4.1] gives $\overrightarrow{\text{grad}}(p_0) = \rho_0 \vec{f_v}$. In the absence of volume forces, the static pressure gradient is zero and equation [4.14] becomes $\partial p/\partial t - c_0^2 \left(\partial \rho/\partial t + \vec{v} \cdot \overrightarrow{\text{grad}}(\rho_0) \right) = 0$. Investigating the difference between the time derivative of the conservation equation of mass [4.12] and the divergence of the Euler equation [4.13], we obtain the relation

$$\frac{\partial^2 \rho}{\partial t^2} + \underbrace{\frac{\partial}{\partial t}\text{div}\,(\rho_0 \vec{v}) - \text{div}\left(\rho_0 \frac{\partial \vec{v}}{\partial t} \right)}_{=0} - \Delta p = 0,$$

in which we can report equation [4.14], which leads to $\partial^2 \rho/\partial t^2 = 1/c^2 \partial^2 p/\partial t^2 - \partial \vec{v}/\partial t \cdot \overrightarrow{\text{grad}}(\rho_0) = 1/c^2 \partial^2 p/\partial t^2 + 1/\rho_0 \overrightarrow{\text{grad}}(p) \cdot \overrightarrow{\text{grad}}(\rho_0)$. From $1/\rho_0 \overrightarrow{\text{grad}}(p) \cdot \overrightarrow{\text{grad}}(\rho_0) - \Delta p = -\rho_0 \text{div} \left(\overrightarrow{\text{grad}}(p)/\rho_0 \right)$, we obtain the wave propagation equation in an anisotropic medium[4]:

$$\frac{1}{c^2(M)} \frac{\partial^2 p(M,t)}{\partial t^2} - \rho_0(M) \text{div} \left(\frac{1}{\rho_0(M)} \overrightarrow{\text{grad}}(p(M,t)) \right) = 0.$$

In a harmonic regime $\exp(-\imath \omega t)$, we obtain the Helmholtz equation with a variable index:

$$\rho_0(M) \text{div} \left(\frac{1}{\rho_0(M)} \overrightarrow{\text{grad}}(p(M)) \right) + k_0^2 n^2(M) p(M) = 0, \qquad [4.15]$$

where $k_0 = \omega/c_0$ is a reference wave number, c_0 is a reference velocity and $n(M) = c_0/c(M)$ is the refractive medium index.

4 We must remark that $\text{div} \left(\vec{A}/q \right) = 1/q \text{div} \vec{A} + \vec{A} \cdot \overrightarrow{\text{grad}}((1/q) = 1/q \text{div} \vec{A} - 1/q^2 \vec{A} \cdot \overrightarrow{\text{grad}}(q)$.

4.7. Medium with a slowly varying celerity

In general, we do not know how to solve the Helmholtz equation with a variable index [4.15] in a simple manner. But with the help of approximations, such as for slowly varying media which are either stratified or whose refractive index varies only in one direction, we can construct numerically efficient solutions for calculating the propagation over long distances or at high frequencies. Some examples of the methods are the parabolic approximation, the ray method or the geometric diffraction theory which takes into consideration the presence of obstacles.

To better understand these ideas, let us consider a medium which only varies in celerity (for example, due to a temperature gradient). Equation [4.15] becomes:

$$\Delta p(M) + k^2(M)p(M) = 0, k(M) = \omega/c(M)$$

We are looking for solutions in the form of $P(M) = A(M)\exp(\imath S(M))$ where $A(M)$ and $S(M)$ are real functions. We substitute this expression in the Helmholtz equation, we equal to zero the real and imaginary parts and obtain two equations:

$$\Delta A(M) + (k^2(M) - \Delta S(M))A(M) = 0, A(M)\Delta S(M) + 2\,\overrightarrow{\mathrm{grad}}A(M)\cdot\overrightarrow{\mathrm{grad}}S(M) = 0.$$

The last equation is the transport equation, which allows us to calculate the amplitude $A(M)$. If we are interested in wave behavior, we should consider the first. It can be written as $1/\left(k^2(M)A(M)\right)\Delta A(M) + 1 - 1/k^2(M)\Delta S(M) = 0$. Assuming that $1/\left(k(M)^2A(M)\right)\Delta A(M) \ll 1$, we obtain the Eikonal equation which allows us to describe the propagation of sound in a medium with a variable index: $\Delta S(M) = k^2(M)$. In this approximation, the wave trajectory is associated with a "sound beam", S is associated with the wave front and $\overrightarrow{\mathrm{grad}}S$ characterizes the direction of the beam. Let us take the example of a medium with revolution symmetry (O, r, z) whose celerity varies only with height ($k(M) = k(r, z) = k(z)$). We define θ as the angle between a beam and the horizontal. The Eikonal equation becomes $(\partial S/\partial r)^2 + (\partial S/\partial z)^2 = k^2(z)$. Let us study in detail the trajectory followed by a beam (locally, the trajectory coincides with the tangent to the curve). We have $dl^2 = dr^2 + dz^2$, with $\sin\theta(z) = dz/dl$ and $\cos\theta(z) = dr/dl$. We thus have $k^2(z) = k^2(z)(\cos^2\theta(z) + \sin^2\theta(z)) = k^2(z)((dr/dl)^2 + (dz/dl)^2)$. Thus, $(\partial S/\partial r)^2 + (\partial S/\partial z)^2 = k^2(z)\left[(dr/dl)^2 + (dz/dl)^2\right]$. Then, by identification, $\partial S/\partial r = k(z)dr/dl$ and $\partial S/\partial z = k(z)dz/dl$. The first relation gives $d/dl\,(\partial S/\partial r) = \partial/\partial r\,(dS/dl) = d/dl\,(k(z)dr/dl)$, but $dS/dl = \partial S/\partial r dr/dl + \partial S/\partial z dz/dl = k(z)\,(dr/dl)^2 + k(z)\,(dz/dl)^2 = k(z)$. That is $d/dl\,(k(z)dr/dl) = \partial/\partial r\,(dS/dl) = \partial/\partial r\,(k(z)) = 0$. Finally, one obtains $d(k(z)\sin\theta(z))/dl = 0$. This shows that $k(z)\cos\theta(z) = C$. This relation is an extension of the Snell–Descartes law for a medium whose characteristics vary continuously. From this law, it is easy to deduce the relationship between the curvature radius and the celerity profile. We start with $c_0 \cos\theta(z) = c(z)\cos\theta_0$, which we can derive with respect to the variable z, that is $-c_0 \sin\theta(z) d\theta(z)/dz = dc(z)/dz \cos\theta_0$. The curvature radius R is given by:

$$R^{-1} = \left|\frac{d\theta(z)}{dz}\right| = \frac{\cos\theta_0}{c_0 \sin\theta(z)}\left|\frac{dc(z)}{dz}\right|.$$

If the celerity gradient is zero, the curvature radius is infinite, the beams propagate in a straight line, if the gradient is constant (the celerity varies linearly with height), the curvature radius is constant and the trajectory is a circular arc.

4.8. Media in motion

4.8.1. *Homogeneous medium in uniform motion*

For a homogeneous ideal gas in uniform motion (parallel to the x axis), we have ρ_0 =cte, p_0 =cte, $\vec{v}_0 = v_0\vec{x}$, v_0 =cte. In the absence of sources, the motion equations are $d\rho/dt + \rho_0 \text{div}\,(\vec{v}) = 0$, $\rho_0 d\vec{v}/dt + \overrightarrow{\text{grad}}(p) = 0$, $ds/dt = 0$ and $p = c_0^2\rho + p/C_v s$. From there, we can deduce the convected wave equation $\Delta p - 1/c_0^2 d^2p/dt^2 = 0$. By introducing into the latter the material derivative, we obtain:

$$\Delta p - \frac{1}{c_0^2}\frac{\partial^2 p}{\partial t^2} - 2\frac{v_0}{c_0^2}\frac{\partial^2 p}{\partial t \partial x} - \frac{v_0^2}{c_0^2}\frac{\partial^2 p}{\partial x^2} = 0. \qquad [4.16]$$

In a harmonic regime $\exp(-\imath\omega t)$, with $M = v_0/c_0$, we obtain $\Delta p + k^2 p + 2\imath k M \partial p/\partial x - M^2 \partial^2 p/\partial x^2 = 0$.

4.8.1.1. *Continuity condition for normal displacements*

Let us take the example of a plane surface Σ whose normal $\vec{n}$ is oriented along the y axis. To establish the condition of continuity of normal displacements, we start from the conservation equation of impulse projected on the normal to the surface $\rho_0 \vec{n} \cdot d\vec{v}/dt + \vec{n} \cdot \overrightarrow{\text{grad}}(p) = 0$, with $\vec{n} \cdot \overrightarrow{\text{grad}}(p) = \partial p/\partial y$, which is coupled with the expression of the material derivative, which, for a flow parallel to the x axis, can be written as:

$$\frac{d}{dt} = \frac{\partial}{\partial t} + c_0 M \frac{\partial}{\partial x} = -\imath\omega + c_0 M \frac{\partial}{\partial x}$$

If $u_\Sigma(Q,\omega)$ is the displacement normal to the surface, and if the normal displacements of the fluid and surface are identical, we have:

$$\rho_0 \left(-\imath\omega u_\Sigma(Q,\omega) + c_0 M \frac{\partial u_\Sigma(Q,\omega)}{\partial x}\right)^2 + \frac{\partial p(Q,\omega)}{\partial y} = 0.$$

4.8.1.2. *Green's kernel*

It is shown that applying a Prandtl–Glauert transformation to Green's kernel in a three-dimensional space at rest, Green's kernel of the convected Helmholtz equation, for a flow parallel to the x axis with the Mach number M, can be expressed as:

$$G_e(M, M') = -\frac{e^{\imath k\tilde{R}}}{4\pi\sqrt{1-M^2}R^*}, \qquad [4.17]$$

$$\text{where } R^* = \sqrt{\frac{(x-x')^2}{1-M^2} + (y-y')^2 + (z-z')^2}$$

$$\text{and } \tilde{R} = \frac{1}{\sqrt{1-M^2}}\left(\frac{x-x'}{\sqrt{1-M^2}}M + R^*\right),$$

where $\sqrt{1-M^2}$ is the Prandtl-Glauert factor. For $M = 0$, we find Green's function in an infinite medium.

4.8.2. *Plane interface between media in motion*

Here, we are interested in the problem of reflection and transmission of a plane wave at the interface (parallel to the y axis) of two media in motion. When the two media are at rest, the boundary conditions are the continuity of both the pressure and acoustic velocity. When there is motion, velocity and frequency change while the displacement and normal direction to the wave front remain unchanged. This indicates that, if we always impose the continuity of acoustic pressure, the continuity of the acoustic velocity must be replaced by a displacement continuity condition.

In addition, a geometric continuity condition for the projection of the wavelength at the interface must be imposed. It is $\lambda_{1y} = \lambda_{2y}$, from where we obtain $k_{1y} = k_{2y}$, where k_{1y} and k_{2y} are the projections along the interface of wave numbers in media 1 and 2.

We assume that the incident field is a plane wave in the form $p_1 = \exp(-\imath\omega t + \imath\vec{k}_1 \cdot \vec{x})$ which can be written as $p_1 = \exp(-\imath\omega t)\exp\imath(k_{1x}x + k_{1y}y)$; we can easily obtain $p_1 = \exp(-\imath\omega t)\exp\imath(k_1\sin\phi_1 x + k_1\cos\phi_1 y)$. To consider the boundary condition, $k_{1y} = k_{2y}$, we must find the wave number expression in the media 1 and 2. To do this, we introduce the expression of pressures in the media 1 and 2 in the convected wave equation $c_i^2\Delta p_i - (\partial/\partial t + \vec{v}_i \cdot \overrightarrow{\text{grad}})^2 p_i = 0$:

$$\left(\frac{\partial}{\partial t} + V_1\frac{\partial}{\partial y}\right)^2 e^{-\imath\omega t}e^{\imath(k_1\sin\phi_1 x + k_1\cos\phi_1 y)} = c_1^2\Delta\left(e^{-\imath\omega t}e^{\imath(k_1\sin\phi_1 x + k_1\cos\phi_1 y)}\right).$$

From this, we obtain $(\omega - V_1 k_1 \cos\phi_1)^2 = c_1^2 k_1^2$. This is $\omega = k_1(c_1 + V_1 \cos\phi_1)$ and thus $k_1 = \omega/(c_1 + V_1 \cos\phi_1)$. The boundary condition $k_{1y} = k_{2y}$ can be written as $k_1 \cos\phi_1 = k_2 \cos\phi_2$. We obtain $V_1 + c_1/cos\phi_1 = V_2 + c_2/\cos\phi_2$ and $\cos\phi_2 = c_2 \cos\phi_1/\left(c_1 + \Delta V \cos\phi_1\right)$, with $\Delta V = V_2 - V_1$.

If the velocity variation observed at the crossing of the discontinuity is positive (for example, if medium 1 is at rest and medium 2 is in motion), then $\phi_2 < \phi_1$ and the critical angle of incidence (the angle below which there are no longer any waves transmitted $\phi_2 = 0$ and thus there is total reflection) is given by $\cos\phi_{1c} = c_1/(c2 + \Delta V)$. Let us now examine the reflection and transmission coefficients of waves.

If the incident and reflected fields are p_i and p_r, the total field in medium 1 is $p_1 = (\exp(\imath(k_{1x}x + k_{1y}y)) + R\exp(\imath(-k_{1x}x + k_{1y}y)))\exp(-\imath\omega t)$. Medium 2 contains a transmitted field in the form of $p_2 = T\exp(\imath(k_{2x}x + k_{2y}y)t)\exp t(-\imath\omega t)$.

The boundary conditions are the conditions of continuity of pressure and normal displacement at the $x = 0$ interface. The first gives $T = 1 - R$. The second is slightly more complicated since it involves the normal displacement u_x which is given, for example in medium 1, by the linearized Euler equation in a harmonic regime $\rho_1\left(\partial/\partial t + V_1 \partial/\partial y\right)^2 u_x^1 = -\partial p_1/\partial x$. This linearized Euler equation allows us to calculate u_x^1. In a harmonic regime, we obtain $\partial/\partial t = -\imath\omega$,

$$\rho_1\left(-\imath\omega + V_1\frac{\partial}{\partial y}\right)^2 u_x^1 = -\imath k_{1x}\left[\mathrm{e}^{\imath(k_{1x}x + k_{1y}y)} - R\mathrm{e}^{\imath(-k_{1x}x + k_{1y}y)}\right]$$

It is obvious that u_x^1 is in the form $u_x^1 = A[\exp(\imath(k_{1x}x + k_{1y}y)) - R\exp(\imath(-k_{1x}x + k_{1y}y))]$. We obtain $\rho_1(-\imath\omega + \imath V_1 k_{1y})^2 A = -\imath k_{1x}$. That is $\rho_1\left(\omega - V_1 k_1 \cos\phi_1\right)^2 A = \imath k_1 \sin\phi_1$, however, we know that $k_1 = \omega/(c_1 + V_1 \cos\phi_1)$, this is $\omega = k_1(c_1 + V_1 \cos\phi_1)$ and thus $\omega - k_1 V_1 \cos\phi_1 = k_1 c_1$. This is $A = \imath k_1 \sin\phi_1/\left(\rho_1 (k_1 c_1)^2\right)$.

Thus, the equation of continuity of normal displacement at $x = 0$ can be written as $u_{1x} = u_{2x}$ which translates into $(\rho_2 c_2^2 k_2 \sin\phi_1)(1 - R) = (\rho_1 c_1^2 k_1 \sin\phi_2)T$. The reflection coefficient is thus given by:

$$R = \frac{\rho_2 c_2^2 k_2 \sin\phi_1 - \rho_1 c_1^2 k_1 \sin\phi_2}{\rho_2 c_2^2 k_2 \sin\phi_1 + \rho_1 c_1^2 k_1 \sin\phi_2}, \cos\phi_2 = \frac{c_2 \cos\phi_1}{c_1 + \Delta V \cos\phi_1},$$

under normal incidence $\phi_1 = \phi_2 = \pi/2$, we obtain a common result $R = (\rho_2 c_2 - \rho_1 c_1)/(\rho_2 c_2 + \rho_1 c_1)$.

In practice, there is obviously never a situation that is as "perfect" as the one considered here, but the results can be useful as a first approximation for more complex propagation problems (such as those related to jets or the guided propagation in the atmosphere). For example, a jet can be considered for the first approximation as a wave guide. The acoustic pressure emitted by the acoustic sources of the jet in a medium at rest presents a "shadow area" in a cone directed forward (in the opposite direction of the jet) with an opening angle of $2\arccos(c_2/(c_1+V))$, where if V is the velocity of the jet, c_1 is the celerity of sound in the jet and c_2 is the celerity of sound at the exterior.

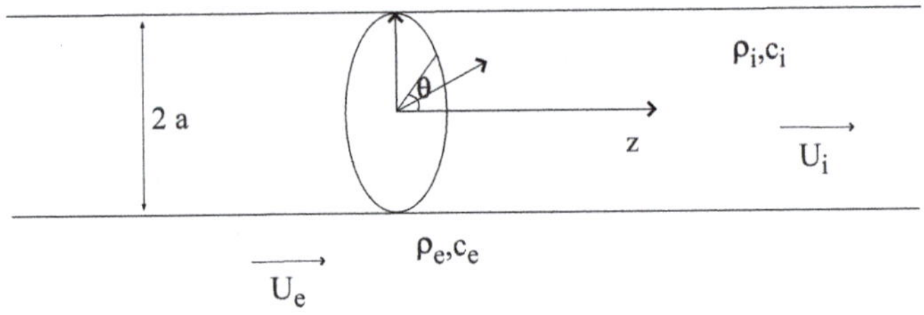

Figure 4.6. *Sheared jet geometry*

4.8.3. *Cylindrical interface between media in motion*

Let us consider an infinite medium marked (see Figure 4.6) by a cylindrical coordinate system (O, r, θ, z). This medium is occupied by a fluid with the density ρ_e and the wave celerity c_e. This fluid is given a uniform motion with a velocity of U_e. Within this medium, a fluid cylinder with a radius a, centered at the origin and extending to infinity, has a fluid density of ρ_i and a wave celerity of c_i. This fluid is given a uniform motion with a velocity of U_i. At the interface of the two media, we impose the continuity of pressure and displacements. The acoustic pressure at the interior and exterior of the cylindrical medium $p_{e,i}(M,t)$, for a medium in uniform motion $U_{e,i}$ parallel to the z axis, contains the solutions to the two wave equations given by [4.16] applied to $U_{e,i}$:

$$\Delta p_{e,i}(M,t) - \frac{1}{c_{e,i}^2}\frac{\partial^2 p_{e,i}(M,t)}{\partial t^2} - 2\frac{U_{e,i}}{c_{e,i}^2}\frac{\partial^2 p_{e,i}(M,t)}{\partial t \partial z} - \frac{U_{e,i}^2}{c_{e,i}^2}\frac{\partial^2 p_{e,i}(M,t)}{\partial z^2}$$
$$= f_{e,i}(M,t).$$

We assume that the following occurs in a harmonic regime. We aim to calculate Green's kernel of this environment. Due to the geometry of this problem, all the functions can be decomposed into an angular Fourier series:

$$f(r,\theta,z,t) = \sum_{m=-\infty}^{m=+\infty} f_m(r,z)\mathrm{e}^{\imath m\theta}\mathrm{e}^{-\imath\omega t}.$$

For each angular harmonic m, Green's kernel $G_m(r,z;r_0,z_0)$ is the solution to the problem:

$$\left(\frac{1}{r}\frac{\partial}{\partial r}r\frac{\partial}{\partial r}+\frac{m^2}{r^2}-\frac{1}{c_i^2}\left(-\imath\omega+U_i\frac{\partial}{\partial z}\right)^2+\frac{\partial^2}{\partial z^2}\right)G_m(r,z;r_0,z_0)$$
$$=\frac{\delta(r-r_0)}{r}\delta(z-z_0),\ \text{for } r<a$$
$$\left(\frac{1}{r}\frac{\partial}{\partial r}r\frac{\partial}{\partial r}+\frac{m^2}{r^2}-\frac{1}{c_e^2}\left(-\imath\omega+U_e\frac{\partial}{\partial z}\right)^2+\frac{\partial^2}{\partial z^2}\right)G_m(r,z;r_0,z_0)$$
$$=0,\ \text{for } r\geq a.$$

$G_m(r,z;r_0,z_0)$ is calculated by a direct and inverse Fourier transform with respect to z: $\tilde{G}_m(r,k_z;r_0,z_0)=\int_{-\infty}^{+\infty}G_m(r,z;r_0,z_0)\exp(-\imath k_z z)dz$. This is:

$$\left(\frac{1}{r}\frac{\partial}{\partial r}r\frac{\partial}{\partial r}+\frac{m^2}{r^2}+\kappa_i^2\right)\tilde{G}_m(r,k_z;r_0,z_0)=\frac{\delta(r-r_0)}{r}\mathrm{e}^{-\imath k_z z_0},\ \text{for } r<a$$
$$\left(\frac{1}{r}\frac{\partial}{\partial r}r\frac{\partial}{\partial r}+\frac{m^2}{r^2}+\kappa_e^2\right)\tilde{G}_m(r,k_z;r_0,z_0)=0,\ \text{for } r\geq a,$$

where $\kappa_i^2=(k_0^i+M_ik_z)^2-k_z^2$ and $\kappa_e^2=(k_0^e+M_ek_z)^2-k_z^2$, with $k_0^{i,e}=\omega/c_{i,e}$ and $\Im(\kappa_{i,e})>0$. The solution of $\left(1/r\partial_r r\partial_r+m^2/r^2+\kappa_i^2\right)\tilde{G}_m(r;r_0)=\delta(r-r_0)/r$ is given by [MOR 68, p. 366]:

$$\tilde{G}_m(r,k_z;r_0,z_0)=-\frac{\imath\pi}{2}\mathrm{H}_m(\kappa_i r_0)\mathrm{J}_m(\kappa_i r),r<r_0$$
$$\tilde{G}_m(r,k_z;r_0,z_0)=-\frac{\imath\pi}{2}\mathrm{H}_m(\kappa_i r)\mathrm{J}_m(\kappa_i r_0),r\geq r_0$$

where $\mathrm{H}_m(z)$ is Hankel's function of the first kind. The presence of the interface in $r=a$ results in the creation of a reflected field $\tilde{G}_{mR}(r;r_0)$ by the interface and a transmitted field $\tilde{G}_{mT}(r;r_0)$ at the exterior. We obtain:

$$\tilde{G}_{mR}(r;r_0)=-\frac{\imath\pi}{2}\mathrm{J}_m(\kappa_i r_0)R_m(k_z)\mathrm{J}_m(\kappa_i r)$$
$$\tilde{G}_{mT}(r;r_0)=-\frac{\imath\pi}{2}\mathrm{J}_m(\kappa_i r_0)T_m(k_z)\mathrm{H}_m(\kappa_e r),$$

where $R_m(k_z)$ and $T_m(k_z)$ are the coefficients of reflection and transmission. They are calculated to satisfy the conditions of continuity. The continuity of the pressure field can be given by $\mathrm{H}_m(\kappa_i a) + \mathrm{J}_m(\kappa_i a)R_m(k_z) = \mathrm{H}_m(\kappa_e a)T_m(k_z)$. Similarly, the continuity of the displacements results in:

$$\mathrm{H}'_m(\kappa_i a) + \mathrm{J}'_m(\kappa_i a)R_m(k_z) = \alpha_m(k_z)\mathrm{H}'_m(\kappa_e a)T_m(k_z) \text{ and } \alpha_m(k_z) = \frac{\kappa_e \rho_i(\omega + U_i k_z)^2}{\kappa_i \rho_e(\omega + U_e k_z)^2}.$$

These relations lead to the expression of the transmission and reflection coefficients:

$$T_m(k_z) = -\frac{2\imath}{\pi\kappa_i a} \frac{1}{\mathrm{J}'_m(\kappa_i a)\mathrm{H}_m(\kappa_e a) - \alpha_m(k_z)\mathrm{H}'_m(\kappa_e a)\mathrm{J}_m(\kappa_i a)},$$

$$R_m(k_z) = \frac{\mathrm{H}_m(\kappa_e a)T_m(k_z) - \mathrm{H}_m(\kappa_i a)}{\mathrm{J}_m(\kappa_i a)}.$$

By a spatially inverse Fourier transform $G_m(r, z; r_0, z_0) = 1/(2\pi)\int_{-\infty}^{+\infty} \tilde{G}_m(r, k_z; r_0, z_0)\exp(\imath k_z z)dk_z$, we obtain:

$$G_m(r, z; r_0, z_0) = \begin{cases} -\frac{\imath}{4}\int_{-\infty}^{+\infty} \mathrm{J}_m(\kappa_i r_0)\mathrm{H}_m(\kappa_e r)T_m(k_z)\mathrm{e}^{\imath k_z(z-z_0)}dk_z, r \geq a \\ -\frac{\imath}{4}\int_{-\infty}^{+\infty} \mathrm{J}_m(\kappa_i r_0)\mathrm{J}_m(\kappa_i r)(1 + R_m(k_z))\mathrm{e}^{\imath k_z(z-z_0)}dk_z, r < a \end{cases} \quad [4.18]$$

4.8.4. *Acoustic radiation of a moving surface*

Here, we establish a model of high-frequency directivity of an acoustic field created by a moving surface. To gain a better understanding of these ideas, let us study the case of a semi-infinite open pipe with zero thickness in the presence of external and internal flow in the pipe. This example describes in a simple way the radiation of a jet engine [MAT 06]. This model is based on the description of radiation from a semi-infinite cylinder using Kirchhoff's approximation. In this approximation, we consider that the radiation depends only on the field at the open end of the cylinder (the diffraction by the exterior of the cylinder is neglected).

The cylinder is immersed in a fluid in subsonic flow. However, for the internal part of the cylinder and in its extension on the outside, it is considered that its flow has characteristics that are different from the external flow: wave celerity, density and flow velocity. If the jet created by a reactor is not circular, it is still reasonable to assume that from an acoustic point of view, the essence of the phenomena of acoustic radiation of a field, which propagate downstream of the turbine, is adequately described by the model.

The chosen radiation model is the result of the work by Cargill [CAR 82]. This theory, developed initially for a sheared jet, remains valid for a medium in uniform flow or at rest. The theory proposed by Cargill is *a priori* a high-frequency theory which is valid only on the forward arc (between 0° and 90°) of the cylinder. Typically, Cargill indicates as the limit a low frequency of $kR \approx 10$ where k is the wave number and R is the radius of the cylinder. Nevertheless, the comparison of the results obtained by this method in a homogeneous medium at rest, and those from an exact method based on the Wiener–Hopf method [HOC 99, HOC 00], has shown an excellent agreement with respect to the forward arc including very low frequency. The main advantage of the method developed here lies in its extremely simple numerical implementation. The precision of the directivity diagram begins to degrade from an angle of about 100° relative to the cylinder axis.

4.8.4.1. *Geometry and notations*

Let us consider an infinite medium marked by a Cartesian coordinate system (O, x, y, z), cylindrical (O, r, θ, z) or spherical (O, r, θ, φ), as required. This medium contains a fluid characterized by its celerity c_e and density ρ_e. This fluid is driven by a uniform motion of velocity U_e parallel to the z axis (at Mach number $M_e = U_e/c_e$). This medium originally contains a semi-infinite rigid cylindrical pipe with a diameter of a. This half-cylinder is extended downstream by a fluid cylinder, in which the fluid celerity is c_i and density ρ_i, driven by a uniform motion of velocity U_i which is parallel to the z axis (at a Mach number $M_i = U_i/c_i$). S_e corresponds to the output section of the wave guide and S corresponds to the external part of the cylinder.

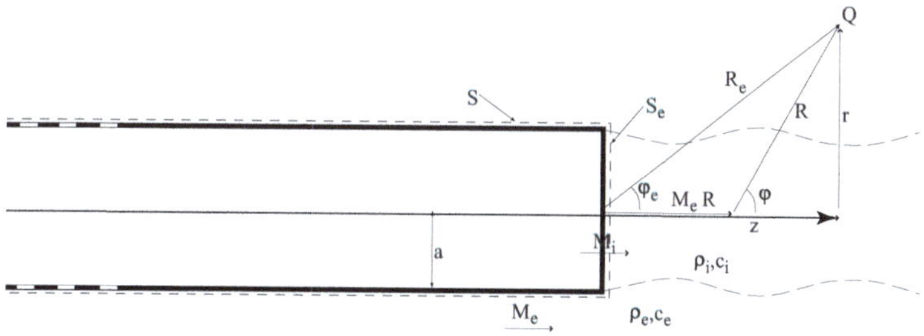

Figure 4.7. *Geometry of a semi-infinite cylinder in the presence of flow*

We consider here the acoustic field $p(M, t)$ received in the point of observation Q located outside of the cylinder and which moves along with the flow. It is equally important to consider that the system, which consists of the cylinder and observer, is described by a fixed observer in a medium at rest (for example, an observer on the ground) and by the cylinder moving according to Mach M_e. This point is identified in the system related to the cylinder by the spherical coordinates R_e, θ, φ_e. For convenience with respect to calculations, as well as to ease the interpretation, let us introduce a coordinate system based on the reception times [CRI 92]. In the reference frame related to the reception time, the coordinates of the point Q are R, θ, φ. We obtain:

$$R_e = \sqrt{R^2 \sin^2 \varphi + R^2(\pm M_e + \cos^2 \varphi)^2} = R\sqrt{1 + M_e^2 - 2M_e \cos^2 \varphi},$$

$$\sin \varphi_e = \frac{R \sin \varphi}{R_e} = \frac{\sin \varphi}{\sqrt{1 + M_e^2 - 2M_e \cos^2 \varphi}}$$

4.8.4.2. *Equation for wave propagation on the outside of the moving surface*

The acoustic pressure, for a medium in uniform motion U parallel to the z axis, is the solution to the equation

$$\left(\Delta - \frac{1}{c^2}\left(\frac{\partial}{\partial t} + U\frac{\partial}{\partial z}\right)^2\right) p(M,t) = 0 \qquad [4.19]$$

We define $\frac{D}{Dt} = \frac{\partial}{\partial t} + U\frac{\partial}{\partial z}$ as the material derivative. We aim to find the acoustic field on the outside of the moving surface (the cylinder) which is defined by $f = 0$ with $f > 0$ on the outside and $f < 0$ on the inside of the surface. To do this, we multiply equation [4.19] by $H(f)$ where $H(x)$ is the Heaviside step. We obtain $H(f)(\Delta p(M,t) - 1/c^2 Dp(M,t)/Dt) = 0$. However, we have $H\Delta p = \Delta(pH) - \text{div}(p\overrightarrow{\text{grad}}H) - \overrightarrow{\text{grad}}p\overrightarrow{\text{grad}}H$ and $HD^2p/Dt^2 = D^2pH/Dt^2 - D/Dt(pDH/Dt) + DH/DtDp/Dt$. From this:

$$\left(\Delta - \frac{1}{c^2}\frac{D}{Dt}\right)(p(M,t)H(f)) = \text{div}(p(M,t)\overrightarrow{\text{grad}}H(f)) + \overrightarrow{\text{grad}}p(M,t)\overrightarrow{\text{grad}}H(f)$$
$$-\frac{1}{c^2}\left(\frac{D}{Dt}\left(p(M,t)\frac{DH(f)}{Dt}\right) + \frac{DH(f)}{Dt}\frac{Dp(M,t)}{Dt}\right) = f^s \qquad [4.20]$$

Equation [4.20] is used to calculate the field radiated by the semi-infinite cylinder in the Kirchhoff approximation. It is assumed that the moving surface consists of the cylinder body S and the exit section S_e. The normal derivative of the pressure field on the surface S is hypothesized to be zero. In the Kirchhoff approximation, it is assumed that the pressure field on this surface is zero. The acoustic source is thus the exit field of the pipe on the S_e surface.

4.8.4.3. *Green's representation for a sheared jet*

To establish Green's representation of pressure, we start from equation [4.20]: $\left(\Delta - \frac{1}{c^2}\frac{D}{Dt}\right)\delta \star (pH) = f^s$, in which $\star$ is the convolution product of space and time. We calculate $G \star \left(\Delta - \frac{1}{c^2}\frac{D}{Dt}\right)\delta \star (pH) = G \star f^s$, where G is Green's function of the sheared jet (note that we only know this function in a harmonic regime). Then we easily obtain $p = G \star f^s$, by expanding the convolution product:

$$\begin{aligned}
p(M,t) = &\int_{V,t} G(M,t;M',t')(\mathrm{div}(p(M',t')\overrightarrow{\mathrm{grad}}H(f)) \\
&+\overrightarrow{\mathrm{grad}}p(M',t')\overrightarrow{\mathrm{grad}}H(f))dM'dt' \\
&-\frac{1}{c^2}\int_{V,t} G(M,t;M',t')\left(\frac{D}{Dt'}\left(p(M',t')\frac{DH(f)}{Dt'}\right)\right. \\
&\left.+\frac{DH(f)}{Dt'}\frac{Dp(M',t')}{Dt'}\right)dM'dt'
\end{aligned}$$

with:

$$\begin{aligned}
&\int_{V,t} G(M,t;M',t')\mathrm{div}(p(M',t')\overrightarrow{\mathrm{grad}}H(f))dM'dt' \\
&= \int_{V,t} \mathrm{div}(G(M;t,M',t')p(M',t')\overrightarrow{\mathrm{grad}}H(f)dM'dt' \\
&- \int_{V,t} p(M',t')\overrightarrow{\mathrm{grad}}_{M'}G(M,t;M',t')\cdot\overrightarrow{\mathrm{grad}}H(f)dM'dt'.
\end{aligned}$$

By integrating by parts, $\int_{V,t}\mathrm{div}(G(M,t;M',t')p(M',t')\overrightarrow{\mathrm{grad}}H(f)dM'dt' = 0$. Finally, we obtain

$$\begin{aligned}
&\int_{V,t} G(M,t;M',t')\mathrm{div}(p(M',t')\overrightarrow{\mathrm{grad}}H(f))dM'dt' \\
&= -\int_{V,t} p(M',t')\overrightarrow{\mathrm{grad}}_{M'}G(M,t;M',t')\cdot\overrightarrow{\mathrm{grad}}H(f)dM'dt'
\end{aligned}$$

Similarly, we obtain:

$$\int_{V,t} G(M,t;M',t')\frac{D\left(p(M',t')\frac{DH(f)}{Dt'}\right)}{Dt'}dM'dt'$$

$$= \int_{V,t}\frac{D}{Dt'}\left(G(M,t;M',t')p(M',t')\frac{DH(f)}{Dt'}\right)dM'dt'$$

$$- \int_{V,t}\frac{DG(M,t;M',t')}{Dt'}p(M',t')\frac{DH(f)}{Dt'}dM'dt'$$

$$= - \int_{V,t}\frac{DG(M,t;M',t')}{Dt'}p(M',t')\frac{DH(f)}{Dt'}dM'dt'$$

because, always integrating by parts, $\int_{V,t}\frac{D}{Dt'}\left(G(M,t;M',t')p(M',t')\frac{DH(f)}{Dt'}\right)$ $dM'dt' = 0$. Finally, from this:

$$p(M,t) = \int_{V,t}(G(M,t;M',t')\overrightarrow{\text{grad}}p(M',t') - p(M',t')\overrightarrow{\text{grad}}_{M'}G(M,t;M',t'))$$

$$\cdot\overrightarrow{\text{grad}}H(f))dM'dt'$$

$$-\frac{1}{c^2}\int_{V,t}\left(G(M,t;M',t')\frac{Dp(M',t')}{Dt'}\right.$$

$$\left.-p(M',t')\frac{DG(M,t;M',t')}{Dt'}\right)\frac{DH(f)}{Dt'}dM'dt'.$$

4.8.4.4. *Acoustic field radiated by the cylinder*

We fix the position of S_e at $z = 0$ and we obtain $\overrightarrow{\text{grad}}H(f)) = \overline{n}\delta(z)$ and $\frac{DH(f)}{Dt} = U_i\delta(z)$ where $\overline{n}$ is the normal to S_e. We obtain Kirchhoff's formula for a moving surface [FAR 96a]:

$$p_e(M,t) = \int_{S_e,t}(G(M,t;M',t')\partial_{\overline{n}}p_e(M',t')$$

$$-p_e(M',t')\partial_{\overline{n}(M')}G(M,t;M',t'))\,dM'dt'$$

$$-\frac{1}{c_i^2}\int_{S_e,t}\left(G(M,t;M',t')\frac{D_i p_e(M',t')}{Dt'}\right.$$
$$\left.-p_e(M',t')\frac{D_i G(M,t;M',t')}{Dt'}\right)U_i dM'dt'.$$

Due to the geometry 2π-periodical in θ of the problem, all the quantities can be decomposed into angular Fourier series $f(r,\theta,z/\varphi) = \sum_{m=-\infty}^{m=+\infty} f_m(r,z/\varphi)\mathrm{e}^{\imath m\theta}$. z/φ indicates that, depending on requirements, we can consider a cylindrical (using variable z) or a spherical (using variable φ) coordinate system. Let us now assume that we adopt a harmonic regime with a time dependence $\mathrm{e}^{-\imath\omega t}$. The acoustic pressure on the outside of the pipe can thus be given by:

$$\begin{aligned}p_{me}(r,z)=&\int_0^a\left(G_m(r,z;r_0,z_0)\frac{\partial p_{me}(r_0,z_0)}{\partial z_0}\right.\\&\left.-p_{me}(r_0,z_0)\frac{\partial G_m(r,z;r_0,z_0)}{\partial z_0}\right)_{z_0=0} r_0 dr_0\\&-\frac{U_i}{c_i^2}\int_0^a\left(G_m(r,z;r_0,z_0)\frac{D_i p_{me}(r_0,z_0)}{Dt_0}\right.\\&\left.-p_{me}(r_0,z_0)\frac{D_i G_m(r,z;r_0,z_0)}{Dt_0}\right)_{z_0=0} r_0 dr_0,\end{aligned} \quad [4.21]$$

where $D_i/Dt_0 = -\imath\omega + U_i\frac{\partial}{\partial z_0}$. We have by continuity on the exit plane of the guide $p_{me}(r_0,z_0) = p_{mi}(r_0,z_0)$ at $z_0 = 0$ and $\partial_{z_0}p_{me}(r_0,z_0) = \partial_{z_0}p_{mi}(r_0,z_0)$ at $z_0 = 0$ where $p_{mi}(r,z)$ is the incident pressure in the pipe. We assume that the incident pressure is given by the mode of propagation of the index mn and the wave number k_{mn}^e [FAR 96b] $p_{mi}(r,z) = \mathrm{J}_m(j'_{mn}r/a)\exp(\imath k_{mn}^e z)$, where $k_{mn}^e = k_0^i/\beta_i^2(-M_i + \sqrt{1-(\beta_i\kappa_{mn}^i/k_0^i)^2})$ where $k_0^i = \omega/c_i$, $\beta_i = \sqrt{1-M_i^2}$ and $\kappa_{mn}^i = j'_{mn}/a$. $\mathrm{J}_m(z)$ is the Bessel function of the first kind (for more details, refer to section 5.9.2). Using the continuity relations and the expressions related to incident pressure and Green's function [4.18], we obtain for the downstream pressure on the outside of the jet:

$$\begin{aligned}p_{me}(r,z) = -\frac{\imath}{4}\int_{-\infty}^{\infty}&\left[T_m(k_z)\left\{-\imath(-k_z - k_{mn}^e)\right.\right.\\&\left.+\imath\frac{U_i}{c_i^2}\left[(\omega - U_i k_{mn}^e) - (\omega + U_i k_z)\right]\right\}\\&\left.\int_0^a \mathrm{J}_m(\kappa_i r_0)\mathrm{J}_m(j'_{mn}\frac{r_0}{a})r_0 dr_0\ \mathrm{H}_m(\kappa_e r)\mathrm{e}^{\imath k_z z}\right] dk_z.\end{aligned}$$

However, we know that $\int(\beta^2 - \alpha^2)z\mathrm{J}_m(z\alpha)\mathrm{J}_m(\beta z)dz = z\left(\alpha\mathrm{J}_m'(\alpha z)\mathrm{J}_m(\beta z)\right.$ $\left.-\beta\mathrm{J}_m(\beta z)\mathrm{J}_m'(\alpha z)\right)$. This is:

$$\int_0^a \mathrm{J}_m(\kappa_i r_0)\mathrm{J}_m(j_{mn}^{'}\frac{r_0}{a})r_0 dr_0 = -\frac{a\kappa_i}{\kappa_i^2 - \left(\frac{j_{mn}^{'}}{a}\right)^2}\mathrm{J}_m(\kappa_i a)\mathrm{J}_m(j_{mn}^{'})$$

Finally, with $\kappa_{i,e} = \kappa_{i,e}(k_z)$, the expression of the pressure radiated on the outside of the jet is:

$$p_{me}(r,z) = -\frac{\imath}{4}\int_{-\infty}^{\infty}\left[T_m(k_z)\left\{-\imath(-k_z - k_{mn}^{e}) + \imath\frac{U_i}{c_i^2}\left[(\omega - U_i k_{mn}^{e}) - (\omega + U_i k_z)\right]\right\}\right.$$
$$\left.\frac{-a\kappa_i}{\kappa_i^2 - \left(\frac{j_{mn}^{'}}{a}\right)^2}\mathrm{J}_m(\kappa_i a)\mathrm{J}_m(j_{mn}^{'})\ \mathrm{H}_m(\kappa_e r)\mathrm{e}^{\imath k_z z}\right] dk_z. \qquad [4.22]$$

4.8.4.5. *Pipe directivity*

Let us consider the far-field $R_e \gg 1$, that is $R \gg 1$. The cylindrical coordinates of the point Q in the reference frame linked to the reception times (see Figure 4.7) are given by $(r,\theta,z) = (R\sin\varphi, \theta, R(M_e + \cos\varphi))$. As long as it is located far enough away from the jet axis, we have $r = R\sin\varphi \gg 1$ and by asymptotic development $\mathrm{H}_m(\kappa_e r) \approx \sqrt{2/(\pi\kappa_e r)}\exp\imath(\kappa_e r - m\pi/2 - \pi/4)$. Equation [4.22] $p_{me}(r,z) = \int_{-\infty}^{\infty} f_m(k_z)\exp(\imath(\kappa_e r - k_z z))dk_z$ with

$$f_m(k_z) = -\frac{\imath}{4}T_m(k_z)\left\{-\imath(-k_z - k_{mn}^{e}) + \imath\frac{U_i}{c_i^2}\left[(\omega - U_i k_{mn}^{e}) - (\omega + U_i k_z)\right]\right\}$$
$$\times\frac{-a\kappa_i}{\kappa_i^2 - \left(\frac{j_{mn}^{'}}{a}\right)^2}\mathrm{J}_m(\kappa_i a)\mathrm{J}_m(j_{mn}^{'})\sqrt{\frac{2}{\pi\kappa_e r}}\mathrm{e}^{-\imath(m\frac{\pi}{2}+\frac{\pi}{4})}.$$

However, the exponential term can be written as $\exp(\imath(\kappa_e r - k_z z)) = \exp(\imath R(\kappa_e(k_z)\sin\varphi - k_z(M_e + \cos\varphi))) = \exp(\imath Rh(k_z))$. We thus have $p_{me}(r,z) = p_{me}(R,\varphi) = \int_{-\infty}^{\infty} f_m(k_z)\exp(\imath Rh(k_z))dk_z$. We approach this integral using the stationary phase approximation [CRI 92] and we obtain:

$$p_{me}(R,\varphi) \approx f_m(k_z^0)\sqrt{\frac{2\pi}{R\left|h''(k_z^0)\right|}}\mathrm{e}^{\imath Rh(k_z^0)+\imath\mathrm{sgn}(h''(k_z^0))\frac{\pi}{4}}$$

if $h'(k_z^0) = 0$ and $h''(k_z^0) \neq 0$. It is easy to show that $k_z^0 = -k_0^e \cos\varphi/(1 + M_e \cos\varphi)$ where $k_0^e = \omega/c_e$. It can be seen that $h(k_z^0) = k_0^e$ and that $h''(k_z^0) = -(1 + M_e \cos\varphi)^3/(k_0^e \sin\varphi)^2 < 0$. Finally, this results in:

$$p_{me}(R,\varphi) \approx f_m(k_z^0)\sqrt{\frac{2\pi(k_0^e \sin\varphi)^2}{R(1 + M_e \cos\varphi)^3}}\mathrm{e}^{\imath R k_0^e}\mathrm{e}^{-\imath\frac{\pi}{4}}.$$

We now define, while following Cargill's notations, $\chi_0(k_z^0) = -k_z^0 + M_i(k_0^i + M_i k_z^0)$, $\chi_{mn}(k_z^0) = -k_{mn}^e - M_i(k_0^i - M_i k_{mn}^e)$ and $D_m(k_z^0) = -\imath a(\chi_0(k_z^0) + \chi_{mn}(k_z^0))$, $I_m(k_z^0) = -2a\kappa_i(k_z^0)/ (\kappa_i^2(k_z^0)a^2 - j_{mn}^{'2})\mathrm{J}_m(\kappa_i(k_z^0)a)\mathrm{J}_m(j_{mn}^{'})$. Finally, this is:

$$\begin{aligned} p_{me}(R,\varphi) &\approx -\frac{\mathrm{e}^{\imath R k_0^e}}{4\pi R}T_m(k_z^0)D_m(k_z^0)I_m(k_z^0)\frac{\pi a}{1 + M_e \cos\varphi}\mathrm{e}^{-\imath m\frac{\pi}{2}} \\ &= -\frac{\mathrm{e}^{\imath k_0^e R}}{4\pi R}\Xi_{mn}^e(k_0^e,\varphi) \end{aligned}$$

where $\Xi_{mn}^e(k_0^e,\varphi)$ is the directivity function (or directivity diagram) of the pipe. This notion is an easy way of representing the spatial distribution of sound pressure. The directivity function is defined by the ratio between the acoustic pressures that radiate from the source in question and an omnidirectional point source located at the origin (Green's kernel) – $\exp(\imath kr)/4\pi r$. This pipe has a maximum directivity of φ_{mn}^{pe}. It is easy to see that the directivity is at its maximum when $\Xi_{mn}^e(k_0^e,\varphi)$ is at its maximum. This occurs when $I_m(k_z^0)$ is at its maximum, that is to say for $\kappa_i^2(k_z^{0pe})a^2 = j_{mn}^{'2}$ with $k_z^{0pe} = -k_0^e \cos\varphi_{mn}^{pe}/((1 + M_e \cos\varphi_{mn}^{pe}))$. From $\kappa_i^2(k_z) = (k_0^i + M_i k_z)^2 - k_z^2$, we use the biquadratic equation in $\cos\varphi_{mn}^{pe}$:

$$(M_\alpha^2 - \alpha_c^2 - \alpha_{mn}^2 M_e^2)\cos^2\varphi_{mn}^{pe} + 2(M_\alpha - \alpha_{mn}^2 M_e)\cos\varphi_{mn}^{pe} + 1 - \alpha_{mn}^2 = 0,$$

with $\alpha_c = c_i/c_e$, $M_\alpha = M_e - \alpha_c M_i$ and $\alpha_{mn} = j_{mn}^{'}/(k_0^i a)$.

We thus obtain

$$\varphi_{mn}^{pe} = \arccos\frac{\alpha_{mn}^2 M_e - M_\alpha + \sqrt{(M_\alpha - \alpha_{mn}^2 M_e)^2 - (1 - \alpha_{mn}^2)(M_\alpha^2 - \alpha_c^2 - \alpha_{mn}^2 M_e^2)}}{M_\alpha^2 - \alpha_c^2 - \alpha_{mn}^2 M_e^2}. \quad [4.23]$$

4.8.4.6. *Results*

Here, we present some results obtained for a cylinder with the radius $a = 0.086\ m$. The inner and outer fluids are air ($\rho_{i,e} = 1.3$, $c_{i,e} = 330\ m/s$). The mode considered is the mode (3.2) whose cutoff frequency is given by $k_0^e a = j'_{32} = 8.015$. Figures 4.8 and 4.9 present the evolution of the maxima of directivity as functions of the Mach numbers and frequency. As shown in the equations, the maximum radiation of the pipe ($\varphi^{pe} \approx 30^o$) depends on the differences between internal and external Mach number. This can be seen in the left curve in Figure 4.8 or at a fixed external Mach number, the maximum radiation of the pipe moves away from its axis along with an increase in the internal Mach number and shows the jet radiation becoming normal to its axis. The effect of this difference is clearly visible in the right curve in Figure 4.8 where there exists an external Mach number from which we can no longer truly define a maximum directivity downstream of the cylinder. Everything happens as if the acoustic field remained confined to the jet. Finally, at fixed Mach numbers, it is easy to see in Figure 4.9 that the maximum directivity of the acoustic radiation of the cylinder moves away from its axis at low frequency.

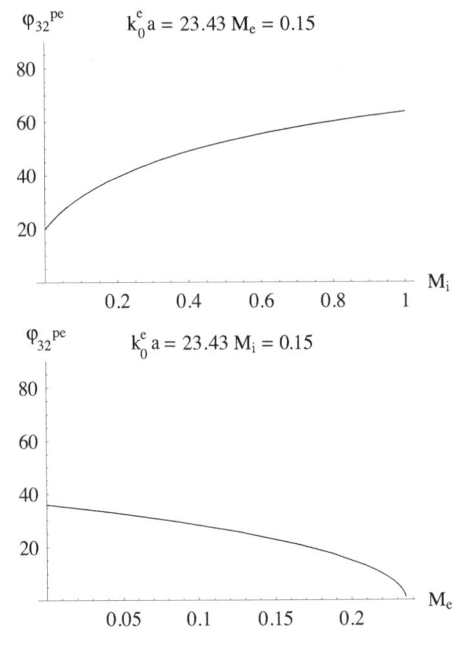

Figure 4.8. *Evolution of the directivity of the cylinder as a function of the internal (left) and external (right) Mach numbers*

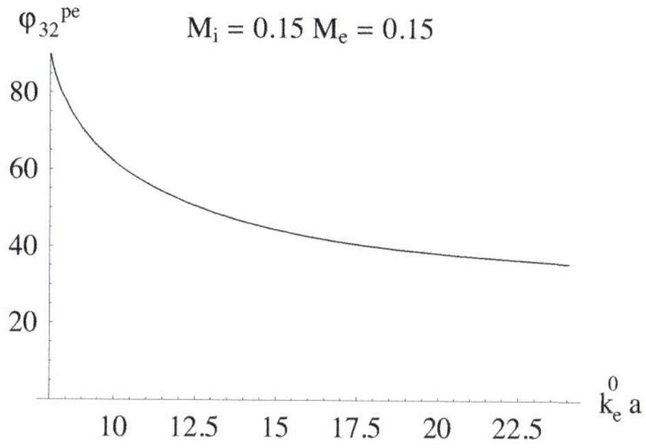

Figure 4.9. *Evolution of the directivity of the cylinder as a function of the frequency*

5

Radiation, Diffraction, Enclosed Space

Here, we present additional concepts, which are the propagation of acoustic waves in a medium in the presence of boundaries and acoustic radiation. We have different approaches for propagation outside of an obstacle, propagation in an enclosed space and in wave guides. We begin by defining acoustic radiation of point sources, then we generalize these concepts to radiation of distributed sources using surface integrals (layer potentials). Then, we arrive at Green's representation of pressure which allows us to introduce the concept of diffraction. We begin by calculating the directivity of a circular baffled piston and then we present several examples of radiation of elementary rectangular structures. For acoustics in an enclosed medium (such as rooms), we distinguish transient (establishment and extinction of sound) and harmonic regimes (time-independent). We start with some aspects of harmonic regimes in an enclosed environment. We then give a brief overview of transient phenomena using Sabine's formula. The properties of guided media can be described partly using the properties of enclosed media, but, they have additional properties, which particularize their study (a preferred direction of propagation and the possible presence of uniform flow). The study of wave propagation in wave guides is of great practical interest. In all ventilation problems, the sound propagates in the ducts for very large distances from the acoustic source. Wind musical instruments strongly rely on particular acoustic phenomena related to their geometry. For the sake of simplicity, we will only study in detail the straight wave guides of a rectangular section. The generalization for different geometries (such as circular and elliptical pipes) requires only little modification of the theoretical developments presented here. We will only deal with a general idea of the propagation in ducts in a non-constant section. This aspect has been the subject of more specific studies in relation to musical instruments [FLE 91, CHA 13].

5.1. Acoustic radiation

5.1.1. *A simple example*

The d'Alembert equation that governs the wave propagation in spherical coordinates in a homogeneous and isotropic medium:

$$\frac{1}{r}\frac{\partial^2 rp}{\partial r^2} - \frac{1}{c_0^2}\frac{\partial^2 p}{\partial t^2} = 0,$$

has a linear combination of outgoing (F) and incoming (G) waves as its solution: $rp(r,t) = F(r - c_0 t) + G(r + c_0 t)$.

Consider the example of a sphere with a radius R, which expands and contracts uniformly with a radial velocity $V(t)$. The surface outflow is $\rho_0 q(t) = 4\pi R^2 \rho_0 V(t)$. Let us calculate the pressure that radiates this sphere. We start with the fundamental law of dynamics $\rho_0 \partial V(t)/\partial t = -\partial p(r,t)/\partial r$, in which the pressure $p(r,t)$ is composed only of an outgoing wave $p(r,t) = P(r - c_0 t)/r$. We have $\partial p(r,t)/\partial r = \partial(P(r - c_0 t)/r)/\partial r = 1/r \partial P(r - c_0 t)/\partial r - 1/r^2 P(r - c_0 t)$. Thus, in $r = R$, we obtain $1/R \partial P/\partial r - P/R^2 = -\rho_0/\left(4\pi R^2\right) \partial q/\partial t$. If the sphere is small and R is small compared to the wavelength (which corresponds to a compact source and/or a large wavelength), we can assume that $P/R \gg \partial P/\partial r$ in $r = R$. Thus, $P \approx (\rho_0/4)\pi dq(t)/dt$ at $r = R$. The radiated pressure is thus:

$$p(r,t) = \frac{\rho_0}{4\pi r} q'(\tau_r), \tau_r = t - r/c_0 \qquad [5.1]$$

The time τ_r, or retarded time, characterizes the time that it takes for the signal to propagate from the source to the receiver. The pressure radiated by a pulsating sphere depends only on the distance from the origin. This is known as monopolar radiation. We must note that, if the flow rate is constant (if the radial velocity of the sphere in the present case is invariable), then the acoustic pressure is zero.

If the flow rate significantly varies, for example, from 0 for $t < 0$ to 1 for $t > 0$, then the flow $q(t) = H(t)$ is a Heaviside step and it is obvious that $dH(t)/dt = \delta(t)$. The wave created will be an impulse modeled by a Dirac distribution $(\rho_0/4\pi r)\ \delta(t - r/c_0)$. We can already specify that the resulting function represents Green's kernel for the d'Alembert equation. It is the function $G(M, t; S, \tau)$, the solution of $1/c_0^2 \partial^2 G/\partial t^2 - \Delta G = \delta(M - S)\delta(t - \tau)$. We have $G(M, t; S, \tau) = 1/(4\pi r)\delta\left(t - \tau - r/c_0\right)$, where $r = d(M, S)$ is the distance between M and S. G represents a spherical impulse which arrives at the observer at time $t = \tau + r/c_0$ and which decreases as $1/r$. The solution of the wave equation for any source $Q(S,t)$ can be given by:

$$p(M,t) = G * Q = \frac{1}{4\pi}\int_{V(S)}\int_{\tau} Q(S,\tau)\delta\left(t-\tau-\frac{r}{c_0}\right)\frac{dS}{r}d\tau$$

$$= \frac{1}{4\pi}\int_{V(S)} Q\left(S,t-\frac{r}{c_0}\right)\frac{dS}{r}. \qquad [5.2]$$

In a harmonic regime, the source can be written as $Q(S,t) = \rho_0 \partial Q_m(S,t)/\partial t = \rho_0 \partial(Q_m \exp(-\imath\omega t)\delta(S))/\partial t$, the acoustic field is given by relation [5.2]:

$$p(M,t) = \frac{\rho_0}{4\pi}\int_{V(S)} Q\left(S,t-\frac{r}{c_0}\right)\frac{dS}{r} = -\imath k\rho_0 c_0 Q_m \mathrm{e}^{-\imath\omega t}\frac{\mathrm{e}^{\imath kr}}{4\pi r},$$

where $k = \omega/c_0$ is the wave number. We get $p(M,t) = \imath k\rho_0 c_0 Q_m \mathrm{e}^{-\imath\omega t}G(M,0)$ where $G(M,S)$ is Green's kernel for the Helmholtz three-dimensional equation.

5.2. Acoustic radiation of point sources

5.2.1. *Multipolar sources in a harmonic regime*

As we have seen, a source that is very small, in comparison to the length of the wave that emits energy in all directions in a uniform manner, presents omnidirectional radiation. It can be described by Green's function $G(M,S)$ and it is called a monopole.

Experience shows that only a few sources present monopolar radiation. It is necessary to consider the radiation of more complex point sources. Let us consider, for example, the system formed by two monopoles. One is located at $S^+(x=\epsilon, y=0, z=0)$ and the other is located at $S^-(x=-\epsilon, y=0, z=0)$. The first one has an amplitude of $-1/2\epsilon$ and the second one has an amplitude of $1/2\epsilon$. In practice, this corresponds to two identical sources in phase opposition ($-1 = \mathrm{e}^{\imath\pi}$). The pressure field created by these two sources is the solution of:

$$(\Delta + k^2)p_\epsilon(M) = -\frac{1}{2\epsilon}\left(\delta_{S^+}(M) - \delta_{S^-}(M)\right).$$

It is then easy to see that we have $p_\epsilon(M) = -\frac{1}{2\epsilon}\left(G(M,S^+) - G(M,S^-)\right)$. This is:

$$p_\epsilon(M) = \frac{1}{2\epsilon}\left(\frac{\mathrm{e}^{\imath kr^+}}{4\pi r^+} - \frac{\mathrm{e}^{\imath kr^-}}{4\pi r^-}\right)$$

with $r^{\pm} = \sqrt{(x \mp \epsilon)^2 + y^2 + z^2}$. We can show that moving toward the limit $\epsilon \to 0$ we obtain:

$$p(M) = -\frac{e^{\imath kr}}{4\pi r}\left(\imath k - \frac{1}{r}\right)\frac{x}{r} = \frac{\partial}{\partial x}\left(-\frac{e^{\imath kr}}{4\pi r}\right).$$

EXERCISE.– Do this.

SOLUTION.–*If ϵ is small, we can perform a perturbation series expansion with respect to ϵ. We set $r^2 = x^2 + y^2 + z^2$, we get $r^{\pm 2} = x^2 \mp 2x\epsilon + \epsilon^2 + y^2 + z^2$, and $r^{\pm 2} \approx r^2 \mp 2x\epsilon$. Finally, it is $r^{\pm} \approx r \mp \epsilon x/r$. We, thus, get:*

$$\frac{e^{\imath kr^{\pm}}}{4\pi r^{\pm}} \approx \frac{e^{\imath k\left(r \mp \epsilon \frac{x}{r}\right)}}{4\pi r\left(1 \mp \epsilon\frac{x}{r^2}\right)} = \frac{e^{\imath kr}}{4\pi r}\frac{e^{\mp \imath k\epsilon \frac{x}{r}}}{1 \mp \epsilon\frac{x}{r^2}}.$$

Expanding $e^{\epsilon} = 1 + \epsilon$ and $1/(1+\epsilon) = 1 - \epsilon$, we obtain:

$$\frac{e^{\imath kr^{\pm}}}{4\pi r^{\pm}} \approx \frac{e^{\imath kr}}{4\pi r}\left(1 \mp \imath k\epsilon\frac{x}{r}\right)\left(1 \pm \epsilon\frac{x}{r^2}\right) \approx \frac{e^{\imath kr}}{4\pi r}\left(1 \mp \epsilon\left(\imath k - \frac{1}{r}\right)\frac{x}{r}\right).$$

From where:

$$p_\epsilon(M) = \frac{1}{2\epsilon}\frac{e^{\imath kr}}{4\pi r}\left(\left(1 - \epsilon\left(\imath k - \frac{1}{r}\right)\frac{x}{r}\right) - \left(1 + \epsilon\left(\imath k - \frac{1}{r}\right)\frac{x}{r}\right)\right) + \mathcal{O}(\epsilon^2)$$

This is $p_\epsilon(M) = -\exp(\imath kr)/(4\pi r)\,(\imath k - 1/r)\,x/r + \mathcal{O}(\epsilon^2)$, we thus obtain the sought result moving toward the limit $\epsilon \to 0$.

This radiation is called the dipole radiation. The corresponding source, the dipole here oriented along the x axis, is represented by the derivative with respect to x of the Dirac distribution. In fact, we get:

$$(\Delta_M + k^2)\frac{\partial}{\partial x}\left(G(M,S)\right) = \frac{\partial}{\partial x}\left((\Delta_M + k^2)G(M,S)\right) = \frac{\partial}{\partial x}\left(\delta_S(M)\right)$$

because in the distribution sense, we can interchange the derivations. On a more general note, let S be the location of a dipole and $\vec{n}_S$ be the unitary vector which defines its orientation, moreover, it radiates throughout the space the pressure $p(M) = \vec{n}_S \cdot \overrightarrow{\text{grad}} G(M,S) = \partial_{n_S} G(M,S)$, where ∂_{n_S} indicates that the normal derivative is with respect to the variable S. The modulus of the pressure radiated by a dipole, for a dipole oriented along the x axis, is given by $|p(M)| = |x|/(4\pi r^2)\sqrt{k^2 - 1/r^2}$. For a fixed r, the pressure is at its maximum along the x axis and zero in the $x = 0$ plane (see Figure 5.1). We say that such a source has two directivity lobes. This type of dipole radiation is such that it can be observed on a single speaker (not embedded within an enclosure) or on a flat vibrating plane (this is similar to doublet source/sink potential in fluid mechanics). When the dipole source is no longer a point, for example, for two sources of flow $\rho_0 Q_m$ and ℓ distance apart, the acoustic pressure depends on the dipole moment $Q_m \ell$. If ϕ is the angle formed by the axis of the dipole and a point in space, in the far-field we get:

$$p(M,t) \approx -\imath k^2 \rho_0 c_0 \mathrm{e}^{-\imath\omega t} Q_m \ell \cos\phi \frac{\mathrm{e}^{\imath kr}}{4\pi r}.$$

Let us recall that an easy way to represent the spatial distribution of sound pressure is to consider the directivity $\Xi(\theta,\phi)$ (sometimes called the shape function), the ratio between acoustic pressures which the considered source radiates and an omnidirectional point source located at the origin. Figure 5.1 presents the directivity $\Xi(\theta,\phi)$ of a monopole and a dipole.

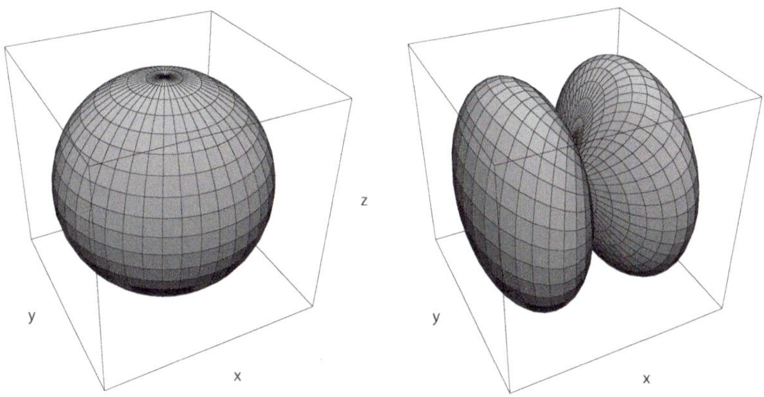

Figure 5.1. *Pressure radiated by a monopole a) with the directivity $\Xi(\theta,\phi) \sim 1$ and a dipole oriented along the x axis b) with the directivity $\Xi(\theta,\phi) \sim \cos\phi$*

Another type of acoustic radiation which is very important, for example, when we are interested in the acoustic radiation of turbulence, is that of the quadrupole. We get

$\Delta p + k^2 p = \partial^2 \delta(M)/\partial x_i \partial x_j$, for which the solution can be obtained by deriving Green's function in the spatial directions considered: $\partial^2 G/\partial x_i \partial x_j$. We get:

$$\frac{\partial^2 G}{\partial x_i \partial x_j} = \frac{e^{\imath kr}}{4\pi r}\left[\frac{k^2 x_i x_j}{r^2} + \frac{\imath k\left(3x_i x_j - r^2 \delta_i^j\right)}{r^3} - \frac{3x_i x_j - r^2\delta_i^j}{r^4}\right],$$

$$x_i = x = r\sin\theta\cos\phi$$
$$x_j = y = r\sin\theta\sin\phi$$

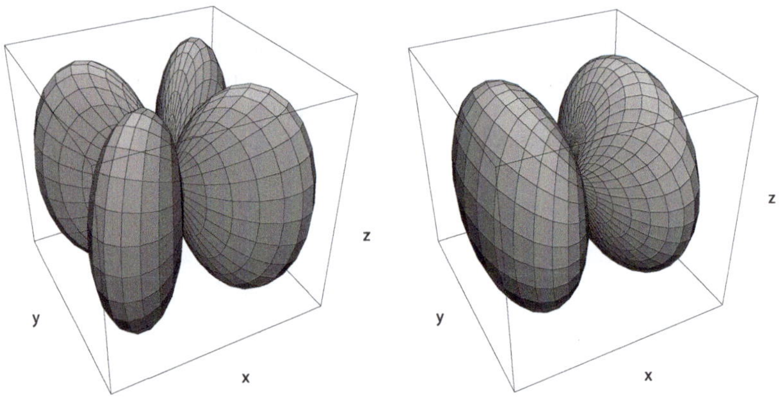

Figure 5.2. *Pressure radiated in the far-field by a quadrupole, lateral with the directivity* $\Xi(\theta,\phi) \sim \cos\phi\sin\phi$ *or longitudinal with the directivity* $\Xi(\theta,\phi) \sim \cos^2\phi$

The acoustic pressure radiated thus strongly depends on the type of the quadrupole. The longitudinal quadrupole can be obtained by $x_i = x_j$ and corresponds to two dipoles placed side by side, while the lateral quadrupole can be obtained by $x_i \neq x_j$ and corresponds to two superimposed dipoles. Figure 5.2 presents the corresponding directivities $\Xi(\theta,\phi)$. In practice, the sources encountered in physics present more complex directivity patterns. The number of lobes (each of these lobes corresponds to a direction in which the radiated energy is significant) can be high. Thus, we must appeal to the concept of a multipolar source. One such source corresponds to a linear combination derived from the Dirac distribution.

$$f(M) = \sum_{l,m,n} A_{lmn}\frac{\partial^{l+m+n}\delta(M)}{\partial x^l \partial y^m \partial z^n},$$

each term radiates a pressure given by:

$$p_{lmn}(M) = A_{lmn} \frac{\partial^{l+m+n}}{\partial x^l \partial y^m \partial z^n} \left(-\frac{e^{\imath kr}}{4\pi r} \right).$$

5.2.2. *Far-field*

For any source $Q(M,t)$, the solution of the wave equation can be given by relation [5.2]:

$$p(M,t) = \frac{1}{4\pi} \int_{V(S)} Q\left(S, t - \frac{r}{c_0}\right) \frac{dS}{r},$$

where $r = d(M,S)$ (see Figure 5.3). We have $r_m = d(M,O)$, $r_S = d(S,O)$, $\overline{M}$ is the vector $\overline{OM}$ and $\overline{S}$ is the vector $\overline{OS}$.

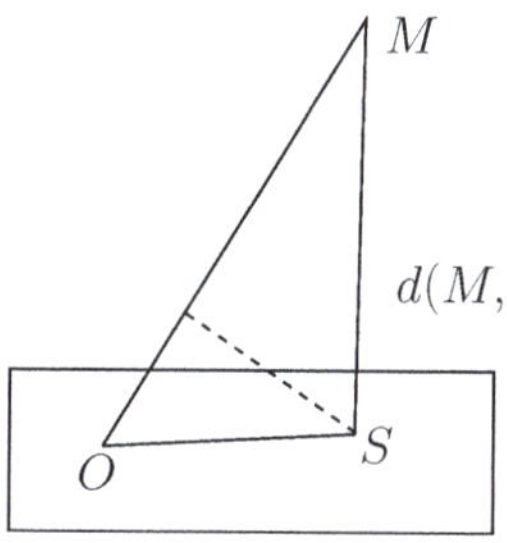

Figure 5.3. *Fraunhofer approximation: geometry and notation*

If the reception point is sufficiently far away from the radiating surface, we have $r^2 = r_M^2 - 2\overline{M} \cdot \overline{S} + r_S^2$, this is with $r_s = \overline{M} \cdot \overline{S} / r_M$ $r^2 = r_M^2 \left(1 - 2r_S/r_M + r_S^2/r_M^2\right)$. Thus, $r \approx r_M - r_S + \mathcal{O}\left(r_S^2/r_M^2\right)$. Hence, the approximation in the far-field is known as the Fraunhofer approximation:

$$p(M,t) \approx \frac{1}{4\pi r_M} \int_{V(S)} Q\left(S, t - \frac{r_M}{c_0} + \frac{r_S}{c_0}\right) dS. \qquad [5.3]$$

5.3. Radiation of distributed sources

5.3.1. *Layer potentials*

When the source cannot be considered as a point, which in acoustics corresponds to the fact that its dimensions are no longer negligible compared to the wave length, we must introduce the concept of a surface source called a layer potential: single layer, double layer and higher order layer.

5.3.1.1. *Simple layer potential*

This is a distribution of an infinite number of monopolar sources distributed over an area Σ with a variable amplitude of $\mu(S)$, $S \in \Sigma$. It is represented by $f(M) = \mu(S) \otimes \delta_\Sigma(M)$, where $\delta_\Sigma(M)$ is the Dirac distribution supported by the surface Σ and $\otimes$ the tensor product. Its acoustic radiation can be given by:

$$\Phi(M) = f(M) * G(M) = (\mu(S) \otimes \delta_\Sigma(M)) * G(M) = \int_\Sigma \mu(S)G(M,S)dS. \quad [5.4]$$

$\Phi(M)$ is the simple layer potential. It verifies the inhomogeneous Helmholtz equation $(\Delta + k^2)\Phi(M) = \mu(S) \otimes \delta_\Sigma(M)$ as well as the Sommerfeld condition. $\Phi(M)$ is an analytic function of the outside coordinates of Σ. $\Phi(M)$ is continuous at the crossing of Σ and its normal derivative is discontinuous at crossing Σ. If $S_0 \in \Sigma$, we have: $\lim_{M\to S_0} \Phi(M) = \Phi(S_0)$ and:

$$\partial_{n(S_0)}\Phi(S_0)^\pm = \pm\frac{\mu(S_0)}{2} + \int_\Sigma \mu(S)\partial_{n(S_0)}G(S_0,S)dS$$

$\partial_{n(S_0)}\Phi(S_0)^\pm$ means that the normal derivative is considered as having a limit of S_0 by parts above or below Σ, the positive value corresponds to a point which is outside the normal.

As an example, let us consider the field created by the potential of a constant simple layer on a circular surface with the radius $R = 0.15$ m. Two types of results are shown. Figure 5.4 shows the pressure level in dB (reference 2×10^{-5} Pa) above the layer at a distance of 3 cm for $kR = 3$ and $kR = 6$. Figure 5.5 shows the directivity at a distance of 1 m from the layer for $kR = 3$ and $kR = 6$.

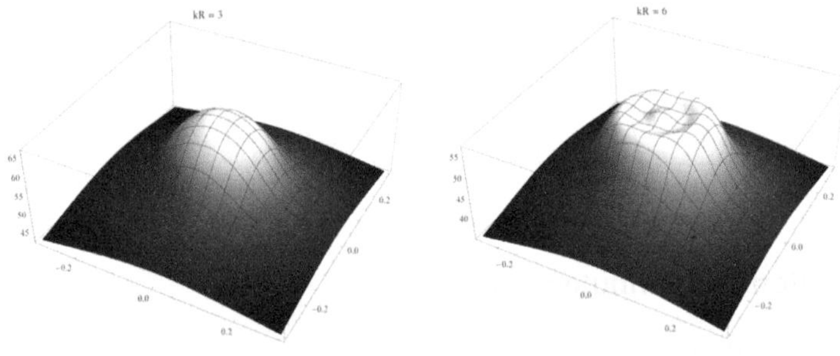

Figure 5.4. *Pressure radiated over a simple layer potential for $kR = 3$ a) and $kR = 6$ b)*

At low frequency, the directivity remains comparable to that of a point source, whereas at high frequency, it is easy to see the marked effects of directivity, which become increasingly pronounced along with the rise in frequency or the complexity of the layer.

5.3.1.2. *Double layer potential*

If, instead of distributing an infinite number of monopoles, we distribute an infinite number of dipoles, with the amplitude $\mu(S)$, oriented along the normal $\vec{n}$ at the surface Σ, we get a double layer represented by the distribution $f(M) = -\vec{n} \cdot \overrightarrow{\text{grad}}(\mu(S) \otimes \delta_\Sigma(M))$. Its radiation can be given by:

$$\Psi(M) = \int_\Sigma \mu(S)\partial_{n(S)}G(M,S)dS.$$

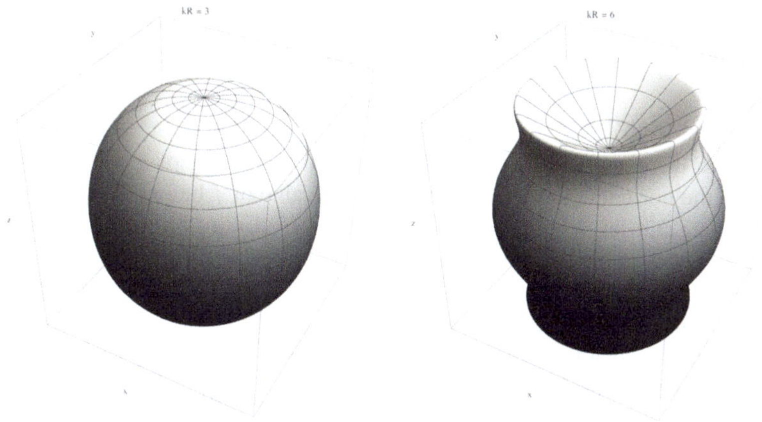

Figure 5.5. *Directivity of a simple layer potential for $kR = 3$ a) and $kR = 6$ b)*

$\Psi(M)$ is the double layer potential, it verifies the inhomogeneous Helmholtz equation $(\Delta + k^2)\Psi(M) = -\vec{n} \cdot \overrightarrow{\text{grad}}(\mu(S) \otimes \delta_\Sigma(M))$. This is an analytic function of the outside coordinates of Σ. The normal derivative of $\Psi(M)$ is continuous at the crossing of Σ and $\Psi(M)$ is discontinuous at the crossing of Σ:

$$\Psi(S_0)^\pm = \pm\frac{\mu(S_0)}{2} + \int_\Sigma \mu(S)\partial_{n(S)}G(S_0,S)dS$$

We can easily define distributed sources of a higher order but this poses a difficulty: when moving toward the limit near the surface which carries the layer, higher order potentials are involved in the expression of the field which they radiate, and the integrals are defined only by Hadamard finite parts and thus complicate the calculation (see equation [3.9]). Moreover, as we will see further on, a linear combination of simple and double layers is enough to describe the acoustic field radiated or diffracted by any source.

5.3.2. *Green's representation of pressure and introduction to the theory of diffraction*

There are numerous ways to present Green's representation of an acoustic field, as its range of application is vast. Green's representation is an extremely powerful tool that transforms a differential boundary problem into an integral problem on the boundaries of the field. This formula, which was initially established for the Helmholtz equation, can be generalized to numerous boundary problems (such as the vibrations of structures such as plates, shells and elastic solids). We begin by examining Green's formula. Then, using an example of diffraction by an obstacle, we will highlight that the diffracting source can be replaced by a superposition of simple and double layers.

5.3.2.1. *Green's formula*

Let σ_f be the jump of a function f of $I\!R^3$ at the crossing of a surface Σ in the direction of the normal (value of f after Σ less the value of f before Σ for an outward facing normal) and let δ_Σ be the Dirac distribution followed by the surface Σ. The partial derivative with respect to the variable x_i can be written in a distribution sense following formula [3.28] $\partial f/\partial x_i = \{\partial f/\partial x_i\} + n_i \sigma_f \delta_\Sigma$ where $\{\cdot\}$ represents the value of the function outside of the discontinuity points. Let us consider a function defined in Σ and zero at the exterior of this surface. We assume that f can be derived twice at the interior of Σ. From this, we get [3.32]:

$$\Delta f = \{\Delta f\} + \sigma_{\partial_n f} \delta_\Sigma + \vec{\nabla} \cdot (\vec{n} \sigma_f \delta_\Sigma) .$$

Let us consider the scalar product between the previous expression and a function ϕ. The second term of the second part of the previous equation (note the similarity between this equation and the simple layer potential) becomes $\langle \sigma_{\partial_n f} \delta_\Sigma, \phi \rangle = -\langle \partial_n f \delta_\Sigma, \phi \rangle$, because f and its partial derivatives are zero at the exterior of the surface. Similarly, the third term of the second part of this equation (note the similarity between this equation and the double layer potential) can be written as $\langle \vec{\nabla} \cdot (\vec{n} \sigma_f \delta_\Sigma), \phi \rangle = -\langle \sigma_f \delta_\Sigma, \vec{\nabla} \cdot (\vec{n}\phi) \rangle$ or, after integration by parts, $\langle \vec{\nabla} \cdot (\vec{n} \sigma_f \delta_\Sigma), \phi \rangle = +\langle f \delta_\Sigma, \partial_n \phi \rangle$. This is:

$$\langle \Delta f, \phi \rangle = \langle \{\Delta f\}, \phi \rangle - \langle \frac{\partial f}{\partial \vec{n}}, \phi \rangle + \langle f \delta_\Sigma, \frac{\partial \phi}{\partial \vec{n}} \rangle = \langle f, \Delta \phi \rangle$$

$\vec{n}$ is the normal exterior to the surface Σ. By clarifying the scalar product and regrouping the volume and surface integrals, we get Green's formula:

$$\int_V f\Delta\phi - \phi\Delta f dV = \int_\Sigma f\frac{\partial\phi}{\partial\vec{n}} - \frac{\partial f}{\partial\vec{n}}\phi dS. \quad [5.5]$$

5.3.2.2. *Green's representation*

Let us consider the following example. We aim to determine the pressure field $p(M)$, solution in the space Ω (bounded or not), using the inhomogeneous Helmholtz equation $(\Delta + k^2)p(M) = f(M)$, where $f(M)$ is any source. We assume further that $p(M)$ verifies the impedance condition $\partial_{\vec{n}}p(S) + \imath k/\zeta p(S) = 0, S \in \Sigma$ on a bounded domain Σ of Ω (see Figure 5.6). If Ω is not bounded, we must add the Sommerfeld conditions to these equations.

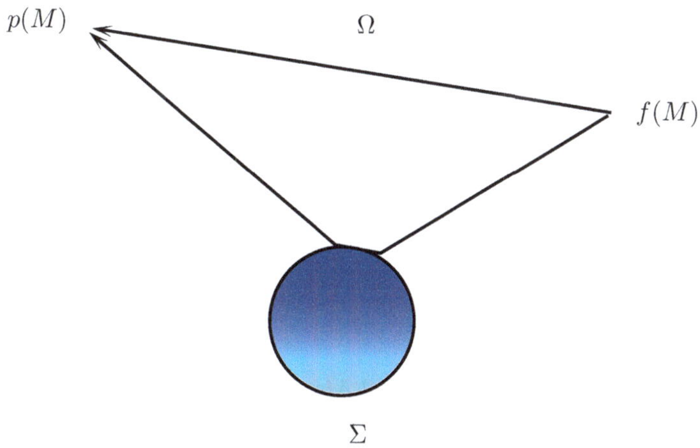

Figure 5.6. *Geometry of the diffraction problem*

Let us recall that Green's kernel $G(M, M')$ is the solution to $(\Delta_M + k^2)G(M, M') = \delta_{M'}(M)$. Let us consider the difference $p(M)\Delta_M G(M, M') - G(M, M')\Delta p(M)$. By adding and subtracting $k^2 p(M)G(M, M)$ we get:

$$p(M)\Delta_M G(M, M') - G(M, M')\Delta p(M) = p(M)(\Delta_M G(M, M') + k^2 G(M, M')) \\ -G(M, M')(\Delta p(M) + k^2 p(M)),$$

then, we integrate this equation across the whole Ω domain and apply it to the first part of Green's formula [5.5]:

$$\int_\Omega (p(Q)\Delta_Q G(Q,M) - G(Q,M)\Delta p(Q))\,dQ = \int_\Sigma \left(p(S)\frac{\partial G(S,M)}{\partial \vec{n}(S)} - G(S,M)\frac{\partial p(S)}{\partial \vec{n}}\right) dS.$$

We thus get:

$$\int_\Omega (p(Q)(\Delta_Q G(Q,M) + k^2 G(Q,M)) - G(Q,M)(\Delta p(Q) + k^2 p(Q)))\,dQ = \int_\Sigma p(S)\frac{\partial G(S,M)}{\partial \vec{n}(S)} dS - \int_\Sigma G(S,M)\frac{\partial p(S)}{\partial \vec{n}} dS.$$

But, hypothetically $\Delta_Q G(Q,M)+k^2 G(Q,M) = \delta_M(Q)$ and $\Delta p(Q)+k^2 p(Q) = f(Q)$. Thus:

$$p(M) - \int_\Omega G(Q,M)f(Q)dQ = \int_\Sigma \left(p(S)\frac{\partial G(S,M)}{\partial \vec{n}(S)} - \frac{\partial p(S)}{\partial \vec{n}} G(S,M)\right) dS.$$

And we know that $p_0(M) = \int_\Omega G(Q,M)f(Q)dQ$ is the response of the medium in the absence of obstacles. Finally, we obtain:

$$p(M) = p_0(M) + \int_\Sigma \left(p(S)\frac{\partial G(S,M)}{\partial \vec{n}(S)} - \frac{\partial p(S)}{\partial \vec{n}} G(S,M)\right) dS. \qquad [5.6]$$

This formula is called Green's representation of pressure. This result shows that the total pressure is made up of two fields. The first field corresponds to the field created by the source in the absence of an obstacle and the second field corresponds to the field created by the diffracting obstacle. The integral over Σ which reflects the effect of the boundary on the total field can be considered as the superposition of two acoustic fields, the first created by a simple layer potential with a density of $-\partial_n p(S)$ and the second created by a double layer potential with a density of $p(S)$. In this case here, to solve this integral equation, we must first note that the boundary condition tells us that $\partial_{\vec{n}} p(S) + \imath k/\zeta p(S) = 0, S \in \Sigma$. Green's representation thus becomes:

$$p(M) = p_0(M) - \int_\Sigma p(S)\left(\frac{\imath k}{\zeta} G(S,M) + \frac{\partial G(S,M)}{\partial \vec{n}(S)}\right) dS,$$

this equation is Green's representation of the sought solution. This integral equation is not easy to solve because the unknown (the acoustic pressure) appears in the left member of the equation as a function of the whole Ω space and in the right member in the form of a function defined on the surface Σ. To solve this, we must find an equation satisfied by the pressure set on the surface Σ. To perform this limiting process for $M \in \Omega \to S \in \Sigma$, we must pay attention to the integral term which contains the normal derivative of $G(S, M)$ which is discontinuous at the crossing of Σ (the double layer potential is discontinuous at the crossing of the surface which bears it). By moving toward the limit, we thus get:

$$\frac{1}{2}p(S) + \int_{\Sigma} p(S') \left(\frac{\imath k}{\zeta} G(S', S) + \frac{\partial G(S', S)}{\partial \vec{n}(S')} \right) dS' = p_0(S), S \in \Sigma$$

This equation is a Fredholm integral equation of the second kind, which in a unique manner determines the unknown $p(S)$. We can also add that the functional space, to which the solution belongs, is the space of square-integrable modulus functions (the finite energy functions).

5.3.2.3. *Solving integral equations*

The previous integral equation can be written in the generic form:

$$\phi(S) + \int_{\Sigma} \phi(S') K(S', S) dS' = \psi(S), S \in \Sigma$$

where $\psi(S)$ and the kernel of the equation $K(S, S')$ are the input quantities of the problem (in general, these integral kernels are singular functions). As we are trying to find square-integrable modulus solutions, it is natural to look for solutions using Galerkin's method.

We divide the surface Σ into N elements Σ_i on which we define a set of N real functions f_i which satisfy two conditions:

– $f_i(S) = 0$ if $S \notin \Sigma_i$, $f_i(S) \neq 0$ if $S \in \Sigma_i$ and $f_i(S_i) = 1$ if S_i is a point which belongs to the sub-domain Σ_i;

– any linear combination of f_i has a regularity which can be fixed: piecewise constant, continuous, continuously derivable, etc. The set of f_i forms the base of the Hilbert space $\mathcal{H}_N$ of dimension N in which we aim to find the solution to the integral function in question.

We are thus looking for $\phi(S) = \sum_i a_i f_i(S)$ where the constants a_i are the new unknowns of the problem. The integral $\int_\Sigma \phi(S')K(S',S)dS'$ can thus be approximated by:

$$\int_\Sigma \phi(S')K(S',S)dS' = \int_\Sigma \sum_i a_i f_i(S')K(S',S)dS'$$
$$= \sum_i a_i \int_{\Sigma_i} f_i(S')K(S',S)dS' \qquad [5.7]$$

The calculation of a_i is chosen so as to minimize in $\mathcal{H}_\mathcal{N}$ the norm of:

$$\sum_i a_i \left(f_i(S) + \int_{\Sigma_i} f_i(S')K(S',S)dS' \right) - \psi(S)$$

This leads to solving the system of N equations with N unknowns:

$$\sum_i a_i \left(\delta_i^j \int_{\Sigma_j} (f_j(S))^2 dS + \int_{\Sigma_j}\int_{\Sigma_i} f_i(S')f_j(S)K(S',S)dS'dS \right)$$
$$= \int_{\Sigma_j} f_j(S)\psi(S)dS,$$

for $j = 1, 2, \cdots, N$. In general, the integrals appearing in these N equations must be calculated numerically. We can proceed in different ways. The first one, which is called *Galerkin's method*, involves numerically estimating the integrations with respect to S and S' with the same precision (by a Gauss or Simpson method). This method is very precise (provided that the quadrature is), but, it is very intensive numerically.

The second method consists of evaluating the integrals with respect to S by a single point rectangle method. In general, we must choose the point S_j of the element belonging to the surface Σ_j for which $f(S_j) = 1$. The previous system thus becomes $\sum_i a_i \left(\delta_i^j + \int_{\Sigma_i} f_i(S')K(S',S_j)dS' \right) = \psi(S_j), j = 1, 2, \cdots, N$. This method, called the *collocation method*, although being the fastest, is not the least precise one. This is one of the most used methods in acoustics because of the simplicity of programming which it requires.

There is an intermediate method, which is called the *collocation-Galerkin method*, which allows us to combine some of the advantages of both methods. It is based on the fact that, with integration kernels being generally singular, the integrals which bear the singularity describe the essence of "the information" and, to increase the quality of the resolution, it is necessary to calculate them with great precision. In practice, the double integrals, which correspond to diagonal terms $\int_{\Sigma_i}\int_{\Sigma_i} f_i(S')f_i(S')K(S',S)dS'dS$, are evaluated by a double numerical integration (Galerkin's method) while the extra-diagonal terms are evaluated by a single point rectangle method (collocation method) $\Sigma_j \int_{\Sigma_i} f_i(S')K(S',S_j)dS'$. This method requires a significantly longer computing time than the collocation method but much less than Galerkin's method, while being more precise than the collocation method (and in practice hardly less precise than Galerkin's method).

From a practical point of view, it should be noted that the convergence of these three methods is assured. The difficulty is to establish a discretization criterion *a priori*. It is generally seen that for a collocation method with a discretization by piecewise constants, a minimum of six points per wavelength are necessary; these correspond to the minimum number of points which are necessary to calculate the integral of a sinusoid over a period by a rectangle method. With a discretization by spline functions, we can settle for a minimum of four points per wavelength. With orthogonal polynomials, such as the Legendre and Chebyshev polynomials (these are referred to as spectral methods), we can go down to π points per wavelength, which appears to be a reasonable minimum. In any event, it is not possible to take less than two points per wavelength due to the Nyquist–Shannon sampling theorem.

5.4. Acoustic radiation of a piston in a plane

Here, we focus on the acoustic pressure created by a circular piston, which is indefinitely extended by a perfectly rigid plane (we often use the word "baffled" with respect to this configuration). This is a good approximation of a speaker, for which a membrane, which oscillates at a constant speed (across its entire surface), is embedded into a rigid enclosure. To get a better understanding of these ideas, we will consider a movement in a harmonic regime. It must be noted that for our study, it is not important to distinguish whether the piston is actually a mechanical piston (e.g. a speaker) or air at the free end of a pipe, such as a wind instrument (the approximation of the latter by a baffled piston is obviously less precise). The effect on the acoustic field created is the same. Moreover, we assume that the medium has no influence on the vibration of the piston.

The problem equations are the following. Let us consider the space $\mathbb{R}^3$, described with the usual Cartesian coordinates (O,x,y,z) and the half-space defined by $z>0$. This domain is occupied by a fluid with a density of ρ_0 and an acoustic wave celerity of c_0. The half-space $z<0$ is occupied by a fluid whose influence we will neglect. The normal to the plane Σ defined by $z=0$ is oriented along the z axis. On the $z=0$ plane, a disk with a radius of R centered at the origin defines the domain S, which is occupied by our piston which is set in motion by a vibratory displacement u given by $u = u_0 \exp(-\imath\omega t)$, where u_0 is a constant. In the rest of the problem, we will omit the time-dependence. We further assume that there is no acoustic source within the fluid. The fluid is assumed to be ideal (without viscosity). Therefore, at the plane-fluid interface, there is continuity of the normal vibration acceleration. The acoustic pressure radiated by the piston $p(M), M \in z > 0$ is the solution to the problem:

$$\begin{aligned}(\Delta + k^2)\, p(M) = 0, M \in z > 0; \partial_n p(Q) = \omega^2 \rho_0 u_0, Q \in S;\\ \partial_n p(Q) = 0, Q \in \Sigma - S\end{aligned}$$

The acoustic pressure solution of the Helmholtz equation can be given by Green's representation in the absence of a source term within the fluid:

$$p(M) = \int_\Sigma p(Q)\partial_{n(Q)}G(M;Q) - \partial_n p(Q)G(M;Q)d\Sigma(Q),$$

the coordinates of the points M and Q are (x, y, z) and $(x', y', z' = 0)$, respectively, because z' belongs to the $z = 0$ plane. The continuity of normal vibration acceleration allows us to describe Green's representation as follows:

$$p(M) = \int_\Sigma p(Q)\partial_{n(Q)}G(M;Q)d\Sigma(S) - \omega^2\rho_0 \int_S u_0 G(M;Q)d\Sigma(Q).$$

Instead of solving this equation (which is a Fredholm integral equation of the second kind), we will benefit from the very specific geometry of the problem to further simplify this equation. We must find Green's kernel $\mathcal{G}$ which satisfies Neumann's problem:

$$(\Delta + k^2)\, \mathcal{G}(M, M') = \delta_{M'}(M), M, M' \in z > 0, \partial_{n(Q)}\mathcal{G}(Q, M) = 0, Q \in \Sigma.$$

To calculate this Green's function, we can use the image method. The field received at point M in space is the superposition of the direct field emitted by the primary source placed at M' and the field reflected by the horizontal plane. The latter is described as the field emitted by a source image, symmetrical to M'' with respect to the horizontal plane of the primary source M' (see Figure 5.7), whose magnitude depends on the boundary condition.

We have:

$$\mathcal{G}(M, M') = \frac{-e^{\imath kd(MM')}}{4\pi d(MM')} + C\frac{-e^{\imath kd(MM'')}}{4\pi d(MM'')},$$

where the coordinates of the points M, M' and M'', M'' is symmetrical to M' with respect to the Σ plane, are given by $M = (x, y, z)$, $M' = (x', y', z')$ and $M'' = (x'' = x', y'' = y', z'' = -z')$. The distances can be written as $d^2(MM') = (x - x')^2 +$

$(y-y')^2+(z-z')^2$ and $d^2(MM'')=(x-x')^2+(y-y')^2+(z+z')^2$. C is a constant, *a priori* a complex, which depends on the boundary condition on the plane. We have:

$$\frac{\partial \mathcal{G}(M,M')}{\partial z} = \frac{-\mathrm{e}^{\imath k d(MM')}}{4\pi d(MM')^2}\left(\imath k - \frac{1}{d(MM')}\right)(z-z') + C\frac{-\mathrm{e}^{\imath k d(MM'')}}{4\pi d(MM'')^2}\left(\imath k - \frac{1}{d(MM'')}\right)(z-z'')$$

At $z=0$, we have $d(MM')=d(MM'')$, it is easy to see that:

$$\frac{\partial}{\partial z}\mathcal{G}(M,M') = 0 \Rightarrow \frac{-\mathrm{e}^{\imath k d(MM')}}{4\pi d(MM')^2}\left(\imath k - \frac{1}{d(MM')}\right)(-z'+Cz') = 0 \text{ if } C=1.$$

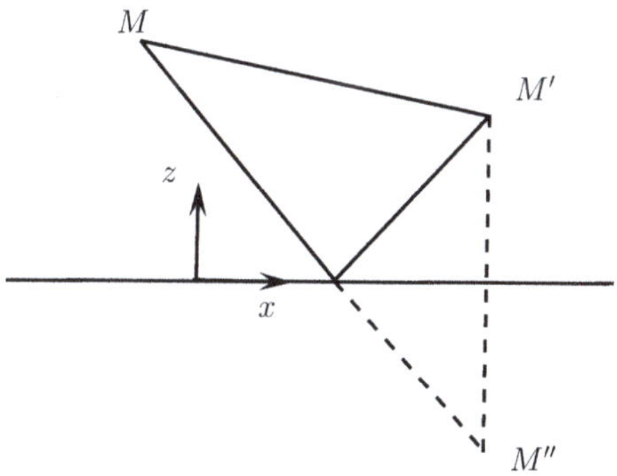

Figure 5.7. *Image method*

Thus, we get:

$$\mathcal{G}(M,M') = -\frac{\mathrm{e}^{\imath k d(MM')}}{4\pi d(MM')} - \frac{\mathrm{e}^{\imath k d(MM'')}}{4\pi d(MM'')}. \quad [5.8]$$

Since the normal derivative of Green's kernel is zero on the Σ plane, the Green representation thus becomes:

$$p(M) = -\omega^2\rho_0 \int_S u_0 \mathcal{G}(M;Q)d\sigma(Q).$$

This integral is often called the Kirchhoff–Helmholtz integral; it states that the acoustic field radiated by a baffled vibrating surface can be approximated by a simple convolution integral between Green's function of the medium and the vibration field of the surface. This formula gives acceptable results in a direction normal to the surface and for a structure which is only slightly influenced by the surrounding fluid (e.g. a speaker or a steel structure in the air). For a very light structure (or a heavy fluid), we must take into consideration the presence of the fluid in order to calculate the displacement of the structure. When we examine the acoustic field near the plane of the structure, the diffraction effects by the boundary become dominating and Kirchhoff's integral gives aberrant results. On the Σ plane, $z' = 0$ and the points M' and M'' coincide, the acoustic pressure radiated by the piston is then given by the following simple result:

$$p(M) = \omega^2 \rho_0 u_0 \int_S \frac{e^{\imath k\left((x-x')^2+(y-y')^2+z^2\right)^{1/2}}}{2\pi\left((x-x')^2+(y-y')^2+z^2\right)^{1/2}} dx'dy'.$$

This result holds for a piston of any form. However, in the case of a circular piston, we can get an (approximate) analytical result.

5.4.1. *Far-field radiation of a circular piston: directivity*

To do this, let us use the spherical coordinates (O, r, θ, ϕ) (see Figure 5.8). The coordinates of the point M can be given by: $x = r\sin\theta\cos\phi$, $y = r\sin\theta\sin\phi$, $z = r\cos\theta$ and for M' by: $x' = r'\sin\theta'\cos\phi'$, $y' = r'\sin\theta'\sin\phi'$, $z' = r'\cos\theta'$. It is easy to see that $(x-x')^2 + (y-y')^2 + (z-z')^2 = r^2 + r'^2 - 2rr'f(\theta,\phi,\theta',\phi')$ with $f(\theta,\phi,\theta',\phi') = \sin\theta\cos\phi\sin\theta'\cos\phi' + \sin\theta\sin\phi\sin\theta'\sin\phi' + \cos\theta\cos\theta'$.

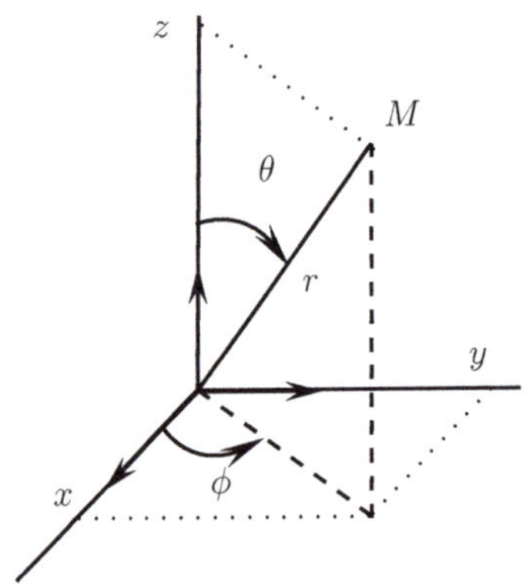

Figure 5.8. *Spherical coordinate system*

Let us recall here that M' is a point on the piston which is close to the origin, thus with $r'/r \ll 1$ and for $kr \gg 1$ and $r'/r \ll 1$ we have the development:

$$\frac{\mathrm{e}^{\imath k\left((x-x')^2+(y-y')^2+(z-z')^2\right)^{1/2}}}{2\pi\left((x-x')^2+(y-y')^2+(z-z')^2\right)^{1/2}} \approx \frac{\mathrm{e}^{\imath kr}}{2\pi r}\mathrm{e}^{-\imath kr' f(\theta,\phi,\theta',\phi')},$$

To demonstrate this result, we must recall that we have, for $r \gg r'$, $r'/r \ll 1$, thus $r^2 + r'^2 - 2rr' f(\theta,\phi,\theta',\phi') = r^2(1 + (r'/r)^2 - 2r'/r f(\theta,\phi,\theta',\phi')) \approx r^2(1 - 2r'/r f(\theta,\phi,\theta',\phi'))$.

In addition we obtain that $\sqrt{r^2 + r'^2 - 2rr' f(\theta,\phi,\theta',\phi')} \approx r(1 - r'/r f(\theta,\phi,\theta',\phi'))$. This expression appears twice in Green's function. Once in the denominator, for which we can approximate it by r. Conversely, at the numerator of the expression, the argument of the complex exponential is a phase term and we must keep the full expression $r - r' f(\theta,\phi,\theta',\phi'))$. Finally, we obtain:

$$\frac{\mathrm{e}^{\imath k\left((x-x')^2+(y-y')^2+(z-z')^2\right)^{1/2}}}{2\pi\left((x-x')^2+(y-y')^2+(z-z')^2\right)^{1/2}} \approx \frac{\mathrm{e}^{\imath kr}}{2\pi r}\mathrm{e}^{-\imath kr' f(\theta,\phi,\theta',\phi')} \qquad [5.9]$$

In fact, this result is not only valid at a high frequency $kr \gg 1$ but also valid for a large source-receiver distance $r \gg r'$. On the Σ plane, we have $z' = r' \cos\theta' = 0 \Rightarrow \theta' = \pi/2$, thus:

$$\frac{\mathrm{e}^{\imath k\left((x-x')^2+(y-y')^2+z^2\right)^{1/2}}}{2\pi\left((x-x')^2+(y-y')^2+z^2\right)^{1/2}} \approx \frac{\mathrm{e}^{\imath kr}}{2\pi r}\mathrm{e}^{-\imath kr' \sin\theta\cos(\phi-\phi')},$$

the acoustic pressure radiated by the piston in the far-field is thus given by the integral:

$$\begin{aligned} p(M) &\approx \omega^2 \rho_0 u_0 \frac{\mathrm{e}^{\imath kr}}{2\pi r}\int_0^R r'\left(\int_0^{2\pi} \mathrm{e}^{-\imath kr'\sin\theta\cos\phi} d\phi\right) dr' \\ &\approx \omega^2 \rho_0 u_0 \frac{\mathrm{e}^{\imath kr}}{2\pi r}\int_0^R r' 2\pi \mathrm{J}_0\left(kr'\sin\theta\right) dr' \text{ because } \mathrm{J}_0(x) = \int_0^{2\pi} \mathrm{e}^{-\imath x\cos\phi} d\phi, \end{aligned}$$

by changing the variable $z = k\sin\theta$, we get:

$$\begin{aligned} p(M) &\approx \omega^2 \rho_0 u_0 R^2 \frac{\mathrm{e}^{\imath kr}}{r}\frac{1}{k^2R^2\sin^2\theta}\int_0^{kR\sin\theta} z\mathrm{J}_0(z)dz \\ &\approx \omega^2 \rho_0 u_0 R^2 \frac{\mathrm{e}^{\imath kr}}{r}\frac{\mathrm{J}_1(kR\sin\theta)}{kR\sin\theta}, \end{aligned}$$

because $\int_0^z t^n \mathrm{J}_{n-1}(t)dt = z^n \mathrm{J}_n(z), n > 0$. Thus, the directivity can be written as $\Xi(\theta, \phi) = -4\pi\omega^2 \rho_0 u_0 R^2 \mathrm{J}_1(x)/x$ with $x = kR\sin\theta$. The appearance of $\mathrm{J}_1(x)/x$ is given in Figure 5.9. We have a function which presents a finite maximum at the origin and then a series of oscillations which dampen quickly.

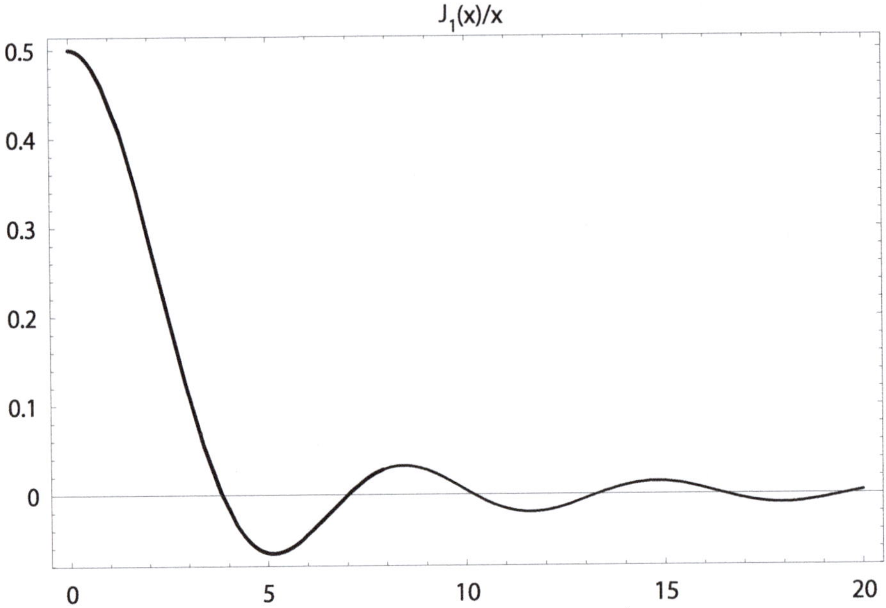

Figure 5.9. *The appearance of the function* $J_1(x)/x$

To better visualize the effect of directivity, we present several examples of directivity diagrams in linear scale at low (Figure 5.10), medium (Figure 5.11) and high (Figure 5.12) frequencies. The piston is located at the origin on the ordinate (which represents the baffle). We have chosen to present these results according to the usual dimensionless number kR. At low frequency $kR \leq 1$, the piston behaves as an omnidirectional source (it does not have a preferred direction). However, as the frequency increases (typically as $kR \geq 5$), "silent" zones start appearing, the pressure radiated by the piston is zero. The space is divided into noise regions (cones) separated by silent zones; the pressure is always at its maximum along the axis of the piston. This (partly) explains why manufacturers mount loudspeakers in several ways; typically, three (for low, medium and high frequencies), which ensures a loudspeaker has an approximately constant directivity over a wide frequency range.

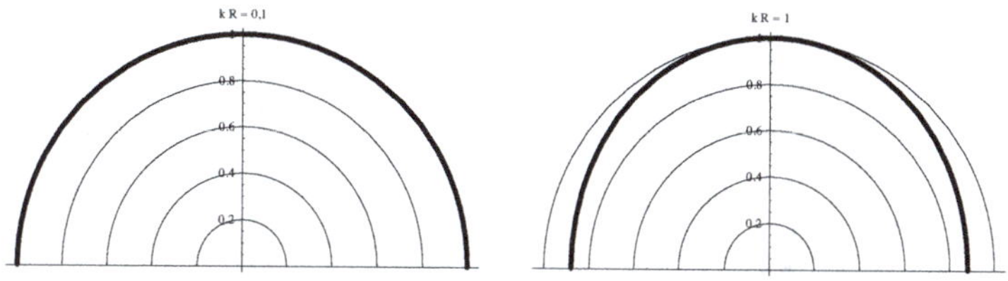

Figure 5.10. *Piston directivity for $kR = 0.1$ a) and $kR = 1$ b)*

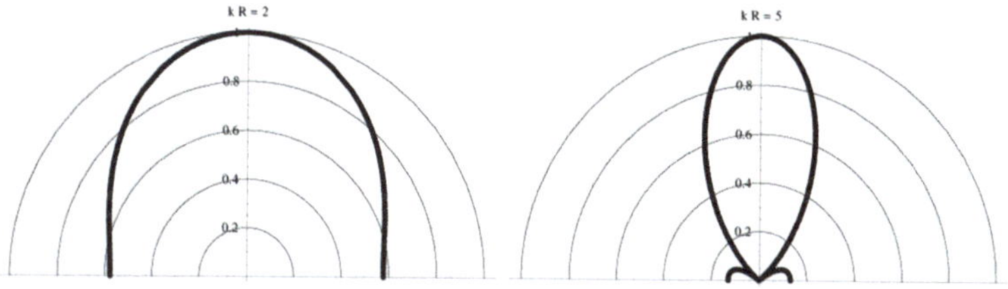

Figure 5.11. *Piston directivity for $kR = 2$ a) and $kR = 5$ b)*

5.4.2. *Radiation along the axis of a circular piston*

In this case, we have $x = 0$, $y = 0$ and $z \neq 0$. The acoustic pressure radiated along the axis can be given by:

$$p(0,0,z) = \omega^2 \rho_0 u_0 \int_0^R \int_0^{2\pi} \frac{\mathrm{e}^{\imath k \left(z^2 + r'^2\right)^{1/2}}}{2\pi \left(z^2 + r'^2\right)^{1/2}} r' dr' d\phi'$$

$$= \omega^2 \rho_0 u_0 \int_0^R \frac{\mathrm{e}^{\imath k \left(z^2 + r'^2\right)^{1/2}}}{\left(z^2 + r'^2\right)^{1/2}} r' dr'.$$

By changing the variable $u = \left(z^2 + r'^2\right)^{1/2}$, we easily obtain:

$$p(0,z) = \omega^2 \rho_0 u_0 \int_z^{(z^2+R^2)^{1/2}} \mathrm{e}^{\imath k u} du = -\frac{\imath \omega^2 \rho_0 u_0}{k} \left[\mathrm{e}^{\imath k (z^2+R^2)^{1/2}} - \mathrm{e}^{\imath k z}\right].$$

In moduli, we obtain $|p(0,z)| = \omega \rho_0 c_0 |u_0| \left|\mathrm{e}^{\imath k (z^2+R^2)^{1/2}} - \mathrm{e}^{\imath k z}\right|$. However, we know that:

$$\left|\mathrm{e}^{\imath \alpha} - \mathrm{e}^{\imath \beta}\right| = \left[(\cos\alpha - \cos\beta)^2 + (\sin\alpha - \sin\beta)^2\right]^{1/2} = 2\left|\sin\frac{1}{2}(\alpha - \beta)\right|.$$

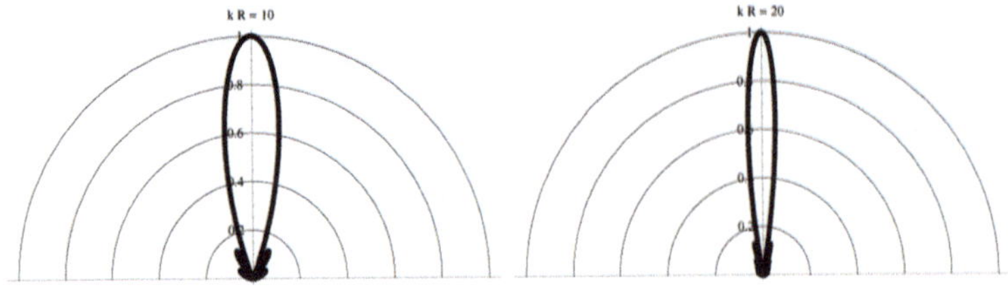

Figure 5.12. *Piston directivity for $kR = 10$ a) and $kR = 20$ b)*

The pressure modulus on the axis is thus $|p(0,z)| = 2\omega\rho_0 c_0 \, |u_0| \, \left|\sin[k/2((z^2+R^2)^{1/2} - z)]\right|$. While for $z \to \infty$, by limited expansion, we get $(z^2 + R^2)^{1/2} - z = z(1 + (R/z)^2)^{1/2} - z = R(1/2 \times (R/z) - 1/8 \times (R/z)^3 + \mathcal{O}((R/z)^5))$. We can thus replace the sine by the first term of the limited development of its argument and we get $|p(0, z \to \infty)| \approx \omega^2 \rho_0 R^2/(2z) \, |u_0|$ which shows a decrease as the inverse of the distance from the source. Figure 5.13 shows the evolution of the pressure radiated by a circular piston along its axis when we move away for three wave numbers ($k = 5$, $k = 10$ and $k = 20$). We can easily notice that in the near field the acoustic field varies considerably, and from a certain distance $z \approx kR^2/\pi = R(2R/\lambda)$, it stabilizes and behaves like a regular monopole in very steady decline of $1/r$. This is usually how we define the transition between the near field and the far-field[1]: the distance to the vibrating body from which the pressure field decreases as $1/r$. We must note that a good approximation of the near field at a medium frequency is to consider that it exists in a domain whose volume is of the order of magnitude of the smallest sphere which contains the source[2]; this volume decreases at low frequency and increases at high frequency, for example, for a piston with a 0.1 m radius, at 200 Hz, the near field is observed for distances less than 1 cm and at 10,000 Hz for distances less than 0.5 m.

5.5. Acoustic radiation of a rectangular baffled structure

To finish off this introduction to radiation, we will examine the acoustic radiation of a baffled vibrating surface. This can be seen as a generalization of a baffled piston or as a first approach to the acoustic radiation of structures. The major difference with what we will study later lies in the fact that here the radiating structure is not

1 There are obviously other definitions that involve the acoustic velocity in addition to pressure, such as the distance from which we have $p\vec{v} = \rho_0 c_0$.

2 At 340 Hz, which is a “commonly used” medium frequency, the wavelength is in the meter range, which is a dimension of the order of structure magnitude commonly encountered in vibroacoustics $a/\lambda \approx 1$.

influenced by the fluid it is in contact with. We can already point out that this is an approximation which is sufficient in many industrial cases (such as anything related to machine radiation).

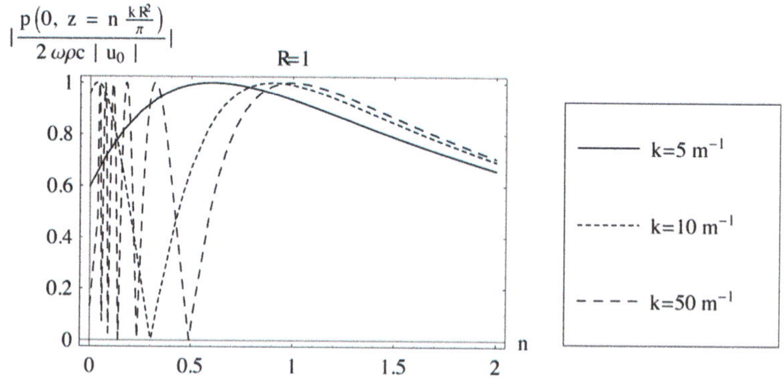

Figure 5.13. *Acoustic pressure radiated along the axis of a piston with a radius of $R = 1m$*

Let us thus consider a very similar problem to the piston problem but in which the vibrating surface is no longer a piston moving with a uniform speed but a rectangular surface $S : (x, y) \in]-a, a[\cup] - b, b[$ animated by a vibratory displacement $u(x, y)$ given by $u(x, y) = u_0(x, y)\mathrm{e}^{-\imath\omega t}$. At the vibrating surface-fluid interface, there is always continuity of normal vibration acceleration. The acoustic pressure $p(M), M \in z > 0$ radiated by the vibrating surface is the solution to the problem:

$$\left(\Delta + k^2\right) p(M) = 0 \in z > 0,$$
$$\partial_n p(Q) = \omega^2 \rho_0 u_0(Q), Q = (x, y) \in S, \partial_n p(Q) = 0, Q \in \Sigma - S$$

The acoustic pressure radiated by the surface is obtained by Green's representation:

$$p(M) = \frac{\omega^2 \rho}{2\pi} \int_{-a}^{a} \int_{-b}^{b} u_0(x', y') \frac{\mathrm{e}^{\imath k\left((x-x')^2+(y-y')^2+z^2\right)^{1/2}}}{\left((x-x')^2 + (y-y')^2 + z^2\right)^{1/2}} dx'dy'.$$

As previously [5.9], we write Green's function in the far-field. On the Σ plane, we have $x' = r' \cos\phi'$, $y' = r' \sin\phi'$, and we obtain:

$$p(r, \theta, \phi) \approx \frac{\omega^2 \rho}{2\pi} \frac{\mathrm{e}^{\imath kr}}{r} \int_{-a}^{a} \int_{-b}^{b} u_0(x', y') \mathrm{e}^{-\imath k \sin\theta\left(x' \cos\phi + y' \sin\phi\right)} dx'dy'.$$

The double integral which appears at the second part can be seen as a partial spatial Fourier transform of the surface displacement:

$$[\tilde{u}_0(\xi,\zeta)]_{\xi=k\sin\theta\cos\phi,\zeta=k\sin\theta\sin\phi} = \int_{-a}^{a}\int_{-b}^{b} u_0(x',y')\mathrm{e}^{-\imath k\sin\theta(x'\cos\phi+y'\sin\phi)}dx'dy'.$$

Finally, this is $p(r,\theta,\phi) \approx \omega^2\rho/(2\pi)\exp(\imath kr)/(r)\tilde{u}_0(k\sin\theta\cos\phi, k\sin\theta\sin\phi)$. To better understand these ideas, let us examine two examples. The first example is a rectangular piston $u_0(x,y) = 1$, we easily obtain that:

$$\tilde{u}_0(k\sin\theta\cos\phi, k\sin\theta\sin\phi) = \frac{2\sin(ak\sin(\theta)\cos(\phi))}{k\sin(\theta)\cos(\phi)} \times \frac{2\sin(bk\sin(\theta)\sin(\phi))}{k\sin(\theta)\sin(\phi)}.$$

Finally, this is:

$$p_0(r,\theta,\phi) \approx \frac{2\omega^2\rho e^{\imath kr}}{\pi r}\frac{\sin(ak\sin(\theta)\cos(\phi))}{k\sin(\theta)\cos(\phi)} \times \frac{\sin(bk\sin(\theta)\sin(\phi))}{k\sin(\theta)\sin(\phi)}.$$

We obtain a behavior similar to that of the circular piston.

The second example is that of a vibrating structure whose deformation is a sinusoidal function of the orders m and n. Two cases stand out: that of a function that is asymmetrical with respect to the center of the structure $u^a_{mn}(x,y) = \sin(m\pi x/a)\sin(n\pi y/b)$ and that of a function which is symmetrical with respect to the center of the structure $u^s_{mn}(x,y) = \cos((m-1/2)\pi x/a)$ $\cos((n-1/2)\pi y/b)$. m and n characterize the number of oscillations of the structure. Thus, we obtain:

$$\tilde{u}^a_{mn}(k\sin\theta\cos\phi, k\sin\theta\sin\phi) = (-1)^{m+n}\frac{2ma\pi\sin(ak\cos(\phi)\sin(\theta))}{(m\pi)^2 - a^2k^2\cos(\phi)^2\sin(\theta)^2}$$

$$\times\frac{2nb\pi\sin(bk\sin(\theta)\sin(\phi))}{(n\pi)^2 - b^2k^2\sin(\theta)^2\sin(\phi)^2}$$

$$\tilde{u}^s_{mn}(k\sin\theta\cos\phi, k\sin\theta\sin\phi) = (-1)^{m+n}\frac{2(m-1/2)a\pi\cos(ak\cos(\phi)\sin(\theta))}{((m-1/2)\pi)^2 - a^2k^2\cos(\phi)^2\sin(\theta)^2}$$

$$\times\frac{2(n-1/2)b\pi\cos(bk\sin(\theta)\sin(\phi))}{((n-1/2)\pi)^2 - b^2k^2\sin(\theta)^2\sin(\phi)^2}.$$

To simplify further study, let us denote $k_m = m\pi/a$ and $k_n = n\pi/b$ (k_m and k_n are wave numbers). We get:

$$p^a_{mn}(r,\theta,\phi) \approx \frac{2\omega^2\rho_0 e^{\imath kr}}{\pi r}(-1)^{m+n}k_m k_n \frac{\sin(ak\cos(\phi)\sin(\theta))}{k_m^2 - k^2\cos(\phi)^2\sin(\theta)^2}$$
$$\times \frac{\sin(bk\sin(\theta)\sin(\phi))}{k_n^2 - k^2\sin(\theta)^2\sin(\phi)^2}$$
$$p^s_{mn}(r,\theta,\phi) \approx \frac{2\omega^2\rho_0 e^{\imath kr}}{\pi r}(-1)^{m+n}k_{m-1/2}k_{n-1/2}\frac{\cos(ak\cos(\phi)\sin(\theta))}{k^2_{m-1/2} - k^2\cos(\phi)^2\sin(\theta)^2}$$
$$\times \frac{\cos(bk\sin(\theta)\sin(\phi))}{k^2_{n-1/2} - k^2\sin(\theta)^2\sin(\phi)^2}.$$

Despite the apparent similarity of the results, we have obtained a behavior which is different from that of the piston. In fact, if we are faced with a very low frequency, the terms in k^2 of the denominators can be neglected. And we can easily see that as a structure presents higher numbers of oscillations, which correspond to higher orders of m and n (and thus higher wave numbers k_m and k_n), the lower the maximum amplitude radiated will become. Along the normal to the surface, we have $p^a_{mn}(r,\theta = 0,\phi) \approx 0$ and $p^s_{mn}(r,\theta = 0,\phi) \approx \omega^2\rho_0\exp(\imath kr)/(4\pi r)(-1)^{m+n}8ab/(\pi^2(m - 1/2)(n-1/2))$, which we compare to the radiation of the rectangular piston along the axis of the piston (attention, the radiation from p^a_{mn} is zero in the center of the surface) $p_0(r,\theta = 0,\phi) \approx \omega^2\rho_0\exp(\imath kr)/(4\pi r)8ab$. The pressure radiated has a maximum whose exact expression can be precisely determined by deriving the function. It is very hard to estimate. However, we can roughly estimate the position and amplitude of these maxima by focusing on the zeros of the denominator. It becomes zero for certain combinations of θ and ϕ if $k_m, k_n < k$. The pressure maxima appear at the angular points θ_0, ϕ_0 given by $k_m = k\cos(\phi_0)\sin(\theta_0)$, $k_n = k\sin(\phi_0)\sin(\theta_0)$. We set $k_s = \sqrt{k_m^2 + k_n^2}$. k_s can be associated with a structurally effective wave length $\lambda_s = 2\pi/k_s$. We obtain the angular coordinates of four points: with $\theta_0 = \sin^{-1} k_s/k$ and $\phi_0 = \tan^{-1} k_n/k_m$, $\phi_1 = -\phi_0$, $\phi_{2,3} = \pi \pm \phi_0$. We call these angles the critical angles or coincidence angles. We can show that the acoustic pressure at the point θ_0 ϕ_0 is independent of the orders m and n for the two types of radiation. Following l'Hospital's rule, we can easily remove the indeterminations (0/0) and we obtain:

$$p^{a,s}_{mn}(r,\theta_0,\phi_p) \approx \omega^2\rho_0\frac{e^{\imath kr}}{4\pi r}2ab \text{ for } k_m, k_n < k, p = 0,1,2,3.$$

Here, we consider the example of a rectangular structure with the dimensions $a = 0.9$ m and $b = 0.7$ m, for which we present the deformations and directivity in several notable configurations. At very low frequency, the radiation is always monopolar for structures with symmetrical deformations and quadrupolar for structures with antisymmetrical deformations. At high frequencies ($k > k_m, k_n$), we can observe that all the modes (with the exception of the first symmetrical mode) exhibit a behavior which can be approximately described as being dipolar for symmetrical modes, such as in Figure 5.16 (sometimes we call them boundary modes), and being quadrupolar for antisymmetrical modes, such as in Figures 5.15 and 5.17 (sometimes referred to as corner modes). This behavior at low and high frequencies is due to the phenomenon of compensation of acoustic radiation in areas adjacent to the structure, which vibrates in phase opposition (see Figure 5.18) and which leads to the almost complete cancelation of acoustic radiation, since then only the peripheral regions radiate efficiently. This compensation phenomenon becomes less pronounced at high frequencies, since, as we can see in the figure (plotted in amplitude and not in levels), there are still many intermediate lobes, as well as a higher maximum level. At medium frequencies, when the mechanic and acoustic wave lengths are comparable, the interference phenomena are very pronounced and the radiation maxima are as high as the vibration amplitude maxima.

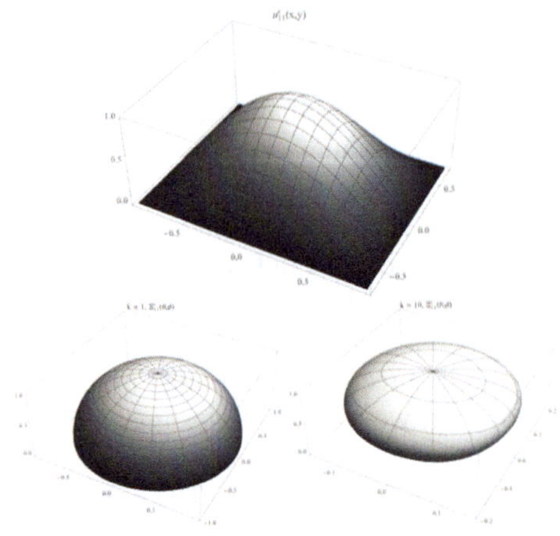

Figure 5.14. *"Symmetrical" deformation and directivity for $m = 1$, $n = 1$ and for $k = 1$ and 10 m^{-1}*

We must note that at very low frequencies, for the modes with symmetrical deformation, which only have one maximum in one direction, there is an obvious monopolar behavior. This is by no means in contradiction with the compensation phenomenon, quite contrary, since in this case the boundaries radiate in phase, because of the significant length of the acoustic waves, this leads to a constructive interference of the two regions that act as a single monopole. At higher frequencies, these zones act as independent sources, as shown by the directivities with two lobes, which can be observed in Figure 5.16.

We must note that the maximum radiation of a structure can very often be observed in the vicinity of its geometrical discontinuities (boundary modes and corner modes), since there the compensation phenomenon is much lower. This is not verified on the boundaries of non-baffled structures, because, in this case, the compensation is maximized by balancing the pressures on either side of the boundary.

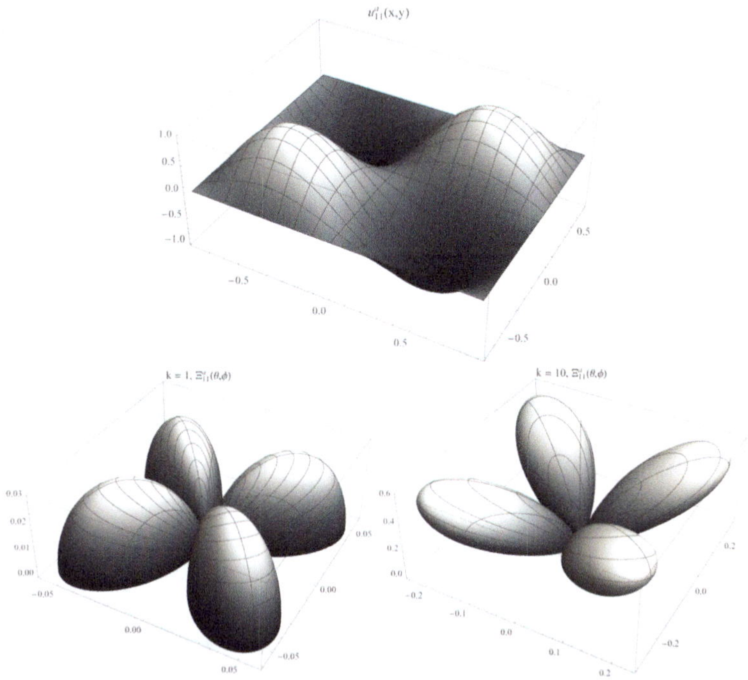

Figure 5.15. *"Antisymmetrical" deformation and directivity for $m = 1$, $n = 1$ and for $k = 1$ and 10 m^{-1}*

5.6. Acoustic radiation of moving sources

5.6.1. *Compact and non-compact sources*

5.6.1.1. *Spatially compact source*

We start out with the Fraunhofer approximation [5.3], by assuming that the source is set in motion at a characteristic velocity v.

$$p(M,t) \approx \frac{1}{4\pi r_M} \int_{V(S)} Q\left(S, t - \frac{r_M}{c_0} + \frac{r_S}{c_0}\right) dS$$

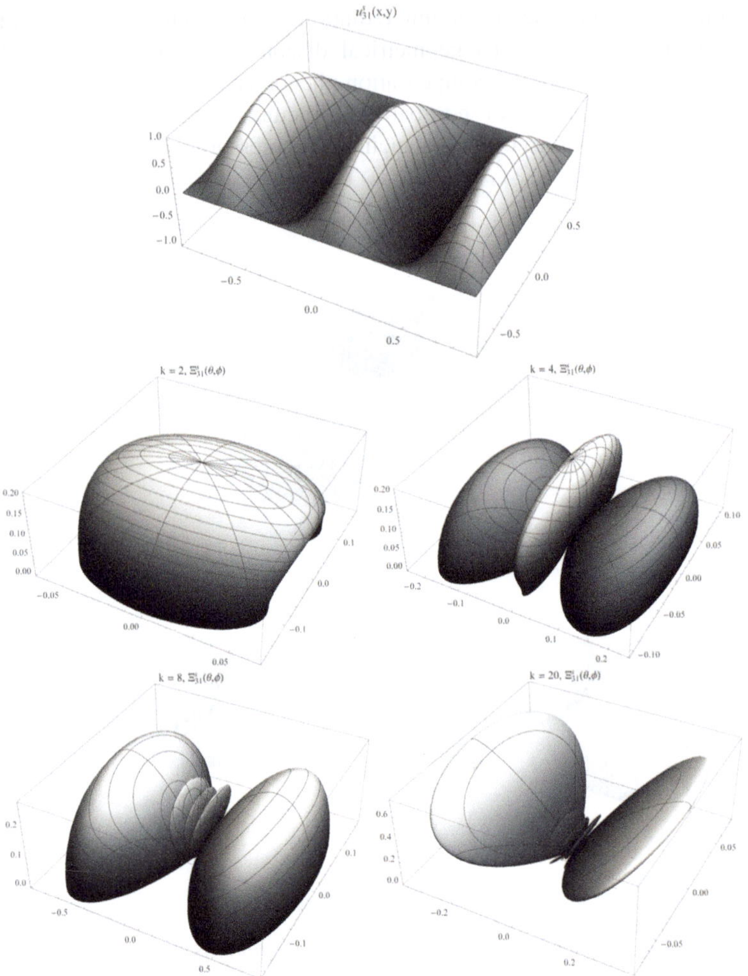

Figure 5.16. *"Symmetrical" deformation and directivity for $m = 3$, $n = 1$ and for $k = 2, 4, 8$ and 20 m^{-1}*

We have $\tau_r \approx t - r_M/c_0 + r_S/c_0 = t - r_M/c_0 + \Delta\tau_r$. The variation of the retarded time is linked to the characteristic dimensions of the source ℓ and we get $\Delta\tau_r = r_S/c_0 \approx \ell/c_0$. For this, we can consider that the source is compact, the retarded time variation on the volume $V \approx \ell^3$ of the source must be small compared to the evolution of the source $\tau_s = \ell/v$, this is $\Delta\tau_r/\tau_s \approx \ell/c_0 v/\ell = M \ll 1$, where M is the Mach number. From this:

$$p(M,t) \approx \frac{1}{4\pi r_M} \int_{V(S)} Q\left(S, t - \frac{r_M}{c_0}\right) dS$$

In the case of a spatially compact source, the wavelength of the signal λ is large compared to the size of the source $\lambda/\ell \approx 1/M >> 1$.

Figure 5.17. *"Antisymmetrical" deformation and directivity for $m = 3$, $n = 1$ and for $k = 2, 4, 8$ and 20 m^{-1}*

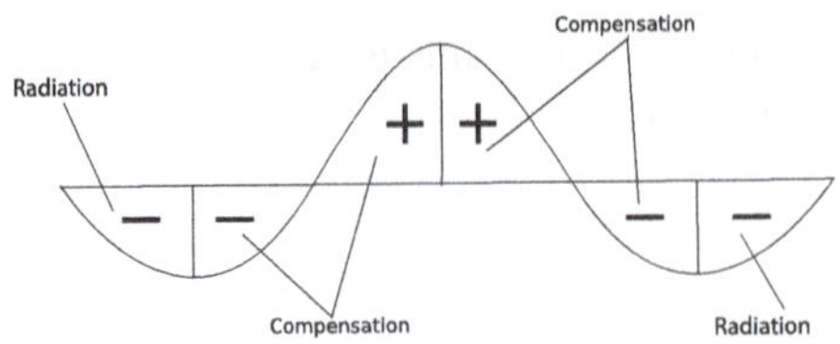

Figure 5.18. *Explanation of the compensation phenomenon*

5.6.1.2. *Spatially non-compact source ($M \gg 1$)*

Let us examine what happens when the source moves at supersonic speed. In the far-field, we always have $r \approx r_M - r_S$. Let $S = (r_S, y_S)$ and let us start out again with the general expression for pressure radiated by any source $Q(S,t)$, which can be given by:

$$
\begin{aligned}
p(M,t) &= \frac{1}{4\pi}\int_{V(S)}\int_{\tau} Q(r_S, y_S, \tau)\delta\left(t-\tau-\frac{r}{c_0}\right)\frac{dr_S dy_S}{r}d\tau \\
&\approx \frac{1}{4\pi r_M}\int_{V(S)}\int_{\tau} Q(r_S, y_S, \tau)\delta\left(t-\tau-\frac{r_M}{c_0}-\frac{r_S}{c_0}\right)dr_S dy_S d\tau \\
&\approx \frac{c_0}{4\pi r_M}\int_{S(S)}\int_{\tau} Q(r_M - c_0(t-\tau), y_S, t)dy_S d\tau
\end{aligned}
$$

Then, we compare the spatial variation of the source term Q in the direction of the observer r_M on the distance $c_0\tau$: $c_0\tau/\ell \approx c_0\ell/u/\ell \approx 1/M \ll 1$. Thus:

$$
p(M,t) = \frac{c_0}{4\pi r_M}\int_{S(S)}\int_{\tau} Q(r_M - c_0 t, y_S, t)dy_S d\tau
$$

5.6.1.3. *The case of the flow source*

For a flow source, we have $Q(S,t) = \rho_0\partial Q_m(S,t)/\partial t \approx \rho_0(v/\ell)/(\ell/v) \approx \rho_0(v/\ell)^2$. If the source is spatially compact:

$$
p(M,t) \approx \frac{1}{4\pi r_M}\int_{V(S)} Q\left(S, t-\frac{r_M}{c_0}\right)dS \approx \frac{1}{r_M}\rho_0\frac{v^2}{\ell^2}\ell^3 \approx \frac{\rho_0 c_0^2 \ell}{r_M}M^2.
$$

The pressure radiated is proportional to $\rho_0 v^2$ and we have:

$$
p(M,t) \approx \frac{1}{4\pi r_M}\frac{\partial}{\partial t}\int_{V(S)} Q_m\left(S, t-\frac{r_M}{c_0}\right)dS.
$$

The radiation is controlled by the time variations of the source. If the source is spatially non-compact, we have:

$$p(M,t) \approx \frac{c_0}{4\pi r_M} \int_S \int_\tau Q(r_M - c_0 t, y_S, t) dy_S d\tau \approx \frac{c_0}{r_M} \rho_0 \frac{v^2}{\ell^2} \ell^2 \frac{\ell}{v} \approx \frac{\rho_0 c_0^2 \ell}{r_M} M.$$

The radiated pressure is proportional to $\rho_0 c_0 v$ and we have with $\partial/\partial t = \partial/\partial r_S \, \partial r_S/\partial t = -c_0 \partial/\partial r_S$:

$$p(M,t) = -\frac{c_0^2}{4\pi r_M} \int_S \int_\tau \frac{\partial}{\partial r_S} Q(r_M - c_0 t, y_S, t) dy_S d\tau,$$

The radiation is controlled by the spatial variations of the source in the direction of the observer.

5.6.2. *Sources in uniform and non-uniform motion*

Let us consider a mass point source localized at the instant t at a point $S(t)$. The pressure that it radiates can be given by:

$$\frac{1}{c_0^2} \frac{\partial^2 p}{\partial t^2} - \Delta p = \frac{\partial}{\partial t} \left(Q(t) \delta(M - S(t)) \right)$$

It is straightforward to deduce that:

$$p(M,t) = \frac{\partial}{\partial t} \int \frac{Q(\tau)\delta(M - S(t))\delta(t - \tau - d(M,M')/c_0)}{4\pi d(M,M')} dM' d\tau$$

From where:

$$p(M,t) = \frac{\partial}{\partial t} \int \frac{Q(\tau)\delta(t - \tau - d(M,S(t))/c_0)}{4\pi d(M,S(t))} d\tau$$

However, we know from formula [3.34] that $\int Q(t)\delta(f(t)) = \sum_{n=1}^{N} Q(\tau_n^*)/|\partial f(\tau)/\partial \tau|$. The sum is extended to N zeros of $f(\tau)$ which are denoted as τ_n^*. Here, $f(\tau) = t - \tau - d(M,S(t))/c_0$. We get:

$$\partial f(\tau)/\partial \tau = -1 + \frac{x_i - x_{si}}{d(M,S(t))c_0} \frac{\partial x_{si}}{\partial \tau} = -1 + M_S$$

where M_S is based on the component of the velocity of the source in the direction of the observer. From where:

$$p(M,t) = \sum_{n=1}^{N} \frac{\partial}{\partial t} \left(\frac{Q(\tau_n^*)}{4\pi d(M,S(t)) \, |1 - M_S|} \right).$$

τ_n^* corresponds to the emission time of the sound received by the observer at time t. At this moment, the source is at $S(\tau_n^*)$ and τ_n^* is the solution of $c_0(t - \tau_n^*) = d(M, S(\tau_n^*))$ and by deriving $c_0\,(1 - \partial\tau_n^*/\partial t) = -(x_i - x_{si})\,/\,(d(M, S(t))c_0) = \partial x_{si}/\partial\tau\partial\tau_n^*/\partial t$. This is $\partial\tau_n^*/\partial t = 1/(1-M_S)$. This again illustrates the relationship between the source time τ_n^* and the observer time t (Doppler effect). Now, let us suppose that the source has a constant amplitude $Q(t) = Q_0$. We thus have:

$$p(M,t) = \frac{Q_0}{4\pi}\sum_{n=1}^{N}\frac{\partial}{\partial t}\left(\frac{1}{d(M,S(t))\,|1 - M_S|}\right)$$

In the far-field ($d(M, S(t)) \gg 1$), we have:

$$p(M,t) \approx \frac{Q_0}{4\pi}\sum_{n=1}^{N}\frac{\partial M_S}{\partial t}\frac{1}{d(M,S(t))\,|1 - M_S|^3}$$

For a source close to the origin, which moves at a low Mach number, we have $p(M,t) \approx Q_0/(4\pi c_0)\partial\overline{V}_S/\partial t{\cdot}\overline{r}_M/r_M^2$. This relation shows that a constant monopole (i.e. without temporal fluctuations of its flow rate) can create a dipole field when it is in non-uniform motion. When the monopole is in uniform movement along the x axis at a velocity V_x, the main contribution in the far-field belongs to the temporal derivative of $Q(\tau_n^*)$. We have:

$$p(M,t) = \sum_{n=1}^{N}\left(\frac{\partial Q(\tau_n^*)/\partial t}{4\pi d(M,S(t))\,|1 - M_x\cos\theta|^2\,\mathrm{sgn}(1 - M_x\cos\theta)}\right) \qquad [5.10]$$

where $\cos\theta = (x - x_S(\tau^*))/d(M, S(\tau^*))$ is the angle between the observer's position and the x axis, measured from the position of the source at the emission time τ^* and $M_x = V_x/c_0$. While a monopole at rest emits an omnidirectional signal, the factor $|1 - M_x\cos\theta|^2$ indicates that a monopole in subsonic movement emits a stronger field toward the front than the back.

The emission time τ^* is then easily calculated by solving the implicit equation:

$$c_0(t - \tau^*) = \sqrt{(x - V_x\tau^*)^2 + y^2 + z^2}$$

which gives:

$$\tau^* = \frac{c_0 t - M_x x \pm \sqrt{(x - V_x t)^2 + (1 - M_x^2)(y^2 + z^2)}}{c_0(1 - M_x^2)}$$

Only solutions for which $\tau^* < t$ and $\tau^* \in I\!R$ make sense. For $M_x < 1$, only the lower sign leads to a physical solution and when $M_x > 1$, there is only the solution in the Mach cone $\sqrt{(1-M_x^2)(y^2+z^2)} < V_x t - x$. It is often useful to introduce a coordinate system R, Θ based on the reception time, with $\cos\Theta = (x - V_x t)/R$ and $R = \sqrt{(x-V_x t)^2 + y^2 + z^2}$. Thus:

$$\tau^* = t - \frac{R}{c_0(1-M_x^2)}\left(M_x \cos\Theta \mp \sqrt{1 - M_x^2 \sin^2\Theta}\right)$$

5.6.2.1. *Doppler effect*

It is a common experience to observe that the sound emitted by an object approaching a receiver is pitched higher than when the object is moving away. This effect also occurs when the wave is reflected on an object in movement with respect to the transmitter/receiver (the principle of speedometers and other radars).

Let us suppose that the wave travels in a medium at rest with a celerity of c_0 (referential R_0), that the velocity of the source is v_s (referential R_s) and that the velocity of the receiver is v_r (referential R_r). For simplicity, let us assume that the emitter and the receiver are moving along a straight line. f_0 is the emitted pulse frequency in the source referential.

In a period of $T_0 = 1/f_0$, the pulses emitted by the source travel a distance $d_0 = c_0 T_0$ in R_0. As during T_0, the source moves over the distance $v_S T_0$, the distance between two pulses in R_0 is $d_s = (c_0 - v_s)T_0$.

Let us calculate the time T_r which is the time between the receiver receiving two pulses. Between the two pulses, the receiver will have traveled the distance $v_r T_r$, the second pulse will have traveled the distance $d_r = d_s + v_r T_r = c_0 T_r$, this is $T_r = d_s/(c_0 - v_r)$ and a frequency of $f_r = 1/T_r = (c_0 - v_r)/d_s$. Thus:

$$f_r = \frac{c_0 - v_r}{c_0 - v_s} f_0 = \frac{1 - v_r/c_0}{1 - v_s/c_0} f_0$$

5.6.2.2. *Shock waves*

The pressure field of a source in motion, which is given by equation [5.10], is clearly singular at a point M such that $M_x \cos\theta = 1$, located on the Mach cone, can be given by the relation $\sqrt{(1-M_x^2)(y^2+z^2)} = V_x t - x$. We can show that the potential of these velocities $\phi(M,t)$ of the Mach wave $(-\rho_0 \partial\phi(M,t)/\partial t = p(M,t))$ in the far-field can be given by a source $Q(S,\tau) = q(x,\tau)\delta(y)\delta(z)$ of finite dimensions and which begins at $t = 0$, by:

$$\phi(M,t) = -\frac{U_x}{4\pi\rho_0 r_M}\int_0^t q(0,\tau)d\tau$$

This means that the whole history of the source is heard by a distant observer in a single "bang". [CRI 92]

5.7. Sound propagation in a bounded medium

The propagation phenomena in bounded media are radically different from sound propagation in free space. The finite nature of the propagation domain highlights the concept of resonance and the possible geometric variety leads to a diversity of situations which are impossible to describe. In order to present the main phenomena, we will introduce and examine some archetypal situations. It must be noted that the problems discussed here concern only ideal media without losses. The reality is quite different because all media present sources of dissipation. In a closed medium, most of the dissipation occurs at the walls, be it because of porous materials or viscothermal phenomena near the walls. We refer the readers to the book by Bruneau [BRU 83] which gives a very complete presentation of this.

5.7.1. *Eigenfrequencies and resonance frequencies*

Any domain of finite dimensions, regardless of the acoustic properties at its boundaries, has eigenfrequencies (real or complex). Let us simply recall that the distinction between eigenfrequencies and resonance frequencies is made when the boundary problem depends on the frequency, such as when using the condition of impedance. Eigenmodes depend on the frequency and verify all the boundary conditions. The resonance modes are independent of frequency and all verify a different limit condition. Eigenfrequencies depend on the geometry of the problem and boundary conditions. They are real in the case of conservative boundary conditions. From a physical point of view, the existence of eigenfrequencies does not mean that there may be a signal in the absence of an excitation source. It simply signifies that, if an acoustic wave is created, the cavity, in which it starts, will have a much stronger response when the excitation frequency corresponds to an eigenfrequency of the system (at least the real part for a damped system).

The following theorem states that the conditions of existence and uniqueness of $p(M)$ can be solved by:

$$\left(\Delta + k_n^2\right) p(M) = f(M), \; M \in \Omega$$
$$\alpha(\omega)\partial_{\vec{n}} p(M) + \beta(\omega) p(M) = 0, \; M \in \partial\Omega,$$

its difficult proof is not given.

1) There exists an enumerable sequence k_n, $n \in \boldsymbol{N}$, of wave numbers such that the previous homogeneous system has solutions that are not zero. These eigen wave numbers correspond to eigenfrequencies f_n.

2) For each wave number k_n, there is a corresponding finite number of solutions ψ_{nm}, $n \in \boldsymbol{N}$, $m \in [1, N_n]$, of the homogeneous problem, which are linearly

independent. These solutions are called eigenmodes of the Helmholtz operator for boundary conditions. The number $N_n < \infty$ of modes which correspond to k_n is called the number of multiplicity of the wave number.

3) Eigen wave numbers k_n are real, if α/β is real, complex if α/β is a complex.

4) If k is equal to one of the eigen wave numbers, the previous inhomogeneous system does not have solutions. If $k \neq k_n$, the solution $p(M)$ exists and is unique for any second part $f(M)$.

5.7.2. *The Helmholtz resonator*

The Helmholtz resonator (an empty bottle is the best example) is an example of the simplest resonant system in acoustics. This is the acoustic equivalent of the mechanical spring-mass system.

The resonator is considered to be a closed cavity with a volume V that communicates with the outside by a tube with a small section of $A = \pi R^2$ and a length of L. All the dimensions are small when compared to the wavelength.

An acoustic wave, which insonifies the resonator, imposes a small displacement of fluid u in the neck. The volume of air displaced in the neck is thus $dV = Au$. If we assume that air is an ideal adiabatic gas (the system is thermally insulated), the Laplace law gives a pressure of $dp = -\gamma p_0/VdV$ in the bottle, where $\gamma = 1.41$ is the ratio of the specific heats. This (variation of) pressure creates a force $dF = Adp = -\gamma p_0 A^2 u/V$.

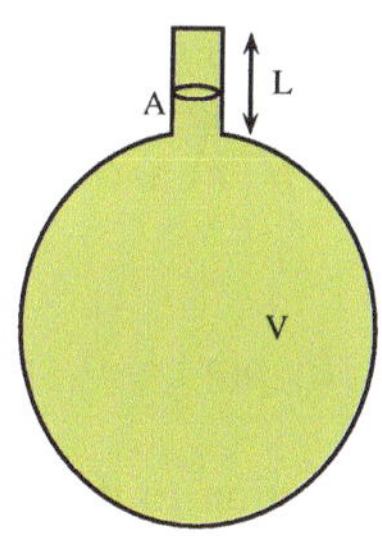

Figure 5.19. *Helmholtz resonator*

The air in the neck satisfies a mass-spring equation whose eigenfrequency depends on the geometry. In fact, the fundamental principle of dynamics leads to $\rho_0 ALd^2u/dt^2 + \gamma p_0 A^2/Vu = 0$, this is $\rho_0 ALd^2u/dt^2 + \rho_0 c_0^2 A^2/Vu = 0$, because $c_0^2 = \gamma p_0/\rho_0$, where $m = \rho_0 AL$ is the mass of air in movement and $r = \rho_0 c_0^2 A^2/V$ is the stiffness associated with the compressible fluid contained in the cavity. This is

$\ddot{u} + \omega_0^2 u = 0$, with $\omega_0 = c_0\sqrt{A/(VL)}$ which is the natural pulsation of the mass (neck) – spring (cavity) system.

In practice, an additional small mass of air is driven to the exterior of the neck. This is accounted for by a length correction which increases the apparent length of the neck $\ell = L + \Delta L$. Specifically, if the resonator leads onto a wall, we have $\Delta L = 0.82R$, if the neck of the resonator is free, we take $\Delta L = 0.61R$. When the section of the neck is non-circular, we take $R = \sqrt{A/\pi}$.

5.7.3. *Example in dimension 1*

Let us consider the example of a Pan flute (or mouth organ) which is composed of a set of independent pipes. Each pipe is considered as a monodimensional domain whose one end is closed and the other end is opened.

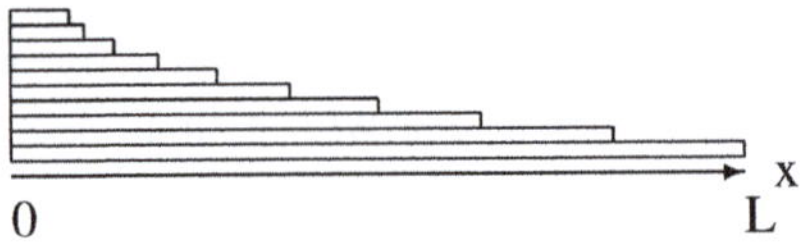

Pan flute. Geometry of the problem.

Each tube is opened at one end ($x = 0$) and is closed at the other end ($x = L$). At the closed end, the tube is enclosed by a rigid surface, well described by the Neumann condition. At the open end, the total pressure is equal to the sum of atmospheric and acoustic pressures, therefore very close to the atmospheric pressure. A good approximation is to impose the Dirichlet boundary condition (acoustic pressure zero). Obviously, the dimensions of the pipe are such that the propagation only depends on the x axis. By choosing the normal exterior to the domain, the Neumann condition gives $\partial_{\vec{n}} p(x) = p'(x)$ and the acoustic pressure is the solution to:

$$p''(x) + k^2 p(x) = 0, \; x \in [0, L], p(x) = 0, \; x = 0, p'(x) = 0, \; x = L.$$

As we have already seen, the solution to the homogeneous Helmholtz equation can be given by $p(x) = A\exp(\imath kx) + B\exp(-\imath kx)$ where A and B are constants. The condition $p(0) = 0$ leads to $A = -B$ and $p'(L) = 0$ to $2\imath A\cos kL = 0$. This is $k = (2n+1)\pi/2L$, $n \in \boldsymbol{N}$. Subsequently, the eigenfrequencies can be given by $f_n = (2n+1)c_0/4L$. The lowest f_0 frequency is the fundamental and is inversely proportional to the length of the pipe. Hence, the "exponential" form of the Pan flute. After normalization, the eigenmodes can be written as $p_n(x) = \sqrt{2/L}\sin(k_n x)$. This shows that the base frequency of a Pan flute with a length of $L = 19.31$ cm is the A_4 of frequency $f = 440$ Hz, for air we take $c_0 = 340$ m/s. As for the Helmholtz resonator, an additional small mass of air is driven into the mouth of the flute. To take it into consideration, we increase the apparent length of the tube $\ell = L + \Delta L$ with $\Delta L = 0.61R$ since a Pan flute never leads to a wall and always has a circular section.

5.7.4. *Example in dimension 3*

Here, we study an example of a rectangular room which limits the domain Ω, defined by $0 < x < \ell_x$, $0 < y < \ell_y$, $0 < z < \ell_z$. The system of equations that needs to be solved is given by:

$$\left(\Delta + k^2\right) p(x,y,z) = 0,\ (x,y,z) \in \Omega, \partial_{\vec{n}} p(x,y,z) = 0,\ (x,y,z) \in \partial\Omega.$$

By convention, let us choose the normal interior to the domain. The boundary conditions can be written as $\partial p(0,y,z)/\partial x = 0$, $\partial p(\ell_x,y,z)/\partial x = 0$, $\partial p(x,0,z)/\partial y = 0$, $\partial p(x,\ell_y,z)/\partial y = 0$, $\partial p(x,y,0)/\partial z = 0$ and $\partial p(x,y,\ell_z)/\partial z = 0$. We must look for the k values for which this system has non-zero values. The room is rectangular. Each side is parallel to a coordinate line. We can thus apply the method of variable separation. We aim to find a solution in the form of a Laplace product $p(x,y,z) = X(x)Y(y)Z(z)$. This can be seen in the Helmholtz equation $\left(\Delta + k^2\right) X(x)Y(y)Z(z) = 0$ which gives $X''(x)/X(x) + Y''(y)/Y(y) + Z''(z)/Z(z) = -k^2$. As k is a constant, it is necessary that each product $X''(x)/X(x) = -\alpha^2$, $Y''(y)/Y(y) = -\beta^2$ and $Z''(z)/Z(z) = -\gamma^2$ where $k^2 = \alpha^2 + \beta^2 + \gamma^2$. Each of the constants α, β and γ is any complex, which does not prejudge its generality. We do the same for the boundary conditions and it is easy to see that $X'(0) = 0$, $X'(\ell_x) = 0$, $Y'(0) = 0$, $Y'(\ell_y) = 0$, $Z'(0) = 0$ and $Z'(\ell_z) = 0$. We solve the three equations in the same way and we obtain:

$$\alpha = \frac{l\pi}{\ell_x}, \beta = \frac{m\pi}{\ell_y}, \gamma = \frac{n\pi}{\ell_z}$$

$$k_{lmn} = \sqrt{\left(\frac{l\pi}{\ell_x}\right)^2 + \left(\frac{m\pi}{\ell_y}\right)^2 + \left(\frac{n\pi}{\ell_z}\right)^2}, (l,m,n) \in \boldsymbol{N}$$

$$\psi_{lmn}(x,y,z) = A_{lmn} \cos\left(\frac{l\pi}{\ell_x} x\right) \cos\left(\frac{m\pi}{\ell_y} y\right) \cos\left(\frac{n\pi}{\ell_z} z\right).$$

k_{lmn} is the series of eigen wave numbers and $\psi_{lmn}(x,y,z)$ are the eigenmodes of the cavity. The constant A_{lmn} can be calculated by imposing on these modes, which are orthogonal, to be of unity norm. The scalar product considered here is, of course, the triple integral over the volume of the cavity:

$$\langle f(x,y,z), g(x,y,z)\rangle = \int_0^{\ell_x} \int_0^{\ell_y} \int_0^{\ell_z} f(x,y,z) g^\star(x,y,z) dx dy dz.$$

To say that two modes are orthonormal is the equivalent of:

$$\begin{aligned}\langle \psi_{lmn}(x,y,z), \psi_{rst}(x,y,z)\rangle &= 0 \text{ if } (l,m,n) \neq (r,s,t)\\ &= 1 \text{ if } (l,m,n) = (r,s,t).\end{aligned}$$

It is simple to show that $A_{lmn} = \sqrt{\epsilon_l \epsilon_m \epsilon_n/(\ell_x \ell_y \ell_z)}$ where $\epsilon_i = 1$, if $i = 0$ and $\epsilon_i = 2$, if $i \geq 1$. The eigenfrequencies can be given by:

$$f_{lmn} = \frac{k_{lmn}c_0}{2\pi} = \frac{c}{2}\sqrt{\left(\frac{l}{\ell_x}\right)^2 + \left(\frac{m}{\ell_y}\right)^2 + \left(\frac{n}{\ell_z}\right)^2}, (l,m,n) \in \boldsymbol{N}.$$

It should be noted that the series of eigen wave numbers is a series of three indices for $V/(6\pi^2)$ which correspond to a countably infinite triple. We show ([COU 53], vol. 1, p. 434, th. 13) that the number N_f of eigenmodes lower than the frequency f, fixed in a room with the volume V, can be given by the high-frequency approximation $N_f \approx V/(6\pi^2)k^3 = 4/3\pi V\,(f/c_0)^3$. We can slightly refine this result ([MOR 48], Chapter 8, p. 394) and get $N_f \approx 4\pi/3V\,(f/c_0)^3 + \pi/4A\,(f/c_0)^2 + L/8f/c_0$, where $A = 2(\ell_x\ell_y + \ell_y\ell_z + \ell_x\ell_z)$ is the total wall area of the room and $L = 4(\ell_x + \ell_y + \ell_z)$ is the total wall length of the room.

The general solution of the system is thus given by the sum of all modes $p(x,y,z) = \sum_{l,m,n=0}^{l,m,n=\infty} E_{lmn}\psi_{lmn}(x,y,z)$, where E_{lmn} is a constant. Let us consider two points of the room $M = (x,y,z)$ and $M' = (x',y',z')$, we can easily demonstrate that Green's function $G(M,M')$ of the room can be solved as $\left(\Delta_M + k^2\right) G(M;M') = \delta_{M'}(M)$, provided that the Neumann boundary conditions for the walls are written as:

$$G(M;M') = c_0^2 \sum_{l,m,n=0}^{l,m,n=\infty} \frac{\psi_{lmn}(x',y',z')\psi_{lmn}(x,y,z)}{\omega^2 - \omega_{lmn}^2},$$

where $\omega_{lmn}^2 = 2\pi f_{lmn}$. It is easy to see that Green's function not only satisfies the reciprocity theorem $G(M;M') = G(M';M)$ but also that $G(M;M') = G(M - M')$. We must also note that the zero-frequency mode, which is present in the domains abiding Neumann's condition on the walls, leads to real difficulty. This mode results in a non-causal temporal solution as the inverse Fourier transform of $p(\omega) \propto 1/\omega^2$ leads[3] to $p(t) \propto t\text{sgn}(t)$. A simple way to solve this difficulty is to introduce slight dissipation in the room walls by using the Helmholtz equation for viscous fluids, which is valid for large cavities (i.e. well above 10μm)

3 As indicated in section 3.4.2.2, it is essential to have modes which come in pairs in anti-Hermitian symmetry to ensure causality.

[BRU 94][4] : $(\Delta + k_a^2)p(M) = 0$, where $k_a^2 = k^2(1 - \imath k\ell_{vh} - k^2\ell'_{vl}\ell_h)^{-1}$, $k = \omega/c_0$, $\ell_{vh} = \ell_v + (\gamma - 1)\ell_h$, $\ell'_{vh} = (\gamma - 1)(\ell_h - \ell_v)$. Where the characteristic lengths are given by $\ell_v = (\mathrm{K} + 4\mu/3)/(\rho_0 c_0)$ and $\ell_h = k_\theta/(\rho_0 c_0 C_p)$, where the shear μ and volume K viscosity coefficients, as well as the heat capacity at constant pressure C_p, are defined in section 2.3.4; k_θ, as we remember, is the thermal conductivity.

5.7.5. *Propagation of pure sound in a circular enclosure*

In this section, we present a comparison of several methods for solving the academic problem of calculating the sound field within a circular enclosure with rigid walls. More precisely, we will compare two methods of resolution based on the direct integration of a differential boundary problem and two methods based on integral representations. As we will see, all these methods result in expressing the displacement in the form of a series of valid functions outside certain frequencies. The objective of this example is to show that the integral equations give identical results to those obtained by the direct solution of the differential problem. This is an indirect way of verifying the equivalence between a system of partial differential equations and the integral equations associated with it. The main difference is that the integral equations are much easier to solve numerically, when the propagation area has boundaries that are not coordinate lines.

Let us consider the following problem. In a two-dimensional space, there is a circular enclosure with the center O and the radius R. It defines a domain S, with a boundary ∂S. In this scenario, $M(r,\theta)$ and $M_0\,(r_0,\theta_0)$ are two points in S, and $Q(R,\theta)$ is a point on the boundary ∂S. We aim to determine the acoustic pressure $P(M)$, which is the solution of the following problem:

$$\left(\Delta + k^2\right) p(M) = \delta_{M_0}(M) \text{ in } S, \partial_n p(Q) = 0 \text{ on } \partial S.$$

This system of partial differential equations reflects the problem of steady radiation of a point source in circular symmetry in $\mathbb{R}^2$, positioned at point M_0 in S whose boundary is assumed to be perfectly reflective. We must recall that, in a circular geometry, we have $\delta_{M_0}(M) = 1/r\delta_{r_0}(r)\delta_{\theta_0}(\theta)$.

5.7.5.1. *Direct integration methods*

5.7.5.1.1. Separation of variables

The previous system is separable with respect to the space variables r and θ. In this simple case, we know how to calculate an exact solution in the form of a series of

4 We must take care when assigning the sign to $\imath k$, which depends on the chosen time convention.

orthogonal functions. These functions are the non-zero solutions of the homogeneous system:

$$\left(\Delta + k_{mn}^2\right) \phi_{mn}(M) = 0 \text{ in } S, \partial_n \phi_{mn}(Q) = 0 \text{ on } \partial S,$$

where the values of k_{mn} are the eigenvalues of the problem. Let us recall that the general solution of the Helmholtz equation (therefore homogeneous) in polar coordinates can be given by:

$$p_g(r,\theta) = \sum_{m=-\infty}^{m=+\infty} A_m \mathrm{J}_m(kr)\mathrm{e}^{im\theta}.$$

To calculate the eigenfrequencies of the domain, we apply the boundary condition. For $\partial_n P(Q) = 0$ over ∂S, it is easy to deduce that $\lim_{r\to R} \partial_r P(r,\theta) = 0$. As this relation is independent of θ, it is easy to obtain that $k\mathrm{J}'_m(kR) = 0$. Because of the properties of the Bessel functions, for each value of the index m, there exists an infinite number of values for k, denoted by $k_{mn}, n \in \boldsymbol{N}^\star$, which satisfy this condition. This double infinity of values forms the series of eigen wave numbers. It is thus easy to obtain the corresponding eigenfrequencies.

The eigenfunctions can be given by $p_{mn}(r,\theta) = c_{mn}\mathrm{J}_m(k_{mn}r)\exp(\imath m\theta)$. We must note that for $m = 0$ and $n = 1$, $k_{01} = 0 \rightarrow p_{01}(r,\theta) = c_{01}$. Due to the properties of Bessel functions and trigonometry, it is possible to show that the eigenmodes satisfy an orthogonality relationship on the section of the guide. Let us calculate c_{mn} for which this base is normalized. With δ_{mn}^{pq}, the Kronecker symbol, we get:

$$c_{mn}c_{pq} \int_0^{2\pi} \left[\int_0^R \mathrm{J}_m(k_{mn}r)\mathrm{J}_p(k_{pq}r)rdr\right] \mathrm{e}^{\imath(m+p)\theta} d\theta = \delta_{mn}^{pq}.$$

For $(mn) = (pq)$, it is obvious that $c_{mn}^2 \int_0^R \mathrm{J}_m^2(k_{mn}r)rdr \int_0^{2\pi} \mathrm{e}^{2\imath m\theta} d\theta = 1$. We show [WAT 44] that if $(m,n) = (0.1)$, then $c_{01}^2 = 1/\left(\pi R^2\right)$ and that for $(m,n) \neq (0,1)$ we have $c_{mn}^2 = 1/[\pi R^2(1 - m^2/(k_{mn}^2 R^2)$ $(\mathrm{J}_m(k_{mn}r))^2)]$.

Let us express $p(r,\theta)$ by a development based on the eigenmodes $p_{mn}(r,\theta)$. We need to find $p(r,\theta)$ in the form of $p(r,\theta) = \sum_{m=-\infty}^{+\infty} \sum_{n=0}^{+\infty} a_{mn} p_{mn}(r,\theta)$ which we incorporate into the Helmholtz equation. We obtain

$\sum_{m=-\infty}^{+\infty}\sum_{n=0}^{+\infty} a_{mn}\left(k^2-k_{mn}^2\right)p_{mn}(r,\theta) = 1/r\delta_{r0}(r)\delta_{\theta_0}(\theta)$. Let us multiply the two parts of this equation by $p_{st}(r,\theta)$ and integrate over the surface:

$$\sum_{m=-\infty}^{+\infty}\sum_{n=0}^{+\infty} a_{mn}\left(k^2-k_{mn}^2\right)\int_0^{2\pi}\int_0^R p_{mn}(r,\theta)p_{st}(r,\theta)rdrd\theta$$
$$= \int_0^{2\pi}\int_0^R \delta_{r0}(r)\delta_{\theta_0}(\theta)p_{st}(r,\theta)drd\theta.$$

The properties of orthogonality of eigenmodes and the Dirac distribution allow us to give the solution in the following form:

$$p(r,\theta) = \sum_{m=-\infty}^{+\infty}\sum_{n=0}^{+\infty}\frac{p_{mn}\left(r_0,\theta_0\right)}{k^2-k_{mn}^2}p_{mn}(r,\theta),$$

and, in expanded form, we get:

$$P(r,\theta) = \sum_{m=-\infty}^{+\infty}\sum_{n=0}^{+\infty}\frac{\mathrm{J}_n\left(k_{mn}r_0\right)}{k^2-k_{mn}^2}\mathrm{J}_n\left(k_{mn}r_0\right)\mathrm{e}^{\imath m(\theta-\theta_0)}. \qquad [5.11]$$

When the current point is near the point source, series [5.11] converges very slowly. We can use another method which leads to a solution in the form of a series that converges faster.

5.7.5.1.2. Direct integration

An efficient integration method is to express the solution in the form of a sum of the general solution of the differential equation and a particular solution: $p(r,\theta) = p_g(r,\theta)+p_0(r,\theta)$. An expression in which $p_g(r,\theta)$ is the solution of the homogeneous Helmholtz equation and $p_0(r,\theta)$ is the solution of the Helmholtz equation with the second part:

$$\left(\Delta+k^2\right)p_0(M) = \delta_{M_0}(M).$$

$p_0(M)$ here equals Green's kernel of the Helmholtz equation, which as we know can be given by $p_0(M) = -\imath/4\mathrm{H}_0\left(kMM_0\right)$. The general solution for $p_g(r,\theta)$ was calculated in the previous section. Thus:

$$p(r,\theta) = -\frac{\imath}{4}\mathrm{H}_0\left(kMM_0\right)+\sum_{m=-\infty}^{+\infty}A_{mn}\mathrm{J}_m(kr)\mathrm{e}^{\imath m\theta}.$$

The values of A_{mn} are calculated by applying the boundary condition to them. In order for us to proceed, we introduce the expansion [WAT 44] of the Hankel function in a series of cylindrical harmonics:

$$\begin{aligned}\mathrm{H}_0\left(kMM_0\right) &= \sum_{m=-\infty}^{+\infty} \mathrm{J}_m\left(kr_0\right)\mathrm{H}_m(kr)\mathrm{e}^{\imath m(\theta-\theta_0)}, \text{ if } r_0 \le r \\ &= \sum_{m=-\infty}^{+\infty} \mathrm{H}_m\left(kr_0\right)\mathrm{J}_m(kr)\mathrm{e}^{\imath m(\theta-\theta_0)}, \text{ if } r_0 \ge r.\end{aligned}$$

When we apply the boundary condition, the current point, identified by its distance from the origin r, is located on the boundary $r = R$. As it is evident that $R \ge r_0$ – the source point is always inside the disk –, we get $\lim_{r\to R} p(r,\theta) = \lim_{r\to R}\sum_{m=-\infty}^{+\infty}\left[-\imath/4\mathrm{H}_m(kr)\mathrm{J}_m\left(kr_0\right)\mathrm{e}^{-\imath m\theta_0}\right]\mathrm{e}^{\imath m\theta}$. The boundary condition gives us that for all m:

$$A_{mn} = \frac{\imath}{4}\frac{\mathrm{H}'_m(kR)\mathrm{J}_m\left(kr_0\right)}{\mathrm{J}'_m(kR)}\mathrm{e}^{\imath m\theta_0} = 0.$$

We must note that this expression is not defined for $k = k_{mn}$ (eigen wave number) as it was defined in the previous section. The acoustic pressure can be written as follows:

$$\begin{aligned}p(r,\theta) &= \sum_{m=-\infty}^{+\infty}\frac{\imath}{4}\left[-\mathrm{H}_m(kr)\mathrm{J}_m\left(kr_0\right) + \frac{\mathrm{H}'_m(kR)\mathrm{J}_m\left(kr_0\right)}{\mathrm{J}'_m(kR)}\mathrm{J}_m(kr)\right]\mathrm{e}^{\imath m(\theta-\theta_0)}, \\ &\quad \text{if } r \ge r_0 \\ &= \sum_{m=-\infty}^{+\infty}\frac{\imath}{4}\left[-\mathrm{J}_m(kr)\mathrm{H}_m\left(kr_0\right) + \frac{\mathrm{H}'_m(kR)\mathrm{J}_m\left(kr_0\right)}{\mathrm{J}'_m(kR)}\mathrm{J}_m(kr)\right]\mathrm{e}^{\imath m(\theta-\theta_0)}, \\ &\quad \text{if } r \le r_0.\end{aligned} \quad [5.12]$$

For comparison, this series converges – approximately – 10 times faster than the series expansion of eigenmodes [5.11]. We must note that it does not seem to be possible to obtain a “more efficient” solution than the latter expression.

5.7.5.2. *Method of integration by integral equations*

The most natural representation (from a mathematical point of view) is Green’s representation which is given for an exterior normal by:

$$p(M) = p_0(M) + \int_{\partial S} p(Q)\partial_{n(Q)}G(M;Q) - \partial_n p(Q)G(M;Q)d\sigma(Q),$$

an expression in which $p_0(M)$ is the incident field in an undefined space. This representation consists of constructing a function defined over the entire $I\!R^n$, equal to the acoustic pressure in the S field, zero outside of it and that satisfies the boundary condition on the boundary of the field ∂S. In fact, there is no obligation to respect this condition. What happens "on the outside" of the area is of little interest to us. We will build on this characteristic to describe the acoustic field in the domain in the form of an integral representation different from Green's representation.

5.7.5.2.1. Green's representation

Let us revisit the previous integral formula and use the fact that at the boundary the normal derivative of the pressure is zero. We get:

$$p(M) = p_0(M) + \int_{\partial S} p(Q)\partial_{n(Q)}G(M,Q)d\sigma(Q),$$

where $p_0(M) = G \star F$ represents the field created by the source $f(M)$ in the absence of the boundary, here $p_0(M) = G(M;M_0) = -\imath/4\mathrm{H}_0\,(kMM_0)$. Let us introduce R and Θ, the coordinates of the point $Q \in \partial S$. Developing in a series of cylindrical harmonics of the Hankel function results in writing the normal derivative on the boundary as $\partial_{n(Q)}\mathrm{H}_0(kMQ) = \sum_{m=-\infty}^{+\infty} k\mathrm{H}'_m(kR)\mathrm{J}_m(kr)\mathrm{e}^{\imath m(\theta-\Theta)}$. With $d\sigma(Q) = Rd\Theta$, the integral can be written as:

$$\int_{\partial S} p(Q)\partial_{n(Q)}G(M;Q)d\sigma(Q) = \sum_{m=-\infty}^{+\infty} k\mathrm{H}'_m(kR)\mathrm{J}_m(kr)\mathrm{e}^{\imath m\theta} \int_0^{2\pi} P(R,\Theta)\mathrm{e}^{-\imath m\Theta}Rd\Theta.$$

Let us develop the pressure $p(R,\Theta)$ into a Fourier series with respect to the angular variable $p(R,\Theta) = \sum_{m=-\infty}^{+\infty} a_m \exp(-\imath m\Theta)$, where $a_m = a_m(R)$ are the Fourier coefficients of acoustic pressure. Thus:

$$\int_{\partial S} p(Q)\partial_{n(Q)}G(M;Q)d\sigma(Q) = 2\pi kR \sum_{m=-\infty}^{+\infty} a_m\mathrm{H}'_m(kR)\mathrm{J}_m(kr)\mathrm{e}^{\imath m\theta}.$$

The development of an incident field in cylindrical harmonics gives, for $r > r_0$, $\mathrm{H}_0(kMM_0) = \sum_{m=-\infty}^{+\infty} \mathrm{J}_m(kr_0)\mathrm{H}_m(kr)\exp(\imath m(\theta-\theta_0))$. Green's representation can thus be written as:

$$p(M) = -\frac{\imath}{4}\sum_{m=-\infty}^{+\infty} \mathrm{J}_m(kr_0)\,\mathrm{H}_m(kr)\mathrm{e}^{\imath m(\theta-\theta_0)} + 2\pi kR\sum_{m=-\infty}^{+\infty} a_m\mathrm{H}'_m(kR)\mathrm{J}_m(kr)\mathrm{e}^{\imath m\theta}.$$

To calculate the coefficients a_m, let us again write the boundary condition on the boundary of the domain. We have:

$$\sum_{m=-\infty}^{+\infty}\left[\mathrm{J}_m(kr_0)\,\mathrm{e}^{-\imath m\theta_0} + 2\pi kRa_m\mathrm{J}'_m(kR)\right]\mathrm{H}'_m(kR)\mathrm{e}^{\imath m\theta} = 0.$$

This expression is zero if each term of the series is $[\mathrm{J}_m(kr_0)\exp(-\imath m\theta_0) + 2\pi kRa_m\mathrm{J}'_m(kR)]\mathrm{H}'_m(kR), \forall m$. If k is real, since all the zeros of the Hankel functions and their derivatives of all orders are complex, we have $\mathrm{H}'_m(kR) \neq 0$. Thus, if $\mathrm{J}'_m(kR)$ is non-zero, which is to say, if k is not a eigen wave number, defined in the previous section, we have:

$$a_m = -\frac{\mathrm{J}_m(kr_0)\,\mathrm{e}^{-\imath m\theta_0}}{2\pi kR\mathrm{J}'_m(kR)}.$$

If we refer to this expression in the expression obtained after integration of Green's representation, Green's representation can be written as:

$$\begin{aligned} p(r,\theta) &= \sum_{m=-\infty}^{+\infty}\frac{\imath}{4}\left[-\mathrm{H}_m(kr)\mathrm{J}_m(kr_0) + \frac{\mathrm{H}'_m(kR)\mathrm{J}_m(kr_0)}{\mathrm{J}'_m(kR)}\mathrm{J}_m(kr)\right]\mathrm{e}^{\imath m(\theta-\theta_0)}, \\ &\quad \text{if } r \geq r_0 \\ &= \sum_{m=-\infty}^{+\infty}\frac{\imath}{4}\left[-\mathrm{J}_m(kr)\mathrm{H}_m(kr_0) + \frac{\mathrm{H}'_m(kR)\mathrm{J}_m(kr_0)}{\mathrm{J}'_m(kR)}\mathrm{J}_m(kr)\right]\mathrm{e}^{\imath m(\theta-\theta_0)}, \\ &\quad \text{if } r \leq r_0. \end{aligned} \quad [5.13]$$

We obtain the same expression as the one by applying the method of direct integration [5.12]. Here, once again, both methods lead to the same result.

5.8. Basics of room acoustics

This section gives a short introduction to room acoustics.

5.8.1. *The concept of acoustic power*

The level of pressure measured around a source depends on the environment, it is often appropriate to define a concept that characterizes the source and which is based on the energy emitted per second (in Watts) by the source (we measure the total energy flux on a surface which surrounds the source). The level L_W is defined as follows:

$$L_W = 10 \log \frac{W}{W_0} = 10 \log \frac{W}{10^{-12}}.$$

We should not confound L_W that is intrinsic to the source and L_p that characterizes the source in its environment. The sound intensity I, at a distance d from the source, can be given by $I = W/(4\pi d^2)$ and the sound level:

$$L_p = 10 \log \frac{I}{10^{-12}} = L_W + 10 \log \frac{1}{4\pi d^2}.$$

EXERCISE.– We measure a sound level of 100 dB at 4 m from a monopole source. What is the power level emitted by this source?

SOLUTION.– *For a monopole source, the level does not depend on the direction of the observation, only the distance. We thus have:*

$$L_W = 100 - 10 \log \frac{1}{64\pi} = 123dB$$

That is a power level of $W = 10^{12.3} * 10^{-12} = 10^{0.3} \approx 2$ W.

5.8.2. *Directivity index*

A simple way to take into consideration the presence of an obstacle in the immediate vicinity of a source is to introduce the directivity index Q[5]. Let us suppose that we have measured the acoustic power level emitted by a source L_W, we can easily calculate the pressure level of the source in an unbounded medium at a

5 The directivity index is a rough extension of the directivity diagram, which is interested only in the environment immediately around the sound source and not the source itself.

distance d from the source in the presence of obstacles by correcting the energy emitted by the source W using the directivity index Q as WQ, or again:

$$L_p = L_W + 10\log Q + 10\log\frac{1}{4\pi d^2}$$

with $Q = 1$ in an open field, $Q = 2$ when the source is located on the ground or against a wall, $Q = 4$ when the source is placed at an angle and $Q = 8$ when the source is placed in a corner. For example, with $Q = 2$, $10\log 2 \approx 3$ and we have $L_p = L_w + 3 - 10\log(4\pi d^2)$.

EXERCISE.– Calculate the sound level measured at 10 m from a source with a power level of $W = 0.01$ W when it is located in a corner.

SOLUTION.– *We have* $L_W = 10\log\frac{0.01}{10^{-12}} = 100$ *dB. This is* $L_p = 100 + 10\log 8 - 10\log(4\pi 100^2) = 78$ *dB.*

It is thus sufficient to add to the power emitted by the source: 3 dB, if the source is located on the ground; 6 dB, if it is at an angle; and 9 dB, if it is in a corner. However, inside a building, we must take into consideration the energy reflected by the other walls, which creates a reverberant field.

5.8.3. *Reverberation duration*

The transitional regimes, which mainly characterize the establishment and extinction of sound, call for more difficult resolution methods than the ones we have seen so far. In particular, with the use of an inverse direct Fourier transform, we can show that the response of the system in a transient regime involves series expansions of resonance modes. Here, we will only introduce the concept of reverberation duration which is widely used in room acoustics.

By definition, the reverberation duration of a room is the length of time necessary for the sound level to decrease to 60 dB in this room.

This reverberation duration (or time) can be measured as follows: a sound source is placed in the room and makes a sufficient level of noise, this level is measured at a point in the room (different from the source point). The power to the source is cutoff and the time it takes for the level to decrease to 60 dB is thus measured. Such a measure is only valid if the sound field is diffuse (independent of the measuring point). This is obviously not the case when the excitation frequency is close to the resonance frequency.

This reverberation time, among other criteria, allows us to characterize the acoustic quality of a room. For example, in a conference room, this duration should be low enough so that during a speech, the sound level of each phoneme must decrease sufficiently before the next is pronounced by the speaker. If this is not the case, the speech becomes unintelligible. Typically, for a conference room or a theater, the reverberation duration should be around 1 s. For a classical music concert hall, this duration should be approximately 1.5 s. In cathedrals, sometimes this duration is approximately 4–5 s. For "rock" concert halls, the problem is different because of the sound systems used. In practice, the technicians require anechoic rooms (with a reverberation duration of zero) in order to be able to create at their convenience (by electronic or digital ways) the optimum reverberation duration. Figure 5.20 presents the reverberation durations which give optimal hearing at the middle and long distances according to the standard DIN 18041. This standard proposes to calculate T_r (expressed in seconds) based on the volume of the room (expressed in m^3) and its destination:

– speech : $T_r = 0.14 + 0.37 \log_{10} V$;

– music : $T_r = 0.07 + 0.45 \log_{10} V$;

– education : $T_r = -0.17 + 0.32 \log_{10} V$;

– sport (2,000 $m^3 < V < 8500$ m^3): $T_r = -2.49 + 1.27 \log_{10} V$.

There are approximate formulas to determine *a priori* the duration of reverberation of a room based on its geometrical characteristics and the materials that cover the walls of this room. The Sabine formula is the simplest and best known. This law relates the slope of the decrease in the sound level to acoustic properties of the walls of this room.

Let us consider a room with a volume V and a total wall area of S. For simplicity, all the walls have the same acoustic properties and are characterized by a single energy absorption coefficient α. By analogy with geometrical optics, we will assume that the sound energy emitted by the source can be decomposed into a set of acoustic "rays" which each contains an identical part of the total energy. Each time a ray hits a wall, it is reflected on it, losing a part α of its initial energy. Between two successive reflections, a ray travels an average distance l, called the mean free path. To estimate this distance, we can assume that the room is a cubic room with the dimensions $L_x \times L_y \times L_z$. Let us consider, for example, that a ray propagates in the direction Ox at an angle θ_x. Each second, it travels a distance $L_x / \cos\theta_x$ and it thus reflects $n_x = c_0/L_x \cos\theta_x$ times. Under the diffuse field hypothesis, each direction is equiprobable and the mean number of reflections will correspond to the mean of all angles, as will the integral over all the angles, weighted by the corresponding solid angle[6]. This is:

$$N_x = \frac{1}{2\pi}\int_0^{2\pi} d\phi \int_0^{\pi/2} \frac{c_0}{L_x}\cos\theta_x \sin\theta_x d\theta_x = \frac{c_0}{2L_x}\int_0^{\pi/2} \sin 2\theta_x d\theta_x = \frac{c_0}{2L_x},$$

6 The term 2π corresponds to a solid angle of $\Omega = \int_0^{2\pi} d\phi \int_0^{\pi/2} \sin\theta d\theta$.

and in the direction x, every second the ray will be reflected $N_x = c_0/(2L_x)$ times[7]. In the directions y and z, in a single second, it will be reflected $N_y = c_0/(2L_y)$ and $N_z = c_0/(2L_z)$ times. Taking into account the three directions, it will be reflected $N = c_0/(2L_x)+c_0/(2L_y)+c_0/(2L_z)$ times per second. This is $N = c_0/2\times(L_yL_z+L_xL_z + L_xL_y)/V$ or again $N = c_0S/(4V)$. The mean free path, which corresponds to the distance traveled between two reflections, is thus $l = c_0/N = 4V/S$.

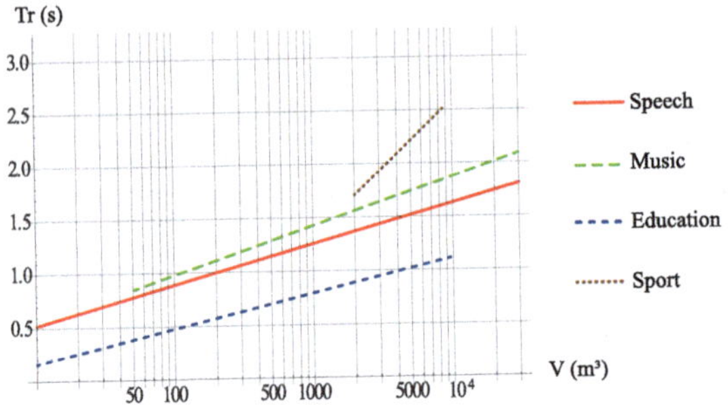

Figure 5.20. *Optimum reverberation duration of rooms (standard DIN 18041). For a color version of the figure, see www.iste.co.uk/anselmet/acoustics.zip*

Since, in 1 s, a ray undergoes $N = c_0/l$ reflections, after 1 s, its energy is equal to $(1-\alpha)^{c_0/l}$ times its initial energy. If we assume that this law is the same for all sound rays that lead to the measurement point, the sound energy decreases as follows:

$$E = E_0(1-\alpha)^{\frac{c_0 t}{l}}. \qquad [5.14]$$

The law of sound decrease can thus be given by $N = N_0 + 10c_0t/l\log_{10}(1-\alpha)$ where N_0 is the sound level before the power cut. The duration of reverberation T_r, which is such that $N_0 - N = 60$, can thus be written as $60 = -10c_0T_r/l\log_{10}(1-\alpha)$. Hence, the Eyring–Millington formula is:

$$T_r = -\frac{24V}{c_0S\log_{10}(1-\alpha)}. \qquad [5.15]$$

If α is small, then $-\log_{10}(1-\alpha) \sim \alpha/\ln 10$. From where, $T_r \sim 4V\ln 10/(c_0S\alpha) \sim 0.16V/(S\alpha)$, where $c_0 = 345$ m/s. This formula is called

7 Another way to view the phenomenon is to assume that each ray will travels in the direction x a mean distance of $2L_x$ before hitting a wall.

the Sabine formula. If the walls do not all have the same absorption, this formula can be generalized as follows:

$$T_r \sim \frac{0.16V}{\sum_i S_i \alpha_i}. \quad [5.16]$$

In fact, the Sabine absorption coefficient, equal to $-\log_{10}(1-\alpha)$, is always greater than α. In some cases, it may be greater than the unit. This is contradictory with the notion of energy loss. A good practice is not to apply the Sabine formula when $\alpha > 0.4$; in practice, this corresponds to an error of the order of about 12% in the approximation $-\ln(1-\alpha) \approx \alpha$, it is 1 dB. We can see that this concept of reverberation duration is vague and requires assumptions that are almost never met. Nevertheless, it is a very useful concept.

5.8.4. *Reverberant fields*

A very important concept in room acoustics is linked to the concept of a reverberant field which corresponds to a field in which the sound level, in a forced regime, does not depend on the measurement point. This concept is slightly difficult to define rigorously[8]. Nevertheless, a good way of introducing it is to assume that it is located in a reverberant field (i.e. in the field produced by successive sound reflections on the walls) from the place where the direct field intensity is smaller than the reverberated field intensity. The energy decrease outside of the sources is given by formula [5.14] and can be written as $E(t) = E_0 \exp(c_0 S/(4V)t \log(1-\alpha))$. In a forced regime, at equilibrium, the power dissipated by the medium is identical to the power P injected by the source and we get $E_0 = P$.

The intensity $I = \overline{p^2}/(\rho_0 c_0)$ of the direct acoustic field can be given by relation [4.10]: $I_d = P/(4\pi r^2)$. The energy density throughout the entire volume can be given by $w = E/V$, where $E = \int_0^\infty E(t)dt$. When examining reverberant fields, we can focus only on the energy of the field which has been reflected at least once. This means replacing the lower bound of integration by l/c_0. This is $E_r = \int_{l/c_0}^\infty E(t)dt$. The density of the reverberated energy can thus be given by $w_r = E_r/V$

$$w_r = \frac{E_r}{V} = -\frac{4P}{c_0 S}\frac{1-\alpha}{\log(1-\alpha)} \approx \frac{4P}{c_0 S}\frac{1-\alpha}{\alpha}, \text{ if } \alpha \ll 1$$

The intensity of the reverberant field, given by $I_r = w_r c_0$, is thus:

$$I_r = \frac{E_r c_0}{V} = \frac{4P}{S}\frac{1-\alpha}{\alpha}. \quad [5.17]$$

8 In practice, the measured sound field always depends on the measurement point, but when the measured variations are not noticeable, we assume that it is located in a reverberant field.

The distance from which we are located in a reverberant field is given by the distance d_r from which we get that $I_r = I_d$, this is for:

$$d_r = \frac{1}{2\sqrt{\pi}}\sqrt{\frac{S\alpha}{4(1-\alpha)}}.$$

This distance, which depends only on the characteristics of the room (as long as the source is smaller than d_c) marks the transition between the classical regime and the statistical regime.

5.8.5. *Pressure level in rooms*

The sound energy I received in any point in the room is the sum of direct energy given by equation [4.10] $I_d = WQ/(4\pi d^2)$ which depends on the distance from the source, as well as the reverberated energy (almost constant), which is reflected by the walls of the room, given by equation [5.17] which we generalize to rooms covered with materials that have different absorption coefficients $I_r = 4WQ(1-\overline{\alpha})(S\overline{\alpha})$, where $\overline{\alpha}$ is the mean absorption coefficient of the room walls and $S\overline{\alpha} = \sum_i \alpha_i S_i$. The indices i mark the i surfaces covered with an absorbent material [9], possibly zero. The total energy is thus:

$$I = I_d + I_r = WQ\left(\frac{1}{4\pi d^2} + 4\frac{1-\overline{\alpha}}{S\overline{\alpha}}\right).$$

The level of sound pressure in the room is thus:

$$L_p = L_W + 10\log Q + 10\log\left(\frac{1}{4\pi d^2} + \frac{4}{A_\alpha}\right),$$

where $A_\alpha = S\overline{\alpha}/(1-\overline{\alpha})$ is the area equivalent to the absorption of the local (surface dimension). When α is small compared to the unity, we confound A_α and S_α. The area equivalent to the absorption of the local is used when we are interested in the reverberation distance d_c. It is easy to see that $d_c = \sqrt{A_\alpha/(16\pi)}$. In a normal living room, we have $d_c \approx 1m$.

EXERCISE.– A device has a power level of $L_W = 80dB$. We place this source in a room of $3 \times 5 \times 10$ m^3. Its directivity index is thus $Q = 3$. The mean absorption coefficient of the room is $\overline{\alpha} = 0.2$. What is the reverberation distance and the pressure level at a distance of 2 m from the source? Similar questions, if $\overline{\alpha} = 0.5$.

---

9 By extension, an open window, which does not reflect acoustic energy, has an absorption coefficient of 1.

SOLUTION.– *The total surface is* $S = 190\ m^2$, *it is* $A_\alpha = S\overline{\alpha}/(1-\overline{\alpha}) = 47.5\ m^2$. *Thus,* $d_c = \sqrt{A_\alpha/(16\pi)} \approx 1\ m$. *We have:*

$$L_p = L_W + 10\log Q + 10\log\left(\frac{1}{4\pi d^2} + \frac{4}{A_\alpha}\right)$$
$$= 80 + 10\log 3 + 10\log\left(\frac{1}{4\pi 2^2} + \frac{4}{47.5}\right) = 75\ dB$$

If $\overline{\alpha} = 0.5$, $A_\alpha = 190\ m^2$, $d_c \approx 2\ m$ *and* $L_p = 71 dB$.

5.8.6. Crossover frequency and the reverberation distance

Let us recall that the eigenfrequencies of a cubic room form a triple sequence defined by $f_{lmn} = \frac{c}{2}\sqrt{(l/\ell_x)^2 + (m/\ell_y)^2 + (n/\ell_z)^2}$. We have seen that at high frequencies the number of modes at a lower frequency f is $N_f = 4/3\pi V(f/c_0)^3$, $V = \ell_x\ell_y\ell_z$. The eigenfrequency number in Hertz at a given frequency f is thus[10] $D_f(f) = 4\pi V f^2/c_0^3$ which implies an inter-resonance spacing $\Delta f = c_0^3/(4\pi V f^2)$. It is considered that two resonances can no longer be distinguished when the bandwidth B at -3 dB of the resonances is equal to three times the inter-resonance spacing. By definition, we have $B = \ln 10^6/(2\pi T_r) = 2.2/T_r$. This is:

$$2.2/T_r = 3c_0^3/(4\pi V f_c^2) \Rightarrow f_c = \sqrt{\frac{3c_0^3}{2.2\times 4\pi}\frac{T_r}{V}} \approx 2000\sqrt{\frac{T_r}{V}},$$

f_c is the crossover or Schroeder frequency. It characterizes the frequency from which we move from a system with clearly identified modes to a statistical system (corresponding to a diffuse field). This formula is valid for any room.

The corresponding wavelength $\lambda_c = c_0/f_c$ can be given by[11]:

$$\lambda_c = \sqrt{\frac{2.2\times 4\pi}{3c_0}\frac{V}{T_r}} = \sqrt{\frac{2.2\times 4\pi}{3\times 0.16c_0}A_\alpha} = \sqrt{\frac{A_\alpha}{6}}.$$

This length is of the same order as the reverberation distance $d_c = \sqrt{A_\alpha/(16\pi)}$. This is not a coincidence since the two lengths describe the transition from a “classical” regime to a “statistical” regime.

Just as we can easily define the number of modes per Hertz and the inter-resonance spacing, we can define the number of reflections (or echos) which arrive after a time

10 It is enough to derive N_f with respect to f.

11 We use the Sabine formula $T_r = 0.16V/A_\alpha$.

t by $N_r = 4/3\pi(t \times c_0)^3/V$, which gives a number of echoes per second $D_r(t) = 4\pi c_0^3 t^2/V$ and an interval between two echoes $\Delta t = V/(4\pi c_0^3 t^2)$.

If this inter-echo interval is smaller than the echo decay time $T_r/\ln 10^6$, a point in space receives many echoes of comparable amplitude, if in addition these are in random directions, the field will diffuse. This results in a "diffusion time", equaling the inter-echo interval and the echo decay time: $t_c = \sqrt{\ln 10^6 V/(4\pi c_0^3 T_r)}$ and a diffusion distance $d_c = c_0 t_c$ which is none other than the reverberation distance $d_c = c_0\sqrt{\ln 10^6 V/(4\pi c_0^3 T_r)} = \sqrt{\ln 10^6 A_\alpha/(4\pi c_0 0.16)} = \sqrt{A_\alpha/(16\pi)}$.

All these results show that these concepts of the Schroeder frequency f_c and reverberation distance d_c describe the same phenomenon of transition from a deterministic regime to a statistical regime but in different spaces.

5.9. Sound propagation in a wave guide

5.9.1. *General solution in a wave guide*

Let us consider a straight conduit with a rectangular section of infinite extension occupying a domain Ω. In the Cartesian coordinate system, Ω is defined by: $x \in]0, \ell_x[$, $y \in]0, \ell_y[$ and $z \in]-\infty, \infty[$. In the absence of sources in the conduit, the equation which governs the wave propagation is the Helmholtz one for pressure $p(x, y, z)$:

$$\Delta p(x, y, z) + k_0^2 p(x, y, z) = 0,$$

where $k_0 = \omega/c_0$ is the wave number and c_0 is the celerity of acoustic waves in an infinite medium. We will assume that the conduit walls are perfectly reflective. This results in a homogeneous Neumann condition on them. By convention, we take the normal oriented toward the outside of Ω. The system of equations for solving it is the following:

$$\Delta p(x, y, z) + k_0^2 p(x, y, z) = 0 \in \Omega,$$
$$\frac{\partial p(x, y, z)}{\partial x} = 0 \qquad \text{for } x = 0, \ell_x, \ \frac{\partial p(x, y, z)}{\partial y} = 0 \text{ for } y = 0, \ell_y.$$

The boundaries of the propagation domain are parallel to the coordinate system axes, as with the cubic room, we can look for the solution in the form of a Laplace product (in separate variables) $p(x, y, z) = X(x)Y(y)Z(z)$, the Helmholtz equation can once more be written as $X''(x)/X(x) + Y''(y)/Y(y) + Z''(z)/Z(z) + k_0^2 = 0$. Here again, the solution of such an equation is obtained by writing that each of these three independent functions is any complex constant (called the separation constant), whose sum must be equal to k_0^2. We thus have $X''(x)/X(x) = -\alpha^2$, $Y''(y)/Y(y) =$

$-\beta^2$, $Z''(z)/Z(z) = -\left(k_0^2 - \alpha^2 - \beta^2\right)$. We can easily show that $\alpha = m\pi/\ell_x$, $m \in$ $\boldsymbol{N}$, $\beta = n\pi/\ell_y$, $n \in \boldsymbol{N}$ and that:

$$X_m(x) = A_m \cos\left(\frac{m\pi}{\ell_x}x\right), m \in \boldsymbol{N}, Y_n(x) = A_n \cos\left(\frac{n\pi}{\ell_y}y\right), n \in \boldsymbol{N}.$$

The general solution of the acoustic pressure in the guide can be written as:

$$p(x,y,z) = \sum_{m=0}^{m=\infty}\sum_{n=0}^{n=\infty}\left(A_{mn}\mathrm{e}^{+\imath k_{mn}z} + B_{mn}\mathrm{e}^{-\imath k_{mn}z}\right) \times \cos\left(\frac{m\pi}{\ell_x}x\right)\cos\left(\frac{n\pi}{\ell_y}y\right), \quad [5.18]$$

where $k_{mn} = \sqrt{k_0^2 - \alpha^2 - \beta^2}$.

EXERCISE.– Show that when the boundary condition on the walls is the Dirichlet condition, we have:

$$p(x,y,z) = \sum_{m=0}^{m=\infty}\sum_{n=0}^{n=\infty}\left(A_{mn}\mathrm{e}^{+\imath k_{mn}z} + B_{mn}\mathrm{e}^{-\imath k_{mn}z}\right)\sin\left(\frac{m\pi}{\ell_x}x\right)\sin\left(\frac{n\pi}{\ell_y}y\right),$$

$$k_{mn} = \sqrt{k_0^2 - (m\pi/\ell_x)^2 - (n\pi/\ell_y)^2}, m \in \boldsymbol{N}^*, n \in \boldsymbol{N}^*.$$

We must note that the solutions $m = 0$ or $n = 0$ are not acceptable because they would lead to a zero solution everywhere.

5.9.2. *Physical interpretation and theory of modes*

Formula [5.18] can be generalized to a large number of cases. It is possible to show that for any straight guide with a constant section whose cross-section is "separable", i.e. such that the walls coincide with the surfaces $u = ctte$, $v = ctte$ of an orthonormal coordinate system (O, u, v, z), the acoustic pressure can be written as:

$$p(x,y,z) = \sum_{m=0}^{m=\infty}\sum_{n=0}^{n=\infty}\left(A_{mn}\mathrm{e}^{+\imath k_{mn}z} + B_{mn}\mathrm{e}^{-\imath k_{mn}z}\right)\Psi_{mn}(u,v), \quad [5.19]$$

where the functions $\Psi_{nm}(u,v)$ are the eigenfunctions of the guide. These functions verify:

$$\Delta\Psi(u,v) + \kappa_{mn}^2\Psi(u,v) = 0, (u,v) \in]0,\ell_u[\times]0,\ell_v[, \partial_{\vec{n}}\Psi(u,v) = 0,$$
$$u = 0, \ell_u, v = 0, \ell_v,$$

and where $k_{mn}^2 = k_0^2 - \kappa_{mn}^2$ is the modal wave number, which depends on k_0, the two parameters m and n and the geometry of the guide.

5.9.2.1. *Modal basis*

The functions $\Psi_{mn}(u,v)$ form a complete orthogonal basis for the section of the guide. In fact, let us consider $\Psi_{mn}(u,v)$ and $\Psi_{pq}(u,v)$, two eigenfunctions of the guide. These functions are the solutions to:

$$\Delta\Psi_{mn}(u,v) + \lambda_{mn}^2\Psi_{mn}(u,v) = 0,\ (u,v) \in S, \partial_{\vec{n}}\Psi_{mn}(u,v) = 0,\ (u,v) \in \partial S,$$

$$\Delta\Psi_{pq}(u,v) + \lambda_{pq}^2\Psi_{pq}(u,v) = 0,\ (u,v) \in S, \partial_{\vec{n}}\Psi_{pq}(u,v) = 0,\ (u,v) \in \partial S.$$

Let us multiply the first equation of the first system by $\Psi_{pq}(u,v)$ and the first equation of the second system by $\Psi_{mn}(u,v)$. Subtract them and integrate the result on the section of the guide. We get:

$$\int_S \Psi_{pq}(u,v)\Delta\Psi_{mn}(u,v) - \Psi_{mn}(u,v)\Delta\Psi_{pq}(u,v)dS$$
$$= \left(\lambda_{pq}^2 - \lambda_{mn}^2\right)\int_S \Psi_{pq}(u,v)\Psi_{mn}(u,v)dS.$$

Let us apply Green's theorem to the first part of this equation:

$$\int_{\partial S} \Psi_{pq}(u,v)\partial_{\vec{n}}\Psi_{mn}(u,v) - \Psi_{mn}(u,v)\partial_{\vec{n}}\Psi_{pq}(u,v)dl$$
$$= \left(\lambda_{pq}^2 - \lambda_{mn}^2\right)\int_S \Psi_{pq}(u,v)\Psi_{mn}(u,v)dS.$$

Now, due to the boundary conditions which satisfy the eigenmodes, it is clear that the integral on the boundary of the section of the guide is zero. Thus, $\left(\lambda_{pq}^2 - \lambda_{mn}^2\right)\int_S \Psi_{pq}(u,v)\Psi_{mn}(u,v)dS = 0$. It is thus obvious that:

$$\int S\Psi_{mn}(u,v)\Psi_{pq}(u,v)dS = S\Lambda_{mn}\delta_{(mn)(pq)} \qquad [5.20]$$

where $\delta_{(mn)(pq)} = 0$ if $m \neq p$ and $n \neq q$. The functions Ψ_{mn} are thus orthogonal. These functions form a complete set of functions for the section of the guide. They are often used for series expansions of eigenfunctions (or modes). This is known as the modal basis.

We must note that it is also possible to define the eigenfunctions $\Psi_{mn}(u,v)$ for the Dirichlet conditions on the walls or for the boundary conditions that combine

Neumann and Dirichlet. For the boundary conditions of the impedance type, for which there is energy loss through the walls, we still get similar results; however, we must note that in this case, the modes and the wave numbers are complexes and generally depend on the excitation frequency and that we should use the scalar product of the complex functions defined by $\int\int_S \Psi_{mn}(u,v)\Psi^{\star}_{pq}(u,v)dS$.

5.9.2.2. *Guide with a circular section*

A practical case, often encountered, is that of a circular guide with a radius R with rigid walls for which the modes ψ_{mn} are defined [MOR 68] by $\psi_{mn}(r,\phi) = \mathrm{J}_m(\kappa_{mn}r)\exp(\imath m\phi)$. The normalization factor Λ_{mn} defined by [5.20] is given by $\Lambda_{mn} = 1/\epsilon_m(1-(m/j'_{m,n})^2)J^2_m(j'_{m,n})$ where $\epsilon_m = 1$ if $m = 0$ and $\epsilon_m = 2$ if $m > 0$ is the Neumann factor. $j'_{m,n}$ is defined by the boundary conditions $\mathrm{J}'_m\left(j'_{m,n}\right) = 0$ and $\kappa_{mn} = j'_{m,n}/R$.

5.9.2.3. *Elements of the modal theory of wave guides*

General formula [5.19] can be interpreted by considering that the propagation in a straight wave guide is formed by the superposition of an infinity of modes (m,n) characterized by their distribution $\Psi_{mn}(u,v)$ in a section of the guide. Each of these modes propagates with a celerity of $c_{mn} = \omega/k_{mn}$, either in the direction of z increasing with an amplitude A_{mn}, or the direction of z decreasing with an amplitude B_{mn}. We must remark that depending on the sign of $k^2_{mn} = k^2_0 - \kappa^2_{mn}$, the modal wave number is either real or pure imaginary. The celerity c_{mn} can be written as:

$$c_{mn} = \frac{\omega}{k_{mn}} = c_0\frac{k_0}{k_{mn}} = c_0\frac{k_0}{\sqrt{k_0^2 - \kappa^2_{mn}}} = \frac{c_0}{\sqrt{1-\left(\frac{f_{mn}}{f}\right)^2}},$$

where $f_{mn} = c_0\kappa_{mn}/2\pi$ is the frequency of the mode Ψ_{mn}. For an excitation frequency f greater than f_{mn}, there is always propagation of the mode corresponding to a celerity greater than the celerity of waves in an infinite medium. Conversely, for a lower excitation frequency, the corresponding mode propagates with an amplitude which decreases exponentially, and it is appropriate to speak of a damped mode. If these modes exist, obeying the discontinuity and source laws, in practice, they can only be observed in their immediate vicinity [BRU 83]. When $f = f_{mn}$, at the cutoff frequency of the mode (m,n), $c_{mn} = \infty$. In fact, at the cutoff frequency of the mode (m,n), due to the interplay of the reflections on the walls, there is a standing wave. The previous result shows that for a frequency higher than the cutoff frequency of the mode (m,n), the phase velocity c_{mn} of the mode (m,n) is greater than the celerity of sound. This apparent contradiction can be waived, if we assume that the nodes of the mode in question propagate and not the energy. To know how fast the signals will move, we need to calculate the group velocity $u_{mn} = d\omega/dk_{mn}$. For this, let us consider $k_{mn} = \sqrt{(\omega/c_0)^2 - \kappa_{mn}^2}$ from which we derive the expression of ω and infer $d\omega/dk_{mn}$. We obtain:

$$u_{mn} = \frac{c_0^2}{\omega}\sqrt{k_0^2 - \kappa_{mn}^2} = c_0 \frac{1}{k_0^2}\sqrt{k_0^2 - \kappa_{mn}^2} = c_0\sqrt{1 - \left(\frac{f_{mn}}{f}\right)^2}.$$

We can see that the group velocity is always less than the sound velocity, which is purely imaginary (which corresponds to a wave which decreases exponentially). We must note that the geometric average of the phase and group velocities is the celerity of sound $u_{mn}c_{mn} = c_0^2$.

In the case of the Neumann condition on all walls, the particular solution $m = n = 0$ can be written as $\Psi_{00}(u, v) = 1p(u, v, z) = A_{00}\exp(+\imath k_{00}z) + B_{00}\exp(-\imath k_{00}z)$, which corresponds to two plane waves (no variations in u and v) that propagate forward and backward. The plane wave cutoff frequency is zero, it is propagating for all frequencies.

5.9.3. *Green's function*

Let us find the function $G(M, M')$ which satisfies:

$$\Delta_M G(M, M') + k_0^2 G(M, M') = \delta_{M'}(M) M, M' \in \Omega \qquad [5.21]$$

$$\text{Boundary conditions on } \partial\Omega, \qquad [5.22]$$

where $M = (u, v, z)$ and $M' = (u', v', z')$. This function can be developed in a series of eigenmodes $\Psi_{mn}(u, v)$, which satisfy the boundary conditions:

$$G(M, M') = \sum_{m=0}^{m=\infty}\sum_{n=0}^{n=\infty} F_{mn}(z, z', u', v')\Psi_{mn}(u, v).$$

Let us put this result into the Helmholtz equation, multiply each part by $\Psi_{mn}(u, v)$ and integrate over the section of the guide. We then have:

$$\left(\frac{d^2}{dz^2} + k_{mn}^2\right) F_{mn}(z, z', u', v') = -\frac{\Psi_{mn}(u', v')}{S\Lambda_{mn}}\delta_{z'}(z).$$

Λ_{mn} is the amplitude of the mode $\Psi_{mn}(u, v)$ and $k_{mn}^2 = k_0^2 - \kappa_{mn}^2$. We know that Green's function of the Helmholtz one-dimensional equation $\left(\frac{d^2}{dz^2} + k^2\right) G(z, z') = \delta_{z'}(z)$ can be given by $G(z, z') = \exp(\imath k|z - z'|/(2\imath k)$. Finally, this is:

$$F_{mn}(z, z', u', v') = \frac{\imath}{2S}\frac{\Psi_{mn}(u', v')}{k_{mn}\Lambda_{mn}} e^{\imath k_{mn}|z-z'|}.$$

Green's function for an infinite guide can be written as:

$$G(M, M') = \frac{\imath}{2S} \sum_{m=0}^{m=\infty} \sum_{n=0}^{n=\infty} \frac{\Psi_{mn}(u, v)\Psi_{mn}(u', v')}{k_{mn}\Lambda_{mn}} \mathrm{e}^{\imath k_{mn}|z-z'|}. \qquad [5.23]$$

We must note that it is necessary to choose a way to determine the radical $k_{mn} = \sqrt{k_0^2 - \kappa_{mn}^2}$. To do this, it is reasonable to assume that the higher order modes are attenuated when $|z - z_0|$ increases. So, we must thus choose $\Im(k_{mn}) > 0$.

5.9.4. *Section change*

5.9.4.1. *Discontinuous variation*

Let us suppose that we are faced with a low-frequency condition, such that only the plane wave can propagate. Let us consider a pipe that has a sudden enlargement of its section. We aim to find the amplitudes of the transmitted and reflected waves. Let s be the upstream section and S, the downstream section. We call 1 the upstream medium and 2 the downstream medium. For simplicity, let us assume that this is a one-dimensional problem. The section change can be found at the abscissa $x = 0$.

It is possible to show [BRU 83] that at the position of a sudden section change of a conduit, an infinity of modes appears. In the low-frequency hypothesis, only the mode $(0, 0)$ can propagate. The other, evanescent, modes rapidly decrease away from the discontinuity. When, in line with the section change, we write the equations of continuity with a single term, the error, which we make, is even lower as its frequency is. Therefore, we can write the continuity of the acoustic pressure $p_1 = p_2$ and the vibration velocity flow $v_1 s = v_2 S$. Only when the plane wave propagates, can we have $p_1 = A_0 \exp(\imath k_0 x) + B_0 \exp(-\imath k_0 x)$ and $v_1 = 1/(\imath\omega_0\rho)dp_1/dx) = 1/(\rho_0 c_0)(A_0 \exp(\imath k_0 x) - B_0 \exp(-\imath k_0 x))$ where A_0 is the amplitude of the incident wave and B_0 is the amplitude of the reflected wave. Similarly, for the pressure in medium 2 $p_2 = C_0 \exp(\imath k_0 x)$ and $v_2 = 1/\rho_0 c_0 C_0 \exp(\imath k_0 x)$ where C_0 is the amplitude of the transmitted wave.

From the continuity equations, in line with the section change, we deduce that $A_0 + B_0 = C_0$ and $(A_0 - B_0)\, s = C_0 S$. Let $m = S/s$ be the ratio between the largest and the smallest sections. We obtain the reflection coefficient $R = B_0/A_0 = (1-m)/(1+m)$ and transmission coefficient $T = C_0/A_0 = 2/(1+m)$ with $1+R = T$. As, according to the hypothesis $m > 1$ there is a part of the energy which is reflected, it becomes even more important for the area ratio to be large. This shows that for an open pipe, at first approximation we have $T \approx 0$ and $R \approx -1$ (which corresponds to the Dirichlet condition). Even though a significant part of the sound energy escapes through the pipe openings, the wave remains largely confined to the pipe. This allows us to establish standing waves in pipes with finite dimensions, such as wind instruments[12] as we have seen in section 5.7.3.

EXERCISE.– Calculate the reflection and transmission coefficients in a narrow section.

SOLUTION.– *Let A_0 be the amplitude of the incident wave; B_0, the reflected wave amplitude and C_0, the transmitted wave amplitude, we get $R = (m-1)/(m+1)$ and $T = 2m/(1+m)$. As $m > 1$, a part of the energy is reflected, especially as the section ratio is large. A way of imagining the phenomenon is to imagine that in the pipe there is a small mass of air that is moving and that, in line with the discontinuity of the section, there is a larger, as large as the section ratio is large, mass of air at rest, whose inertia opposes the movement. At the extreme limit, if the section ratio approaches infinity, the inertia of the external fluid is such that it remains at rest and the waves are completely reflected in the pipe (Dirichlet condition). In practice, this is clearly not the case, and there is always a part of the wave which is transmitted (especially when the frequency is high).*

EXERCISE.– The same question in the case of a chamber expansion for which there is a narrowing and then a widening section, so as to regain an identical section to the one before the enlargement.

SOLUTION.– *Similarly, as before, we assume that A is the amplitude of the incident wave, B is the amplitude of the reflected wave and C is the transmitted wave. m is always the ratio between the largest section and the smallest section and L is the length of the expansion chamber. We obtain $R = ((1-m^2) + \exp(2\imath k_0 L)(m^2 - 1))/D$, where $D = (1+m)^2 - \exp(2\imath k_0 L)(m-1)^2$ and $T = 4m\exp(2\imath k_0 L)/D$.*

It should be noted that in this case, the transmission coefficient depends on the product $k_0 L$. In practice, for a certain length of the expansion chamber, the transmission coefficient passes through a succession of maxima and minima when the frequency varies. Conversely, by adapting the length of this chamber, it is possible to reduce the sound levels transmitted by the chamber for certain frequencies.

5.9.4.2. *Continuous variation: pavilions*

A pavilion is a wave guide whose cross-section continuously varies. Let us assume that the wave guide in perfectly rigid walls (the Neumann condition) is of infinite length and with a small section with respect to the wavelength (only the plane wave propagates), which corresponds to a uniform pressure in a small transverse section of the guide.

12 For the study of wind instrument radiation, see [CHA 13, FLE 91].

Let us consider a point on the x axis. At this point, the straight section of the guide is $A(x)$, at $x+\Delta x$, the straight section is $A(x+\Delta x)$. We integrate the wave equation over the entire volume of the guide between x and $x+\Delta x$. We have:

$$\iiint_V \Delta p dV - \frac{1}{c_0^2}\iiint_V \frac{\partial^2 p}{\partial t^2} dV = 0$$

but Ostrogradsky's theorem tells us that $\iiint_V \mathrm{div}\vec{u} dV = \iint_S \vec{u}\cdot\vec{n} dS$. Hence, with $\Delta p = \mathrm{div}\cdot\overrightarrow{\mathrm{grad}}p$,

$$\iint_S \overrightarrow{\mathrm{grad}}p\cdot\vec{n} dS - \frac{1}{c_0^2}\iiint_V \frac{\partial^2 p}{\partial t^2} dV = 0$$

We know that the normal pressure gradient is zero on the walls of the guide (the Neumann condition) so that $\overrightarrow{\mathrm{grad}}p\cdot\vec{n} = \frac{\partial p}{\partial x}$ on the straight sections of the guide and that the pressure is uniform over a section of the guide (low-frequency hypothesis). Furthermore, if we assume that Δx is sufficiently small, then $\partial^2 p/\partial t^2$ can be considered constant over the integration volume. We thus have:

$$\iint_{A(x+\Delta x)} \frac{\partial p}{\partial x} dS - \iint_{A(x)} \frac{\partial p}{\partial x} dS - \frac{1}{c_0^2}\iint_A \frac{\partial^2 p}{\partial t^2} dS \int dx = 0$$

with $\int dx = \Delta x$. We divide this equation by Δx and take the limit $\Delta x \to 0$:

$$\begin{aligned}\lim_{\Delta x\to 0} & \frac{\iint_{A(x+\Delta x)} \frac{\partial p}{\partial x} dS - \iint_{A(x)} \frac{\partial p}{\partial x} dS}{\Delta x} - \frac{1}{c_0^2}\iint_A \frac{\partial^2 p}{\partial t^2} dS \\ & = \frac{\partial}{\partial x}\iint_A \frac{\partial p}{\partial x} dS - \frac{1}{c_0^2}\frac{\partial^2}{\partial t^2}\iint_A p dS = 0\end{aligned}$$

If the pressure is uniform over the straight section of the guide $1/A(x)\partial\left(A(x)\partial p/\partial x\right)/\partial x - 1/c_0^2\partial^2 p/\partial t^2 = 0$, this equation can also be written as:

$$\frac{\partial^2 p}{\partial x^2} + \frac{A'(x)}{A(x)}\frac{\partial p}{\partial x} - \frac{1}{c_0^2}\frac{\partial^2 p}{\partial t^2} = 0$$

which is called the Webster equation. Let us solve this equation when the ratio $A'(x)/A(x)$ is constant. For example, let us take the elementary case of an exponential pavilion, for which $S(x) = S_0 \exp(2\tau x)$. We then solve $\partial^2 p/\partial x^2 + 2\tau\partial p/\partial x - 1/c_0^2\partial^2 p/\partial t^2 = 0$. In a harmonic regime $\exp(-\imath\omega t)$, it becomes a differential equation with constant coefficients $\partial^2 p/\partial x^2 + 2\tau\partial p/\partial x + k^2 p = 0$, from which we aim to find the general solution using the Laplace method $p(x) = A\exp(\alpha x)$.

The corresponding characteristic polynomial is $\alpha^2 + 2\tau\alpha + k^2$ whose roots are $\alpha = -\tau\left(1 \pm \sqrt{1-(k/\tau)^2}\right)$. Three cases are considered:

1) $k = \tau$ thus $p(x,t) = A\exp(-\tau x)\exp(-\imath\omega t)$, the wave does not propagate;

2) $k < \tau$ the wave does not propagate;

3) $k > \tau$ thus $p(x,t) = A\exp(-\tau x)\exp(-\imath(\omega t \pm k'x))$, where $k' = \sqrt{k^2-\tau^2}$. The wave propagates with an amplitude which decreases with the enlargement of the pavilion.

The interesting phenomenon is that there is a cutoff frequency $f_c = \tau c_0/2\pi$ below which waves do not propagate. We must note that the slower the section variation, the lower the cutoff frequency; at the limit for a constant section, we find that the cutoff frequency is zero for the rigid guide. The exponential pavilions allow us to increase the effectiveness of the loudspeakers. Conversely, the pavilions can be used to amplify the sound perceived, as in the case of ear trumpets.

5.9.5. *Propagation in a conduit in the presence of flow*

Here, we are interested in calculating the propagation in an infinitely rigid two-dimensional conduit in the presence of uniform flow with a constant subsonic velocity v_0 parallel to the x axis. The height ℓ of the conduit is assumed as constant. The propagation equation, in a harmonic regime $\exp(-\imath\omega t)$, with $M = v_0/c_0$, can be given by :

$$\Delta p(x,y) + k^2 p(x,y) + 2\imath k M \frac{\partial p(x,y)}{\partial x} - M^2 \frac{\partial^2 p(x,y)}{\partial x^2} = 0$$

$$\frac{\partial^2 p(x,y)}{\partial y^2} + k^2 p(x,y) + 2\imath k M \frac{\partial p(x,y)}{\partial x} + (1-M^2)\frac{\partial^2 p(x,y)}{\partial x^2} = 0$$

The boundary conditions are $\partial_n p(x,y) = 0$, at $y = 0$ and $y = \ell$, where $\vec{n}$ is the vector normal to the walls of the conduit. This equation can be solved by the variable separation method, which consists of searching a solution in the form of a product $p(x,y) = X(x)Y(y)$. The boundary conditions can thus be written as $Y'(0) = 0$ and $Y'(\ell) = 0$. We obtain:

$$\frac{Y''(y)}{Y(y)} + k^2 + 2\imath k M \frac{X'(x)}{X(x)} + (1-M^2)\frac{X''(x)}{X(x)} = 0$$

The solution in separate variables of this equation can be written as $Y''(y)/Y(y) = -k_y^2$, $2\imath k M X'(x)/X(x) + (1-M^2)X''(x)/X(x) + k^2 - k_y^2 = 0$. The first of these equations is a Helmholtz equation in one dimension, thus the solution is $Y(y) = A\cos(k_y y) + B\sin(k_y y)$. The boundary conditions give $B = 0$ and $k_y(m) = m\pi/\ell$. If we normalize the modes $Y_m(y)$ ($\langle Y_m, Y_n\rangle = \delta_m^n$) we obtain, with $\langle f, g\rangle = \int_0^\ell f(y)g^*(y)dy$, $Y_m(y) = \sqrt{2/\ell}\cos k_y(m)y$. The solution to the second equation $(1-M^2)X''(x) + 2\imath kMX'(x) + (k^2 - k_y^2)X(x) = 0$ is obtained by the Laplace method, which just like any linear differential equation with constant coefficients can be expressed by a linear combination of complex exponentials: $X(x) = \exp(\imath k_x x)$. We obtain the bisquared equation $-(1-M^2)k_x^2 - 2kMk_x + (k^2 - k_y^2) = 0$ whose discriminant is $\Delta' = k^2 - k_y^2\beta^2$, with $\beta^2 = 1 - M^2$. This leads to two values of $k_x = k_x^\pm$:

$$k_x^\pm = \frac{-kM \pm \sqrt{k^2 - k_y^2\beta^2}}{\beta^2} = \frac{k}{\beta^2}\left(-M \pm \sqrt{1 - \left(\frac{k_y\beta}{k}\right)^2}\right)$$

where we have chosen as the way of determining the complex root $\sqrt{z} = \sqrt{|z|}\exp(\imath\arg(z)/2), 0 \le \arg(z) < 2\pi$. If $k^2 - k_y^2\beta^2 > 0$, the wave numbers $k_x^\pm$ are purely real and there is propagation. If $k^2 - k_y^2\beta^2 < 0$, the wave numbers $k_x^\pm$ are complexes with a non-zero imaginary part and the wave propagates with an exponential decrease, this is called an evanescent wave (when $M = 0$, there is no propagation). The frequency for which $k^2 - k_y^2\beta^2 = 0$, given by $f_m^c = k_y c_0/(2\pi)\beta = mc_0/(2\ell)\beta$, is the cutoff frequency mode. If an excitation frequency is higher than the cutoff frequency mode, the latter propagates or it is attenuated. The phase velocity $c_m^\pm = \omega/k_x^\pm$ is:

$$c_m^+ = \frac{c_0\beta^2}{-M + \sqrt{1 - (f_m^c/f)^2}},$$

the phase velocity can be higher than the wave celerity but it can also be negative. The group velocity $v_m = \partial\omega/\partial k_x$ is:

$$v_m^+ = \frac{c_0\beta^2\sqrt{1 - (f_m^c/f)^2}}{1 - M\sqrt{1 - (f_m^c/f)^2}}.$$

The group velocity (and thus the energy) is lower than the wave celerity, the Mach number $M = 0$, $\beta = 1$ and we find the geometric average of the velocities $c_m^+ v_m^+ = c_0^2$. The axial wave numbers k_x^+ for which the group and phase velocities are positive are called the downstream modes, the category of modes of axial wave numbers k_x^+ that have a positive group velocity and a negative phase velocity are called inverse downstream modes because the phase planes move against the current while the energy follows the flow. Similarly, modes associated with wave numbers k_x^- are called upstream modes.

6

Wave Propagation in Elastic Media

Wave propagation in elastic media significantly differs from wave propagation in fluid media. For example, while in a fluid there is only one type of wave that exists, in a homogeneous and isotropic infinite elastic solid, in addition to the longitudinal waves – particle movement occurring in the direction of the propagation, the transverse waves can propagate independently of the first – the particle movement is in an orthogonal direction to the direction of propagation. The coupling between these two types of waves occurs in the vicinity of the discontinuities (geometric or mechanical). Most internal source mechanisms produce the two wave types simultaneously.

In addition, when one or more dimensions are smaller than others, as we will see in the last two sections dedicated to thin structures, we can significantly simplify the three-dimensional equations obtained reducing them to surface or line problems.

A thin body is a solid which occupies in space a volume limited by two surfaces, such that the distance between them, as well as the inverse of the mechanical wave length (low-frequency hypothesis), is small compared to the other dimensions of the structure. For example, a window is a thin plate, the body of a guitar or a violin is composed of thin plates. We assume from here that the displacements and the deformations are small. We define the mean surface as the location of points that are equidistant to the two surfaces which limit the body (the medium plane in our window). This concept of a mean surface (geometric property) is often confounded with the neutral surface (mechanical property). In fact, let us imagine that we bend this plate. The lower side of the plate will be compressed, while the upper side will undergo an extension. The region of the plate, where the effort it undergoes is canceled and the sign changes, is the neutral surface. Usually, the neutral and mean surfaces are the same. In the case of structures such as beams (we will assume that they are rectangular), this notion of a surface degenerates into the notion of a mean (or neutral) line.

6.1. Equation of mechanical wave propagation

Let us consider a homogeneous and isotropic solid. As we have already seen, this solid can be characterized by a density ρ (which we here assume is invariable in time) and two coefficients; (E, ν), which are named Young's modulus (E) and Poisson's ratio (ν). We may also use Lame's coefficient (λ, μ), which relate to the previous by two equations [2.14] $\lambda = \mathrm{E}\nu / ((1-2\nu)(1+\nu)), \mu = \mathrm{E}/(2(1+\nu))$. Similarly as for the fluids, we are interested in the propagation of a perturbation within the solid. This assumption in most cases allows us to focus only on the "linear" part of the movement and to assume that the displacements and deformations that exist within the body are sufficiently small. In this case, we know that the law that links the deformation undergone by each point of the solid to the strain suffered by these points is given by Hooke's law [2.16], which in the homogeneous and isotropic case can be written as $\sigma_{ij} = \lambda d_{ll}\delta_{ij} + 2\mu d_{ij}$.

Let us return to the equation of impulse conservation [2.11] (with Einstein's convention of summation of repeated indices), in which we have introduced the invariability of density over time:

$$\rho\frac{\partial^2 u_i}{\partial t^2} + \frac{\partial\left(\rho v_i v_j - \sigma_{ij}\right)}{\partial x_j} = F_i \qquad [6.1]$$

where u_i are the components of the displacement field of a solid, σ_{ij} are the components of the stress tensor and F_i are the components of the applied force. Under the small deformation hypothesis, we linearize this equation. This leads to neglecting the term $v_i v_j$. Hooke's law gives us:

$$\frac{\partial \sigma_{ij}}{\partial x_j} = \frac{\partial \lambda d_{ll}\delta_{ij} + 2\mu d_{ij}}{\partial x_j} = \lambda\frac{\partial d_{ll}}{\partial x_i} + 2\mu\frac{\partial d_{ij}}{\partial x_j}. \qquad [6.2]$$

because $\partial\left(d_{ll}\delta_{ij}\right)/\partial x_j = \partial d_{ll}/\partial x_i$. But the linearized deformation stress relations can be given by [2.7] $d_{ij} = \frac{1}{2}\left(\frac{\partial u_i}{\partial x_j} + \frac{\partial u_j}{\partial x_i}\right)$. This is:

$$\frac{\partial \sigma_{ij}}{\partial x_j} = \lambda\frac{\partial^2 u_l}{\partial x_i \partial x_l} + \mu\left(\frac{\partial^2 u_i}{\partial x_j^2} + \frac{\partial^2 u_j}{\partial x_i \partial x_j}\right) = (\lambda+\mu)\frac{\partial \mathrm{div}\vec{u}}{\partial x_i} + \mu\Delta u_i. \qquad [6.3]$$

Applying the law of impulse conservation to this expression, each displacement component is the solution of:

$$\rho\frac{\partial^2 u_i}{\partial t^2} - (\lambda+\mu)\frac{\partial \mathrm{div}\vec{u}}{\partial x_i} - \mu\Delta u_i = F_i. \qquad [6.4]$$

In vector form, this equation can be written as:

$$\rho\frac{\partial^2\vec{u}}{\partial t^2} - (\lambda+\mu)\overrightarrow{\text{grad}}\,(\text{div}\vec{u}) - \mu\Delta\vec{u} = \vec{F}. \qquad [6.5]$$

This equation is the linearized equation of wave propagation in elasticity, sometimes called Navier's equation.

6.2. Free waves

6.2.1. *Volumic waves*

For general solutions of this equation, let us study the propagation of free waves. The Helmholtz–Hodge theorem allows us to decompose the velocity field in a unique way, in potential and rotational parts:

$$\vec{u} = \overrightarrow{\text{grad}}\Psi^L + \overrightarrow{\text{curl}}\vec{\Psi}^T. \qquad [6.6]$$

Introducing this expression into wave equation [6.5] in which the second member, for the search of free waves, is taken as zero:

$$\rho\left(\overrightarrow{\text{grad}}\ddot{\Psi}^L + \overrightarrow{\text{curl}}\ddot{\vec{\Psi}}^T\right) - (\lambda+\mu)\overrightarrow{\text{grad}}\left(\text{div}\left(\overrightarrow{\text{grad}}\Psi^L + \overrightarrow{\text{curl}}\vec{\Psi}^T\right)\right) - \mu\Delta\left(\overrightarrow{\text{grad}}\Psi^L + \overrightarrow{\text{curl}}\vec{\Psi}^T\right) = \vec{0}$$

where $\ddot{\Psi}^L = d^2\Psi^L/dt^2$ and $\ddot{\vec{\Psi}}^T = d^2\vec{\Psi}^T/dt^2$. The Laplacian operator commutes with the gradient ($\overrightarrow{\text{grad}}$), divergence (div) and rotational ($\overrightarrow{\text{curl}}$) operators. On the other hand, we know that $\text{div}\overrightarrow{\text{curl}} = 0$ and that $\text{div}\overrightarrow{\text{grad}} = \Delta$. We deduce that:

$$-\rho\left(\overrightarrow{\text{grad}}\ddot{\Psi}^L - \overrightarrow{\text{curl}}\ddot{\vec{\Psi}}^T\right) + \mu\Delta\overrightarrow{\text{grad}}\Psi^L + \mu\Delta\overrightarrow{\text{curl}}\vec{\Psi}^T + (\lambda+\mu)\overrightarrow{\text{grad}}\Delta\Psi^L = \vec{0},$$

or again:

$$\overrightarrow{\text{grad}}\left\{(\lambda+2\mu)\Delta\Psi^L - \rho\ddot{\Psi}^L\right\} + \overrightarrow{\text{curl}}\left\{\mu\Delta\vec{\Psi}^T - \rho\ddot{\vec{\Psi}}^T\right\} = \vec{0}. \qquad [6.7]$$

For this equation to be satisfied, it is sufficient for the scalar and vector fields to be constant. We can easily choose these constants as zeros. We obtain two wave equations, a scalar one $(\lambda+2\mu)\Delta\Psi^L - \rho\ddot{\Psi}^L = 0$ and a vector one $\mu\Delta\vec{\Psi}^T - \rho\ddot{\vec{\Psi}}^T = \vec{0}$.

Let us define c_L and c_T as the celerities of the longitudinal and transverse waves, respectively:

$$c_L = \sqrt{\frac{\lambda + 2\mu}{\rho}} = \sqrt{\frac{E(1-\nu)}{\rho(1+\nu)(1-2\nu)}} \quad [6.8]$$

$$c_T = \sqrt{\frac{\mu}{\rho}} = \sqrt{\frac{E}{2\rho(1+\nu)}} = c_L\sqrt{\frac{1-2\nu}{2(1-\nu)}} \approx c_L/2 \text{ for } \nu = 1/3. \quad [6.9]$$

The previous two equations then take the most common form, written as follows:

$$\Delta\Psi^L - \frac{1}{c_L^2}\ddot{\Psi}^L = 0, \Delta\vec{\Psi}^T - \frac{1}{c_T^2}\ddot{\vec{\Psi}}^T = \vec{0}. \quad [6.10]$$

These two decoupled wave operators indicate the existence of two types of waves, which propagate independently in the medium at different velocities. As $0 \leq \nu \leq 1/2$, propagation velocities are linked by $c_L \geq \sqrt{2}c_T$. For example, for steel $c_L =$ 5900 m/s and $c_T =$ 3200 m/s; for aluminum $c_L =$ 6300 m/s and $c_T =$ 3100 m/s and for glass $c_L =$ 5800 m/s and $c_T =$ 3400 m/s.

In a harmonic regime with a time dependence $\Psi^{L-T}(\vec{x}, t) = \psi^{L-T}(\vec{x})$ $\exp(-\imath\omega t)$, these equations become:

$$\Delta\psi^L + \frac{\omega^2}{c_L^2}\ddot{\psi}^L = 0, \Delta\vec{\psi}^T + \frac{\omega^2}{c_T^2}\ddot{\vec{\psi}}^T = \vec{0}. \quad [6.11]$$

When the material has an internal dissipation (for example, viscoelasticity), we have already seen that to describe the dissipation we use a complex Young's modulus $\hat{E} = E_0(1 - \imath\eta)$, with $\eta > 0$ and $E_0 \in \mathbb{R}$. For a temporal dependence of the type $\exp(+\imath\omega t)$, we would have $\hat{E} = E_0(1 + \imath\eta)$, with $\eta > 0$. This allows us to easily introduce complex celerities which can be written as $\hat{c}_{L,T} = c_{L,T}\sqrt{1 - \imath\eta} \approx c_{L,T}(1 - \imath\eta/2)$, if η is small.

6.2.2. *Plane wave case*

Let us, for example, use Cartesian coordinates and assume that $\vec{u} = \vec{u}(x)$ only depends on the variable x. We get: $\vec{u} = (u_x, u_y, u_z)$, $\text{div}\vec{u} = u_x'$ and $\overrightarrow{\text{curl}}\vec{u} = (0, u_z', u_y')$. We obtain three wave propagation equations:

$$\Delta u_x - \frac{1}{c_L^2}\frac{\partial^2 u_x}{\partial t^2} = 0, \Delta u_y - \frac{1}{c_T^2}\frac{\partial^2 u_y}{\partial t^2} = 0, \Delta u_z - \frac{1}{c_T^2}\frac{\partial^2 u_z}{\partial t^2} = 0. \quad [6.12]$$

For a wave that propagates in the direction x, the displacement is parallel to the direction of propagation (longitudinal polarization). It is appropriate to talk about longitudinal waves or compression waves. For the others (which propagate along y and z), the displacement is directed in a plane perpendicular to the direction of propagation (transverse polarization), such a wave is called a transverse or shear wave. In seismics, a longitudinal wave is a primary wave since, because of its celerity being almost twice that of a transverse wave, it arrives first. Let us recall that in fluid media, there are no shear waves – fluid acoustics are irrational.

6.2.3. *Surface waves*

Many types of waves exist near the boundaries of the elastic media. In particular, for semi-infinite media, we can show that there is a particular type of waves, which propagate near the free boundary with an exponential decrease with depth. These waves, of which there are two types, correspond to movements recorded by seismographs during earthquakes.

The Rayleigh waves are very similar to the waves at the surface of the water; the displacement that they induce is complex: a retrograde movement at the surface of an elliptical surface both horizontally and vertically. During the seismic recordings, the three components of the seismograph capture them. These waves are the ones responsible for the perceived rumble during earthquakes. When the surface of the elastic medium is in contact with a fluid (air or water), to the Rayleigh wave (which becomes complex due to the attenuation by acoustic radiation), we can add another surface wave whose energy is mainly localized in the fluid, the Scholte–Stoneley wave.

Love waves induce a transverse movement of the ground in a horizontal plane, perpendicular to its direction of propagation. The displacement is essentially the same as that of the shear waves without vertical movement. It is recorded only in the horizontal components of the seismograph. In contrast, the celerity of Love waves is frequency-dependent. These waves are said to be dispersive, because for a given time signal, the amplitudes of different frequencies contained in the signal will propagate at different velocities, thus causing a gradual dispersion of the original signal during the propagation.

6.2.3.1. *Rayleigh waves*

Let us examine a simplified case of a two-dimensional semi-infinite problem, defined in Cartesian coordinates (O, x, z), by $-\infty < x < \infty$ and $0 < z < \infty$ such that the surface $z = 0$ is not subject to any excitation. The free wave equation (without the forcing term) can be given by equation [6.7] $(\lambda + \mu)\overrightarrow{\text{grad}}\,(\text{div}\vec{u}) + \mu\Delta\vec{u} = \rho\partial^2\vec{u}/\partial t^2$, which we can write in the following form

$(c_L^2 - c_T^2)\overrightarrow{\text{grad}}\,(\text{div}\vec{u}) + c_T^2\Delta\vec{u} = \partial^2\vec{u}/\partial t^2$. We aim to find, whether the free waves can propagate along the x direction while exponentially attenuating along z:

$$u_x = A_x \mathrm{e}^{-Bz}\mathrm{e}^{\imath k(x-ct)}, u_z = A_z \mathrm{e}^{-Bz}\mathrm{e}^{\imath k(x-ct)}, u_z = 0,$$

where A_x and A_z are complex constants, B is a positive real number and k is a real number. c is the wave propagation velocity to be determined. Substituting these three expressions in the three wave propagation equations, we obtain the following system:

$$\begin{aligned}&\left((c^2 - c_L^2)k^2 + c_T^2 B^2\right) A_x - \imath(c_L^2 - c_T^2)BkA_z = 0,\\ -&\imath(c_L^2 - c_T^2)BkA_x + \left((c^2 - c_T^2)k^2 + c_L^2 B^2\right) A_z = 0.\end{aligned}$$

In order to have the amplitudes A_x and A_y not identically zero, the determinant associated with the linear system of the unknowns A_x and A_z must be zero. We obtain $\left((c^2 - c_T^2)k^2 + c_T^2 B^2]\right)\left((c^2 - c_L^2)k^2 + c_L^2 B^2\right)$, which we solve for the unknown B. We obtain two solutions $B' = k\sqrt{1 - c^2/c_T^2}$ and $B'' = k\sqrt{1 - c^2/c_L^2}$. For B to be real, we must have $c < c_T < c_L$. We deduce the amplitude ratio, which corresponds to B' and B'' : $(A_z/A_x)' = \imath k/B'$ and $(A_z/A_x)'' = -B''/(\imath k)$. With $u_z = 0$, we obtain:

$$\begin{aligned}u_x &= A'_x \mathrm{e}^{-B'z}\mathrm{e}^{\imath k(x-ct)} + A''_x \mathrm{e}^{-B''z}\mathrm{e}^{\imath k(x-ct)},\\ u_z &= \frac{\imath k}{B'}A'_x \mathrm{e}^{-B'z}\mathrm{e}^{\imath k(x-ct)} - \frac{B''}{\imath k}A''_x \mathrm{e}^{-B''z}\mathrm{e}^{\imath k(x-ct)}.\end{aligned}$$

The unknowns A'_x, A''_x and c can be determined by applying the free surface conditions: $\sigma_{zz} = \sigma_{zx} = \sigma_{yz} = 0$ at $z = 0$. Hooke's law allows us to write these conditions in the form of:

$$\frac{\partial u_x}{\partial z} + \frac{\partial u_z}{\partial x} = 0, \text{ at } z = 0, \lambda\left(\frac{\partial u_x}{\partial x} + \frac{\partial u_z}{\partial z}\right) + 2\mu\frac{\partial u_z}{\partial z} = 0, \text{ at } z = 0.$$

With $\mu = \rho c_T^2$ and $\lambda = \rho\left(c_L^2 - 2c_T^2\right)$, the boundry conditions become $-\left(B' + k^2/B'\right)A'_x - 2A''_x B'' = 0$ and $\left[(c_L^2 - 2c_T^2) - c_L^2\frac{B''^2}{k^2}\right]A''_x - 2c_T^2 A'_x = 0$. However, we know that $B'^2 = k^2(1 - c^2/c_T^2)$ and $B''^2 = k^2(1 - c^2/c_L^2)$, thus $2A''_x B'' + (2 - c^2/c_T^2)k^2\frac{A'_x}{B'} = 0$ and $(2 - c^2/c_T^2)A''_x + 2A'_x = 0$. Once more, we obtain a linear system whose unknowns A'_x and A''_x are not identically zero, if the determinant of the system is zero, that is to say, if c satisfies the characteristic equation:

$$\left(2 - \frac{c^2}{c_T^2}\right)^2 = 4\sqrt{1 - \frac{c^2}{c_L^2}}\sqrt{1 - \frac{c^2}{c_T^2}}, \qquad [6.13]$$

Let $p = c^2/c_T^2$ and $q = c_T^2/c_L^2 = \mu/(\lambda + 2\mu) = (1 - 2\nu)/(2(1 - \nu))$. The characteristic equation can thus be written as $(2-p)^2 = 4\sqrt{1-p}\sqrt{1-qp}$. We square and factorize by p. We obtain $p\left(p^3 - 8p^2 + 8(3 - 2q)p - 16(1 - q)\right) = 0$. The case $p = 0$ is not of interest to us, so we let $f(p, q) = p^3 - 8p^2 + 8(3 - 2q)p - 16(1 - q)$. The characteristic equation of Rayleigh waves is thus $f(p, q) = 0$. The coefficients of the equation are real, the cubic equation $f(p, q) = 0$ has one or three real roots. As $c_T < c_L$ we get $q < 1$. It is easy to see that $f(p, q)$ is negative for $p = 0$, positive for $p = 1$ and that $\partial f(p, q)/\partial p > 0$ for $p \leq 1$. Thus, there is always a real root in the interval $0 < p < 1$. This shows that at the interface there can exist waves, which propagate with a velocity c_R which is less than the celerity of the transverse waves. If $\nu = 1/4$, then $\lambda = \mu$, that is $c_L = \sqrt{3}c_T$ or $q = 1/3$. The Rayleigh equation can be reduced to $p^3 - 8p^2 + 56/3p - 32/3 = 0$, and gives the three real roots $p = 4$, $p = 2 + 2/\sqrt{3}$ and $p = 2 - 2/\sqrt{3}$. Only the latter is less than 1 and corresponds to a wave that attenuates with depth (real B' and B'') and which propagates with a velocity of $c_R = 0.9194c_T$ independent of the frequency (the wave is not dispersive). The two other roots are roots that can be regarded as parasites[1]. The calculation of the celerity of Rayleigh waves has been the subject of numerous works that have led to several approximately equivalent expressions:

$$\frac{c_R}{c_T} \approx \frac{0.862 + 1.14\nu}{1 + \nu} \text{ or } \frac{c_R}{c_T} \approx 1 - \frac{1}{2}\delta - \frac{5}{8}\delta^2 + \frac{27}{16}\delta^3, \qquad [6.14]$$

with $\delta = 1/(4(3 - 4q)) = (1 - \nu)/(4(1 + \nu)) \leq 1/4$. The extreme values are given by $\nu = 0$, $\delta = 1/4$ and $c_R \approx 0.862c_T$ (the exact value is $c_R = 0.8741c_T$) and for $\nu = 0.5$, $\delta = 1/12$ and $c_R \approx 0.955c_T$ (the exact value is $c_R = 0.9554c_T$).

6.2.3.2. *Scholte–Stoneley waves*

When the elastic medium is in contact with a fluid medium, with a density ρ_f and the celerity of the volumic waves is c_0 and, if we take into account the continuity conditions of stress and displacement normal to the interface, the Rayleigh equation [6.13] is modified in the following way:

$$4\sqrt{1-p}\sqrt{1-qp} - (2-p)^2 = mp^2\sqrt{\frac{1-qp}{1-Qp}}, \qquad [6.15]$$

where we set $Q = c_T^2/c_0^2$ and $m = \rho_f/\rho_s$ if ρ_s is the density of the solid. We can show that this equation has two roots. The first root, with an imaginary non-zero part (even higher than the ratio ρ_f/ρ_s), corresponds to a wave which radiates across the interface and attenuates during propagation. This is the generalized Rayleigh wave. The second root, purely real, is associated with a surface wave located in the fluid, which propagates without attenuation. This is the Scholte–Stoneley wave, which

1 Alternatively, B' and B'' would be pure imaginary and would propagate according to depth.

propagates at a velocity c_S, which is slightly lower than the velocity of the volumic waves in the fluid. As $c_S < c_0$, we get $\sin\theta_S = c_0/c_S > 1$, it is not possible to generate a Scholte–Stoneley wave on a planar interface from an incident wave in a fluid (θ_S would be complex). We can estimate, by asymptotic calculations, the celerity c_S when the fluid is a gas, such as air. We thus get $m \ll 1$ and $Q \gg 1$, the modified Rayleigh equation [6.15] can be written as:

$$4\sqrt{1-p} - \frac{(2-p)^2}{\sqrt{1-qp}} = mp^2\frac{1}{\sqrt{1-Qp}}, \qquad [6.16]$$

as $c_S < c_0$ then $p = c_S^2/c_T^2 \ll 1$. By using the Taylor expansion around $p = 0$ limited to the first order of the left side of equation [6.16], after squaring we obtain $1 - Qp \approx m^2p^2/(4(1-q)^2)$. If we assume that the right side is small compared to the unit, as a first approximation, we obtain $p \approx 1/Q < 1$; as a second approximation, we introduce this result in the part on the right and we obtain $p \approx \frac{1}{Q}(1 - m^2/(4Q^2(1-q)^2))$, from where it is easy to extract, since $m^2/(4Q^2(1-q)^2) \ll 1$, that:

$$c_S \approx c_0\left(1 - \frac{m^2}{8Q^2(1-q)^2}\right), \qquad [6.17]$$

which is slightly smaller than c_0.

6.2.3.3. *Love waves*

The Rayleigh and the Scholte–Stoneley surface waves are characterized by a movement of particles, which occurs only in the propagation plane. In the example that we just discussed, there is no propagation in the direction y normal to the plane $(0, x, z)$. The surface waves whose movement is perpendicular to the propagation plane cannot exist in semi-infinite elastic media. Nevertheless, if the elastic medium is composed of a homogeneous layer superposed on a semi-infinite elastic medium with different characteristics (see Figure 6.1), we can show that these transverse waves, with exponential attenuation with depth, (or Love waves) exist. These waves are particularly destructive for buildings, and have an amplitude comparable to the Rayleigh waves during earthquakes.

We aim to find, whether it is possible for free waves to propagate in the direction x while exponentially attenuating along z and moving along the y direction:

$$u_x = 0, u_y = \left[A_{1z}e^{-kz\sqrt{1-\frac{c^2}{c_{1T}^2}}} + B_{1z}e^{kz\sqrt{1-\frac{c^2}{c_{1T}^2}}}\right] e^{\imath k(x-ct)} \text{ in } M_1$$

$$u_z = 0, u_y = A_{2z}e^{-kz\sqrt{1-\frac{c^2}{c_{2T}^2}}} e^{\imath k(x-ct)} \text{ in } M_2$$

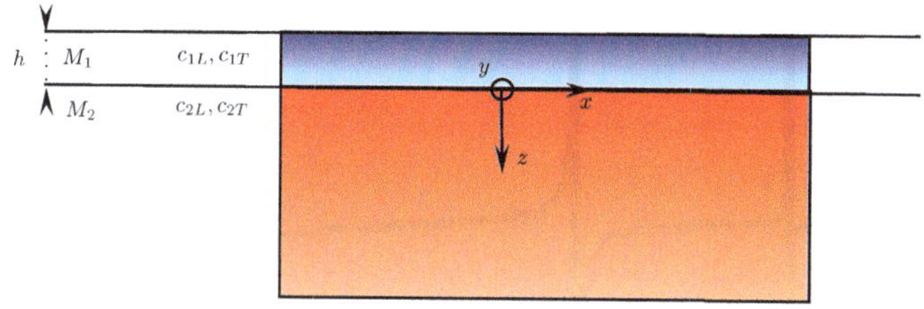

Figure 6.1. *The geometry of a bilayer medium*

The boundary conditions imply that u_y and σ_{yz} are continuous when crossing the $z = 0$ surface and that $\sigma_{yz} = \mu\left(\partial u_z/\partial y + \partial u_y/\partial z\right) = 0$ at $z = -h$. We thus obtain:

$$A_{2z} = A_{1z} + B_{1z}, \mu_2 A_{2z}\sqrt{1 - \frac{c^2}{c_{2T}^2}} = \mu_1\left(A_{1z} - B_{1z}\right)\sqrt{1 - \frac{c^2}{c_{1T}^2}},$$

$$A_{1z}\mathrm{e}^{kh\sqrt{1-\frac{c^2}{c_{1T}^2}}} = B_{1z}\mathrm{e}^{-kh\sqrt{1-\frac{c^2}{c_{1T}^2}}}.$$

The first two relations give $(A_{1z} - B_{1z})/(A_{1z} + B_{1z}) = \mu_2/\mu_1\sqrt{1 - c^2/c_{2T}^2}/\sqrt{1 - c^2/c_{1T}^2}$. We must note that $\tanh x = (\exp(x) - \exp(-x))/(\exp(x) + \exp(-x))$. This is:

$$\tanh\left(kh\sqrt{1 - \frac{c^2}{c_{1T}^2}}\right) = \frac{\mathrm{e}^{kh\sqrt{1-\frac{c^2}{c_{1T}^2}}} - \mathrm{e}^{-kh\sqrt{1-\frac{c^2}{c_{1T}^2}}}}{\mathrm{e}^{kh\sqrt{1-\frac{c^2}{c_{1T}^2}}} + \mathrm{e}^{-kh\sqrt{1-\frac{c^2}{c_{1T}^2}}}} = -\frac{A_{1z} - B_{1z}}{A_{1z} + B_{1z}}.$$

We thus obtain, with $\mu_i = \rho_i c_{iT}^2$, the equation that governs the wave propagation velocity in a bilayer medium:

$$\frac{\rho_2 c_{2T}^2\sqrt{1 - \frac{c^2}{c_{2T}^2}}}{\rho_1 c_{1T}^2\sqrt{\frac{c^2}{c_{1T}^2} - 1}} = \tan\left(kh\sqrt{\frac{c^2}{c_{1T}^2} - 1}\right). \qquad [6.18]$$

If $c_{1T} > c_{2T}$, it is easy to see that this equation does not have a real solution[2]. If $c_{1T} < c_{2T}$, we can show in the same way that this equation has at least one real solution $c_{1T} < c < c_{2T}$ for all the values of kh, the Love wave. The displacement is located in the top layer which acts as the wave guide.

2 If $c > c_{1T}$, the right side is purely imaginary and the left side is real; if $c < c_{2T}$, the right side is negative and the left side is positive; and if $c_{1T} > c > c_{2T}$, the left side is purely imaginary and the right side is real.

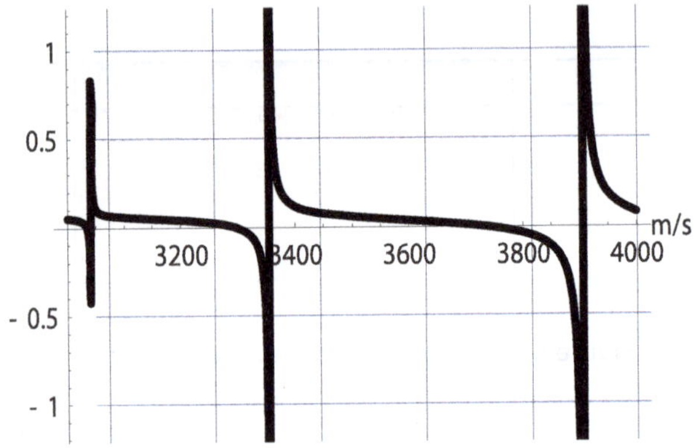

Figure 6.2. *Dispersion equation of Love waves*

The velocity of the Love waves is generally larger than the velocity of Rayleigh waves and depends on the frequency[3]. If kh is very small, then the wave propagation equation in a bilayer medium becomes $\rho_2 c_{2T}^2 \sqrt{1 - c^2/c_{2T}^2} \approx 0$ and the Love wave velocity approaches that of shear waves in a half space. Figure 6.2 shows an example of a plot with a normalized amplitude of the dispersion relation for $kh = 3\pi$ for media with identical densities $\rho_1 = \rho_2$ and the celerity $c_1 = 3$ km/s and $c_2 = 4$ km/s. For a fixed k, there is at least one solution, and, in practice, n, the number of roots of the dispersion relation depends on the number of tangent branches in the interval. We get $n = k_n h/\pi\sqrt{1 - c_{1T}^2/c_{2T}^2} + 1$, rounded to the nearest integer. For example, consider Figure 6.2, we have $n = 2.98 \approx 3$. Conversely, with each Love mode, we can associate a cutoff wave number (and thus a cutoff frequency) equal to $k_n = \pi(n-1)/\left(h\sqrt{1 - c_{1T}^2/c_{2T}^2}\right)$. At this frequency, the associated velocity corresponds to the velocity of the half-space c_{2T} since it decreases when the frequency increases to meet that of the layer.

These waves are very useful for studying the Earth's mantle, especially for measuring its thickness. For example, on the surface of the Earth, compression waves have a velocity of about 6 km/s, shear waves of about 4.1 km/s, Love waves of about 4 km/s and Rayleigh waves of about 3.8 km/s.

6.3. Green's kernels in a harmonic regime

Similarly as for d'Alembert's formula, we can assume that the time dependence is of the type $\mathrm{e}^{-\imath\omega t}$. Navier's equation thus becomes $(\lambda+\mu)\overrightarrow{\text{grad}}\,(\text{div}\vec{u})+\mu\Delta\vec{u}+\omega^2\rho\vec{u} = -\vec{F}$.

3 Love waves are, therefore, dispersive.

The concept of Green's kernel can be generalized without difficulty to vector differential equations. We will call $\vec{\vec{G}}$, Green's tensor, the solution to the following equation $(\lambda + \mu)\overrightarrow{\text{grad}}\left(\text{div}\vec{\vec{G}}\right) + \mu\Delta\vec{\vec{G}} + \omega^2\rho\vec{\vec{G}} = \delta\vec{\vec{I}}$, where $\vec{\vec{I}}$ is the identity matrix and δ is the Dirac distribution. Schematically, $\vec{\vec{G}}$ is a squared matrix 3×3 whose components are Green's kernels. Let $k_L = \omega/c_L$ and $k_T = \omega/c_T$, the wave numbers of longitudinal and transverse waves, and R, the Euclidean distance between the source point and the current point. We show [ACH 82] that the components G_{ij} of Green's tensor can be given by:

$$G_{ij} = \frac{1}{\mu k_T^2}\frac{\partial^2}{\partial x_i \partial x_j}\left[-G\left(k_L R\right) + G\left(k_T R\right)\right] + G\left(k_T R\right)\delta_i^j, \qquad [6.19]$$

where δ_i^j is Kronecker's symbol with $G\left(k_\alpha R\right) = -\exp(\imath k_\alpha R)/\left(4\pi R\right)$ in $I\!R^3$ and $G\left(k_\alpha R\right) = -\imath/4\text{H}_0\left(k_\alpha R\right)$ in $I\!R^2$ with $\alpha = L, T$. In the two-dimensional case, the indices i and j vary between 1 and 2 and $\vec{\vec{G}}$ is a square matrix 2×2.

6.4. Thin body approximation for plannar structures

Under the previous assumptions, the proposed method consists of expanding the displacement field into a Taylor series with respect to thickness. This series is truncated to the first order. This imposes a linear variation of the stresses in the thickness. If we further assume that the sides of the structure are free of any constraints, we can show that the terms of the first order of the Taylor series of displacement are expressed as functions of the terms of the order zero, which are defined on the mean surface. The displacement fields obtained thus express all the kinematics of the medium from the quantities defined on the mean surface. The three-dimensional problem is thus reduced to a two-dimensional problem or even a one-dimensional problem in the case of beams. The application of Hamilton's principle on the Lagrangian of a structure allows us to obtain the operator of the structure, as well as the associated boundary conditions.

We examine the assumptions behind the Kirchhoff–Love framework:

1) low thickness compared to the other dimensions;

2) displacements and deformations are small enough for neglecting the second-order terms;

3) the transverse shear is low compared to the normal forces;

4) a part of the material, transverse to the non-deformed mean surface, behaves as an undeformable solid.

To these assumptions, we add that the thickness must be small compared to the wave lengths (low-frequency hypothesis). Moreover, we assume that the material which forms the shells is homogeneous and isotropic and, finally, that the stress–strain relation can be given by Hooke's law. The kinematic relations of Kirchhoff–Love are both intended to approach the three-dimensional elasticity equations and eliminate the internal behavior of the material. These approximations impose that a part of the material transverse to the mean surface must behave as an undeformable solid. The movement of the structure is then deduced from the rotation and translation fields. These fields reflect the two phenomena that govern the equilibrium of thin bodies. The rotation field reflects the bending action and the translation field reflects the effect of the membrane which is linked to specific movements on the mean surface. An important consequence of the thin body approximation is that the bending effect is governed by an operator of the fourth order. This operator (usually a bi-Laplacian) governs the behavior of the displacement field in the direction in which the structure is thin (referring to thickness), while the membrane effect is reflected by an operator of the second order (similar to the three-dimensional elasticity equations). Thus, we see that, if the approximation of the thin body simplifies the problem by reducing the number of unknowns, a level of complexity is introduced to the equations *via* higher order operators.

In this section, we present only the results for plane structures (thin plates and beams). The following section is dedicated to the study of thin cylindrical shells.

6.4.1. *Straight beams*

Let us examine a problem with a thin beam whose rectangular section is constant. Let us use the usual Cartesian coordinates (O, x, y, z). This coordinate system is centered at the middle line of the beam (see Figure 6.3). Let us define x as the longitudinal coordinate, y and z, as the transverse coordinates. The dimensions of the beam can be given by: $x \in]0, \ell[$, $y \in] - e/2, +e/2[$ and $z \in] - h/2, +h/2[$. e is the width of the beam and h is its thickness. In the following, we assume $e \ll \ell$ and $h \ll \ell$. The material, which constitutes the beam, is assumed to be homogeneous and isotropic, and the stress–strain relationship is given by Hooke's law.

6.4.1.1. *Displacement field*

Since the structure is thin along the y and z directions, expanding the displacement field $\vec{U}$ into a Taylor series with respect to these variables gives:

$$\vec{U}(x, y, z, t) = \vec{u}^0(x, t) + y\vec{u}^1(x, t) + z\vec{u}^2(x, t) \quad [6.20]$$

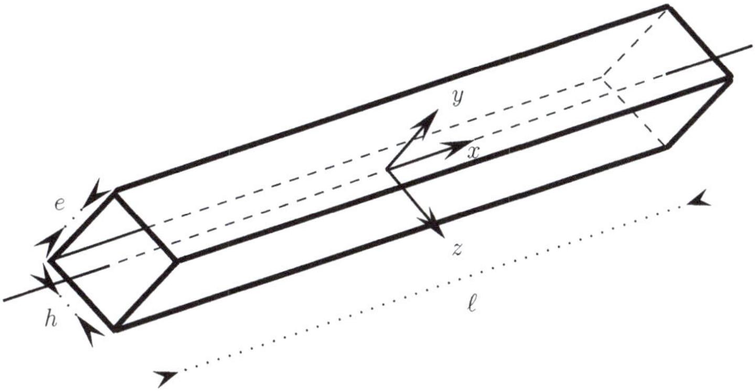

Figure 6.3. *The geometry of a straight beam*

$\vec{u}^0(x,t)$ is, of course, the value of the displacement field on the neutral line: $\vec{u}^0(x,t) = \vec{U}(x,0,0,t)$. In addition, we have $\vec{u}^1(x,t) = \vec{U}_{,y}(x,0,0,t)$ and $\vec{u}^2(x,t) = \vec{U}_{,z}(x,0,0,t)$.

In each component of the stress-strain tensor, we introduce the previous Taylor expansion. To simplify, the dependencies on x and on t are omitted. We have:

$$\begin{aligned} d_{xx} &= U_{x,x} = u^0_{x,x} + yu^1_{x,x} + zu^2_{x,x}, d_{yy} = U_{y,y} = u^1_y, d_{zz} = U_{z,z} = u^2_z \\ d_{xy} &= \tfrac{1}{2}\left(U_{x,y} + U_{y,x}\right) = \tfrac{1}{2}\left(u^1_x + u^0_{y,x} + yu^1_{y,x} + zu^2_{y,x}\right) \\ d_{xz} &= \tfrac{1}{2}\left(U_{x,z} + U_{z,x}\right) = \tfrac{1}{2}\left(u^2_x + u^0_{z,x} + yu^1_{z,x} + zu^2_{z,x}\right) \\ d_{yz} &= \tfrac{1}{2}\left(U_{y,z} + U_{z,y}\right) = \tfrac{1}{2}\left(u^2_y + u^1_z\right). \end{aligned} \qquad [6.21]$$

The free side hypothesis implies that all the side faces of the beam are free from constraints. So for $y = \pm e/2$ and $z = \pm h/2$, we get $\vec{\vec{S}} \cdot \vec{n}_\alpha = 0$ for $\alpha = y, z$. This last relation is equivalent to $\sigma_{\alpha y} = 0$ and $\sigma_{\alpha z} = 0$ for $\alpha = x, y$; if we expand the stress tensor into a Taylor series with respect to thickness, it is easy to see that these components are zero throughout the thickness. The stress–strain relationship [2.17] results in $\sigma_{xy} = 0 \Rightarrow d_{xy} = 0$, $\sigma_{xz} = 0 \Rightarrow d_{xz} = 0$ and $\sigma_{yz} = 0 \Rightarrow d_{yz} = 0$. Similarly, from $\sigma_{zz} = 0$ and $\sigma_{yy} = 0$, we get $d_{yy} + d_{zz} = -2\nu d_{xx}$ from where $\sigma_{xx} = Ed_{xx}$. From the relations $d_{xy} = 0$, $d_{xz} = 0$ and $d_{yz} = 0$ we deduce $u^1_x + u^0_{y,x} + yu^1_{y,x} + zu^2_{y,x} = 0$, $u^2_x + u^0_{z,x} + yu^1_{z,x} + zu^2_{z,x} = 0$ and $u^2_y + u^1_z = 0$. These relations are true for $y = \pm e/2$ and $z = \pm h/2$, it is thus evident that each coefficient for y and z must be zero. Thus, we need to verify the following relationships $u^1_x + u^0_{y,x} = 0$, $u^1_{y,x} = 0$, $u^2_{y,x} = 0$, $u^2_x + u^0_{z,x} = 0$, $u^1_{z,x} = 0$, $u^2_{z,x} = 0$ and $u^2_y + u^1_z = 0$. That is $u^1_x = -u^0_{y,x}$, $u^1_y = C_1$, $u^2_y = C_2$, $u^2_x = -u^0_{z,x}$, $u^1_z = C_3$, $u^2_z = C_4$, $u^2_y = -u^1_z \Rightarrow C_2 = -C_3$, where $C_i, i = 1, \cdots 4$ are constants. To determine them, we use the fourth Kirchhoff–Love hypothesis which indicates that a part of the material, transverse to the non-deformed mean surface, behaves as an undeformable solid. In practice, the movements of the beam in the direction where the structure is thin are rigid. Thus, the displacement of the beam along the y direction, given by U_y, does not depend on y and thus $u_y^1 = C_1 = 0$; similarly, the displacement of the beam along the z direction, given by U_z, does not depend on z and thus $u_z^2 = C_4 = 0$. We note that brutally imposing $u_y^1 = 0$ and $u_z^2 = 0$ in the equations would lead [6.21] to imposing $d_{yy} = 0$ and $d_{zz} = 0$ and as $d_{yy} + d_{zz} = -2\nu d_{xx}$ we would have $d_{xx} = 0$. In practice, the conditions $u_y^1 = 0$ and $u_z^2 = 0$ are written strictly as $yu_y^1 \ll u_y^0$ and $zu_z^2 \ll u_z^0$. With these reservations, we thus have $C_1 \equiv 0$ and $C_4 \equiv 0$, as well as the transverse components of the displacement field, whose appearance is presented [6.4], become $U_y = u_y^0 + zC_2$ and $U_z = u_z^0 - yC_2$. Except for the displacement given by u_y^0 and u_z^0, the movement of each section of the beam is a rotation, to which a low shear is added, if the section of the beam is rectangular. This induces a transverse shear along the length of the beam. However, the third Kirchhoff–Love assumption indicates that the transverse shear is low compared with the normal forces and thus in order to address it, we should take $C_2 = 0$.

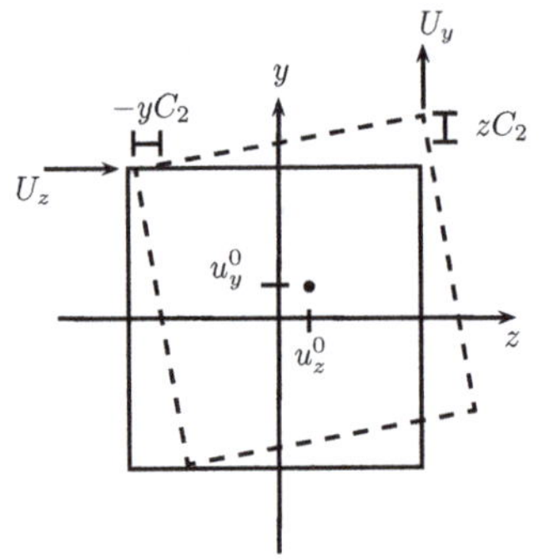

Figure 6.4. *Transverse displacement induced by $U_y = u_y^0 + zC_2$ and $U_z = u_z^0 - yC_2$ for a beam with a square section*

Displacement field [6.20] of the beam can, finally, be written as follows:

$$U_x = u_x^0 - yu_{y,x}^0 - zu_{z,x}^0, U_y = u_y^0, U_z = u_z^0. \quad [6.22]$$

The displacement field that we obtain (denoted by U_α), valid throughout the volume of the beam, is expressed from values (denoted by u_α^0) defined on the middle line of the beam. For convenience with respect to calculations, further on, we will remove the exponent "0". It is important to note that in further calculations, the exponent "2" represents squaring of the quantity and not the term of the second order of the Taylor series.

6.4.1.2. *Beam operator*

To calculate the beam operator, we apply Hamilton's theorem, which states that the variation of the total energy of the system (given by the Lagrangian) is equal to the work of the external forces, which are applied to it. Let us calculate the Lagrangian for the beam which is given by the difference between the kinetic energy $\mathcal{T}$ and the potential energy $\mathcal{P}$ of the beam. We know that:

$$\mathcal{T}(\vec{U}) = \frac{1}{2}\int_\Omega \rho U_{\alpha,t} U_{\alpha,t} dV,$$

where Ω is the volume occupied by the beam. The density of the elastic potential energy can be written as:

$$\mathcal{P}(\vec{U}) = \frac{1}{2}\int_\Omega \sigma_{\alpha\beta} d_{\alpha\beta} dV.$$

Let us calculate the density of the elastic potential energy $dP(\vec{U}) = \frac{1}{2}\sigma_{\alpha\beta} d_{\alpha\beta}$. We get:

$$dP(\vec{U}) = \frac{1}{2}\sigma_{xx} d_{xx} = \frac{E}{2} d_{xx}^2 = \frac{E}{2} U_{x,x}^2. \quad [6.23]$$

We must note that this density of the elastic potential energy is independent of ν. This corresponds to the fact that an elongation is not accompanied by a lateral contraction because all the kinematics of the beam have been reduced to the neutral line which, by definition, cannot contract. This remark naturally leads to the assumption that the longitudinal and transverse displacements of low amplitude of a rectilinear beam are decoupled, since none of the longitudinal elongations (transverse, respectively) are accompanied by a transverse contraction (longitudinal, respectively).

6.4.1.2.1. Longitudinal vibrations (compression)

In this case, the neutral line is not subjected to displacements along the directions orthogonal to it. This is $u_x^0 \neq 0$, $u_y^0 = u_z^0 = 0$. The displacement field of the beam [6.20] is thus simplified to $U_x = u_x^0$, $U_y = 0$ and $U_z = 0$. For simplicity, further on, we note that $u_x^0 = u_x$. The density of the internal elastic potential energy [6.23] can thus be written as $dP(\vec{U}) = E/2\,(u_{x,x})^2$. Similarly, the density of internal kinetic energy is given by $dT(\vec{U}) = \rho/2\,(u_{x,t})^2$. To simplify, let us assume that only a force with the density $\vec{F}$, constant over a straight section (area A) and whose only component is F_x, is applied to the beam. The work of this force density in a virtual displacement $\delta\vec{U}$ is given by the scalar product $\vec{F}\cdot\delta\vec{U}$. We get:

$$\delta\left(\mathcal{T}(\vec{U}) - \mathcal{P}(\vec{U})\right) = \int_V \int_0^t \vec{F}\cdot\delta\vec{U} dV dt = \int_0^\ell \int_0^t A F_x \delta u_x dx dt.$$

After integration in the straight section of the beam, the kinetic energy $\mathcal{T}(\vec{U}) = \int_V \int_0^t dT(\vec{U})dVdt$ and potential energy become $\mathcal{P}(\vec{U}) = \int_V \int_0^t dP(\vec{U})dVdt$:

$$\mathcal{T}(\vec{U}) = \int_0^\ell \int_0^t \frac{\rho A}{2} (u_{x,t})^2 \, dxdt \text{ and } \mathcal{P}(\vec{U}) = \int_0^\ell \int_0^t \frac{EA}{2} (u_{x,x})^2 \, dxdt.$$

Hamilton's principle is then written as:

$$\int_0^\ell \int_0^t \left[\rho A u_{x,t} \left(\delta u_x\right)_{,t} - EAu_{x,x} \left(\delta u_x\right)_{,x}\right] dxdt = \int_0^\ell \int_0^t AF_x \delta u_x dxdt,$$

which we integrate by parts with respect to x and t. If Young's modulus and the density are constant over the entire length of the beam and over time, we get:

$$\int_0^\ell \int_0^t \left[Eu_{x,xx} - \rho u_{x,tt} - F_x\right] \delta u_x dxdt + \int_0^\ell \left[\rho u_{x,t}\right]_t \delta u_x dx - \int_0^\ell \left[Eu_{x,x}\right]_0^\ell \delta u_x dt = 0.$$

We must note that, rather than Young's modulus, the product EA (and ρA) must be constant. From this, we deduce:

$$\int_0^\ell \int_0^t \left[Eu_{x,xx} - \rho u_{x,tt} - F_x\right] \delta u_x dxdt = 0, \int_0^\ell \left[\rho u_{x,t}\right]_t \delta u_x dx = 0,$$
$$\int_0^t \left[Eu_{x,x}\right]_0^\ell \delta u_x dt = 0, \quad [6.24]$$

as the virtual displacement δu_x is arbitrary, the integrands are all zero. We obtain:

$$Eu_{x,xx} - \rho u_{x,tt} = F_x, \quad \forall x \in]0, \ell[, \forall t, \quad [6.25]$$

$$\rho u_{x,t} = 0 \text{ and } u_x = 0, \quad \forall x \in]0, \ell[\text{ at } t = 0, \quad [6.26]$$

$$Eu_{x,x} = 0 \text{ or } u_x = 0, \quad \text{at } x = 0, \ell, \forall t. \quad [6.27]$$

Equation [6.25] is the equation of beams that vibrate at compression, [6.26] giving the temporal boundary conditions (the "and" is linked to the principle of causality) and [6.27] giving the spatial boundary conditions. We have two possibilities: the clamped end $u_x = 0$ or the free end $EAu_{x,x} = 0$.

6.4.1.2.2. Weak formulation of the problem

Hamilton's principle simply gives us access to the weak formulation (or the variational formulation) of the problem. This writing is the one which is used when we are interested in solving the problem by finite elements. Let us call $a(u_x, \delta u_x)$, the symmetric bilinear form given by $a(u_x, \delta u_x) = \int_0^\ell \left[u_{x,x} \left(\delta u_x\right)_{,x} \right] dx$ and let us define the scalar product $\langle u_x, \delta u_x \rangle$ by $\langle u_x, \delta u_x \rangle = \int_0^\ell u_x \delta u_x dx$. We must note that $\mathcal{A}(u_x, \delta u_x) = \int_0^t a(u_x, \delta u_x) dt$ is also the symmetric bilinear form and that $\int_0^t \langle u_x, \delta u_x \rangle dt$ is yet again a scalar product. We must note that $EA\mathcal{A}(u_x, u_x)$ represents the internal potential energy of the beam. Hamilton's principle can be written as:

$$-EA\mathcal{A}(u_x, \delta u_x) + \rho A \int_0^t \langle u_{x,t}, \delta u_{x,t} \rangle dt = A \int_0^t \langle F_x, \delta u_x \rangle dt.$$

We integrate by parts with respect to time considering that the completely integrated term is equal to zero (the displacements u_x and δu_x satisfy the temporal boundary conditions). We get $-E\mathcal{A}(u_x, \delta u_x) - \rho \int_0^t \langle u_{x,tt}, \delta u_x \rangle dt = \int_0^t \langle F_x, \delta u_x \rangle dt$. The weak formulation of the problem consists of searching for u_x which satisfies the boundary conditions of the problem so that for any displacement v_x which also satisfies these boundary conditions, we get:

$$-E\mathcal{A}(u_x, v_x) - \rho A \int_0^t \langle u_{x,tt}, v_x \rangle dt = \int_0^t \langle F_x, v_x \rangle dt. \qquad [6.28]$$

In a harmonic regime, we get $u_{x,tt} = -\omega^2 u_x$ this is:

$$-Ea(u_x, v_x) + \rho A \omega^2 \langle u_x, v_x \rangle = \langle F_x, v_x \rangle. \qquad [6.29]$$

$-Ea(u_x, v_x)$ is the stiffness term (this is the analogue to a spring in the spring-mass system) and $\rho A \langle u_x, v_x \rangle$ is the mass term.

6.4.1.2.3. Transverse vibrations (bending)

Here, we focus on the movements that occur along a normal to the neutral line, for example along z. The method remains the same, if we focus on the movements along y. We, therefore, impose that $u_x = 0$, $u_y = 0$ and $u_z \neq 0$. Displacement [6.20] is thus simplified as $U_x = -z u^0_{z,x}$, $U_y = 0$ and $U_z = 0$. Thus, for the potential energy density (to simplify u^0_z is denoted as u_z), we have $\mathcal{P}(\vec{U}) = E/2z^2 \left(u_{z,xx}\right)^2$. Similarly, the internal kinetic energy density is given by $\mathcal{T}(\vec{U}) = \rho / \left[z^2 \left(u_{z,xt}\right)^2 + \left(u_{z,t}\right)^2 \right]$. Let us appoint, for a beam with a rectangular section, $I = \int_S z^2 dS = eh^3/12$ the moment of inertia of the beam[4]. If the beam is subjected to a force with the density $\vec{F} = F_z$, Hamilton's theorem can be written as:

$$\delta \int_0^\ell \int_0^t \left[\frac{\rho}{2} \left[I \left(u_{z,xt}\right)^2 + A \left(u_{z,t}\right)^2 \right] - \frac{EI}{2} \left(u_{z,xx}\right)^2 \right] dx dt = \int_0^\ell \int_0^t AF_z \delta u_x dx dt,$$

before proceeding with the integration by parts, let us simplify the previous writing. For this, let us assume that $I \left(u_{z,xt}\right)^2 \ll A \left(u_{z,t}\right)^2$. For this assumption to be satisfied, as in the case of a thin beam $I \ll A$, it is necessary for the deformation velocity term $u_{z,xt}$ to not be too large compared to the velocity term $u_{z,t}$. We must note that in the case of beams, small transverse deformations can induce large displacements (such as a sailboat mast and a bended plate), thus justifying our assumption. We thus have:

$$\int_0^\ell \int_0^t \left[\rho A u_{z,t} \left(\delta u_z\right)_{,t} - EI u_{z,xx} \left(\delta u_z\right)_{,xx} \right] dx dt = \int_0^\ell \int_0^t AF_z \delta u_x dx dt,$$

integrating by parts, once with respect to t and twice with respect to x, we obtain:

$$\begin{aligned} &\int_0^\ell \int_0^t \left[-\rho A u_{z,tt} - EI u_{z,xxxx} \right] \delta u_z dx dt \\ &+ \int_0^\ell \left[\rho A u_{z,t} \delta u_z \right]_t dx + \int_0^t \left[-EI u_{z,xx} \delta u_{z,x} + EI u_{z,xxx} \delta u_z \right]_0^\ell dt \\ &= \int_0^\ell \int_0^t AF_z \delta u_x dx dt. \end{aligned} \qquad [6.30]$$

By canceling out the integrands of different integrals, we obtain:

$$-\rho A u_{z,tt} - EI u_{z,xxxx} = AF_z, \forall x \in]0, \ell[, \forall t, \qquad [6.31]$$

$$\rho A u_{z,t} = 0 \text{ and } u_z = 0, \forall x \in]0, \ell[, t = 0, \qquad [6.32]$$

$$EI u_{z,xx} = 0 \text{ or } u_{z,x} = 0 \text{ and } EI u_{z,xxx} = 0 \text{ and } u_z = 0, \text{ at } x = 0, \ell, \forall t. \qquad [6.33]$$

Equation [6.31] is the equation for the bending movement of thin beams, which is called Euler's equation, [6.32] gives the temporal boundary conditions and [6.33] gives the spatial boundary conditions. We must note that, of the four possible, only three boundary condition couples make sense: the clamped end, $u_z = 0$ and $u_{z,x} = 0$, the simply supported end, $u_z = 0$ and $EI u_{z,xx} = 0$ and finally the free end, $IE u_{z,xxx} = 0$ and $EI u_{z,xx} = 0$; the fourth couple of boundary conditions, $u_{z,x} = 0$ and $EI u_{z,xxx} = 0$, is called guided clamping in the literature and its physical significance is unclear [FIL 08].

4 The term "moment of inertia" is misleading because it is not a matter of mass; nevertheless, we will use it, as it is referred to in this way in the literature.

6.4.1.2.4. Weak formulation of the problem

As previously, we can define the variational formulation of the problem by introducing symmetric bilinear forms $a(u_z, v_z)$ and $\mathcal{A}(u_z, v_z)$ by:

$$a(u_z, v_z) = \int_0^\ell u_{z,xx} v_{z,xx} dx,$$
$$\mathcal{A}(u_z, v_z) = \int_0^t a(u_z, v_z) dt,$$

the weak formulation of the problem consists of searching for u_z which satisfies the boundary conditions of the problem so that for any displacement v_z that also satisfies these boundary conditions, we get:

$$-EI\mathcal{A}(u_z, v_z) - \rho A \int_0^t \langle u_{z,tt}, v_z \rangle dt = \int_0^t \langle F_z, v_z \rangle dt. \quad [6.34]$$

In a harmonic regime:

$$-EIa(u_z, v_z) + \rho A \omega^2 \langle u_z, v_z \rangle = \langle F_z, v_z \rangle. \quad [6.35]$$

Let us note that in the directions where the structure is thin, its behavior is controlled by an operator of the fourth order, while in the longitudinal direction we have an operator of the second order. We appoint $D_{y-z} = EI_{y-z}$, the stiffness modulus when bending the beam, $k_l^2 = \omega^2 \rho / E$ the longitudinal wave number, $k_{ty-z}^4 = \omega^2 \rho / D_{y-z}$ the transverse wave number. The three mouvement equations can be written as:

$$u_x'' + k_l^2 u_x = \frac{F_x}{E}, u_y^{(iv)} - k_{ty}^4 u_y = -\frac{AF_y}{EI_z}, u_z^{(iv)} - k_{tz}^4 u_z = -\frac{AF_z}{EI_y}.$$

In the first of these equations, we can recognize a Helmholtz equation. The following two are Euler equations (in a harmonic regime). Let us briefly examine the celerities of the corresponding waves. We know that the wave celerity in a medium, dispersive or not, is given by the relation $c = \partial \omega / \partial k$. In the present case, we have:

$$c_l = \sqrt{\frac{E}{\rho}}, c_{ty-z} = \sqrt{\frac{D_{y-z}}{\rho} \frac{\partial k_{ty-z}^2}{\partial k_{ty-z}}} = 2k_{ty-z} \sqrt{\frac{D_{y-z}}{\rho}}.$$

The celerity of longitudinal waves is analogous to that obtained for compression waves in an infinite medium, apart from Poisson's ratio. This is due to the fact that a beam is considered here as a one-dimensional medium, which is to say a medium which can extend but not contract during elongation, since its section is already zero. The celerity of transverse waves increases with frequency. This can be explained by the fact that the approximations, which we developed, are only valid at low frequencies.

6.4.2. *Plane plates*

Let us examine the problem of a thin plate (see Figure 6.5). A plate is considered thin when its thickness is sufficiently low compared to its other dimensions. Similarly, the deformations are assumed to be sufficiently small; in this case, the "smallness" criterion is given by the displacements of the plate, which must be small compared to the thickness of the plate. Let us make use of the usual Cartesian coordinates (O, x, y, z). The origin is centered at any point on the mean surface of the plate. Let us define x and y as the plane coordinates of the non-deformed plate. z is the coordinate normal to the plane of the plate. The thickness of the plate, small compared to the other dimensions, is defined by $z \in]-h/2, +h/2[$. Similarly as for the beam, the material which constitutes the plate is homogeneous and isotropic, and the stress–strain relation is given by Hooke's law.

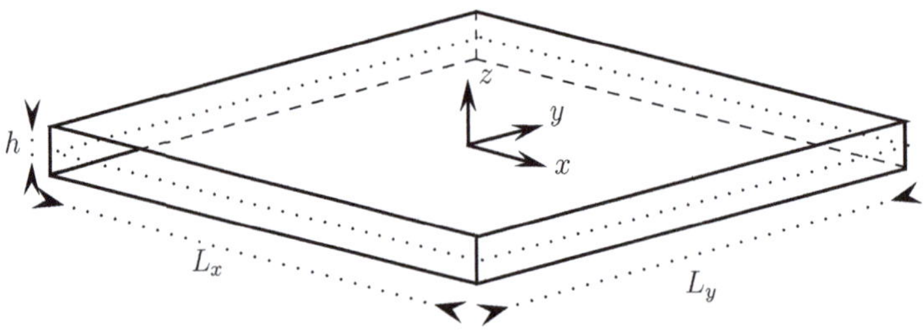

Figure 6.5. *The geometry of a plane plate*

6.4.2.1. *Displacement field*

Since the structure is thin along the z direction, we expand the displacement field $\vec{U}$ into a Taylor series with respect to thickness:

$$\vec{U}(x, y, z, t) = \vec{u}^0(x, y, t) + z\vec{u}^1(x, y, t)$$

$\vec{u}^0(x, y, t)$ is, of course, the value of the displacement field on the mean surface: $\vec{u}^0(x, y, t) = \vec{U}(x, y, 0, t)$. And we get $\vec{u}^1(x, y, t) = \vec{U}_{,z}(x, y, 0, t)$. For each component of the strain-stress tensor, we refer to the previous development. Once more, to simplify writing, the dependencies on x, y and t are omitted. We get:

$$\begin{aligned} d_{xx} = U_{x,x} = u^0_{x,x} + zu^1_{x,x}, d_{yy} = U_{y,y} = u^0_{y,y} + zu^1_{y,y}, d_{zz} = U_{z,z} = u^1_z \\ d_{xy} = \frac{1}{2}(U_{x,y} + U_{y,x}) = \frac{1}{2}\left(u^0_{x,y} + zu^1_{x,y} + u^0_{y,x} + zu^1_{y,x}\right) \\ d_{xz} = \frac{1}{2}(U_{x,z} + U_{z,x}) = \frac{1}{2}\left(u^1_x + u^0_{z,x} + zu^1_{z,x}\right) d_{yz} = \frac{1}{2}(U_{y,z} + U_{z,y}) \\ = \frac{1}{2}\left(u^1_y + u^0_{z,y} + zu^1_{z,y}\right) \end{aligned}$$

The free side hypothesis implies that the side faces of the plate are free from any constraints. Thus, for $z = \pm h/2$, we have $\vec{\vec{S}} \cdot \vec{n}_3 = 0$. This last equation is equivalent to $\sigma_{xz} = 0$, $\sigma_{yz} = 0$ and $\sigma_{zz} = 0$. Based on the stress–strain relations [2.17], from $\sigma_{xz} = 0$ and $\sigma_{yz} = 0$, we get $d_{xz} = 0$ and $d_{yz} = 0$. The strain–displacement relations [2.18] lead to $u_x^1 + u_{z,x}^0 + zu_{z,x}^1 = 0$ and $u_y^1 + u_{z,y}^0 + zu_{z,y}^1 = 0$. As these relations are true for $z = \pm h/2$, it is necessary for each z coefficient to be zero. We obtain $u_x^1 = -u_{z,x}^0$, $u_{z,x}^1 = 0$, $u_y^1 = -u_{z,y}^0$ and $u_{z,y}^1 = 0$. This is $u_x^1 = -u_{z,x}^0$ and $u_y^1 = -u_{z,y}^0$. It is then obvious that u_z^1 is constant, to determine this constant, we must impose the fourth Kirchhoff–Love assumption. It imposes that the straight section of the plate does not undergo deformation during movement. Physically, a "slice" of the material taken in the thickness of the plate locally behaves as an undeformable solid. The displacement in the z direction is thus constant. We thus obtain $u_z^1 = 0$. The displacement field (denoted by U_α) of the plate can then be written as:

$$U_x = u_x^0 - zu_{z,x}^0, U_y = u_y^0 - zu_{z,y}^0, U_z = u_z^0. \qquad [6.36]$$

Similarly as for the beams, the field is expressed in quantities (denoted by u_α^0) defined on the mean line. In all the following calculations, we remove the exponent "0".

6.4.2.2. *Plate operator*

Similarly as for beams, we apply Hamilton's theorem. We always have:

$$\mathcal{T}(\vec{U}) = \frac{1}{2}\int_0^t \int_\Omega \rho U_{\alpha,t} U_{\alpha,t} dV dt,$$

where Ω is the volume occupied by the plate. The elastic potential energy density can be written as:

$$\mathcal{P}(\vec{U}) = \frac{1}{2}\int_0^t \int_\Omega \sigma_{\alpha\beta} d_{\alpha\beta} dV dt.$$

Let us calculate the elastic potential energy density. We begin with $dP(\vec{U}) = 1/2\sigma_{\alpha\beta}d_{\alpha\beta}$. With the "free side" conditions which lead to $\sigma_{\alpha 3} = 0$, $\alpha = 1, 2, 3$, throughout the thickness of the plate, we obtain $dP(\vec{U}) = 1/2(\sigma_{xx}d_{xx} + \sigma_{yy}d_{yy} + 2\sigma_{xy}d_{xy})$. In addition to $\sigma_{zz} = 0$, we get $d_{zz} = -\nu/(1-\nu)(d_{xx} + d_{yy})$. We obtain $\sigma_{xx} = E/(1-\nu^2)(d_{xx} + \nu d_{yy})$, $\sigma_{yy} = E/(1-\nu^2)(d_{yy} + \nu d_{xx})$ and $\sigma_{xy} = E/(1+\nu)d_{xy}$. If we introduce these results in the internal potential energy expression, it can be written that $dP(\vec{U}) = E/(2(1-\nu^2)) - ((d_{xx} + d_{yy})^2 + 2(1-\nu)(d_{xy}^2 - d_{xx}d_{yy}))$. The various components of the strain tensor are given by relations [2.18]. We get $d_{xx} = u_{x,x} - zu_{z,xx}$, $d_{yy} = u_{y,y} - zu_{z,yy}$ and $d_{xy} = 1/2\left(u_{x,y} + u_{y,x} - 2zu_{z,xy}\right)$. This is:

$$\begin{aligned} dP(\vec{U}) = \tfrac{E}{2(1-\nu^2)} \Big\{ & \left(u_{x,x} - zu_{z,xx} + u_{y,y} - zu_{z,yy}\right)^2 \\ & +2(1-\nu)\left(\tfrac{1}{4}\left(u_{x,y} + u_{y,x} - 2zu_{z,xy}\right)^2 \right. \\ & \left. - \left(u_{x,x} - zu_{z,xx}\right)\left(u_{y,y} - zu_{z,yy}\right)\right)\Big\}. \end{aligned}$$

Similarly as for the beams, we need to integrate this energy density in the thickness of the plate. It is thus obvious that only the terms with an even power at z are contributing to the result. The internal potential energy density is thus composed of two terms:

$$
\begin{aligned}
dP(\vec{U}) \equiv & \tfrac{E}{2(1-\nu^2)} \left[\left(u_{x,x} + u_{y,y}\right)^2 + \tfrac{1-\nu}{2} \left(\left(u_{x,y} + u_{y,x}\right)^2 - 4u_{x,x}u_{y,y} \right) \right] \\
& + z^2 \tfrac{E}{2(1-\nu^2)} \left[\left(u_{z,xx} + u_{z,yy}\right)^2 + 2(1-\nu) \left(u_{z,xy}^2 - u_{z,xx}u_{z,yy}\right) \right].
\end{aligned}
$$

Now, we calculate the internal kinetic energy density. Similarly as for the potential energy density, after integration in thickness, only the coefficients with an even power at z contribute to the integral. We then have, based on the expression of the displacement field, $dT(\vec{U}) = \rho/2 \left(U_{1,t}^2 + U_{2,t}^2 + U_{3,t}^2\right]$. This is keeping only the even terms of z:

$$
dT(\vec{U}) \equiv \frac{1}{2}\rho \left[u_{x,t}^2 + u_{y,t}^2 + u_{z,t}^2 + z^2 \left(u_{z,xt}^2 + u_{z,yt}^2 \right) \right].
$$

We calculate the variation of potential energy integrated in the thickness of the plate. We get:

$$
\begin{aligned}
\delta \int_{\Omega} \int_0^t dP(\vec{U}) dV dt = & \delta \frac{Eh}{2\left(1-\nu^2\right)} \int_S \int_0^t \left(u_{x,x} + u_{y,y}\right)^2 \\
& + \frac{1-\nu}{2} \left(\left(u_{x,y} + u_{y,x}\right)^2 - 4u_{x,x}u_{y,y} \right) dSdt \\
& + \delta \frac{Eh^3}{24\left(1-\nu^2\right)} \int_S \int_0^t \left(u_{z,xx} + u_{z,yy}\right)^2 \\
& + 2(1-\nu) \left(u_{z,xy}^2 - u_{z,xx}u_{z,yy}\right) dSdt,
\end{aligned}
$$

where dS is the integration element on the surface of the plate. Let us consider separately the two integrals J_{P1} and J_{P2}:

$$
\begin{aligned}
J_{P1} = & \frac{Eh}{2\left(1-\nu^2\right)} \delta \int_S \int_0^t \left(u_{x,x} + u_{y,y}\right)^2 \\
& + \frac{1-\nu}{2} \left(\left(u_{x,y} + u_{y,x}\right)^2 - 4u_{x,x}u_{y,y} \right) dSdt \\
J_{P2} = & \frac{Eh^3}{24\left(1-\nu^2\right)} \delta \int_S \int_0^t \left(u_{z,xx} + u_{z,yy}\right)^2 \\
& + 2(1-\nu) \left(u_{z,xy}^2 - u_{z,xx}u_{z,yy}\right) dSdt.
\end{aligned}
$$

J_{P1} and J_{P2} represent the elastic potential energy density of compression and bending of the plate.

First, we examine the variation of J_{P1}. We obtain:

$$
\begin{aligned}
J_{P1} = \mathcal{D}\int_S\int_0^t & \left(u_{x,x}+u_{y,y}\right)\left(\delta u_{x,x}+\delta u_{y,y}\right) \\
& +\frac{1-\nu}{2}\left(u_{x,y}+u_{y,x}\right)\left(\delta u_{x,y}+\delta u_{y,x}\right) \\
& -\left(1-\nu\right)\left(\delta u_{x,x}u_{y,y}+u_{x,x}\delta u_{y,y}\right)dSdt.
\end{aligned}
$$

$\mathcal{D} = Eh/\left(1-\nu^2\right)$ is the rigidity of the plate membrane. This integral is composed of three terms $J_{P1} = \mathcal{D}\left(j_1+\frac{1-\nu}{2}j_2-\left(1-\nu\right)j_3\right)$, with $j_1 = \int_S\int_0^t\left(u_{x,x}+u_{y,y}\right)\left(\delta u_{x,x}+\delta u_{y,y}\right)dSdt$, $j_2 = \int_S\int_0^t\left(u_{x,y}+u_{y,x}\right)$ $\left(\delta u_{x,y}+\delta u_{y,x}\right)dSdt$ and $j_3 = \int_S\int_0^t\delta u_{x,x}u_{y,y}+u_{x,x}\delta u_{y,y}dSdt$. We introduce the vectors $\vec{A} = \left(u_x,u_y\right)$, $\vec{B} = \left(u_y,u_x\right)$, $\vec{C} = \left(u_{x,x}\delta u_x,u_{y,y}\delta u_y\right)$ and $\vec{D} = \left(u_{x,xx},u_{y,yy}\right)$. We get:

$$
\begin{aligned}
\left(u_{x,x}+u_{y,y}\right)\left(\delta u_{x,x}+\delta u_{y,y}\right) &= \text{div}\left(\vec{A}\right)\text{div}\left(\delta\vec{A}\right) \\
\left(u_{x,y}+u_{y,x}\right)\left(\delta u_{x,y}+\delta u_{y,x}\right) &= \text{div}\left(\vec{B}\right)\text{div}\left(\delta\vec{B}\right) \\
\delta u_{x,x}u_{y,y}+u_{x,x}\delta u_{y,y} &= \text{div}\left(\vec{A}\right)\text{div}\left(\delta\vec{A}\right)-\text{div}\left(\vec{C}\right)+\vec{D}\cdot\delta\vec{A}.
\end{aligned}
$$

We know that if p is a scalar and $\vec{q}$ is a vector $\text{div}(p\vec{q}) = p\text{div}(\vec{q})+\overrightarrow{\text{grad}}p\cdot\vec{q}$, then $\text{div}(\vec{A})\text{div}(\delta\vec{A}) = \text{div}\left(\text{div}(\vec{A})\delta\vec{A}\right)-\overrightarrow{\text{grad}}\left(\text{div}(\vec{A})\right)\cdot\delta\vec{A}$. We apply this result to the integrands of j_1, j_2 and j_3 and then we apply Ostrogradsky's theorem. We obtain:

$$
\begin{aligned}
j_1 &= \int_{\partial S}\int_0^t\text{div}(\vec{A})\delta\vec{A}\cdot\vec{n}\,dldt-\int_S\int_0^t\overrightarrow{\text{grad}}\left(\text{div}(\vec{A})\right)\cdot\delta\vec{A}dSdt, \\
j_2 &= \int_{\partial S}\int_0^t\text{div}(\vec{B})\delta\vec{B}\cdot\vec{n}\,dldt-\int_S\int_0^t\overrightarrow{\text{grad}}\left(\text{div}(\vec{B})\right)\cdot\delta\vec{B}dSdt, \\
j_3 &= \int_{\partial S}\int_0^t\text{div}(\vec{A})\delta\vec{A}\cdot\vec{n}-\vec{C}\cdot\vec{n}dldt-\int_S\int_0^t\overrightarrow{\text{grad}}\left(\text{div}(\vec{A})\right)\cdot\delta\vec{A}-\vec{D}\cdot\delta\vec{A}dSdt,
\end{aligned}
$$

where $\vec{n}$ is the vector normal to the edge of the plate, contained in the plate itself; ∂S is the edge of the plate. dl is the integration element defined on the edge of the plate.

Finally, if we group the results, this is:

$$J_{P1} = \mathcal{D}\left\{\int_{\partial S}\int_0^t \left[\nu\text{div}(\vec{A})\delta\vec{A} + \frac{1-\nu}{2}\text{div}(\vec{B})\delta\vec{B} + (1-\nu)\vec{C}\right]\cdot\vec{n}dldt\right.$$
$$-\int_S\int_0^t \nu\overrightarrow{\text{grad}}\left(\text{div}(\vec{A})\right)\cdot\delta\vec{A} + \frac{1-\nu}{2}\overrightarrow{\text{grad}}\left(\text{div}(\vec{B})\right)\cdot\delta\vec{B}]$$
$$\left.+(1-\nu)\vec{D}\cdot\delta\vec{A}dSdt\right\}.$$

Now, we examine the variation of the integral J_{P2}. We get:

$$J_{P2} = \frac{Eh^3}{24\left(1-\nu^2\right)}\delta\int_S\int_0^t \left(\Delta u_z\right)^2 + 2(1-\nu)\left(u_{z,xy}^2 - u_{z,xx}u_{z,yy}\right)dSdt.$$

where $\Delta u_z = u_{z,xx}+u_{z,yy}$ is the Laplacian. We appoint $D = \frac{Eh^3}{12(1-\nu^2)}$ as the bending stiffness coefficient of the plate. We get:

$$J_{P2} = D\int_S\int_0^t \Delta u_z\Delta\delta u_z + (1-\nu)$$
$$\times\left(2u_{z,xy}\delta u_{z,xy} - \delta u_{z,xx}u_{z,yy} - u_{z,xx}\delta u_{z,yy}\right)dSdt. \qquad [6.37]$$

We once again decompose this integral into two parts. We set $J_{P2} = Dj_1 + D(1-\nu)j_2$, where:

$$j_1 = \int_S\int_0^t \Delta u_z\Delta\delta u_z dSdt \text{ and}$$
$$j_2 = \int_S\int_0^t u_{z,xy}\delta u_{z,xy} - \delta u_{z,xx}u_{z,yy} - u_{z,xx}\delta u_{z,yy}dSdt. \qquad [6.38]$$

Let us consider j_1. We know that $\Delta u = \text{div}\left(\overrightarrow{\text{grad}}u\right)$. Thus, $j_1 = \int_S\int_0^t \Delta u_z$ $\text{div}\left(\overrightarrow{\text{grad}}\delta u_z\right)dSdt$. We apply the formula $\text{div}(p\vec{q}) = p\text{div}(\vec{q}) + \overrightarrow{\text{grad}}p\cdot\vec{q}$ twice: the first with $p = \Delta u_z$ and $\vec{q} = \overrightarrow{\text{grad}}\left(u_z\right)$ and the second with $p = \delta u_z$ and $\vec{q} = \overrightarrow{\text{grad}}\left(\Delta u_z\right)$. Therefore:

$$j_1 = \int_S\int_0^t \text{div}\left(\Delta u_z\overrightarrow{\text{grad}}\delta u_z - \overrightarrow{\text{grad}}\left(\Delta u_z\right)\delta u_z\right)dSdt + \int_S\int_0^t \Delta^2 u_z\delta u_z\, dSdt,$$

where Δ^2 is the bi-Laplacian and $\Delta^2 u_z = u_{z,xxxx} + 2u_{z,xxyy} + u_{z_y yyy}$. We apply Ostrogradsky's theorem to the first integral, with $\overrightarrow{\text{grad}} u \cdot \vec{n} = \partial_n u$, we obtain:

$$j_1 = \int_{\partial S} \int_0^t \Delta u_z \partial_n \delta u_z - \partial_n \Delta u_z \delta u_z \, dldt + \int_S \int_0^t \Delta^2 u_z \delta u_z \, dSdt.$$

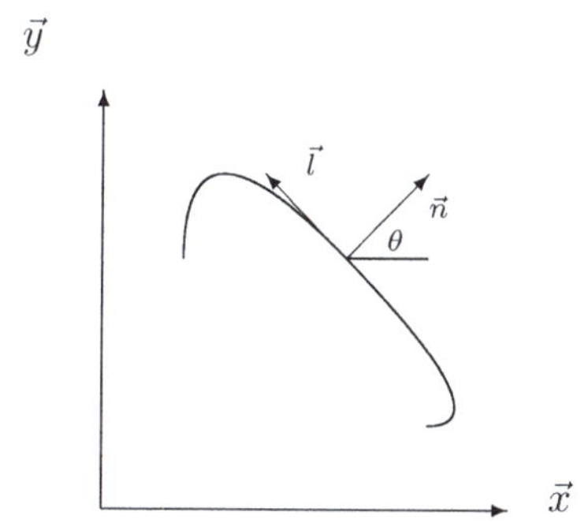

Figure 6.6. *Geometry of the plate edge*

Now, we consider j_2. The integrand is written as the divergence of a vector $\vec{\Psi}$, with the coordinates $\Psi_1 = u_{z,xy}\delta u_{z,y} - \delta u_{z,x} u_{z,yy}$ and $\Psi_2 = u_{z,xy}\delta u_{z,x} - \delta u_{z,y} u_{z,yy}$ which we apply to Ostrogradsky's theorem $j_2 = \int_{\partial S} \int_0^t \vec{\Psi} \cdot \vec{n} \, dldt$. In order for us to proceed, we must explain the scalar product $\vec{\Psi} \cdot \vec{n}$. We define (see Figure 6.6) $\vec{x}$ and $\vec{y}$ as the unit vectors of the x and y axes. Thus, the normal vector can be written as $\vec{n} = \cos\theta \vec{x} + \sin\theta \vec{y}$ and the tangent vector $\vec{l} = -\sin\theta\vec{x} + \cos\theta\vec{y}$. The normal derivative is thus $\partial_n = \cos\theta\partial_x + \sin\theta\partial_y$, while the tangent derivative is $\partial_l = -\sin\theta\partial_x + \cos\theta\partial_y$. It is thus easy to see that $\partial_x = \cos\theta\partial_n - \sin\theta\partial_l$ and that $\partial_y = \sin\theta\partial_n + \cos\theta\partial_l$. We show that $\vec{\Psi} \cdot \vec{n} = \cos\theta\Psi_1 + \sin\theta\Psi_2 = -\partial_n \delta u_z \partial_l^2 u_z + \partial_l \delta u_z \partial_n \partial_l u_z$. Hence, the expression $j_2 = \int_{\partial S} \int_0^t -\partial_n \delta u_z \partial_l^2 u_z + \partial_l \delta u_z \partial_n \partial_l u_z \, dldt$. We integrate by parts the second term of the second integral. We get:

$$\int_{\partial S} \int_0^t \partial_l \delta u_z \partial_n \partial_l u_z \, dldt = [\delta u_z \partial_n \partial_l u_z]_{\partial S} - \int_{\partial S} \int_0^t \delta u_z \partial_l \left(\partial_n \partial_l u_z\right) \, dldt.$$

Since the plate is of finite dimensions, its borders ∂S form a closed contour. It is obvious that we have $[\delta u_z \partial_n \partial_l u_z]_{\partial S} = 0$. We obtain:

$$j_2 = \int_{\partial S} \int_0^t -\partial_n \delta u_z \partial_l^2 u_z - \delta u_z \partial_l \partial_n \left(\partial_l u_z\right) \, dldt. \qquad [6.39]$$

We regroup the already calculated terms. We get:

$$\begin{aligned}J_{P2} = D &\int_S \int_0^t \Delta^2 u_z \delta u_z \, dSdt \\ &+ D \int_{\partial S} \int_0^t \left(\Delta u_z - (1-\nu)\partial_l^2 u_z \right) \partial_n \delta u_z \\ &- \left(\partial_n \Delta u_z + (1-\nu)\partial_l \left(\partial_n \partial_l u_z \right) \right) \delta u_z \, dldt.\end{aligned}$$

The potential energy variation can thus be written as:

$$\begin{aligned}\delta \int_\Omega \int_0^t \mathcal{P}(\vec{u}) \, dV dt = -\mathcal{D} &\int_S \int_0^t \nu \overrightarrow{\text{grad}} \left(\text{div}(\vec{A}) \right) \cdot \delta \vec{A} \\ &+ \frac{1-\nu}{2} \left\{ \overrightarrow{\text{grad}} \left(\text{div}(\vec{B}) \right) \cdot \delta \vec{B} + 2\vec{D} \cdot \delta \vec{A} \right\} dSdt \\ &+ D \int_S \int_0^t \Delta^2 u_z \delta u_z \, dSdt \\ &+ D \int_{\partial S} \int_0^t \left(\Delta u_z - (1-\nu)\partial_l^2 u_z \right) \partial_n \delta u_z \\ &- \left(\partial_n \Delta u_z + (1-\nu)\partial_l \left(\partial_n \partial_l u_z \right) \right) \delta u_z \, dldt \\ &+ \mathcal{D} \int_{\partial S} \int_0^t \left[\nu \text{div}(\vec{A}) \delta \vec{A} + \frac{1-\nu}{2} \text{div}(\vec{B}) \delta \vec{B} + (1-\nu)\vec{C} \right] \cdot \vec{n} \, dldt.\end{aligned}$$

It is not necessary to regroup the surface and border integrals, resulting from the effects of compression and bending, since they concern independent variables. We now calculate the kinetic energy variation of the plate. After integration in the thickness of the plate of the internal kinetic energy density, we get:

$$\begin{aligned}\delta \int_\Omega \int_0^t \mathcal{T}(\vec{u}) \, dV dt = &\frac{\delta}{2} \rho h \int_S \int_0^t u_{x,t}^2 + u_{y,t}^2 + u_{z,t}^2 \, dSdt \\ &+ \frac{\delta}{2} \rho h \chi \int_S \int_0^t u_{z,xt}^2 + u_{z,yt}^2 \, dSdt,\end{aligned}$$

where $\chi = h^2/12$. As previously, we set $j_1 = \frac{\delta}{2}\int_S \int_0^t u_{x,t}^2 + u_{y,t}^2 + u_{z,t}^2 \, dSdt$ and $j_2 = \frac{\delta}{2}\int_S \int_0^t u_{z,xt}^2 + u_{z,yt}^2 \, dSdt$. The calculation of j_1 is trivial, we obtain:

$$\begin{aligned} j_1 = &-\int_S \int_0^t u_{x,tt}\delta u_x + u_{y,tt}\delta u_y + u_{z,tt}\delta u_z \, dSdt \\ &+ \int_S \left[u_{x,t}\delta u_x + u_{y,t}\delta u_y + u_{z,t}\delta u_z\right]_0^t \, ds. \end{aligned}$$

The calculation of j_2 is carried out as previously:

$$\begin{aligned} j_2 &= \int_S \int_0^t u_{z,xt}\delta u_{z,xt} + u_{z,yt}\delta u_{z,yt} \, dSdt \\ &= \int_S \int_0^t \operatorname{div}\left(\vec{\Phi}\right) - \vec{Y} \cdot \delta\vec{A} \, dSdt, \end{aligned} \qquad [6.40]$$

where $\vec{\Phi} = (u_{z,xt}\delta u_{z,t}, u_{z,yt}\delta u_{z,t})$ and $\vec{Y} = (u_{z,xxt}, u_{z,yyt})$. We apply Ostrogradsky's theorem $j_2 = -\int_S \int_0^t \vec{y} \cdot \delta\vec{u} \, dSdt + \int_{\partial S} \int_0^t \vec{\Phi} \cdot \vec{n} \, dldt$. We get $\vec{\Phi} \cdot \vec{n} = \partial_n u_{z,t}\delta u_{z,t}$. Hence, if we introduce this result in the expression for j_2 and integrate by parts with respect to time:

$$j_2 = -\int_S \int_0^t \vec{y} \cdot \delta\vec{u} \, dSdt + \int_{\partial S} \left[\partial_n u_{z,t}\delta u_z\right]_0^t \, dl - \int_{\partial S} \int_0^t \partial_n u_{z,tt}\delta u_z \, dldt.$$

Finally, the kinetic energy variation can be written as:

$$\begin{aligned} \delta \int_\Omega \int_0^t dT(\vec{u}) \, dVdt = &-\rho h \int_S \int_0^t u_{x,tt}\delta u_x + u_{y,tt}\delta u_y + u_{z,tt}\delta u_z \\ &+\chi\vec{Y} \cdot \delta\vec{A} \, dSdt \\ &-\rho h\chi \int_{\partial S} \int_0^t \partial_n u_{z,tt}\delta u_z \, dldt \\ &+\rho h\chi \int_{\partial S} \left[\partial_n u_{z,t}\delta u_z\right]_0^t \, dl \\ &+\rho h \int_S \left[u_{x,t}\delta u_x + u_{y,t}\delta u_y + u_{z,t}\delta u_z\right]_0^t \, ds. \end{aligned}$$

Applying Hamilton's theorem, after grouping together the integrals of the same type, leads us to the following result:

$$
\begin{aligned}
0 = &-\int_S \int_0^t \left(\rho h \left(u_{x,tt}\delta u_x + u_{y,tt}\delta u_y + u_{z,tt}\delta u_z + \chi \vec{Y}\cdot\delta\vec{A}\right)\right.\\
&-\mathcal{D}\left\{\nu\overrightarrow{\text{grad}}\left(\text{div}(\vec{A})\right)\cdot\delta\vec{A} + \frac{1-\nu}{2}\overrightarrow{\text{grad}}\left(\text{div}(\vec{B})\right)\cdot\delta\vec{B} + (1-\nu)\vec{D}\cdot\delta\vec{A}\right\}\\
&\left.+D\Delta^2 u_z\delta u_z\right)\,dSdt\\
&-\int_{\partial S}\int_0^t \left(\rho h\chi\partial_n u_{z,tt}\delta u_z + D\left\{\left(\Delta u_z - (1-\nu)\partial_l^2 u_z\right)\partial_n\delta u_z\right.\right.\\
&\left.-\left(\partial_n\Delta u_z + (1-\nu)\partial_l\left(\partial_n\partial_l u_z\right)\right)\delta u_z\right\}\\
&\left.+\mathcal{D}\left[\nu\text{div}(\vec{A})\delta\vec{A} + \frac{1-\nu}{2}\text{div}(\vec{B})\delta\vec{B} + (1-\nu)\vec{C}\right]\cdot\vec{n}\right)\,dldt\\
&+\int_S \rho h\left[u_{x,t}\delta u_x + u_{y,t}\delta u_y + u_{z,t}\delta u_z\right]_0^t\,ds + \int_{\partial S}\rho h\chi\left[\partial_n u_{z,t}\delta u_z\right]_0^t\,dl.
\end{aligned}
$$

For this sum of integrals to be zero, each of them must be zero, since the integration areas are different. We thus obtain:

$$
\begin{aligned}
&\int_S\int_0^t \mathcal{D}\left\{\nu\overrightarrow{\text{grad}}\left(\text{div}(\vec{A})\right)\cdot\delta\vec{A} + \frac{1-\nu}{2}\overrightarrow{\text{grad}}\left(\text{div}(\vec{B})\right)\cdot\delta\vec{B} + (1-\nu)\vec{D}\cdot\delta\vec{A}\right\}\\
&-\rho h\left(u_{x,tt}\delta u_x + u_{y,tt}\delta u_y + u_{z,tt}\delta u_z + \chi\vec{Y}\cdot\delta\vec{A}\right) - D\Delta^2 u_z\delta u_z\,dSdt = 0.
\end{aligned}
$$

$$
\begin{aligned}
&\int_{\partial S}\int_0^t -\rho h\chi\partial_n u_{z,tt}\delta u_z - \mathcal{D}\left[\nu\text{div}(\vec{A})\delta\vec{A} + \frac{1-\nu}{2}\text{div}(\vec{B})\delta\vec{B} + (1-\nu)\vec{C}\right]\cdot\vec{n}\\
&-D\left\{\left(\Delta u_z - (1-\nu)\partial_l^2 u_z\right)\partial_n\delta u_z - \left(\partial_n\Delta u_z + (1-\nu)\partial_l\left(\partial_n\partial_l u_z\right)\right)\delta u_z\right\}\,dldt = 0.
\end{aligned}
$$

$$
\int_S \rho h\left[u_{x,t}\delta u_x + u_{y,t}\delta u_y + u_{z,t}\delta u_z\right]_0^t\,ds = 0.
$$

$$
\int_{\partial S}\rho h\chi\left[\partial_n u_{z,t}\delta u_z\right]_0^t\,dl = 0.
$$

The first equation gives us the movement equation of the plate, the second equation leads to the spatial boundary conditions, the third equation results in the temporal boundary conditions, while the last equation is both a spatial and temporal boundary condition. Since the virtual displacements are arbitrary, all the integrands are zero. Taking into consideration that the terms affected by the coefficient $\chi = h^2/12$ are negligible compared to the others, we obtain the relations:

– For S, $\forall t$:

$$\begin{aligned}&\mathcal{D}\left\{\nu\overrightarrow{\text{grad}}\left(\text{div}(\vec{A})\right)\cdot\delta\vec{A}+\tfrac{1-\nu}{2}\overrightarrow{\text{grad}}\left(\text{div}(\vec{B})\right)\cdot\delta\vec{B}+(1-\nu)\vec{D}\cdot\delta\vec{A}\right\}\\&\quad-\rho h\left(u_{x,tt}\delta u_x+u_{y,tt}\delta u_y+u_{z,tt}\delta u_z\right)-D\Delta^2u_z\delta u_z=0.\end{aligned}$$

– For ∂S, $\forall t$:

$$\begin{aligned}&-\mathcal{D}\left[\nu\text{div}(\vec{A})\delta\vec{A}+\tfrac{1-\nu}{2}\text{div}(\vec{B})\delta\vec{B}+(1-\nu)\vec{C}\right]\cdot\vec{n}\\&-D\left\{\left(\Delta u_z-(1-\nu)\partial_l^2u_z\right)\partial_n\delta u_z\right.\\&\left.-\left(\partial_n\Delta u_z+(1-\nu)\partial_l\left(\partial_n\partial_lu_z\right)\right)\delta u_z\right\}=0.\end{aligned}$$

– For S, at $t=0$:

$$\rho h\left(u_{x,t}\delta u_x+u_{y,t}\delta u_y+u_{z,t}\delta u_z\right)=0.$$

PLATE MOVEMENT EQUATIONS.– In Cartesian coordinates, we have:

$$\begin{aligned}\overrightarrow{\text{grad}}\left(\text{div}(\vec{A})\right)\cdot\delta\vec{A}&=\left(u_{x,xx}+u_{y,xy}\right)\delta u_x+\left(u_{x,xy}+u_{y,yy}\right)\delta u_y\\\overrightarrow{\text{grad}}\left(\text{div}(\vec{B})\right)\cdot\delta\vec{B}&=\left(u_{y,xx}+u_{x,xy}\right)\delta u_y+\left(u_{y,xy}+u_{x,yy}\right)\delta u_x\\\vec{D}\cdot\delta\vec{A}&=u_{x,xx}\delta u_x+u_{y,yy}\delta u_y.\end{aligned}$$

We introduce these results in the first equations and group the coefficients of the various virtual displacements. We thus get:

$$\begin{aligned}&\left(\mathcal{D}\left\{\nu\left(u_{x,xx}+u_{y,xy}\right)+\tfrac{1-\nu}{2}\left(u_{y,xy}+u_{x,yy}\right)+(1-\nu)u_{x,xx}\right\}-\rho hu_{x,tt}\right)\delta u_x\\&+\left(\mathcal{D}\left\{\nu\left(u_{x,xy}+u_{y,yy}\right)+\tfrac{1-\nu}{2}\left(u_{y,xx}+u_{x,xy}\right)+(1-\nu)u_{y,yy}\right\}-\rho hu_{y,tt}\right)\delta u_y\\&+\left(-D\Delta^2u_z-\rho hu_{z,tt}\right)\delta u_z=0.\end{aligned}$$

The virtual displacements δu_α, $\alpha=x,y,z$ are arbitrary. Any of their coefficients are, therefore, zero. We thus obtain the movement equations, if the plate is subjected to a force $\vec{F}=(f_x,f_y,-f_z)$:

$$\begin{aligned}\mathcal{D}\left\{u_{x,xx}+\tfrac{1+\nu}{2}u_{y,xy}+\tfrac{1-\nu}{2}u_{x,yy}\right\}-\rho hu_{x,tt}&=f_x\\\mathcal{D}\left\{u_{y,yy}+\tfrac{1+\nu}{2}u_{x,xy}+\tfrac{1-\nu}{2}u_{y,xx}\right\}-\rho hu_{y,tt}&=f_y\\D\Delta^2u_z+\rho hu_{z,tt}&=f_z.\end{aligned}\qquad [6.41]$$

We must note that, similarly as for the beams, in the direction where the plate is thin, the movement is governed by an operator of the fourth order, while in the plane of the plate, we always have an operator of the second degree. We must note that, if we take the impulse conservation equation, in which we neglect the components σ_{i3}, we find, without difficulty, the first two of the previous equations. For example, let us consider in the equations for plane – or longitudinal – movement of our plate that the strain depends only on x. We thus have $\mathcal{D}u_{x,xx}-\rho hu_{x,tt}=f_x$ and $\mathcal{D}(1-\nu)/2u_{y,xx}-\rho hu_{y,tt}=f_y$. The first equation characterizes the longitudinal wave propagation at a celerity of $c_l = \sqrt{E/\left(\rho\left(1-\nu^2\right)\right)}$ and the second equation characterizes the transverse waves which propagate at a celerity of $c_t = \sqrt{E/\left(2\rho(1+\nu)\right)}$. We must note that the transverse waves have a lower celerity than longitudinal waves, as in an infinite medium. It is interesting to note that the bending movement of the plate is completely decoupled from the latter plane movements. Usually, when we talk about the plate equation (or Sophie Germain's equation or Kirchhoff's equation), we must understand the equation of the bending movement of the plate $-D\Delta^2 u_z - \rho h u_{z,tt} = f_z$.

TEMPORAL BOUNDARY CONDITIONS.– The temporal boundary conditions are obvious, for all S, we get:

$$\begin{aligned} \rho h u_{x,t} &= 0 \text{ and } u_x = 0, \text{ at } t=0, \rho h u_{y,t} = 0 \text{ and } u_y = 0, \text{ at } t=0, \\ \rho h u_{z,t} &= 0 \text{ and } u_z = 0, \text{ at } t=0. \end{aligned} \quad [6.42]$$

SPATIAL BOUNDARY CONDITIONS.– Finally, let us clarify the spatial boundary conditions. By grouping the coefficients of various virtual displacements, we have:

$$\begin{aligned} &-\mathcal{D}\delta u_x\left[u_{x,x}\cos\theta + \nu u_{y,y}\cos\theta + \tfrac{1-\nu}{2}\sin\theta\left(u_{y,x}+u_{x,y}\right)\right] \\ &\quad -\mathcal{D}\delta u_y\left[u_{y,y}\sin\theta + \nu u_{x,x}\sin\theta + \tfrac{1-\nu}{2}\cos\theta\left(u_{y,x}+u_{x,y}\right)\right] \\ &\quad -D\left\{\left(\Delta u_z - (1-\nu)\partial_l^2 u_z\right)\partial_n\delta u_z - \left(\partial_n\Delta u_z + (1-\nu)\partial_l\left(\partial_n\partial_l u_z\right)\right)\delta u_z\right\} = 0. \end{aligned}$$

Thus, it is obvious that for ∂S, $\forall t$:

$$\begin{aligned} &-\mathcal{D}\delta u_x\left[u_{x,x}\cos\theta + \nu u_{y,y}\cos\theta + \tfrac{1-\nu}{2}\sin\theta\left(u_{y,x}+u_{x,y}\right)\right] = 0. \\ &-\mathcal{D}\delta u_y\left[u_{y,y}\sin\theta + \nu u_{x,x}\sin\theta + \tfrac{1-\nu}{2}\cos\theta\left(u_{y,x}+u_{x,y}\right)\right] = 0. \\ &-D\left\{\left(\Delta u_z - (1-\nu)\partial_l^2 u_z\right)\partial_n\delta u_z - \left(\partial_n\Delta u_z + (1-\nu)\partial_l\left(\partial_n\partial_l u_z\right)\right)\delta u_z\right\} = 0. \end{aligned}$$

The boundary conditions for the plane movements of the plate over ∂S, $\forall t$ can be given by:

$$\begin{aligned} -\mathcal{D}\left[u_{x,x}\cos\theta + \nu u_{y,y}\cos\theta + \tfrac{1-\nu}{2}\sin\theta\left(u_{y,x}+u_{x,y}\right)\right] &= 0 \text{ or } u_x = 0. \\ -\mathcal{D}\left[u_{y,y}\sin\theta + \nu u_{x,x}\sin\theta + \tfrac{1-\nu}{2}\cos\theta\left(u_{y,x}+u_{x,y}\right)\right] &= 0 \text{ or } u_y = 0. \end{aligned} \quad [6.43]$$

If the first condition is satisfied, the displacement of the plate is arbitrary, the plate is free. If the second condition is satisfied, the plate is attached to the edge – we then talk of a clamped plate. We must note that these equations are not expressed in local coordinates of the contour.

The boundary conditions for bending movements can be given by four possible pairs of boundary conditions:

$$\begin{aligned} -D\left(\Delta u_z - (1-\nu)\partial_l^2 u_z\right) &= 0 \text{ or } \partial_n u_z = 0 \text{ and} \\ -D\left(\partial_n\Delta u_z + (1-\nu)\partial_l\left(\partial_n\partial_l u_z\right)\right) &= 0 \text{ or } \partial_n u_z = 0. \end{aligned} \quad [6.44]$$

These boundary conditions, unlike the boundary conditions of longitudinal movements, are expressed in local coordinates at the contour that allows us to overcome the form of it. Among these four pairs of, mathematically acceptable, conditions, only three have physical significance which can be easily interpreted:

– Free plate

$$-D\left(\Delta u_z - (1-\nu)\partial_l^2 u_z\right) = 0 \text{ and}$$
$$-D\left(\partial_n \Delta u_z + (1-\nu)\partial_l\left(\partial_n \partial_l u_z\right)\right) = 0 \qquad [6.45]$$

– Supported plate

$$u_z = 0 \text{ and } -D\left(\Delta u_z - (1-\nu)\partial_l^2 u_z\right) = 0 \qquad [6.46]$$

– Clamped plate

$$u_z = 0 \text{ and } \partial_n u_z = 0 \qquad [6.47]$$

The pair of conditions $\partial_n u_z = 0$ and $-D\left(\partial_n \Delta u_z + (1-\nu)\partial_l\left(\partial_n \partial_l u_z\right)\right) = 0$, sometimes referred to in the literature as guided clamping, does not have a clear physical signification [FIL 08].

6.4.2.3. *Harmonic regime*

In a harmonic regime with a time dependence $\exp(-\imath\omega t)$, the temporal boundary conditions are equally satisfied. The spatial boundary conditions do not change and the equations of motion become:

$$\begin{aligned} \mathcal{D}\left\{u_{x,xx} + \tfrac{1+\nu}{2}u_{y,xy} + \tfrac{1-\nu}{2}u_{x,yy}\right\} + \omega^2\rho h u_x &= f_x \\ \mathcal{D}\left\{u_{y,yy} + \tfrac{1+\nu}{2}u_{x,xy} + \tfrac{1-\nu}{2}u_{y,xx}\right\} + \omega^2\rho h u_y &= f_y \\ D\Delta^2 u_z - \omega^2\rho h u_z &= f_z. \end{aligned} \qquad [6.48]$$

Similarly as for the transverse movements of a beam, the bending movements of an infinite plate occur with a wave celerity which depends on the frequency. In fact, we aim to find whether there exists a free monochromatic elastic bending wave with a pulsation ω, a wave vector $\vec{k} = (k_x, k_y)$ which can propagate in the plate in the absence of the second part. We aim to find a solution to the equation $D\Delta^2 u_z - \omega^2\rho h u_z = 0$ in the form of $u_z = \exp(\imath\vec{k}\cdot\vec{r} - \imath\omega t)$. We thus obtain the dispersion relation, which, as we remember, determines the variation of the frequency as a function of the wave vector $Dk^4 - \omega^2\rho h = 0$, where $k^4 = k_x^4 + 2k_x^2k_y^2 + k_y^4$. The dispersion relation can be written in the following form $\omega = k^2\sqrt{D/(\rho h)}$ or $k = \sqrt{\omega}\sqrt[4]{\rho h/D}$. From this relation, we can deduce the phase velocity $v = \omega/k = k\sqrt{D/(\rho h)} = \sqrt{\omega}\sqrt[4]{D/(\rho h)}$. The group velocity[5] of the bending waves depends on the frequency. It is given by $c = \partial\omega/\partial k$:

$$c = 2k\sqrt{\frac{D}{\rho h}} = 2\sqrt{\omega}\sqrt[4]{\frac{D}{\rho_p h}}. \qquad [6.49]$$

This shows that the thin plates are dispersive media, in which the velocity of wave propagation depends on the frequency. As we will see later, this has very important implications on everything that relates to radiation.

6.5. Thin body approximation for cylindrical structures

In this section, we present a theory on thin cylindrical shells. We will establish the operator expression that governs the dynamic behavior of cylindrical shells and rings. Due to the curvature, the operators (the square matrices 3×3 of differential operators), which govern this type of structures, are more complex than the plate or beam operators. In particular, the three components of displacement are all interrelated.

Once more, we expand the displacement field in a Taylor series of the first order in thickness, we assume that the sides of the structure are free from any constraints. This once again allows us to express all the kinematics of the medium from quantities defined on the mean surface. To this, we add the Kirchhoff–Love assumptions. To these assumptions, we add that the thickness must be small compared to the wave length (low-frequency hypothesis) and that the width must be small compared to the radius of the rings. In the case of cylindrical shells, the overall equilibrium which minimizes the elastic potential energy is maintained by the membrane effect and thus by the stiffness of the mean surface (which itself depends on the curvature of the shell). In addition to the resistance due to the material, the shells present a resistance due to their shape[6].

6.5.1. *Cylinder*

Several levels of approximations exist for this operator. In particular, we can show that the different operators can be expressed as a modification of the Donnell–Mushtari operator (the simplest one) by modified operator that depends on

5 We must recall that the group velocity corresponds to the velocity of the energy propagation.

6 Planar structures may be considered as curved structures with curvatures with an infinite radius; the membrane effect thus becomes negligible compared to the bending effect which depends only on the mechanical properties (Young's modulus and Poisson's ratio) and the local geometry, such as thickness.

the ratio $\chi = h^2/(12R^2) \ll 1$, where h is the thickness of the shell and R is its radius. We present here the two most common operators: the Donnell–Mushtari operator and Flügge's operator (one of the most complete). For a general theory on thin shells (ellipsoidal and spherical), we can turn to classic books by Leissa [LEI 93b] and Flügge [FLU 73].

6.5.1.1. *Displacement field*

Let us consider a cylindrical coordinate system defined on the mean surface of the shell (O, z, ϕ, r) (see Figure 6.7). In this system, the shell occupies the domain $-L < z < +L$, $0 \leq \phi < 2\pi$ and $-h/2 < r < +h/2$. In comparison with the usual cylindrical coordinate system, we have simply carried out the transformation $r \to R + r$.

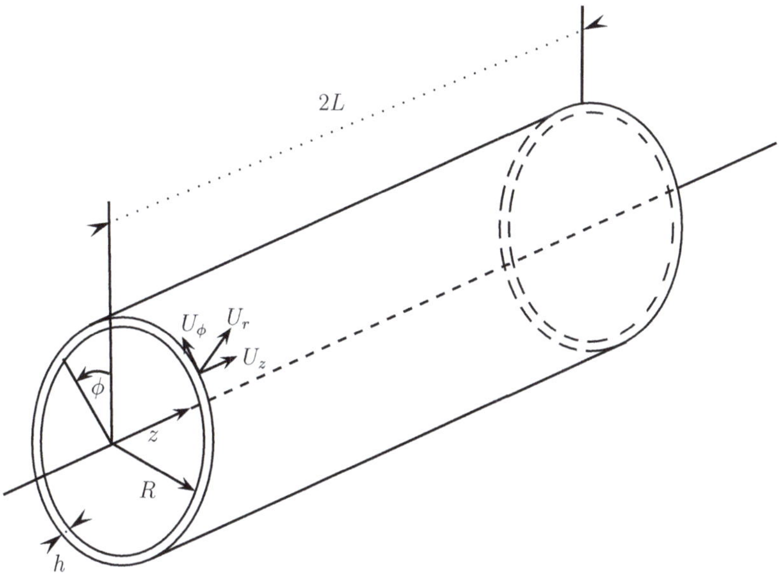

Figure 6.7. *The geometry of a thin cylindrical shell*

The three components of the displacement field of the shell $\vec{U}(z, \phi, r, t)$ are given by $U_z(z, \phi, r, t)$, $U_\phi(z, \phi, r, t)$ and $U_r(z, \phi, r, t)$. Since the structure is thin along the r direction, we can expand each component of this displacement field in a Taylor series around $r = 0$ at the first order: $U_\alpha(r, \phi, z, t) = u^0_\alpha(\phi, z, t) + r u^1_\alpha(\phi, z, t)$, $\alpha = r, \phi, z$. $u^0_\alpha(\phi, z, t)$ represents the displacement of the shell on the mean surface. We aim to express $u^1_\alpha(\phi, z, t)$ as a function of $u^0_\alpha(\phi, z, t)$. To do this, we will need the strain tensor, which is given in cylindrical coordinates (taking into account the change of origin on the neutral line) by relations [2.19]:

$$\begin{aligned}
&d_{zz} = U_{z,z}(z,\phi,r,t), d_{\phi\phi} = \tfrac{1}{R+r}\left(U_{\phi,\phi}(z,\phi,r,t) + U_r(z,\phi,r,t)\right)\\
&d_{rr} = U_{r,r}(z,\phi,r,t)\\
&d_{\phi,z} = d_{z,\phi} = \tfrac{1}{2}\left(U_{\phi,z}(z,\phi,r,t) + \tfrac{1}{R+r}U_{z,\phi}(z,\phi,r,t)\right)\\
&d_{rz} = d_{zr} = \tfrac{1}{2}\left(U_{r,z}(z,\phi,r,t) + U_{z,r}(z,\phi,r,t)\right)\\
&d_{r\phi} = d_{\phi r} = \tfrac{1}{2}\left(U_{\phi,r}(z,\phi,r,t) - \tfrac{1}{R+r}U_{\phi}(z,\phi,r,t) + \tfrac{1}{R+r}U_{r,\phi}(z,\phi,r,t)\right).
\end{aligned}$$

The free side assumption implies that all the lateral sides of the shell are free from any constraints. At $r = \pm h/2$, the stress tensor components[7] $\sigma_{\alpha r}(z,\phi,r,t) = 0$. From Hooke's law, given by [2.16], $\sigma_{\alpha\beta} = E/\left(2(1+\nu)\right)$ $(d_{\alpha\beta} + \nu/(1-2\nu)d_{\gamma\gamma}d_{\alpha\beta})$, we deduce that the following three relations must be verified for $r = \pm h/2$, $d_{rr} = -\nu/(1-\nu)(d_{zz} + d_{\phi\phi})$, $d_{rz} = 0$ and $d_{r\phi} = 0$. From the second relation $d_{rz} = 0$, we get $U_{r,z}(z,\phi,r,t) + U_{z,r}(z,\phi,r,t) = 0$ at $r = \pm h/2$ this is $u^0_{r,z}(z,\phi,t) + ru^1_{r,z}(z,\phi,t) + u^1_{z,r}(z,\phi,t) = 0, r = \pm h/2$, which is written as $u^0_{r,z}(z,\phi,t) + u^1_{z,r}(z,\phi,t) + \frac{h}{2}u^1_{r,z}(z,\phi,t) = 0$ and $u^0_{r,z}(z,\phi,t) + u^1_{z,r}(z,\phi,t) - \frac{h}{2}u^1_{r,z}(z,\phi,t) = 0$. These two relations are satisfied, if $u^0_{r,z}(z,\phi,t) + u^1_{z,r}(z,\phi,t) = 0$ and $u^1_{r,z}(z,\phi,t) = 0$. This is $u^1_{z,r}(z,\phi,t) = u^0_{r,z}(z,\phi,t)$ and $u^1_r(z,\phi,t) = f(\phi)$.

The relation $d_{r\phi} = 0$ can be written as $U_{\phi,r}(z,\phi,r,t) - 1/(R+r)$ $U_\phi(z,\phi,r,t) + 1/(R+r)U_{r,\phi}(z,\phi,r,t) = 0$, this is $u^1_\phi(z,\phi,t) - 1/(R+r)(u^0_\phi(z,\phi,t) + ru^1_\phi(z,\phi,t)) + 1/(R+r)(u^0_{r,\phi}(z,\phi,t) + ru^1_{r,\phi}(z,\phi,t)) = 0$. But $r \ll R$ this is $1/(R+r) \approx 1/R(1-r/R)$. We thus obtain, at the first order in r/R the relation $u^1_\phi(z,\phi,t) + (u^0_{r,\phi}(z,\phi,t) - u^0_\phi(z,\phi,t))/R + r/R(-1/R(u^0_{r,\phi}(z,\phi,t) - u^0_\phi(z,\phi,t)) + (u^1_{r,\phi}(z,\phi,t) - u^1_\phi(z,\phi,t))) = 0$ which is written at $r = \pm h/2$. As previously, we obtain two relations $u^1_\phi(z,\phi,t) + \left(u^0_{r,\phi}(z,\phi,t) - u^0_\phi(z,\phi,t)\right)/R = 0$ and $-1/R(u^0_{r,\phi}(z,\phi,t) - u^0_\phi(z,\phi,t)) + u^1_{r,\phi}(z,\phi,t) - u^1_\phi(z,\phi,t) = 0$; from there, it is easy to deduce that $u^1_{r,\phi}(z,\phi,t) = 0$, yet we have seen that $u^1_r(z,\phi,t) = f(\phi)$, therefore, $u^1_r(z,\phi,t)$ is constant. From the fourth Kirchhoff–Love assumption, we can deduce that, if a slice of the material transverse to the mean surface does not undergo deformation, then the normal displacement of this surface must be constant in the thickness and therefore that $u^1_r(z,\phi,t) = Cte = 0$. Finally, the three components at the first order of the displacement field are given by $u^1_r(z,\phi,t) = 0$, $u^1_z(z,\phi,t) = -u^0_{r,z}$ and

---

7 We must note that based on the second Kirchhoff–Love assumption, we can expand in a Taylor series around $r = 0$ each component $\sigma_{\alpha\beta}$. Thus, for $r = \pm h/2$, we have $\sigma_{\alpha\beta}(z,\phi,r,t) = \sigma^0_{\alpha r}(z,\phi,t) \pm h/2\sigma^1_{\alpha r}(z,\phi,t) = 0$ for $\alpha = z,\phi,r$. It is easy to see that we thus have $\sigma^0_{\alpha r}(z,\phi,t) = 0$ and $\sigma^1_{\alpha r}(z,\phi,t) = 0$ and that the three components $\sigma_{\alpha r}(z,\phi,r,t)$ are zero in the whole thickness apart from the factor r^2.

$u_\phi^1(z,\phi,t) = -1/R\left(u_{r,\phi}^0(z,\phi,t) - u_\phi^0(z,\phi,t)\right)$. As we have shown for thin plates, the displacement field of the shell, denoted by the Flügge displacement field $\vec{U}^F = (U_z^F, U_\phi^F, U_r^F)^t$, is expressed only in quantities defined on the mean surface. It is given by:

$$\begin{aligned} U_z^F(z,\phi,r,t) &= u_z^0(z,\phi,t) - ru_{r,z}^0(z,\phi,t) \\ U_\phi^F(z,\phi,r,t) &= u_\phi^0(z,\phi,t) - \tfrac{r}{R}\left(u_{r,\phi}^0(z,\phi,t) - u_\phi^0(z,\phi,t)\right) \\ U_r^F(z,\phi,r,t) &= u_r^0(z,\phi,t). \end{aligned} \quad [6.50]$$

We can simplify this displacement field a bit, if we consider that $r/R \ll 1$ and neglect $r/R\ u_\phi^0(z,\phi,t)$ compared to $u_\phi^0(z,\phi,t)$ in the second component of the Flügge displacement field[8]. We thus obtain the Donnell–Mushtari displacement field $\vec{U}^{DM} = (U_z^{DM}, U_\phi^{DM}, U_r^{DM})^t$ given by:

$$\begin{aligned} U_z^{DM}(z,\phi,r,t) &= u_z^0(z,\phi,t) - ru_{r,z}^0(z,\phi,t) \\ U_\phi^{DM}(z,\phi,r,t) &= u_\phi^0(z,\phi,t) - \tfrac{r}{R}u_{r,\phi}^0(z,\phi,t) \\ U_r^{DM}(z,\phi,r,t) &= u_r^0(z,\phi,t). \end{aligned} \quad [6.51]$$

In practice, we have $\vec{U}^F = \vec{U}^{DM} + \vec{U}^{mo}$, where $\vec{U}^{mo}$ is a modified displacement field given by:

$$\begin{aligned} U_z^{mo}(z,\phi,r,t) &= 0 \\ U_\phi^{mo}(z,\phi,r,t) &= \tfrac{r}{R}u_\phi^0(z,\phi,t) \\ U_r^{mo}(z,\phi,r,t) &= 0. \end{aligned} \quad [6.52]$$

In order to simplify writing, further on, when it does not lead to confusion, we denote $u_\alpha^0(z,\phi,t) = u_\alpha$ as $\alpha = z,\phi,r$.

6.5.1.2. *Thin shell operators*

As we have seen for the displacement field, the differences between the two operators of interest, here, are solely related to the refinements of the displacements and deformations. To establish the expressions for these operators, we once more apply Hamilton's theorem. Minimizing the Lagrangian action $L = T - P$ once more requires knowledge of kinetic $\mathcal{T} = \int_V dTdV$ and potential $\mathcal{P} = \int_V dPdV$ (elastic) energy of the shell, to which we must eventually add the external potential energy. It must be noted that the integration element in the modified cylindrical coordinate system is given by $dV = (R+r)dzd\phi dr$. The kinetic and potential energy densities are given by $dT = \rho/2\sum_\alpha u_{\alpha,t}^2$ and $dP = 1/2\sum_{\alpha\beta}\sigma_{\alpha\beta}d_{\alpha\beta}$, but we have seen that $\sigma_{\alpha r} = 0, \forall\alpha$ in the entire shell. Thus, the elastic potential energy comes down to

8 We must note that for the Donnell–Mushtari operator, in the expansions of the type $1 + r/R$, we retain only the terms of the zero order, whereas for Flügge's operator, we "push" the order of the expansion up by one, keeping the terms at the first order.

$dP = 1/2(\sigma_{zz}d_{zz} + \sigma_{\phi\phi}d_{\phi\phi} + \sigma_{z\phi}d_{z\phi})$. By Hooke's law, we deduce that $dP = 1/2E/(1-\nu^2)\left(d_{zz}^2 + d_{\phi\phi}^2 + 2\nu d_{zz}d_{\phi\phi} + 2(1-\nu)d_{z\phi}^2\right)$, which is an analogue result to the one obtained for plane plates.

6.5.1.3. *Elastic potential energy*

We assume that the shell is truncated perpendicularly to its generator at $z = \pm L$, the potential energy is given by the following integral:

$$\mathcal{P} = \frac{ER}{2(1-\nu^2)} \int_0^T \int_{\frac{-h}{2}}^{\frac{+h}{2}} \int_0^{2\pi} \int_{-L}^{+L} (d_{zz}^2 + d_{\phi\phi}^2 + 2\nu d_{zz}d_{\phi\phi} + 2(1-\nu)d_{z\phi}^2) \times \left(1 + \frac{r}{R}\right) dz d\phi dr dt.$$

We calculate d_{zz}, $d_{\phi\phi}$ and $d_{z\phi}$ for the Donnell–Mushtari and the Flügge displacement models. The idea is to express the components of the strain tensor for the Flügge model, as a correction of those derived from the Donnell–Mushtari model. By omitting the dependence on the space variables, for the Donnell–Mushtari model, we have:

$$\begin{aligned} d_{zz}^{DM} &= U_{z,z}^{DM} = u_{z,z} - r u_{r,zz} \\ d_{\phi\phi}^{DM} &= \tfrac{1}{R+r}\left(U_{\phi,\phi}^{DM} + U_r^{DM}\right) = \tfrac{1}{R}u_{\phi,\phi} - \tfrac{r}{R^2}u_{r,\phi\phi} + \tfrac{1}{R}u_r \\ d_{z\phi}^{DM} &= \tfrac{1}{2}\left(U_{\phi,z}^{DM} + \tfrac{1}{R+r}U_{z,\phi}^{DM}\right) = \tfrac{1}{2R}\left(u_{z,\phi} + RU_{\phi,z} - 2ru_{r,z\phi}\right). \end{aligned}$$

and for the Flügge model:

$$\begin{aligned} d_{zz}^{F} &= U_{z,z}^{F} = u_{z,z} - r u_{r,zz} \\ d_{\phi\phi}^{F} &= \tfrac{1}{R+r}\left(U_{\phi,\phi}^{F} + U_r^{F}\right) = \tfrac{1}{R}u_{\phi,\phi} - \tfrac{r}{R(R+r)}u_{r,\phi\phi} + \tfrac{1}{R+r}u_r \\ d_{z\phi}^{F} &= \tfrac{1}{2}\left(U_{\phi,z}^{F} + \tfrac{1}{R+r}U_{z,\phi}^{F}\right) = \tfrac{1}{2R}\left(\tfrac{1}{R+r}u_{z,\phi} + \tfrac{R+r}{R}u_{\phi,z} - \left(\tfrac{r}{R} + \tfrac{r}{R+r}\right)u_{r,z\phi}\right). \end{aligned}$$

Before calculating the various terms which compose the potential energies, we must note that in the elastic potential energy expression, the deformation terms appear either squared or multiplied two by two. And thus, even if we have only kept in the expansions in series the powers of r/R whose terms are of the first order, in the various expressions of the deformation terms, we need to do the expansions of the second order and to neglect the terms of the third order for r/R. We thus have $1/(R+r) \approx 1/R\left(1 - r/R + (r/R)^2\right)$, $(R+r)/R = 1 + r/R$, $r/(R(R+r)) \approx 1/R\left(r/R - (r/R)^2\right)$ and $r/R + r/(R+r) \approx 2r/R - (r/R)^2$. If we introduce these expansions in each term of the strain tensor and write $d^F_{\alpha\beta} = d^{DM}_{\alpha\beta} + d^{mo}_{\alpha\beta}$, it is easy to obtain:

$$\begin{aligned} d^F_{zz} &= u_{z,z} - r u_{r,zz} \\ d^F_{\phi\phi} &= \tfrac{1}{R} u_{\phi,\phi} - \tfrac{1}{R}\left(\tfrac{r}{R} - \left(\tfrac{r}{R}\right)^2\right) u_{r,\phi\phi} + \tfrac{1}{R}\left(1 - \tfrac{r}{R} + \left(\tfrac{r}{R}\right)^2\right) u_r \\ d^F_{z\phi} &= \tfrac{1}{2R}\left(\tfrac{1}{R}\left(1 - \tfrac{r}{R} + \left(\tfrac{r}{R}\right)^2\right) u_{z,\phi} + \left(1 + \tfrac{r}{R}\right) u_{\phi,z} \right. \\ &\quad \left. - \left(\tfrac{r}{R} + \tfrac{1}{R}\left(\tfrac{r}{R} - \left(\tfrac{r}{R}\right)^2\right)\right) u_{r,z\phi}\right). \end{aligned}$$

From this, we deduce:

$$\begin{aligned} d^{mo}_{zz} &= 0 \\ d^{mo}_{\phi\phi} &= \tfrac{1}{R}\tfrac{r}{R}\left(\left(-1 + \tfrac{r}{R}\right) u_r + \tfrac{r}{R} u_{r,\phi\phi}\right) \\ d^{mo}_{z\phi} &= \tfrac{1}{2}\tfrac{r}{R}\left(\tfrac{1}{R}\left(-1 + \tfrac{r}{R}\right) u_{z,\phi} + u_{\phi,z} \tfrac{r}{R} u_{r,z\phi}\right). \end{aligned}$$

We express the elastic potential energy. To do this, with $A \equiv F$ or $A \equiv DM$, we define:

$$\begin{aligned} dP^A &= \tfrac{E}{2(1-\nu^2)}[(d^A_{zz})^2 + (d^A_{\phi\phi})^2 + 2\nu d^A_{zz} d^A_{\phi\phi} + 2(1-\nu)(d^A_{z\phi})^2] \\ dP^{mo} &= \tfrac{E}{2(1-\nu^2)}[(d^{mo}_{\phi\phi})^2 + 2 d^{DM}_{\phi\phi} d^{mo}_{\phi\phi} + 2\nu d^{DM}_{zz} d^{mo}_{\phi\phi} + 2(1-\nu) \\ &\quad \times \left(2 d^{DM}_{z\phi} d^{mo}_{z\phi} + (d^{mo}_{z\phi})^2\right)]. \end{aligned}$$

We thus have simply $dP^F = dP^{DM} + dP^{mo}$ and thus, with $\mathcal{P}^F = \mathcal{P}^{DM} + \mathcal{P}^{mo}$, the elastic potential energies for the Donnell–Mushtari and Flügge models are given by:

$$\mathcal{P}^{DM} = R\int_0^T \int_{\frac{-h}{2}}^{\frac{+h}{2}} \int_0^{2\pi} \int_{-L}^{+L} dP^{DM} dz d\phi dr dt$$

$$\mathcal{P}^{F} = R\int_0^T \int_{\frac{-h}{2}}^{\frac{+h}{2}} \int_0^{2\pi} \int_{-L}^{+L} dP^{F}\left(1 + \frac{r}{R}\right) dz d\phi dr dt$$

$$\mathcal{P}^{mo} = R\int_0^T \int_{\frac{-h}{2}}^{\frac{+h}{2}} \int_0^{2\pi} \int_{-L}^{+L} \left(\frac{r}{R} dP^{DM} + dP^{mo}\left(1 + \frac{r}{R}\right)\right) dz d\phi dr dt.$$

We first calculate the elastic potential energy for the Donnell–Mushtari model. We know that $dP^{DM} = 1/2E/(1-\nu^2)(d^{DM}_{zz})^2 + (d^{DM}_{\phi\phi})^2 + 2\nu d^{DM}_{zz} d^{DM}_{\phi\phi} + 2(1-\nu)(d^{DM}_{z\phi})^2$. Since we need to integrate this expression according to the thickness between $-h/2$ and $h/2$, in the energy density expression we can keep only the even powers of r. We obtain[9]:

$$\begin{aligned}
&\left(d_{zz}^{DM}\right)^2 \equiv u_{z,z}^2 + r^2 u_{r,zz}^2, \left(d_{\phi\phi}^{DM}\right)^2 \equiv \tfrac{1}{R^2}\left(u_{\phi,\phi}+u_r\right)^2 + \tfrac{r^2}{R^4}u_{r,\phi\phi}^2 \\
&\left(d_{z\phi}^{DM}\right)^2 \equiv \tfrac{1}{4R^2}\left(\left(u_{z,\phi}+Ru_{\phi,z}\right)^2 + 4r^2 u_{r,z\phi}^2\right) \\
&d_{zz}^{DM} d_{\phi\phi}^{DM} \equiv \tfrac{1}{R}u_{z,z}\left(u_{\phi,\phi}+u_r\right) + \tfrac{r^2}{R^2}u_{r,\phi\phi}u_{r,zz}.
\end{aligned}$$

This is:

$$\begin{aligned}
dP^{DM} \equiv\ & \tfrac{E}{2(1-\nu^2)}\Big[u_{z,z}^2 + \tfrac{1}{R^2}\left(u_{\phi,\phi}+u_r\right)^2 + 2\nu\tfrac{1}{R}u_{z,z}\left(u_{\phi,\phi}+u_r\right) \\
& + \tfrac{1-\nu^2}{2R^2}\left(u_{z,\phi}+Ru_{\phi,z}\right)^2 \\
& + r^2\left(u_{r,zz}^2 + \tfrac{1}{R^4}u_{r,\phi\phi}^2 + \tfrac{2\nu}{R^2}u_{r,\phi\phi}u_{r,zz} + \tfrac{2(1-\nu)}{R^2}u_{r,z\phi}^2\right)\Big].
\end{aligned}$$

$\mathcal{D} = Eh/(1-\nu^2)$ is always the membrane stiffness; we define $\chi = h^2/12R^2$, and we obtain after integration with respect to r:

$$\begin{aligned}
\mathcal{P}^{DM} =\ & \frac{R}{2}\mathcal{D}\int_0^T\int_0^{2\pi}\int_{-L}^{+L}\Bigg[u_{z,z}^2 + \frac{1}{R^2}\left(u_{\phi,\phi}+u_r\right)^2 \\
& +2\nu\frac{1}{R}u_{z,z}\left(u_{\phi,\phi}+u_r\right) + \frac{1-\nu^2}{2R^2}\left(u_{z,\phi}+Ru_{\phi,z}\right)^2 \\
& +\chi\Bigg(\left(Ru_{r,zz}\right)^2 + \left(\frac{u_{r,\phi\phi}}{R}\right)^2 \\
& +2\nu u_{r,\phi\phi}u_{r,zz} + 2(1-\nu)u_{r,z\phi}^2\Bigg)\Bigg]\, dz d\phi dt.
\end{aligned}$$

We then calculate the elastic potential energy of Flügge's shell model, by calculating the term that modifies the elastic potential energy of the Donnell–Mushtari shell model. We get $\mathcal{P}^{mo} = R\int_{-\frac{h}{2}}^{\frac{h}{2}}\int_0^{2\pi}\int_{-L}^{+L}$ $(r/RdP^{DM} + (1+r/R)dP^{mo})dzd\phi dr$. We must, therefore, first calculate the terms with even power in r for $r/RdP^{DM} + (1+r/R)dP^{mo}$. The coefficients of r/RdP^{DM} are (here again, for the four terms, there is not a strict equality):

$$\begin{aligned}
&\tfrac{r}{R}\left(d_{zz}^{DM}\right)^2 \equiv -2\tfrac{r^2}{R}u_{z,z}u_{r,zz}, \tfrac{r}{R}\left(d_{\phi\phi}^{DM}\right)^2 \equiv -\tfrac{2}{R^2}\tfrac{r^2}{R^2}\left(u_{\phi,\phi}+u_r\right)u_{r,\phi\phi} \\
&\tfrac{r}{R}\left(d_{z\phi}^{DM}\right)^2 \equiv -\tfrac{1}{R}\tfrac{r^2}{R^2}u_{r,z\phi}\left(u_{z,\phi}+Ru_{\phi,z}\right) \\
&\tfrac{r}{R}d_{zz}^{DM}d_{\phi\phi}^{DM} \equiv \tfrac{r^2}{R^2}\left(-\tfrac{1}{R}u_{z,z}u_{r,\phi\phi} - u_{r,zz}(u_{\phi,\phi}+u_r)\right).
\end{aligned}$$

9 There is no equality, since, to simplify the writing, which is already quite complex, we have stopped writing r of odd powers, whose contribution to the integral is zero.

This is:

$$\frac{r}{R}dP^{DM} \equiv \frac{E}{2(1-\nu^2)}\frac{r^2}{R^2}\left[-2Ru_{z,z}u_{r,zz} - \frac{2}{R}\left(u_{\phi,\phi}+u_r\right)u_{r,\phi,\phi}\right.$$
$$+2\nu\left(-\frac{1}{R}u_{z,z}u_{r,\phi\phi} - u_{r,zz}\left(u_{\phi,\phi}+u_r\right)\right)$$
$$\left.+2(1-\nu)\left(-\frac{1}{R}u_{r,z\phi}\left(u_{z,\phi}+Ru_{\phi,z}\right)\right)\right].$$

$$\left(1+\frac{r}{R}\right)\left(d_{\phi\phi}^{mo}\right)^2 \equiv \frac{1}{R^2}\frac{r^2}{R^2}u_r^2, \left(1+\frac{r}{R}\right)d_{\phi\phi}^{DM}d_{\phi\phi}^{mo}$$
$$\equiv \frac{1}{R}\frac{r^2}{R^2}\left(\frac{1}{R}u_{\phi,\phi}u_{r,\phi\phi} + \frac{2}{R}u_r u_{r,\phi\phi}\right)$$
$$\left(1+\frac{r}{R}\right)d_{zz}^{DM}d_{\phi\phi}^{mo} \equiv \frac{r^2}{R^2}\left(\frac{1}{R}u_{z,z}u_{r,\phi\phi} + u_r u_{r,zz}\right), \left(1+\frac{r}{R}\right)\left(d_{z\phi}^{mo}\right)^2$$
$$\equiv \frac{r^2}{4R^2}\left(\frac{1}{R^2}u_{z,\phi}^2 + u_{\phi,z}^2 - \frac{2}{R}u_{z,\phi}u_{\phi,z}\right)$$
$$2\left(1+\frac{r}{R}\right)d_{z\phi}^{DM}d_{z\phi}^{mo} \equiv \frac{1}{2R}\frac{r^2}{R^2}\left(u_{z,\phi}u_{\phi,z} + 3u_{z,\phi}u_{r,z\phi} - Ru_{\phi,z}u_{r,z\phi} + Ru_{\phi,z}^2\right)$$

This is:

$$(1+r/R)dP^{mo} \equiv \frac{E}{2(1-\nu^2)}\frac{r^2}{R^2}\left[\frac{1}{R^2}u_r^2 + \frac{2}{R}\left(\frac{1}{R}u_{\phi,\phi}u_{r,\phi\phi} + \frac{2}{R}u_r u_{r,\phi\phi}\right)\right.$$
$$+2\nu\left(\frac{1}{R}u_{z,z}u_{r,\phi\phi} + u_r u_{r,zz}\right)$$
$$\left.+\frac{1-\nu}{2}\left(\frac{3}{R^2}u_{z,\phi}^2 + u_{\phi,z}^2\right) + (1-\nu)\left(\frac{3}{R}u_{z,\phi}u_{r,z\phi} - u_{\phi,z}u_{r,z\phi}\right)\right].$$

We group the terms of the expression $r/RdP^{DM} + (1+r/R)dP^{mo}$. After simplifying and integrating with respect to thickness, we finally obtain:

$$\mathcal{P}^{mo} = \frac{R}{2}\mathcal{D}\chi\int_0^T\int_0^{2\pi}\int_{-L}^{+L}\left[\frac{1-\nu}{2}\left(\frac{u_{z,\phi}}{R}\right)^2 + (1-\nu)\frac{u_{z,\phi}}{R}u_{r,z\phi}\right.$$
$$-2Ru_{z,z}u_{r,zz} + \frac{3(1-\nu)}{2}u_{\phi,z}^2$$
$$\left.-3(1-\nu)u_{\phi,z}u_{r,z\phi} - 2\nu u_{\phi,\phi}u_{r,zz} + \left(\frac{u_r}{R}\right)^2 + \frac{2}{R^2}u_r u_{r,\phi\phi}\right]dzd\phi dt.$$

6.5.1.4. *Kinetic energy*

The shell is always truncated perpendicularly to its generator at $z = \pm L$. We have the following expression for kinetic energy:

$$\mathcal{T} = \frac{\rho R}{2} \int_0^T \int_{\frac{-h}{2}}^{\frac{+h}{2}} \int_0^{2\pi} \int_{-L}^{+L} \left(U_{z,t}^2 + U_{\phi,t}^2 + U_{r,t}^2\right) \left(1 + \frac{r}{R}\right) dz d\phi dr dt.$$

We use the definitions of each displacement field and we obtain:

$$\begin{aligned}\mathcal{T}^{DM} &= \frac{\rho R}{2} \int_0^T \int_{\frac{-h}{2}}^{\frac{+h}{2}} \int_0^{2\pi} \int_{-L}^{+L} \Big[\left(u_{z,t} - r u_{r,zt}\right)^2 + \left(u_{\phi,t} - \frac{r}{R} u_{r,\phi t}\right)^2 \\ &\quad + \left(u_{r,t}\right)^2\Big] dz d\phi dr dt, \\ \mathcal{T}^F &= \frac{\rho R}{2} \int_0^T \int_{\frac{-h}{2}}^{\frac{+h}{2}} \int_0^{2\pi} \int_{-L}^{+L} \Big[\left(u_{z,t} - r u_{r,zt}\right)^2 + \left(u_{\phi,t} - \frac{r}{R}\left(-u_{\phi,t} + u_{r,\phi t}\right)\right)^2 \\ &\quad + \left(u_{r,t}\right)^2 \left(1 + \frac{r}{R}\right)\Big] dz d\phi dr dt.\end{aligned}$$

After expanding each expression and integrating according to the thickness, we get:

$$\begin{aligned}\mathcal{T}^{DM} &= \frac{\rho h R}{2} \int_0^T \int_0^{2\pi} \int_{-L}^{+L} \left[\left(u_{z,t}^2 + u_{\phi,t}^2 + u_{r,t}^2\right) + \chi\left(R^2 u_{r,zt}^2 + u_{r,\phi t}^2\right)\right] dz d\phi dt, \\ \mathcal{T}^F &= \frac{\rho h R}{2} \int_0^T \int_0^{2\pi} \int_{-L}^{+L} \Big[\left(u_{z,t}^2 + u_{\phi,t}^2 + u_{r,t}^2\right) + \chi\Big(R^2 u_{r,zt}^2 + \left(u_{r,\phi t} - u_{\phi,t}\right)^2 \\ &\quad -2R u_{z,t} u_{r,zt} - 2u_{\phi,t}(u_{r,\phi t} - u_{\phi,t})\Big)\Big] dz d\phi dt.\end{aligned}$$

We assume here that the kinetic energies of rotation and shear (all terms affected by the χ coefficient) are negligible. And we obtain:

$$\mathcal{T}^{DM} = \mathcal{T}^F = \frac{\rho h R}{2} \int_0^T \int_0^{2\pi} \int_{-L}^{+L} \left(u_{z,t}^2 + u_{\phi,t}^2 + u_{r,t}^2\right) dz d\phi dt \qquad [6.53]$$

6.5.1.5. *Variational equations: operators*

To establish the Donnell–Mushtari and Flügge's thin shell operators, we calculate the variation of potential and kinetic energies. $\delta\mathcal{P}^{DM}$ can be written by factoring in the virtual displacements and their derivatives:

$$\begin{aligned}\delta\mathcal{P}^{DM} &= R\mathcal{D}\int_0^T\int_0^{2\pi}\int_{-L}^{+L}\Big[\Big(u_{z,z}+\frac{\nu}{R}\left(u_{\phi,\phi}+u_r\right)\Big)\left(\delta u_z\right)_{,z}\\ &+\frac{1-\nu}{2R}\left(u_{z,\phi}+Ru_{\phi,z}\right)\left(\delta u_\phi\right)_{,z}\\ &+\left(\frac{1}{R^2}\left(u_{\phi,\phi}+u_r\right)+\frac{\nu}{R}u_{z,z}\right)\left(\delta u_\phi\right)_{,\phi}+\frac{1-\nu}{2R^2}\left(u_{z,\phi}+Ru_{\phi,z}\right)\left(\delta u_z\right)_{,\phi}\end{aligned}$$

$$\begin{aligned}&+\left(\frac{1}{R^2}\left(u_{\phi,\phi}+u_r\right)+\frac{\nu}{R}u_{z,z}\right)\delta u_r+\chi\Big\{\left(R^2u_{r,zz}+\nu u_{r,\phi\phi}\right)\left(\delta u_r\right)_{,zz}\\ &+\left(\frac{1}{R^2}u_{r,\phi\phi}+\nu u_{r,zz}\right)\left(\delta u_r\right)_{,\phi\phi}+2(1-\nu)u_{r,z\phi}\left(\delta u_r\right)_{,z\phi}\Big\}\Big]\,dzd\phi dt.\end{aligned}$$

We integrate by parts with respect to z and ϕ one or two times (the multiplicative terms of χ) to obtain a term integrated over the volume of the shell, which depends only on the virtual displacements. As the shell is truncated at $z=\pm L$ and $\phi\in[0,2\pi[$, the part integrated over ϕ will be zero. Only the terms integrated over z (which will lead to the boundary conditions) will remain. We obtain:

$$\begin{aligned}\delta\mathcal{P}^{DM} &= -R\mathcal{D}\int_0^T\int_0^{2\pi}\int_{-L}^{+L}\Big[\Big(u_{z,zz}+\frac{\nu}{R}\left(u_{\phi,z\phi}+u_{r,z}\right)\\ &+\frac{1-\nu}{2R^2}\left(u_{z,\phi\phi}+Ru_{\phi,z\phi}\right)\Big)\delta u_z\\ &+\left(\frac{1}{R^2}\left(u_{\phi,\phi\phi}+u_{r,\phi}\right)+\frac{\nu}{R}u_{z,z\phi}+\frac{1-\nu}{2R}\left(u_{z,\phi z}+Ru_{\phi,zz}\right)\right)\delta u_\phi\\ &-\Big(\frac{1}{R^2}\left(u_{\phi,\phi}+u_r\right)+\frac{\nu}{R}u_{z,z}\\ &+\chi\left(R^2u_{r,zzzz}+2u_{r,zz\phi\phi}\frac{1}{R^2}u_{r,\phi\phi\phi\phi}\right)\Big)\delta u_r\Big]\,dzd\phi dt\\ &+R\mathcal{D}\int_0^T\int_0^{2\pi}\Big[\Big(u_{z,z}+\frac{\nu}{R}\left(u_{\phi,\phi}+u_r\right)\Big)\delta u_z+\frac{1-\nu}{2R}\left(u_{z,\phi}+Ru_{\phi,z}\right)\delta u_\phi\end{aligned}$$

$$+\chi\left(-R^2u_{r,zzz}-(2-\nu)u_{r,\phi\phi z}\right)\delta u_r$$
$$+\chi\left(R^2u_{r,zz}+\nu u_{r,\phi\phi}\right)(\delta u_r)_{,z}\Big]_{z=-L}^{z=+L}d\phi dt.$$

We proceed in the same way for the variation of $\mathcal{P}^{mo}$. We get:

$$\delta\mathcal{P}^{mo}=R\mathcal{D}\chi\int_0^T\int_0^{2\pi}\int_L^{+L}\left[\frac{1-\nu}{2R^2}u_{z,\phi}(\delta u_z)_{,\phi}+\frac{1-\nu}{2R}\left(u_{z,\phi}(\delta u_r)_{,z\phi}\right.\right.$$
$$\left.+\,u_{r,z\phi}(\delta u_z)_{,\phi}\right)$$
$$-R\left(u_{z,z}(\delta u_r)_{,zz}+u_{z,zz}(\delta u_r)_{,z}\right)+\frac{3(1-\nu)}{2}u_{\phi,z}(\delta u_\phi)_{,z}$$
$$-\frac{3(1-\nu)}{2}\left(u_{\phi,z}(\delta u_r)_{,z\phi}+u_{r,z\phi}(\delta u_\phi)_{,z}\right)-\nu\left(u_{\phi,\phi}(\delta u_r)_{,zz}\right.$$
$$\left.\left.+u_{r,zz}(\delta u_\phi)_{,\phi}\right)+\frac{1}{R^2}u_r\delta u_r+\frac{1}{R^2}\left(u_r(\delta u_r)_{,\phi\phi}+u_{r,\phi\phi}\delta u_r\right)\right]dzd\phi dt.$$

After integrating by parts, we obtain:

$$\delta\mathcal{P}^{mo}=R\mathcal{D}\chi\int_0^T\int_0^{2\pi}\int_L^{+L}-\left[\left(\frac{1-\nu}{2R^2}u_{z,\phi\phi}+\frac{1-\nu}{2R}u_{r,z\phi\phi}-Ru_{r,zzz}\right)\delta u_z\right.$$
$$+\left(3\frac{1-\nu}{2}u_{\phi,zz}-\frac{3-\nu}{2}u_{r,zz\phi}\right)\delta u_\phi+\left(-\frac{u_r}{R^2}-\frac{2}{R^2}u_{r,\phi\phi}\right.$$
$$\left.\left.+Ru_{z,zzz}+\frac{3-\nu}{2}u_{\phi,zz\phi}-\frac{1-\nu}{2R}u_{z,z\phi\phi}\right)\delta u_r\right]dzd\phi dt$$
$$+R\mathcal{D}\chi\int_0^T\int_0^{2\pi}\left[-Ru_{r,zz}\delta u_z+\left(3\frac{1-\nu}{2}u_{\phi,z}-3\frac{1-\nu}{2}u_{r,z\phi}\right)\delta u_\phi\right.$$
$$+\left(-\frac{1-\nu}{2R}u_{z,\phi\phi}+\frac{3(1-\nu)}{2}u_{\phi,z\phi}+Ru_{z,zz}+\nu u_{\phi,z\phi}\right)\delta u_r$$
$$-\left(Ru_{z,z}+\nu u_{\phi,\phi}\right)(\delta u_r)_{,z}\Big]_{z=-L}^{z=+L}d\phi dt.$$

Similarly, it is easy to show that the kinetic energy variation can be given by:

$$\delta\mathcal{T}^{DM/F}=-\rho hR\int_0^T\int_0^{2\pi}\int_L^{+L}\left(u_{z,tt}\delta u_z+u_{\phi,tt}\delta u_\phi+u_{r,tt}\delta u_r\right)dzd\phi dt$$
$$+\rho hR\int_0^{2\pi}\int_L^{+L}\left[u_{z,t}\delta u_z+u_{\phi,t}\delta u_\phi+u_{r,t}\delta u_r\right]_0^T dzd\phi \qquad [6.54]$$

Hamilton's theorem states that among the paths allowed by the system, subject to restrictive conditions $\delta\vec{U}(t=0)=\delta\vec{U}(t=T)=0$ at the ends of the time interval, the real trajectory is the one that minimizes the Lagrangian action $\mathcal{L}=\mathcal{T}-\mathcal{U}$. $\mathcal{U}$. The potential energy is composed of two terms: the internal potential energy $\mathcal{P}$ which we have already calculated and the external potential energy $\mathcal{P}_e$ which only depends on the position and not the strain. We have $\delta\mathcal{P}_e=-\int_0^T\int_V\vec{G}\cdot\delta\vec{U}dVdt-\int_0^T\int_S\vec{F}\cdot\delta\vec{U}dSdt$, where V is the volume occupied by the body and S is its surface. $\vec{G}$ represents the volume forces and $\vec{f}=(F_z,F_\phi F_r)^t$ represents the surface forces (equivalent to a pressure), $\int_S\vec{f}\cdot\delta\vec{U}dS$ is the work carried out by external forces. In the absence of volume forces and if the edges of the shell are subjected to conditions such that the integrals of the edges are at $z=\pm L$, for the Donnell–Mushtari displacement field, we obtain:

$$
\begin{aligned}
&-R\mathcal{D}\int_0^T\int_0^{2\pi}\int_L^{+L}\left[\left(u_{z,zz}+\frac{\nu}{R}\left(u_{\phi,z\phi}+u_{r,z}\right)\right.\right.\\
&+\left.\frac{1-\nu}{2R^2}\left(u_{z,\phi\phi}+Ru_{\phi,z\phi}\right)+\frac{\rho h}{\mathcal{D}}u_{z,tt}\right)\delta u_z\\
&+\left(\frac{1}{R^2}\left(u_{\phi,\phi\phi}+u_{r,\phi}\right)+\frac{\nu}{R}u_{z,z\phi}+\frac{1-\nu}{2R}\left(u_{z,\phi z}+Ru_{\phi,zz}\right)+\frac{\rho h}{\mathcal{D}}u_{\phi,tt}\right)\delta u_\phi\\
&-\left(\frac{1}{R^2}\left(u_{\phi,\phi}+u_r\right)+\frac{\nu}{R}u_{z,z}\right.\\
&\left.\left.+\chi\left(R^2u_{r,zzzz}+2u_{r,zz\phi\phi}+\frac{1}{R^2}u_{r,\phi\phi\phi\phi}\right)-\frac{\rho h}{\mathcal{D}}u_{r,tt}\right)\delta u_r\right]dzd\phi dt\\
&=R\int_0^T\int_0^{2\pi}\int_L^{+L}\left(F_z\delta u_z+F_\phi\delta u_\phi+F_r\delta u_r\right)dzd\phi dt.
\end{aligned}
$$

As this relation is true for any virtual displacement $\vec{U}$, we must have:

$$
\begin{aligned}
&-\mathcal{D}\left[u_{z,zz}+\frac{\nu}{R}\left(u_{\phi,z\phi}+u_{r,z}\right)+\frac{1-\nu}{2R^2}\left(u_{z,\phi\phi}+Ru_{\phi,z\phi}\right)\right]+\rho h u_{z,tt}=F_z\\
&-\mathcal{D}\left[\frac{1}{R^2}\left(u_{\phi,\phi\phi}+u_{r,\phi}\right)+\frac{\nu}{R}u_{z,z\phi}+\frac{1-\nu}{2R}\left(u_{z,\phi z}+Ru_{\phi,zz}\right)\right]+\rho h u_{\phi,tt}=F_\phi\\
&\mathcal{D}\left[\frac{1}{R^2}\left(u_{\phi,\phi}+u_r\right)+\frac{\nu}{R}u_{z,z}+\chi\left(R^2u_{r,zzzz}+2u_{r,zz\phi\phi}+\frac{1}{R^2}u_{r,\phi\phi\phi\phi}\right)\right]\\
&-\rho h u_{r,tt}=F_r.
\end{aligned}
$$

Generally, we set $\vec{F} = (-F_z, -F_\phi F_r)^t$ to obtain a symmetrical operator[10] and we get $\mathcal{D}\overline{\overline{C}}^{DM}\vec{U} = \vec{F}$, where $\overline{\overline{C}}^{DM}$, the Donnell–Mushtari thin shell operator, is given by the matrix:

$$\overline{\overline{C}}^{DM} = \begin{pmatrix} \frac{\partial^2}{\partial z^2} + \frac{1-\nu}{2R^2}\frac{\partial^2}{\partial \phi^2} - \frac{\rho h}{\mathcal{D}}\frac{\partial^2}{\partial t^2} & \frac{1+\nu}{2R}\frac{\partial^2}{\partial \phi \partial z} & \frac{\nu}{R}\frac{\partial}{\partial z} \\ \frac{1+\nu}{2R}\frac{\partial^2}{\partial \phi \partial z} & \frac{1}{R^2}\frac{\partial^2}{\partial \phi^2} + \frac{1-\nu}{2}\frac{\partial^2}{\partial z^2} - \frac{\rho h}{\mathcal{D}}\frac{\partial^2}{\partial t^2} & \frac{1}{R^2}\frac{\partial}{\partial \phi} \\ \frac{\nu}{R}\frac{\partial}{\partial z} & \frac{1}{R^2}\frac{\partial}{\partial \phi} & \frac{1}{R^2} + \chi\Delta^2 + \frac{\rho h}{\mathcal{D}}\frac{\partial^2}{\partial t^2} \end{pmatrix} \quad [6.55]$$

where we have defined $\Delta^2 = R^2\partial^4/\partial z^4 + 2\partial^4/\partial z^2\partial\phi^2 + 1/R^2\partial^4/\partial\phi^4$ as the bi-Laplacian. The thin shell operator involves two terms: a membrane term, and, if we note that $\mathcal{D}\chi = D$, a bending term analogous to the thin plate operator given by $\mathcal{D}\chi\Delta^2$.

For the Flügge operator, the procedure is similar and we get $\mathcal{D}\left(\overline{\overline{C}}^{DM} + \chi\overline{\overline{C}}^{mo}\right)\vec{U} = \mathcal{D}\overline{\overline{C}}^{F}\vec{U} = \vec{F}$, with:

$$\overline{\overline{C}}^{mo} = \begin{pmatrix} \frac{1-\nu}{2R^2}\frac{\partial^2}{\partial \phi^2} & 0 & -R\frac{\partial^3}{\partial z^3} + \frac{1-\nu}{2R}\frac{\partial^3}{\partial z \partial \phi^2} \\ 0 & 3\frac{1-\nu}{2}\frac{\partial^2}{\partial z^2} & -\frac{3-\nu}{2}\frac{\partial^3}{\partial z^2 \partial \phi} \\ -R\frac{\partial^3}{\partial z^3} + \frac{1-\nu}{2R}\frac{\partial^3}{\partial z \partial \phi^2} & -\frac{3-\nu}{2}\frac{\partial^3}{\partial z^2 \partial \phi} & \frac{1}{R^2} + \frac{2}{R^2}\frac{\partial^2}{\partial \phi^2} \end{pmatrix}. \quad [6.56]$$

6.5.1.6. *Boundary conditions*

Obviously, we need to reflect the fact that the shell is truncated at its ends by $z = \pm L$. If no energy is supplied or dissipated by the borders of the shell, we must cancel the integrals at the edge of the domain. We can easily obtain the boundary conditions in space (at each end of the shell) for the Donnell–Mushtari model:

$$\begin{cases} \mathcal{D}\left(u_{z,z} + \frac{\nu}{R}(u_{\phi,\phi} + u_r)\right) = 0 & \text{or } u_z = 0 \\ \mathcal{D}\left(\frac{1-\nu}{2R}(u_{z,\phi} + Ru_{\phi,z})\right) = 0 & \text{or } u_\phi = 0 \\ \mathcal{D}\chi\left(-R^2u_{r,zzz} - (2-\nu)u_{r,\phi\phi z}\right) = 0 & \text{or } u_r = 0 \\ \mathcal{D}\chi\left(R^2u_{r,zz} + \nu u_{r,\phi\phi}\right) = 0 & \text{or } u_{r,z} = 0 \end{cases}. \quad [6.57]$$

The first column corresponds to the mechanical conditions and represents the four free edge conditions. The last column represents the geometric conditions and reflects the clamped edge condition. Supported shell conditions are shown in the first and last

10 It must be noted that this manipulation renders the shell operator non-self-adjoint.

equations of the first column, and the second and third equations of the last column $\mathcal{D}\left(u_{z,z} + \nu/R\left(u_{\phi,\phi} + u_r\right)\right) = 0$, $u_\phi = 0$, $u_r = 0$ and $\mathcal{D}\chi\left(R^2 u_{r,zz} + \nu u_{r,\phi\phi}\right) = 0$.

For the Flügge model, we obtain:

$$\begin{cases} \mathcal{D}\left(u_{z,z} + \frac{\nu}{R}\left(u_{\phi,\phi} + u_r\right) - \chi R u_{r,zz}\right) = 0 & \text{or } u_z = 0 \\ \mathcal{D}\left(\frac{1-\nu}{2R}\left(u_{z,\phi} + R u_{\phi,z}\right) + \chi\left(3\frac{1-\nu}{2}u_{\phi,z} - 3\frac{1-\nu}{2}u_{r,z\phi}\right)\right) = 0 & \text{or } u_\phi = 0 \\ \mathcal{D}\chi\left(-R^2 u_{r,zzz} - (2-\nu)u_{r,\phi\phi z} - \frac{1-\nu}{2R}u_{z,\phi\phi}\right. & \\ \qquad \left. + \frac{3(1-\nu)}{2}u_{\phi,z\phi} + R u_{z,zz} + \nu u_{\phi,z\phi}\right) = 0 & \text{or } u_r = 0 \\ \mathcal{D}\chi\left(R^2 u_{r,zz} + \nu u_{r,\phi\phi} - R u_{z,z} - \nu u_{\phi,\phi}\right) = 0 & \text{or } u_{r,z} = 0 \end{cases} \quad [6.58]$$

We must note that these operators give correct results as long as the ratio L/R is sufficiently large (to be more precise, we must have a shell length of at least three or four times the radius) because these operators poorly describe edge effects. The temporal boundary conditions are analogous to the Cauchy conditions. We have for $t = 0$, $u_z = u_{z,t} = 0$, $u_\phi = u_{\phi,t} = 0$ and $u_r = u_{r,t} = 0$, or in more compact form $\vec{U} = \vec{U}_{,t} = \vec{0}$.

6.5.1.7. *Harmonic regime*

In a harmonic regime with the dependence $\exp(-\imath\omega t)$, the temporal boundary conditions are identically satisfied. The spatial boundary conditions do not change and the movement equations for the Donnell–Mushtari model become:

$$\begin{cases} \mathcal{D}\left[u_{z,zz} + \frac{1-\nu}{2R^2}u_{z,\phi\phi} + \frac{\rho h\omega^2}{\mathcal{D}}u_z + \frac{1+\nu}{2R}u_{\phi,z\phi} + \frac{\nu}{R}u_{r,z}\right] = F_z \\ \mathcal{D}\left[\frac{1+\nu}{2R}u_{z,z\phi} + \frac{1-\nu}{2}u_{\phi,zz} + \frac{1}{R^2}u_{\phi,\phi\phi} + \frac{\rho h\omega^2}{\mathcal{D}}u_\phi + \frac{1}{R^2}u_{r,\phi}\right] = F_\phi \\ \mathcal{D}\left[\frac{\nu}{R}u_{z,z} + \frac{1}{R^2}u_{\phi,\phi} + \frac{1}{R^2}u_r + \chi\Delta^2 u_r - \frac{\rho h\omega^2}{\mathcal{D}}u_r\right] = F_r. \end{cases} \quad [6.59]$$

6.5.1.8. *Angular Fourier series*

Because of the 2π-periodicity of the geometry of the shell, it is easy to introduce a Fourier series decomposition of the variable ϕ (we often talk of a series of angular harmonics). Each component of the displacement $u_z(z,\phi)$, $u_\phi(z,\phi)$ and $u_r(z,\phi)$ is decomposed into a Fourier series with respect to the variable ϕ:

$$u_\alpha(z,\phi) = \sum_{m=-\infty}^{m=+\infty} u_{\alpha m}(z)\mathrm{e}^{\imath m\phi}, u_{\alpha m}(z) = \int_0^{2\pi} u_\alpha(z,\phi)\mathrm{e}^{-\imath m\phi}d\phi, \alpha = z,\phi,r. \quad [6.60]$$

For each angular component m, with $F_\alpha(z,\phi) = \sum_{m=-\infty}^{m=+\infty} F_{\alpha m}(z) e^{\imath m\phi}$ for $\alpha = z, \phi, r$, the movement equations for the Donnell–Mushtari model become:

$$\mathcal{D}\left[u_{zm,zz} - m^2\frac{1-\nu}{2R^2}u_{zm} + \frac{\rho h\omega^2}{\mathcal{D}}u_{zm} + \imath m\frac{1+\nu}{2R}u_{\phi m,z} + \frac{\nu}{R}u_{rm,z}\right] = F_{zm}$$

$$\mathcal{D}\left[\imath m\frac{1+\nu}{2R}u_{zm,z} + \frac{1-\nu}{2}u_{\phi m,zz} - \frac{m^2}{R^2}u_{\phi m} + \frac{\rho h\omega^2}{\mathcal{D}}u_{\phi m} + \frac{\imath m}{R^2}u_{rm}\right] = F_{\phi m}$$

$$\mathcal{D}\left[\frac{\nu}{R}u_{zm,z} + \frac{\imath m}{R^2}u_{\phi m} + \frac{1}{R^2}u_{rm} + \chi\Delta_m^2 u_{rm} - \frac{\rho h\omega^2}{\mathcal{D}}u_{rm}\right] = F_{rm},$$

where we have defined $\Delta_m^2 = (R^2\partial^4/\partial z^4 - 2m^2\partial^2/\partial z^2 + m^4/R^2)$. It must be noted that the order of the derivations with respect to ϕ in each term is such that we could simply search the partial Fourier series for each component of the following form:

$$u_z(z,\phi) = \sum_{m=0}^{m=+\infty} u_{zm}(z)\cos(m\phi),$$

$$u_\phi(z,\phi) = \sum_{m=0}^{m=+\infty} u_{\phi m}(z)\sin(m\phi) u_r(z,\phi) = \sum_{m=0}^{m=+\infty} u_{rm}(z)\cos(m\phi).$$

Each component of the angular Fourier series [6.60] is the solution of $\mathcal{D}\overline{\overline{C}}_m^{DM}\vec{U}_m = \vec{F}_m$, where the operator $\overline{\overline{C}}_m^{DM}$ represents the Fourier component $\overline{\overline{C}}^{DM}$, which is given by:

$$\overline{\overline{C}}_m^{DM} = \begin{pmatrix} \frac{\partial^2}{\partial z^2} - m^2\frac{1-\nu}{2R^2} + \frac{\rho h\omega^2}{\mathcal{D}} & \imath m\frac{1+\nu}{2R}\frac{\partial}{\partial z} & \frac{\nu}{R}\frac{\partial}{\partial z} \\ \imath m\frac{1+\nu}{2R}\frac{\partial}{\partial z} & -\frac{m^2}{R^2} + \frac{1-\nu}{2}\frac{\partial^2}{\partial z^2} + \frac{\rho h\omega^2}{\mathcal{D}} & \frac{\imath m}{R^2} \\ \frac{\nu}{R}\frac{\partial}{\partial z} & \frac{\imath m}{R^2} & \frac{1}{R^2} + \chi\Delta_m^2 - \frac{\rho h\omega^2}{\mathcal{D}} \end{pmatrix}, \quad [6.61]$$

and similarly for Flügge's operator:

$$\overline{\overline{C}}_m^{mo} = \begin{pmatrix} -m^2\frac{1-\nu}{2R^2} & 0 & -R\frac{\partial^3}{\partial z^3} - m^2\frac{1-\nu}{2R}\frac{\partial}{\partial z} \\ 0 & 3\frac{1-\nu}{2}\frac{\partial^2}{\partial z^2} & -\imath m\frac{3-\nu}{2}\frac{\partial^2}{\partial z^2} \\ -R\frac{\partial^3}{\partial z^3} - m^2\frac{1-\nu}{2R}\frac{\partial}{\partial z} & -\imath m\frac{3-\nu}{2}\frac{\partial^2}{\partial z^2} & \frac{1}{R^2} - \frac{2m^2}{R^2} \end{pmatrix}. \quad [6.62]$$

6.5.2. *Ring*

Similarly as for planar structures, for which the beam operator cannot be obtained by simplifying the plate operator, we cannot obtain a ring operator by simplifying the cylinder operator. We will thus establish a ring operator following a displacement model of the Donnell–Mushtari type.

6.5.2.1. *Displacement field*

We once more use a cylindrical coordinate system defined on the mean surface of the shell (O, z, ϕ, r) (see Figure 6.8). In this coordinate system, the ring occupies the domain $-e/2 < z < +e/2$, $0 \leq \phi < 2\pi$ and $-h/2 < r < +h/2$. Here, in addition to assuming that $h/R \ll 1$, we also assume that $e/R \ll 1$. We call $\vec{U}(z, \phi, r, t)$ the displacement field of the ring.

The three components of the displacement field $\vec{U}(z, \phi, r, t)$, $U_z(z, \phi, r, t)$, $U_\phi(z, \phi, r, t)$, $U_r(z, \phi, r, t)$ are expanded in a double Taylor series around $r = 0$ and $z = 0$ at the first order: $U_\alpha(r, \phi, z) = u_\alpha^0(\phi, t) + ru_\alpha^1(\phi, t) + zu_\alpha^2(\phi, t)$, $\alpha = r, \phi, z$. $u_\alpha^0(\phi, t)$ represents the displacement of the middle line of the ring. We aim to express $u_\alpha^1(\phi, t)$ and $u_\alpha^2(\phi, t)$ as a function of $u_\alpha^0(\phi, t)$.

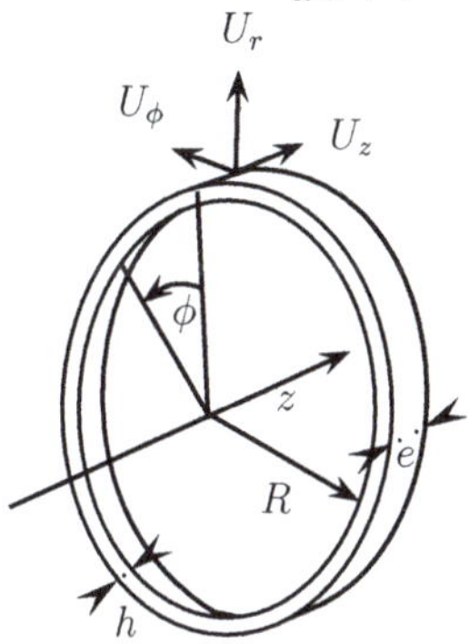

Figure 6.8. *The geometry of a thin cylindrical ring*

We introduce into the strain tensor expression the expressions obtained for the displacement field:

$$
\begin{aligned}
d_{zz} &= u_z^2(\phi,t), d_{rr} = u_r^1(\phi,t), d_{rz} = \tfrac{1}{2}\left(u_r^2(\phi,t) + u_z^1(\phi,t)\right) \\
d_{\phi\phi} &= \tfrac{1}{R+r}\left(u_{\phi,\phi}^0(\phi,t) + ru_{\phi,\phi}^1(\phi,t) + zu_{\phi,\phi}^2(\phi,t) + u_r^0(\phi,t)\right. \\
&\quad \left. + ru_r^1(\phi,t) + zu_r^2(\phi,t)\right) \\
d_{\phi z} &= \tfrac{1}{2}\left(u_\phi^2(\phi,t) + \tfrac{1}{R+r}\left(u_{z,\phi}^0(\phi,t) + ru_{z,\phi}^1(\phi,t) + zu_{z,\phi}^2(\phi,t)\right)\right) \\
d_{r\phi} &= \tfrac{1}{2}\left(u_\phi^1(\phi,t) - \tfrac{1}{R+r}\left(u_\phi^0(\phi,t) + ru_\phi^1(\phi,t) + zu_\phi^2(\phi,t)\right)\right. \\
&\quad \left. + \tfrac{1}{R+r}\left(u_{r,\phi}^0(\phi,t) + ru_{r,\phi}^1(\phi,t) + zu_{r,\phi}^2(\phi,t)\right)\right).
\end{aligned}
$$

Thus, we expand the terms in a series limited to the first order of the small parameters z/R and r/R:

$$d_{zz} = u_z^2(\phi,t), d_{rr} = u_r^1(\phi,t),$$
$$d_{\phi\phi} = \frac{u_{\phi,\phi}^0(\phi,t)}{R} + \frac{u_r^0(\phi,t)}{R} + \frac{r}{R}\left(u_{\phi,\phi}^1(\phi,t) + u_r^1(\phi,t) - \frac{u_{\phi,\phi}^0(\phi,t)}{R} - \frac{u_r^0(\phi,t)}{R}\right)$$
$$+\frac{z}{R}\left(u_{\phi,\phi}^2(\phi,t) + u_r^2(\phi,t)\right)$$
$$d_{\phi z} = \frac{1}{2}\left(u_\phi^2(\phi,t) + \frac{u_{z,\phi}^0(\phi,t)}{R} + \frac{r}{R}\left(u_{z,\phi}^1(\phi,t) - \frac{u_{z,\phi}^0(\phi,t)}{R}\right) + \frac{z}{R}u_{z,\phi}^2(\phi,t)\right)$$
$$d_{r\phi} \approx \frac{1}{2}\left(u_\phi^1(\phi,t) - \frac{u_\phi^0(\phi,t)}{R} + \frac{u_{r,\phi}^0(\phi,t)}{R}\right.$$
$$+\frac{r}{R}\left(-u_\phi^1(\phi,t) + u_{r,\phi}^1(\phi,t) + \frac{u_\phi^0(\phi,t)}{R} - \frac{u_{r,\phi}^0(\phi,t)}{R}\right)$$
$$\left.+\frac{z}{r}\left(-u_\phi^2(\phi,t) + u_{r,\phi}^2(\phi,t)\right)\right).$$

The free side assumption implies that all the side faces of the ring are free from any constraints. From here, it is easy to see that the only non-zero component of the stress tensor is $\sigma_{\phi\phi}$. We thus have $z = \pm e/2$ and $r = \pm h/2$ that verify the relation $d_{z\phi} = 0$, $d_{zr} = 0$, $d_{r\phi} = 0$, $d_{zz} = -\nu/(1-\nu)(d_{\phi\phi} + d_{rr})$ and $d_{rr} = \nu/(1-\nu)(d_{zz} + d_{\phi\phi})$. The two latter relations lead to $d_{rr} = -\nu d_{\phi\phi}$ and $d_{zz} = -\nu d_{\phi\phi}$ and thus[11] we obtain $\sigma_{\phi\phi} = E d_{\phi\phi}$. The first three relations lead us to the following seven conditions $u_\phi^2(\phi,t) + u_{z,\phi}^0(\phi,t)/R = 0$, $u_{z,\phi}^1(\phi,t) - u_{z,\phi}^0(\phi,t)/R = 0$, $u_{z,\phi}^2(\phi,t) = 0$, $u_r^2(\phi,t) + u_z^1(\phi,t) = 0$, $u_\phi^1(\phi,t) - u_\phi^0(\phi,t)/R + u_{r,\phi}^0(\phi,t)/R = 0$, $-u_\phi^1(\phi,t) + u_{r,\phi}^1(\phi,t) + u_\phi^0(\phi,t)/R - u_{r,\phi}^0(\phi,t)/R = 0$ and $-u_\phi^2(\phi,t) + u_{r,\phi}^2(\phi,t) = 0$. From here, we deduce the expressions of the terms of the first and second order of each component of the displacement $u_z^1(\phi,t) = u_z^0(\phi,t)/R + C_1$, $u_z^2(\phi,t) = C_2$, $u_\phi^1(\phi,t) = u_\phi^0(\phi,t)/R - u_{r,\phi}^0(\phi,t)/R$, $u_\phi^2(\phi,t) = -u_{z,\phi}^0(\phi,t)/R$, $u_r^1(\phi,t) = C_3$ and $u_r^2(\phi,t) = -u_z^0(\phi,t)/R + C_4$ where $C_{1..4}$ are constants. The fourth Kirchhoff–Love assumption imposes on the material to behave as an undeformable solid in the direction where the structure is thin, allowing us to once again set the integration constants C_2 and C_3 to zero. From $u_r^2(\phi,t) + u_z^1(\phi,t) = 0$ and the relations $u_z^1(\phi,t) = u_z^0(\phi,t)/R + C_1$ and $u_r^2(\phi,t) = -u_z^0(\phi,t)/R + C_4$, we obtain $C_1 = -C_4 = C$. We thus have $U_z(z,\phi,r,t) = u_z^0(\phi,t) + \frac{r}{R}u_z^0(\phi,t) + rC$ and $U_r(z,\phi,r,t) = u_r^0(\phi,t) - \frac{z}{R}u_z^0(\phi,t) - z$. We approach the first equation $(1 + r/R \approx 1)$ by $U_z(z,\phi,r,t) = u_z^0(\phi,t) + rC$, which corresponds to a transverse shear of the ring. Similarly as for the beams, we will take into consideration the third Kirchhoff–Love assumption, which indicates that the ring is not subject to transverse shear, and set $C = 0$. We must note that the displacement U_r contains two terms, whereas for beams, the two components of the transverse displacement behave as a

11 Similarly as with the beam, this relation implies that the strain affects the neutral line and occurs without lateral contraction. Poisson's ratio, therefore, does not intervene.

single term. We finally obtain the displacement field of the ring following the Donnell–Mushtari model:

$$\begin{array}{l} U_z(z,\phi,r,t) = u_z^0(\phi,t) \\ U_\phi(z,\phi,r,t) = u_\phi^0(\phi,t) - \frac{r}{R}u_{r,\phi}^0(\phi,t) - \frac{z}{R}u_{z,\phi}^0 \\ U_r(z,\phi,r,t) = u_r^0(\phi,t) - \frac{z}{R}u_z^0(\phi,t) \end{array}$$

Further on, we remove the exponent "0".

6.5.2.2. *Ring operator*

We calculate the internal potential energy density $dP = E/2d_{\phi\phi}^2$. We easily get[12]:

$$d_{\phi\phi} = \frac{1}{R}\left(u_{\phi,\phi}(\phi,t) + u_r(\phi,t) - \frac{r}{R}u_{r,\phi\phi}(\phi,t) - \frac{z}{R}\left(u_{z,\phi\phi}(\phi,t) + u_z(\phi,t)\right)\right).$$

We integrate over r between $-h/2$ and $h/2$ and over z between $-e/2$ and $e/2$, we only keep r and z of even powers. The internal potential energy density can be written as:

$$dP \equiv \frac{E}{2R^2}\left((u_{\phi,\phi}(\phi,t) + u_r(\phi,t))^2 + \left(\frac{r}{R}\right)^2 u_{r,\phi\phi}(\phi,t)^2 \right.$$
$$\left. + \left(\frac{z}{R}\right)^2 (u_{z,\phi\phi}(\phi,t) + u_z(\phi,t))^2\right).$$

After integration, we obtain:

$$\mathcal{P} = R\frac{E}{2R^2}\int_0^T\int_0^{2\pi}\Big(A\left(u_{\phi,\phi}(\phi,t) + u_r(\phi,t)\right)^2$$
$$+ I_z u_{r,\phi\phi}(\phi,t)^2 + I_r\left(u_{z,\phi\phi}(\phi,t) + u_z(\phi,t)\right)^2\Big)\, d\phi dt,$$

where $A = eh$ is the area of a cross-section of the ring and $I_z = eh^3/12R^2$ and $I_r = e^3h/12R^2$ are the main inertia moments.

The kinetic energy can be given by:

$$\mathcal{T} = \frac{\rho AR}{2}\int_0^T\int_0^{2\pi}\left(u_{z,t}^2 + u_{\phi,t}^2 + u_{r,t}^2\right) d\phi dt.$$

---

12 Let us recall that under the Donnell–Mushtari assumptions, we take $(1 - r/R) \approx 1$ in the energy density terms.

The calculations of the partial derivative equations, which govern the dynamic behavior of the ring, are once more obtained by applying Hamilton's theorem. Rather than proposing a complete new calculation based on integration by parts, we use, here, a method based on the calculus of variations [MOR 53]. To better understand these ideas, we consider the integral $I(y) = \int_{x_1}^{x_2} \mathcal{F}(x, y, y_{,x})\, dx$. We show that the condition for which $I(y)$ goes through an extreme, such as $\delta I(y) = 0$, is that $\mathcal{F}$ is the solution to Euler's equation $\partial\mathcal{F}/\partial y - d/dx\,(\partial\mathcal{F}/\partial y_{,x}) = 0$. This result can be generalized as follows: the condition for which

$$\int_{x_1}^{x_2} \mathcal{F}\left(x, y, y_{,x}, y_{,xx}, y_{,\underbrace{x\cdots x}_{n}}\right) dx$$

passes through a minimum such as $\delta I(y) = 0$ is that $\mathcal{F}$ is the solution of:

$$\frac{\partial \mathcal{F}}{\partial y} - \frac{d}{dx}\left(\frac{\partial \mathcal{F}}{\partial y_{,x}}\right) + \frac{d^2}{dx^2}\left(\frac{\partial \mathcal{F}}{\partial y_{,xx}}\right) + \cdots + (-1)^n \frac{d^n}{dx^n}\left(\frac{\partial \mathcal{F}}{\partial y_{,x\cdots x}}\right) = 0.$$

If $\vec{F} = (-F_z, -F_\phi, F_r)^t$ represents the surface force applied to the ring, we can easily obtain the equilibrium equations:

$$\left\{\begin{array}{ll} -\frac{\partial \mathcal{P}}{\partial u_z} - \left(\frac{\partial \mathcal{P}}{\partial u_{z,\phi\phi}}\right)_{,\phi\phi} - \left(\frac{\partial \mathcal{T}}{\partial u_{z,t}}\right)_{,t} & = F_z \\ \left(\frac{\partial \mathcal{P}}{\partial u_{\phi,\phi}}\right)_{,\phi} - \left(\frac{\partial \mathcal{T}}{\partial u_{\phi,t}}\right)_{,t} & = F_\phi \\ \frac{\partial \mathcal{P}}{\partial u_r} + \left(\frac{\partial \mathcal{P}}{\partial u_{r,\phi\phi}}\right)_{,\phi\phi} + \left(\frac{\partial \mathcal{T}}{\partial u_{r,t}}\right)_{,t} & = F_r \end{array}\right. \quad [6.63]$$

The equilibrium equations are thus:

$$\left\{\begin{array}{ll} -\frac{E}{R^2} I_r \left(u_{z,\phi\phi\phi\phi} + 2u_{z,\phi\phi} + u_z\right) - \rho A u_{z,tt} & = F_z \\ \frac{E}{R^2} A \left(u_{\phi,\phi\phi} + u_{r,\phi}\right) - \rho A u_{\phi,tt} & = F_\phi \\ \frac{E}{R^2} \left(A\left(u_{\phi,\phi} + u_r\right) + I_z u_{r,\phi\phi\phi\phi}\right) + \rho A u_{r,tt} & = F_r \end{array}\right. \quad [6.64]$$

A common effect for all these structures can be noted: the operator, which governs the component of displacement in the directions where the structure is thin (here in z and r), is an operator of the fourth order. We also note that the axial component of the displacement is decoupled from the other components, while the latter two are coupled: typically, a bending movement in the ring plane induces neither its rotation nor expansion.

6.5.2.3. *Harmonic regime: solution in angular harmonics*

It must be noted that in a harmonic temporal regime for each angular harmonic m, the previous differential equations can be reduced to an algebraic system[13]:

$$\begin{cases} -\frac{E}{R^2} I_r (m^4 - 2m^2 + 1) u_{zm} + \rho A \omega^2 u_{zm} & = F_{zm} \\ \frac{E}{R^2} A \left(-m^2 u_{\phi m} + \imath m u_{rm} \right) + \rho A \omega^2 u_{\phi m} & = F_{\phi m} \\ \frac{E}{R^2} \left(A \left(\imath m u_{\phi m} + u_{rm} \right) + m^4 I_z u_{rm} \right) - \omega^2 \rho A u_{rm} & = F_{rm}, \end{cases} \qquad [6.65]$$

where $u_\alpha(\phi) = \sum_{m=-\infty}^{m=+\infty} u_{\alpha m} \exp(\imath m \phi)$, $\alpha = z, \phi, r$. Solving the first equation of system [6.65] is trivial. We obtain:

$$u_{zm} = -\frac{F_{zm}}{\frac{E}{R^2} I_r (m^2 - 1)^2 - \rho A \omega^2}.$$

When $m \neq 1$, the solution u_{zm} is obviously not defined for the eigenfrequencies:

$$\omega_{zm} R = 2\pi R f_{zm} = \sqrt{\frac{E}{\rho}} \frac{e|m^2 - 1|}{\sqrt{12}}.$$

We know that the longitudinal wave celerity in the thin beam approximation does not depend on Poisson's ratio and if we set $c_L = \sqrt{E/\rho}$, the eigenfrequencies thus correspond to the wave lengths $\lambda_{zm} = c_L / f_{zm} \propto 2\pi R$ proportional to the perimeter of the ring.

The equations for the components $u_{\phi m}$ and u_{rm} of the ring, corresponding to the last two equations of system [6.65], can be solved easily. It is trivial to show that we get:

$$\begin{pmatrix} u_{\phi m} \\ u_{rm} \end{pmatrix} = \frac{1}{\Delta} \begin{pmatrix} \frac{E}{R^2} I_z m^4 + \frac{EA}{R^2} - \rho A \omega^2 & -\imath m \frac{EA}{R^2} \\ -\imath m \frac{EA}{R^2} & -\frac{EA}{R^2} m^2 + \rho A \omega^2 \end{pmatrix} \begin{pmatrix} F_{\phi m} \\ F_{rm} \end{pmatrix}, \qquad [6.66]$$

where the determinant Δ is given by:

$$\Delta = -\rho^2 A^2 \omega^4 + \frac{E}{R^2} \left(A + A m^2 + I_z m^4 \right) \rho A \omega^2 - \left(\frac{E\rho}{R^2} \right)^2 A I_z m^6.$$

13 It is enough to replace in the equations $u_\alpha(z, \phi, r, t)$ by $u_\alpha(z, r, \omega) \exp(im\phi) \exp(-\imath\omega t)$, and to eliminate the exponential terms.

This polynomial is of the second degree in $\rho A\omega^2$. We can easily obtain the four roots of the polynomial, which correspond to four natural pulsations (or free waves) of $u_{\phi m}$ and u_{rm}. We get:

$$\omega R = \pm\sqrt{\frac{E}{\rho}\frac{1}{\sqrt{2A}}\left(A + Am^2 + I_z m^4 \pm\sqrt{\left(A + Am^2 + I_z m^4\right)^2 - 4AI_z m^6}\right)}. \quad [6.67]$$

We aim to find whether the evanescent waves (with a non-zero imaginary part) can exist. This case can only occur for $\left(A + Am^2 + I_z m^4\right)^2 < 4AI_z m^6$. It is easy to see that we have $A + Am^2 + I_z m^4 - 2\sqrt{AI_z}m^3 = A\left(1 + m^2 - 2\sqrt{\chi}m^3 + \chi m^4\right)$, where $\chi = h^2/(12R^2)$. The four zeros of the fourth-order polynomial $1 + x^2 - 2\sqrt{\chi}x^3 + \chi x^4$ are given by $x = 12\sqrt{\chi}\left(1 \pm \sqrt{1 \pm 4\imath\sqrt{\chi}}\right)$. The polynomial $1 + x^2 - 2\sqrt{\chi}x^3 + \chi x^4$ is never canceled for $x \in I\!R$. It never changes its sign and its value is always positive. Thus, $1 + m^2 + \chi m^4 > 2\sqrt{\chi}m^3$ for all m. The free waves given by [6.67] are always propagating.

7

Vibrations of Thin Structures

This chapter is dedicated to studying vibration phenomena in thin media. We will thoroughly investigate the beams and plates, and, on a slightly more superficial level, we will also address the vibration problems of thin shells.

7.1. Beam vibrations

In this section, we focus on the vibrations of beams of finite dimensions. We assume, further on, that the time dependence is monochromatic in the form of $e^{-\imath\omega t}$. To make writing easier, we simplify this dependence in the equations. As we saw in Chapter 6, the longitudinal and transverse movements are completely decoupled. We can thus study the compression and bending movements of our beam independently.

To simplify writing, we denote the longitudinal coordinate as x.

7.1.1. *Beam compression vibrations*

We consider the equation that governs the longitudinal movements of a beam in a harmonic regime. If u_1 is the longitudinal displacement of the beam, we have $EA\left(u_1'' + k_l^2 u_1\right) = f$, where f is the excitation force, $k_l = \omega/c_l$ is the longitudinal wave number and $c_l = \sqrt{E/\rho}$ is the celerity of the compression waves. It is necessary to add to this equation the boundary conditions: for the free end $EAu_{1,1}$ and for the clamped end $u_1 = 0$. We must note that $\sigma_{11} = Eu_{1,1}$ is the axial stress and $A\sigma_{11}$ is the axial force. This equation and the associated boundary conditions are exactly the same as the Helmholtz equation subjected to the Neumann (normal derivative of a displacement is zero) or Dirichlet (displacement is zero) type boundary conditions. We must note that $D = EA$ is called the stiffness modulus in compression.

The general solution of the equation for beams that vibrate in compression $u_g(x)$, solution of the homogeneous equation $u_g''(x) + k_l^2 u_g(x) = 0$, is obviously obtained by the Laplace method. It is a linear combination of complex exponentials, given by:

$$u_g(x) = A\mathrm{e}^{\imath kx} + B\mathrm{e}^{-\imath kx}. \qquad [7.1]$$

The calculation of Green's kernel can be carried out as proposed in section 4.3.2 by taking into consideration the stiffness D. The solution to the inhomogeneous equation:

$$\Gamma''(x, x_0) + k_l^2 \Gamma(x, x_0) = \frac{1}{D}\delta_{x_0}(x)$$

is obtained by a direct and inverse Fourier transform of this equation. It is easy to show that we get a solution identical to [4.11]:

$$\Gamma(x, x_0) = \frac{1}{2\imath k_l D}\mathrm{e}^{\imath k_l |x-x_0|}. \qquad [7.2]$$

Here again, just as in the general solution, Green's kernel is constituted by a propagating wave (with a sinusoidal behavior).

7.1.1.1. *Clamped beam and several solution methods*

In this section, we focus on solving the problem of a clamped beam, excited by a point force, using three different methods. This is intended to show that there are several resolution methods – which is obvious – for an inhomogeneous boundary problem but especially – which is a good reminder – that some methods are more effective than others. Let us consider the following problem:

$$u''(x, x_0) + k_l^2 u(x, x_0) = \frac{1}{D}\delta_{x_0}(x), u(0, x_0) = 0, u(\ell, x_0) = 0. \qquad [7.3]$$

7.1.1.2. *Expansion based on eigenmodes*

We begin by calculating the eigenmodes, which are the non-zero solutions of the homogeneous boundary problem associated with the previous inhomogeneous problem. We proceed just as in section 3.4.1. We thus aim to find $u_m(x)$ and k_m such that:

$$u_m''(x) + k_m^2 u_m(x) = 0, u_m(0) = 0, u_m(\ell) = 0,$$

where m is an integer parameter. We know that the eigenmodes can be written as $u_m(x) = A \sin m\pi x/\ell$ and the eigen wave numbers as $k_m = m\pi/\ell$. We must note that the solution $m = 0$ corresponds to an eigenmode identical to zero and has no physical significance. These eigenmodes are orthogonal with respect to the scalar product $\langle f, g \rangle = \int_0^\ell f(x)g(x)dx$. The constant A is calculated by imposing these modes to be of norm unity. We obtain $\langle u_m(x), u_m(x) \rangle = 1$. It is easy to show that $A = \sqrt{2/\ell}$. The expansion based on the eigenmodes of the problem [7.3] is then easy to obtain. We find:

$$u(x, x_0) = \frac{1}{D} \sum_{n=1}^{\infty} \frac{u_m(x_0)u_m(x)}{k_m^2 - k_l^2}, k_l \neq k_m. \quad [7.4]$$

7.1.1.3. *Solution using Green's representation*

Writing the beam equation in terms of distributions results in:

$$u''(x, x_0) + k_l^2 u(x, x_0) = \{u''(x, x_0) + k_l^2 u(x, x_0)\}$$
$$+\sigma_{u'(0)}\delta(x) + \sigma_{u'(\ell)}\delta_\ell(x) + \sigma_{u(0)}\delta'(x) + \sigma_{u(\ell)}\delta'_\ell(x),$$

where $\sigma_{u(0)} = u(0^-) - u(0^+)$ is the jump of the function $u(x)$ at $x = 0$, $\sigma_{u'(\ell)} = u'(\ell^+) - u'(\ell^-)$ is the derivative of the displacement at $x = \ell$. As the interior of the domain is occupied by the beam, the boundary conditions give $u(0^+) = 0$ and $u(\ell^-) = 0$ and as on the outside we can assume that $u(0^-) = 0$ and $u(\ell^+) = 0$, we obtain $\sigma_{u(0)} = 0$ and $\sigma_{u(\ell)} = 0$. Thus:

$$u''(x, x_0) + k_l^2 u(x, x_0) = \{u''(x, x_0) + k_l^2 u(x, x_0)\} + \sigma_{u'(0)}\delta(x) + \sigma_{u'(\ell)}\delta_\ell(x).$$

Green's kernel of the equation for beams which vibrate in compression is given by [7.2] $\Gamma(x - x_0) = \exp(\imath k_l|x - x_0|)/(2ik_l D)$. Using Green's representation, we obtain:

$$u(x, x_0) = \Gamma(x - x_0) + \sigma_{u'(0)}\Gamma(x) + \sigma_{u'(\ell)}\Gamma(x - \ell).$$

We can thus see that the solution of the problem depends only on the two unknown constants $\sigma_{u'(0)}$ and $\sigma_{u'(\ell)}$. To calculate them, we apply the boundary conditions. We obtain:

$$\sigma_{u'(0)}\Gamma(0) + \sigma_{u'(\ell)}\Gamma(-\ell) = -\Gamma(x_0), \sigma_{u'(0)}\Gamma(\ell) + \sigma_{u'(\ell)}\Gamma(0) = -\Gamma(\ell - x_0).$$

We obtain a linear system of two equations with two unknowns. The determinant of this system is given by $\Delta = -\exp(\imath k_l \ell) \sin k_l \ell/(2\imath kD)^2$. Of course, we find the same condition of existence for the solution, as with the expansion based on eigenmodes: $k_l \neq k_m = m\pi/\ell$. Reserving the right to not cancel the determinant, we can easily show that $\sigma_{u'(0)} = -\sin k_l(\ell - x_0)/\sin k_l\ell$ and $\sigma_{u'(\ell)} = -\sin k_l x_0/\sin k_l\ell$. The solution can thus be written as:

$$u(x, x_0) = \frac{1}{2\imath k_l D}\left\{\mathrm{e}^{\imath k_l|x-x_0|} - \frac{1}{\sin k_l\ell}\left[\sin k_l(\ell - x_0)\mathrm{e}^{\imath k_l x} + \sin k_l x_0 \mathrm{e}^{\imath k_l(\ell-x)}\right]\right\}. \qquad [7.5]$$

This result is interesting because it involves only a finite (and low) number of terms, while the modal expansion [7.4] consists of an infinite number of terms. It is thus beneficial to use as much as possible this parsimonious representation.

7.1.1.4. *General integration method*

We aim to find the displacement, i.e. the solution of an inhomogeneous differential equation, as the sum of the general solution and the particular solution: $u = u_0 + u_g$. In this case, the particular solution is known, since it is Green's kernel. In the case of a more complicated source term, we could use the product of the convolution of the source term and Green's kernel. The constants, which appear in the general solution, are calculated so as to satisfy the boundary conditions. We obtain:

$$u(x, x_0) = \frac{1}{2\imath k_l D}\mathrm{e}^{\imath k_l|x-x_0|} + A\mathrm{e}^{\imath k_l x} + B\mathrm{e}^{-\imath k_l x}.$$

When we write the boundary conditions, we obtain a linear system of two equations with two unknowns. The condition to solve this system is the same as previously. The constants A and B can be written as:

$$A = -\frac{1}{2\imath k_l D}\frac{\sin k_l(\ell - x_0)}{\sin k_l\ell}, B = -\frac{1}{2\imath k_l D}\frac{\sin k_l x_0}{\sin k_l\ell}\mathrm{e}^{\imath k\ell},$$

the solution obtained is identical to the one obtained using Green's representation. We must note that when Green's representation can be calculated (even numerically), it is often the most efficient way of calculating the solution.

7.1.1.5. *Beam excited at one end*

We consider here the problem of a beam with a length ℓ which vibrates in compression. This beam is free at the $x = \ell$ end. It is excited by a harmonic force ($f = EAF$) located at $x = 0$. The boundary condition at $x = 0$ tells us that the axial force (which is the product of the axial stress and the surface) $A\sigma_{11}$ is imposed. We thus have to solve the system of equations:

$$EA\left(u_1'' + k_l^2 u_1\right) = 0, \text{ for } x \in]0, \ell[, A\sigma_{11} = EAF \text{ at } x = 0, A\sigma_{11} = 0 \text{ at } x = \ell.$$

This is $u_1'' + k_l^2 u_1 = 0$ for $x \in]0, \ell[$, with the boundary conditions $u_1' = F_1$ at $x = 0$ and $u_1' = 0$ at $x = \ell$. It is evident that we obtain $u_1 = Ae^{\imath k_l x_1} + Be^{-\imath k_l x_1}$. The boundary conditions result in:

$$\begin{aligned} A\imath k_l - B\imath k_l &= F_1 \\ A\imath k_l e^{\imath k_l \ell} - B\imath k_l e^{-\imath k_l \ell} &= 0 \end{aligned}$$

We solve this system using Cramer's rule. Its determinant can be written as $D = 2k_l \sin k_l \ell$. If the excitation frequency is not an eigenfrequency of the system (for which the determinant cancels), such that $f_m = mc_l/(2\ell)$, $m \in \boldsymbol{N}$, the constants A and B are given by $A = -F\exp(-\imath k_l \ell)/D$ and $B = -F\exp(\imath k_l \ell)/D$. The displacement is thus given by $u_1(x) = -2F\cos\left(k_l(x-\ell)\right)/D$. We must note that the solution of the problem is purely real.

7.1.2. *Beam bending vibrations*

In this section, we focus on studying the beam bending vibrations. Due to the similarity of the two equations which govern the transverse movements of the displacements, we do not need to particularize the displacements. We thus consider the equation:

$$EIu^{(4)} - \rho A\omega^2 u(x) = f$$

in which the displacement u is any of the transverse displacements. The moment of inertia, of course, corresponds to the displacement considered. To this equation, we must add the boundary conditions given by: for the clamped end, $u(x) = 0$ and $u'(x) = 0$, for the simply supported end, $u(x) = 0$ and $EIu''(x) = 0$, or for the free end, $EIu''(x) = 0$ and $EIu'''(x) = 0$.

7.1.2.1. *General solution*

We set $k_t^4 = \rho A\omega^2/(EI)$. We aim to find $u_g(x)$, the solution of the homogeneous equation $u_g^{(4)}(x) - k_t^4 u_g(x) = 0$. This equation can be written as $\left(d^2/dx^2 - k_t^2\right)\left(d^2/dx^2 + k_t^2\right) u_g(x) = 0$. $u_g(x)$ is thus a linear combination of two functions $u_1(x)$ and $u_2(x)$, such that $\left(d^2/dx^2 + k_t^2\right) u_1(x) = 0$ and $\left(d^2/dx^2 - k_t^2\right) u_2(x) = 0$. u_1 and u_2 are solutions of a Helmholtz one-dimensional equation for a real wave number k_t for u_1 and a pure imaginary wave number $\imath k_t$ for u_2. It is, thus, obvious that we have $u_1(x) = Ae^{\imath k_t x} + Be^{-\imath k_t x}$ and $u_2(x) = Ce^{k_t x} + De^{-k_t x}$. The general solution for Euler's equation is thus given by any of the following formulations:

$$\begin{aligned} u_g(x) &= Ae^{\imath k_t x} + Be^{-\imath k_t x} + Ce^{k_t x} + De^{-k_t x} \\ u_g(x) &= A'\sin k_t x + B'\cos k_t x + C'\sinh k_t x + D'\cosh k_t x. \end{aligned} \qquad [7.6]$$

7.1.2.2. *Green's kernels*

Now, we aim to find the solution to the following inhomogeneous equation $\Gamma^{(4)}(x, x_0) - k_t^4 \Gamma(x, x_0) = 1/D\delta_{x_0}(x)$, where $D = EI$ is the stiffness modulus in bending. The solution to this equation is obtained by the direct and inverse Fourier transforms of this equation. Green's kernel is an elementary kernel that satisfies an energy conservation condition, such as the principle of limit absorption. We aim to find a solution for $\Gamma_\epsilon(x, x_0)$ such that:

$$\Gamma_\epsilon^{(4)}(x, x_0) - (k_t + \imath\epsilon)^4 \Gamma_\epsilon(x, x_0) = \frac{1}{D}\delta_{x_0}(x),$$

where ϵ is a real positive number which represents a loss of energy in the material and D is the stiffness modulus in bending. The solution Γ is obtained by approaching the limit $\epsilon \to 0$. $\hat{\Gamma_\epsilon}(\zeta)$ is the Fourier transform of the displacement. Using the Fourier transform, Euler's equation becomes $(16\pi^4\zeta^4 - (k_t + \imath\epsilon)^4)\hat{\Gamma_\epsilon}(\zeta, x_0) = 1/D\exp(-2\imath\pi x_0\zeta)$. It is obvious that the solution $\hat{\Gamma_\epsilon}(\zeta)$ is given by:

$$\hat{\Gamma_\epsilon}(\zeta, x_0) = \frac{1}{2(k_t + \imath\epsilon)^2 D}\left\{\frac{1}{\left(4\pi^2\zeta^2 - (k_t + \imath\epsilon)^2\right)} - \frac{1}{\left(4\pi^2\zeta^2 + (k_t + \imath\epsilon)^2\right)}\right\} e^{-2\imath\pi x_0\zeta}.$$

We thus have:

$$\Gamma_\epsilon(x, x_0) = \int_{-\infty}^{+\infty} \frac{e^{2\imath\pi(x - x_0)\zeta}}{2(k_t + \imath\epsilon)^2 D}\left\{\frac{1}{\left(4\pi^2\zeta^2 - (k_t + \imath\epsilon)^2\right)} - \frac{1}{\left(4\pi^2\zeta^2 + (k_t + \imath\epsilon)^2\right)}\right\} d\zeta.$$

The calculation of this integral is done using the residue method. Because of the limit absorption principle, the poles have a non-zero positive imaginary part, which allows us to apply the residue theorem once more. After approaching the limit, we obtain:

$$\Gamma(x, x_0) = \frac{1}{4k_t^3 D}\left\{\imath e^{\imath k_t|x - x_0|} - e^{-k_t|x - x_0|}\right\}. \qquad [7.7]$$

We can see that Green's kernel, as the general solution [7.6], is a superposition of a propagating wave (with sinusoidal behavior) and an evanescent wave (exponentially decreasing).

7.1.2.3. *Beams of finite length*

Let us now examine some practical examples of solving problems related to the finite dimension of a beam. In particular, we will study some simple examples of basic problems with beam eigenfrequencies and the expansion based on eigenmodes.

7.1.2.4. *Supported beam*

We assume that the beam, with a length of ℓ, is simply supported at both of its ends. The boundary conditions are then given by:

$$u(0) = u(\ell) = u''(0) = u''(\ell) = 0.$$

EIGENVALUES.– We aim to find the eigenvalues of this beam. If we start from a general solution $u(x) = A\sin k_t x + B\cos k_t x + C\sinh k_t x + D\cosh k_t x$ and if we write the four boundary conditions, we obtain the system:

$$\begin{pmatrix} 0 & 1 & 0 & 1 \\ \sin k_t\ell & \cos k_t\ell & \sinh k_t\ell & \cosh k_t\ell \\ 0 & -1 & 0 & 1 \\ -\sin k_t\ell & -\cos k_t\ell & \sinh k_t\ell & \cosh k_t\ell \end{pmatrix} \begin{pmatrix} A \\ B \\ C \\ D \end{pmatrix} = \begin{pmatrix} 0 \\ 0 \\ 0 \\ 0 \end{pmatrix}.$$

The system only has a non-zero solution, if its determinant $\Delta = 4\sin k_t\ell \sinh k_t\ell$ is zero. This relation is satisfied, if $\sinh k_t\ell = 0$ or $\sin k_t\ell = 0$. In the first condition, we deduce that it is necessary and sufficient that $k_t = 0$. The wave numbers k_t that satisfy the second relation form a sequence $k_{tm} = m\pi/\ell$, $m \in \boldsymbol{N}$. From here, we deduce that the sequence of eigenfrequencies of the beam can be given by:

$$f_m = \frac{m^2\pi}{2\ell^2}\sqrt{\frac{EI}{\rho A}}. \qquad [7.8]$$

EIGENMODES.– Let us now calculate the eigenmode expression. The boundary conditions at $x = 0$ give us $B + D = 0$ and $B - D = 0$. From here, it is obvious that $B = D = 0$. Thus, we obtain the displacement expression $u(x) = A\sin k_t x + C\sinh k_t x$. We use, for example, the fact that $u(\ell) = 0$. We thus have $C = -A\sin k_t\ell/\sinh k_t\ell$, however, we know that $\sin k_t\ell = 0$ and $\sinh k_t\ell \neq 0$. Therefore, $C = 0$. From this, the eigenmode expression is $u_m(x) = A\sin\frac{m\pi}{\ell}x$. We consider the scalar product $\langle f, g\rangle = \int_0^\ell f(x)g^\star(x)dx$, where $g^\star(x)$ is the conjugate of $g(x)$. It is evident that these eigenmodes are orthogonal, two by two: $\langle u_m, u_n\rangle = 0$, if $m \neq n$. The constant A is arbitrary. To determine it, we must impose on the eigenmodes to be normalized to one. This is $\|u_m(x)\|^2 = 1$ which leads to $A = \sqrt{2/\ell}$. The eigenmodes can be given by the expression:

$$u_m(x) = \sqrt{\frac{2}{\ell}} \sin \frac{m\pi}{\ell} x. \qquad [7.9]$$

We note that imposing $m = 0$ leads to an eigenmode identical to zero and must be rejected. We, thus, obtain $m \in \boldsymbol{N}^{\star}$.

FORCED REGIME.– We now solve the inhomogeneous problem $u^{(4)}(x) - k_t^4 u(x) = f(x)$ for $x \in]0, \ell[$, $u(x) = u''(x) = 0$, for $x = 0, \ell$, where $f(x)$ is a given variable of the problem. We aim to find $u(x)$ in the form of an expansion based on eigenmodes. We obtain $u(x) = \sum_{m=1}^{m=\infty} a_m u_m(x)$, where the coefficients a_m are the new unknowns of the problem. We expand the exciting force based on the eigenmodes $f(x) = \sum_{m=1}^{m=\infty} f_m u_m(x)$, where the coefficients $f_m = \int_0^\ell f(x) u_m(x) dx$ are known. Thus, $\sum_{m=1}^{m=\infty} a_m u_m^{(4)}(x) - k_t^4 \sum_{m=1}^{m=\infty} a_m u_m(x) = \sum_{m=1}^{m=\infty} f_m u_m(x)$. If $k_{tm} \neq k_t$, which corresponds to a different excitation frequency than the eigenfrequency of the beam, hence $a_m = f_m/(k_{tm}^4 - k_t^4)$. The displacement can thus be written as:

$$u(x) = \sum_{m=1}^{m=\infty} \frac{f_m}{k_{tm}^4 - k_t^4} u_m(x). \qquad [7.10]$$

If $f(x) = \delta_{x_0}(x)$, it is obvious that $f_m = u_m(x_0)$. The previous expansion, which includes the expression f_m, is the expansion based on the eigenmodes of Green's kernel of the simply supported beam.

7.1.2.5. *Clamped beam*

Let us assume now that the beam, always with a length of ℓ, is clamped at its two ends. The boundary conditions are given by:

$$u(0) = u(\ell) = u'(0) = u'(\ell) = 0.$$

EIGENVALUES.– We aim to find the eigenvalues of this beam. We begin with the general solution $u(x) = A \sin k_t x + B \cos k_t x + C \sinh k_t x + D \cosh k_t x$ and we write the four boundary conditions. We obtain the system:

$$\begin{aligned} B + D &= 0 \\ A \sin k_t\ell + B \cos k_t\ell + C \sinh k_t\ell + D \cosh k_t\ell &= 0 \\ k_t A + k_t C &= 0 \\ A k_t \cos k_t\ell - B k_t \sin k_t\ell + C k_t \cosh k_t\ell + D k_t \sinh k_t\ell &= 0 \end{aligned}$$

This system has a non-zero solution only if its determinant is zero. This determinant can be easily calculated and its cancelation leads to the eigenvalue equation $2\,(1 - \cos k_t\ell \cosh k_t\ell) = 0$. There is only one numerical solution possible for this equation. However, an approximate calculation gives us a very good approximation of the solutions for this equation. We must note that this calculation is required, even if we aim to find only one numerical solution to the eigenvalue equation. It is well known that numerically finding the zeros of a function is only possible if we know an approximation of the sought value. The graphical method, which we propose, allows us to find these values with good precision. We have to solve $\cos k_t\ell \cosh k_t\ell = 1$. As such, we can look for the intersection points of $\cos k_t\ell$ and $1/\cosh k_t\ell$. These two curves are shown in Figure 7.1.

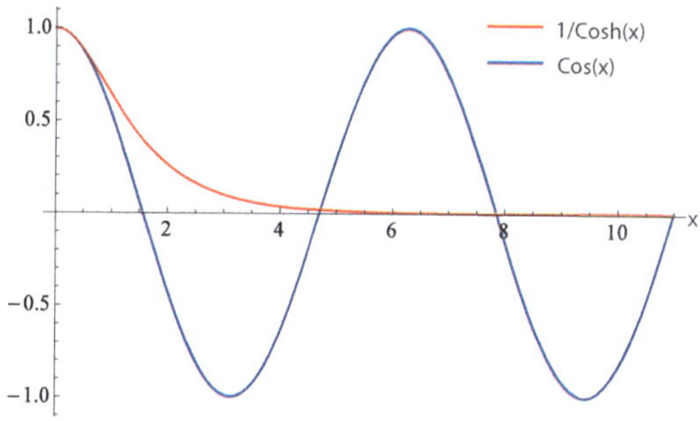

Figure 7.1. *A graphic solution of the eigenvalue equation. For a color version of the figure, see www.iste.co.uk/anselmet/acoustics.zip*

The first intersection is at $3\pi/2 + \epsilon_1$, the second intersection is at $5\pi/2 + \epsilon_2$. From them, we can deduce the approximate values of the following eigen wave numbers:

$$k_{tm} \sim \frac{2m+1}{2\ell}\pi + \frac{\epsilon_m}{\ell}, \qquad [7.11]$$

where $\lim_{m\to\infty} \epsilon_m = 0$. The corresponding eigenfrequencies can be deduced easily. We must note that ϵ_m becomes negligible very quickly, for example $|\epsilon_2| < 0.001$.

It is possible to give an approximate expression of the corrective term ϵ_m using an asymptotic expansion. Let us consider $k_{tm}\ell \sim (2m+1)\pi/2 + \epsilon_m$. We introduce these results into the eigenvalue equation. We obtain $\cos\left((2m+1)\pi/2 + \epsilon_m\right)$ $\cosh\left((2m+1)\pi/2 + \epsilon_m\right) = 1$, if we expand the cosine $-\sin(2m + 1)\pi/2 \sin\epsilon_m \left\{\cosh(2m+1)\pi/2 \cosh\epsilon_m - \sinh(2m+1)\pi/2 \sinh\epsilon_m\right\} = 1$, or $\sin\epsilon_m \sim \epsilon_m$, $\sinh\epsilon_m \sim \epsilon_m$ and $\cosh\epsilon_m \sim 1$, from this we deduce that $\epsilon_m \sim -(-1)^m\left(\cosh(2m+1)\pi/2\right)$. For example, $\epsilon_1 \sim 1.79\ 10^{-2}$ (instead of $1.76\ 10^{-2}$), $\epsilon_2 \sim -7.76\ 10^{-4}$ (instead of $-7.77\ 10^{-4}$).

EIGENMODES.– We return once more to the general solution $u(x) = A\sin k_t x + B\cos k_t x + C\sinh k_t x + D\cosh k_t x$. From the conditions $u(0) = 0$ and $u'(0) = 0$, we obtain $A = -C$ and $B = -D$. With $u(\ell) = 0$, we find:

$$B = -A\frac{\sin k_t\ell - \sinh k_t\ell}{\cos k_t\ell - \cosh k_t\ell}.$$

The eigenmodes can, thus, be given by:

$$u_m(x) = A\left\{\sin k_{tm}x - \sinh k_{tm}x - \frac{\sin k_{tm}\ell - \sinh k_{tm}\ell}{\cos k_{tm}\ell - \cosh k_{tm}\ell}\right. \\ \left. \times (\cos k_{tm}x - \cosh k_{tm}x)\right\} \qquad [7.12]$$

A synthetic notation is to introduce the Duncan functions:

$$s_1(x) = \sinh x + \sin x, c_1(x) = \cosh x + \cos x,$$
$$s_2(x) = \sinh x - \sin x, c_2(x) = \cosh x - \cos x,$$

these functions can be deduced from each other by deriving $s_1'''(x) = c_1''(x) = s_2'(x) = c_2(x)$. We thus have:

$$u_m(x) = \frac{A}{-c_2(k_{tm}\ell)}\left(s_2(k_{tm}x)\, c_2(k_{tm}\ell) - c_2(k_{tm}x)\, s_2(k_{tm}\ell)\right).$$

To show the orthogonality of these eigenmodes, it is necessary to use Green's formula of Euler's equation. We recall that Green's formula of the beam equation actually consists of three independent formulas (one for each displacement). To establish Green's formula for a structure (beam, plate or shell) vibration equation, the simplest method consists of establishing it based on the variational formulation associated with the expression of the energy of the system. We consider, for example, the weak form of the beam bending problem, given by equation [6.30] which we apply to the displacement u_2 in a harmonic regime. If the beam is subjected to a force density f_2, the weak form of the problem can be written as:

$$\int_0^\ell A\left(\rho\omega^2 u_2 - \frac{EI_3}{A}u_2^{(iv)}\right)\delta u_2 dx - EI_3\left[u_2'''\delta u_2 + u_2''\delta u_2'\right]_0^\ell = \int_0^\ell f_2\delta u_2 dx$$

this is $\rho A\omega^2\int_0^\ell u_2\delta u_2 dx - EI_3\int_0^\ell u_2^{(iv)}\delta u_2 dx - EI_3\left[u_2'''\delta u_2 + u_2''\delta u_2'\right]_0^\ell = \int_0^\ell f_2\delta u_2 dx$. If we introduce the scalar product $\langle u, v\rangle = \int_0^\ell u(x)v(x)dx$, we obtain the weak formulations which consist of finding the functions u_2 satisfying the boundary conditions, such that for any function δu_2 satisfying the boundary conditions, we have $\rho A\omega^2\langle u_2, \delta u_2\rangle - a(u_2, \delta u_2) = \langle f_2, \delta u_2\rangle$. $a(u,v) = EI_3 \int_0^\ell u^{(iv)} v dx - EI_3 [u'''v + u''v']_0^\ell$ is a symmetric bilinear form. $a(u_2, u_2)$ represents the internal potential energy of the displacement u_2. To establish Green's representation of the displacement u_2, we write $a(u_2, v_2) - a(v_2, u_2) = 0$. We obtain:

$$\int_0^\ell u_2^{(iv)} v_2 - v_2^{(iv)} u_2 dx = [(u_2''' v_2 - v_2''' u_2) + (u_2'' v_2' - v_2'' u_2')]_0^\ell . \quad [7.13]$$

The eigenmode u_m satisfies the system of equations $u_m^{(4)} - k_{tm}^4 u_m = 0$, for $]0, \ell[$ provided that the boundary conditions are $u_m = u_m' = 0$ at 0 and ℓ. The mode u_n satisfies the system of equations $u_n^{(4)} - k_{tn}^4 u_n = 0$, for $]0, \ell[$ provided that the boundary conditions are $u_n = u_n' = 0$ at 0 and ℓ. We multiply the first equation of the first system by $u_n(x)$, the second equation of the second system by $u_m(x)$. By subtracting them and integrating along the length of the beam, we obtain $\int_0^\ell u_m^{(4)} u_n - u_n^{(4)} u_m dx = \left(k_{tm}^4 - k_{tn}^4\right) \int_0^\ell u_n u_m dx$. Green's theorem [7.13] applied to the first part leads to $[EI(u_m''' v_n - v_n''' u_m) - EI(u_m'' v_n' - v_n'' u_m')]_0^\ell = \left(k_{tm}^4 - k_{tn}^4\right) \int_0^\ell u_n u_m dx$. According to the boundary conditions, it is obvious that the first part of this equation is zero. Thus:

$$\left(k_{tm}^4 - k_{tn}^4\right) \int_0^\ell u_n u_m dx = 0.$$

If $k_{tm} \neq k_{tn}$ then $\int_0^\ell u_n u_m dx = 0$. On the other hand, we have $k_{tm} = k_{tn}$ if $m = n$. In this case, $\int_0^\ell u_n^2 dx$ is arbitrary. It is convenient to fix a value equal to one for this integral, which allows us to obtain an orthonormal system. We calculate this normalization constant and we get:

$$u_m(x) = A_m \left\{ \sin k_{tm} x - \sinh k_{tm} x - \frac{\sin k_{tm}\ell - \sinh k_{tm}\ell}{\cos k_{tm}\ell - \cosh k_{tm}\ell} \times (\cos k_{tm} x - \cosh k_{tm} x) \right\},$$

thus:

$$\int_0^\ell |u_m(x)|^2 dx = \frac{A_m^2}{k_{tm}} \int_0^{k_{tm}\ell} \left\{ \sin x - \sinh x - \frac{\sin k_{tm}\ell - \sinh k_{tm}\ell}{\cos k_{tm}\ell - \cosh k_{tm}\ell} \times (\cos x - \cosh x) \right\}^2 dx = 1.$$

Let $\mu_m = k_{tm}\ell$. We show that:

$$1 = -\frac{A_m^2}{k_{tm}} \{2(\cos\mu_m - 2\mu_m \sin\mu_m)\sinh\mu_m \\ -2\cosh\mu_m\left(\sin\mu_m + \cos^2\mu_m \sinh\mu_m\right) \\ -\mu_m\cos 2\mu_m + \mu_m\cosh 2\mu_m + \cosh^2\mu_m \sin 2\mu_m\} / 2\left(\cos\mu_m - \cosh\mu_m\right)^2$$

At first approximation, we know that $\mu_m \approx (2m+1)/2\pi$. Thus, we have $\cos\mu_m \approx 0$, $\sin\mu_m \approx (-1)^m$, $\cos 2\mu_m \approx -1$, $\sin 2\mu_m \approx 0$. From these results, we deduce:

$$1 \approx \frac{A_m^2}{k_{tm}} \frac{\mu_m(1 + \mu_m\cosh 2\mu_m) - 2(-1)^m(\cosh\mu_m + 2\mu_m\sinh\mu_m)}{2\cosh^2\mu_m}.$$

Moreover, if $\mu_m \gg 1$, $\cosh 2\mu_m \approx 2\cosh^2\mu_m \gg \cosh\mu_m$, $\cosh 2\mu_m \gg \sinh\mu_m$ and $\cosh 2\mu_m \gg \mu_m$. We obtain $1 \approx A_m^2\mu_m/k_{tm}$. From here, it is easy to deduce that $A_m \approx \sqrt{1/\ell}$. We must note that this approximation, precise to 2% of the first eigenvalue, is almost exact for higher orders. For $m = 2$, we already obtain three significant figures. We must note that the numerical calculation of this integral is unrealistic for high values of m, i.e. it becomes impossible from $m = 10$. The analytic approximations developed here become more meaningful.

7.1.2.6. *Other boundary conditions*

Table 7.1 gives the eigenvalues obtained for all the combinations of boundary conditions with the exception of guided clamped, which, as we can recall, has no physical meaning.

Boundary conditions	$k_{t1}\ell$	$k_{t2}\ell$	$k_{t3}\ell$	$k_{tm}\ell, n > 3$
free–free	0	4.73	7.853	$(2m-1)\frac{\pi}{2}$ approximated
free–supported	0	3.927	7.069	$(4m-3)\frac{\pi}{4}$ approximated
free–clamped	1.875	4.694	7.855	$(2m-1)\frac{\pi}{2}$ approximated
supported–supported	3.142	6.283	9.425	$m\pi$ exact
supported–clamped	3.927	7.069	10.21	$(4m+1)\frac{\pi}{4}$ approximated
clamped–clamped	4.73	7.853	10.996	$(2m+1)\frac{\pi}{2}$ approximated

Table 7.1. *Bending beam eigenvalues*

7.1.2.7. *Two cantilever beams coupled with a spring*

Let us consider two beams of identical length, clamped on one end and free on the other (cantilevers) coupled with a spring of rigidity r_c placed on their free end. We can consider that the coupling of the two beams equates to a rigidity force, dependent on each extremity's relative movement, applied to the free end of each beam. Let us establish the eigenvalue equations.

In the absence of any excitation force, the movement of the first beam, $u_1(x)$, satisfies equation $EIu_1^{(4)} - \rho A\omega^2 u_1(x) = 0$, while the movement of the second satisfies equation $EIu_2^{(4)} - \rho A\omega^2 u_2(x) = 0$. To these equations we add eight boundary conditions $u_1(0) = 0$, $u_2(0) = 0$, $u_1'(0) = 0$, $u_2'(0) = 0$, $u_1''(\ell) = 0$, $u_2''(\ell) = 0$, $EIu_1'''(\ell) = r_c(u_1(\ell) - u_2(\ell))$ and $EIu_2'''(\ell) = r_c(u_2(\ell) - u_1(\ell))$. Take the general solution [7.6], written for $i = 1, 2$, as $u_i(x) = A_i \cos k_p x + B_i \sin k_p x + C_i \cosh k_p x + D_i \sinh k_p x$ with $k_p = \rho A\omega^2/(EI)$. We apply the eight boundary conditions and obtain a homogeneous system of eight linear equations with eight unknowns (constants A_i, B_i, C_i and D_i for $i = 1,2$). In order for this homogeneous system to allow a solution non-identically zero, we must cancel out its determinant $\mathcal{D}(\mu)$ where we have assumed that $\mu = k_p\ell$. We can therefore demonstrate the following:

$$\mathcal{D}(\mu) = 4\frac{EI}{\ell^6}\mu^3(1 + \cos\mu\cosh\mu)(EI\mu^3 + \cosh\mu(EI\mu^3\cos\mu + 2\ell^3 r_c \sin\mu)$$
$$-2\ell^3 r_c \cos\mu \sinh\mu).$$

$\mathcal{D}(\mu)$ cancels itself out for two sets of eigenvalues. The first, which is given by values of μ_1^j that satisfy $1 + \cos\mu_1^j \cosh\mu_1^j = 0$, $j = 1, 2, 3, \cdots$, depends only on the characteristics of each beam (as they are identical, they possess the same values). The second is given by values of μ_2^j that satisfy $(\mu_2^j)^3(1 + \cosh\mu_2^j \cos\mu_2^j) + \kappa_c(\cosh\mu_2^j \sin\mu_2^j - \sinh\mu_2^j \cos\mu_2^j) = 0$, $j = 1, 2, 3, \cdots$, where we have written $\kappa_c = 2\ell^3 r_c/(EI)$; these eigenvalues correspond to the vibrations in phase opposition of the beams. We can demonstrate that each time $r_c \neq 0$, the eigenvalues satisfy $\mu_1^1 < \mu_1^2 < \mu_2^1 < \mu_2^2 < \mu_3^1 < \mu_3^2 < \cdots$; the two sets of eigenvalues are overlayed. We note that for each value of j, $\mu_1^j < \mu_2^j$ is in accord with the principle of additional links that states a movement of the spectrum towards treble frequencies for every additional stress, i.e. the rigidity of the coupling, to the system.

7.1.2.8. *Identification of mechanical properties*

One of the most used devices in vibration dynamics for accessing the mechanical characteristics of a material is the cantilever beam (a free–clamped beam). The classic procedure is described by the American Society of Testing and Materials under the reference ASTM E756-05: "Standard Test Method for Measuring Vibration-Damping Properties of Materials" available at [AST 10]. This method requires a dedicated experimental setup: a contactless exciter, delicate clamping for implementation with different beam geometries, measurements using a laser vibrometer and modal analysis. Moreover, when we examine the phenomena for damping purposes, the clamping and excitation are sometimes difficult to control, because of the loss of energy that they can introduce [WOJ 04]. However, the acoustic energy lost by radiations is very low and most of the dissipation (and thus the decrease in vibrations or acoustic field measured in the near field) is due to the material which makes up the beam.

We have seen that the precise calculation of the resonances of a clamped-free beam leads to the first eigenvalues which are $k_1L = 1.19373\pi/2$, $k_2L = 2.98835\pi/2$ and $k_3L = 5.00049\pi/2$. For higher values, we can choose with great precision $k_mL = (2m-1)\pi/2$. Subsequently, the eigenfrequencies are then given by:

$$f_m = \frac{\pi h}{16\sqrt{3}L^2}\sqrt{\frac{E}{\rho}}a_m, \text{ with } a_1 = (1.194)^2, a_2 = (2.988)^2,$$

$$a_m = (2m-1)^2, m \geq 3$$

The measurement of Young's modulus and the damping of a material, based on the resonances of the cantilever beam, can be easily carried out. For continuous systems, we saw that one of the simplest models for describing viscous-type damping is to consider that for each frequency, Young's modulus is a complex variable in the form of $E = E_0(1 - \imath\eta)$, where the loss factor η is positive, if the time dependence is in the form of $\exp(-\imath\omega t)$. We recall that in the relevant literature there is also mention of the damping factor $\zeta = \eta/2$. We can then easily see that the first eigen pulsations of the beam can be represented as complex variables, which can be written in the form of $\omega_m = \dot{\omega}_m - \imath\hat{\omega}_m$, $\dot{\omega}_m = 2\pi a_m \pi h/(16\sqrt{3}L^2)\sqrt{E_0/\rho}$ and $\hat{\omega}_m = \dot{\omega}_m\eta/2$, if $0 < \eta \ll 1$. We recall that the logarithmic decrement of a signal $s(t)$, which is defined by $\delta = \log s(t)/s(t + 2\pi/\dot{\omega}_1)$ [CRA 70], can be written as $\delta = \log\exp(-\imath\omega_1 t)\exp(-\hat{\omega}_1 t)/(\exp(-\imath\omega_1(t + 2\pi/\dot{\omega}_1))\exp(-\hat{\omega}_1(t + 2\pi/\dot{\omega}_1)))$. This is $\delta = \hat{\omega}_1(2\pi/\dot{\omega}_1) = \pi\eta = 2\pi\zeta$ for $\eta \ll 1$.

The measurement of the first resonance frequencies and the logarithmic decrements or decrease times very simply leads to an estimation of Young's modulus and the loss factor or the damping factor for the beam resonances. In practice, we are often limited to a few frequencies (typically 3 or 4) because the experimental results diverge from the theoretical model as soon as we go beyond these frequencies (because of geometric deviations and approximations of the thin beam, which are only valid at low frequency). A very simple measurement setup consists of exciting the beam by an impact and measuring the fundamental frequency and the decrease time of the pseudo-periodic pressure signal emitted at the pulsation $\dot{\omega}_1 = 2\pi(1.194)^2\pi h/\left(16\sqrt{3}L^2\right)\sqrt{E/\rho}$, where, for a viscous-type damping, the acoustic pressure signal is in the form $x(t) = A\exp(-\imath\dot{\omega}_1 t)\exp(-\gamma_1 t)$, where $\gamma_1 = \hat{\omega}_1$ is the decrease time of the fundamental mode. Of course, this result is only valid at a single frequency. In order to be able to access a broader frequency range, it is sufficient to repeat the measurement for beams of different lengths. We must note, however, that for the usual frequency ranges in acoustics, the damping related to the material varies only slightly.

Another measurement technique consists of measuring the bandwidth at –3 dB [BLE 90b] which gives direct access to the damping factor[1]: $\Delta f_3/f_r = \eta$. The difficulty of application in the case of the beam is exciting it at a given frequency without modifying its characteristics, which requires a contactless excitation, for example, by a sound wave (which is not very efficient, since the beam is weakly coupled to the fluid), and a contactless vibration measurement, for example, using laser vibrometry.

7.2. Plate vibrations

In this section, we are only interested in bending vibrations of plates of infinite and finite dimensions in a harmonic regime with a dependence of $\exp(-\imath\omega t)$. The longitudinal movements, due to the complexity of the system of coupled equations which describe them, are beyond the scope of this chapter; in fact, the methods used are very close to those used for solving three-dimensional elasticity problems.

We begin by studying the vibrations of an infinite plate by means of calculating the general solution and Green's kernel. Then, we examine the problem of a plate with finite dimensions. After acquiring a theorem on the existence and uniqueness of the solution of the boundary problem, we present some analytic methods in simple cases of rectangular and circular plates. Finally, we give a solution to the problem of a plate, of any shape or subject to non-uniform boundary conditions, using an integral method based on Green's representation of the plate equation.

7.2.1. *Infinite plate*

We consider the usual Cartesian coordinates (O, x, y, z). The plate occupies the plane $z = 0$. $u(x, y)$ is the normal displacement of this plate. As the material which makes up the plate is homogeneous and isotropic, it is obvious that the properties of the plate have a circular symmetry with respect to the origin of the coordinates. The displacement of the plate obviously has the same properties. The solution depends only on the distance from the origin.

The geometry of the problem is presented in Figure 7.2. In polar coordinates (O, r, ϕ), the Laplacian is given by $\Delta = \partial^2/\partial r^2 + \partial/(r\partial r) + \partial^2/\left(r^2\partial\phi^2\right)$. The plate equation $\Delta^2 u(r,\phi) - \left(\omega^2\rho_p h\right)/Du(r,\phi) = 0$ can be written as:

$$\left(\frac{\partial^2}{\partial r^2} + \frac{\partial}{r\partial r} + \frac{\partial^2}{r^2\partial\phi^2} + (\imath k_p)^2\right)\left(\frac{\partial^2}{\partial r^2} + \frac{\partial}{r\partial r} + \frac{\partial^2}{r^2\partial\phi^2} + k_p^2\right) u(r,\phi) = 0$$

where $k_p = \sqrt[4]{\left(\omega^2\rho_p h\right)/D}$ is the wave number in the plate.

---

1 This formula is only a particular case of the result $1/\sqrt{x^2 - 1}\Delta f_n/f_r = \eta$, where Δf_n is the bandwidth at $-n$ dB, $x = 10^{n/20}$ and f_r is the resonance frequency.

7.2.1.1. *General solution*

Similarly as with beams, the plate equation can be broken down into a product of two Helmholtz operators, one for real wave numbers and the other one for complex wave numbers. It is obvious that the general solution of this equation is a linear combination of two general solutions of the corresponding Helmholtz equations.

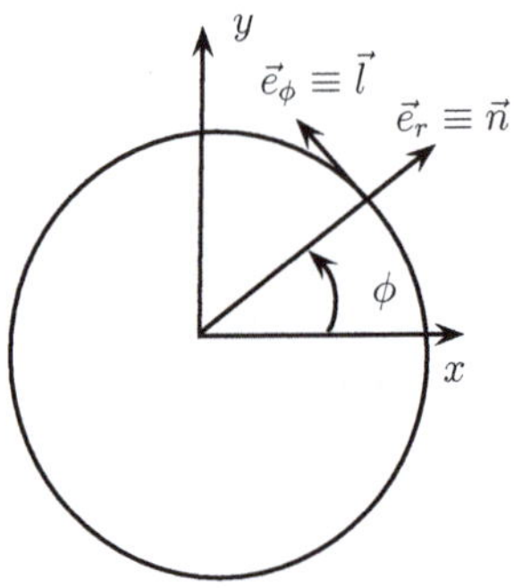

Figure 7.2. *Plate geometry in polar coordinates. $\vec{l}$ is the tangent vector and $\vec{n}$ is the normal vector*

7.2.1.2. *Polar coordinates*

We aim to find the solution u_1 of the equation for real wave numbers $\partial^2 u_1(r,\phi)/\partial r^2 + \partial u_1(r,\phi)/(r\partial r) + \partial^2 u_1(r,\phi)\left(r^2\partial\phi^2\right) + k_p^2 u_1(r,\phi) = 0$ in the form of a Laplace product $u_1(r,\phi) = R(r)\Theta(\phi)$. We proceed as for the Helmholtz equation in polar coordinates whose solution is a linear combination of the Bessel function given by [4.8]. We thus obtain $u_1(r,\phi) = \sum_{n=-\infty}^{+\infty} A_n \mathrm{J}_n(k_p r)\mathrm{e}^{\imath n\phi}$.

The second solution family is obtained if we replace the wave number k_p by $\imath k_p$. We must note that $\mathrm{J}_n(\imath z) = \imath^{-n}\mathrm{I}_n(z)$, where $\mathrm{I}_n(z)$ which is the modified Bessel function with an integer index [WAT 44] (an exponentially increasing function for $z \to \infty$). We obtain $u_2(r,\phi) = \sum_{n=-\infty}^{+\infty} B_n \imath^{-n}\mathrm{I}_n(k_p r)\mathrm{e}^{\imath n\phi}$. The general solution of the plate equation is thus given by:

$$u(r,\phi) = \sum_{n=-\infty}^{+\infty} \left\{A_n \mathrm{J}_n(k_p r) + B_n \mathrm{I}_n(k_p r)\right\} \mathrm{e}^{\imath n\phi}. \qquad [7.14]$$

We must note that the behavior of the general solution for plates is analogous to the beam one whose general solution is the sum of an oscillating function (the role played by the Bessel function $\mathrm{J}_n(z)$) and an exponentially growing function (the role played by the Bessel function $\mathrm{I}_n(z)$).

7.2.1.3. *Cartesian coordinates*

It is easy to express the general solution in Cartesian coordinates (O, x, y). As previously, it is necessary to break down the plate operator into two Helmholtz operators. We obtain $\left(\Delta + (\imath k_p)^2\right)\left(\Delta + k_p^2\right) u(x, y) = 0$. Thus, $u(x, y) = Au_1(x, y) + Bu_2(x, y)$, where $\left(\Delta + k_p^2\right) u_1 = 0$ and $\left(\Delta + (\imath k_p)^2\right) u_2 = 0$, and A and B are arbitrary constants. Let us calculate u_1. To do this, we break down u_1 into the form of a Laplace product: $u_1(x, y) = X(x)Y(y)$. We obtain:

$$\frac{X''(x)}{X(x)} + \frac{Y''(y)}{Y(y)} + k_p^2 = 0,$$

since k is a constant independent of x and y, it is evident that we have:

$$\frac{X''(x)}{X(x)} = -\alpha^2, \alpha \in \mathbb{C}, \frac{Y''(y)}{Y(y)} = -\beta^2, \beta \in \mathbb{C}, \alpha^2 + \beta^2 = k_p^2.$$

$X(x)$ and $Y(y)$ are the solutions of a one-dimensional Helmholtz problem. It is thus obvious that we have $X(x) = Ae^{\imath\alpha x} + Be^{-\imath\alpha x}$ and $Y(y) = Ce^{\imath\alpha y} + De^{-\imath\alpha y}$, with $\alpha^2 + \beta^2 = k_p^2$. It is obvious that u_2 is calculated in the same way by replacing k_p with $\imath k_p$. We, therefore, obtain:

$$\begin{aligned} u(x, y) &= A_1 e^{\imath(\alpha x + \beta y)} + B_1 e^{\imath(\alpha x - \beta y)} + C_1 e^{-\imath(\alpha x + \beta y)} + D_1 e^{-\imath(\alpha x - \beta y)} \\ &\quad + A_2 e^{\alpha x + \beta y} + B_2 e^{\alpha x - \beta y} + C_2 e^{-\alpha x + \beta y} + D_2 e^{-\alpha x - \beta y} \\ &= A_1 \sin \alpha x \sin \beta y + B_1 \sin \alpha x \cos \beta y + C_1 \cos \alpha x \sin \beta y \\ &\quad + D_1 \cos \alpha x \cos \beta y + A_2 \sinh \alpha x \sinh \beta y + B_2 \sinh \alpha x \cosh \beta y \\ &\quad + C_2 \cosh \alpha x \sinh \beta y + D_2 \cosh \alpha x \cosh \beta y, \end{aligned} \qquad [7.15]$$

where $\alpha^2 + \beta^2 = k_p^2$. We must note that again the general solution is composed of oscillating functions and functions of exponential behavior.

7.2.1.4. *Dispersion relation*

We aim to find the free waves which propagate in the plate. For this, it is enough to examine the case of a wave which propagates along the direction of the x axis: $u(x, y) = \exp(\imath kx)$. By rotating the axes, we can access all the waves. Mathematically, this amounts to finding the general solution of the problem using the Laplace method. The plate equation $\Delta^2 u(x, y) - k_p^4 u(x, y) = 0$ can thus be written as $k^4 \exp(\imath kx) - k_p^4 \exp(\imath kx) = 0$. From this, it is simple to deduce that $k = \sqrt[4]{1} k_p$. This is $k = \pm k_p$ and $k = \pm \imath k_p$. We note the presence of four waves: two propagating waves and two waves with exponential behavior.

7.2.1.5. *Green's kernel*

We now aim to find $G(M, M_0)$ such that $D\Delta_M^2 G(M, M_0) - Dk_p^4 G(M, M_0) = \delta_{M_0}(M)$, where Δ_M^2 indicates that the bi-Laplacian acts on the variable M, and which satisfies a condition of the conservation of energy, the limit absorption principle. We aim to find $G(M, M_0)$, with a limit of $G_\epsilon(M, M_0)$ solution of the plate equation for a complex wave number $(k_p + \imath\epsilon)$. We must note from now that, as has been observed for the general solution, Green's kernel of the plate equation must be a particular linear combination of Green's kernels which correspond to two Helmholtz equations for a real wave number and pure imaginary wave number. Green's function is calculated once again using the direct and inverse Fourier transforms. The corresponding kernel is deduced simply by replacing the distance from the origin M by the distance MM_0.

The direct and inverse Fourier transforms of a purely radial two-dimensional function $f(r)$ are given by:

$$\hat{f}(\rho) = 2\pi \int_0^{+\infty} f(r)\mathrm{J}_0(2\pi\rho r)r dr, f(r) = \pi \int_{e^{\imath\pi}\infty}^{+\infty} \hat{f}(\rho)\mathrm{H}_0(2\pi\rho r)\rho d\rho, \quad [7.16]$$

where $\mathrm{H}_0(z)$ is Hankel's function of the first kind of the order zero. Thus, if r is the distance between M and the origin, we get $16\pi^4\rho^4\hat{G}_\epsilon(\rho)v\left(k_p + \imath\epsilon\right)^4 \hat{G}_\epsilon(\rho) = 1/D$. This is:

$$\begin{aligned} D\hat{G}_\epsilon(\rho) &= \frac{1}{2\left(k_p + \imath\epsilon\right)^2}\left(\frac{1}{4\pi^2\rho^2 - \left(k_p + \imath\epsilon\right)^2} + \frac{-1}{4\pi^2\rho^2 + \left(k_p + \imath\epsilon\right)^2}\right) \\ &= D\hat{G}_1(\rho) + D\hat{G}_2(\rho). \end{aligned}$$

We calculate the inverse of $\hat{G}_1(\rho)$ using an inverse Fourier transform, $\hat{G}_2(\rho)$ can be deduced immediately.

$$G_1(r) = \frac{\pi}{2\left(k_p + \imath\epsilon\right)^2} \int_{e^{\imath\pi}\infty}^{+\infty} \frac{1}{4\pi^2\rho^2 - \left(k_p + \imath\epsilon\right)^2} \mathrm{H}_0(2\pi\rho r)\rho d\rho.$$

We calculate this integral using the method of residues on the following contour.

The integrand has only two simple poles $2\pi\rho = \pm\left(k_p + \imath\epsilon\right)$ with a non-zero imaginary part. The contour chosen contains only the pole $2\pi\rho = k_p + \imath\epsilon$. The residue theorem applies without any particular difficulties and we obtain $G_1(r) = \imath/(8(k_p + \imath\epsilon)^2)\ \mathrm{H}_0\left(\left(k_p + \imath\epsilon\right)r\right)$. Similarly, we obtain $G_2(r) = -\imath/\left(8\left(k_p + \imath\epsilon\right)^2\right) \mathrm{H}_0\left(\imath\left(k_p + \imath\epsilon\right)r\right)$. By approaching the limit $\epsilon \to 0$, we finally obtain $G(r) = \imath/\left(8Dk_p^2\right)\left(\mathrm{H}_0\left(k_p r\right) - \mathrm{H}_0\left(\imath k_p r\right)\right)$. With $\mathrm{H}_0(\imath z) = -2\imath/\pi \mathrm{K}_0(z)$, where $\mathrm{K}_0(z)$ is modified Bessel's function of the second kind [WAT 44] or Kelvin's function of order zero (exponentially decreasing function for $z \to \infty$), the expression of Green's kernel of the plate equation is thus given by:

$$G(M, M_0) = \frac{\imath}{8Dk_p^2}\left(\mathrm{H}_0(k_p MM_0) + \frac{2\imath}{\pi}\mathrm{K}_0(k_p MM_0)\right) \qquad [7.17]$$

which highlights the sum of an oscillating function and an exponentially decreasing function. We study the behavior of this kernel in the vicinity of the origin. We know that $-\imath\mathrm{H}_0(z) \approx 2/\pi \ln(z/2)$ and that $\mathrm{K}_0(z) \approx -\ln(z/2)$ [WAT 44]. We obtain:

$$r \to 0 : G(r) \approx \frac{\imath}{8Dk_p^2}, G'(r) \approx 0, G''(r) \approx 1/(4D\pi k_p^2)\ln r,$$

$$G'''(r) \approx C\, 1/(4D\pi k_p^2) 1/r.$$

The amplitude of the plate is limited and its slope at the origin is zero. This result reflects the symmetry of revolution of the plate.

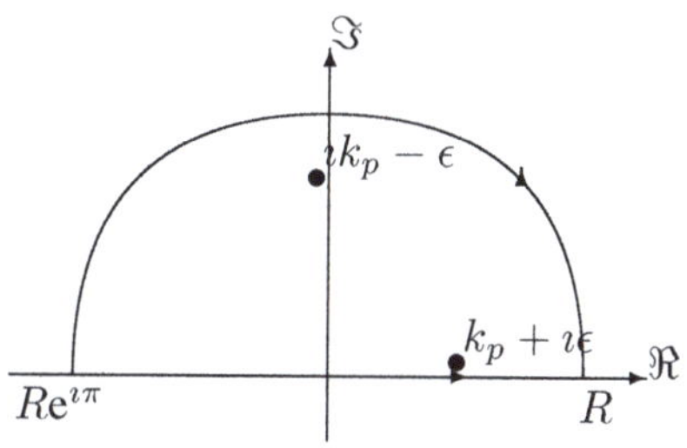

Figure 7.3. *Integration contour*

7.2.1.6. *Thick plate*

We know that in a thin plate the phase velocity (which reflects the wave propagation) and the group velocity (which reflects the energy propagation velocity) increase with frequency and can thus *a priori* propagate faster than the mechanical waves (longitudinal or transverse). To correct this issue, we must take into consideration that at high frequencies, the wave lengths have a dimension comparable to the thickness and that the transverse shear, neglected to establish the Kirchhoff equation, must be taken into account (which is equivalent of considering a thick "thin plate"). To do this, we use the homogeneous Timoshenko–Mindlin equation, which adds a shearing effect to the bending effect described by Kirchhoff. The starting equations form a coupled system of two equations which involve the displacement $u(x,t)$ and the rotation $\partial_x u(x,t)$. This system, in the absence of the second member, can be reduced to a single equation on the displacement:

$$\left(D\Delta - \frac{\rho_p h^3}{12}\frac{\partial^2}{\partial t^2}\right)\left(\Delta - \frac{\rho_p}{\kappa^2 G}\frac{\partial^2}{\partial t^2}\right)u(x,t) + \rho_p h \frac{\partial^2 u(x,t)}{\partial t^2} = 0$$

where $G = E/2(1+\nu)$ is the shear modulus and κ is the shear coefficient with $\kappa^2 = \pi^2/12$ in the dynamic case or $5/6$ in the static case. In a harmonic regime, we obtain:

$$\left(D\Delta + \frac{\rho_p h^3}{12}\omega^2\right)\left(\Delta + \frac{\rho_p}{\kappa^2 G}\omega^2\right)u(x) - \rho_p h\omega^2 u(x) = 0. \qquad [7.18]$$

We aim to find the solution to equation [7.18] using the Laplace method, which comes down to searching whether the free waves can propagate in a medium which can be described by the considered equation, $u(x) = \exp(\imath kx)$, and we solve it for k. We obtain $k^4 - k^2\left((1-\nu^2)/E + 1/(\kappa^2 G)\right)\rho_p\omega^2 + \rho_p^2\omega^4(1-\nu^2)/(\kappa^2 EG) - \rho_p h\omega^2/D = 0$ which is a bisquare equation in k^2 which can be solved very easily. We show that:

$$2\left(k^{\pm}\right)^2 = \left(\frac{1-\nu^2}{E} + \frac{1}{\kappa^2 G}\right)\rho_p\omega^2 \pm \sqrt{\rho_p^2\omega^4\left(\frac{1-\nu^2}{\kappa^2 EG} - \frac{1}{\kappa^2 G}\right)^2 + 4\frac{\rho_p h\omega^2}{D}},$$

and we see that for ω, there is at least one real root k^+ and that for a sufficiently large ω, there is a second real root k^-. The transition for ω_0 is such that $\left(k^-\right)^2 = 0$, this is $\omega_0 = 12\kappa^2/h^2 E/(2\rho_p h(1+\nu))$. However, we know that $c_T^2 = E/(2\rho_p(1-\nu^2)) = G/\rho_p$ and $c_L^2 = (1-\nu^2)/(2\rho_p(1+\nu)(1-2\nu))$. This is $\omega_0 = 12/h^2\kappa^2 c_T^2$. We define the dimensionless frequency $\Omega = \omega h/c_T$, and we have $\Omega_0^2 = 12\kappa^2 = \pi^2$. This is $\omega_0 h/c_T = \pi$, with $c_T \approx 3000$ m/s (steel, glass and aluminum), we have $f_0 = c_T/2h \approx 1500/h$ Hz. At low frequency, $\omega \to 0$, we recover the results obtained from Kirchhoff's equation $\left(k^{\pm}\right)^4 \approx \pm\rho_p h\omega^2/D$. At high frequency, $\omega \to \infty$, we obtain $\left(k^+\right)^2 \approx \rho_p\omega^2\left(1-\nu^2\right)/E = \omega^2/c_l^2(1-\nu)^2/(1-2\nu)$ and $\left(k^-\right)^2 \approx \rho_p\omega^2 2(1+\nu)/\left(\kappa^2 E\right) = \omega^2/\left(c_t^2\right)1/\kappa^2$. If $\nu \to 0$, k^+ is the wave number of waves in compression in an unbounded medium k_L, if $\nu \approx 1/6$, $k^+ \approx k_L$. For all ν, we have $k^- \approx 1,2k_T$. At high frequency, the Timoshenko–Mindlin waves propagate at finite velocities, which are always smaller than or equal to the velocities of the compression waves and the shear waves. Obviously, at very high frequency, the plate cannot be regarded as a two-dimensional medium and the waves which propagate in the plate are thus the compression waves and the shear waves.

In Figures 7.4, 7.5 and 7.6, we show the shape of the wave numbers, the phase velocities and the group velocities for the various types of waves. These figures show that schematically Kirchhoff's equation is valid up to about $f_0/10$, which, to give a better idea, for a steel plate with a thickness of 5 cm is around 3 kHz. This corresponds to a wavelength $\lambda = c_T/f \approx 0.1$ m, thus a half wave-length comparable to the thickness. The Timoshenko–Mindlin equation is very precise and describes the volume effects with the appearance of two natural wave types for elastic solids, whose observed velocities asymptotically approach the corresponding velocities in infinite media.

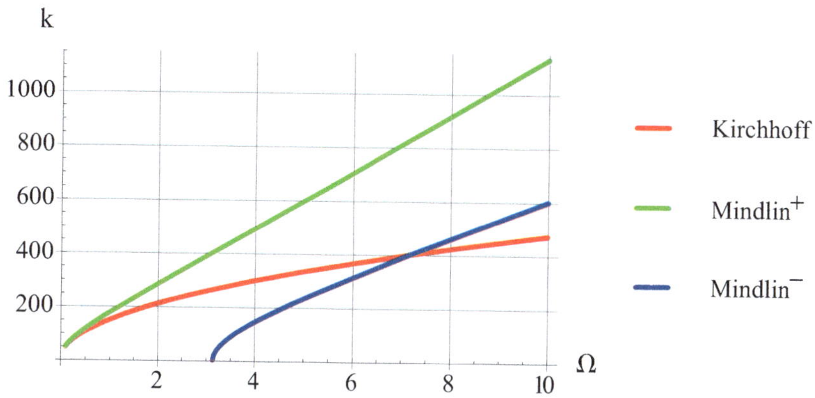

Figure 7.4. *Wave numbers for the Kirchhoff and the Timoshenko–Mindlin thin plates. For a color version of the figure, see www.iste.co.uk/anselmet/acoustics.zip*

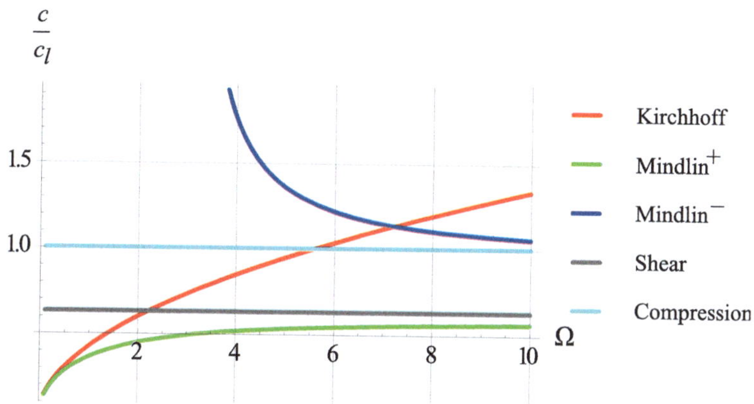

Figure 7.5. *Phase celerities for the Kirchhoff and Timoshenko–Mindlin thin plates. In this figure, we focus on the phase celerities of the compression waves and the shear waves. For a color version of the figure, see www.iste.co.uk/anselmet/acoustics.zip*

7.2.2. Finite plate

Because of the notable variety in possible forms, there are only a few configurations for which it is possible to express, in a simple manner, the eigenmodes and eigenfrequencies of a plate. We thus focus on those practical cases, for which it is possible to give an analytic solution and we then propose a more general method that can be applied to any geometry.

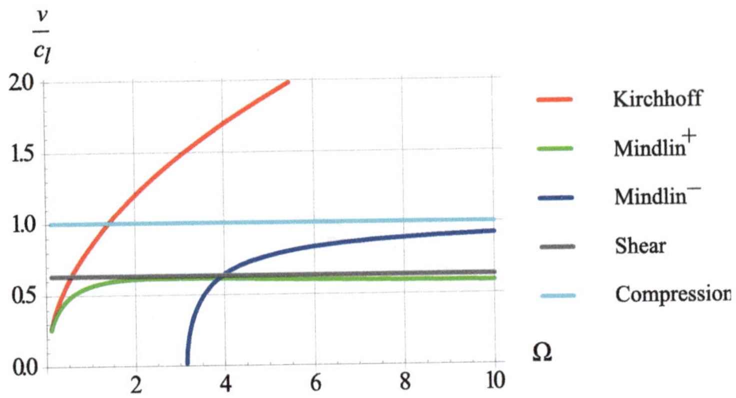

Figure 7.6. *Group celerities for the Kirchhoff and Timoshenko–Mindlin thin plates. In this figure, we focus on the group celerities of the compression waves and the shear waves. For a color version of the figure, see www.iste.co.uk/anselmet/acoustics.zip*

We consider the boundary problem governed by the plate equation and a couple of permissible boundary conditions at the edge of the field. We have the theorem (its demonstration, which is too difficult, is not given):

1) There exists a countable double sequence k_{mn} of values k_p such that the homogeneous problem possesses a finite number of linearly independent solutions u_{mn}. k_{mn} is the eigen wave number and u_{mn} is the corresponding eigenmode. If the order of multiplicity of k_{mn} is greater than 1 (multiple eigen wave number), we obtain similar results.

2) If $k_p = k_{mn}$, the inhomogeneous problem does not have solutions. If $k_p \neq k_{mn}$, there exists a unique solution regardless of the excitation.

3) The double sequence u_{nm} can always be made orthonormal and constitutes a basis on which the solution of the inhomogeneous problem can be developed in a convergent series.

7.2.2.1. *Rectangular plate with simply supported edges*

Consider a plate which occupies the area of the plane (O, x, y) defined by $0 \leq x \leq \ell_x \cup 0 \leq y \leq \ell_y$. This plate is simply supported at its contour. The boundary conditions $u = 0$ and $\Delta u - (1-\nu)\partial^2_{l^2} u = 0$ can be written in Cartesian coordinates:

$$u(x,y) = 0, \frac{\partial^2 u(x,y)}{\partial x^2} + \nu \frac{\partial^2 u(x,y)}{\partial y^2} = 0 \quad \text{at } x = 0, \ell_x, \forall y \in]0, +\ell_y[$$

$$u(x,y) = 0, \frac{\partial^2 u(x,y)}{\partial y^2} + \nu \frac{\partial^2 u(x,y)}{\partial x^2} = 0 \quad \text{at } y = 0, \ell_y, \forall x \in]0, +\ell_x[.$$

Let us clarify these boundary conditions. The sought displacement is in the form of a Laplace product $u(x,y) = U(x)V(y)$. As the coordinate lines are parallel to the axis of the coordinate system, this product allows us to simplify the boundary conditions. We consider, for example, the boundary conditions at $x = 0$. We have $U(0)V(y) = 0$, $\forall y \in]0, +\ell_y[$ and $\partial^2 U(0)V(y)/\partial x^2 + \nu \partial^2 U(0)V(y)/\partial y^2 = 0$, $\forall y \in]0, +\ell_y[$. The first relation is satisfied if $U(0) = 0$ or if $V(y) = 0$. As the second possibility implies that the displacement is zero at any point on the plate, we must exclude it. Thus, the first equation gives us $U(0) = 0$. The second relation gives $U''(0)V(y) + \nu U(0)V''(y) = 0$ $\forall y \in]0, +\ell_y[$, taking into account the relation $U(0) = 0$, we obtain $U''(0)V(y) = 0$ $\forall y \in]0, +\ell_y[$. The same reasoning leads us to impose as a boundary condition $U''(0) = 0$. If we do the same for the other boundary conditions, we finally obtain $u(x,y) = 0, \partial^2 u(x,y)/\partial x^2 = 0$ at $x = 0, \ell_x, \forall y \in]0, +\ell_y[$ and $u(x,y) = 0, \partial^2 u(x,y)/\partial y^2 = 0$ at $y = 0, \ell_y, \forall x \in]0, +\ell_x[$.

7.2.2.2. *Modal basis*

To evaluate the expression of the eigenmodes, we apply the boundary conditions to the general solution [7.15]. We have:

$$\begin{aligned} u(x,y) = {} & A_1 \sin \alpha x \sin \beta y + B_1 \sin \alpha x \cos \beta y + C_1 \cos \alpha x \sin \beta y \\ & + D_1 \cos \alpha x \cos \beta y + A_2 \sinh \alpha x \sinh \beta y + B_2 \sinh \alpha x \cosh \beta y \\ & + C_2 \cosh \alpha x \sinh \beta y + D_2 \cosh \alpha x \cosh \beta y. \end{aligned}$$

Let us now examine the first couple of boundary conditions. The conditions at $x = 0$ lead us to $C_1 \sin \beta y = -D_1 \cos \beta y$ and $C_2 \sinh \beta y = -D_2 \cosh \beta y$. From this, if we introduce these results into the expression of the general solution $u(x,y)$:

$$\begin{aligned} u(x,y) = {} & A_1 \sin \alpha x \sin \beta y + B_1 \sin \alpha x \cos \beta y + A_2 \sinh \alpha x \sinh \beta y \\ & + B_2 \sinh \alpha x \cosh \beta y. \end{aligned}$$

Similarly, the conditions at $y = 0$ lead us to $B_1 = B_2 = 0$. This is $u(x,y) = A_1 \sin \alpha x \sin \beta y + A_2 \sinh \alpha x \sinh \beta y$. Applying the boundary conditions at $x = \ell_x$ leads to $\sin \alpha \ell_x \sin \beta y = 0$ and to $\sinh \alpha \ell_x \sinh \beta y = 0$, $\forall y \in]0, +\ell_y[$. From there, it is obvious that we have $\sin \alpha \ell_x = 0$ or $\sinh \alpha \ell_x = 0$. The boundary conditions at $y =$

ℓ_y lead to a similar result $\sin \beta\ell_y = 0$ or $\sinh \beta\ell_y = 0$. These relations are satisfied if the sine terms are zero. The hyperbolic sine can only be zero if $\alpha = 0$, which corresponds to a solution $u(x, y)$ with zero everywhere and is obviously undesirable. Thus, $\alpha\ell_x = m\pi$, $m \in \boldsymbol{N}^\star$ and $\beta\ell_y = n\pi$, $n \in \boldsymbol{N}^\star$. The eigen wave numbers of the rectangular plate with simply supported edges are thus given by:

$$k_{mn} = \sqrt{\left(\frac{m\pi}{\ell_x}\right)^2 + \left(\frac{n\pi}{\ell_y}\right)^2}.$$

The corresponding eigenfrequencies, with $\omega_{mn} = \sqrt{D/\rho_p h}k_{mn}^2$, are given by:

$$f_{mn} = \frac{\pi}{2}\sqrt{\frac{D}{\rho_p h}}\left(\left(\frac{m}{\ell_x}\right)^2 + \left(\frac{n}{\ell_y}\right)^2\right).$$

The expression of the eigenmodes is given by $u_{mn}(x, y) = A_{mn} \sin(m\pi x/\ell_x) \sin(n\pi y/\ell_y)$. The mode, being defined to a multiplicative constant, is fixed by normalizing the modes. We examine the scalar product $\langle f, g\rangle = \int_0^{\ell_x} \int_0^{\ell_y} f(x, y)g^\star(x, y)dxdy$. The scalar product of the two eigenmodes $\langle u_{mn}, u_{pq}\rangle$ is, therefore:

$$\begin{aligned}\langle u_{mn}, u_{pq}\rangle &= A_{mn}^2 \int_0^{\ell_x} \sin\left(\frac{m\pi x}{\ell_x}\right) \sin\left(\frac{p\pi x}{\ell_x}\right) dx \int_0^{\ell_y} \sin\left(\frac{n\pi y}{\ell_y}\right) \sin\left(\frac{q\pi y}{\ell_y}\right) dy \\ &= A_1^2 \frac{\ell_x}{2} \delta_m^p \frac{\ell_y}{2} \delta_n^q\end{aligned}$$

if we choose $A_{mn} = 2/\sqrt{\ell_x \ell_y}$, it is obvious that the eigenmodes are orthonormal. We finally obtain: $u_{mn}(x, y) = 2\sqrt{\ell_x \ell_y} \sin(m\pi x/\ell_x) \sin(n\pi y/\ell_y)$.

When the plate is square, the couples (mn) and (nm) correspond to identical eigenfrequencies. Such a situation is called the degenerescence condition. In this case, the plate vibrates at a frequency of f_{mn} but its shape is given by any linear combination of the corresponding eigenmodes. Another case of degenerescence is where the transverse dimensions of the plate are multiple. We must note, however, that in this case, the notion of the basis of eigenmodes is preserved.

7.2.2.3. *Green's kernel*

We calculate Green's kernel of the rectangular plate with simply supported edges using the expansion based on eigenmodes. For this, we solve the problem:

$$\Delta_M^2 G(M, M') - k_p^4 G(M, M') = \frac{1}{D}\delta_{M'}(M), \text{ on the plate,}$$

$$G(M, M') = 0 \text{ and } \Delta_M G(M, M') - (1 - \nu)\partial_{\ell^2}^2 G(M, M') = 0, \text{ at the edge,}$$

using the expansion based on eigenmodes. If $M = (x, y)$ and $M' = (x', y')$, it is obvious that we can write:

$$G(M, M') = \frac{1}{D} \sum_{n=1}^{\infty} \sum_{m=1}^{\infty} \frac{f_{mn} u_{mn}(x, y)}{k_{mn}^4 - k_p^4}, \text{ if } k_p \neq k_{pm}$$

where $f_{mn} = \langle \delta_{x',y'}(x, y), u_{mn}(x, y) \rangle$. Because of the properties of the Dirac measure, it is clear that we obtain $f_{mn} = u_{mn}(x', y')$ and thus $G(M, M') = G(M', M)$ and $G(M, M') = G(M - M')$.

7.2.2.4. *Clamped or free rectangular plate*

In contrast with the boundary conditions of simply supported plates (which are simply supported either on the four sides or two opposite ones), which allow us to obtain an analytic solution, we do not know how to find an exact simple solution for a separate variable, when a rectangular plate is clamped or free on more than three sides or more than two adjacent sides[2]. Nevertheless, we propose approximate analytic solutions for the eigenmodes and frequencies, allowing us to calculate modal shape with great precision and eigenfrequencies with a more than acceptable precision (around 1%). To do this, we approximate the modes of a plate by a product of beam modes that satisfy the boundary conditions imposed on two opposite edges [LEI 93a].

7.2.2.5. *Clamped plate*

Without going into detail about the calculations, the orthonormal modes of a clamped rectangular plate with the dimensions $[0, \ell_x] \times [0, \ell_x]$ are given by a product of beam eigenmodes that satisfy the boundary conditions $u_{mn}(x, y) = 2/\sqrt{\ell_x \ell_y} X_m^{\ell_x}(x) X_n^{\ell_y}(y)$, $m, n = 1, 2, 3, \cdots$ where

$$m = 1, X_m^{\ell_x}(x) = 0,991289 \cos \gamma_1 (\frac{x}{\ell_x} - \frac{1}{2})\pi + \frac{\sin \frac{\gamma_1}{2}\pi}{\sinh \frac{\gamma_1}{2}\pi} \cosh \gamma_1 (\frac{x}{\ell_x} - \frac{1}{2})\pi, \tag{7.19}$$

$$m \text{ odd}, X_m^{\ell_x}(x) = \cos \gamma_m (\frac{x}{\ell_x} - \frac{1}{2})\pi + \frac{\sin \gamma_m \frac{\pi}{2}}{\sinh \gamma_m \frac{\pi}{2}} \cosh \gamma_m (\frac{x}{\ell_x} - \frac{1}{2})\pi, \tag{7.20}$$

$$m \text{ even}, X_m^{\ell_x}(x) = \sin \gamma_m (\frac{x}{\ell_x} - \frac{1}{2})\pi - \frac{\sin \gamma_m \frac{\pi}{2}}{\sinh \gamma_m \frac{\pi}{2}} \sinh \gamma_m (\frac{x}{\ell_x} - \frac{1}{2})\pi, \tag{7.21}$$

2 This is due to the fact that the system formed by the plate equation and the boundary conditions, free or clamped, cannot be separated. To see this, it is enough, for example, to assume that the displacement of a completely clamped plate is separable, and to apply the eight boundary conditions to determine the eight constants and to see that the last two boundary conditions lead to an impossible solution. This shows, by contradiction, that the base assumption, namely the separability of the solution, was false.

with $\gamma_1 \approx 1.505618$ and $\gamma_m = m + 1/2$. The eigenfrequencies are given by:

$$f_{mn} = \frac{\pi}{2}\sqrt{\frac{D}{\rho_p h}}\sqrt{\left(\frac{G(m)}{\ell_x}\right)^4 + \left(\frac{G(n)}{\ell_y}\right)^4 + 2\frac{H(m)}{\ell_x^2}\frac{H(n)}{\ell_y^2}}, \qquad [7.22]$$

with $G(m) = \gamma_m$ and $H(m) = 1.248$ if $m = 1$ and $H(m) = (m+1/2)^2 (1 - 2/(m+1/2)\pi)$ if $m > 1$.

7.2.2.6. *Free plate*

The orthonormal modes of a free rectangular plate with the dimensions $[0, \ell_x] \times [0, \ell_x]$ are given by $u_{mn}(x, y) = 2/\sqrt{\ell_x \ell_y} X_m^{\ell_x}(x) X_n^{\ell_y}(y)$, $m, n = -1, 0, 1, 2, \cdots$ where :

$$m = -1, X_m^{\ell_x}(x) = \frac{1}{\sqrt{2}}, m = 0, X_m^{\ell_x}(x) = \sqrt{\frac{3}{2}}\left(1 - \frac{2x}{\ell_x}\right)$$

$$m = 1, X_m^{\ell_x}(x) = 0.991289 \cos\gamma_1(\frac{x}{\ell_x} - \frac{1}{2})\pi - \frac{\sin\frac{\gamma_1}{2}\pi}{\sinh\frac{\gamma_1}{2}\pi}\cosh\gamma_1(\frac{x}{\ell_x} - \frac{1}{2})\pi,$$

$$m \text{ odd }, X_m^{\ell_x}(x) = \cos\gamma_2(\frac{x}{\ell_x} - \frac{1}{2})\pi - \frac{\sin\gamma_2\frac{\pi}{2}}{\sinh\gamma_2\frac{\pi}{2}}\cosh\gamma_2(\frac{x}{\ell_x} - \frac{1}{2})\pi,$$

$$m \text{ even }, X_m^{\ell_x}(x) = \sin\gamma_2(\frac{x}{\ell_x} - \frac{1}{2})\pi + \frac{\sin\gamma_2\frac{\pi}{2}}{\sinh\gamma_2\frac{\pi}{2}}\sinh\gamma_2(\frac{x}{\ell_x} - \frac{1}{2})\pi,$$

with $\gamma_1 \approx 1.505618$ and $\gamma_m = m + \frac{1}{2}$. We must note that, even if the modal shapes of a free plate seem very similar to those of a clamped plate, with the exception of the additional modes $m = -1$ and $m = 0$, this is misleading (note the sign change of each second term), because, for a free plate, the displacement is always at its maximum at the edge, whereas, for a clamped plate, it is zero at the edge. The eigenfrequencies are given by:

$$f_{mn} = \frac{\pi}{2}\sqrt{\frac{D}{\rho_p h}}\sqrt{\left(\frac{G(m)}{\ell_x}\right)^4 + \left(\frac{G(n)}{\ell_y}\right)^4 + \frac{2}{\ell_x^2 \ell_y^2}\left(\nu H(m)H(n) + (1-\nu)J(m)J(n)\right)} \qquad [7.23]$$

with $G(-1) = 0$, $G(0) = 0$, $G(m) = \gamma_m$ if $m > 1$, $H(-1) = 0$, $H(0) = 0$, $H(1) = 1.248$ and $H(m) = (m+1/2)^2 (1 - 2/(m+1/2)\pi)$ if $m > 1$, $J(-1) = 0$, $J(0) = 12/\pi^2$, $J(1) = 5.017$ and $J(m) = (m+1/2)^2 (1 + 6/(m+1/2)\pi)$ if $m > 1$.

An example of the first eigenfrequencies calculated and measured for a free plate is proposed in Table 7.2. The experiment was conducted using a rectangular plate with the dimensions 29.1 cm $\times$ 47.5 cm $\times$ 0.15 cm made of aluminum $E_0 = 75$ GPa, $\nu = 0.3$ and $\rho_p = 2.850$ kg/m^3. The plate was hung by two fishing wires fixed in the thickness of the plate. In order to avoid the loss of energy by the excitation and measurement system, the plate was set in vibration by a sound wave emitted from a loudspeaker placed nearby and its vibration velocity was measured using a laser vibrometer. The identification of the complex resonance frequencies was carried out by comparing the response of the plate with a Lorentzian[3] near each resonance. In order to account for the energy losses within the plate, we take into consideration three energy losses. The first is induced by viscous dissipation, the value of the viscous damping coefficient is fixed at $\eta_v = 0.0002$. The second is the outcome of the thermoelastic dissipation using the Zener model, the damping coefficient is the one given by equation [2.30]:

$$\eta_z = \frac{E_0 \alpha^2 T}{c_v \rho_p} \frac{\frac{\omega}{\omega_\tau}}{1 - \left(\frac{\omega}{\omega_\tau}\right)^2}, \text{ where } \omega_\tau = \frac{k_\theta}{c_v \rho_p \left(\frac{h}{\pi}\right)^2}, \qquad [7.24]$$

with $\alpha = 23 \cdot 10^{-6}$, $c_v = 877$ J/kg/K, $k_\theta = 200$ W/m/K at an ambient temperature of $T_0 = 300\,\mathrm{deg\,K}$. Finally, the last source of damping is the energy loss during acoustic radiation, the calculation of the resonance frequencies of the plate coupled to the fluid $\tilde{f}_{mn}$, using a perturbation method, is given by formula [8.26] $\tilde{f}_{mn} \approx f_{mn}/\sqrt{1 - \epsilon\beta_{mn}}$, where f_m is given by equation [7.23], $\epsilon = \rho_f/\rho_p h$, if ρ_f is the density of air and β_{mn} represents, without the factor $\omega^2 \rho_f$, the energy lost by radiation of the mode mn. The resonance frequencies $\breve{f}_{mn}$ of the dissipative plate are thus given by the equation $\breve{f}_{mn} = \tilde{f}_{mn} \times \sqrt{(1 - \imath\eta_v)} \times \sqrt{(1 - \imath\eta_z)}$.

7.2.2.7. *Identification of experimental resonance frequencies*

The experimental complex resonance frequencies can be estimated from the frequency response of a mechanical system measured in some points (less than 10 are normally enough) around the considered resonance. To do this, we begin with the modal representation [3.70] and we assume that around each resonance frequency $\breve{f}_{mn}$, the complete series can be written by a single term corresponding to the considered resonance $H_{mn}(f) \approx A_{mn}/\left(\left(f - \breve{f}_{mn}\right)\left(f + \breve{f}^*_{mn}\right)\right)$. The calculations of the amplitudes and corresponding frequencies can be obtained by a fairly simple iterative procedure. At each iteration, we aim to find the frequency corresponding to the maximum response, then, around this frequency, we identify the amplitude A_{mn} and the resonance f_{mn}. Then, we subtract from the entire frequency response the

3 This leads to the assumption that around each resonance, the response of the plate can be described as $A_{mn}/(f^2 - f^2_{mn})$ where the amplitude A_{mn}, which depends on the measuring point, and the complex frequency f_{mn} need to be identified.

contribution of the estimated mode, and we restart the procedure once again. So constructed, the frequency response can be written in the form of a finite sum of Lorentzian functions by f:

$$H(f) \approx \sum_{m,n} \frac{A_{mn}}{\left(f - \check{f}_{mn}\right)\left(f + \check{f}^*_{mn}\right)} \qquad [7.25]$$

where $2\pi \check{f}_{mn} = \Omega_{mn} - \imath\alpha_{mn}$ is the resonance pulsation and $\check{f}^*_{mn}$ is the conjugate of $\check{f}_{mn}$. Figure 7.7 shows an example of such an identification carried out on a free rectangular plate with the dimensions 35 cm $\times$ 40 cm $\times$ 0.2 cm made from aluminum. If we ignore the measurement noise, we can reconstruct the behavior of the plate with great accuracy.

The results show that the observed damping can only be described properly by taking into account all the damping components (viscoelastic, thermoelastic and acoustic). Although in this particular example, the thermoelastic damping is the predominant one and acoustic damping is negligible[4], the different dissipation sources are comparable (as for a steel plate for clamped or simply supported boundary conditions). We should bear in mind that in this case with non-free boundary conditions, a damping source, which is difficult to quantify, comes from the dissipation of vibrations at the edge (by vibration transfer or friction).

We must note here that the thermoelastic damping according to the Zener model is only an approximation and that, strictly speaking, we should address the coupled thermoelastic problem given by the Duhamel–Neumann equations [2.21] or [2.22] and the linearized thermal conduction equation [2.24]. To solve these equations, we proceed iteratively [ZOG 06]. We begin by solving the dynamic equation with the constitutive equation $\sigma_{ij} = E/(1+\nu)\left(d_{ij} + \nu/(1-2\nu)d_{ll}\delta_{ij}\right)$, then, with the data of the constraint σ_{ll}, we solve the heat equation with the second part $\rho c_v dT/dt - k_\theta \Delta T = -\alpha T_0 d\sigma_{ll}/dt$. Then, we calculate a thermal stress matrix $\sigma^{th}_{ij} = -3K\alpha(T - T_0)\delta_{ij}$ as well as the energy $E_{th} = \int_V d_{ii}\sigma^{th}_{kk}$ of the thermoelastic strain. This energy allows us to deduce, using perturbation and for each mode of the structure, the modification to the resonances $\rho_p h\omega^2_{mn,th} = \rho_p h\omega^2_{th} + E_{th}$. The results presented in Figure 7.8 show a comparison of the thermoelastic loss factor modeled for different mechanical boundary conditions using the iterative method described previously and the Zener damping given by relation [7.24]. The Zener model (whose maximum thermoelastic damping here is in the vicinity of $\omega_\tau/2\pi \approx 35$ Hz) gives outstandingly accurate results for the clamped and supported cases and it is sufficient for most practical cases [ZOG 15].

4 A free plate is rarely coupled with a fluid, because the acoustic pressure difference on either side of the plate is negligible due to the pressure balance on the edge of the plate [ZOG 06]; the main effect here is the added mass effect (and thus a decrease in the real part of the resonance frequency).

We also show that the frequency shift does no exceed 0.5% and can be ignored[5]. By contrast, the Zener model overestimates the thermoelastic damping when the plate is free. This is explained by the fact that the compression strain energy, which governs the loss of thermoelastic energy as shown by the Duhamel–Neumann law, heavily depends on the vibration mode when the plate is free and not at all for the other boundary conditions (thus, the use of a homogeneous model is justified).

$\tilde{f}_{mn}$		$m = -1$	$m = 0$	$m = 1$	$m = 2$
$n = -1$	vis.	0	0	$94.64 - \imath 0.009$	$260.81 - \imath 0.026$
	the.	0	0	$94.64 - \imath 0.09$	$260.81 - \imath 0.12$
	aco.	0	0	$94.10 - \imath 5.10^{-6}$	$259.16 - \imath 0.043$
	sum.	0	0	$94.10 - \imath 0.10$	$259.16 - \imath 0.19$
	exp.	–	–	$99.64 - \imath 0.18$	$276.90 - \imath 0.20$
$n = 0$	vis.	0	$36.78 - \imath 0.004$	$120.58 - \imath 0.012$	$283.35 - \imath 0.028$
	the.	0	$36.78 - \imath 0.037$	$120.58 - \imath 0.10$	$283.35 - \imath 0.12$
	aco.	0	$36.52 - \imath 4.10^{-5}$	$119.99 - \imath 4.10^{-6}$	$281.86 - \imath 0.006$
	sum.	0	$36.53 - \imath 0.04$	$119.99 - \imath 0.11$	$281.86 - \imath 0.15$
	exp.	0	–	$125.76 - \imath 0.15$	$290.65 - \imath 0.13$
$n = 1$	vis.	$35.52 - \imath 0.004$	$82.72 - \imath 0.008$	$184.01 - \imath 0.018$	$349.53 - \imath 0.035$
	the.	$35.52 - \imath 0.036$	$82.72 - \imath 0.085$	$184.01 - \imath 0.11$	$349.53 - \imath 0.12$
	aco.	$35.23 - \imath 5.10^{-7}$	$82.26 - \imath 3.10^{-6}$	$183.21 - \imath 9.10^{-6}$	$347.91 - \imath 0.001$
	sum.	$35.23 - \imath 0.04$	$82.25 - \imath 0.09$	$183.21 - \imath 0.13$	$347.91 - \imath 0.16$
	exp.	$41.03 - \imath 0.04$	$78.44 - \imath 0.27$	$177.15 - \imath 0.13$	$353.81 - \imath 0.23$
$n = 2$	vis.	$97.89 - \imath 0.010$	$147.81 - \imath 0.015$	$267.26 - \imath 0.027$	$444.21 - \imath 0.044$
	the.	$97.89 - \imath 0.094$	$147.81 - \imath 0.11$	$267.26 - \imath 0.12$	$444.21 - \imath 0.12$
	aco.	$97.04 - \imath 0.002$	$146.93 - \imath 1.10^{-4}$	$266.04 - \imath 5.10^{-5}$	$442.02 - \imath 0.018$
	sum.	$97.04 - \imath 0.11$	$146.93 - \imath 0.12$	$266.04 - \imath 0.15$	$442.02 - \imath 0.18$
	exp.	–	$146.43 - \imath 0.10$	$262.23 - \imath 1.20$	$447.59 - \imath 0.2$
$n = 3$	vis.	$191.86 - \imath 0.019$	$240.57 - \imath 0.024$	$370.93 - \imath 0.037$	$560.48 - \imath 0.056$
	the.	$191.86 - \imath 0.11$	$240.57 - \imath 0.12$	$370.93 - \imath 0.12$	$560.48 - \imath 0.12$
	aco.	$190.88 - \imath 1.10^{-5}$	$239.56 - \imath 1.10^{-4}$	$369.59 - \imath 2.10^{-5}$	$558.34 - \imath 4.10^{-6}$
	sum.	$190.88 - \imath 0.13$	$239.56 - \imath 0.14$	$369.59 - \imath 0.16$	$558.34 - \imath 0.18$
	exp.	$207.96 - \imath 0.16$	$231.12 - \imath 0.07$	$370.94 - \imath 0.36$	$569.00 - \imath 0.24$
$n = 4$	vis.	$317.15 - \imath 0.032$	$364.24 - \imath 0.036$	$500.46 - \imath 0.05$	$700.94 - \imath 0.07$
	the.	$317.15 - \imath 0.12$	$364.24 - \imath 0.12$	$500.46 - \imath 0.12$	$700.94 - \imath 0.5$
	aco.	$315.44 - \imath 0.085$	$362.70 - \imath 0.09$	$498.64 - \imath 2.10^{-3}$	$698.22 - \imath 0.049$
	sum.	$315.44 - \imath 0.24$	$362.70 - \imath 0.16$	$498.64 - \imath 0.17$	$698.22 - \imath 0.24$
	exp.	$335.73 - \imath 0.21$	$375.49 - \imath 0.53$	$495.79 - \imath 0.12$	$698.73 - \imath 0.25$

Table 7.2. *A comparison between the calculations and experimental results of the resonance frequencies of a free plate (with the dimensions 29.1 cm × 47.5 cm × 0.15 cm). vis.: only viscous damping; the.: only thermoelastic damping; aco.: only acoustic damping; sum.: viscous, thermoelastic and acoustic damping; exp.: experimental damping*

5 In practice, the experimental uncertainties of the material (homogeneity and isotropy), the geometry and the boundary conditions limit the estimation of resonance frequencies to a relative precision of a few percent at best.

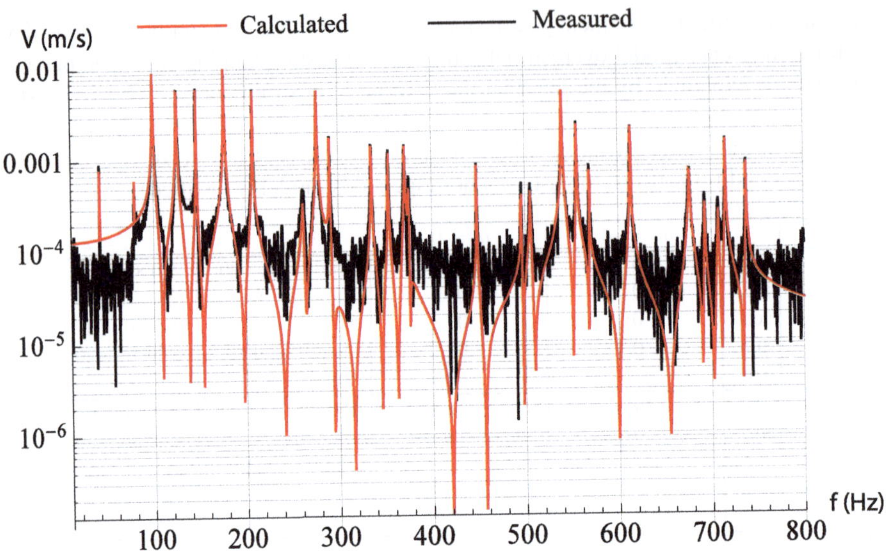

Figure 7.7. *Comparison between the measured vibration velocities and the ones reconstructed by a sum of Lorentzians of 25 terms for a free aluminum plate (with the dimensions 35 cm × 40 cm × 0.2 cm). For a color version of the figure, see www.iste.co.uk/anselmet/acoustics.zip*

7.2.2.8. *Clamped circular plate*

Unlike for the rectangular plates, for which only the simply supported boundary conditions allowed us to obtain an analytic solution, the circular plate problem can be solved analytically, regardless of being clamped, supported or free. We will explore the two extreme cases of clamped and free plates, which show the main characteristics of the associated problem. Finally, we present some results for a supported plate.

We consider a circular plate, which occupies a domain defined in the polar coordinates (O, r, ϕ) by $0 \leq r < R$, $0 \leq \phi < 2\pi$. This plate satisfies the equation $\Delta^2 u(r,\phi) - k_p^4 u(r,\phi) = 0, r \in [0, R[, \phi \in [0, 2\pi[$. It is assumed to be clamped along its entire contour. The boundary conditions are given by the cancelation of the displacement $u(r,\phi) = 0, r = R, \phi \in [0, 2\pi[$ and its normal derivative at the edge $\partial_n u(r,\phi) = 0, r = R, \phi \in [0, 2\pi[$.

Because of the geometry of the problem, it is obvious that we have $\partial_n u(r,\phi) = \partial_r u(r,\phi)$. We use the expression of the general solution in polar coordinates [7.14] $u(r,\phi) = \sum_{n=-\infty}^{+\infty} (A_n \mathrm{J}_n(k_p r) + B_n \mathrm{I}_n(k_p r)) \exp(\imath n\phi)$. Applying the boundary conditions leads to satisfy for all $\phi \in [0, 2\pi[$, and therefore for all n, the two relations $A_n \mathrm{J}_n(k_p R) + B_n \mathrm{I}_n(k_p R) = 0$ and $A_n \mathrm{J}_n'(k_p R) + B_n \mathrm{I}_n'(k_p R) = 0$. It is obvious that this system has a non-zero solution only if its

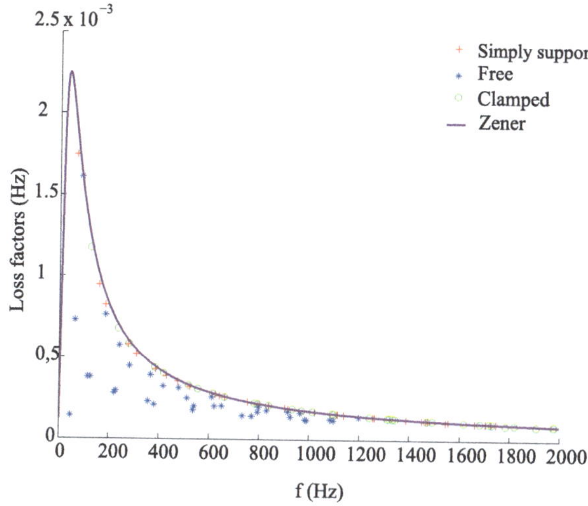

Figure 7.8. *Comparison of thermoelastic loss factors obtained by different mechanical boundary conditions and described by the homogeneous Zener model for a plate made out of aluminum, simply supported, clamped or free (with the dimensions 35 cm × 40 cm × 0.2 cm). For a color version of the figure, see www.iste.co.uk/anselmet/acoustics.zip*

determinant is zero. We use the properties of the Bessel functions $z\mathrm{J}_n'(z) = n\mathrm{J}_n(z) - z\mathrm{J}_{n+1}(z)$ and $z\mathrm{I}_n'(z) = n\mathrm{I}_n(z) + z\mathrm{I}_{n+1}(z)$, from which we obtain the eigenvalue equations $\mathrm{J}_n(k_pR)\mathrm{I}_{n+1}(k_pR) + \mathrm{I}_n(k_pR)\mathrm{J}_{n+1}(k_pR) = 0$. For each value of n, this equation has an infinite number of real roots. The first ones are given in Table 7.3.

$k_{mn}R$	$m=0$	$m=1$	$m=2$	$m=3$	$m=4$	$m=5$
$n=0$	3.1962	6.3064	9.4395	12.5771	15.7164	18.8565
$n=1$	4.6109	7.7993	10.9581	14.1086	17.2557	20.401
$n=2$	5.9057	9.1969	12.4022	15.5795	18.744	21.9015
$n=3$	7.1435	10.5367	13.7951	17.0053	20.1923	23.3663
$n=4$	8.3466	11.8367	15.1499	18.396	21.6084	24.8015
$n=5$	9.5257	13.1074	16.4751	19,7583	22.9979	26.2117

Table 7.3. *Non-dimensionalized eigen wave numbers for a clamped circular plate*

The eigenvalues form a countable double infinity, which corresponds well to the physics of this two-dimensional problem. For each (mn) couple, we have a unique shape. The number of nodal diameters corresponds to the order n and the number of nodal circles corresponds to m. We recall that a nodal line is the analogue in two dimensions of a vibration node of a string or a beam. The modes which correspond to $n = 0$ are called axisymmetric. We calculate the expression of eigenmodes. From the first boundary condition, we deduce $A_n/B_n = -\mathrm{I}_n(k_pR)/\mathrm{J}_n(k_pR)$. The eigenmodes can thus be written as (for $k_p = k_{mn}$):

$$u_{mn}(r,\phi) = \frac{A_{mn}}{\mathrm{I}_n(k_{mn}R)}\left\{\mathrm{I}_n(k_{mn}R)\,\mathrm{J}_n(k_{mn}r) - \mathrm{J}_n(k_{mn}R)\,\mathrm{I}_n(k_{mn}r)\right\}\mathrm{e}^{\imath n\phi}.$$

The orthonormality of these eigenmodes can be demonstrated by using the properties of the Bessel functions. We recall that the scalar product is defined, on a circular domain, for complex functions by:

$$\langle u_{mn}, u_{pq}\rangle = \int_0^{2\pi}\int_0^R u_{mn}(r,\phi)u^\star_{pq}(r,\phi)rdrd\phi,$$

where $f^\star$ is the conjugate of f. The value of the coefficient A_{mn} is fixed by normalizing the modes. For example, we show that for axisymmetric modes (for $n = 0$, the modes do not depend on ϕ) the normalization leads to $A_{m0} = 1/\left(R\mathrm{J}_0\left(k_{m0}R\right)\right)$.

7.2.2.9. *Forced regime*

We now suppose that this plate is subjected to a unitary point force located at the origin. The equation satisfied by this plate is $D(\Delta^2 u(r,\phi) - k_p^4 u(r,\phi)) = 1/r\delta(r)\delta(\phi), r \in [0,R[, \phi \in [0,2\pi[$. It is easy to show that the development on the basis of eigenmodes is given by:

$$u(r,\phi) = \frac{1}{D}\sum_{m=1}^{m=+\infty}\sum_{n=-\infty}^{n=+\infty}\frac{f_{mn}}{k^4_{mn} - k^4_p}u_{mn}(r,\phi),$$

where $f_{mn} = \int_0^{2\pi}\int_0^R \delta(r)\delta(\phi)u^\star_{mn}(r,\phi)drd\phi = u_{mn}(0,0)$. However, we know that $\mathrm{J}_n(0) = \mathrm{I}_n(0) = 0$, if $n \geq 1$ and $\mathrm{J}_0(0) = \mathrm{I}_0(0) = 1$. Thus, $u_{mn}(0,0) = 0$ if $n \geq 1$, this is:

$$u(r,\phi) = \frac{1}{D}\sum_{m=1}^{m=+\infty}\frac{1}{k^4_{m0} - k^4_p}\frac{A_{m0}}{\mathrm{I}_0(k_{m0}R)}\left\{\mathrm{I}_0(k_{m0}R) - \mathrm{J}_0(k_{m0}R)\right\}u_{m0}(r,\phi),$$

we note that only the axisymmetric modes are excited since $u_{m0}(r,\phi) = u_{m0}(r)$ does not depend on ϕ. The solution has revolution symmetry. This corresponds well to the physics of the problem because the excitation, located at the center of the plate, is at the revolution symmetry.

The previous series converges very slowly near the excitation point. We aim to find a solution that would be better adapted to numerical calculations. To do this, we aim to find the solution of the initial problem as the sum of the vibration field in an infinite medium, which satisfies the inhomogeneous plate equation (subjected to the same excitation force), and the diffracted vibration field, the homogeneous plate equation, and such that this sum satisfies the boundary conditions. Here, we use the method of integration of inhomogeneous differential equations, which consists of searching for a solution to the equation as the sum of the general solution and the particular solution (here, Green's kernel of the plate equation since the second part is a Dirac function). The expression of the general solution is sought in a way to satisfy the boundary condition. Thus, this is $u = u_0 + u_g$. As the excitation is a Dirac distribution, u_0 which is Green's kernel of the plate equation is given by [7.17]:

$$u_g = \frac{1}{D} \sum_{n=-\infty}^{+\infty} \left\{ A_n \mathrm{J}_n(k_p r) + B_n \mathrm{I}_n(k_p r) \right\} \mathrm{e}^{\imath n \phi}, \text{ general solution}$$

$$u_0 = \frac{\imath}{8Dk_p^2} \left(\mathrm{H}_0\left(k_p M M_0\right) - \mathrm{H}_0\left(\imath k_p M M_0\right) \right), \text{ particular solution}$$

where $M = (r, \phi)$, and $M_0 = (r_0, \phi_0)$. The boundary conditions give us $u_g = -u_0$ at $r = R$ and $u_g' = -u_0'$ at $r = R$. Recall the expansion into cylindrical harmonic series of the Hankel function [MOR 53] $\mathrm{H}_0(k_p M M_0) = \sum_{n=-\infty}^{+\infty} \mathrm{J}_n(k_p r)\mathrm{H}_n(k_p r_0) \exp(\imath n(\phi - \phi_0))$, if $r \leq r_0$ and $\mathrm{H}_0(k_p M M_0) = \sum_{n=-\infty}^{+\infty} \mathrm{H}_n(k_p r)\mathrm{J}_n(k_p r_0) \exp(\imath n(\phi - \phi_0))$, if $r \geq r_0$. By passing to the limit, we find $R \geq r_0$, thus:

$$\lim_{r \to R} u_0 = \lim_{r \to R} \frac{\imath}{8Dk_p^2} \left(\sum_{n=-\infty}^{+\infty} \left\{ \mathrm{H}_n(k_p r)\mathrm{J}_n(k_p r_0) \right.\right.$$
$$\left.\left. - \frac{2}{\imath\pi} \mathrm{K}_n(k_p r)\mathrm{I}_n(k_p r_0) \right\} \mathrm{e}^{\imath n(\phi - \phi_0)} \right).$$

The boundary conditions for each term of the series n can be written as:

$$A_n \mathrm{J}_n(k_p R) + B_n \mathrm{I}_n(k_p R) = -\frac{\imath}{8Dk_p^2} \left(\mathrm{H}_n(k_p R)\mathrm{J}_n(k_p r_0) \right.$$
$$\left. - \frac{2}{\imath\pi} \mathrm{K}_n(k_p R)\mathrm{I}_n(k_p r_0) \right) \mathrm{e}^{-\imath n \phi_0}$$

$$A_n \mathrm{J}_n'(k_p R) + B_n \mathrm{I}_n'(k_p R) = -\frac{\imath}{8Dk_p^2} \left(\mathrm{H}_n'(k_p R)\mathrm{J}_n(k_p r_0) \right.$$
$$\left. - \frac{2}{\imath\pi} \mathrm{K}_n'(k_p R)\mathrm{I}_n(k_p r_0) \right) \mathrm{e}^{-\imath n \phi_0}.$$

We obtain a linear system with the determinant $\Delta_n = \mathrm{J}_n(k_pR)\mathrm{I}'_n(k_pR) + \mathrm{J}'_n(k_pR)\mathrm{I}_n(k_pR)$, and normally find the same eigenfrequencies as before. The constants A_m and B_m are given by:

$$A_n = \frac{\left(C_n\left(R, r_0\right)\mathrm{I}'_n(k_pR) - D_n\left(R, r_0\right)\mathrm{J}_n(k_pR)\right)\mathrm{e}^{-\imath n\phi_0}}{\Delta_n}$$

$$B_n = \frac{\left(D_n\left(R, r_0\right)\mathrm{J}_n(k_pR) - C_n\left(R, r_0\right)\mathrm{I}'_n(k_pR)\right)\mathrm{e}^{-\imath n\phi_0}}{\Delta_n},$$

where we have set $C_n\left(R, r_0\right) = -\imath/\left(8Dk_p^2\right)\left(\mathrm{H}_n(k_pR)\mathrm{J}_n\left(k_pr_0\right) + 2\imath/\pi\mathrm{K}_n(k_pR)\right.$ $\left.\mathrm{I}_n\left(k_pr_0\right)\right)$ and $D_n\left(R, r_0\right) = -\imath/\left(8Dk_p^2\right)\left(\mathrm{H}'_n(k_pR)\mathrm{J}_n\left(k_pr_0\right) + 2\imath/\pi\mathrm{K}'_n(k_pR)\right.$ $\left.\mathrm{I}_n\left(k_pr_0\right)\right)$. We get a series which converges very quickly (outside the eigenfrequencies). For a given precision, the number of necessary terms is practically 10-fold smaller with respect to the series of eigenmodes.

7.2.2.10. *Free circular plate*

The case of a circular plate with a free edge is more difficult to treat than that of a clamped plate due to the boundary conditions. In fact, the boundary conditions for the plates are trace operators on the edge. The evoked differential operators are oblique derivatives related to the fact that an elongation involves a contraction, which results in a non-zero Poisson's ratio, and thus all the differential operators of the surface, which act on the boundary conditions, must in a certain way reflect a volume effect. As the plate is infinitely thin, this is called an oblique derivative. This effect does not exist for beams, which are described as "lines" and thus, unlike surfaces, cannot contract. For example, $\Delta u - (1-\nu)\partial_l^2 u = 0$ at $r = R$ should be understood as the limit for a point on the plate, tending toward the edge $r \to R$. In practice, this does not really change the writing. This is more difficult for the term $\partial_n \Delta u + (1-\nu)\partial_l\left(\partial_n\partial_l u\right) = 0$ at $r = R$ which should be understood as $\lim_{r\to R}\left(\partial_n\Delta u + (1-\nu)\partial_l \lim_{r\to R}\left(\partial_n\partial_l u\right)\right) = 0$. When the edge is regular, one can easily invert the operators of the derivation and the limit passage and return to a more classic writing. However, in the case of a geometric singularity (as with a polygonal plate), we must be careful when calculating ∂_l and its limit passage. We know that $\Delta = \partial_l^2 + \partial_n^2$ and we have $\partial_l = 1/r\partial/\partial\phi$, $\partial_n = \partial/\partial r$, $\partial_l^2 = 1/r\partial/\partial r + 1/r^2\partial^2/\partial\phi^2$, $\partial_n^2 = \partial^2/\partial r^2$ and $\partial_l\partial_n = -1/r^2\partial/\partial\phi + 1/r\partial^2/\partial\phi\partial r$. The boundary conditions can be written for all $\phi \in [0, 2\pi[$: $\Delta u - (1-\nu)\partial_l^2 u = 0, r = R$, this is $\Rightarrow \lim_{r\to R}(\Delta u - (1-\nu)\partial_l^2 u) = 0$, and $\partial_n\Delta u + (1-\nu)\partial_l\left(\partial_n\partial_l u\right) = 0, r = R$, this is $\lim_{r\to R}(\partial_r\Delta u + (1-\nu)1/r\partial_\phi(\lim_{r\to R}(-1/r^2\partial_\phi u + 1/r\partial^2_{\phi r}u)) = 0$ or $\lim_{r\to R}(\partial_r\Delta u + (1-\nu)(-1/r^3\partial^2_{\phi^2}u + 1/r^2\partial^3_{\phi^2 r}u)) = 0$.

We must note that in these relations, we have not expanded the Laplace operator. In fact, these equations are applied to the functions, which are the products of the Bessel functions and the periodical complex exponentials $2-\pi$. The Bessel equation of integer order n is $f''(z) + 1/zf'(z) + \left(1 - n^2/z^2\right)f(z) = 0$, which allows us to write

$\Delta\left(\mathrm{J}_n(k_p r)\exp(-\imath n\phi)\right) = -k_p^2 \mathrm{J}_n(k_p r)\exp(-\imath n\phi)$ and $\Delta\left(\mathrm{I}_n(k_p r)\exp(-\imath n\phi)\right) = k_p^2 \mathrm{I}_n(kr)\exp(-\imath n\phi)$.

To calculate the eigenfrequencies, we should, of course, apply the boundary conditions to the general solution in polar coordinates [7.14]. After simplification, we obtain the two relations:

$$A_n\left(-k_p^2\mathrm{J}_n(k_pR) - \frac{1-\nu}{R}\left(k\mathrm{J}'_n(k_pR) - n^2\frac{\mathrm{J}_n(k_pR)}{R}\right)\right)$$
$$+B_n\left(k_p^2\mathrm{I}_n(k_pR) - \frac{1-\nu}{R}\left(k\mathrm{I}'_n(k_pR) - n^2\frac{\mathrm{I}_n(k_pR)}{R}\right)\right) = 0$$
$$A_n\left(-k_p^3\mathrm{J}'_n(k_pR) - \frac{(1-\nu)n^2}{R^2}\left(-\frac{\mathrm{J}_n(k_pR)}{R} + k\mathrm{J}'_n(k_pR)\right)\right)$$
$$+B_n\left(-k_p^3\mathrm{I}'_n(k_pR) - \frac{(1-\nu)n^2}{R^2}\left(-\frac{\mathrm{I}_n(k_pR)}{R} + k\mathrm{I}'_n(k_pR)\right)\right) = 0.$$

As previously, the values of k_pR which cancel the determinant of this system are the eigenvalues of the plate. The eigenvalue equation is given by:

$$\frac{\lambda^2\mathrm{J}_n(\lambda) + (1-\nu)\left(\lambda\mathrm{J}'_n(\lambda) - n^2\mathrm{J}_n(\lambda)\right)}{\lambda^3\mathrm{J}'_n(\lambda) + (1-\nu)n^2\left(\lambda\mathrm{J}'_n(\lambda) - \mathrm{J}_n(\lambda)\right)} = \frac{\lambda^2\mathrm{I}_n(\lambda) - (1-\nu)\left(\lambda\mathrm{I}'_n(\lambda) - n^2\mathrm{I}_n(\lambda)\right)}{\lambda^3\mathrm{I}'_n(\lambda) - (1-\nu)n^2\left(\lambda\mathrm{I}'_n(\lambda) - \mathrm{I}_n(\lambda)\right)},$$

where $\lambda = k_pR$. It should be noted that the eigenvalues depend on Poisson's ratio, whereas the values were independent of it for the clamped plate. The first eigenvalues are given in Table 7.4 (for $\nu = 1/3$).

$k_{mn}R$	$m=0$	$m=1$	$m=2$	$m=3$	$m=4$
$n=0$	0	3.0139	6.2089	9.3702	12.53
$n=1$	0	4.5299	7.7175	10.9087	14.0784
$n=2$	2.2919	5.9372	9.1597	12.4097	15.5788
$n=3$	3.4971	7.2739	10.5499	13.86	17.0499
$n=4$	4.6476	8.5499	11.9499	15.2413	18.4499

Table 7.4. *Non-dimensionalized eigen wave numbers for a free circular plate*

We show that the roots λ are composed of the zeros of $\mathrm{J}'_n(z)$ and $\mathrm{J}_n(z)$. The higher order roots can be calculated by the asymptotic expansion $k_{mn}R \approx \alpha_{mn}(m+1)(8\alpha_{mn}) - 4/3(7s^2+22s+11)/(8\alpha_{mn})^2 + \cdots$ where $\alpha_{mn} = (2m+n)\pi/2$ and $s = 4n^2$. Asymptotically, we have for $m \to \infty$, $k_{mn}R \approx (2m+n)\pi/2$. We remark that the two "first" eigenvalues have a frequency of zero. These are the rigid body movements, which correspond to the movements of translation and rotation of the entire plate, which are characteristics of free structures. In addition, the eigenvalues are lower than those obtained for the clamped plate. This is consistent with the principle of the additional connections which impose a shift of the spectrum toward the treble for all additional stress added on the system.

From an experimental point of view, the totally free plate case has little meaning. However, the circular plates free at the edge and clamped at the center are of significant practical interest. This is particularly useful for determining the characteristics of the material of the plate for identifying Young's modulus and Poisson's ratio. The technique consists of searching the vibration modes of an annular plate, free at the edges and fixed at its center to a vibrating pot, thus the approximation of clamping is generally acceptable. An example of such an implementation is presented in Figure 7.9. In this case, we can show that for the modes that have a vibration node at the center (namely the modes which have a nodal diameter or more, for which $n \geq 1$), the results slightly differ from the free plate ones. We can show [LEI 93a] that for such a plate with a ratio of the inner diameter over the outer diameter of α, we must calculate the roots of the following determinants, for $n = 0$ and with $A = 2(1-\nu)/(\lambda)$:

$$\begin{vmatrix} \mathrm{J}_0(\lambda) & \mathrm{Y}_0(\lambda) & -\mathrm{I}_0(\lambda) + A\mathrm{I}_1(\lambda) & -\mathrm{K}_0(\lambda) - A\mathrm{K}_1(\lambda) \\ \mathrm{J}_1(\lambda) & \mathrm{Y}_1(\lambda) & \mathrm{I}_1(\lambda) & -\mathrm{K}_1(\lambda) \\ \mathrm{J}_0(\alpha\lambda) & \mathrm{Y}_0(\alpha\lambda) & \mathrm{I}_0(\alpha\lambda) & \mathrm{K}_0(\alpha\lambda) \\ \mathrm{J}_1(\alpha\lambda) & \mathrm{Y}_1(\alpha\lambda) & -\mathrm{I}_1(\alpha\lambda) & \mathrm{K}_1(\alpha\lambda) \end{vmatrix} = 0.$$

For $n = 1$, with $A = 2(1-\nu)/(\lambda)$, $B = -1+4(1-\nu)/(\lambda)^2$ and $C = 8(1-\nu)\lambda)^3$, we have:

$$\begin{vmatrix} \mathrm{J}_0(\lambda) & \mathrm{Y}_0(\lambda) & B\mathrm{I}_0(\lambda) - C\mathrm{I}_1(\lambda) & -\mathrm{K}_0(\lambda) - A\mathrm{K}_1(\lambda) \\ \mathrm{J}_1(\lambda) & \mathrm{Y}_1(\lambda) & \mathrm{I}_1(\lambda) & -\mathrm{K}_1(\lambda) \\ \mathrm{J}_0(\alpha\lambda) & \mathrm{Y}_0(\alpha\lambda) & \mathrm{I}_0(\alpha\lambda) & \mathrm{K}_0(\alpha\lambda) \\ \mathrm{J}_1(\alpha\lambda) & \mathrm{Y}_1(\alpha\lambda) & \mathrm{I}_1(\alpha\lambda) & \mathrm{K}_1(\alpha\lambda) \end{vmatrix} = 0.$$

For $n = 2$, with $A = \lambda/4 - (3+\nu)(2\lambda)$, $B = 12(1-\nu)\lambda/(12(1-\nu^2) - (\lambda)^4)$, $C = (2(1-\nu)(7+\nu+(\lambda)^2) - (\lambda)^4)/\left(12(1-\nu^2) - (\lambda)^4\right)$ and $D = \lambda/4 + (3+\nu)/(2\lambda) - AC$, we have:

$$\begin{vmatrix} \mathrm{J}_0(\lambda) & \mathrm{Y}_0(\lambda) & (1-4AB)\mathrm{I}_0(\lambda) - D\mathrm{I}_1(\lambda) & (1-4AB)\mathrm{K}_0(\lambda) + D\mathrm{K}_1(\lambda) \\ \mathrm{J}_1(\lambda) & \mathrm{Y}_1(\lambda) & 4B\mathrm{I}_0(\lambda) - C\mathrm{I}_1(\lambda) & 4B\mathrm{K}_0(\lambda) + C\mathrm{K}_1(\lambda) \\ \mathrm{J}_0(\alpha\lambda) & \mathrm{Y}_0(\alpha\lambda) & -\mathrm{I}_0(\alpha\lambda) + \frac{4}{\alpha\lambda}\mathrm{I}_1(\alpha\lambda) & -\mathrm{K}_0(\alpha\lambda) - \frac{4}{\alpha\lambda}\mathrm{K}_1(\alpha\lambda) \\ \mathrm{J}_1(\alpha\lambda) & \mathrm{Y}_1(\alpha\lambda) & \mathrm{I}_1(\alpha\lambda) & -\mathrm{K}_1(\alpha\lambda) \end{vmatrix} = 0.$$

For example, for $\alpha = 0.05$, which corresponds to a plate with a 10 cm radius and clamped at its center by a bolt/washer system with a diameter of 1 cm, the first eigenfrequencies have the values shown in Table 7.5 (always for $\nu = 1/3$).

$k_{mn}R$	$m=0$	$m=1$	$m=2$	$m=3$	$m=4$
$n=0$	1.986	4.766	8.139	11.463	14.783
$n=1$	2.052	4.871	8.206	11.514	14.824
$n=2$	2.286	5.975	9.269	12.527	15.789

Table 7.5. *Non-dimensionalized eigen wave numbers for a free circular plate clamped at its center*

We note that in practice, the values for $n = 0$ and $n = 1$ are confounded. The values obtained for $n \geq 2$ are almost identical to those obtained for the completely free plate; they become even closer as the interior radius decreases. This is understandable, since all the modes $n \geq 1$ have a vibration node at the center which corresponds to the clamped boundary condition. Figure 7.9 shows the two examples of nodal lines observed by accumulation of sand, or Chladni figures, on such a plate when it is excited in a harmonic regime at the frequency of the corresponding mode.

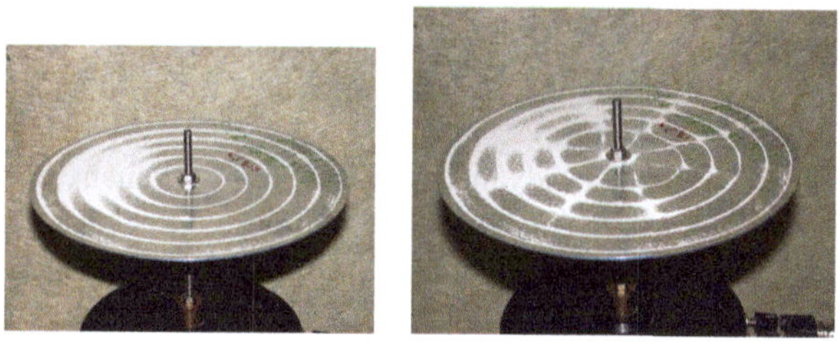

Figure 7.9. *Chladni figures (modes 0-6 and 4-5) of a clamped–free circular plate*

7.2.2.11. *Supported circular plate*

We are limiting this section to the essential results. The methodology employed is not different from the one used previously. The boundary conditions lead to the system:

$$A_n \mathrm{J}_n(\lambda) + B_n \mathrm{I}_n(\lambda) = 0$$
$$A_n \left[\mathrm{J}_n''(\lambda) + \frac{\nu}{\lambda}\mathrm{J}_n'(\lambda)\right] + B_n \left[\mathrm{I}_n''(\lambda) + \frac{\nu}{\lambda}\mathrm{I}_n'(\lambda)\right] = 0.$$

The cancelation of the determinant of this system, using the properties of the Bessel functions, leads to the following eigenvalue equation:

$$\frac{\mathrm{J}_{n+1}(\lambda)}{\mathrm{J}_n(\lambda)} + \frac{\mathrm{I}_{n+1}(\lambda)}{\mathrm{I}_n(\lambda)} = \frac{2\lambda}{1-\nu}.$$

The first eigenvalues are presented in Table 7.6 (always for $\nu = 1/3$).

$k_{mn}R$	$m = 0$	$m = 1$	$m = 2$	$m = 3$	$m = 4$
$n = 0$	2.2314	5.4548	8.6133	11.7622	14.9079
$n = 1$	3.7330	6.9651	10.1393	13.2979	16.4498
$n = 2$	5.0645	8.3756	11.5901	14.7727	17.9408
$n = 3$	6.3239	9.7253	12.9887	16.2023	19.3918
$n = 4$	7.5416	11.0333	14.3486	17.5966	20.8106

Table 7.6. *Non-dimensionalized eigen wave numbers for a supported circular plate*

7.2.3. *Plate of arbitrary shape*

When the plate shape is arbitrary or subjected to non-uniform boundary conditions, the previous methods are no longer applicable. It becomes necessary to use more refined methods. The latter are only applicable with the help of a computer. There are two types of these methods. The Rayleigh–Ritz method (and its corollary the finite element method) and the boundary integral equation method [VIV 74, FIL 08]. The advantage of the finite element method is that it applies to any type of geometry (such as plates of variable thickness) and all types of mechanical behavior. Its inconvenience arises precisely from its general nature in the sense that it is very cumbersome to implement and that its precision is sometimes difficult to evaluate. The other method, based on Green's representation of the plate, consists of establishing an integral representation of the displacement. The main advantage is that the initial differential system is transformed into an integral system (better suited for digital calculation) defined at the edge of the plate. The dimension of the problem is reduced by an order of magnitude. The new unknowns are defined on the edge of the domain. The drawback is that this method is only applicable if we can use Green's kernel of the plate equation. This limits its scope to homogeneous and isotropic plates.

7.2.3.1. *Green's formula*

Let Ω be the set of points on the plate and u and v be the two zero functions on the outside of Ω. We consider the bilinear form of $a(u, v)$ defined in Cartesian coordinates, by the integral [6.37] taking u for u_z and v for δu_z which we calculate in a harmonic regime:

$$a(u, v) = \int_\Omega \Delta u \Delta v + (1 - \nu)\left(2u_{,xy}v_{,xy} - u_{,xx}v_{,yy} - u_{,yy}v_{,xx}\right) d\Omega.$$

$Da(u, u)$ represents the internal potential energy density of the plate in bending. A first integration by parts leads to (see section 6.4.2.2) $a(u, v) = \int_\Omega \left(\Delta^2 u\right) v+$

$\int_{\partial\Omega} (\ell_1 u)\, \partial_n v - (\ell_2 u)\, v$, where $\ell_1 u = \Delta u - (1-\nu)\partial_l^2 u$ and $\ell_2 u = \partial_n \Delta u + (1-\nu)\partial_l (\partial_n \partial_l u)$. A second integration by parts leads to $a(u,v) = \int_\Omega (\Delta^2 v)\, d\Omega + \int_{\partial\Omega} (\ell_1 v)\, \partial_n u - (\ell_2 v)\, d\partial\Omega$, using the difference between the two expressions, we obtain Green's formula for the plate operator:

$$\int_\Omega (\Delta^2 v)\, u - (\Delta^2 u)\, v d\Omega = + \int_{\partial\Omega} (\ell_1 v)\, \partial_n u - (\ell_1 u)\, \partial_n v$$
$$- [(\ell_2 v)\, u - (\ell_2 u)\, v]\, d\partial\Omega.$$

It must be noted that this formulation is valid for any type of geometry, assuming that we know how to express the tangent and normal derivatives that appear in the boundary operators $\partial_n u$, $\ell_1 u$ and $\ell_2 u$.

7.2.3.2. *Green's representation of the displacement of the plate*

We define $G(M, M')$ as the Green kernel of the plate equation. We construct the set of solutions for:

$$D\left(\Delta^2 - k_p^4\right) u(M) = F(M), m \in \Omega.$$

Let us extend u and G by zero on the outside of Ω and apply Green's formula to $u(M')$ and $G(M, M')$. It becomes:

$$\int_\Omega (\Delta^2 u(M'))\, G(M,M') - (\Delta_{M'}^2 G(M,M'))\, u(M') dM'$$
$$= \int_{\partial\Omega} (\ell_1 u(M'))\, \partial_{n(M')} G(M,M') dM' - \int_{\partial\Omega} (\ell_{1M'} G(M,M'))\, \partial_n u(M') dM'$$
$$- \int_{\partial\Omega} (\ell_2 u(M'))\, G(M,M') dM' + \int_{\partial\Omega} (\ell_{2M'} G(M,M')\, u(M') dM',$$

where $\ell_{1M'}$ indicates that the operator ℓ_1 addresses the variable M'. We add to and subtract from the first part of this equation $k_p^4 u(M') G(M,M')$. We group the terms and we obtain:

$$D \int_\Omega (\Delta^2 u(M') - k_p^4 u(M'))\, G(M,M') dM' = \int_\Omega F(M') G(M,M') dM'$$

$$D \int_\Omega (\Delta_{M'}^2 G(M,M') - k_p^4 G(M,M'))\, u(M') dM' = \int_\Omega \delta_M(M') u(M') dM'$$
$$= u(M).$$

The integral representation of the displacement is thus given by:

$$u(M) = u_0(M) + D\int_{\partial\Omega} (\ell_1 u(M'))\, \partial_{n(M')} G(M, M') - (\ell_2 u(M'))\, G(M, M')dM'$$
$$-D\int_{\partial\Omega} (\ell_{1M'} G(M, M'))\, \partial_n u(M') - (\ell_{2M'} G(M, M'))\, u(M')dM', \quad [7.26]$$

where $u_0(M) = \int_\Omega F(M')G(M, M')dM'$ is the source term. This integral relation is Green's representation of the displacement $u(M)$. It is possible to show that the boundary integral is the general solution to the homogeneous equation, which regulates the vibration movements of a thin plate in bending.

7.2.3.3. *Boundary integral equations*

In Green's representation of the displacement u, there appear to be four unknown functions $\ell_1 u$, $\ell_2 u$, $\partial_n u$ and u at the edge of the domain that we must determine. If we apply the boundary conditions imposed by the problem, two of these unknown functions are zeros. It then remains to determine the other two. To do this, it is enough to simply express two of the functions $\ell_1 u$, $\ell_2 u$, $\partial_n u$ or u based on the integral representation and once more apply the boundary conditions to obtain a system of two boundary integral equations with two unknowns. For example, let us consider the case of a clamped plate on $\partial\Omega$. We get $\partial_n u = 0$ and $u = 0$ on $\partial\Omega$ and thus $\ell_1 u$, $\ell_2 u$ are unknowns. The integral representation of u can be written as:

$$u(M) = u_0(M, M') + D\int_{\partial\Omega} (\ell_1 u(M')\, \partial_{n(M')} G(M, M')$$
$$- (\ell_2 u(M'))\, G(M, M')dM'.$$

To calculate $\ell_1 u$ and $\ell_2 u$, we can write that u and its normal derivative are zero at the boundary of the plate. Since Green's kernel of the plate and its derivative are regular, it is not necessary to take particular precautions for the calculation of this integral and its normal derivative at the boundary of the domain. However, in the case of a supported or a free plate, the integrals must be calculated as the main Cauchy values, if there are any derivatives of the second order or finite parts in the Hadamard sense for the derivatives of a higher order kernel or equal to three. The limit (for a point of Ω which tends toward a point of $\partial\Omega$) of each integral is the limit integral. We obtain the following system:

$$D\int_{\partial\Omega} (\ell_1 u(M'))\, \partial_{n(M')} G(Q, M') - (\ell_2 u(M'))\, G(Q, M')dM' = u_0(Q)$$

$$D\int_{\partial\Omega} (\ell_1 u(M')\, \partial_{n(M')}\partial_{n(Q)} G(Q, M') - (\ell_2 u(M'))\, \partial_{n(Q)} G(Q, M')dM'$$
$$= \partial_{n(Q)} u_0(Q),$$

for $Q \in \partial\Omega$. We obtain a linear system of two equations with two unknowns. It is not our objective to solve these equations; this cannot be done without a computer. The basic idea consists of transforming this integral system into a linear algebraic system. To do this, we aim to find the unknowns of the problem in the form of a series expansion. The coefficients of this series become the new unknowns of the problem. Let us, for example, calculate $I = \int_{\partial\Omega} (\ell_2 u(M')) G(Q, M') dM'$. We set $\ell_2 u(M') = \sum_{i=1}^{i=N} a_i \gamma_i(M')$, where the $\gamma_i(M')$ are the chosen approximation functions. We obtain $I = \sum_{i=1}^{i=N} a_i \int_{\partial\Omega} \gamma_i(M') G(Q, M') dM'$, an expression in which the integral can be digitally calculated without difficulty. A very significant number of choices for approximation functions exist as piecewise constant functions or orthogonal polynomials. We finally obtain a system of $2N$ unknowns with $2N$ coefficients. The principle of these methods consists of calculating the integrals using more or less fine digital quadratures. Approximating the function using piecewise constants leads to calculating the integral using the rectangle method. The approximation by orthogonal polynomials leads to estimating the integral in question using Gauss' method. The two main resolution methods are the Galerkin method (the most precise) and the collocation method (the easiest to implement). We broadly outline these.

For the Galerkin method, we consider H a Hilbert space in which the scalar product $< \cdot, \cdot >$ is defined and let H^N be a sub-space with finite dimensions: $\dim\left(H^N\right) = N$. An approximation $f^N \in H^N$ of f is the solution of the equation $Af = g$ in H, where A is a linear operator (in our case, an integral operator) of the Galerkin method that satisfies $\left\langle Af^N, \Phi \right\rangle =< g, \Phi >$, for each $\Phi \in H^N$. This is $f^N = \sum_{k=1}^{k=N} c_k f_k \Phi_{k-1}$, where $\{\Phi_k\}_k$ is a basis of H. Thus, we have $\sum_{k=1}^{k=N} c_k f_k \left\langle A\Phi_{k-1}, \Phi \right\rangle =< g, \Phi >$, this equation is valid for all Φ of H^N. In particular, for each $\Phi_{j-1}, j = 1..N$. Thus, $\sum_{k=1}^{k=N} c_k f_k \left\langle A\Phi_{k-1}, \Phi_{j-1} \right\rangle = \left\langle g, \Phi_{j-1} \right\rangle, j = 1..N$. The coefficients f_k are evaluated by solving a linear algebraic system $N \times N$ in which each coefficient is given by a scalar product. We must note that this method is generally applied to polynomial approximation functions. For example, the Legendre or Chebyshev polynomials. We will present this method in more detail when studying the radiation of a thin plate under flow in section 8.6.1.

For the collocation method, f is approximated by $\breve{f}^N = \sum_{k=1}^{k=N} c_k \breve{f}_k \Phi_{k-1}$. This equation is satisfied in a finite number of points $\sum_{k=1}^{k=N} c_k \breve{f}_k A\Phi_{k-1} = g$ at $x = x_i, i = 1..N$, the points $x_i, i = 1..N$ are the collocation points. The coefficients f_k are once more evaluated solving a linear algebraic system $N \times N$. The advantage of this method is that the coefficients $\breve{f}_k$ are easier to calculate than in the Galerkin method (a simple integral instead of a double integral). The precision is obviously lower. In the collocation method, we aim to find the solution by interpolating the unknown functions – the function and its interpolated function are equal in a finite number of points. In the Galerkin method, we aim to find the solution by constructing an approximation of the unknown function – the distance between the function and its approximation is minimized in the sense of the chosen scalar product. We will not develop these numerical methods any further. There are numerous books available devoted to them.

We enunciate the following existence theorem (which also shows the equivalence between the starting differential system and the final integral system) [VIV 74, FIL 08]:

1) there exists a countable – double – sequence with a value of k_{mn} of k_p for which the homogeneous system of the boundary integral equations has a finite number of non-zero solutions. The k_{mn} are the eigen wave numbers of the plate;

2) to each k_{mn}, there corresponds a finite number of functions u_{mn}, defined by the integral representation and which are the solutions to the homogeneous boundary problem for $k = k_{mn}$;

3) if $k = k_{mn}$, the inhomogeneous boundary problem does not have solutions. If $k_p \neq k_{mn}$, the inhomogeneous problem has only one solution, regardless of the second part. The initial solution to the boundary problem was given by the integral representation previously.

7.3. Cylindrical shell vibrations

In this section, we only focus on thin cylindrical shell vibrations with infinite and finite dimensions in a harmonic regime. Given the complexity of the equations and the solutions derived, we cannot address the problem with the same level of detail as for the beams and plates. We begin by studying the vibrations of an infinite shell by calculating the general solution and Green's kernel. Then, we examine the problem of a shell of finite dimensions. We present some of the results obtained using a boundary integral equation method in the case of a shell excited by an internal turbulent flow [DUR 00].

7.3.1. *Infinite shell*

We consider an elastic thin cylindrical shell, with a finite length $2L$ or not, which occupies the domain Σ. In the usual cylindrical coordinate system (z, ϕ, r), Σ occupies the domain $z \in]-L, +L[$, where L may be finite or infinite in this section, and $\phi \in [0, 2\pi[$, $r = R$. ρ_c is the density of the shell, E is its Young's modulus and ν is its Poisson's ratio. In the following calculations, we sometimes use the change of variables $s = z/R$ (this change is often used in the literature). With this variable change, the Donnell–Mushtari operator given by the system [6.59] can be written as:

$$
\begin{aligned}
&u_{s,ss} + \tfrac{1-\nu}{2} u_{s,\phi\phi} + \tfrac{\rho h R^2 \omega^2}{D} u_s + \tfrac{1+\nu}{2} u_{\phi,s\phi} + \nu u_{r,s} = \tfrac{R^2}{D} F_s \\
&\tfrac{1+\nu}{2} u_{s,s\phi} + \tfrac{1-\nu}{2} u_{\phi,ss} + u_{\phi,\phi\phi} + \tfrac{\rho h R^2 \omega^2}{D} u_\phi + u_{r,\phi} = \tfrac{R^2}{D} F_\phi \\
&\nu u_{s,s} + u_{\phi,\phi} + u_r + \chi \left(u_{r,ssss} + 2u_{r,ss\phi\phi} + u_{r,\phi\phi\phi\phi} \right) - \tfrac{\rho h R^2 \omega^2}{D} u_r = \tfrac{R^2}{D} F_r.
\end{aligned}
\qquad [7.27]
$$

7.3.1.1. *General solution*

We aim to calculate the solution of the previous system for a shell with an infinite length, when the second part is zero. First, we aim to find a displacement field in the following form:

$$\begin{aligned} u_s(s,\phi) &= A\cos(\lambda s)\cos(m\phi) \\ u_\phi(s,\phi) &= B\sin(\lambda s)\sin(m\phi) \\ u_r(s,\phi) &= C\sin(\lambda s)\cos(m\phi) \end{aligned} \qquad [7.28]$$

It is easy to see that, because of the different orders of derivation, this choice, after factoring in the functions cos and sin, allows us to eliminate the dependence on s and ϕ. We thus obtain:

$$\begin{aligned} &\left(-\lambda^2 - \tfrac{1-\nu}{2}m^2 + \tfrac{\rho h R^2\omega^2}{\mathcal{D}}\right)A + \tfrac{1+\nu}{2}\lambda m B + \nu\lambda C = 0 \\ &\tfrac{1+\nu}{2}\lambda m + \left(-\tfrac{1-\nu}{2}\lambda^2 - m^2 + \tfrac{\rho h R^2\omega^2}{\mathcal{D}}\right)B - mC = 0. \\ &-\nu\lambda A + mB + \left(1 + \chi\left(\lambda^2+m^2\right)^2 - \tfrac{\rho h R^2\omega^2}{\mathcal{D}}\right)C = 0 \end{aligned} \qquad [7.29]$$

Since this system has non-zero solutions $u_\alpha(s,\phi)$ (and thus to have A, B and C not all zeros), we must set either ω, in which case it is enough that there exist such non-zero λ values that the determinant of this system becomes zero, or we must set λ and find the values of ω which cancel this determinant. Since $s = z/R$, the half-length of the wave of the displacement field in the direction z is ℓ, if we take $\lambda = \pi R/\ell$, the eigenfrequencies of the free vibrations can be found easily. We set $\omega_a = \sqrt{\mathcal{D}/(\rho h)}/R = c_L/R$ as the ring pulsation where c_L is the longitudinal wave celerity[6]. The ring frequency $f_a = \omega_a/(2\pi) = c_L/(2\pi R)$ is such that the wave length of the longitudinal waves is equal to the perimeter of the shell.

We set $\Omega = \omega/\omega_a$ and we easily obtain the dispersion relation $\Omega^6 - K_2\Omega^4 + K_1\Omega^2 - K_0 = 0$, with:

$$\begin{aligned} K_0 &= \tfrac{1-\nu}{2}\left((1-\nu^2)\lambda^4 + \chi(m^2+\lambda^2)^4\right) \\ K_1 &= \tfrac{1-\nu}{2}\left((3+2\nu)\lambda^2 + m^2 + (m^2+\lambda^2)^2 + \tfrac{3-\nu}{1-\nu}\chi(m^2+\lambda^2)^3\right) \\ K_2 &= 1 + \tfrac{3-\nu}{2}(m^2+\lambda^2) + \chi(m^2+\lambda^2)^2. \end{aligned}$$

The dispersion relation is a relation of the third degree for Ω^2, it has three roots (and their opposites), three real numbers or one real and two complex conjugates when the coefficients are real.

6 Another way of defining this ring frequency is to set $m = \rho h$ as the surface mass of the shell and $r_m = \mathcal{D}/R^2$ as the membrane stiffness, obviously with $\omega_a^2 = r_m/m$; which shows that this ring frequency is a concept closely linked to the membrane stiffness of the shells. We can anticipate a little bit about the coupling effect, to indicate that, at this frequency, similarly as for the plates, the shells are completely transparent to the acoustic fields.

7.3.1.2. *Green's kernel*

By definition, if $\overline{\overline{C}}$ is the Donnell–Mushtari operator of cylindrical shells, Green's tensor $\overline{\overline{\Gamma}}_m(z,\phi;z',\phi')$ of this operator is defined by $\mathcal{D}\overline{\overline{C}}\overline{\overline{\Gamma}}(z,\phi;z',\phi') = \overline{\overline{I}}\delta_{z'}(z)\delta_{\phi'}(\phi)$, where $\overline{\overline{I}}$ is the identity matrix. It is obvious that $\overline{\overline{\Gamma}}(z,\phi;z',\phi') = \overline{\overline{\Gamma}}(z-z',\phi-\phi')$. We expand this operator into an angular Fourier series $\overline{\overline{\Gamma}}(z-z',\phi-\phi') = \sum_{m=-\infty}^{+\infty} \overline{\overline{\Gamma}}_m(z-z')\exp(\imath m(\phi-\phi'))$. From there, it is easy to obtain for each angular harmonic the equation:

$$\mathcal{D}\overline{\overline{C}}_m\overline{\overline{\Gamma}}_m(z-z') = \overline{\overline{I}}\delta_{z'}(z), \qquad [7.30]$$

where $\overline{\overline{C}}_m$ are the angular components of $\overline{\overline{C}}$ given by relation [6.61]. To calculate $\overline{\overline{\Gamma}}_m(z-z')$, we proceed with a direct and inverse Fourier transform with respect to the variable z. Let $f(z)$ be a function of the axial variable, its Fourier transform will be $\hat{f}(\xi) = \int_{-\infty}^{+\infty} f(z)\exp(-2\imath\pi\xi z)dz$, the inversion formula is $f(z) = \int_{-\infty}^{+\infty} \hat{f}(\xi)\exp(2\imath\pi\xi z)d\xi$. If we call $\hat{\hat{C}}_m(\xi)$ the transform matrix of $\overline{\overline{C}}_m$, it is easy to show that the Fourier transform of this equation [7.30] can be written as $\mathcal{D}\hat{\hat{C}}_m(\xi)\hat{\hat{\Gamma}}_m(\xi,\xi') = \overline{\overline{I}}\mathrm{e}^{-2\imath\pi\xi z'}$, with:

$$\hat{\hat{C}}_m(\xi) = \begin{pmatrix} C^z_{mz}(\xi) & C^\phi_{mz}(\xi) & C^r_{mz}(\xi) \\ C^z_{m\phi}(\xi) & C^\phi_{m\phi}(\xi) & C^r_{m\phi}(\xi) \\ C^z_{mr}(\xi) & C^\phi_{mr}(\xi) & C^r_{mr}(\xi) \end{pmatrix},$$

where the matrix coefficients are:

$$\begin{array}{l} C^z_{mz}(\xi) = -4\pi^2\xi^2 - m^2\frac{1-\nu}{2R^2} + \frac{\rho h\omega^2}{\mathcal{D}}, C^\phi_{m\phi}(\xi) = -\frac{m^2}{R^2} - 4\pi^2\xi^2\frac{1-\nu}{2} + \frac{\rho h\omega^2}{\mathcal{D}} \\ C^r_{mr}(\xi) = \frac{1}{R^2} + \chi\left(R^2 16\pi^4\xi^4 + 8m^2\pi^2\xi^2 + \frac{m^4}{R^2}\right) - \frac{\rho h\omega^2}{\mathcal{D}} \\ C^z_{mr}(\xi) = C^r_{mz}(\xi) = 2\imath\pi\xi\frac{\nu}{R}, C^\phi_{mr}(\xi) = C^r_{m\phi}(\xi) = \frac{\imath m}{R^2}, \\ C^\phi_{mz}(\xi) = C^z_{m\phi}(\xi) = -2\pi\xi m\frac{1+\nu}{2R}. \end{array} \qquad [7.31]$$

From where, using the matrix inversion, $\hat{\hat{\Gamma}}_m(\xi,\xi') = 1/\mathcal{D}\hat{\hat{C}}_m(\xi)^{-1}\mathrm{e}^{-2\imath\pi\xi z'}$. By an inverse Fourier transform, $\overline{\overline{\Gamma}}_m(z,z') = \overline{\overline{\Gamma}}_m(z-z') = \int_{-\infty}^{\infty}\hat{\hat{\Gamma}}_m(\xi)\mathrm{e}^{2\imath\pi\xi(z-z')}d\xi$, with $\hat{\hat{\Gamma}}_m(\xi) = \hat{\hat{C}}_m(\xi)^{-1}/\mathcal{D}$. Since $\overline{\overline{C}}_m$ is a symmetric matrix, $\hat{\hat{\Gamma}}_m(\xi)$ is a symmetric matrix too. We call $\hat{\Gamma}^\beta_{m\alpha}(\xi)$ the components of the matrix $\hat{\hat{\Gamma}}_m(\xi)$. We call $D_m(\xi^2)$ the determinant of the matrix $\hat{\hat{C}}_m(\xi)$; this is a fourth-degree polynomial with respect to ξ^2. If we assume that the roots ξ_{ml}^2 of the determinant are all simple[7], $D_m(\xi^2) = \gamma_m \prod_{l=1}^{4}(\xi^2 - \xi_{ml}^2)$, γ_m is a constant. Without going into detail, we can show that:

$$\hat{\Gamma}_{m\alpha}^{\beta}(\xi) = \frac{N_{m\alpha}^{\beta}(\xi^2)}{D_m(\xi^2)}, \text{ if } (\alpha,\beta) \neq (z,\phi),(z,r), \hat{\Gamma}_{m\alpha}^{\beta}(\xi) = \xi \frac{N_{m\alpha}^{\beta}(\xi^2)}{D_m(\xi^2)}, \\ \text{if } (\alpha,\beta) = (z,\phi),(z,r),$$

where $N_{m\alpha}^{\beta}(\xi^2)$ are the minors of the matrix $\hat{\hat{C}}_m(\xi)$ (for example, we have $N_{mz}^{z}(\xi^2) = C_{m\phi}^{z}(\xi)C_{mr}^{r}(\xi) - \left(C_{m\phi}^{r}(\xi)\right)^2$), where the coefficients $C_{m\alpha}^{\beta}$ are given by relations [7.31]. To calculate these diverse integrals, we apply the residue theorem on either of the contours $\mathcal{C}^+$ or $\mathcal{C}^-$.

In order to keep a finite value to the exponential on the semi-circles, we must choose the integration contour according to the sign of $z - z'$. If $z - z' > 0$, we must integrate over $\mathcal{C}^+$, it is thus easy to see that the upper half-plane contains the four roots of $D_m(\xi^2)$ with positive imaginary parts. We thus have:

$$\Gamma_{m\alpha}^{\beta}(z-z') = \frac{\imath\pi}{\gamma_m}\sum_{k=1}^{k=4}\frac{N_{m\alpha}^{\beta}(\xi_{mk}^2)}{\xi_{mk}\prod_{\substack{l=1\\l\neq k}}^{4}(\xi_{mk}^2-\xi_{ml}^2)}\mathrm{e}^{2\imath\pi\xi_{ml}(z-z')}, \\ \text{if } (\alpha,\beta) \neq (z,\phi),(z,r)$$

$$\Gamma_{m\alpha}^{\beta}(z-z') = \frac{\imath\pi}{\gamma_m}\sum_{k=1}^{k=4}\frac{N_{m\alpha}^{\beta}(\xi_{mk}^2)}{\prod_{\substack{l=1\\l\neq k}}^{4}(\xi_{mk}^2-\xi_{ml}^2)}\mathrm{e}^{2\imath\pi\xi_{ml}(z-z')}, \text{ if } (\alpha,\beta) = (z,\phi),(z,r).$$

If $z - z' < 0$, we must integrate over $\mathcal{C}^-$, it is thus easy to see that the lower half-plane contains the four roots of $D_m(\xi^2)$ with negative imaginary parts. We obtain:

$$\Gamma_{m\alpha}^{\beta}(z-z') = \frac{\imath\pi}{\gamma_m}\sum_{k=1}^{k=4}\frac{N_{m\alpha}^{\beta}(\xi_{mk}^2)}{\xi_{mk}\prod_{\substack{l=1\\l\neq k}}^{4}(\xi_{mk}^2-\xi_{ml}^2)}\mathrm{e}^{-2\imath\pi\xi_{ml}(z-z')}, \\ \text{if } (\alpha,\beta) \neq (z,\phi),(z,r)$$

$$\Gamma_{m\alpha}^{\beta}(z-z') = -\frac{\imath\pi}{\gamma_m}\sum_{k=1}^{k=4}\frac{N_{m\alpha}^{\beta}(\xi_{mk}^2)}{\prod_{\substack{l=1\\l\neq k}}^{4}(\xi_{mk}^2-\xi_{ml}^2)}\mathrm{e}^{-2\imath\pi\xi_{ml}(z-z')}, \\ \text{if } (\alpha,\beta) = (z,\phi),(z,r).$$

7 For double roots, the following modifications are minor.

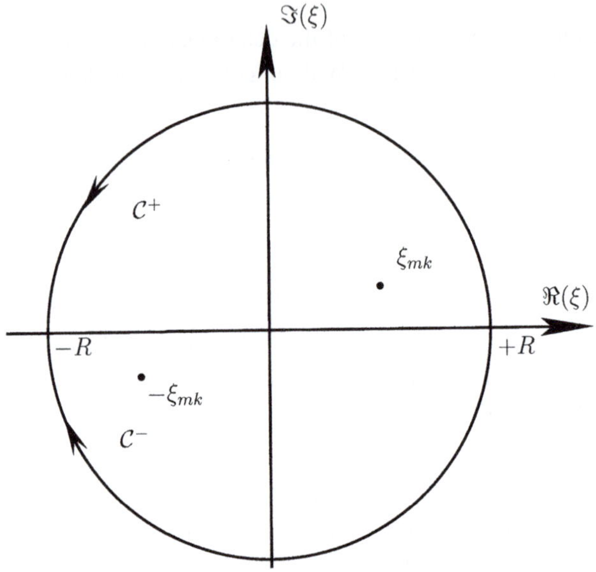

Figure 7.10. *Integration contour*

From this, the result:

$$\Gamma^{\beta}_{m\alpha}(z-z') = \frac{\imath\pi}{\gamma_m}\sum_{k=1}^{k=4}\frac{N^{\beta}_{m\alpha}(\xi^2_{mk})}{\xi_{mk}\prod_{\substack{l=1\\l\neq k}}^{4}(\xi^2_{mk}-\xi^2_{ml})}\mathrm{e}^{2\imath\pi\xi_{ml}|z-z'|},$$

$$\text{if } (\alpha,\beta)\neq(z,\phi),(z,r)$$

$$\Gamma^{\beta}_{m\alpha}(z-z') = \frac{\imath\pi}{\gamma_m}\mathrm{sgn}(z-z')\sum_{k=1}^{k=4}\frac{N^{\beta}_{m\alpha}(\xi^2_{mk})}{\prod_{\substack{l=1\\l\neq k}}^{4}(\xi^2_{mk}-\xi^2_{ml})}\mathrm{e}^{2\imath\pi\xi_{ml}|z-z'|},$$

$$\text{if } (\alpha,\beta)=(z,\phi),(z,r).$$

Each angular component of Green's tensor of the shell operator is a simple linear combination of complex exponentials.

7.3.2. *Finite shell*

7.3.2.1. *Special case of the supported shell*

We consider a shell with a length L supported at $s=0$ and $s=+L/R$. We have seen that the boundary conditions $\mathcal{D}\left(u_{s,s}+\nu\left(u_{\phi,\phi}+u_r\right)\right)=0$, $u_\phi=0$, $u_r=0$ and

$\mathcal{D}\chi\left(u_{r,ss}+\nu u_{r,\phi\phi}\right)=0$ correspond to a simply supported shell[8]. The angular components of these boundary conditions become $u'_{sm}+\nu\left(\imath m u_{\phi m}+u_{rm}\right)=0$, $u_{\phi m}=0$, $u_{rm}=0$ and $u''_{rm}-\nu m^2 u_{rm}=0$; after simplifying, this is $u'_{sm}=0$, $u_{\phi m}=0$, $u_{rm}=0$ and $u''_{rm}=0$. The general solution [7.28] also satisfies the boundary conditions. If the shell is supported at $s=0$ and $s=L/R$, it becomes $\lambda_n=n\pi R/L=k_nR$, $n=0,1,\cdots$. The eigen pulsations are obtained by replacing λ_n in the dispersion relation and solving it; each m,n couple has three corresponding eigen pulsations. The lowest of the three frequencies generally corresponds to a predominantly radial mode.

7.3.2.2. *Other boundary conditions*

By combining the different boundary conditions possible at each end of the cylinder, we arrive at a very high number of different cases for finite cylinders (there are a total of 136). Although computationally difficult, it is possible to analytically obtain the eigenmodes for all these cases. We will obviously not go into detail but will simply indicate the major steps (this is very similar to what we have examined already). We begin by searching for a displacement field in the following form:

$$\begin{aligned}u_s(s,\phi)&=A\exp(\lambda s)\cos(m\phi)\\u_\phi(s,\phi)&=B\exp(\lambda s)\sin(m\phi)\\u_r(s,\phi)&=C\exp(\lambda s)\cos(m\phi),\end{aligned}$$

that we introduce into the system [7.27] in which the second part has been rendered to zero. We thus obtain a linear system for λ and m given by [7.29]. Canceling the determinant of this system leads to an algebraic relation of the fourth degree with respect to λ^2 given by $\lambda^8+l_6\lambda^6+l_4\lambda^4+l_2\lambda^2+l_0=0$, where:

$$\begin{aligned}l_6&=-4m^2+\tfrac{3-\nu}{1-\nu}\Omega^2, l_4=6m^4-3\tfrac{3-\nu}{1-\nu}m^2\Omega^2+\tfrac{2}{1-\nu}\Omega^4+\tfrac{1}{\chi}(1-\nu^2-\Omega^2)\\l_2&=-4m^6+\tfrac{m^2\Omega^2}{1-\nu}\left(3(3-\nu)m^2-4\Omega^2\right)+\tfrac{\Omega^2}{\chi}\left(3+2\nu+2m^2-\tfrac{3-\nu}{1-\nu}\Omega^2\right)\\l_0&=m^8-\tfrac{m^2\Omega^2}{\chi}\left(1+\chi m^4+m^2-\Omega^2\right)-\tfrac{2\Omega^2}{1-\nu}\left(m^6-\tfrac{\Omega^2}{\chi}(1+\chi m^4+m^2-\Omega^2)\right).\end{aligned}$$

In practice, for the usual parameters of the shell [LEI 93b] and $m\geq 1$, we get $\lambda=\pm\lambda_1$, $\pm\imath\lambda_2$, $\pm(\lambda_3\pm\lambda_4)$, with $\lambda_i\in\mathbb{R}^+, i=1,\cdots 4$. Thus, the various components of the displacement field $u_\alpha(s,\phi)$ can be expressed as a function of eight unknown constants. By applying the boundary conditions, we can calculate seven of these constants and the frequency parameter Ω. The final constant can be calculated by normalizing the modes.

---

8 More specifically, these conditions describe a shell closed by two thin circular plates fixed – for example, by a weld – to the shell.

7.3.2.3. *Green's formula*

Here, we establish, in a formal manner, Green's representation of a shell described by the Donnell–Mushtari operator $\overline{\overline{C}}$. We recall that Green's tensor of the shell $\overline{\overline{\Gamma}}$ which is associated with this operator is the solution of $\overline{\overline{C\Gamma}} = \overline{\overline{I}}\delta$. We define by $\overline{\overline{C}}^a$ the adjoint of $\overline{\overline{C}}$[9]. We define the scalar product $\langle f, g^\star\rangle = \int_\Sigma fgdS$ where Σ is the surface occupied by the shell.

We construct the scalar product $\int_\Sigma \overline{\overline{\Gamma}}\left(\overline{\overline{C}}\vec{U}\right)dS = \langle\overline{\overline{\Gamma}},\left(\overline{\overline{C}}\vec{U}\right)^\star\rangle$. Based on the definition of an adjoint, we have $\langle\overline{\overline{\Gamma}},\left(\overline{\overline{C}}\vec{U}\right)^\star\rangle = \langle\overline{\overline{C}}^{\star a}\overline{\overline{\Gamma}},\vec{U}^\star\rangle$, this is $\langle\overline{\overline{\Gamma}},\left(\overline{\overline{C}}\vec{U}\right)^\star\rangle = \langle\left(\overline{\overline{C}}^a\overline{\overline{\Gamma}}^\star\right)^\star,\vec{U}^\star\rangle$. Moreover, we know by definition that $\overline{\overline{\Gamma}}$ is symmetric, thus $\overline{\overline{\Gamma}}^T = \overline{\overline{\Gamma}}$, and that $\overline{\overline{\Gamma}}^{T\star} = \overline{\overline{\Gamma}}^a$. Thus, $\langle\overline{\overline{\Gamma}},\left(\overline{\overline{C}}\vec{U}\right)^\star\rangle = \langle\left(\overline{\overline{C}}^a\overline{\overline{\Gamma}}^a\right)^\star,\vec{U}^\star\rangle$. Finally, this is $\langle\overline{\overline{\Gamma}},\left(\overline{\overline{C}}\vec{U}\right)^\star\rangle = \langle\left(\overline{\overline{I}}\delta\right)^a,\vec{U}^\star\rangle = \langle\left(\overline{\overline{I}}\delta\right),\vec{U}^\star\rangle = \vec{U}$.

We notice that, in general, for two operators A and B, we have $(AB)^a = B^a A^a$, but here $\overline{\overline{C}}$ is a symmetric operator and $\overline{\overline{\Gamma}}$ is a symmetric matrix, therefore $\overline{\overline{C}}^a\overline{\overline{\Gamma}}^a = \left(\overline{\overline{C\Gamma}}\right)^a$ (which is clearly false, if neither is symmetric); moreover, *a posteriori* this result is obvious, when we remember that, when establishing the operator, we changed the sign of the first two lines of the operator. Thus, since the initial operator was self-adjoint (so that $\overline{\overline{C}}^a = \overline{\overline{C}}$), the newly obtained operator is also symmetric.

The solution of $\overline{\overline{C}}\vec{U} = \vec{f}$ can be obtained easily. Let us examine the difference between $\overline{\overline{J}}_1 = \int_\Sigma \overline{\overline{\Gamma}}\left(\overline{\overline{C}}\vec{U}\right)dS$ and $\overline{\overline{J}}_2 = \int_\Sigma \overline{\overline{C}}^a\overline{\overline{\Gamma}}\vec{U}dS$. It is easy to see that $-\int_\Sigma \overline{\overline{\Gamma}}\left(\overline{\overline{C}}\vec{U}\right)dS + \int_\Sigma \overline{\overline{C}}^a\overline{\overline{\Gamma}}\vec{U}dS = -\int_\Sigma \overline{\overline{\Gamma}}\vec{F}dS + \int_\Sigma \overline{\overline{I}}\delta\vec{U}dS$. Thus, with $\overline{\overline{C}}\vec{U} = \vec{F}$ and $\overline{\overline{C}}^a\overline{\overline{\Gamma}} = \overline{\overline{I}}\delta$, we easily obtain $\vec{U} = \int_\Sigma \overline{\overline{\Gamma}}\vec{F}dS - \int_\Sigma \overline{\overline{\Gamma}}\left(\overline{\overline{C}}\vec{U}\right)dS + \int_\Sigma \overline{\overline{C}}^a\overline{\overline{\Gamma}}\vec{U}dS$.

---

9 We recall that based on the definition of the adjoint of a differential operator, we get: $a\partial(\cdot)/\partial x = -\partial(a\cdot)/\partial x$. The adjoint $\overline{\overline{C}}^a$ of a matrix $\overline{\overline{C}}$ is given by: $\overline{\overline{C}}^a = \overline{\overline{C}}^{t\star}$ where $\overline{\overline{C}}^t$ is the transpose and $\overline{\overline{C}}^\star$ is the conjugate. If C is an operator, then $\langle u, Cv\rangle = \langle C^a u, v\rangle$.

The idea is to integrate by parts $\overline{\overline{J}}_1$ and $\overline{\overline{J}}_2$ to show, by contrast, the integrals at the edge of the shell. Each component α of $\overline{\overline{J}}_1$ can be given by:

$$J_{1\alpha} = \mathcal{D}\int_\Sigma \Bigg[\Gamma_{\alpha z}\left(u_{z,zz} + \frac{1-\nu}{2R^2}u_{z,\phi\phi} + \frac{\rho h\omega^2}{\mathcal{D}}u_z + \frac{1+\nu}{2R}u_{\phi,z\phi} + \frac{\nu}{R}u_{r,z}\right)$$
$$+\Gamma_{\alpha\phi}\left(\frac{1+\nu}{2R}u_{z,z\phi} + \frac{1-\nu}{2}u_{\phi,zz} + \frac{1}{R^2}u_{\phi,\phi\phi} + \frac{\rho h\omega^2}{\mathcal{D}}u_\phi + \frac{1}{R^2}u_{r,\phi}\right)$$
$$+\Gamma_{\alpha r}\left(\frac{\nu}{R}u_{z,z} + \frac{1}{R^2}u_{\phi,\phi} + \frac{1}{R^2}u_r\right.$$
$$\left.+ \chi\left(R^2u_{r,zzzz} + 2u_{r,zz\phi\phi} + \frac{1}{R^2}u_{r,\phi\phi\phi\phi}\right) - \frac{\rho h\omega^2}{\mathcal{D}}u_r\right)dS\Bigg].$$

We integrate by parts this integral (and twice the terms in χ)[10]. If $\partial\Sigma$ is the edge of the cylinder ($z = \pm L, \phi \in [0, 2\pi[$), we can show that we have:

$$J_{1\alpha} = -\mathcal{D}\int_\Sigma \Bigg[\Gamma_{\alpha z,z}\left(u_{z,z} + \frac{\nu}{R}\left(u_{\phi,\phi} + u_r\right)\right) + \Gamma_{\alpha z,\phi}\frac{1-\nu}{2R}\left(\frac{u_{z,\phi}}{R} + u_{\phi,r}\right)$$
$$+\Gamma_{\alpha\phi,\phi}\left(\frac{\nu}{R}u_{z,z} + \frac{1}{R^2}\left(u_{\phi,\phi} + u_r\right)\right) + \Gamma_{\alpha\phi,z}\frac{1-\nu}{2}\left(\frac{u_{z,\phi}}{R} + u_{\phi,z}\right)$$
$$-\Gamma_{\alpha r}\left(\frac{\nu}{R}u_{z,z} + \frac{1}{R^2}\left(u_{\phi,\phi} + u_r\right)\right)$$
$$-\Gamma_{\alpha z}\frac{\rho h\omega^2}{D}u_z - \Gamma_{\alpha\phi}\frac{\rho h\omega^2}{D}u_\phi + \Gamma_{\alpha r}\frac{\rho h\omega^2}{D}u_r\Bigg]dS$$
$$+\mathcal{D}\chi\int_\Sigma \Bigg[\Gamma_{\alpha r,zz}\left(R^2u_{r,zz} + \nu u_{r,zz}\right) + \Gamma_{\alpha r,\phi\phi}\left(\frac{1}{R^2}u_{r,\phi\phi} + \nu u_{r,zz}\right)$$
$$+\Gamma_{\alpha r,z\phi}2(1-\nu)u_{r,z\phi}\Bigg]dS$$
$$+\mathcal{D}\int_{\partial\Sigma}\Bigg[\Gamma_{\alpha z}\left(u_{z,z} + \frac{\nu}{R}\left(u_{\phi,\phi} + u_r\right)\right) + \Gamma_{\alpha\phi}\frac{1-\nu}{2}\left(\frac{u_{z,\phi}}{R} + u_{\phi,z}\right)$$
$$+\chi R^2\left(\Gamma_{\alpha r}\left(u_{r,zzz} + \frac{2-\nu}{R^2}u_{r,\phi\phi z}\right) - \Gamma_{\alpha r,z}\left(u_{r,zz} + \frac{\nu}{R^2}u_{r,\phi\phi}\right)\right)\Bigg]dS.$$

10 In fact, this leads to carrying out the inverse operation of the one which we carried out when establishing the shell equation.

Similarly, we show that each component α of $\overline{\overline{J}}_2$ can be given by (we must pay attention to the sign change for the simple derivations which occur due to the adjoint):

$$J_{2\alpha} = \mathcal{D}\int_{\Sigma}\left[u_z\left(\Gamma_{\alpha z,zz} + \frac{1-\nu}{2R^2}\Gamma_{\alpha z,\phi\phi} + \frac{\rho h\omega^2}{\mathcal{D}}\Gamma_{\alpha z} + \frac{1+\nu}{2R}\Gamma_{\alpha\phi,z\phi} - \frac{\nu}{R}\Gamma_{\alpha r,z}\right)\right.$$
$$+u_\phi\left(\frac{1+\nu}{2R}\Gamma_{\alpha z,z\phi} + \frac{1-\nu}{2}\Gamma_{\alpha\phi,zz} + \frac{1}{R^2}\Gamma_{\alpha\phi,\phi\phi} + \frac{\rho h\omega^2}{\mathcal{D}}\Gamma_{\alpha\phi} - \frac{1}{R^2}\Gamma_{\alpha r,\phi}\right)$$
$$+u_r\left(-\frac{\nu}{R}\Gamma_{\alpha z,z} - \frac{1}{R^2}\Gamma_{\alpha\phi,\phi} + \frac{1}{R^2}\Gamma_{\alpha r}\right.$$
$$\left.\left.+\chi\left(R^2\Gamma_{\alpha r,zzzz} + 2\Gamma_{\alpha r,zz\phi\phi} + \frac{1}{R^2}\Gamma_{\alpha r,\phi\phi\phi\phi}\right) - \frac{\rho h\omega^2}{\mathcal{D}}\Gamma_{\alpha r}\right)dS\right].$$

We integrate by parts once more:

$$J_{2\alpha} = -\mathcal{D}\int_{\Sigma}\left[u_{z,z}\left(\Gamma_{\alpha z,z} + \frac{\nu}{R}\left(\Gamma_{\alpha\phi,\phi} - \Gamma_{\alpha r}\right)\right) + u_{z,\phi}\frac{1-\nu}{2R}\left(\frac{\Gamma_{\alpha z,\phi}}{R} + \Gamma_{\alpha\phi,r}\right)\right.$$
$$+u_{\phi,\phi}\left(\frac{\nu}{R}\Gamma_{\alpha z,z} + \frac{1}{R^2}\left(\Gamma_{\alpha\phi,\phi} - \Gamma_{\alpha r}\right)\right) + u_{\phi,z}\frac{1-\nu}{2}\left(\frac{\Gamma_{\alpha z,\phi}}{R} + \Gamma_{\alpha\phi,z}\right)$$
$$-u_r\left(-\frac{\nu}{R}\Gamma_{\alpha z,z} + \frac{1}{R^2}\left(-\Gamma_{\alpha\phi,\phi} + \Gamma_{\alpha r}\right)\right)$$
$$\left.-u_z\frac{\rho h\omega^2}{D}\Gamma_{\alpha z} - u_\phi\frac{\rho h\omega^2}{D}\Gamma_{\alpha\phi} + u_r\frac{\rho h\omega^2}{D}\Gamma_{\alpha r}\right]dS$$
$$+\mathcal{D}\chi\int_{\Sigma}\left[u_{r,zz}\left(R^2\Gamma_{\alpha r,zz} + \nu\Gamma_{\alpha r,zz}\right) + u_{r,\phi\phi}\left(\frac{1}{R^2}\Gamma_{\alpha r,\phi\phi} + \nu\Gamma_{\alpha r,zz}\right)\right.$$
$$\left.+u_{r,z\phi}2(1-\nu)\Gamma_{\alpha r,z\phi}\right]dS$$
$$+\mathcal{D}\int_{\partial\Sigma}\left[u_z\left(\Gamma_{\alpha z,z} + \frac{\nu}{R}\left(\Gamma_{\alpha\phi,\phi} - \Gamma_{\alpha r}\right)\right) + u_\phi\frac{1-\nu}{2}\left(\frac{\Gamma_{\alpha z,\phi}}{R} + \Gamma_{\alpha\phi,z}\right)\right.$$
$$\left.+\chi R^2\left(u_r\left(\Gamma_{\alpha r,zzz} + \frac{2-\nu}{R^2}\Gamma_{\alpha r,\phi\phi z}\right) - u_{r,z}\left(\Gamma_{\alpha r,zz} + \frac{\nu}{R^2}\Gamma_{\alpha r,\phi\phi}\right)\right)\right]dS.$$

It is easy to see that the integrals related to Σ are identical, by difference of $\overline{\overline{J}}_1$ and $\overline{\overline{J}}_2$, only the integrals at the edge of the shell remain. We obtain Green's representation of the shell:

$$\vec{U} = \int_{\Sigma} \overline{\overline{\Gamma}} \vec{F} dS - \mathcal{D} \int_{\partial\Sigma} \left[\Gamma_{\alpha z} \left(u_{z,z} + \frac{\nu}{R} (u_{\phi,\phi} + u_r) \right) \right.$$
$$+ \Gamma_{\alpha z,z}(-u_z) + \Gamma_{\alpha z,\phi} \left(-\frac{1-\nu}{2R} u_\phi \right)$$
$$+ \Gamma_{\alpha\phi} \frac{1-\nu}{2} \left(\frac{u_{z,\phi}}{2} + u_{\phi,z} \right) + \Gamma_{\alpha\phi,z} \left(-\frac{1-\nu}{2} u_\phi \right) + \Gamma_{\alpha\phi,\phi} \left(-\frac{\nu}{R} u_z \right)$$
$$+ \Gamma_{\alpha r} \left[\chi R^2 \left(u_{r,zzz} + \frac{2-\nu}{R^2} u_{r,\phi\phi z} \right) + \frac{\nu}{R} u_z \right] + \Gamma_{\alpha r,zz} \left(\chi R^2 u_{r,z} \right)$$
$$- \Gamma_{\alpha r,z} \chi R^2 \left(u_{r,zz} + \frac{\nu}{R^2} u_{r,\phi\phi} \right) + \Gamma_{\alpha r,\phi\phi} \left(\chi R^2 \frac{\nu}{R^2} u_{r,z} \right)$$
$$\left. + \Gamma_{\alpha r,\phi\phi z} \left(-\chi R^2 \frac{2-\nu}{R^2} u_r \right) + \Gamma_{\alpha r,zzz} \left(-\chi R^2 u_r \right) \right] dS.$$

In a general way, as $\int_{\partial\Sigma} (\cdot) dS = \int_0^{2\pi} [(\cdot)]_{-L}^{+L} R d\phi$, we can write this Green's representation in a more compact form:

$$\vec{U} = \overline{\overline{\Gamma}} \star \vec{F} + \overline{\overline{\Gamma}} \star \left(\vec{S}^{+L} - \vec{S}^{-L} \right), \qquad [7.32]$$

where $\vec{S}^{\pm L}$ are the boundary sources defined by:

$$S_z^{\pm L} = -\left(\sigma_{z,z}^{\pm L} + \frac{\nu}{R} \left(\sigma_{\phi,\phi}^{\pm L} + \sigma_r^{\pm L} \right) \right) \delta^{\pm L}(z)\delta(\phi) + \sigma_z^{\pm L} \delta_{,z}^{\pm L}(z)\delta(\phi)$$
$$+ \frac{1-\nu}{2R} \sigma_\phi^{\pm L} \delta^{\pm L}(z) \delta_{,\phi}(\phi)$$
$$S_\phi^{\pm L} = -\frac{1-\nu}{2} \left(\frac{1}{R} \sigma_{z,z}^{\pm L} + \sigma_{\phi,z}^{\pm L} \right) \delta^{\pm L}(z)\delta(\phi) + \frac{1-\nu}{2} \sigma_\phi^{\pm L} \delta_{,z}^{\pm L}(z)\delta(\phi)$$
$$+ \frac{\nu}{R} \sigma_z^{\pm L} \delta^{\pm L}(z) \delta_{,\phi}(\phi)$$
$$S_r^{\pm L} = \left[-\chi R^2 \left(\sigma_{r,zzz}^{\pm L} + \frac{2-\nu}{R^2} \sigma_{r,\phi\phi z}^{\pm L} \right) - \frac{\nu}{R} \sigma_r^{\pm L} \right] \delta^{\pm L}(z)\delta(\phi)$$
$$- \chi R^2 \left[\left(\sigma_{r,zz}^{\pm L} + \frac{\nu}{R^2} \sigma_{r,\phi\phi}^{\pm L} \right) \delta_{,z}^{\pm L}(z)\delta(\phi) - \sigma_{r,z}^{\pm L} \delta_{,zz}^{\pm L}(z)\delta(\phi) \right]$$

$$+\chi R^2\left(-\frac{\nu}{R}\sigma_{r,z}^{\pm L}\delta^{\pm L}(z)\delta_{,\phi\phi}(\phi)+\frac{2-\nu}{R^2}\sigma_r^{\pm L}\delta_{,z}^{\pm L}(z)\delta_{,\phi\phi}(\phi)\right.$$
$$\left.+\sigma_r^{\pm L}\delta_{,zzz}^{\pm L}(z)\delta(\phi)\right),$$

where $\sigma_{\alpha,\beta}^{\pm L} = u_{\alpha,\beta}^{\pm L^+} - u_{\alpha,\beta}^{\pm L^-}$ is the jump in the value $u_{\alpha,\beta}$ when crossing the discontinuity $\pm L$ taken in the direction of the normal. $\delta^{\pm L}(z)\delta(\phi)$ is the Dirac distribution for $z = \pm L, \forall\phi$. To gain a better understanding of these ideas, we consider a shell clamped at $z = \pm L$. The values on the inside of the domain at $z = \pm L$ are zero: $u_z = 0$, $u_\phi = 0$, $u_r = 0$ and $u_{r,z} = 0$. We assume that the same is true for the values outside of the domain occupied by the shell and therefore the jumps become zero: $\sigma_z^{\pm L} = 0$, $\sigma_\phi^{\pm L} = 0$, $\sigma_r^{\pm L} = 0$ and $\sigma_{r,z}^{\pm L} = 0$. Thus, the boundary sources $\vec{S}^{\pm L}$ in this case become:

$$S_z^{\pm L} = -\left(\sigma_{z,z}^{\pm L} + \frac{\nu}{R}\left(\sigma_{\phi,\phi}^{\pm L}\right)\right)\delta^{\pm L}(z)\delta(\phi) = s_1^{\pm L}\delta^{\pm L}(z)\delta(\phi)$$

$$S_\phi^{\pm L} = -\frac{1-\nu}{2}\left(\frac{1}{R}\sigma_{z,z}^{\pm L} + \sigma_{\phi,z}^{\pm L}\right)\delta^{\pm L}(z)\delta(\phi) = s_2^{\pm L}\delta^{\pm L}(z)\delta(\phi)$$

$$S_r^{\pm L} = -\chi R^2\left(\sigma_{r,zzz}^{\pm L} + \frac{2-\nu}{R^2}\sigma_{r,\phi\phi z}^{\pm L}\right)\delta^{\pm L}(z)\delta(\phi)$$
$$-\chi R^2\left(\sigma_{r,zz}^{\pm L} + \frac{\nu}{R^2}\sigma_{r,\phi\phi}^{\pm L}\right)\delta_{,z}^{\pm L}(z)\delta(\phi)$$
$$= s_3^{\pm L}\delta^{\pm L}(z)\delta(\phi) + s_4^{\pm L}\delta_{,z}^{\pm L}(z)\delta(\phi).$$

Green's representation of the shell thus introduces four unknowns $s_i^{\pm L}$ at each end of the shell. We can easily calculate them by writing the four boundary conditions at each end of the shell. We can easily show that for each angular harmonic m, this is reduced to solving a system of simultaneous equations of eight unknowns, which can be solved trivially[11].

7.3.2.4. *Response of a shell excited by a turbulent boundary layer*

Let us consider the example of a cylindrical shell of a finite length, clamped on both ends in vacuum, which is set in vibration by a random process in the space and time representative of an internal turbulent boundary layer of air[12]. This process can

11 We obtain a similar result for the Flügge operator, which due to the cumbersome calculations, we will not discuss in detail.

12 It seems a bit surprising to describe the movement of a shell in vacuum when it is excited by a turbulent boundary layer, nevertheless, be aware that for a steel shell in contact with air, the effect of the presence of a fluid on the dynamic behavior of the shell is negligible. Moreover, the turbulent excitation can be fully described as a distributed force since it is considered that the vibration of the shell does not change the structure of the turbulent boundary layer.

be represented by its interspectral power density $\Phi_p(Q;Q',\omega)$ where ω is the pulsation. We have chosen to represent this density according to the Corcos model (given in section 8.3.9). In this case, $\Phi_p(Q;Q',\omega)$ can be expressed in a separable form. For a more complete presentation of these concepts, the readers are invited to refer to sections 9.6 and 10.2. Let (z,ϕ) be the coordinates of Q and (z',ϕ') such that Q', $\Phi_p(Q;Q',\omega)$ can be written as $\Phi_p(Q;Q',\omega) = \Phi(f)\exp(- \mid \zeta \mid /L_\zeta(f))\exp(- \mid \eta \mid /L_\eta(f))\exp(-2\imath\pi f\zeta/U_c(\zeta,f))$, where $\zeta = z - z'$ and $\eta = \phi - \phi'$ are the longitudinal and circumferential separations. The parameters of the model are given by interpolations constructed on the basis of experimental data measured in an anechoic wind tunnel. These parameters are the spectral density $\Phi(f)$, given by the Durant–Robert model [8.119], and the longitudinal $L_\zeta(f)$ and circumferential $L_\eta(f)$ correlation lengths and the convection velocity $U_c(\zeta,f)$, which are given by the following interpolations [DUR 99]:

$$\frac{L_\zeta(f)}{R} = 0.762 + 16.528 f_e - 37.646 f_e^2 + 34.115 f_e^3 - 15.329 f_e^4 \\ +3.393 f_e^5 - 0.295 f_e^6$$

$$\frac{L_\eta(f)}{R} = 0.1 + 0.594 f_e - 1.169 f_e^2 + 0.952 f_e^3 - 0.399 f_e^4 \\ +0.084 f_e^5 - 0.007 f_e^6$$

$$\frac{U_c(f)}{U_0} = 0.394 + 1.656 f_e - 2.866 f_e^2 + 2.373 f_e^3 - 1.032 f_e^4 \\ +0.226 f_e^5 - 0.019 f_e^6$$

where $f^e = fR/U_0$ represents the frequency measured in the scale system in external flow variable. The time scale is determined by the flow velocity $U_0 = 100$ m/s and the radius R.

If $\vec{u}(Q;M,\omega)$ is the response of a system at point M to a normal unitary exciting point force $(0,0,\delta_Q)$ applied at Q, the interspectral densities of each component of the displacement u_z, u_ϕ and u_r are given by:

$$S_{U_j}(M;M',\omega) = \int_\Sigma\int_\Sigma u_j(Q;M,\omega)\Phi_p(Q;Q',\omega)u_j^\star(Q';M',\omega)dQdQ',$$

where $u_j^\star$ is the conjugate of u_j. The initial random problem is thus reduced to solving a series of deterministic problems in a harmonic regime $\mathrm{e}^{-\imath\omega t}$ for which we can show that there exists a unique solution for any real excitation frequency, if the shell is damped. When we focus on the spectral density of the vibration velocity, it is enough to calculate $S_{V_j}(M;M',\omega) = -\omega^2 S_{U_j}(M;M',\omega)$.

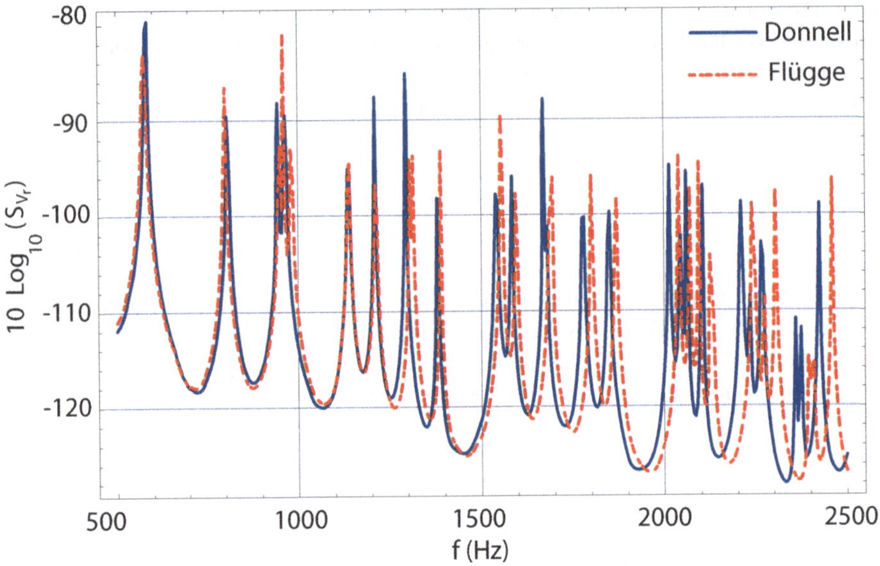

Figure 7.11. *Comparison of power spectral densities of radial vibration velocity for the Donnell–Mushtari and Flügge shell operators. For a color version of the figure, see ww.iste.co.uk/anselmet/acoustics.zip*

The Donnell–Mushtari operator gives satisfactory results, however, its precision decreases as the frequency increases. In order to refine the calculations, we can opt for a Flügge shell model which is one of the most accurate thin shell operators. The results have shown a good agreement between the calculations and measurements over a wide frequency range. Figure 7.11 presents a comparison of radial vibration velocities (thus, the normal velocities of the shell), at a point located on the axis of the shell at 20 cm from the upstream end of the shell, obtained by the two thin shell models for excitation by a turbulent boundary later. In the two cases, the shell, slightly dampened by a viscosity model with a complex Young's modulus, is made out of steel ($\rho_c = 7800$ Kg/m^3, $E = 200(1 - \imath\eta)$ Gpa, $\nu = 0.3$, $\eta = 5.10^{-4}$) 46 cm in length, 12.5 cm in diameter and 0.5 mm in thickness. Despite the small ratio $h/R = 0.8\%$, there is a significant difference between the results. The Flügge operator is a Donnell–Mushtari operator, which is refined to better account for the mechanical and geometrical effects in the thickness. We can see it as a slightly thicker thin shell operator. At low frequency, due to the added mass effect, the Flügge operator has a spectrum that is slightly offset toward the low frequencies by approximately 10 Hz (approximately 2% difference deviation from eigenfrequencies), while at high frequencies, due to the added stiffness, the Flügge operator has a spectrum shifted toward the treble up to 35 Hz (approximately 1.5%).

Table 7.7 includes the first shell resonance frequencies described by a Flügge model[13], and the values measured by identification of the parameters of the frequency response, based on the frequency response measured by a shock hammer. The calculated frequencies are extremely close to the measured ones; the imaginary part is slightly underestimated due to not taking into account the dissipation by the edges of the shell.

f_{mn}	$n = 1$	$n = 2$	$n = 3$	$n = 4$
$m = 1$ c.	$2034.4 - \imath 0.50$			
m.	$2024.8 - \imath 1.03$			
$m = 2$ c.	$973.1 - \imath 0.70$	$2238.8 - \imath 1.0$		
$m = 3$ c.	$569.9 - \imath 0.26$	$1298.8 - \imath 0.9$	$2238.4 - \imath 0.34$	$3336 - \imath 1.5$
m.	$565.2 - \imath 0.63$	$1306.2 - \imath 1.0$		
$m = 4$ c.	$573.9 - \imath 0.41$	$945.3 - \imath 0.65$	$1551.8 - \imath 0.97$	$2296.9 - \imath 0.46$
m.	$570.7 - \imath 1.10$	$944.8 - \imath 0.51$	$1560.5 - \imath 0.31$	$2323 - \imath 1.3$
$m = 5$ c.	$795.9 - \imath 0.35$	$957.9 - \imath 0.42$	$1307.5 - \imath 0.45$	$1796.6 - \imath 0.91$
m.	$794.6 - \imath 0.61$	$959.7 - \imath 0.56$	$1310.3 - \imath 0.81$	$1815.3 - \imath 0.51$
$m = 6$ c.	$1134.3 - \imath 0.25$	$1209.6 - \imath 0.20$	$1383.9 - \imath 0.50$	$1684.4 - \imath 0.81$
m.	$1131.5 - \imath 1.02$	$1206.9 - \imath 1.2$	$1387.1 - \imath 0.90$	$1692.6 - \imath 0.70$
$m = 7$ c.	$1550.1 - \imath 0.41$	$1593.7 - \imath 0.27$	$1688.1 - \imath 0.95$	$1865.1 - \imath 0.43$
m.	$1544.7 - \imath 1.7$	$1588.8 - \imath 0.89$	$1687.1 - \imath 1.35$	$1873.6 - \imath 0.53$

Table 7.7. *Resonance frequencies of a cylindrical shell (c.: calculated, m.: measured)*

Figure 7.12 shows a comparison between the calculated radial vibration velocity power spectral densities and those measured by a laser vibrometer at a point located 5 cm from the center of the shell in the direction of the flow. At the center of the flow, the air velocity is $U_0 = 100$ m/s. Apart from the low-frequency noise levels that the model cannot describe, the main cause of dispersion is related to the limited dynamic range of the measuring device which digitizes the signal using a 12-bit analog-digital converter. Such a range corresponds to a maximum dynamic range of $20 \log_{10} 2^{12} =$ 72 dB; as, in practice, the last two bits are not important, the dynamic is limited to approximately $72 - 12 = 60$ dB. On the curve, we can observe a dynamic measured at approximately 56 dB, which is in line with the previous analysis. An analysis of other octaves shows differences of more than 1.5 dB between 500 and 3,000 Hz. It is only when the modal density becomes too high (greater than 50 modes) that this gap becomes significant. We clearly see that the turbulent boundary level has the ability to excite all the vibroacoustic modes of a structure. It is, therefore, very important to understand all the phenomena involved, which can be found in many areas, such as sonar sampling, fluid flow and, more generally, high-speed transport.

13 These frequencies are complex because the shell is damped by viscosity.

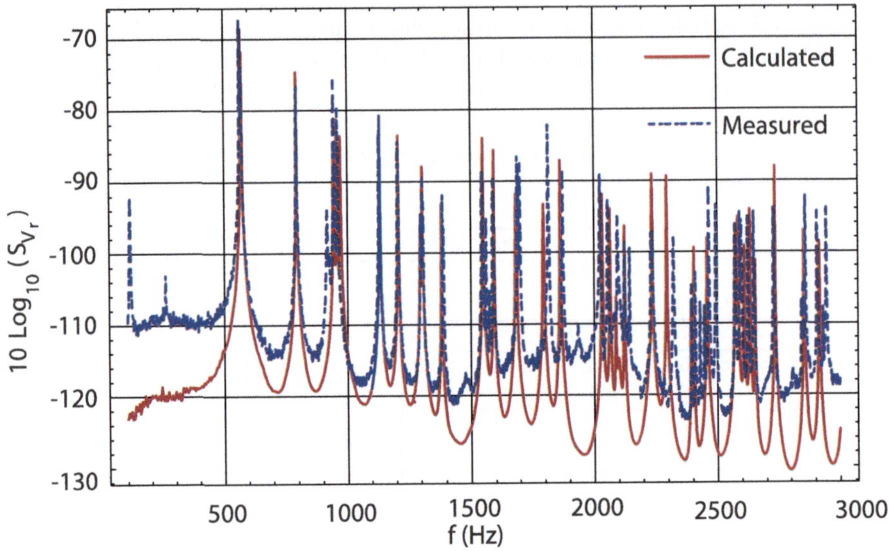

Figure 7.12. *Radial velocity power spectral density of the shell (U_0 = 100 m/s). For a color version of the figure, see www.iste.co.uk/anselmet/acoustics.zip*

8

Acoustic Radiation of Thin Plates

The appearance of an acoustic wave in a fluid is often caused by the vibration of an elastic solid. This is the case of many musical instruments (except for the wind instruments) and most machines. For example, for string instruments, the vibrating string transfers its movement to the body of the instrument, which in turn begins to vibrate, transmitting its vibration energy into the air surrounding the instrument. Most of the energy radiated by a violin is by its body and virtually nothing is radiated by the string in vibration. Apart from being a pleasant (or irritating) aspect, it is important to note that the noise radiated by an object can be a good way of controlling it. A cracked plate has a different acoustic "signature" than that of a healthy one. The opposite aspect of this phenomenon is setting in vibration a structure by a sound field and the sound transmission through it. Hence, we can hear what is happening on the other side of a wall or window. A classic example is that a singer singing a particularly high note can break a piece of glass. The frequency of the sung note corresponds to the eigenfrequency of glass, which then vibrates with such an amplitude that it is destroyed. We must note that the vibration of a structure can also be induced by a fluid in motion, either by a turbulent boundary layer or by regular vortex shedding. In acoustics of building, due to the large wall surfaces, a very small vibration of one of these surfaces is enough (due to the high flow rate) to create significant sound waves. For example, a room with 3 m long walls, for which one of the walls has a 1m^2 window that vibrates with an amplitude of 0.1 mm, induces a relative volume variation $\Delta V/V$ and an acoustic field in the cavity (based on the barotropic state law, $p = \gamma p_0 \Delta V/V = 1.4 \times 10^5 \times 0.0001/3^3$) of around 90 dB.

Here, we will discuss several simple examples of thin plates, immersed in an ideal fluid at rest and in motion, and excited by external forces.

First, we study a particularly simple one-dimensional example. This purely academic example provides us with complete analytic calculations and identifies several essential ideas. We begin with an infinite plate and we examine the conditions under which a bending plane wave can propagate in a plate and eventually radiate energy. Then, we examine the response of the plate-fluid system when a point source excites the structure. We will see in particular that the pressure radiated in the far field can be calculated very easily. In both cases, we focus on the case where the fluid has little influence on the behavior of the structure (such as a steel plate in air). This hypothesis is known as the "light fluid" approximation although gravity does not play a role in this case. This hypothesis allows us to develop several analytic approximations that are useful both for understanding the underlying physics and for validating the calculation codes. We then examine a slightly less simplistic case, where the structure is of finite dimensions, prolongated by a perfectly reflecting plane which extends to infinity. We often talk of baffled structures. The presence of the baffle is somewhat unrealistic but makes it possible to carry out simple calculations, while disregarding the essential ideas that are only relevant in the case of a non-baffled structure. In this latter case, in fact, there is involvement of the effects of diffraction by the edges of the plate which are difficult to address theoretically and numerically. We show how the coupled differential equation boundary problem can be reduced to an integro-differential boundary problem which is easier to address. We thus examine three "exact" methods for solving this type of problem. In the first method, we transform the initial integral problem into two coupled integral equations for which we propose a numerical solution. The advantage of using this type of method is that the dimension of the problem is reduced by an order of magnitude and that the function approximations are on the values at the edge of the surface. The wave propagation equation is perfectly respected. In the second method, we express the displacement of the coupled plate in a series of eigenmodes of the plate *in vacuo*. Then, we define the eigenmodes of the coupled plate which form a natural basis of the problem and we develop approximate methods based on the light fluid approximation. We present a simple method which allows us to extend to heavy fluid coupling the asymptotic methods developed for light fluid coupling, as well presenting an unexpected consequence of heavy fluid coupling on the spectral analysis. We will conclude this long chapter with the description of some models and effects related to the integration of fluid flow around a plate.

8.1. First notions of vibroacoustics: a simple example

We consider an infinite wave guide with a constant unit section (see Figure 8.1). At $x = 0$, the guide has a "partition" consisting of a thin wall (compared to the wavelength) with a mass m and suspended from a string with the stiffness r. On both sides of the partition, the guide is filled with an ideal fluid with the density ρ_0 and with the sound celerity c_0. This wall separates the guide into two "isolated" regions (the fluid cannot cross the partition). Moreover, we assume that only the plane waves can propagate parallel to the guide axis.

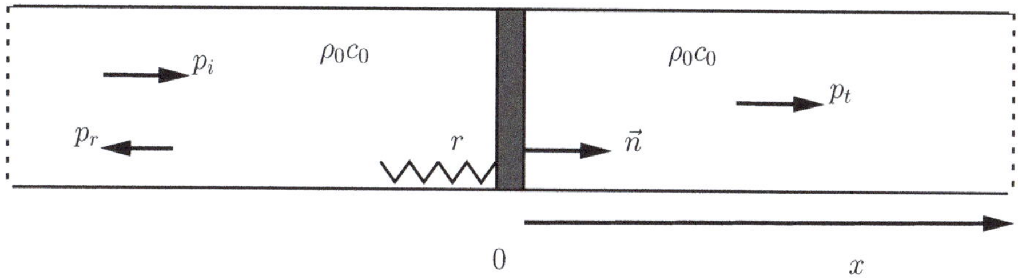

Figure 8.1. *1D wave guide geometry*

8.1.1. *Motion equations*

Let F be the force acting on the partition wall and U be its displacement. U is the solution to the (classic) mass-spring differential equation:

$$m\frac{d^2U(t)}{dt^2} + rU(t) = F.$$

The force F consists of two terms. The first term is an external force F_e and the second term F_p is the difference between acoustic pressures on either side of the partition wall. We find $F = F_e - F_p$. The choice of the sign of the excitation force is guided by physical considerations. The force F_e is a force injected into the system, thus it brings energy and its sign is positive. The pressure force F_p is a force which is lost by the partition, its sign is thus negative. Let S^+ and S^- be acoustic sources placed in the half-guides $x > 0$ and $x < 0$. P^+ and P^- are the acoustic pressures prevailing in the two half-guides. We thus obtain:

$$\frac{\partial^2 P^{\pm}(x,t)}{\partial x^2} - \frac{1}{c^2}\frac{\partial^2 P^{\pm}(x,t)}{\partial t^2} = S^{\pm}(x,t).$$

As the fluid is assumed to be ideal, there is continuity of the normal velocities at the wall. We have $dU(t)/dt = V^+(0,t) = V^-(0,t)$ where $= V^{\pm}(0,t)$ are the velocities of fluid particles present within $x < 0$ and $x > 0$. Thus, based on the conservation equation of impulse, we get the continuity condition of normal acceleration $\rho_0 d^2U/dt^2 = -\partial_n p(x=0) = -\partial p(x=0,t)/\partial x$ where ρ_0 represents the mass per unit of length of the fluid (one-dimensional problem). To this, we add that the waves emitted and reflected by the partition propagate toward $x > 0$ for the half-guide $x > 0$ (and toward $x < 0$ for the half-guide $x < 0$).

In a harmonic regime, when this does not cause confusion, the dependence on the pulsation ω is omitted. We thus set $f(\omega)$, $u(\omega)$, $p^{\pm}(x,\omega)$ and $s^{\pm}(x,\omega)$ as the Fourier transforms of the excitation term, the displacement of the wall, the pressures and the acoustic sources. We find $F_e(t) = f_e e^{-\imath\omega t}$, $F_p(t) = f_p \exp(-\imath\omega t)$, $S^{\pm}(x,t) = s^{\pm}(x)\exp(-\imath\omega t)$, $U(t) = u\exp(-\imath\omega t)$, $P^{\pm}(x,t) = p^{\pm}(x)\exp(-\imath\omega t)$. The system of equations that need to be solved can be written, if $f_p = p^+(0) - p^-(0)$:

$$-m\omega^2 u + ru = f_e - p^+(0) + p^-(0), \frac{d^2p^{\pm}(x)}{dx^2} + k^2p^{\pm}(x) = s^{\pm}(x).$$

To these equations, we can add the equation of conservation of momentum $\omega^2\rho_0 u = p'(x)|_{x=0}$. We must note that the first equation above shows a specific pulsation ω_0, defined by $\omega_0^2 = r/m$. ω_0 is the eigen pulsation of the partition wall. We consider two cases. In the first case, the partition wall is subjected to an external force $f_e \neq 0$. Moreover, we assume that there is no acoustic source in $x > 0, S^+ = 0$. This is the phenomenon of acoustic radiation from a vibrating structure. In the second case, we assume that the external force is zero $f_e = 0$ and that there is an acoustic source located in $x < 0$ and it emits an acoustic field; the wall is thus excited through the air and we find ourselves faced with the case of acoustic transmission.

8.1.2. *Acoustic radiation*

We thus assume that there is no energy source within the fluid. Only the partition wall is subjected to an external force of amplitude f_e. The acoustic waves generated by the partition wall, which vibrates like a piston, thus propagate toward $x > 0$ and $x < 0$ in the semi-guides $x > 0$ and $x < 0$. The pressure fields $p^{\pm}(x)$ are the solutions to the homogeneous one-dimensional Helmholtz equations $(p''(x) + k^2p(x) = 0)$. The principle of energy conservation (or the Sommerfeld conditions) indicates that only the waves, which are moving away, have a physical meaning. This implies that the acoustic pressures are in the form of $p^{\pm}(x) = A^{\pm}\exp(\pm kx)$.

The continuity equations of normal velocities can be written as $\omega^2\rho_0 u = dp^{\pm}(0)/dx$, that is $\omega^2\rho_0 u = p^{+\prime}(0)$ and $\omega^2\rho_0 u = p^{-\prime}(0)$. We thus obtain a system of three equations $-m(\omega^2 - \omega_0^2)u = f_e - p^+(0) + p^-(0)$, $\omega^2\rho_0 u = p^{+\prime}(0)$ and $\omega^2\rho_0 u = p^{-\prime}(0)$. As $p^{\pm}(x) = A^{\pm}\exp(\pm kx)$, the unknowns are thus u, A^+ and A^- which are the solutions to a linear system of three equations and three unknowns:

$$\begin{pmatrix} 1 & -1 & m\left(\omega^2 - \omega_0^2\right) \\ 1 & 0 & +\imath\omega\rho_0 c_0 \\ 0 & 1 & -\imath\omega\rho_0 c_0 \end{pmatrix} \begin{pmatrix} A^+ \\ A^- \\ u \end{pmatrix} = \begin{pmatrix} f_e \\ 0 \\ 0 \end{pmatrix}$$

The determinant of this system can be given by $D = -2m(\imath\omega\rho_0 c_0/m + \omega^2 - \omega_0^2)$. It is easy to see that due to the damping by coupling with the fluid, the determinant of the system is never zero and that there always exists a solution for it. If ρ_0 is zero, D is zero at $\omega = \omega_0$ which leads to an undetermination in the system. We, thus, obtain:

$$u = -f_e \frac{1}{m\left(2\imath\omega\frac{\rho_0 c_0}{m} + \omega^2 - \omega_0^2\right)}, A^{\pm} = \pm f_e \frac{\imath\omega\frac{\rho_0 c_0}{m}}{2\imath\omega\frac{\rho_0 c_0}{m} + \omega^2 - \omega_0^2}.$$

The eigen pulsation ω_0 of the partition wall has a specific role. If $\omega = \omega_0$, the displacement of the wall and the pressures radiated are at their maximum. In particular, we find that $A^{\pm} = \pm f_e/2$, which means that all the excitation energy is transmitted to the fluid. The pressures $p^{\pm}$ have the same modulus, which is obvious because of the symmetry of the system with respect to x. In addition, these pressures have opposite signs. We obtain $p^+(x) = -p^-(x) = \exp(\imath\pi)p^-(x)$. The pressures are in phase opposition. The result is also evident since, during its movement, when the wall moves toward $x > 0$, it creates an overpressure at $x > 0$ and a depression at $x < 0$.

We calculate P_t, the total mean power (over a period T) supplied to the system. By definition, we obtain:

$$P_t = \frac{1}{T}\int_0^T \Re\left(F_e(t)\right)\Re\left(v(t)\right)dt = \frac{1}{T}\int_0^T \Re\left(f_e \mathrm{e}^{-\imath\omega t}\right)\Re\left(-\imath\omega u \mathrm{e}^{-\imath\omega t}\right)dt.$$

We assume that $F_e \in I\!R$. We find:

$$\begin{aligned} P_t &= \frac{1}{T}\int_0^T f_e^2\cos(\omega t)\Re\left(\imath\omega\frac{-2\imath\omega\frac{\rho_0 c_0}{m} + \omega^2 - \omega_0^2}{m\left(4\omega^2\left(\frac{\rho_0 c_0}{m}\right)^2 + \left(\omega^2 - \omega_0^2\right)^2\right)}\mathrm{e}^{-\imath\omega t}\right)dt \\ &= \frac{f_e^2}{m\left(4\omega^2\left(\frac{\rho_0 c_0}{m}\right)^2 + \left(\omega^2 - \omega_0^2\right)^2\right)}\omega^2\frac{\rho_0 c_0}{m}. \end{aligned}$$

It is easy to see that $P_t > 0$. If $\omega = \omega_0$, the total permissible power is at its maximum and is $P_t = F_e^2/\left(4\rho_0 c_0\right)$. Similarly, we can define the mean power transmitted in each semi-guide $P^{\pm}$ as the mean over a period of the instantaneous power $P_i^{\pm}$ which crosses a section of the guide at the abscissa x. $P_i^{\pm}$ is given by $P_i^{\pm}(t) = Re\left(p^{\pm}(x,\omega)\mathrm{e}^{-\imath\omega t}\right)\Re\left(v^{\pm}(x,\omega)\mathrm{e}^{-\imath\omega t}\right)$, where $v^{\pm}(x,\omega) = 1/(\imath\omega\rho_0)$ $dp^{\pm}(x,\omega)/dx$ is the instantaneous vibration velocity of the fluid.

$$P^{\pm} = \frac{1}{T}\int_0^T P_i^{\pm}(x,t)dt = \frac{1}{T}\int_0^T \Re\left(p^{\pm}(x,t)\right)\Re\left(v^{\pm}(x,t)\right)dt.$$

We can easily show that $P^{\pm} = P_t/2$.

8.1.3. *"Light fluid" approximation*

We introduce the parameter $\epsilon = \rho_0 c_0/m$. If the density of the fluid is low compared to the mass of the partition wall, we find $\epsilon \ll 1$. We expand u and $A^{\pm}$ into a series of increasing powers of ϵ. It becomes:

$$u = u_0 + \epsilon u_1 + \epsilon^2 u_2 + +\mathcal{O}(\epsilon^3), A^{\pm} = A_0^{\pm} + \epsilon A_1^{\pm} + \epsilon^2 A_2^{\pm} + +\mathcal{O}(\epsilon^3).$$

We start again with the three initial equations:

$$A^{+} - A^{-} - m\left(\omega^2 - \omega_0^2\right) u = f_e, -\imath k A^{\pm} \pm \omega^2 \rho_0 u = 0,$$

in which, we show that $\epsilon = \rho_0 c_0/m$. For example, we consider $\imath k A^{+} - \omega^2 \rho_0 u = 0$. We obtain $A^{+} + \imath \rho_0 c_0 \omega u = 0$ that is $A^{+}/m + \imath \epsilon \omega u = 0$. Hence the system:

$$\frac{A^{+}}{m} - \frac{A^{-}}{m} - \left(\omega^2 - \omega_0^2\right) u = \frac{f_e}{m}, \frac{A^{\pm}}{m} \pm \imath \epsilon \omega u = 0.$$

In this system, we carry out the expansions into a series of perturbations limited to the order 1. We obtain:

$$\frac{A_0^{+} + \epsilon A_1^{+}}{m} - \frac{A_0^{-} + \epsilon A_1^{-}}{m} - \left(\omega^2 - \omega_0^2\right)\left(u_0 + \epsilon u_1\right) + \mathcal{O}(\epsilon^2) = \frac{f_e}{m},$$
$$\frac{A_0^{\pm-} + \epsilon A_1^{\pm}}{m} \pm \imath \epsilon \omega (u_0 + \epsilon u_1) + \mathcal{O}(\epsilon^2) = 0.$$

We regroup the terms with the same power of ϵ. We obtain at the order zero for ϵ:

$$\frac{A_0^{+}}{m} - \frac{A_0^{-}}{m} - \left(\omega^2 - \omega_0^2\right) u_0 = \frac{f_e}{m}, \frac{A_0^{\pm}}{m} = 0.$$

Similarly, we obtain the system at the order one for ϵ.

$$\frac{A_1^{+}}{m} - \frac{A_1^{-}}{m} - \left(\omega^2 - \omega_0^2\right) u_1 = 0, \frac{A_1^{\pm}}{m} \pm \imath \omega u_0 = 0.$$

The system at the order zero is very easy to solve. We find $A_0^{\pm} = 0$ and $u_0 = -f_e/m(\omega^2 - \omega_0^2)$ which is the solution that we obtain in the absence of a fluid (mass-spring *in vacuo* and a null acoustic pressure). The system at the order one can be solved based on the solution at the order zero. This is general practice for the perturbation methods in which the solution at a given order depends on the solution at the lower order (or orders). We find:

$$A_1^{\pm} = \mp \imath m \omega u_0 = \pm \frac{\imath \omega f_e}{\omega^2 - \omega_0^2}, u_1 = \frac{A_1^{+} - A_1^{-}}{m(\omega^2 - \omega_0^2)} = -\frac{2\imath \omega u_0}{\omega^2 - \omega_0^2} = \frac{2\imath \omega f_e}{m(\omega^2 - \omega_0^2)^2}.$$

This result can be easily interpreted, if we note that the amplitude of the pressure at the order one corresponds to that which the spring-mass system *in vacuo* radiates. Similarly, the displacement at the order one corresponds to the displacement of the spring-mass system when the exciting force corresponds to the one due to the pressure difference created by the displacement of the partition wall in the absence of a fluid on both sides of the partition wall.

We show that at any order m, we obtain $u_m = -2\imath\omega/(\omega^2 - \omega_0^2)u_{m-1}$. It is thus obvious that we can write u_m in the following form $u_m = (-2\imath\omega/(\omega^2 - \omega_0^2))^m u_0$. We thus obtain the expansion:

$$u = \sum_m \epsilon^m u_m = u_0 \sum_m \left(-2\epsilon\imath\omega/(\omega^2 - \omega_0^2)\right)^m, \text{ or } \sum_m (-x)^m = \frac{1}{1+x}, \text{ that is}$$

$$u = \frac{u_0}{1 + 2\epsilon\imath\omega/(\omega^2 - \omega_0^2)} = -\frac{f_e}{m(\omega^2 - \omega_0^2 + 2\epsilon\imath\omega)}.$$

This shows that the expansion into a perturbation series, when it exists and when it can be continued as far as we want, can lead to an exact solution. However, it is very rare for the obtained series to converge. In general, we obtain asymptotic series. For a fixed value of the parameters, these series are divergent and for each problem there exists a single number of terms which lead to an optimal approximation (see, for example, [NAY 93, pp. 18–22]).

These results require several comments. We must note that rather than ϵ, these expansions involve the ratio $\eta = \epsilon/\left(\omega^2 - \omega_0^2\right)$. In fact, the real convergence condition of these expansions focuses on η and not on ϵ. The distinction is important because in the vicinity of the eigen pulsation of the wall, η can be very large as ϵ can be very small. These calculations are, therefore, valid "outside" of the eigenfrequency of the partition wall. We must note that the method which consists of solving a fluid-structure coupling problem, neglecting the influence of the fluid on the displacement and calculating the pressure radiated by the structure based on its displacement *in vacuo*, is commonly used.

8.1.4. *Sound transmission*

To simplify the explanation, we can assume that there are no acoustic sources in $x < 0$ but that the acoustic pressure $p^-(x)$ is the sum of a plane wave with a unity amplitude $p_i(x) = \exp(\imath kx)$ (which leads to rejecting the source s^- to infinity) and a wave reflected by the wall to determine $p_r(x)$. We thus obtain $p^-(x) = \exp(\imath kx) + p_r(x)$. The acoustic pressures $p^\pm(x)$ both satisfy homogeneous Helmholtz equations. The principle of conservation of energy implies that we have $p_r(x,\omega) = R(\omega)\mathrm{e}^{-\imath kx}$ and $p^+(x,\omega) = p_t(x,\omega) = T(\omega)\mathrm{e}^{\imath kx}$. This means a wave reflected by the partition wall which moves toward $x < 0$ and a transmitted wave which moves toward $x > 0$. We determine $u(\omega)$, $R(\omega)$ the reflection coefficient and $T(\omega)$ the transmission coefficient. The equation of the spring-mass system can be written as $-m\omega^2 u + ru = -p^+(0) + p^-(0)$, that is $m\left(\omega^2 - \omega_0^2\right)u + R - T = -1$. The continuity equations of normal velocities can be written as:

$$\omega^2 \rho_0 u = \frac{dp^\pm(0)}{dx} \Rightarrow \begin{cases} \omega^2 \rho_0 u - \imath k T = 0 \\ \omega^2 \rho_0 u - \imath k(1 - R) = 0. \end{cases}$$

Again, we obtain a linear system of three equations with three unknowns:

$$\begin{pmatrix} 1 & -1 & m\left(\omega^2 - \omega_0^2\right) \\ 1 & 0 & -\imath\omega\rho_0 c_0 \\ 0 & -1 & -\imath\omega\rho_0 c_0 \end{pmatrix} \begin{pmatrix} R \\ T \\ u \end{pmatrix} = \begin{pmatrix} -1 \\ 1 \\ 0 \end{pmatrix}$$

thus, the determinant Δ is given by $\Delta = -m(2\imath\omega\rho_0 c_0/m + \omega^2 - \omega_0^2)$. It is easy to see that Δ is never zero for a real excitation frequency. From here, it is easy to deduce that:

$$R = \frac{\omega^2 - \omega_0^2}{2\imath\omega\frac{\rho_0 c_0}{m} + \omega^2 - \omega_0^2}, T = \frac{2\imath\omega\rho_0 c_0}{2\imath\omega\rho_0 c_0 + m\left(\omega^2 - \omega_0^2\right)},$$

$$u = \frac{-2}{2\imath\omega\rho_0 c_0 + m\left(\omega^2 - \omega_0^2\right)}.$$

If $\omega = \omega_0$ then $R = 0$, $T = 1$, $u = 1/\left(\imath\omega_0\rho_0 c_0\right)$. The partition wall transmits all the energy of the incident wave and becomes acoustically transparent to acoustic waves. We use τ as the ratio between the mean power (over a period) of the energy transmitted by the partition wall and the mean incident power (here unity). We obtain $\tau = TT^\star$. This is:

$$\tau = \frac{4\omega^2\left(\frac{\rho_0 c_0}{m}\right)^2}{4\omega^2\left(\frac{\rho_0 c_0}{m}\right)^2 + \left(\omega^2 - \omega_0^2\right)^2}.$$

Let i be the insertion loss index, which is the difference between the incident wave and the transmitted wave. We obtain $i = 10\log_{10}(1/\tau)$. If $\epsilon = \rho_0 c_0/m \ll 1$ (such as a concrete wall in contact with air), we obtain:

$$\tau = \frac{4\omega^2\epsilon^2}{4\omega^2\epsilon^2 + \left(\omega^2 - \omega_0^2\right)^2} = \frac{4\epsilon^2}{\omega^2\left(1 - \frac{\omega_0^2}{\omega^2}\right)^2} + \mathcal{O}(\epsilon^4).$$

Moreover, we assume that the excitation frequency is sufficiently high to have $\frac{\omega_0^2}{\omega^2} \ll 1$. We thus obtain:

$$\tau \simeq \frac{4\rho_0^2 c^2}{m^2\omega^2}, i = 20 \log_{10}\left(\frac{m\omega}{2\rho_0 c}\right).$$

This law, valid "outside" of the eigenfrequency of the wall (high frequency), is called the mass law and states that the amplitude of the transmitted wave is inversely proportional to the mass of the partition wall. The insertion loss index thus presents a slope of 6 dB per octave: $i(2\omega) = i(\omega) + 20\log_{10}(2) = i(\omega) + 6$.

EXERCISES.–

1) We consider a one-dimensional problem (see Figure 8.2) which consists of a wave guide with a length ℓ, closed at the end $x = 0$ by a spring-mass partition wall and at $x = \ell$ by a perfectly rigid wall (Neumann condition). This guide contains a fluid with a density per unit length of ρ_0 and a wave celerity of c_0. This system is a good approximation of the loud-speaker enclosure at low frequency:

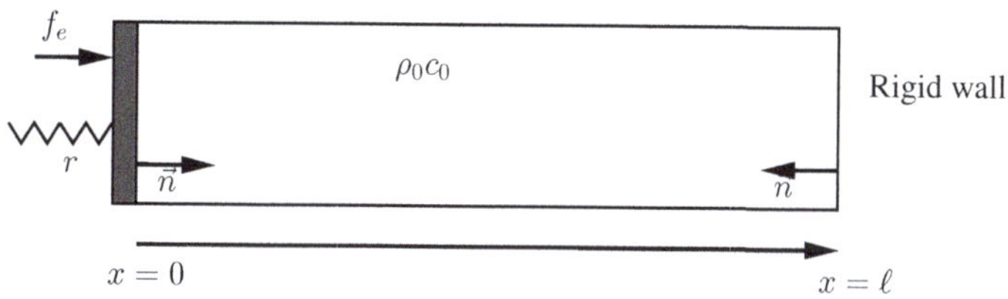

Figure 8.2. *Guide geometry closed by a rigid termination*

a) Show that the fluid mass-spring system has an infinite number of eigenfrequencies.

b) Calculate the displacement at low frequency, assuming that $\omega\ell/c_0 \ll 1$. Show that in this case the presence of the air layer can be reduced to an added stiffness effect.

2) The same problem as before but we replace the rigid wall at $x = \ell$ by a wall described by a Dirichlet condition which is a first approximation of a pipe open into air:

a) Show that the fluid spring-mass system has an infinite number of eigenfrequencies.

b) Calculate the displacement at low frequency, assuming that $\omega\ell/c_0 \ll 1$. Show that in this case the presence of the air layer can be reduced to an added mass effect.

3) Here, we look at the problem of transmission through a double-leaf wall (see Figure 8.3).

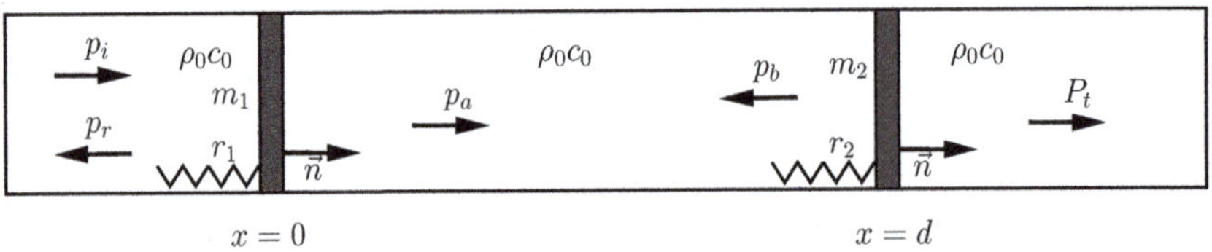

Figure 8.3. *Double-leaf wall geometry*

We call u_1 the displacement of the wall at $x = 0$ (wall 1) and u_2 the displacement of the wall at $x = d$ (wall 2). We assume that the incident field is a unitary plane wave $p_i(x) = \exp(\imath kx)$. It is obvious that we have: $p_r(x) = R\exp(-\imath kx)$, $p_a(x) = A\exp(\imath kx)$, $p_b(x) = B\exp(-\imath kx)$ and $p_t(x) = T\exp(\imath kx)$.

a) Write the six motion equations. To simplify, we use $p_1 = p_i(0^-) + p_r(0^-)$, $p_2 = p_a(0^+) + p_b(0^+)$, $p_3 = p_a(d^-) + p_b(d^-)$ and $p_4 = p_t(d^+)$.

b) Further on, we assume that $kd \ll 1$.

i) Show that the pressure in the cavity is uniform;

ii) Calculate the pressure in the cavity as a function of the difference $u_1 - u_2$ and show that the air acts as a spring with a stiffness $r_c = \rho_0 c_0^2/d$;

iii) Calculate the ratio u_1/u_2 and show that when $d \to 0$, the wall movements are identical;

iv) Calculate the transmission coefficient T. Show that the external fluid introduces a damping;

v) If we neglect the mechanical stiffness and the acoustic damping, we can show that the transmission coefficient is at its maximum for the mass-air-mass pulsation given by $\omega_0 = \sqrt{r_c(m_1 + m_2)/m_1 m_2}$. Calculate the insertion loss indices for $\omega \ll \omega_0$ and $\omega \gg \omega_0$.

SOLUTIONS.–

1) The equations of the problem are given by:

$$\begin{aligned} &\text{at } x=0 \quad -m\omega^2 u + ru = f_e - p(0)\\ &\qquad\qquad p'(0) = \omega^2 \rho_0 u\\ &\text{at } x=\ell \quad p'(\ell)=0 \qquad (\partial_n = 0 \Leftrightarrow -dp/dx = 0)\\ &x \in]0,\ell[\quad p''(x) + k^2 p(x) = 0 \end{aligned}$$

a) The Helmholtz equation leads to $p(x) = A\exp(\imath kx) + B\exp(-\imath kx)$. *The first three equations of the previous system can thus be written as:*

$$-m(\omega^2 - \omega_0^2)u + A + B = f_e \qquad [8.1]$$

$$\imath\omega\rho_0 c_0 u + A - B = 0 \qquad [8.2]$$

$$Ae^{\imath k\ell} - Be^{-\imath k\ell} = 0. \qquad [8.3]$$

To show that the fluid spring-mass system has an infinite number of eigenfrequencies, it is enough to show that there exists an infinite number of real frequencies that allow non-zero solutions for an excitation force of zero. This means that there are frequencies for which the determinant of the matrix associated with the linear system of simultaneous equations [8.1–8.3] is zero. This is given by $D = -2\imath[m(\omega^2 - \omega_0^2)\sin k\ell - \omega\rho_0 c_0 \cos k\ell]$. *To solve* $D = 0$, *it is enough to solve* $m(\omega^2 - \omega_0^2)\sin k\ell = \omega\rho_0 c_0 \cos k\ell$, *that is* $\tan \omega\ell/c_0 = \omega\rho_0 c_0/m(\omega^2 - \omega_0^2)$. *Solving this transcendental equation is particularly difficult. Nevertheless, we have a very simple graphical way to estimate the roots (and thus the eigenfrequencies) of the determinant. We aim to find the common points of the curves* $\tan \omega\ell/c_0$ *and* $\omega\rho_0 c_0/m(\omega^2 - \omega_0^2)$. *We can clearly see in the example (*ω_0 *is represented by a gray dot) in Figure 8.4, where the points of intersection are represented by black dot, that we have an infinite number of zeros and thus eigenfrequencies. If we carry out detailed analysis of these points, those which correspond to zeros of* $\tan \omega\ell/c$ *are the zeros that are slightly shifted toward the trebles of the system consisting of a rigid cavity at both ends; this can be easily explained by the fact that at the origin the mass-spring system adds some stiffness to the rigid cavity system due to the presence of the spring. We can also observe that the eigenfrequency of the spring-mass system is slightly shifted toward the trebles, this is due to the fact that the fluid adds some stiffness to the spring-mass system.*

b) The displacement is given by:

$$u = -\frac{f_e}{(m(\omega^2 - \omega_0^2) - \omega\rho_0 c_0 \cot \frac{\omega\ell}{c_0})}.$$

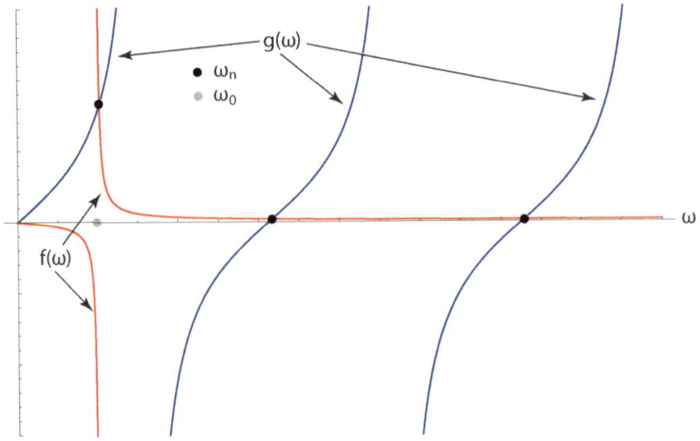

Figure 8.4. *The zeros of the system determinant with infinitely rigid termination with $f(\omega) = \omega\rho_0 c_0/m(\omega^2 - \omega_0^2)$, $g(\omega) = \tan \omega\ell/c_0$. For a color version of the figure, see www.iste.co.uk/anselmet/acoustics.zip*

For $\omega\ell/c_0 \ll 1$, we obtain $\cot \omega\ell/c_0 \approx 1/(\omega\ell/c_0)$. This is:

$$u \approx -\frac{f_e}{(m(\omega^2 - \omega_0^2) - \frac{\rho_0 c_0^2}{\ell})}.$$

But $r_c = \rho_0 c_0^2/\ell$ has a stiffness dimension $[Kg]/[s]^2$. We then set $\omega_1^2 = \omega_0^2 + r_c/m > \omega_0^2$, and we obtain the displacement of a spring-mass system with increased stiffness:

$$u \approx -\frac{f_e}{m(\omega^2 - \omega_1^2)}.$$

2) With the same notation, only changing the condition at $x = \ell$, given by $p(\ell) = 0$, the equations of the problem are given by:

$$\begin{aligned}
&\textit{at } x = 0 \quad -m\omega^2 u + ru = f_e - p(0) \\
&\qquad\qquad p'(0) = \omega^2 \rho_0 u \\
&\textit{at } x = \ell \quad p(\ell) = 0 \\
&x \in]0, \ell[\quad p''(x) + k^2 p(x) = 0
\end{aligned}$$

a) We obtain the system:

$$-m(\omega^2 - \omega_0^2)u + A + B = f_e$$

$$\imath\omega\rho_0 c_0 u + A - B = 0$$

$$Ae^{\imath k\ell} + Be^{-\imath k\ell} = 0.$$

The determinant of this system is given by $D = -2\imath[m(\omega^2 - \omega_0^2)\cos k\ell + \omega\rho_0 c_0 \sin k\ell]$. *To solve* $D = 0$, *it is enough to solve* $\cot \omega\ell/c_0 = -\omega\rho_0 c_0/m(\omega^2 - \omega_0^2)$. *As previously, we proceed graphically. We can easily see in Figure 8.5 that we have an infinite number of eigenfrequencies. Detailed analysis of these points shows that those which correspond to zeros of* $\cot \omega\ell/c_0$ *are the zeros that are slightly offset toward the trebles of the system consisting of a rigid cavity at the origin and a soft cavity at the other end; this can be easily explained by the fact that the spring-mass system at the origin always adds some stiffness to the rigid/soft cavity system due to the presence of the spring. However, we can see that the eigenfrequency of the spring-mass system is slightly shifted toward the low frequencies, this is due to the mass of the fluid which affects the wall with its movement.*

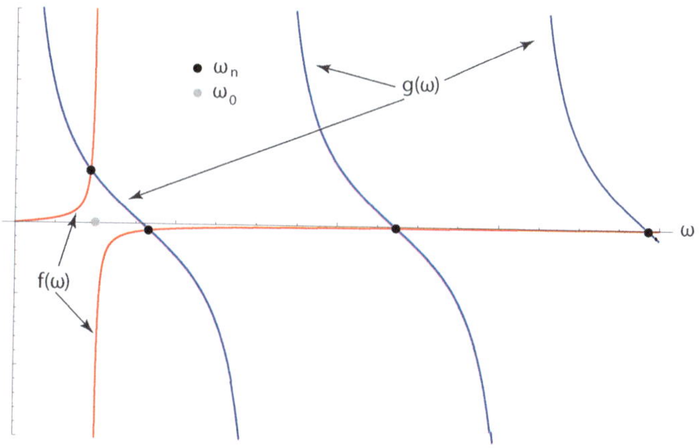

Figure 8.5. *Zeros of the system determinant with infinitely soft termination with* $f(\omega) = -\omega\rho_0 c_0/m(\omega^2 - \omega_0^2)$, $g(\omega) = \cot \omega\ell/c_0$

b) The displacement is given by:

$$u = -\frac{f_e}{(m(\omega^2 - \omega_0^2) + \omega\rho_0 c_0 \tan\frac{\omega\ell}{c_0})}. \qquad [8.4]$$

For $\omega\ell/c_0 \ll 1$, *we have* $\tan\omega\ell/c_0 \approx \omega\ell/c_0$. *This is:*

$$u \approx -\frac{f_e}{(m(\omega^2-\omega_0^2)+\omega^2\rho_0\ell)}.$$

But $\rho_0\ell$ *has a dimension of mass. We then set* $m_f = \rho_0\ell$, *and we obtain the displacement of a mass-spring system with increased mass:*

$$u \approx -\frac{f_e}{((m+m_f)\omega^2 - m\omega_0^2)},$$

which reflects the mass effect added by the presence of the fluid.

3) We set $\omega_1^2 = r_1/m_1$ *and* $\omega_2^2 = r_2/m_2$, $p_1 = 1+R$, $p_2 = A+B$, $p_3 = A\exp(\imath kd) + B\exp(-\imath kd)$ *and* $p_4 = T\exp(\imath kd)$*;*

a) For a normal interior to the system (directed toward $x > 0$ *at* $x = 0$ *and toward* $x < 0$ *at* $x = d$*), we obtain the system:*

$$\begin{aligned}
&-m_1(\omega^2-\omega_1^2)u_1 = p_2 - p_1\\
&-m_2(\omega^2-\omega_2^2)u_2 = p_3 - p_4\\
&\omega^2\rho_0 u_1 = p_i'(0) + p_r'(0) \Rightarrow -\imath\omega\rho_0 c_0 u_1 = 1 - R\\
&\omega^2\rho_0 u_1 = p_a'(0) + p_b'(0) \Rightarrow -\imath\omega\rho_0 c_0 u_1 = A - B\\
&\omega^2\rho_0 u_2 = p_a'(d) + p_b'(d) \Rightarrow -\imath\omega\rho_0 c_0 u_2 = A\exp(\imath kd) - B\exp(-\imath kd)\\
&\omega^2\rho_0 u_2 = p_t'(d) \Rightarrow -\imath\omega\rho_0 c_0 u_2 = T\exp(\imath kd)
\end{aligned}$$

b) From here, we assume that $kd \ll 1$.

i) The pressure in the cavity can be given by $p_c = A\exp(\imath kx) + B\exp(-\imath kx)$. *If* $kd \ll 1$, *then for all* $x \in]0, d[$ *we find* $kx \ll 1$ *and* $\exp(\pm\imath kx) \approx 1$, *as well as* $p_c \approx A + B$ *which is indeed uniform,*

ii) We start with $-\imath\omega\rho_0 c_0 u_2 A\exp(\imath kd) + B\exp(-\imath kd)$ *and we expand each term into a Taylor series at the order 1:* $-\imath\omega\rho_0 c_0 u_2 A(1+\imath kd) + B(1-\imath kd) = A - B + \imath kd(A+B)$. *However, we know that* $A+B = p_c$ *and that* $-\imath\omega\rho_0 c_0 u_1 = A - B$. *From this, we take* $-\imath\omega\rho_0 c_0(u_2 - u_1) = \imath kdp_c$. *From this, the expression for the pressure in the cavity is* $p_c = \rho_0 c_0^2/d(u_2 - u_1)$. *It is thus obvious to see that the air acts as a spring with a stiffness of* $r_c = \rho_0 c^2/d$,

iii) We start with $-m_2(\omega^2 - \omega_2^2)u_2 = p_3 - p4 = p_c - p_4$, *which becomes, according to the continuity equation,* $p_4 = -\imath\omega\rho_0 c_0 u_2$. *From this, we deduce that* $-m_2(\omega^2 - \omega_2^2)u_2 = p_c + \imath\omega\rho_0 c_0 u_2 = r_c(u_1 - u_2) + \imath\omega\rho_0 c_0 u_2$. *Thus, we obtain* $-m_2(\omega^2 - \omega_2^2) = r_c(u_1/u_2 - 1) + \imath\omega\rho_0 c$. *After several basic transformations, we can deduce the expression for the wall displacement ratio,* $u_1/u_2 = (r_c - m_2(\omega^2 -$

$\omega_2^2) - \imath\omega\rho_0 c_0)/r_c$, *if* $d \to 0$, $r_c \to \infty$ *and* $u_1/u_2 \to 1$. *Thus, when d is very small, the air behaves like a spring with an infinite stiffness, i.e. as a rigid connection, and thus* $u_1 = u_2$,

iv) We start with $-\imath\omega\rho_0 c_0 u_1 = 1 - R$. *As* $p_1 = 1 + R$, *therefore* $-\imath\omega\rho_0 c_0 u_1 = 2 - p_1$, *that is* $p_1 = 2 + \imath\omega\rho_0 c_0 u_1$. *On the other hand, the movement equation of wall 1 gives* $p_1 = p_2 - m_1(\omega^2 - \omega_1^2)u_1$, *that is* $p_1 = p_c - m_1(\omega^2 - \omega_1^2)u_1 = r_c(u1 - u2) - m_1(\omega^2 - \omega_1^2)u_1$, *that is* $2 + \imath\omega\rho_0 c_0 u_1 = r_c(u1 - u2) - m_1(\omega^2 - \omega_1^2)u_1$, *which we divide by* u_2: $2/u_2 + \imath\omega\rho_0 c_0 u_1/u_2 = r_c(u1/u_2 - 1) - m_1(\omega^2 - \omega_1^2)u_1/u_2$. *We rearrange this expression:* $2/u_2 = u_1/u_2(r_c - m_1(\omega^2 - \omega_1^2) - \imath\omega\rho_0 c_0) - r_c$. *But, we know that* $-\imath\omega\rho_0 c_0 u_2 = T\exp(\imath kd) = T$ *if* $d \ll 1$, *then* $2/u_2 = -2\imath\omega\rho_0 c_0 u_2/T$. *This leads to:*

$$-\frac{2\imath\omega\rho_0 c_0}{T} = \frac{u_1}{u_2}(r_c - m_1(\omega^2 - \omega_1^2) - \imath\omega\rho_0 c_0) - r_c$$
$$= \frac{r_c - m_2(\omega^2 - \omega_2^2) - \imath\omega\rho_0 c_0}{r_c}(r_c - m_1(\omega^2 - \omega_1^2) - \imath\omega\rho_0 c_0) - r_c.$$

Finally, that is:

$$T = -\frac{2\imath\omega\rho_0 c_0 r_c}{(r_c - m_2(\omega^2 - \omega_2^2) - \imath\omega\rho_0 c_0)(r_c - m_1(\omega^2 - \omega_1^2) - \imath\omega\rho_0 c_0) - r_c^2}.$$

We can clearly see the effect of the external fluid, such as damping $\imath\omega\rho_0 c_0$,

v) We ignore $\imath\omega\rho_0 c_0$ *and the stiffness values* r_1 *and* r_2 *(which impose that* ω_1 *and* ω_2 *must be zero) with respect to the stiffness introduced by the internal fluid. We thus have:*

$$T = -\frac{2\imath\omega\rho_0 c_0 r_c}{(r_c - m_2\omega^2)(r_c - m_1\omega^2) - r_c^2}$$
$$T = -\frac{2\imath\omega\rho_0 c_0 r_c}{\omega^2(m_1 m_2\omega^2 - (m_1 + m_2)r_c)}.$$

It is obvious that T *is at its maximum for* $\omega = \sqrt{(m_1 + m_2)r_c/m_1 m_2}$, ω_0 *is the mass-air-mass pulsation. It is obvious that when* $\omega = \omega_0$, *to ensure that* T *remains finite, we cannot neglect* $\imath\omega\rho_0 c_0$. T *can also be written as* $T = -2\imath\omega\rho_0 c_0 r_c/(m_1 m_2\omega^2(\omega^2 - \omega_0^2))$ *and we get the following limit cases:*

$$\text{if } \omega \ll \omega_0, \quad T \approx \frac{2\imath\rho_0 c_0 r_c}{\omega m_1 m_2 \omega_0^2}$$
$$T \approx \frac{2\imath\rho_0 c_0}{\omega(m_1 + m_2)}$$
$$\text{if } \omega \gg \omega_0, \quad T = -\frac{2\imath\rho_0 c_0 r_c}{\omega^3 m_1 m_2}.$$

Because we know that $m_1 m_2 \omega_0^2 = r_c(m_1+m_2)$, *we recall that* $i = -20\log_{10}|T|$:

$$\textit{if}\ \omega \ll \omega_0,\ \ i = -20\log_{10}\frac{2\rho_0 c_0}{\omega(m_1+m_2)}$$

$$: \textit{mass law}\ m_1+m_2\ \textit{at -6dB per octave}$$

$$\textit{if}\ \omega \gg \omega_0,\ \ i = -20\log_{10}\frac{2\rho_0 c_0 r_c}{\omega^3 m_1 m_2}$$

$$: \textit{stiffness law}\ r_c\ \textit{at -18dB per octave.}$$

At low frequencies, the additional attenuation due to double glazing (when compared with single glazing) is mainly due to the mass added by the second glass thus, at high frequencies, the gain is provided by the stiffness of air between the two glasses, which act as a spring, and is inversely proportional to the distance between the glasses, and this, of course, outside the eigenfrequencies of the walls.

8.1.5. *Transient regime*

We assume that there are no sources within the fluid and that $F_e(t)$ begins at $t = 0$. The displacement of the wall is thus given by the inverse Fourier transform of the displacement in a harmonic regime, i.e. using the Fourier integral:

$$U(t) = -\frac{1}{2\pi}\int_{-\infty}^{\infty}\frac{f_e(\omega)}{m\left(2\imath\omega\frac{\rho_0 c_0}{m}+\omega^2-\omega_0^2\right)}\mathrm{e}^{-\imath\omega t}d\omega.$$

To slightly simplify the calculations, we can assume that $F_e(t) = \delta(t) \Rightarrow f_e(\omega) = 1$, and then we only have to calculate the momentum response of the wall $U_\delta(t)$:

$$U_\delta(t) = -\frac{1}{2\pi}\int_{-\infty}^{\infty}\frac{\mathrm{e}^{-\imath\omega t}}{m\left(2\imath\omega\frac{\rho_0 c_0}{m}+\omega^2-\omega_0^2\right)}d\omega.$$

The integrand is a meromorphic function and the explicit calculation of this integral can be carried out using the method of residues on either contour $C^{\pm}$ consisting of a real line limited by $\pm R$ and the half-circles located in the upper and lower half-planes (see Figure 8.6).

A priori, we are allowed to integrate over any of the defined contours, or even over any contour which contains the real axis. We thus have:

$$-\frac{1}{2\pi}\int_{C^{\pm}}\frac{\mathrm{e}^{-\imath\omega t}}{m\left(2\imath\omega\frac{\rho_0 c_0}{m}+\omega^2-\omega_0^2\right)}d\omega = -\frac{\pm 2\imath\pi}{2\pi}\sum_{res(C^{\pm})}\frac{\mathrm{e}^{-\imath\omega t}}{m\left(2\imath\omega\frac{\rho_0 c_0}{m}+\omega^2-\omega_0^2\right)},$$

where $res(C^{\pm})$ are the residues of the integrand in the contour considered. We have a $+$ sign when we integrate over C^+ and a $-$ sign when we integrate over C^- because we turn in the opposite direction in a trigonometric sense. As we have already noted, we should choose the contour which, using Jordan's lemma, would allow us to cancel the integral over the semi-circle when we pass to the limit $R \to \infty$. We must integrate over C^+ for $t < 0$ and over C^- for $t > 0$. Now, we examine the zeros of the denominator of the integrand. It is evident that we have a polynomial of degree 2 with respect to ω. We assume that the discriminant $\Delta' = \omega_0^2 - (\rho_0 c_0/m)^2$ is strictly positive. It is easy to see that the roots have an anti-Hermitian symmetry:

$$\Omega^+ = -\imath\frac{\rho_0 c_0}{m} + \sqrt{\omega_0^2 - \left(\frac{\rho_0 c_0}{m}\right)^2} = \dot{\Omega} - \imath\hat{\Omega}$$

$$\Omega^- = -\imath\frac{\rho_0 c_0}{m} - \sqrt{\omega_0^2 - \left(\frac{\rho_0 c_0}{m}\right)^2} = -\dot{\Omega} - \imath\hat{\Omega} = -(\Omega^+)^*$$

with $\dot{\Omega} > 0$ and $\hat{\Omega} > 0$. As these roots both have a negative imaginary part, the contour C^+ does not contain a pole and thus, for $t < 0$, we have $U_\delta(t) = 0$ which satisfies the principle of causality. For $t > 0$, we obtain with $z - z^* = 2\Im[z]$:

$$U_\delta(t) = \frac{\imath}{2m}\left[\frac{e^{-\imath\Omega^+ t}}{\Omega^+ + \imath\frac{\rho_0 c_0}{m}} + \frac{e^{-\imath\Omega^- t}}{\Omega^- + \imath\frac{\rho_0 c_0}{m}}\right]$$

$$= \frac{1}{m}\Im\left[\frac{e^{-\imath\dot{\Omega}t}}{\Omega^+ + \imath\frac{\rho_0 c_0}{m}}\right] e^{-\hat{\Omega}t} = \frac{\sin\dot{\Omega}t}{m\dot{\Omega}} e^{-\hat{\Omega}t}.$$

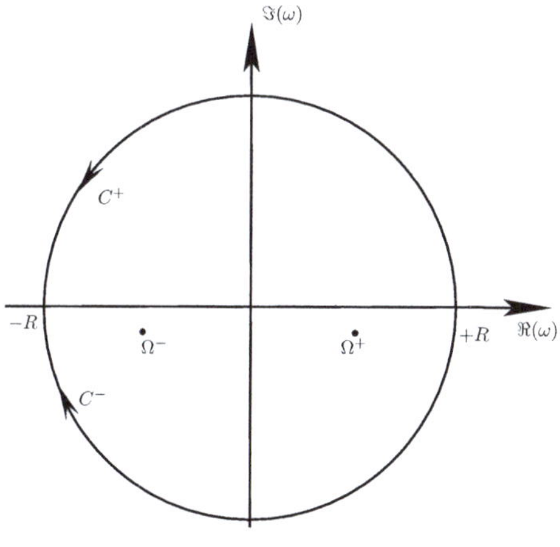

Figure 8.6. *Integration contour*

As $\dot{\Omega}$ and $\hat{\Omega}$ are both positive, $U_\delta(t)$ is a damped oscillation, especially since $\hat{\Omega} = \rho_0 c_0/m$ is large, meaning that the fluid has a high density or that the mass of the wall is small, or that the parameter ϵ is high.

With $A^+(\omega)$ calculated for $f_e = 1$, the acoustic pressure in the guide $x > 0$ is given by:

$$P_\delta^+(x,t) = \frac{1}{2\pi}\int_{-\infty}^{\infty} A^+(\omega)\mathrm{e}^{+\imath kx}\mathrm{e}^{-\imath\omega t}d\omega$$

$$= -\frac{1}{2\pi}\int_{-\infty}^{\infty} \frac{\imath\omega\rho_0 c_0}{m\left(2\imath\omega\frac{\rho_0 c_0}{m} + \omega^2 - \omega_0^2\right)}\mathrm{e}^{-\imath\omega\left(t-\frac{x}{c}\right)}d\omega.$$

Because of the term $\exp(-\imath\omega(t - x/c_0))$, it is necessary to integrate over C^+ for $t < x/c_0$. Thus, the pressure is zero for $t < x/c_0$, which corresponds to the time it takes for the sound wave to reach the point of abscissa x. We obtain, with $z + z^* = 2\Re[z]$, for $t > \frac{x}{c_0}$:

$$P_\delta^+(x,t) = -\frac{\rho_0 c_0}{m}\Re\left[\frac{\Omega^+}{\Omega^+ + \imath\frac{\rho_0 c_0}{m}}\mathrm{e}^{-\imath\dot{\Omega}\left(t-\frac{x}{c_0}\right)}\right]\mathrm{e}^{-\hat{\Omega}\left(t-\frac{x}{c_0}\right)}$$

$$= -\hat{\Omega}\left(\cos\dot{\Omega}\left(t - \frac{x}{c_0}\right) - \frac{\hat{\Omega}}{\dot{\Omega}}\sin\dot{\Omega}\left(t - \frac{x}{c_0}\right)\right)\mathrm{e}^{-\hat{\Omega}\left(t-\frac{x}{c_0}\right)}.$$

If we focus on temporal excitation, other than a Dirac momentum, there are two methods. We either directly calculate the Fourier transform of $u(\omega)$ and we obtain:

$$U(t) = -\frac{1}{m}\Im\left[f_e(\Omega^+)\frac{\mathrm{e}^{-\imath\dot{\Omega}t}}{\Omega^+ + \imath\frac{\rho_0 c_0}{m}}\right]\mathrm{e}^{-\hat{\Omega}t},$$

if $f_e(\Omega^{+*}) = -f_e^*(\Omega^+)$, or we calculate the response of the system by temporal convolution between the momentum response and the excitation. For example, if $F_e(t) \neq 0$ for $t \in [0, T]$, we obtain:

$$U(t) = \int_0^t U_\delta(t-\tau)F_e(\tau)d\tau, t < T \text{ and } U(t) = \int_0^T U_\delta(t-\tau)F_e(\tau)d\tau, t \geq T.$$

EXERCISE.– Calculate the temporal response of the piston when $\rho_0 c_0/m > \omega_0$.

SOLUTION.– *The roots of the denominator can always be written as* $\Omega^\pm = -\imath\rho_0 c_0/m \pm \sqrt{\omega_0^2 - (\rho_0 c_0/m)^2}$. *For* $\rho_0 c_0 > m\omega_0$, *we set* $\beta = \omega_0 m/\rho_0 c_0, 0 < \beta < 1$. *We thus have* $\Omega^\pm = -\imath\rho_0 c_0/m(1 \mp \sqrt{1-\beta^2})$. *Since the roots have an imaginary negative part, the response of the piston is always causal and for* $t > 0$ *we have:*

$$U_\delta(t) = \frac{\imath}{2m}\left[\frac{e^{-\imath\Omega^+ t}}{\Omega^+ + \imath\frac{\rho_0 c_0}{m}} + \frac{e^{-\imath\Omega^- t}}{\Omega^- + \imath\frac{\rho_0 c_0}{m}}\right] = \frac{1}{m}\frac{\sinh\left(\frac{\rho_0 c_0}{m}\sqrt{1-\beta^2 t}\right)}{\frac{\rho_0 c_0}{m}\sqrt{1-\beta^2}}e^{-\frac{\rho_0 c_0}{m}t}.$$

We set $\epsilon = \rho_0 c_0/m$ and $\alpha = \epsilon\sqrt{1-\beta^2} = \sqrt{\epsilon^2 - \omega_0^2}$. We have $U_\delta(t) = 1/m\alpha \, \sinh(\alpha t)\exp(-\epsilon t)$. The displacement of the wall is a non-oscillatory function of time. We refer to supercritical damping. The displacement increases up to the time $t_m = 1/2\alpha\ln((\epsilon+\alpha)/(\epsilon-\alpha)) = 1/\sqrt{\epsilon^2-\omega_0^2}\text{arccosh}(\epsilon/\omega_0)$ up to a maximum amplitude $A_m = \exp(-t_m\epsilon)/m\omega_0$ and then it decreases exponentially. We give an example of the movement of a wall in Figure 8.7 for a mass of $m = 1$, $\omega_0 = 10$, $\epsilon = 10$ for the continuous curve and $\epsilon = 20$ and $\epsilon = 50$ for the discontinuous curves. In practice, this case may be present for a very light weight associated with a spring with a low stiffness; in air, this is almost never observed (even for a loudspeaker membrane); we can only observe this phenomenon of supercritical damping for coupling with a high-density fluid, such as water.

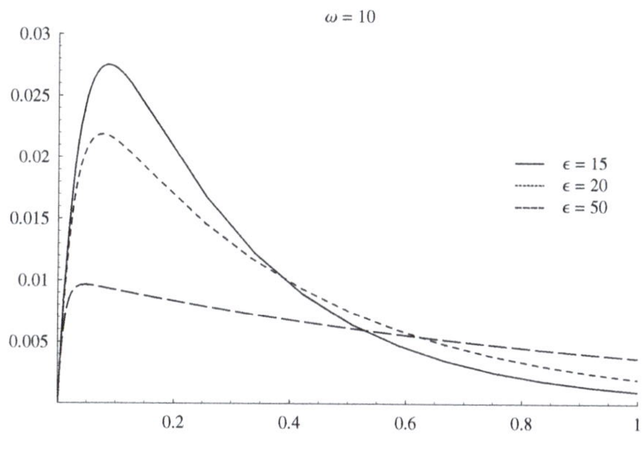

Figure 8.7. *An example of wall movements in supercritical damping for several values of* ϵ

We can also note that when ϵ, the damping by radiation, increases, the maximum amplitude of the wall decreases along with the time taken to reach this value. However, the temporal decrease in the amplitude is less rapid. This leads to a slightly paradoxical result in which increasing the damping does not systematically lead to a lower movement amplitude. In practice, we have an optimal damping value (optimal in the sense of the fastest decline) which is such that $\epsilon = \omega_0 = \sqrt{r/m}$ that is $\rho_0 c_0 = \sqrt{rm}$[1].

EXERCISE.– In the absence of a fluid, a very simple way to introduce damping into a mechanical system is to translate the energy loss into the form of a complex stiffness $r = r_0(1 - \imath\eta)$, where $0 < \eta \ll 1$ is the damping factor. Show that this method leads to a non-causal response.

SOLUTION.– *The movement of the damped wall in the absence of a fluid in a harmonic regime is given by* $u(\omega) = -f_e/m(\omega^2 - \omega_0^2(1 - i\eta))$. *The resonance pulsations* $\Omega^{\pm} = \pm\omega_0(1 - i\eta/2)$ *if* $\eta \ll 1$ *have their real and imaginary parts with opposite signs. As we have two roots with imaginary parts with opposite signs, we cannot have a causal response.*

8.2. Free waves in an infinite plate immersed in a fluid

We consider the plane $z = 0$ which is occupied by a homogeneous and isotropic thin plate with a stiffness of $D = Eh^3/12(1-\nu^2)$ and a surface mass of $\rho_p h$. The two half-spaces, $z > 0$, and $z < 0$, are occupied by an ideal fluid at rest characterized by a density ρ_0 and an acoustic wave propagation velocity c_0. We denote the displacement of the plate by $u(x, y)$ and we denote the acoustic pressure in the two half-spaces by $p^{\pm}(x, y, z)$. Finally, we assume that the movements are harmonic, with a time dependence (neglected in the following) $\mathrm{e}^{-\imath\omega t}$. The system of equations that needs to be solved in the absence of external excitation (mechanical force) is as follows:

$$D\left(\Delta^2 - k_p^4\right) u(x, y) = -\left(p^+(x, y, z = 0) - p^-(x, y, z = 0)\right) \forall x, y \qquad [8.5]$$

$$\left(\Delta + k_f^2\right) p^{\pm}(x, y, z) = 0 \text{ in } z \gtrless 0 \qquad [8.6]$$

$$\omega^2 \rho_0 u(x, y) = \partial_z p^{\pm}(x, y, z = 0), \qquad [8.7]$$

with $k_p^4 = \rho_p h\omega^2/D$ and $k_f^2 = \omega^2/c_0^2$. We note that the normal is directed toward $z > 0$. The third equation expresses the continuity of the accelerations normal to the plate and fluid. To these equations, we must add an energy conservation condition, such as the Sommerfeld condition or the limit absorption principle. In the latter case, the wave number $k_{p,f} = \lim_{\epsilon\to 0} k_{p,f}(1 + \imath\epsilon)$, $\epsilon > 0$. We must note that if we had chosen a time dependence in the form of $\exp(+\imath\omega t)$, the limit absorption principle should have been written as $k_{p,f} = \lim_{\epsilon\to 0} k_{p,f}(1 - \imath\epsilon)$.

We examine whether it is possible for a plane wave to appear in the plate. To do this, we recall that it is enough to examine the case of a wave which propagates in the

1 Here, we find a classic result which says that the optimal damping of a mass-spring-damper system is obtained by a damping value of two times the square root of the product of the mass and the stiffness.

direction of the x axis: $u(x, y) = \exp(\imath kx)$. By rotating the axes, we can access all the waves. Mathematically, this leads to finding the general solution of the problems [8.5], [8.6] and [8.7] in the form of a complex exponential (Laplace's method) $\exp(\imath kx)$. The continuity equation of normal vibration velocities gives $\omega^2 \rho_0 e^{\imath kx} = \partial_z p^+(x, y, z = 0) = \partial p^-(x, y, z = 0)/\partial z$ [FIL 08], as in addition acoustic pressures satisfy a Helmholtz equation. If $\partial_z p^\pm(x, y, z)_{z=0} = Ae^{\imath kx}$, it is necessary that we have $p^\pm(x, y, z) = A/(\pm\imath\zeta)\exp(\imath(kx \pm \zeta z))$. Once again that is $p^\pm(x, z) = \omega^2\rho_0/(\pm\imath\zeta)\exp(\imath(kx \pm \zeta z))$, the sign of the coefficient ζ is chosen in a way so as to satisfy the energy conservation principle ($+$ in the domain $z > 0$, and $-$ in $z < 0$). As $p^\pm(x, z)$ satisfies the Helmholtz equation, we have $k_f^2 = k^2 + \zeta^2$, with $\zeta > 0$ if $\zeta^2 > 0$ or $\Im\zeta > 0$ if $\Im\zeta \neq 0$. Once more, this choice is justified by the energy conservation principle, such that the waves moving away from the plate should be damped when $|z|$ increases. We thereby obtain:

$$u(x, y) = e^{\imath kx}, p^\pm(x, y, z) = \frac{\omega^2\rho_0}{\pm\imath\zeta} e^{\imath(kx \pm \zeta z)}, k_f^2 = k^2 + \zeta^2.$$

The equation satisfied by the displacement of the plate leads to the algebraic relation $\imath D\zeta(k^4 - k_p^4) = -2\omega^2\rho_0$, with $\zeta^2 = k_f^2 - k^2$, $\Im\zeta > 0$ if $\Im\zeta \neq 0$ and $\zeta > 0$ if $\Im\zeta = 0$. The relation $\mathcal{D}(k) = \imath\sqrt{k_f^2 - k^2}(k^4 - k_p^4) + 2\omega^2\rho_0/D = 0$ is known as the dispersion relation. This name comes from the plate equation that models a dispersive medium in which the wave celerity (the phase velocity) depends on the frequency.

8.2.1. *Roots of the dispersion equation*

The dispersion relation can also be written as $(k_f^2 - k^2)(k^4 - k_p^4)^2 + 4\omega^2\rho_0^2/D^2 = 0$. This is a polynomial of degree 5 with respect to k^2. This polynomial has five roots. As it has real coefficients (for real physical parameters), it can have five real roots or three real roots and one pair of conjugate complex roots or one real root and two pairs of conjugate complex roots. It is evident that if k is a root, $-k$ is also a root. We must distinguish between two cases, $k_f > k_p$ and $k_f < k_p$. When $k_f = k_p$, the mechanical and acoustic wavelengths are identical, as well as the wave propagation velocities. We recall that the bending wave propagation celerity in the plate is dispersive (it increases along with the frequency). The pulsation for which $k_f = k_p$ is called the critical pulsation (or the coincidence pulsation). It is denoted with ω_c. From $k_f = k_p$, we can easily derive that:

$$\omega_c = c_0^2\sqrt{\frac{\rho_p h}{D}}. \qquad [8.8]$$

For a steel plate with a thickness of 1 cm in air, the critical frequency value is $f_c = 1,200$ Hz, for lead 5,500 Hz, for steel, aluminum or glass 1,200 Hz, for copper

1,800 Hz, for concrete 2,000 Hz, for plexiglass 4,000 Hz and approximately 3,000 Hz for a 13 mm thick gypsum plate.

At low frequency, for $\omega < \omega_c$, we find $k_f < k_p$, the phase velocities given by $c_{0,p} = \omega/k_{f,p}$ are such that $c_0 > c_p$: the bending waves in the plate propagate at a velocity that is lower than the sound propagation velocity, this is referred to as a subsonic regime. Conversely, at high frequency $\omega > \omega_c$, we find $k_f > k_p$ and the phase velocities are such that $c_0 < c_p$ and this is referred to as a supersonic regime. Sometimes in the literature, the ratio $M = c_p/c_0 = k_f/k_p = \sqrt{\omega/\omega_c}$ is just called the Mach number of the phase velocities. We can graphically prove that there is at least one real root $k_3 > k_{f,p}$. To do this, we aim to find whether there exist any intersection points for the curves $(k_f^2 - k^2)(k^4 - k_p^4)^2$ and $4\omega^2\rho_0^2/D^2$, in the subsonic and supersonic cases.

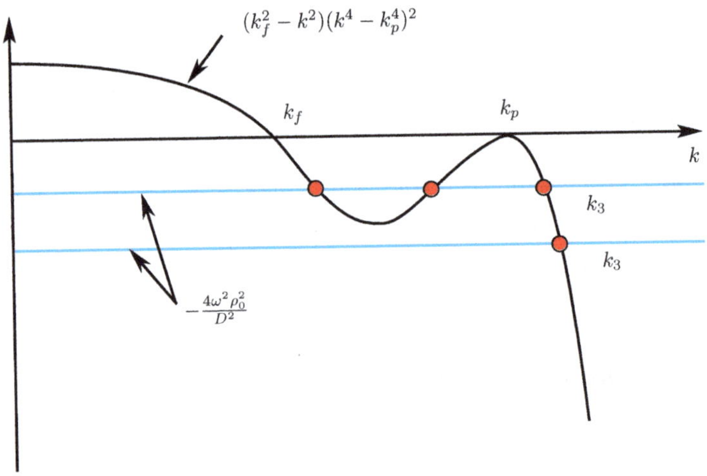

Figure 8.8. *Graphical solution of the dispersion relation $k_f < k_p$ (subsonic regime)*

In a subsonic regime, according to Figure 8.8, it is evident that there always exists at least one real solution $k_3 > k_p > k_f$. In a supersonic regime, we can see that according to Figure 8.9, there always exists a real solution $k_3 > k_f > k_p$. Similarly, we can show that the dispersion relation has one or two other pairs of conjugate complex roots in the form of $k_+ = \kappa + \imath\tau$ and $k_- = -\kappa + \imath\tau$, as well as an imaginary part $\tau > 0$ and equal and opposite real parts $\kappa > 0$. The real root k_3 corresponds to a subsonic plane wave which propagates in the plate without damping toward $x > 0$ as $u_0 = \exp(\imath k_3 x)$. The real root k_3 of the dispersion equation corresponds to the acoustic pressure $p_0^+ = \omega^2\rho_0/(\imath\zeta)\exp(\imath(k_3 x + \zeta z))$ and, as in all the cases $k_3 > k_f$, therefore $\zeta^2 = k_f^2 - k_3^2 < 0$, and thus $\zeta = \imath\sqrt{k_3^2 - k_f^2}$. The pressure field radiated by the wave k_3 is an evanescent plane wave. It decreases exponentially with z and does not transport any energy in the fluid. This result is *a priori* obvious since in the plate-fluid system this bending wave propagates at constant amplitude and therefore the plate cannot radiate energy.

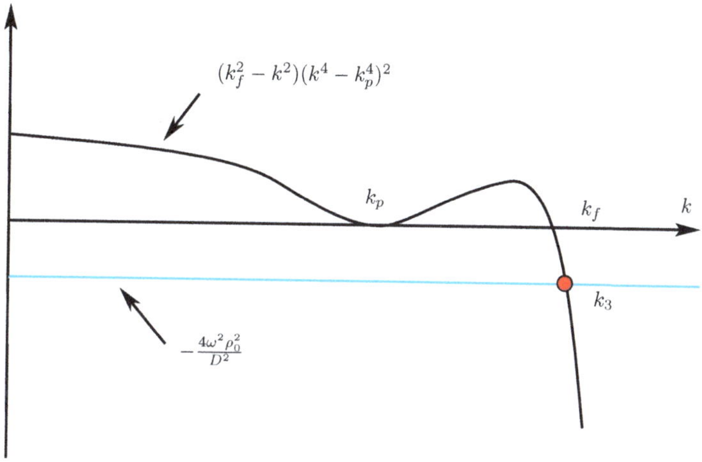

Figure 8.9. *Graphical solution of the dispersion relation $k_f > k_p$ (supersonic regime)*

8.2.2. *Light fluid approximation*

We set $\epsilon = \rho_0/(\rho_p h)$. The dispersion relation can thus be written as $(k_f^2-k^2)(k^4-k_p^4)^2 + 4k_p^8\epsilon^2 = 0$, and, when ϵ approaches zero, the dispersion equation can be reduced to $(k_f^2 - k^2)(k^4 - k_p^4)^2 = 0$, the roots of which are naturally $k = \pm k_p$, $k = \pm \imath k_p$ and $k = \pm k_f$. When the influence of the fluid on the plate is very small, it is reasonable to assume that the roots of the dispersion equation of the coupled plate are very similar to those *in vacuo*. We thus aim to find the roots in the form of $k_1 = k_p(1 + \epsilon\gamma_1)$, $k_2 = \imath k_p(1 + \epsilon\gamma_2)$ and $k_3 = k_f(1 + \epsilon^2\gamma_3)$. Thus, the unknowns are γ_i. We begin with $k_1 = k_p(1 + \epsilon\gamma_1)$ which we introduce in the dispersion relation $(k_f^2 - (k_p(1+\epsilon\gamma_1))^2)((k_p(1+\epsilon\gamma_1))^4 - k_p^4)^2 + 4k_p^8\epsilon^2 = 0$, which leads for a small ϵ to $(k_f^2 - k_p^2(1 + 2\epsilon\gamma_1))(k_p^4(1 + 4\epsilon\gamma_1) - k_p^4)^2 + 4k_p^8\epsilon^2 \approx 0$. If k_f and k_p are different enough, we can write that $k_f^2 - k_p^2(1 + 2\epsilon\gamma_1) = k_f^2 - k_p^2$. The dispersion relation thus becomes $(k_f^2 - k_p^2)(4\epsilon\gamma_1 k_p^4)^2 + 4k_p^8\epsilon^2 = 0$, that is $(k_f^2 - k_p^2)4\gamma_1^2 = -1$. So that $\gamma_1 \approx \pm 1/\left(2\imath\sqrt{k_f^2 - k_p^2}\right)$. Similarly, we take $k_2 = \imath k_p(1 + \epsilon\gamma_2)$, which we introduce in the dispersion relation $(k_f^2 + k_p^2(1 + 2\epsilon\gamma_2))(k_p^4(1 + 4\epsilon\gamma_2) - k_p^4)^2 + 4k_p^8\epsilon^2 = 0$. The dispersion relation thus becomes $(k_f^2 + k_p^2)(4\epsilon\gamma_2 k_p^4)^2 + 4k_p^8\epsilon^2 \approx 0$. From this, we can easily deduce that $\gamma_2 \approx \pm 1/\left(2\imath\sqrt{k_f^2+k_p^2}\right)$. Finally, we take $k_3 = \imath k_p(1+\epsilon^2\gamma_3)$, which we introduce in the dispersion relation. We obtain $(-2\epsilon^2\gamma_3 k_f^2)(k_f^4(1+4\epsilon^2\gamma_3)-k_p^4)^2+4k_p^8\epsilon^2=0$, that is $(-2\epsilon^2\gamma_3 k_f^2)(k_f^4-k_p^4)^2+4k_p^8\epsilon^2 \approx 0$ if k_f and k_p are different enough. From here, we take $\gamma_3 \approx 2k_p^8/\left(k_f^2(k_f^4-k_p^4)^2\right)$. We must note that all these results are valid far away from the coincidence frequency. Let us now examine the subsonic and supersonic cases.

8.2.2.1. *Subsonic regime*

Below the subsonic frequency, we find $k_p > k_f$. This is $\sqrt{k_f^2-k_p^2} = \imath\sqrt{k_p^2-k_f^2}$. Thus, $\gamma_1 = \pm 1/\left(2\sqrt{k_p^2-k_f^2}\right)$, $\gamma_2 = \pm\imath/\left(2\sqrt{k_p^2+k_f^2}\right)$ and $\gamma_3 = 2k_p^8/\left(k_f^2(k_f^4-k_p^4)^2\right)$. We obtain the roots:

$$k_1^+ = k_p\left(1+\frac{\epsilon}{2\sqrt{k_p^2-k_f^2}}\right), k_1^- = k_p\left(1-\frac{\epsilon}{2\sqrt{k_p^2-k_f^2}}\right)$$

$$k_2^+ = \imath k_p\left(1+\frac{\imath\epsilon}{2\sqrt{k_f^2+k_p^2}}\right), k_2^- = \imath k_p\left(1-\frac{\imath\epsilon}{2\sqrt{k_f^2+k_p^2}}\right)$$

$$k_3 = k_f\left(1+\frac{2\epsilon^2 k_p^8}{k_f^2(k_f^4-k_p^4)^2}\right).$$

To these roots, we add those with the opposite sign. It is obvious that we obtain a pair of conjugate complex roots and three real roots; this corresponds to the result presented in Figure 8.8 where we can see a root that is slightly greater than k_f (here called k_3) and two roots on either sides of k_p (here called $k_1^{\pm}$).

8.2.2.2. *Supersonic regime*

Above the coincidence frequency, we find $k_f > k_p$. Thus, $\gamma_1 = \pm 1/2(k_p^2 - k_f^2)^{-1/2}$, $\gamma_2 = \pm\imath/\left(2\sqrt{k_f^2+k_p^2}\right)$ and $\gamma_3 = 2k_p^8/\left(k_f^2(k_f^4-k_p^4)^2\right)$. We obtain the roots:

$$k_1^+ = k_p\left(1+\frac{\imath\epsilon}{2\sqrt{k_f^2-k_p^2}}\right), k_1^- = k_p\left(1-\frac{\imath\epsilon}{2\sqrt{k_f^2-k_p^2}}\right)$$

$$k_2^+ = \imath k_p \left(1 + \frac{\imath\epsilon}{2\sqrt{k_f^2 + k_p^2}}\right), k_2^- = \imath k_p \left(1 - \frac{\imath\epsilon}{2\sqrt{k_f^2 + k_p^2}}\right)$$

$$k_3 = k_f \left(1 + \frac{2\epsilon^2 k_p^8}{k_f^2(k_f^4 - k_p^4)^2}\right).$$

To these roots, we add those with opposite signs. We can see that we obtain a real root greater than k_f and k_p, and two pairs of conjugate complex roots.

EXERCISE.– Study the sonic regime. Show that the roots k_2 do not change and that the roots k_1 and k_3 are merged. We set $k_1 = k_p(1+\epsilon^{2/3}\gamma_1)$ and $k_3 = k_f(1+\epsilon^{2/3}\gamma_3)$.

SOLUTION.– *In a sonic regime, $k_f = k_p$. We can easily see from the previous results that the roots $k_2^\pm$ are identical in subsonic and supersonic regimes. However, the roots k_1 and k_3 are not defined in a sonic regime.*

We start with $k_1 = k_p(1 + \epsilon^{2/3}\gamma_1)$ which we introduce in the dispersion relation $(k_f^2 - k_p^2(1+2\epsilon^{2/3}\gamma_1))(k_p^4(1+4\epsilon^{2/3}\gamma_1) - k_p^4)^2 + 4k_p^8\epsilon^2 \approx 0$. As $k_f = k_p$, we obtain $-2k_p^2\epsilon^{2/3}\gamma_1(4\epsilon^{2/3}\gamma_1 k_p^4)^2 + 4k_p^8\epsilon^2 = 0$. This is $8k_p^2\epsilon^{6/3}\gamma_1^3 = -\epsilon^2$. Finally, that is $\gamma_1 = 1/(2k_p^2)^{1/3}$. We can also show that $\gamma_3 = 1/(2k_p^2)^{1/3}$. We obtain a single real root.

8.3. Transmission of a plane wave by a thin plate

Here, we consider the two-dimensional problem of a plate with an infinite extension which separates two semi-infinite domains occupied by the same ideal fluid (see Figure 8.10):

The plate is insonified by a plane wave whose angle of incidence with respect to the normal is given by θ_i. In $x < 0$, the pressure p^- is composed of two terms p_i and p_r. In $x > 0$, the pressure p^+ is composed of the only transmitted pressure p_t. Additionally, we can simplify this problem by considering the problem as one-dimensional for the plate $u(y,z) = u(z)$ and two-dimensional for the pressures $p^+(x,y,z) = p^+(x,z)$ and $p^-(x,y,z) = p^-(x,z)$. The equations of the problem are:

$$\left(\frac{d^4}{dz^4} - k_p^4\right) u(z) = -\frac{1}{D}\left(p^+(0,z) - p^-(0,z)\right),$$

$$\left(\Delta + k_f^2\right) p^\pm(x,z) = 0, \omega^2\rho_0 u(z) = \partial_x p^\pm, \forall z.$$

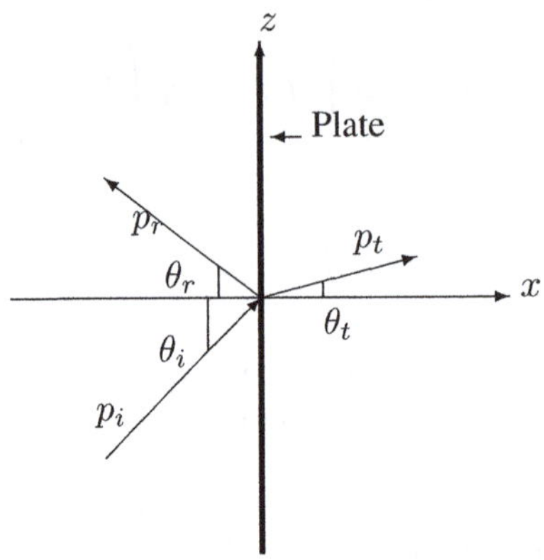

Figure 8.10. *Problem geometry*

To further simplify the problem, we can assume that the incident pressure is unitary. We thus have:

$$p_i(x,z) = e^{\imath k_f(z\sin\theta_i + x\cos\theta_i)}, p_r(x,z) = Re^{\imath k_f(z\sin\theta_r - x\cos\theta_r)},$$
$$p_t(x,z) = Te^{\imath k_f(z\sin\theta_t + x\cos\theta_t)}.$$

The condition of continuity of normal velocities gives:

$$\omega^2\rho_0 u(z) = \imath k_f\cos\theta_i e^{\imath k_f z\sin\theta_i} - R\imath k_f\cos\theta_r e^{\imath k_f z\sin\theta_r} = T\imath k_f\cos\theta_t e^{\imath k_f z\sin\theta_t}.$$

As these relations are valid for all x, it is necessary that $\imath k_f\sin\theta_i = \imath k_f\sin\theta_r = \imath k_f\sin\theta_t$, from where $\theta_i = \theta_r = \theta_t = \theta$. The second continuity equation gives $u(z) = 1/(\omega^2\rho_0)T\imath k_f\cos\theta_t\exp(\imath k_f z\sin\theta)$. If we introduce this displacement expression in the equation which it satisfies, we obtain:

$$\left(\frac{d^4}{dz^4} - k_p^4\right)u(z) = \frac{1}{\omega^2\rho_0}T\imath k_f\cos\theta_t\left(k_f^4\sin^4\theta - k_p^4\right)e^{\imath k_f z\sin\theta}$$
$$= -\frac{1}{D}\left(p^+(0,z) - p^-(0,z)\right),$$

from where:

$$\frac{1}{\omega^2\rho_0}T\imath k_f\cos\theta_t\left(k_f^4\sin^4\theta - k_p^4\right)e^{\imath k_f z\sin\theta}$$
$$= -\frac{1}{D}\left(Te^{\imath k_f z\sin\theta} - (1+R)e^{\imath k_f z\sin\theta}\right).$$

We can simplify this by $\exp(\imath k_f z \sin\theta)$: $1/\left(\omega^2\rho_0\right) T\imath k_f \cos\theta \left(k_f^4 \sin^4\theta - k_p^4\right) = -1/D(T-(1+R))$. To this equation, we add the outcome of the continuity condition of normal velocities, $T = 1 - R$. From these two equations, we can easily obtain the reflection and transmission coefficients:

$$R = \frac{\imath k_f \cos\theta \left(k_f^4 \sin^4\theta - k_p^4\right)}{\mathcal{D}\left(k_f \sin\theta\right)} T = \frac{\frac{2\omega^2\rho_0}{D}}{\mathcal{D}\left(k_f \sin\theta\right)},$$

where $\mathcal{D}(k)$ is the dispersion relation of the coupled plates. As $k_f \sin\theta < k_f$, there are no real roots, since they are all greater than k_f (see Figures 8.8 and 8.9) thus $\mathcal{D}\left(k_f \sin\theta\right) \neq 0$. If $k_f \sin\theta = k_p$, it is easy to see that $R = 0$ and $T = 1$: there is coincidence and the plate is completely transparent. Let us now consider under normal incidence that $\theta = 0$. We thus have:

$$R = \frac{\imath k_f k_p^4}{\imath k_f k_p^4 - \frac{2\omega^2\rho_0}{D}}, T = \frac{\frac{2\omega^2\rho_0}{D}}{\frac{2\omega^2\rho_0}{D} - \imath k_f k_p^4}.$$

As, $k_p^4 = m\omega^2/D$ and $k_f = \omega/c_0$, thus:

$$R = \frac{\omega m}{\omega m - 2\imath\rho_0 c_0}, T = \frac{2\rho_0 c_0}{2\rho_0 c_0 - \omega m}.$$

We recall the definition of τ which is the ratio between the average power (over a period) of the transmitted energy and the incident energy (here unitary). We show that $e_T \propto TT^\star$, $e_I = 1$. We thus obtain $\tau = TT^\star = 4\rho_0^2 c_0^2/(4\rho_0^2 c_0^2 + \omega^2 m^2)$. We define i as the insertion loss index. We obtain $i = -10\log_{10}\tau$. As $1/\tau$ is proportional to the square of the frequency, it is easy to verify that i presents a slope of 6 dB per octave (which corresponds to the classic mass law).

Similarly, if $\theta \neq 0$, we can show that:

$$\tau = \frac{4\omega^4\rho_0^2}{4\omega^4\rho_0^2 + D^2 k_f^2 \cos^2\theta \left(k_f^4 \sin^4\theta - k_p^4\right)^2}.$$

The insertion loss index has three different zones. Below the coincidence frequency[2], the index has a slope of 6 dB per octave which corresponds to the mass law; at the coincidence frequency, the plate is transparent, beyond it, we have a slope, typical for damping by stiffness, of 18 dB per octave. In Figure 8.11 we present the

2 We recall that the coincidence frequency depends on the angle of incidence: it is minimal at grazing incidence and rejected at infinity at normal incidence.

insertion loss index curves for a steel plate in air. The mechanical and geometrical characteristics are, for the plate $E = 2\ 10^{11}$ Pa, $\nu = 0.3$, $\rho_p = 8,000\ \mathrm{Kg/m^3}$, $h = 1$ cm, and for air: $\rho_0 = 1.3\ \mathrm{Kg/m^3}$, $c_0 = 340\ \mathrm{m/s}$. The coincidence frequency of this plate coupled to air is $\omega_c = 7,640$ rd/s that is $f_c = 1,216$ Hz. It should be noted that under grazing incidence (for $\theta = \pi/2$), the insertion loss index is always zero under the critical frequency.

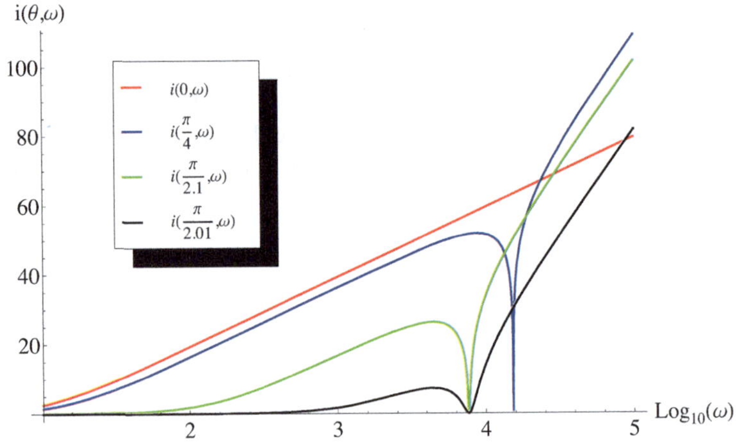

Figure 8.11. *The insertion loss index of a steel plate in air. For a color version of the figure, see www.iste.co.uk/anselmet/acoustics.zip*

To avoid this transparency phenomenon near the coincidence frequency, it is advantageous to have the highest possible critical frequency[3]. Therefore, we aim to use the heaviest walls and those with the lowest possible stiffness (such as lead). This goes against the technical evolution, where the walls must participate in the solidity of the whole structure and thus must be light and stiff (such as the epoxy carbon fiber composites).

8.4. Radiation of an infinite plate under point excitation

We now consider a mechanical excitation of the plate by a point force (modeled by a Dirac distribution) $\delta(x, y)$ located at the origin. The movement equations are equations [8.5], [8.6] and [8.7] in which only the second part of the first one changes:

3 We recall that ω_c decreases along with the stiffness E and thickness h, and increases with the mass ρ_p.

$$D\left(\Delta^2 - k_p^4\right) u(x,y) = \delta(x,y) - \left(p^+(x,y,z=0) - p^-(x,y,z=0)\right) \text{ over } z=0$$

$$\left(\Delta + k_f^2\right) p^{\pm}(x,y,z) = 0 \text{ in } z \gtrless 0$$

$$\omega^2 \rho_0 u(x,y) = \partial_z p^{\pm}(x,y,z=0).$$

To this, we add an energy conservation condition.

8.4.1. *Integro-differential equation with respect to* u

Let $\mathcal{G}(M,M')$ be Green's kernel of the Helmholtz equation in the half space $z > 0$ which satisfies a Neumann condition on the $z = 0$ plane. This kernel is calculated very simply using the image method and is given by [5.8]. Using Green's representation, the acoustic pressure in the half-space $z > 0$ can be given by:

$$p^+(M) = -\int_{z=0} p^+(Q)\partial_n \mathcal{G}(M,Q) - \partial_n p^+(Q)\mathcal{G}(M,Q)dQ,$$

according to the hypothesis, if $Q \in [z=0]$: $\partial_n \mathcal{G}(M,Q) = 0$ and $\partial_n p^+(Q) = \omega^2 \rho_0 u(Q)$, we then simply obtain:

$$p^{\pm}(x,y,z) = \mp 2\omega^2 \rho_0 \int_{z=0} u(x',y') \frac{e^{\imath k_f \sqrt{(x-x')^2+(y-y')^2+z^2}}}{4\pi\sqrt{(x-x')^2+(y-y')^2+z^2}} dx'dy'.$$

We obtain an integro-differential equation whose only unknown is the displacement of the plate $u(x,y)$, which can be given by:

$$D\left(\Delta^2 - k_p^4\right) u(x,y) - 4\omega^2 \rho_0 \int_{z=0} u(x',y') \frac{e^{\imath k_f \sqrt{(x-x')^2+(y-y')^2}}}{4\pi\sqrt{(x-x')^2+(y-y')^2}} dx'dy' = \delta(x,y).$$

8.4.2. *Fourier transform of* u

We must note that the integro-differential operator and the second part of the previous equation have a rotational symmetry with respect to the origin. Thus, $u(x,y) = u(r)$ only depends on the distance from the origin r. The direct and inverse Fourier transforms of a radial function are defined by [7.16]. It might be easy to see that the Fourier transform of the bi-Laplacian can be given by $16\pi^4\rho^4$, however, the calculation of the Fourier transform of the integral is more difficult. First, we must note that we can write:

$$- \int_{z=0} u(x', y') \frac{e^{\imath k_f \sqrt{(x-x')^2+(y-y')^2}}}{4\pi\sqrt{(x-x')^2+(y-y')^2}} dx'dy'$$

$$= \lim_{z\to 0} \left\{ \left(\frac{-e^{\imath k_f d(x,y,z)}}{4\pi d(x,y,z)} \star u(x,y) \right) \otimes \delta_z) \right\},$$

where $f(x, y) \otimes g(z)$ is the tensor product of $f(x, y)$ and $g(z)$, and $\star$ is the convolution product with respect to the spatial variables x and y. We know that the Fourier transform of a convolution product is simply the product of the Fourier transforms of each of the terms. To calculate the Fourier transform of the convolution kernel, we begin with:

$$(\Delta + k_f^2) \frac{-e^{\imath k_f d(x,y,z)}}{4\pi d(x,y,z)} = \delta_x \otimes \delta_y \otimes \delta_z,$$

because of the radial symmetry of the problem we can write that $d(x, y, z) = d(r, z)$. By using a radial Fourier transform, we obtain:

$$\left(\frac{d^2}{dz^2} + k_f^2 - 4\pi^2\rho^2 \right) \hat{G}(\rho, z) = \delta_z, \hat{G}(\rho, z) = \left(\widehat{\frac{-e^{\imath k_f d(r,z)}}{4\pi d(r,z)}} \right),$$

where $\hat{G}(\rho, z)$ is the radial Fourier transform of $-\exp(\imath k_f d(r, z))/(4\pi d(r, z))$. If we set $k_f^2 - 4\pi^2\rho^2 = K^2(\rho)$ with $\Im K(\rho) > 0$, we can see that the sought function is Green's kernel of a one-dimensional Helmholtz problem whose wave number is $K(\rho)$. We can, thus, very simply obtain that:

$$\hat{G}(\rho, z) = \frac{e^{\imath K(\rho)|z|}}{2\imath K(\rho)},$$

from where, by passing to the limit at $z = 0$:

$$\mathcal{F}\left(\lim_{z\to 0} \left\{ \left(\frac{e^{\imath k_f d(x,y,z)}}{4\pi d(x,y,z)} \right) \star u \otimes \delta_z \right\} \right) = \hat{u}(\rho) \frac{1}{2\imath K(\rho)}.$$

The Fourier transform of the displacement of the coupled plate can be given by:

$$\hat{u}(\rho) = \frac{1}{D} \frac{\imath K(\rho) D}{(16\pi^4\rho^4 - k_p^4)\, \imath K(\rho) D + 2\omega^2 \rho_0}, K^2(\rho) = k_f^2 - 4\pi^2\rho^2, \Im K(\rho) > 0 \quad [8.9]$$

8.4.3. *Calculation of* $u(r)$

The inversion formula [7.16] leads to:

$$u(r) = \pi \frac{1}{D} \int_{\infty e^{\imath\pi}}^{\infty} \frac{\imath K(\rho) D}{\left(16\pi^4\rho^4 - k_p^4\right) \imath K(\rho) D + 2\omega^2 \rho} \mathrm{H}_0^1(2\pi\rho r)\rho d\rho.$$

We calculate this integral using the method of residues over the following contour:

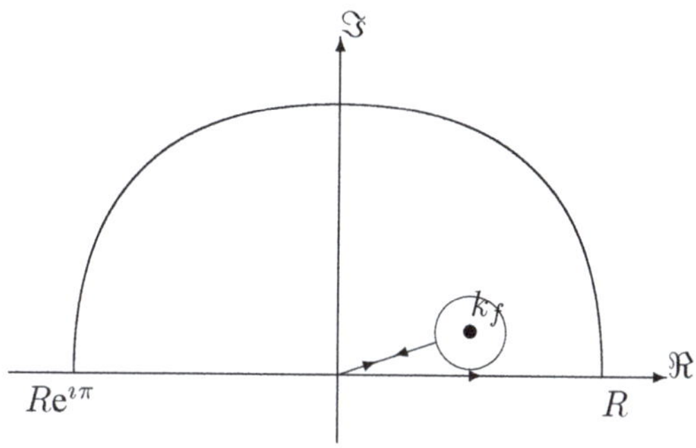

Figure 8.12. *Integration contour*

Using Jordan's lemma, we can show that for $R \rightarrow \infty$, the integral over the half-circles approaches zero. The one over the real segment $]Re^{\imath\pi}, R[$ approaches the sought integral. However, the function $K(\rho)$ is a multiform function, and we must turn around and avoid the point $2\pi\rho_0 = k_f$. The way of avoiding this point has been correctly defined using the limit absorption principle which imposes on $K(\rho)$ to have a non-zero positive imaginary part. The displacement can thus be expressed as the sum of the residues contained in the integration contour supplemented by a branch integral. We will not examine here the calculations of this branch integral; nevertheless, we need to know that for $R \gg 1$, it decreases exponentially [FIL 08]. The displacement is thus correctly described by:

$$u(r) = 2\imath\pi^2 \frac{1}{D} \sum_{n=1}^{3} A_n \mathrm{H}_0^1(k_n r),$$

where k_n are the roots of the dispersion equation. The coefficients A_n are given by $A_n = \imath K_n D(k_n/2\pi)/\mathcal{D}'(k_n/2\pi)$, where $K_n = k_f^2 - k_n^2, \Im K_n > 0$ if $K_n \in \mathbb{C}$ and $K_n > 0$ if $K_n \in \mathbb{R}$. $\mathcal{D}'(\rho)$ is the derivative with respect to ρ of the denominator given by:

$$\mathcal{D}'(\rho) = \imath K'(\rho) D(16\pi^4\rho^4 - k_p^4) + \imath K(\rho) D \, 64\pi^4\rho^3.$$

We must note that near the origin $u(r)$ has a behavior analogous to that of a plate *in vacuo*:

$$u(0) \propto 1, u'(0) = 0, u''(0) \propto \lim_{\epsilon \to 0} \ln \epsilon, u'''(0) \propto \lim_{\epsilon \to 0} \frac{1}{\epsilon}.$$

8.4.4. *Radiated acoustic pressure*

We can calculate the Fourier transforms of $p^{\pm}(x, y, z)$ without difficulty. The calculation of the inverse transforms, which uses the same computational techniques, poses the same level of difficulty. Therefore, we will not develop them here. However, it is very easy to calculate the asymptotic behavior – at a great distance – of the pressure radiated by the plate. For this, we establish a general result for the asymptotic behavior of radiation from a source with an arbitrary density of $f(M)$. We find:

$$\tilde{p}(M) = \int_{I\!R^3} f(M') \frac{e^{\imath k_f r(M,M')}}{4\pi r(M,M')} dM'.$$

We define by (r, θ, ϕ) the spherical coordinates of M and by (r', θ', ϕ') those of M'. We use the expansion, valid for $k_f r \gg 1$, of the kernel of the integral given by [5.9]:

$$-\frac{e^{\imath k_f r(M,M')}}{4\pi r(M,M')}$$
$$= -\frac{e^{\imath k_f r}}{4\pi r} e^{-\imath k_f r' \left(\sin\theta \cos\phi \sin\theta' \cos\phi' + \sin\theta \sin\phi \sin\theta' \sin\phi' + \cos\theta \cos\theta' \right)} + \mathcal{O}(k_f r)^{-2},$$

where $x' = r' \sin\theta' \cos\phi'$, $y' = r' \sin\theta' \sin\phi'$ and $z' = r' \cos\theta'$. We obtain:

$$\tilde{p}(M) \approx \frac{e^{\imath k_f r}}{4\pi r} \int_{I\!R^3} f(M') e^{-\imath k_f \left(x' \sin\theta \cos\phi + y' \sin\theta \sin\phi + z' \cos\theta \right)} dx' dy' dz'.$$

We can recognize the Fourier transform of $f(x', y', z')$ taken at points:

$$2\pi\xi_1 = k_f \sin\theta \cos\phi, 2\pi\xi_2 = k_f \sin\theta \sin\phi, 2\pi\xi_3 = k_f \cos\theta,$$

from where the final result:

$$\tilde{p}(M) \approx \frac{e^{\imath k_f r}}{4\pi r} \hat{f} \left(\frac{k_f \sin\theta \cos\phi}{2\pi}, \frac{k_f \sin\theta \sin\phi}{2\pi}, \frac{k_f \cos\theta}{2\pi} \right),$$

where $\hat{f}(\xi_1, \xi_2, \xi_3) = \int_{\mathbb{R}^3} f(x, y, z) \exp(-2\imath\pi(x\xi_1 + y\xi_2 + z\xi_3))dxdydz$. With this result, it is easy to calculate the asymptotic behavior of $p^+(x, y, z)$. As we previously noted, $p^+(x, y, z)$ is the radiation of a source with a density $f(x, y, z) = -2\omega^2\rho_0 u(x, y) \otimes \delta_z$ whose radial Fourier transform can be given by $\mathcal{F}(u(x, y) \otimes \delta_z) = \hat{u}(\rho)$, taken at $\rho = \sqrt{\xi_1^2 + \xi_2^2}$. With $2\pi\xi_1 = k_f \sin\theta \cos\phi$ and $2\pi\xi_2 = k_f \sin\theta \sin\phi$, it is easy to see $\rho = k_f \sin\theta / 2\pi$. With $\hat{u}(\rho)$ given by [8.9], we obtain:

$$p^+(M) \approx -2\omega^2\rho_0 \frac{e^{\imath k_f r}}{4\pi r} \hat{u}\left(\frac{k_f \sin\theta}{2\pi}\right).$$

We obtain a similar result for $p^-(x, y, z)$. The calculation of a directivity diagram (or a form or directivity function) $\Theta(\theta, \phi)$ does not require Fourier inversion and thus can be done without difficulty. We obtain:

$$p^+(M) \approx \frac{-2\omega^2\rho_0}{D} \frac{e^{\imath k_f r}}{4\pi r} \frac{\imath D k_f \cos\theta}{\left(k_f^4 \sin^4\theta - k_p^4\right) \imath D k_f \cos\theta + 2\omega^2\rho_0},$$

$$\Theta^\pm(\theta, \phi) = \frac{-2\omega^2\rho_0}{D} \frac{\imath D k_f \cos\theta}{\left(k_f^4 \sin^4\theta - k_p^4\right) \imath D k_f \cos\theta + 2\omega^2\rho_0}.$$

This function presents a marked maximum when the denominator is at its minimum, i.e. for $k_f^4 \sin^4\theta = k_p^4$. This corresponds to the angle $\theta = \arcsin k_p/k_f = \arcsin 1/M$. Thus, in a supersonic regime, we obtain an angle (called the Mach angle) for which the plate has a maximum directivity.

8.5. Acoustic radiation and vibration of finite plates

8.5.1. *Statement of the problem*

Unless stated otherwise, in what follows we assume that we are dealing with a one-dimensional plate which is in contact with the same ideal fluid on both sides. The plate, with a finite length of $2L$, is simply supported at its ends. It is prolonged by an infinitely rigid screen, which separates the two fluid media (see Figure 8.13).

The equations of the problem are the following:

$$\left(\frac{d^4}{dx^4} - k_p^4\right) u(x) = \frac{f(x)}{D} - \frac{1}{D}\left(p^+(x, 0) - p^-(x, 0)\right),$$

$$\left(\Delta + k_f^2\right) p^\pm(x, z) = 0 \text{ in } z \gtrless 0, \omega^2\rho_0 u(x) = \partial_z p^\pm(x, z = 0), \forall x,$$

to which we add the simply supported boundary conditions $u(x)$ given by $u(\pm L) = u''(\pm L) = 0$. By using Green's kernel of the Helmholtz equation for the Neumann problem on the surface $z = 0$ which occupies the plate and baffle, Green's representation of acoustic pressure allows us to write the latter in the following form:

$$p^{\pm}(x,z) = \text{sgn}z\, \omega^2 \rho_0 \int_{-L}^{+L} \frac{-\imath}{2} \text{H}_0 \left(k_f \sqrt{(x-x')^2 + z^2} \right) u(x') dx'.$$

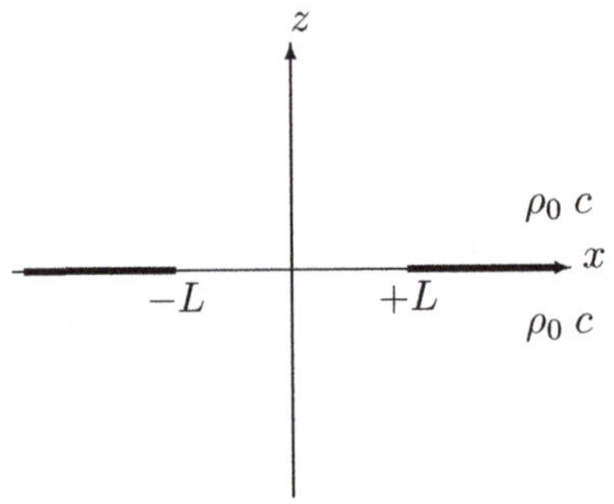

Figure 8.13. *Geometry of the problem*

The initial differential system transforms into an integro-differential system:

$$\left(\frac{d^4}{dx^4} - k_p^4 \right) u(x) = \frac{f(x)}{D} - P(x),$$

$$P(x) = \frac{2\omega^2 \rho_0}{D} \int_{-L}^{+L} \frac{-\imath}{2} \text{H}_0 \left(k_f |x - x'| \right) u(x') dx',$$

to which we add the simply supported boundary conditions $u(\pm L) = u''(\pm L) = 0$. There are many methods to solve this problem. We will study some of them. However, all of them require numerical calculation.

8.5.2. *Exact methods*

There does not exist an exact method. All that we can hope to find is an approximation of the solution of the problem that is as accurate as possible. Certain methods, such as approximations by a series of orthogonal polynomials, can (theoretically) offer infinite precision. It is important to understand that attempting to get more than two or three significant figures is very difficult numerically. Moreover, we only rarely have access to the mechanical and physical characteristics of the materials with a precision of less than a few percent[4]; in practice, we rarely know the spectrum of a structure with a precision better than a few percent.

8.5.2.1. *Solution using integral equations*

We consider G_p the Green kernel of the one-dimensional plate equation, given by:

$$G_p(x;x') = G_p(x-x') = \frac{1}{4k_p^3 D}\left(\imath e^{\imath k_p|x-x'|} - e^{-k_p|x-x'|}\right).$$

The Green representation of displacement $u(x)$ can be written as:

$$u(x) = u_0(x) + S_1(x-L) + S_2(x-L) + S_1(x+L) + S_2(x+L) + \int_{-L}^{+L} P(x')G_p(x-x')dx',$$

where $u_0(x) = 1/Df(x) \star G_p(x)$. The functions $S_{1,2}$ are sources located on the boundaries of the plate at $\pm L$. They allow us to take into account the discontinuity of the displacement at the boundaries of the plate. We obtain:

$$S_1(x \pm L) = \alpha_1^{\pm L} G_p(x - \pm L), S_2(x \pm L) = \alpha_2^{\pm L} G_p''(x - \pm L),$$

the four constants $\alpha_1^{\pm L}$, which can be written as a function of the displacement derivative at the boundaries, are the unknowns which we can calculate, for example, by writing the four boundary conditions. Finally, we obtain two coupled integral equations:

$$u(x) = G_p + S_1(x-L) + S_2(x-L) + S_1(x+L) + S_2(x+L) + \int_{-L}^{+L} P(x')G_p(x-x')dx'$$

$$P(x) = \frac{4\omega^2\rho_0}{D}\int_{-L}^{+L} \frac{-\imath}{4} \mathrm{H}_0\left(k_f|x-x'|\right) u(x')dx'.$$

We must note that the displacement can also be written as:

$$u(x) = G_p + S_{1,2}(x \pm L) + \frac{2\omega^2\rho_0}{D}\int_{-L}^{+L}\int_{-L}^{+L} \frac{-\imath}{2}\mathrm{H}_0\left(k_f|x''-x'|\right) u(x')G_p(x-x'')dx'dx''.$$

4 We must take into consideration that the eigenfrequencies of a plate directly depend on its thickness and it is unrealistic to believe that a metal sheet or a glass with a thickness of 2 mm is manufactured with a significantly better tolerance than 20 microns.

The only way to calculate $u(x)$ and $P(x)$ consists of finding a numerical approximation to the displacement and the pressure jump in the form of a series expansion of known functions, the new unknowns become the coefficients of these series:

$$u(x) = \sum_{m=1}^{M} u_m \gamma_m^u(x), P(x) = \sum_{n=1}^{N} P_n \gamma_n^P(x).$$

The new unknowns are u_m and P_n. They can be calculated using the Galerkin or collocation method. The approximation functions $\gamma_{m,n}^{u,P}$ can be piecewise constant, quadratic or polynomial. The number of functions that need to be taken into consideration (the series approximation order) depends on the excitation frequency, the geometrical and mechanical characteristics of the structure or the fluid. In practice, if we use piecewise constant functions, we assume that the minimum of six points per wavelength, which was required in the diffraction problems, is no longer sufficient, and that 10 or 20 points per wavelength are essential near the resonance frequencies of the coupled plate. Moreover, for this type of problem, for which the given wavelengths (mechanical and acoustic) are different, we must consider the shortest (and, thus, the most costly) wavelengths. We finally obtain a linear system of simultaneous equations with the dimension $M + N + 4$ whose size increases along with frequency. We can show that a solution exists for any real frequency due to the loss of energy by radiation. The zeros of the determinant of the associated linear system are complex. These are the resonance frequencies of the system.

8.5.2.2. *Series expansion of eigenmodes in vacuo*

If we consider the eigenmode-eigen wave number couple $(u_n(x), k_{pn})$, this is the non-zero solution of the homogeneous system:

$$\left(\frac{d^4}{dx^4} - k_{pn}^4\right) u_n(x) = 0, u_n(\pm L) = u_n''(\pm L) = 0,$$

where $u_n(x) = 1/\sqrt{L} \sin k_{pn}(x + L), k_{pn} = n\pi/(2L)$, are the eigenmodes, which are orthonormal with respect to the scalar product $\langle f, g\rangle = \int_{-L}^{+L} f(x)g^\star(x)dx$. We aim to find $u(x)$ in the form of a series expansion of these eigenmodes $u(x) = \sum_{m=1}^{\infty} \alpha_m u_m(x), \alpha_m = \langle u(x), u_m(x)\rangle$. We introduce this expression in the integro-differential equation satisfied by the displacement. We obtain:

$$\sum_{m=1}^{\infty} \alpha_m \left\{ (k_{pm}^4 - k_p^4) u_m(x) + \frac{2\omega^2 \rho_0}{D} \int_{-L}^{+L} \frac{-\imath}{2} \mathrm{H}_0\left(k_f |x - x'|\right) u_m(x') dx' \right\} = \frac{f(x)}{D}.$$

With $f_n = 1/D \int_{-L}^{+L} f(x) u_n^\star(x) dx$, we obtain:

$$\sum_{m=1}^{\infty} \alpha_m \left\{ \left(k_{pm}^4 - k_p^4\right) \delta_m^n \right.$$
$$\left. + \frac{2\omega^2 \rho_0}{D} \int_{-L}^{+L} \int_{-L}^{+L} \frac{-\imath}{2} \mathrm{H}_0 \left(k_f |x - x'|\right) u_m(x') u_n^\star(x) dx dx' \right\} = f_n,$$

where δ_m^n is Kronecker's symbol. We can then see that the coefficients α_n are the solutions of a full infinite linear system. We must note that the zeros of the determinant of this system all have a non-zero imaginary part. The difference between the latter and that obtained *in vacuo* arises from the double integral, which reflects the effect of the fluid on the plate. The fluid introduces intermodal coupling. When the influence of the fluid on the structure is weak (for example, a steel plate in air), this intermodal coupling is weak and we can ignore the extra-diagonal terms:

$$\frac{4\omega^2 \rho_0}{D} \int_{-L}^{+L} \int_{-L}^{+L} \frac{-\imath}{4} \mathrm{H}_0 \left(k_f |x - x'|\right) u_m(x') u_n^\star(x) dx dx', m \neq n$$

with regard to the diagonal terms, and obtain an infinite linear system, which is diagonal, and can be analytically inverted to give:

$$\alpha_n = \frac{f_n}{\left(k_{pn}^4 - k_p^4\right) + \frac{2\omega^2 \rho_0}{D} \int_{-L}^{+L} \int_{-L}^{+L} \frac{-\imath}{2} \mathrm{H}_0 \left(k_f |x - x'|\right) u_n(x') u_n^\star(x) dx dx'}.$$

When the fluid density is not negligible, we must retain the non-diagonal terms. In order to be able to calculate the coefficients α_n, it is necessary to truncate the modal series (and obtain a finite linear system). The difficulty lies in the choice of truncation. Quite often, the only method consists of choosing a mode number N, inverting the system and calculating $u^N(x)$, then we start once again with a mode number $N' > N$ and calculate $u^{N'}(x)$ hoping to get the desired precision.

8.5.2.3. *Series expansion of coupled eigenmodes*

Similarly as for the eigenmodes *in vacuo*, it is possible to define a eigenmode-eigen wave number couple $(\tilde{u}_n(x), \tilde{k}_{pn})$, the non-zero solution of the homogeneous integro-differential equation:

$$\left(\frac{d^4}{dx^4} - \tilde{k}_{pn}^4\right) \tilde{u}_n(x) + \frac{2\rho_0}{\rho_p h} \tilde{k}_{pn}^4 \int_{-L}^{+L} \frac{-\imath}{2} \mathrm{H}_0 \left(k_f |x - x'|\right) \tilde{u}_m(x') dx' = 0,$$

given the boundary conditions $\tilde{u}_n(\pm L) = \tilde{u}_n''(\pm L) = 0$. These eigenmodes depend on the excitation frequency since the kernel $\mathrm{H}_0 \left(k_f |x - x'|\right)$ depends on frequency. Similarly, the eigenfrequencies depend on frequency.

To establish the properties of these eigenmodes, it is useful to introduce the weak (or energetic) formulation of the problem. We must first note that, if $f \in L^2[-L,+L]$ (square-integrable function class), then the solution $u(x)$ belongs to the space of square-integrable functions over]-L,+L[as well as their derivatives up to the order 2, which are zero, and their second derivatives at $x = \pm L$. The eigenmodes *in vacuo* constitute a basis of this set. This means that any solution of a coupled plate problem can be expanded in a convergent series of this mode. Once again, we define as a scalar product $\langle u, v\rangle = \int_{-L}^{+L} u(x)v^\star(x)dx$. We define the bilinear forms:

$$a(u,v) = D\int_{-L}^{+L} u''(x)v''^\star(x)dx,$$

$$\beta_\omega(u,v) = \frac{-\imath}{2}\int_{-L}^{+L}\int_{-L}^{+L} u(x')\mathrm{H}_0\left(k_f|x-x'|\right)v^\star(x')dxdx'. \quad [8.10]$$

We note that $a(u,v)$ is the energy of deformation of the plate. $\omega^2\rho_0\beta_\omega(u,v)$ corresponds to the energy lost by radiation. The initial boundary problem can be written in the weak form as follows: find the function u which satisfies the boundary conditions, such that for any function, which also satisfies these conditions, we obtain:

$$a(u,v) - \rho_p h\omega^2\left(\langle u,v\rangle - \frac{2\rho_0}{\rho_p h}\beta_\omega(u,v)\right) = \langle f,v\rangle. \quad [8.11]$$

We must note that the problem has one and only one solution for any real frequency. It has a countable sequence of eigenvalues with non-zero imaginary parts. We can associate a finite number of eigen functions to each of them.

By definition, the eigenmodes $\tilde{u}_n(x)$ and the eigenvalues $\tilde{\Lambda}_n = \rho_p h\omega^2$ are the non-zero solutions of the following problem:

$$a(\tilde{u}_n,v) = \tilde{\Lambda}_n\left(\langle\tilde{u}_n,v\rangle - \epsilon\beta_\omega(\tilde{u}_n,v)\right), \text{ for all } v, \quad [8.12]$$

where $\epsilon = 2\rho_0/(\rho_p h)$ is the coupling parameter. We assume that the multiplicity order of eigenmodes is arbitrary. We set $v = \tilde{u}^\star_m$. We obtain[5]:

$$a(\tilde{u}_n,\tilde{u}^\star_m) = \tilde{\Lambda}_n\left(\langle\tilde{u}_n,\tilde{u}^\star_m\rangle - \epsilon\beta_\omega(\tilde{u}_n,\tilde{u}^\star_m)\right),$$

where we have the following symmetries $a(\tilde{u}_n,\tilde{u}^\star_m) = a(\tilde{u}_m,\tilde{u}^\star_n)$, $\langle\tilde{u}_n,\tilde{u}^\star_m\rangle = \langle\tilde{u}_m,\tilde{u}^\star_n\rangle$ and $\beta_\omega(\tilde{u}_n,\tilde{u}^\star_m) = \beta_\omega(\tilde{u}_m,\tilde{u}^\star_n)$. Thus, from the first relation,

5 We must note that the term $\omega^2\rho_0\beta_\omega(\tilde{u}_n,\tilde{u}^\star_m)$ is the radiation mode impedance of the mode mn.

we can deduce that $\tilde{\Lambda}_n(\langle \tilde{u}_n, \tilde{u}_m^\star\rangle - \epsilon\beta_\omega(\tilde{u}_n, \tilde{u}_m^\star)) = \tilde{\Lambda}_m(\langle \tilde{u}_m, \tilde{u}_n^\star\rangle - \epsilon\beta_\omega(\tilde{u}_m, \tilde{u}_n^\star))$, that is if $m \neq n$ and $\tilde{\Lambda}_m \neq \tilde{\Lambda}_n$, $\langle \tilde{u}_n, \tilde{u}_m^\star\rangle - \epsilon\beta_\omega(\tilde{u}_n, \tilde{u}_m^\star) = 0$, or $a(\tilde{u}_n, \tilde{u}_m^\star) = 0$. This relation is an orthogonality relation of coupled eigenmodes. It is analogous to the one satisfied by the eigenmodes *in vacuo*; it is enough to take $\epsilon = 0$ to find it. The norm associated with this relation is given by $a(\tilde{u}_n, \tilde{u}_n^\star) = \tilde{\Lambda}_n\left[\langle \tilde{u}_n, \tilde{u}_n^\star\rangle - \epsilon\beta_\omega(\tilde{u}_n, \tilde{u}_n^\star)\right]$. This expression, which connects the eigenvalues to the eigenmodes, shows that for the common vibroacoustic problems the eigenvalues depend on the frequency via the coupling term. It is the same for eigenmodes. For each frequency, we must thus determine the resulting eigenfrequencies and eigenmodes.

Now, we aim to find the solution to our problem by using a series expansion of these coupled eigenmodes. Therefore, we aim to find $u(x) = \sum_{j=1}^{\infty} \alpha_j \tilde{u}_j(x)$. We introduce this development in the variational equation:

$$\sum_{j=1}^{\infty} \alpha_j \left\{a(\tilde{u}_j, \tilde{u}_i^\star) - \rho_p h\omega^2 \left(\langle \tilde{u}_j, \tilde{u}_i^\star\rangle - \epsilon\beta_\omega(\tilde{u}_j, \tilde{u}_i^\star)\right)\right\} = \langle f, \tilde{u}_i^\star\rangle,$$

where we have set $v = \tilde{u}_i^\star$. The orthogonality relation, satisfied by the eigenmodes, leads us to:

$$\alpha_i \left\{a(\tilde{u}_i, \tilde{u}_i^\star) - \rho_p h\omega^2 \left(\langle \tilde{u}_i, \tilde{u}_i^\star\rangle - \epsilon\beta_\omega(\tilde{u}_i, \tilde{u}_i^\star)\right)\right\} = \langle f, \tilde{u}_i^\star\rangle.$$

Finally, that is:

$$\alpha_i \left\{\frac{\tilde{\Lambda}_i - \rho_p h\omega^2}{\tilde{\Lambda}_i} a(\tilde{u}_i, \tilde{u}_i^\star)\right\} = \langle f, u_i^\star\rangle.$$

Thus, we obtain the series expansion of coupled modes which is defined for any real excitation frequency:

$$u(x) = \sum_{j=1}^{\infty} \frac{\tilde{\Lambda}_i}{\tilde{\Lambda}_i - \rho_p h\omega^2} \frac{\langle f, u_i^\star\rangle}{a(\tilde{u}_i, \tilde{u}_i^\star)} \tilde{u}_j(x).$$

The difficulty related to this method lies in the calculation of eigenmodes, which can only be performed numerically.

8.5.3. *Light fluid approximation*

When the fluid density is low in comparison to the structure, its influence is weak. The displacement of the plate is very similar to that of a plate *in vacuo*. We will see that the perturbation methods (series expansion of powers of the small parameter ϵ) allow us to reduce the problem to solving a set of simpler *in vacuo* plate problems. However, this technique has the significant disadvantage of not keeping the existence and uniqueness of the solution. We then apply these techniques to coupled eigenmode calculations. The latter approach is the most satisfactory one, since it allows us to retain, in the approximate solution, the main properties of the solution of the initial problem: existence and unity of the solution for any excitation frequency. Finally, we propose some analytic approximations, which allow us to link the mathematical solution to the physics of the problem.

8.5.3.1. *Solution using a series expansion of eigenmodes in vacuo*

Once again, we start with the initial integro-differential boundary problem:

$$Du''''(x) - \rho_p h\omega^2 \left(u(x) - \epsilon \int_{-L}^{+L} \frac{-\imath}{2} \mathrm{H}_0 \left(k_f |x - x'|\right) u(x) dx \right) = f(x)$$

given the boundary conditions $u(\pm L) = u''(\pm L) = 0$; if ϵ is sufficiently small ($\epsilon = 0.033\ \mathrm{m}^{-1}$ for a steel plate with a 1 cm thickness in air), we assume that the expansion of $u(x)$ in a series of increasing powers of ϵ converges: $u(x) = u^0(x) + \epsilon u^1(x) + \epsilon^2 u^2(x) + \cdots$. We introduce this expansion in the integro-differential problem that we need to solve. We obtain:

$$D(u^{0iv}(x) + \epsilon u^{1iv}(x)) - \rho_p h\omega^2 \left(u^0(x) + \epsilon u^1(x) - \epsilon \int_{-L}^{+L} \frac{-\imath}{2} \mathrm{H}_0 \left(k_f |x - x'|\right) \right.$$
$$\left. \times \left(u^0(x) + \epsilon u^1(x)\right) dx \right) = f(x),$$

and, for the boundary conditions, $u^0(\pm L) + \epsilon u^1(\pm L) = u^{0''}(\pm L) + \epsilon u^{1''}(\pm L) = 0$. We regroup the terms of power of ϵ. At the order zero, with respect to ϵ, we obtain $Du^{0iv}(x) - \rho_p h\omega^2 u^0(x) = f(x)$ and $u^0(\pm L) = u^{0''}(\pm L) = 0$. It is obvious that the solution can be calculated very easily by an expansion on the basis of eigenmodes *in vacuo*. The main problem lies in the fact that this solution is not defined for the frequencies of the plate *in vacuo*, whereas the solution of the initial problem is defined for any frequency.

At the order one, with respect to ϵ, we obtain $Du^{1iv}(x) - \rho_p h\omega^2 u^1(x) = -\rho_p h\omega^2 \int_{-L}^{+L} -\imath/2\mathrm{H}_0 \left(k_f |x - x'|\right) u^0(x) dx$, and for the boundary conditions $u^1(\pm L) = u^{1''}(\pm L) = 0$. Once again, the solution to this problem can be easily calculated on the basis of eigenmodes *in vacuo*, always with the problem of a non-existing solution for the eigenfrequencies of the plate *in vacuo*. Finally, we obtain, for $i = 0$ and $i = 1$, $u^i(x) = \sum_{n=1}^{\infty} f_n^i \left(k_{pn}^4 - k_p^4\right) u_n(x)$, where:

$$
\begin{aligned}
f_n^0 &= \frac{1}{D}\int_{-L}^{+L} f(x)u_n(x)dx \\
f_n^1 &= \frac{-\rho_p h\omega^2}{D}\int_{-L}^{+L}\int_{-L}^{+L}\frac{-\imath}{2}\mathrm{H}_0\left(k_f|x-x'|\right)u^0(x)u_n(x')dxdx' \\
&= \frac{-\rho_p h\omega^2}{D}\sum_{m=1}^{\infty}\frac{f_n^0}{k_{pn}^4-k_p^4}\beta_\omega(u_m,u_n^\star).
\end{aligned}
$$

This latter expression involves double integrals. The calculation of higher terms of the perturbation series always requires even higher order integration. At the order two, we must calculate a triple integral. The calculation quickly becomes impractical. This is where we find a general property of the perturbation series; these are very powerful methods as long as it is enough to examine only the first two terms, beyond that, it is better to solve the exact problem.

8.5.3.2. *Calculation of eigenmodes and eigenfrequencies of a coupled plate*

In order to avoid this problem of indeterminacy of the solution near eigenfrequencies *in vacuo*, we must use the series expansion of coupled eigenmodes. We calculate them using the perturbation method. We must note that since the eigenvalues and eigenfrequencies depend on the coupling with the fluid, we must try to find them both in the form of a perturbation series. This method is called the Rayleigh–Schrödinger method [NAY 73]. We obtain:

$$
\tilde{u}_n(x) = \tilde{u}_n^0(x) + \epsilon\tilde{u}_n^1(x) + \epsilon^2\tilde{u}_n^2(x) + \cdots, \tilde{\Lambda}_n = \tilde{\Lambda}_n^0 + \epsilon\tilde{\Lambda}_n^1 + \epsilon^2\tilde{\Lambda}_n^2 + \cdots,
$$

and we then introduce these expressions in the equation satisfied by the eigenmodes and eigenvalues given by [8.12]:

$$
a(\tilde{u}_n^0 + \epsilon\tilde{u}_n^1, v) = \left(\tilde{\Lambda}_n^0 + \epsilon\tilde{\Lambda}_n^1\right)\left(\langle\tilde{u}_n^0 + \epsilon\tilde{u}_n^1, v\rangle - \epsilon\beta_\omega(\tilde{u}_n^0 + \epsilon\tilde{u}_n^1, v)\right), \text{ for all } v.
$$

Due to the linearity of $a(u,v)$, $\langle u,v\rangle$ and $\beta_\omega(u,v)$, it is obvious that, for any v, we obtain:

$$
\begin{aligned}
a(\tilde{u}_n^0, v) + \epsilon a(\tilde{u}_n^1, v) &= \tilde{\Lambda}_n^0\langle\tilde{u}_n^0, v\rangle \\
&+\epsilon\left\{\tilde{\Lambda}_n^1\langle\tilde{u}_n^0, v\rangle + \tilde{\Lambda}_n^0\left(\langle\tilde{u}_n^1, v\rangle - \beta_\omega(\tilde{u}_n^0, v)\right)\right\} + \cdots.
\end{aligned}
$$

At the order zero, we obtain $a(\tilde{u}_n^0, v) = \tilde{\Lambda}_n^0 \langle \tilde{u}_n^0, v\rangle$. This equation is the equation of eigenmodes and eigenvalues *in vacuo*, for which we know an analytic solution, for the particular case considered here. Whereas at the order one with respect to ϵ:

$$a(\tilde{u}_n^1, v) = \tilde{\Lambda}_n^1 \langle \tilde{u}_n^0, v\rangle + \tilde{\Lambda}_n^0 \left(\langle \tilde{u}_n^1, v\rangle - \beta_\omega(\tilde{u}_n^0, v)\right).$$

In the latter, we set $v = \tilde{u}_n^{0\star}$. We obtain:

$$a(\tilde{u}_n^1, \tilde{u}_n^{0\star}) = \tilde{\Lambda}_n^1 \langle \tilde{u}_n^0, \tilde{u}_n^{0\star}\rangle + \tilde{\Lambda}_n^0 \left(\langle \tilde{u}_n^1, \tilde{u}_n^{0\star}\rangle - \beta_\omega(\tilde{u}_n^0, \tilde{u}_n^{0\star})\right),$$

now $a(\tilde{u}_n^1, \tilde{u}_n^{0\star}) = a(\tilde{u}_n^0, \tilde{u}_n^{1\star})$ and, due to the definition of eigenmodes *in vacuo*, we get $a(\tilde{u}_n^0, \tilde{u}_n^{1\star}) = \tilde{\Lambda}_n^0 \langle \tilde{u}_n^0, \tilde{u}_n^{1\star}\rangle$. If we introduce this result in the previous equation, we obtain:

$$\tilde{\Lambda}_n^0 \langle \tilde{u}_n^0, \tilde{u}_n^{1\star}\rangle = \tilde{\Lambda}_n^1 \langle \tilde{u}_n^0, \tilde{u}_n^{0\star}\rangle + \tilde{\Lambda}_n^0 \left(\langle \tilde{u}_n^1, \tilde{u}_n^{0\star}\rangle - \beta_\omega(\tilde{u}_n^0, \tilde{u}_n^{0\star})\right),$$

that is $\tilde{\Lambda}_n^1 = \tilde{\Lambda}_n^0 \beta_\omega(\tilde{u}_n^0, \tilde{u}_n^{0\star}) / \langle \tilde{u}_n^0, \tilde{u}_n^{0\star}\rangle$. We must recall that the eigenmodes *in vacuo* are of norm unity, $\langle \tilde{u}_n^0, \tilde{u}_n^{0\star}\rangle = 1$. Finally, $\tilde{\Lambda}_n^1 = \tilde{\Lambda}_n^0 \beta_\omega(\tilde{u}_n^0, \tilde{u}_n^{0\star})$.

We expand $\tilde{u}_n^1$ in a series of eigenmodes *in vacuo* $\tilde{u}_n^1(x) = \sum_{m=1}^{\infty} \alpha_n^m \tilde{u}_m^0(x)$. We introduce this relation in the mode equation $\tilde{u}_m^1(x)$ (equation of the order one with respect to ϵ), taken for $v = \tilde{u}_j^{0\star}$. We obtain:

$$\sum_{m=1}^{\infty} \alpha_n^m a(\tilde{u}_m^0, \tilde{u}_j^{0\star}) = \tilde{\Lambda}_n^1 \langle \tilde{u}_n^0, \tilde{u}_j^{0\star}\rangle + \tilde{\Lambda}_n^0 \left(\sum_{m=1}^{\infty} \alpha_n^m \langle \tilde{u}_m^0, \tilde{u}_j^{0\star}\rangle - \beta_\omega(\tilde{u}_n^0, \tilde{u}_j^{0\star}) \right).$$

By regrouping the terms we obtain:

$$\sum_{m=1}^{\infty} \alpha_n^m \left(a(\tilde{u}_m^0, \tilde{u}_j^{0\star}) - \tilde{\Lambda}_n^0 \langle \tilde{u}_m^0, \tilde{u}_j^{0\star}\rangle \right) = \tilde{\Lambda}_n^1 \langle \tilde{u}_n^0, \tilde{u}_j^{0\star}\rangle - \tilde{\Lambda}_n^0 \beta_\omega(\tilde{u}_n^0, \tilde{u}_j^{0\star}).$$

Based on the properties of eigenmodes *in vacuo*, we obtain $a(\tilde{u}_m^0, \tilde{u}_j^{0\star}) = \tilde{\Lambda}_j^0 \langle \tilde{u}_j^0, \tilde{u}_j^{0\star}\rangle \delta_m^j = \tilde{\Lambda}_j^0 \delta_m^j$, $\langle \tilde{u}_m^0, \tilde{u}_j^{0\star}\rangle = \delta_m^j$ and evidently $\langle \tilde{u}_n^0, \tilde{u}_j^{0\star}\rangle = \delta_n^j$. Thus, if $n \neq j$, we obtain $\alpha_n^j (\tilde{\Lambda}_j^0 - \tilde{\Lambda}_n^0) \langle \tilde{u}_j^0, \tilde{u}_j^{0\star}\rangle = -\tilde{\Lambda}_n^0 \beta_\omega(\tilde{u}_n^0, \tilde{u}_j^{0\star})$. This is $\alpha_n^j (\tilde{\Lambda}_j^0 - \Lambda_n^0) = -\tilde{\Lambda}_n^0 \beta_\omega(\tilde{u}_n^0, \tilde{u}_j^{0\star})$. Finally, that is:

$$\alpha_n^j = -\frac{\tilde{\Lambda}_n^0}{\tilde{\Lambda}_j^0 - \tilde{\Lambda}_n^0} \beta_\omega(\tilde{u}_n^0, \tilde{u}_j^{0\star}).$$

The coefficients α_n^j are determined in a unique way. However, if $n = j$, we obtain $\alpha_j^j (\tilde{\Lambda}_j^0 - \tilde{\Lambda}_j^0) = \tilde{\Lambda}_j^1 - \tilde{\Lambda}_j^0 \beta_\omega(\tilde{u}_j^0, \tilde{u}_j^{0\star})$, now $\tilde{\Lambda}_j^1 = \tilde{\Lambda}_j^0 \beta_\omega(\tilde{u}_j^0, \tilde{u}_j^{0\star})$, thus, we get $\alpha_j^j \times 0 =$

$0 \times \beta_\omega(\tilde{u}_j^0, \tilde{u}_j^{0\star})$ which is in an indeterminate form. We can show that the only possible choice is $\alpha_j^j = 0$. In order to do this, we impose that the orthogonality relation of coupled modes is satisfied up to ϵ^2 that is $\langle \tilde{u}_n, \tilde{u}_n^\star \rangle = \langle \tilde{u}_n^0 + \epsilon \tilde{u}_n^1, \tilde{u}_n^{0\star} + \epsilon \tilde{u}_n^{1\star} \rangle = 1$. After expansion, that is $\langle \tilde{u}_n^0, \tilde{u}_n^{0\star} \rangle + 2\epsilon \langle \tilde{u}_n^1, \tilde{u}_n^{0\star} \rangle + \mathcal{O}(\epsilon^2) = 1$. Obviously, at the order 0 with respect to ϵ: $\langle \tilde{u}_n^0, \tilde{u}_n^{0\star} \rangle = 1$ and at the order 1 with respect to ϵ: $2\langle \tilde{u}_n^1, \tilde{u}_n^{0\star} \rangle = 0$. The first relation is satisfied due to the properties of eigenmodes *in vacuo*. The second gives:

$$\begin{aligned}\langle \tilde{u}_n^1, \tilde{u}_n^{0\star} \rangle &= \langle \sum_{m=1}^{\infty} \alpha_n^m \tilde{u}_m^0(x), \tilde{u}_n^{0\star} \rangle = \sum_{m=1}^{\infty} \alpha_n^m \langle \tilde{u}_m^0(x), \tilde{u}_n^{0\star} \rangle \\ &= \alpha_m^m \langle \tilde{u}_m^0(x), \tilde{u}_m^{0\star} \rangle = \alpha_m^m = 0.\end{aligned}$$

Finally, we obtain:

$$\tilde{u}_n^1(x) = \sum_{m=1, m\neq n}^{\infty} \frac{-\tilde{\Lambda}_n^0}{\tilde{\Lambda}_m^0 - \tilde{\Lambda}_n^0} \beta_\omega(\tilde{u}_n^0, \tilde{u}_m^{0\star}) \tilde{u}_m^0(x).$$

We can note that the corrective term of each modal deformation is orthogonal to the considered mode and that the sound intensity is governed by intermodal coupling (term $\beta_\omega(\tilde{u}_n^0, \tilde{u}_m^{0\star})$). We recognize here a classic result established by Rayleigh [STR 45, vol. I, pp. 115–118]. Obviously, this result is not valid if the modes are degenerate (if there exist two modes m and n such that $\tilde{\Lambda}_m^0 = \tilde{\Lambda}_n^0$).

The series expansion of coupled modes can, finally, be written as:

$$u(x) = \sum_{m=1}^{\infty} \frac{\tilde{\Lambda}_n^0 + \epsilon \tilde{\Lambda}_n^1}{\tilde{\Lambda}_n^0 + \epsilon \tilde{\Lambda}_n^1 - \rho_p h \omega^2} \frac{\langle f, \tilde{u}_m^{0\star} + \epsilon \tilde{u}_m^{1\star} \rangle}{a(\tilde{u}_m^0 + \epsilon \tilde{u}_m^1, \tilde{u}_m^{0\star} + \epsilon \tilde{u}_m^{1\star})} (\tilde{u}_m^0(x) + \epsilon \tilde{u}_m^1(x)).$$

We must note that the terms $\tilde{\Lambda}_n^1$ have a non-zero imaginary part and thus allow us to obtain a solution defined for any real frequency. The response of the plate is, of course, at its maximum when the excitation frequency is equal to the real part of the eigenfrequency.

The eigenvalues $\tilde{\Lambda}_n$ can be broken down into real and imaginary parts $\tilde{\Lambda}_n = \tilde{\Lambda}_n^0(1 + \epsilon/2 i_n(k_f) - \imath\epsilon/2 j_n(k_f))$, where $i_n(k_f) = \int_{-L}^{+L}\int_{-L}^{+L} \tilde{u}_n^0(x)$ $\mathrm{J}_0(k_f|x - x'|)\tilde{u}_n^0(x')dx\,dx'$ and $j_n(k_f) = \int_{-L}^{+L}\int_{-L}^{+L} \tilde{u}_n^0(x)\mathrm{J}_0(k_f|x - x'|)$ $\tilde{u}_n^0(x')dx\,dx'$. At low frequency, $k_f \ll 1$, the Bessel functions behave as $\mathrm{Y}_0(k_f x) \approx \frac{2}{\pi} \ln k_f x < -\eta$ and $\mathrm{J}_0(k_f x) \approx 1$. If we replace these approximations in the previous integrals, we obtain $i_n(k_f) \approx \left(\int_{-L}^{+L} \tilde{u}_n^0(x)dx\right)^2 > 0$ and

$j_n(k_f) < -\eta \left(\int_{-L}^{+L} \tilde{u}_n^0(x)dx\right)^2 < 0$. Therefore, we can see that, at low frequency, the eigenvalues have a negative imaginary part, which corresponds to damping by radiation. This result remains valid at high frequencies. The principle of energy conservation is satisfied. Moreover, we can see, by the shift of the real part of the eigenvalues toward the low frequencies, that an effect appears which is known as the "added mass" effect. This effect reflects the fact that the presence of the fluid increases the bulk density of the structure, which introduces a spectrum shift toward the bass. We must note this effect diminishes when the excitation frequency increases. Similarly, we can see that $\int_{-L}^{+L} \tilde{u}_n^0(x)dx = 1/\sqrt{L}\int_{-L}^{+L} \sin\left(n\pi(x+L)/(2L)\right)dx = 2\sqrt{L}/(n\pi)(1 - (-1)^n)$. Everything happens as if the even modes (or antisymmetric modes, as they have a node in the middle) were not influenced by the fluid for the approximation carried out here, since both the damping and added mass are zero. This result, even though it is the result of two approximations (light fluid and low frequency), describes the vibroacoustic behavior of simply supported plates quite well.

8.5.3.3. *Estimating the energy loss by radiation of a coupled plate*

When a plate is of finite dimensions, regardless of its shape, it has an infinite number of eigenmodes and above the critical frequency, the radiation efficiency of the plate is at its maximum since the power radiated by a mode is directly proportional to the radiation factor [LES 88]. The plate behaves as an infinite plate and all modes radiate approximately in the same way. An example of this behavior is presented in Figure 8.14 where we plot the evolution of the acoustic damping $\alpha_{ac}(\omega/\omega_c)$ as a function of the ratio ω/ω_c following the approximation proposed in [CHA 01]. For a baffled plate[6], we get:

$$\alpha_{ac}\left(\frac{\omega}{\omega_c}\right) = -\frac{\omega}{\omega_c}\frac{2\rho_0 c_0}{\rho_p h}\Im\left(\frac{0.0620\left(\imath\frac{\omega}{\omega_c}\right)+0.5950\left(\imath\frac{\omega}{\omega_c}\right)^2+1.0272\left(\imath\frac{\omega}{\omega_c}\right)^3}{1.1669+1.6574\left(\imath\frac{\omega}{\omega_c}\right)+1.5528\left(\imath\frac{\omega}{\omega_c}\right)^2+\left(\imath\frac{\omega}{\omega_c}\right)^3}\right),$$

where c_0 is the wave celerity in the fluid, ρ_0 is the density of the fluid, ρ_p is the density of the plate and h is its thickness. We recall that this damping α_{ac} is the imaginary part of the resonance frequencies of the structure. In this approximation, ω is the pulsation and ω_c is the critical pulsation, $\omega_c = c_0^2\sqrt{\frac{\rho_p h}{D}}$. For an aluminum plate with a thickness of 5 cm, for example, we get $f_c \approx 256$ Hz. We can show without difficulty that we get $\lim_{\omega\to\infty}\alpha_{ac}\left(\frac{\omega}{\omega_c}\right) = -1/2\rho_0 c_0/(\rho_p h)$. For this non-baffled aluminum plate with a thickness of 5 cm in air, we get $\alpha_{ac}(\omega/\omega_c) \approx -1.5$. This relation shows that above the critical frequency, the acoustic damping is fairly constant.

6 For a non-baffled plate, a first approximation consists of dividing the fluid effect by 4.

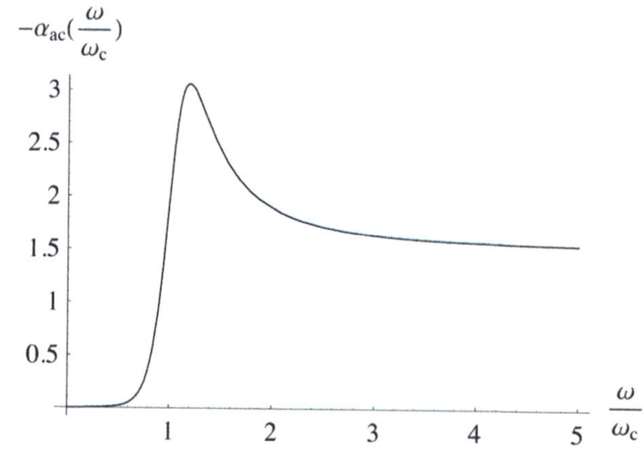

Figure 8.14. *Acoustic damping for a non-baffled 5 cm thick aluminumm plate in air*

8.5.4. *Higher order approximations*

The numerical efficiency of these analytic methods is such that we aim to use them even beyond their natural limits. Typically, we want to try to calculate the spectrum of a coupled plate in a high-density fluid such as water. Of course, in this case, using the perturbation methods in a rough way does not work at all and the numerical results reveal a divergence from the very first term. However, there exists a series resummation technique which gives remarkable results. We begin by examining its principle using a trivial example with our simply supported plate, when it is damped, not by radiation, but by viscosity, and only then we move on to its application in the notably more difficult case, i.e. when the plate is damped by radiation.

8.5.4.1. *Higher order perturbation method for a viscous plate*

In this section, we present several classic results [LAN 02, WOO 98] of the eigenmodes and eigenvalues for a plate damped by viscosity. $u(x, \omega)$ is always the normal displacement of the plate in a harmonic regime with a time dependence $\exp(-\imath\omega t)$. It is shown in section 3.4.2.3 that, in the temporal domain, the damping by viscosity can be written in the form of a differential operator given by $\eta_p \partial/\partial t$, where η_p is the damping constant. If we introduce the parameter $\epsilon_p = \eta_p/\rho_p h$, then $u(x, \omega)$ satisfies the damped Kirchhoff equation $D\Delta^2 u(x, \omega) - \rho_p h \omega^2 \left(1 - \epsilon_p \frac{-\imath}{\omega}\right) u(x, \omega) = F(x, \omega)$, given the boundary conditions $u(0, \omega) = 0$, $u''(0, \omega) = 0$, $(l_x, \omega) = 0$, $u''(L_x, \omega) = 0$.

As we saw for a viscous beam in section 3.4.2, it is easy to establish the exact expression of the eigenmodes and eigenvalues. Let us recall a broad outline of the steps. We set $\rho_p^{'} = \rho_p\left(1 - \epsilon_p \frac{-\imath}{\omega}\right)$. The eigenmodes $\tilde{u}_m(x,\omega)$ and the eigen pulsations $\tilde{\omega}_m(\omega)$ are the non-zero solutions of $D\Delta^2 \tilde{u}_m(x,\omega) - \rho_p^{'}(\omega) h \tilde{\omega}_m^2(\omega)\tilde{u}_m(x,\omega) = 0$ given the boundary conditions $\tilde{u}_m(0,\omega) = 0$, $\tilde{u}_m^{''}(0,\omega) = 0$, $\tilde{u}_m(l_x,\omega) = 0$, $\tilde{u}_m^{''}(L_x,\omega) = 0$. As a one-dimensional plate only differs from a beam in terms of coefficients, we thus obtain a result comparable to [3.71]:

$$\tilde{u}_m(x,\omega) = \sqrt{\frac{2}{L_x}} \sin \frac{m\pi}{L_x} x, \tilde{\omega}_m(\omega) = \left(\frac{m\pi}{L_x}\right)^2 \sqrt{\frac{D}{\rho_p h}} \frac{1}{\sqrt{1 + \epsilon_p \frac{\imath}{\omega}}}.$$

The eigen pulsations depend on the frequency. The eigenmodes do not depend on it; this result is absolutely classic and arises from the fact that the chosen viscous damping is very specific, since it is uniformly distributed throughout the whole material (and therefore does not depend on the geometry of the considered mode). The calculation of the resonance modes $\hat{u}_m(x)$ and the resonance pulsations $\hat{\omega}_m$ is trivial and, since the eigenmodes do not depend on the frequency, they coincide with the resonance modes. The resonance frequencies can be given by relation [3.72] in which the term E/ρ is replaced by its analogue for the plate, $D/\rho_p h$, and we obtain the two resonance pulsations of the mode m:

$$\hat{\omega}_m = -\frac{\imath\epsilon_p}{2} \pm \sqrt{\frac{D}{\rho_p h}\left(\frac{m\pi}{L_x}\right)^4 - \left(\frac{\epsilon_p}{2}\right)^2}.$$

We find the anti-Hermitian symmetry of the resonances with respect to the imaginary axis, which ensures the existence and uniqueness of the temporal solution.

8.5.4.1.1. Higher order perturbation method for eigenmodes and series resummation

At this point, it is very easy to introduce the calculations for a weak formulation of the problem, which is analogous to the one introduced for establishing the plate and beam operators, in particular for our one-dimensional problem. It is enough to revisit equation [6.35], setting:

$$\langle u, v\rangle = \int_0^{L_x} u(x,\omega)v(x,\omega)dx, a(u,v) = D\langle u_{,xx} v_{,xx}\rangle, \beta_\omega(u,v) = -\imath/\omega\langle u, v\rangle.$$

The calculation of eigenmodes can then be written to find $\tilde{u}_m(x,\omega)$ and $\tilde{\Lambda}_m(\omega)$, where $\tilde{\Lambda}_m = \rho_p h \tilde{\omega}_m^2(\omega)$ satisfies the boundary conditions, so that for each $v(x,\omega)$ satisfying the boundary conditions, the following equation is satisfied:

$$a(\tilde{u}_m, v) - \tilde{\Lambda}_m\left(\langle \tilde{u}_m, v\rangle - \epsilon_p \beta_\omega(\tilde{u}_m, v)\right) = 0. \qquad [8.13]$$

We introduce the Rayleigh–Schrödinger series expansion of eigenmodes $\tilde{u}_m$ and eigenvalues $\tilde{\Lambda}_m$ with respect to ϵ:

$$\begin{aligned}\tilde{u}_m(M) &= \tilde{u}_m^{(0)}(M) + \epsilon\tilde{u}_m^{(1)}(M) + \cdots + \epsilon^s\tilde{u}_m^{(s)}(M) + \cdots \tilde{\Lambda}_m \\ &= \tilde{\Lambda}_m^{(0)} + \epsilon\tilde{\Lambda}_m^{(1)} + \cdots + \epsilon^s\tilde{\Lambda}_m^{(s)} + \cdots\end{aligned}$$

We substitute these two expansions in equation [8.13], we collect and set to zero the coefficients of the same power with respect to ϵ:

$$\epsilon^0 : a(\tilde{u}_m^{(0)}, v) - \tilde{\Lambda}_m^{(0)}\langle\tilde{u}_m^{(0)}, v\rangle = 0, \quad [8.14]$$

$$\epsilon^1 : a(\tilde{u}_m^{(1)}, v) - \tilde{\Lambda}_m^{(0)}\langle\tilde{u}_m^{(1)}, v\rangle = \tilde{\Lambda}_m^{(1)}\langle\tilde{u}_m^{(0)}, v\rangle - \tilde{\Lambda}_m^{(0)}\beta\omega(\tilde{u}_m^{(0)}, v), \quad [8.15]$$

$$\begin{aligned}\epsilon^2 : a(\tilde{u}_m^{(2)}, v) - \tilde{\Lambda}_m^{(0)}\langle\tilde{u}_m^{(2)}, v\rangle &= \tilde{\Lambda}_m^{(2)}\langle\tilde{u}_m^{(0)}, v\rangle + \tilde{\Lambda}_m^{(1)}\langle\tilde{u}_m^{(1)}, v\rangle \\ &-\tilde{\Lambda}_m^{(0)}\beta_\omega(\tilde{u}_m^{(1)}, v) - \tilde{\Lambda}_m^{(1)}\beta_\omega(\tilde{u}_m^{(0)}, v),\end{aligned} \quad [8.16]$$

$$\vdots$$

$$\begin{aligned}\epsilon^s : a(\tilde{u}_m^{(s)}, v) - \tilde{\Lambda}_m^{(0)}\langle\tilde{u}_m^{(s)}, v\rangle &= \sum_{l=1}^{l=s}\tilde{\Lambda}_m^{(l)}\langle\tilde{u}_m^{(s-l)}, v\rangle \\ &- \sum_{l=0}^{l=s-1}\tilde{\Lambda}_m^{(l)}\beta_\omega(\tilde{u}_m^{(s-l-1)}, v)\end{aligned} \quad [8.17]$$

The solution at the order zero with respect to ϵ can be obtained by solving the eigenmode equation [8.14] which gives with $\tilde{\Lambda}_m^{(0)} = \rho_p h(\tilde{\omega}_m^{(0)})^2$:

$$\tilde{u}_m^{(0)}(x) = \sqrt{\frac{2}{L_x}}\sin\frac{m\pi}{L_x}x, \langle\tilde{u}_m^{(0)}, \tilde{u}_m^{(0)*}\rangle = 1, \tilde{\omega}_m^{(0)} = \left(\frac{m\pi}{L_x}\right)^2\sqrt{\frac{D}{\rho_p h}}.$$

The calculation of the solution at the order one with respect to ϵ can be obtained by solving equation [8.15]. In order for a solution to this equation to exist, its second member must be orthogonal to the considered mode. This leads to the condition (in the literature, it is referred to as the solvability condition): $\tilde{\Lambda}_m^{(1)}\langle\tilde{u}_m^{(0)}, \tilde{u}_m^{(0)*}\rangle - \tilde{\Lambda}_m^{(0)}\beta_\omega(\tilde{u}_m^{(0)}, \tilde{u}_m^{(0)*}) = 0$. As the eigenmodes $\tilde{u}_m^{(0)}(M)$ of the elastic plate *in vacuo* are normalized, it is easy to find that $\tilde{\Lambda}_m^{(1)}/\tilde{\Lambda}_m^{(0)} = \beta_\omega^{mm}$, where, to simplify writing, we have written $\beta_\omega^{mn} = \beta_\omega(\tilde{u}_m^{(0)}, \tilde{u}_n^{(0)*})$. As previously, we expand $\tilde{u}_m^{(1)}(M)$ in a series of modes at the order zero: $\tilde{u}_m^{(1)}(M) = \sum_n \alpha_m^{1n}\tilde{u}_n^{(0)}(M)$ which we introduce in equation [8.15] by taking $v(M) = \tilde{u}_j^{(0)*}(M)$, and we obtain:

$$\sum_n \alpha_m^{1n}\left(a(\tilde{u}_n^{(0)}, \tilde{u}_j^{(0)*}) - \tilde{\Lambda}_m^{(0)}\langle\tilde{u}_n^{(0)}, \tilde{u}_j^{(0)*}\rangle\right) = \tilde{\Lambda}_m^{(1)}\langle\tilde{u}_m^{(0)}, \tilde{u}_j^{(0)*}\rangle - \tilde{\Lambda}_m^{(0)}\beta_\omega^{mj}.$$

If $j \neq m$ and for simple eigenvalues, it is obvious that $\alpha_m^{1j} = \tilde{\Lambda}_m^{(0)} / \left(\tilde{\Lambda}_m^{(0)} - \tilde{\Lambda}_j^{(0)}\right) \beta_\omega^{mj}$. For $m = j$, α_m^{1m} is undetermined. As previously, we calculate it by imposing on $\tilde{u}_m^{(0)}(M) + \epsilon \tilde{u}_m^{(1)}(M)$ to be of norm unity. We then obtain $\alpha_m^{1m} = 0$. Now, $\beta_\omega^{mj} = \beta_\omega(\tilde{u}_m^{(0)}, \tilde{u}_j^{(0)*}) = -\imath/\omega\delta_m^j$. Thus, all the coefficients α_m^{1j} are zero and thus $\tilde{u}_m^{(1)}(M)$ is identically zero.

The calculation of the solution at the order two with respect to ϵ requires solving equation [8.16]. In order for a solution to this equation to exist, its second member must, once again, satisfy a solvability condition: $\tilde{\Lambda}_m^{(2)} \langle \tilde{u}_m^{(0)}, \tilde{u}_m^{(0)*} \rangle + \tilde{\Lambda}_m^{(1)} \langle \tilde{u}_m^{(1)}, \tilde{u}_m^{(0)*} \rangle - \tilde{\Lambda}_m^{(0)} \beta_\omega(\tilde{u}_m^{(1)}, \tilde{u}_m^{(0)*}) - \tilde{\Lambda}_m^{(1)} \beta_\omega(\tilde{u}_m^{(0)}, \tilde{u}_m^{(0)*}) = 0$. The modes *in vacuo* are normalized, it is then evident that:

$$\tilde{\Lambda}_m^{(2)} = \tilde{\Lambda}_m^{(0)} \beta_\omega^{mm} (\tilde{u}_m^{(1)}, \tilde{u}_m^{(0)*}) + \tilde{\Lambda}_m^{(1)} \left[\beta_\omega(\tilde{u}_m^{(0)}, \tilde{u}_m^{(0)*}) - \langle \tilde{u}_m^{(1)}, \tilde{u}_m^{(0)*} \rangle\right].$$

Introducing the results of the expansion at the order one in the previous equality leads to:

$$\tilde{\Lambda}_m^{(2)} = \tilde{\Lambda}_m^{(0)} \sum_n \alpha_m^{1n} \beta_\omega^{mn} + \tilde{\Lambda}_m^{(0)} \beta_\omega^{mm} \left[\beta_\omega^{mm} - \underbrace{\sum_n \alpha_m^{1n} \overbrace{\langle \tilde{u}_n^{(0)}, \tilde{u}_m^{(0)*} \rangle}^{\delta_m^n}}_{=\alpha_m^{1m}=0}\right].$$

From where, finally, $\tilde{\Lambda}_m^{(2)} / \tilde{\Lambda}_m^{(0)} = (\beta_\omega^{mm})^2$. Once again, we expand $\tilde{u}_m^{(2)}(M)$ in a series of modes at the order zero: $\tilde{u}_m^{(2)}(M) = \sum_n \alpha_m^{2n} \tilde{u}_n^{(0)}(M)$. If we introduce this expansion in equation [8.16] with $v(M) = \tilde{u}_j^{(0)*}(M)$, after calculations for $m \neq j$, we obtain:

$$\alpha_m^{2j} = \frac{\tilde{\Lambda}_m^{(0)}}{\tilde{\Lambda}_m^{(0)} - \tilde{\Lambda}_j^{(0)}} \left[\sum_{i \neq m} \alpha_m^{1i} \beta_\omega^{mi} + \beta_\omega^{mm} \left(\beta_\omega^{mj} - \alpha_m^{1j}\right)\right] = 0.$$

The term α_m^{2m} is still undetermined, we calculate it by once again imposing the mode normalization condition (this condition will be satisfied at ϵ^3):

$$\begin{aligned}
\langle \tilde{u}_m, \tilde{u}_m * \rangle &= \langle \tilde{u}_m^{(0)} + \epsilon \tilde{u}_m^{(1)} + \epsilon^2 \tilde{u}_m^{(2)}, \tilde{u}_m^{(0)*} + \epsilon \tilde{u}_m^{(1)*} + \epsilon^2 \tilde{u}_m^{(2)*} \rangle \\
&= \langle \tilde{u}_m^{(0)}, \tilde{u}_m^{(0)*} \rangle + 2\epsilon \langle \tilde{u}_m^{(0)}, \tilde{u}_m^{(1)*} \rangle + \epsilon^2 \left(2 \langle \tilde{u}_m^{(0)}, \tilde{u}_m^{(2)*} \rangle + \langle \tilde{u}_m^{(1)}, \tilde{u}_m^{(1)*} \rangle\right) \\
&= 1 + 0 \times \epsilon + 0 \times \epsilon^2.
\end{aligned}$$

Thus, that is $\langle\tilde{u}_m^{(0)}, \tilde{u}_m^{(0)*}\rangle = 1$, $\langle\tilde{u}_m^{(0)}, \tilde{u}_m^{(1)*}\rangle = 0$ and $2\langle\tilde{u}_m^{(0)}, \tilde{u}_m^{(2)*}\rangle + \langle\tilde{u}_m^{(1)}, \tilde{u}_m^{(1)*}\rangle = 0$. The first two relations are already satisfied and the third relation leads to the result $\alpha_m^{2m} = -\frac{1}{2}\sum_p \left(\alpha_m^{1p}\right)^2 = 0$. Thus, $\tilde{u}_m^{(2)}(M)$ is identically zero.

We can similarly show that, at any order i, we get $\tilde{\Lambda}_m^{(i)} = \tilde{\Lambda}_m^{(0)}\left(\beta_\omega^{mm}\right)^i$ and $\tilde{u}_m^{(i)}(M) = 0$. Therefore, the Rayleigh–Schrödinger expansion of eigenvalues can be written as:

$$\tilde{\Lambda}_m(\omega) = \tilde{\Lambda}_m^{(0)}\left(1 + \epsilon\beta_\omega^{mm} + \epsilon^2\left(\beta_\omega^{mm}\right)^2 + \cdots + \epsilon^i\left(\beta_\omega^{mm}\right)^i + \cdots\right). \quad [8.18]$$

We recognize in equation [8.18] the expansion around the origin of the function $1/(1-x)$ for $x = \epsilon\beta_\omega^{mm}$. By identification, we obtain:

$$\tilde{\Lambda}_m(\omega) = \tilde{\Lambda}_m^{(0)}\frac{1}{1-\epsilon\beta_\omega^{mm}} = \tilde{\Lambda}_m^{(0)}\frac{1}{1+\frac{\imath\epsilon_p}{\omega}},$$

and finally, we obtain a result comparable to the one obtained in section 8.1.3:

$$\tilde{u}_m(x,\omega) = \tilde{u}_m^{(0)}(x), \tilde{\omega}_m(\omega) = \tilde{\omega}_m^{(0)}\frac{1}{\sqrt{1+\frac{\imath\epsilon_p}{\omega}}},$$

which shows that the Rayleigh–Schrödinger method coupled to a series resummation in this case allows us to find the exact solution. We now apply this technique to the case where a plate is coupled to a fluid.

8.5.4.2. *Higher order perturbation method for a finite coupled plate*

We start again with the system [8.5], [8.6] and [8.7] which characterizes the dynamic behavior of a coupled baffled plate. We assume that this plate has finite dimensions and mechanical boundary conditions at the edge. For this purpose, it is not necessary to clarify here the type of conditions imposed on the plate, it is only important to know that the plate has finite dimensions. In a harmonic regime, using Green's representation of pressure, we can easily show that solving this problem requires solving the integro-differential equation:

$$\begin{aligned} D\Delta^2 U(M,\omega) - \rho_p h\omega^2\left(U(M,\omega) - \epsilon\int_\Sigma U(M',\omega)\mathcal{G}(M-M',\omega)dM'\right) \\ = F(M,\omega) \end{aligned} \quad [8.19]$$

with the boundary conditions, in which $\epsilon = 2\rho_0/(\rho_p h)$ and $\mathcal{G}(M-M',\omega)$ is Green's kernel of the Neumann problem for the Helmholtz equation written for the points M and M' on the plate. As previously, this system can be written in the weak form equivalent to that given by equation [8.11]. $\langle U, V\rangle = \int_\Sigma U(M)V^*(M)dM$ is the usual scalar product. We aim to find a $U(M,\omega)$ that would satisfy the boundary conditions so that, for each $V(M,\omega)$ satisfying the boundary conditions, the following equation would be verified:

$$a(U,V) - \Lambda\left(\langle U,V\rangle - \epsilon\beta_\omega(U,V)\right) = \langle F,V\rangle, \qquad [8.20]$$

where $\Lambda = \rho_p h\omega^2$, $a(U,V)$ corresponds to the elastic (or potential) deformation energy of the structure. $\rho_0\omega^2\beta_\omega(U,V) = \rho_0\omega^2 \int_\Sigma\int_\Sigma U(M)\mathcal{G}(M - M',\omega) V^*(M')dMdM'$ is analogous to the radiation impedance and characterizes the loss of energy and the mass added by radiation. $\rho_p h\omega^2\langle U,V\rangle$ is the kinetic energy. Since in this case $\mathcal{G}(M - M',\omega)$ depends on frequency, $\beta_\omega(U,V)$ obviously depends on frequency and we must distinguish eigenmodes from resonance modes.

We examine the Rayleigh–Schrödinger series expansion of eigenmodes $\tilde{u}_m$ and eigenvalues $\tilde{\Lambda}_m$ with respect to ϵ:

$$\tilde{u}_m(M) = \tilde{u}_m^{(0)}(M) + \epsilon\tilde{u}_m^{(1)}(M) + \cdots + \epsilon^s\tilde{u}_m^{(s)}(M) + \cdots$$
$$\tilde{\Lambda}_m = \tilde{\Lambda}_m^{(0)} + \epsilon\tilde{\Lambda}_m^{(1)} + \cdots + \epsilon^s\tilde{\Lambda}_m^{(s)} + \cdots$$

As previously, after introducing this in the weak formulation [8.20] of these two expansions, we obtain, collect and set to zero the diverse powers of ϵ, and we find:

$$\begin{aligned}
&\epsilon^0 : a(\tilde{u}_m^{(0)}, v) - \tilde{\Lambda}_m^{(0)}\langle\tilde{u}_m^{(0)}, v\rangle = 0,\\
&\epsilon^1 : a(\tilde{u}_m^{(1)}, v) - \tilde{\Lambda}_m^{(0)}\langle\tilde{u}_m^{(1)}, v\rangle = \tilde{\Lambda}_m^{(1)}\langle\tilde{u}_m^{(0)}, v\rangle - \tilde{\Lambda}_m^{(0)}\beta\omega(\tilde{u}_m^{(0)}, v),\\
&\epsilon^2 : a(\tilde{u}_m^{(2)}, v) - \tilde{\Lambda}_m^{(0)}\langle\tilde{u}_m^{(2)}, v\rangle = \tilde{\Lambda}_m^{(2)}\langle\tilde{u}_m^{(0)}, v\rangle + \tilde{\Lambda}_m^{(1)}\langle\tilde{u}_m^{(1)}, v\rangle\\
&-\tilde{\Lambda}_m^{(0)}\beta_\omega(\tilde{u}_m^{(1)}, v) - \tilde{\Lambda}_m^{(1)}\beta_\omega(\tilde{u}_m^{(0)}, v), \qquad [8.21]\\
&\vdots\\
&\epsilon^s : a(\tilde{u}_m^{(s)}, v) - \tilde{\Lambda}_m^{(0)}\langle\tilde{u}_m^{(s)}, v\rangle = \sum_{l=1}^{l=s}\tilde{\Lambda}_m^{(l)}\langle\tilde{u}_m^{(s-l)}, v\rangle - \sum_{l=0}^{l=s-1}\tilde{\Lambda}_m^{(l)}\beta_\omega(\tilde{u}_m^{(s-l-1)}, v),
\end{aligned}$$

What differentiates this system from the system consisting of equations [8.14], [8.15], [8.16] and [8.17] is only the expression of $\beta_\omega(u,v)$. Similarly as for a viscous plate, the first equation [8.21] leads to eigenmodes and eigenvalues. For the following, it requires the second members to be orthogonal to the mode $\tilde{u}_m$, and requires solving the compatibility (or solvability) conditions:

$$\sum_{l=1}^{l=s} \tilde{\Lambda}_m^{(l)} \langle \tilde{u}_m^{(s-l)}, \tilde{u}_m^* \rangle - \sum_{l=0}^{l=s-1} \tilde{\Lambda}_m^{(l)} \beta_\omega (\tilde{u}_m^{(s-l-1)}, \tilde{u}_m^*) = 0. \quad [8.22]$$

To simplify writing, we note that $\beta_\omega^{mn} = \beta_\omega(\tilde{u}_m, \tilde{u}_n^*)$. Solving these compatibility conditions leads to the expansion of eigenvalues at the order one with respect to $\tilde{\Lambda}_m^{(1)}/\tilde{\Lambda}_m^{(0)} = \beta_\omega^{mm}$, at the order two with respect to $\tilde{\Lambda}_m^{(2)}/\tilde{\Lambda}_m^{(0)} = (\beta_\omega^{mm})^2 + \sum_{p\neq m} \alpha_m^{1p} \beta_\omega^{mp}$ and, finally, at the order three with respect to $\tilde{\Lambda}_m^{(3)}/\tilde{\Lambda}_m^{(0)} = (\beta_\omega^{mm})^3 + 2\beta_\omega^{mm} \sum_{p\neq m} \alpha_m^{1p} \beta_\omega^{mp} + \sum_{p\neq m} \alpha_m^{2p} \beta_\omega^{mp}$. Here, once again, we expand each term of the Rayleigh–Schrödinger series of eigenmodes in a series of modes of the order zero:

$$\tilde{u}_m^{(s)}(M) = \sum_n \alpha_m^{sn} \tilde{u}_n^{(0)}(M). \quad [8.23]$$

We obtain at the order 1:

$$\alpha_m^{1j} = \frac{\tilde{\Lambda}_m^{(0)}}{\tilde{\Lambda}_m^{(0)} - \tilde{\Lambda}_j^{(0)}} \beta_\omega^{mj}, m \neq j \text{ and } \alpha_m^{1m} = 0,$$

at the order 2:

$$\alpha_m^{2j} = \frac{\tilde{\Lambda}_m^{(0)}}{\tilde{\Lambda}_m^{(0)} - \tilde{\Lambda}_j^{(0)}} \left[\sum_{i\neq m} \alpha_m^{1i} \beta_\omega^{mi} + \beta_\omega^{mm} \left(\beta_\omega^{mj} - \alpha_m^{1j} \right) \right], m \neq j$$

$$\text{and } \alpha_m^{2m} = -\frac{1}{2} \sum_p \left(\alpha_m^{1p} \right)^2,$$

and at the order 3:

$$\alpha_m^{3j} = \frac{\tilde{\Lambda}_m^{(0)}}{\tilde{\Lambda}_m^{(0)} - \tilde{\Lambda}_j^{(0)}} \left[\sum_{i\neq m} \alpha_m^{2i} \beta_\omega^{mi} \right.$$

$$+ (\beta_\omega^{mm})^2 \left(\beta_\omega^{mj} - \alpha_m^{1j} \right) + \left(\sum_{i\neq m} \alpha_m^{1i} \beta_\omega^{mi} \right) \left(\beta_\omega^{mj} - \alpha_m^{1j} \right)$$

$$\left. + \beta_\omega^{mm} \left(\sum_{i\neq m} \alpha_m^{1i} \beta_\omega^{mi} + \alpha_m^{2m} - \alpha_m^{2j} \right) \right], m \neq j \text{ and } \alpha_m^{3m} = - \sum_p \alpha_m^{1p} \alpha_m^{2p}.$$

Now, we calculate the expansions of eigenvalues up to the order 3:

$$\frac{\tilde{\Lambda}_m}{\tilde{\Lambda}_m^{(0)}} = \sum_{l=0}^{l=3} \epsilon^l \frac{\tilde{\Lambda}_m^{(l)}}{\tilde{\Lambda}_m^{(0)}} + \cdots .$$

After rearranging the terms, we arrive at:

$$\begin{aligned}\frac{\tilde{\Lambda}_m}{\tilde{\Lambda}_m^{(0)}} &= \left[1 + \epsilon\beta_\omega^{mm} + \epsilon^2(\beta_\omega^{mm})^2 + \epsilon^3(\beta_\omega^{mm})^3 + \epsilon^4(\beta_\omega^{mm})^4 + \epsilon^5(\beta_\omega^{mm})^5 + \cdots\right] \\ &+\epsilon^2 \sum_{p\neq m} \alpha_m^{1p}\beta_\omega^{mp}\left[1 + 2\epsilon\beta_\omega^{mm} + 3\epsilon^2(\beta_\omega^{mm})^2 + 4\epsilon^3(\beta_\omega^{mm})^3 + \cdots\right] \\ &+\epsilon^3 \sum_{p\neq m} \alpha_m^{2p}\beta_\omega^{mp}\left[1 + 2\epsilon\beta_\omega^{mm} + 3\epsilon^2(\beta_\omega^{mm})^2 + \cdots\right] + \cdots \end{aligned} \quad [8.24]$$

For $\epsilon\beta_\omega^{mm} < 1$, we can recognize in each square bracket in [8.24] the expansion of $1/(1-\epsilon\beta_\omega^{mm})$ and its successive powers. By identification of the expansion with the original functions, we obtain:

$$\frac{\tilde{\Lambda}_m}{\tilde{\Lambda}_m^{(0)}} = \frac{1}{(1-\epsilon\beta_\omega^{mm})} + \epsilon^2\frac{\sum_{p\neq m}\alpha_m^{1p}\beta_\omega^{mp}}{(1-\epsilon\beta_\omega^{mm})^2} + \epsilon^3\frac{\sum_{p\neq m}\alpha_m^{2p}\beta_\omega^{mp}}{(1-\epsilon\beta_\omega^{mm})^2} + \mathcal{O}(\epsilon^4). \quad [8.25]$$

In practice, if we can limit our focus solely on the first term, we can obtain an approximate expression of the eignevalues, referred to as Warburton's approximation [FIL 08]:

$$\tilde{\Lambda}_m \approx \frac{1}{(1-\epsilon\beta_\omega^{mm})}\tilde{\Lambda}_m^{(0)}, \quad [8.26]$$

which is particularly precise for relatively high values of the ϵ parameter [MAT 07]. We recall that, since $\tilde{\Lambda}_m = \rho_p h\tilde{\omega}_m^2$, the eigenvalue equation [8.26] can be written in the form of $\tilde{\omega}_m^2(1-\epsilon\beta_\omega^{mm}) - \left(\tilde{\omega}_m^{(0)}\right)^2 = 0$. We must note that the pulsations $\tilde{\omega}_m(\omega)$, which satisfy this equation, are eigen pulsations and that, by definition, the resonance pulsations are the pulsations $\hat{\omega}_m$ which satisfy the relation $\hat{\omega}_m^2(1-\epsilon\beta_{\hat{\omega}_m}^{mm}) - \left(\tilde{\omega}_m^{(0)}\right)^2 = 0$. Many numerical experiments carried out on plates have shown that the spectrum of a metal plate in water can be estimated with a precision comparable to that obtained by solving the exact equations [FIL 08, MAT 07]. This is a particularly useful result to know, since, as we will see in the next section, it allows us to fairly precisely estimate the eigenvalues of a structure coupled to a fluid with any density at a low computational cost, which is barely higher than the one required for the calculations *in vacuo*.

8.6. Heavy fluid coupling: resonance estimation

As shown by Sanchez [SAN 89], in a harmonic regime, the coupling with a low-density fluid reveals two relatively distinct families of frequencies in the spectrum. These frequencies are all complex with negative imaginary parts, if the time dependence belongs to the $\exp(-\imath\omega t)$ type, to ensure an amplitude which decreases over time. The first family corresponds to the eigenfrequencies of the structure *in vacuo* (real in the absence of coupling for purely elastic structures), whose real part is slightly reduced (due to the added mass effect) and whose imaginary part (generally a small fraction of the real part[7]) corresponds to the loss of energy by radiation. The second set corresponds to the diffraction spectrum of the Helmholtz equation for the rigid body whose geometry is that of a vibrating structure; this spectrum has a discrete set of highly damped frequencies (whose imaginary part is very important) and whose trace is made invisible to the response of structures in light fluid coupling. When a structure is coupled to a high-density fluid, as we will see further on, the spectral families can mix, making the calculations and interpretation of results particularly difficult. We begin by presenting a numerical method for solving exact equations, in which, after transforming the starting differential system of integral equations at the edge, we solve these integral equations using a collocation-Galerkin method in which the unknowns of the problem are expanded in a series of orthogonal polynomials. We will see that calculating the determinant of the resulting linear system allows us to estimate the spectrum of a plate coupled with a heavy fluid and highlights the mix of spectral families. To better understand their origin, we will carry out an analytic study of resonances based on the Warburton approximation [8.26]. We will show, using a theorem which is based on the location of zeros of entire functions, that the coupling of a vibrating structure with a fluid with a comparable density radically changes the nature of its spectrum, making the use of modal methods difficult.

8.6.1. *Clamped rectangular plate coupled with a heavy fluid*

8.6.1.1. *Numerical solution of exact equations*

We consider a rectangular flat steel plate in contact with water. The plate is composed of a purely elastic material with a Young's modulus of $E = 200$ GPa, a Poisson's ratio of $\nu = 0.3$ and a density of $\rho_p = 7,800$ kg/m^3 . The thickness of the plate h (the thickness, which is an important parameter here, will be specified later) occupies the domain Σ which is defined in Cartesian coordinates (O, x, y, z) by $-L_x/2 < x < +L_x/2, -L_y/2 < y < +L_y/2, z = 0$ with $L_x = 0.7$ m and $L_y = 1$ m. $\partial\Sigma$ is the boundary of Σ with an outer normal. $\Gamma_i, i = 1.4$ are the four segments that make up the edge of the plate $\partial\Sigma = \bigcup_{i=1}^{4} \Gamma_i$ (see Figure 8.15). Σ' is

7 The numerical value of this imaginary part generally varies between 5% for the first modes and up to less than 0.1% of the real part for the higher order modes.

the complement to Σ in the $z = 0$ plane. $D = Eh^3/\left(12(1-\nu^2)\right)$ is the bending stiffness of the plate. The plate is clamped on all four edges along $\partial\Sigma$ and prolonged by an infinitely rigid flat screen, which occupies Σ'. The half-spaces $\Omega^{\pm} = \{x, y, z \gtrless 0\}$ contain water, with a density of $\rho_0 = 1,000$ Kg/m^3 and an acoustic wave celerity of $c_0 = 1,500$ m/s .

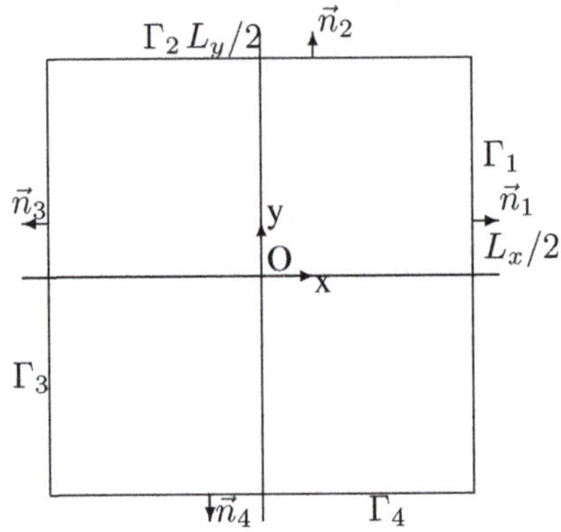

Figure 8.15. *Geometry of the problem in the* $(O, x, y, z = 0)$ *plane*

Here, we generalize the result of Green's representation of the plate *in vacuo*, given by equation [7.26]. Using Green's representation, the system of differential equations with partial derivatives is transformed into a system of coupled integral equations. The unknowns of the problem, the displacement of the plate, the acoustic pressure at the surface of the plate and the boundary sources are expanded in a series of Chebyshev polynomials of the first kind. These polynomials have important properties: an explicit formulation in terms of trigonometric functions, which enables easy calculations of the zeros, a discrete orthogonality relation (useful for digital processing) and a stable recurrence relation, which allows us to calculate higher degree polynomials without accumulating rounding errors. Moreover, just as any polynomial approximation, they are well suited for approximating functions on domains with a separable geometry.

The plate is excited by a harmonic unitary point force $F(x, y)$. Let $M = (x, y)$ be a point of the plate and $Q = (x, y, z)$ be a point of the half-space Ω^+. The displacement of the plate is denoted as $\mathcal{U}(M, t) = \Re\left(u(M)e^{-\imath\omega t}\right)$. $\mathcal{P}(Q, t) = \Re\left(p(Q)e^{-\imath\omega t}\right)$ is the instantaneous acoustic pressure. The functions $u(M)$ and $p(Q)$ satisfy the following system of differential equations:

$$\left(\Delta^2 - \frac{\rho_p h\omega^2}{D}\right) u(M) = \frac{1}{D}\left(F(M) - P(M)\right),$$

$$M \in \Sigma \text{ where } P(M) = \lim_{Q \to M \in \Sigma} \left(p^+(Q) - p^-(Q)\right)$$

$$\left(\Delta + \frac{\omega}{c_0}^2\right) p^{\pm}(Q) = 0, Q \in \Omega^{\pm}, \text{with } \frac{\partial p^{\pm}(M)}{\partial \vec{n}} = \begin{cases} \omega^2 \rho_0 u(M), M \in \Sigma \\ 0, M \in \Sigma' \end{cases}$$

To these equations, we must add the boundary conditions $u(S) = \partial u(S)/\partial \vec{n} = 0, S \in \partial\Sigma$ as well as a principle of conservation of energy such as the Sommerfeld conditions or the principle of limit absorption. Let $k_p = \sqrt[4]{\rho_p h\omega^2/D}$ be the wave number of the infinite plate *in vacuo* and $k_f = \omega/c_0$ be the acoustic wave number. The wavelengths in the plate *in vacuo* and in the fluid are given by $\lambda_{p,f} = 2\pi/k_{p,f}$. The latter equation represents the continuity of normal accelerations of the plate (and the rigid baffle) and the fluid in contact with it.

Green's kernel $\mathcal{G}^{\pm}$ of the Helmholtz equation for the half-space $\Omega^{\pm}$, which satisfies the Neumann condition on the $z = 0$ plane, as well as the condition of conservation of energy at infinity, is always:

$$\mathcal{G}^{\pm}\left(Q, Q'\right) = -\frac{e^{\imath k_f r(Q,Q')}}{4\pi r\left(Q, Q'\right)} - \frac{e^{\imath k_f r(Q,Q'')}}{4\pi r\left(Q, Q''\right)},$$

where Q and Q' are the points of $\Omega^{\pm}$. Q'' is the image of Q' with respect to the $z = 0$ plane. $r\left(Q, Q'\right)$ is the distance between the points Q and Q'. In this case, of vibrating structures without sources within the fluid, Q'' always coincides with Q' at $z = 0$. The Green representation of acoustic pressure is, therefore:

$$p^{\pm}(Q) = - \pm \omega^2 \rho_0 \int_{\Sigma} u(M) \frac{e^{\imath k_f r(Q,M)}}{2\pi r(Q, M)} dM. \qquad [8.27]$$

Now, we consider G_p the Green kernel of the plate *in vacuo*, which satisfies the limit absorption principle, which as we know is given by:

$$G_p\left(M, M'\right) = \frac{\imath}{8k_p^2 D}\left[\mathrm{H}_0\left(k_p r\left(M, M'\right)\right) + \frac{2\imath}{\pi}\mathrm{K}_0\left(k_p r\left(M, M'\right)\right)\right],$$

where $\mathrm{H}_0(z)$ is the Hankel function of the first kind at the order zero and $\mathrm{K}_0(z)$ is the modified Bessel function of the first kind at the order zero. Using Green's representation, the displacement of the plate can be written as:

$$u(M) = \int_{\Sigma} F(M') \, G_p(M, M') \, dM' - \int_{\Sigma} P(M') \, G_p(r(M, M')) \, dM' + \int_{\partial\Sigma} \sigma_0(S) \, G_p(r(M, S)) \, dS + \int_{\partial\Sigma} \sigma_1(S) \frac{\partial G_p(r(M, S))}{\partial \vec{n_S}} dS, \quad [8.28]$$

where $\partial \vec{n_{S'}}$ is the normal derivative at the point S'. In equation [8.28], the two simple integrals are layer potentials introduced to account for the boundary conditions. The layer potentials introduce new unknowns, which are the edge sources $\sigma_0(S)$ and $\sigma_1(S)$, $S \in \partial\Sigma$.

When the exciting force is a Dirac momentum located at (x_0, y_0), the equations [8.27] and [8.28] and the boundary conditions result in the following system:

$$\begin{aligned}
u(x, y) = {} & G_p(x - x_0, y - y_0) - \int_{-\frac{L_x}{2}}^{\frac{L_x}{2}} \int_{-\frac{L_y}{2}}^{\frac{L_y}{2}} P(x', y') \, G_p(x - x', y - y') \, dx'dy' \\
& + \int_{-\frac{L_y}{2}}^{\frac{L_y}{2}} \left[\sigma_0^1(y') \, G_p(x - x', y - y') + \sigma_1^1(y') \frac{\partial G_p(x - x', y - y')}{\partial x'} \right]_{x' = \frac{L_x}{2}} dy' \\
& + \int_{-\frac{L_x}{2}}^{\frac{L_x}{2}} \left[\sigma_0^2(x') \, G_p(x - x', y - y') + \sigma_1^2(x') \frac{\partial G_p(x - x', y - y')}{\partial y'} \right]_{y' = \frac{L_y}{2}} dx' \\
& + \int_{-\frac{L_y}{2}}^{\frac{L_y}{2}} \left[\sigma_0^3(y') \, G_p(x + x', y - y') + \sigma_1^3(y') \frac{\partial G_p(x - x', y - y')}{\partial x'} \right]_{x' = -\frac{L_x}{2}} dy' \\
& + \int_{-\frac{L_x}{2}}^{\frac{L_x}{2}} \left[\sigma_0^4(x') \, G_p(x - x', y + y') + \sigma_1^4(x') \frac{\partial G_p(x - x', y - y')}{\partial y'} \right]_{y' = -\frac{L_y}{2}} dx',
\end{aligned}$$

[8.29]

$$P(x,y) = -2\omega^2 \rho_0 \int_{-\frac{L_x}{2}}^{\frac{L_x}{2}} \int_{-\frac{L_y}{2}}^{\frac{L_y}{2}} u(x', y') \frac{e^{\imath k_f \sqrt{(x-x')^2+(y-y')^2}}}{2\pi\sqrt{(x-x')^2+(y-y')^2}} dx'dy' \qquad [8.30]$$

to which we add the boundary conditions:

$$u\left(x, \pm\frac{L_y}{2}\right) = 0, \left[\frac{\partial u(x,y)}{\partial y}\right]_{y=\pm\frac{L_y}{2}} = 0 \ \ \forall x \in \left[-\frac{L_x}{2}, \frac{L_x}{2}\right] \qquad [8.31]$$

$$u\left(\pm\frac{L_x}{2}, y\right) = 0, \left[\frac{\partial u(x,y)}{\partial x}\right]_{x=\pm\frac{L_x}{2}} = 0 \ \ \forall y \in \left[-\frac{L_y}{2}, \frac{L_y}{2}\right] \qquad [8.32]$$

The two coupled integral equations [8.29] and [8.30] and the eight boundary conditions [8.31]–[8.32] result in a system consisting of 10 coupled integral equations with 10 unknowns (pressure at the surface of the plate, displacement and eight edge sources). The only way to calculate these unknowns is to find their numerical approximations.

In this section, we present only the general aspects of the numerical solution of the integral equations. For a more detailed analysis, see [MAT 96]. The principle of spectral approximation using Chebyshev polynomials is to expand the unknowns in a truncated series of Chebyshev polynomials [ABR 65, BER 91]. The new unknowns then become the coefficients of the truncated series. When we calculate the unknowns using the Chebyshev collocation points, we can show that the method is equivalent to a Galerkin method [MAT 96]. This allows us to combine the efficiency of the collocation method with the precision of the Galerkin method. We can show that a minimum of π points per wavelength is necessary. In practice, imposing four points per wavelength ensures precision of less than a decibel. Moreover, the simple analytic expressions which we use to describe the Chebyshev polynomials and their zeros lead to simple numerical expansions. For example, the integrable function $f(x)$, $L^2_{[-1,1]}$, defined in $[-1,1]$, can be expanded in a series of Chebyshev polynomials $f(x) = \sum_{n=1}^{\infty} c_n f_n T_{n-1}(x)$, where $c_1 = 1/2$ and $c_n = 1, n \geq 2$. The series coefficients are given by the scalar product, weighted in $L^2_{[-1,1]}$, between the function and the corresponding polynomial $f_n = \langle f, T_n \rangle_{L^2_{[-1,1]}} = 2/\pi \int_{-1}^{+1} f(x)$ $T_{n-1}(x)/\sqrt{1-x^2}dx$. Thus, an approximation of f can be given by $f^N(x) = \sum_{n=1}^{N} c_n f_n T_{n-1}(x)$. This is $x_k = \cos((2k-1)\pi/2N)$, the N zeros of the Chebyshev polynomial of degree N. For a given N, the set $E_T^N = \{x_k, k \in [1,N]\}$ is the Chebyshev support. It is well known that these polynomials satisfy a discrete orthogonality relation. This relation, equivalent to the continuous relation defined in $L^2_{[-1,1]}$, is given by:

$$\langle T_i, T_j \rangle_{E_T^N} = \sum_{k=1}^{N} T_i(x_k) T_j(x_k) = \begin{cases} 0 & i \neq j \\ N/2 & i = j \neq 0 \\ N & i = j = 0 \end{cases} .$$

With this relation, it is easy to obtain the approximation [PRE 92] $\tilde{f}^N(x) = \sum_{n=1}^{N} c_n \tilde{f}_n T_{n-1}(x)$, where the coefficients are given by the scalar product $\tilde{f}_n = 2/N \langle f, T_n \rangle_{E_T^N} = 2/N \sum_{k=1}^{N} f(x_k) T_{n-1}(x_k)$. In the two-dimensional case, a function $f(x, y)$ defined in $[-1, 1] \times [-1, 1]$ is approximated by:

$$\tilde{f}^{N_x N_y}(x, y) = \sum_{n_x=1}^{N_x} \sum_{n_y=1}^{N_y} c_{n_x} c_{n_y} \tilde{f}_{n_x n_y} T_{n_x-1}(x) T_{n_y-1}(y),$$

where, if $x_j \in E_T^{N_x}$ and $y_k \in E_T^{N_y}$, the coefficients of the series $\tilde{f}_{n_x n_y}$ are given by:

$$\tilde{f}_{n_x n_y} = \frac{4}{N_x N_y} \sum_{j=1}^{N_x} \sum_{k=1}^{N_y} f(x_j, y_k) T_{n_x-1}(x_j) T_{n_y-1}(y_k) .$$

The displacement of the plate, the pressure at the surface and the edge sources are the solutions of equations [8.30], [8.29], [8.31] and [8.32], which can be approximated using:

$$u(x, y) \approx u^{N_x N_y}(x, y) = \sum_{n_x=1}^{N_x} \sum_{n_y=1}^{N_y} c_{n_x} c_{n_y} u_{n_x n_y} T_{n_x-1}\left(\frac{2x}{L_x}\right) T_{n_y-1}\left(\frac{y}{\frac{L_y}{2}}\right)$$

$$P(x, y) \approx P^{M_x M_y}(x, y) = \sum_{m_x=1}^{M_x} \sum_{m_y=1}^{M_y} c_{m_x} c_{m_y} p_{m_x m_y} T_{m_x-1}\left(\frac{2x}{L_x}\right) T_{m_y-1}\left(\frac{2y}{L_y}\right)$$

$$\sigma_i^j(y) \approx \sigma_i^{jS_y}(y) = \pm \sum_{s_y=1}^{S_y} c_{s_y} \sigma_{is_y}^j T_{s_y-1}\left(\frac{2y}{L_y}\right) \begin{array}{l} + \text{ if } (i, j) = (0, 1), (0, 3), (1, 1) \\ - \text{ if } (i, j) = (1, 3) \end{array}$$

$$\sigma_i^j(x) \approx \sigma_i^{jS_x}(x) = \pm \sum_{s_x=1}^{S_x} c_{s_x} \sigma_{is_x}^j T_{s_x-1}\left(\frac{2x}{L_x}\right) \begin{array}{l} + \text{ if } (i, j) = (0, 2), (0, 4), (1, 2) \\ - \text{ if } (i, j) = (1, 4) \end{array}$$

By introducing these expressions in [8.30], [8.29], [8.31] and [8.32], we obtain:

$$\sum_{m_x=1}^{M_x}\sum_{m_y=1}^{M_y} c_{m_x}c_{m_y}p_{m_xm_y}T_{m_x-1}\left(\frac{2x}{L_x}\right)T_{m_y-1}\left(\frac{2y}{L_y}\right)+2\omega^2\rho_0$$
$$\times\sum_{n_x=1}^{N_x}\sum_{n_y=1}^{N_y} c_{n_x}c_{n_y}u_{n_xn_y}I_{n_xn_y}(x,y)=0 \qquad [8.33]$$

for Green's representation of pressure where:

$$I_{n_xn_y}(x,y)=\int_{-\frac{L_x}{2}}^{\frac{L_x}{2}}\int_{-\frac{L_y}{2}}^{\frac{L_y}{2}} T_{n_x-1}\left(\frac{2x'}{L_x}\right)T_{n_y-1}\left(\frac{2y'}{L_y}\right)$$
$$\times\frac{e^{\imath k_f\sqrt{(x-x')^2+(y-y')^2}}}{2\pi\sqrt{(x-x')^2+(y-y')^2}}dx'dy', \qquad [8.34]$$

and for Green's representation of displacement:

$$\sum_{n_x=1}^{N_x}\sum_{n_y=1}^{N_y} c_{n_x}c_{n_y}u_{n_xn_y}T_{n_x-1}\left(\frac{2x}{L_x}\right)T_{n_y-1}\left(\frac{2y}{L_y}\right)$$
$$+\sum_{m_x=1}^{M_x}\sum_{m_y=1}^{M_y} c_{m_x}c_{m_y}p_{m_xm_y}J_{m_xm_y}(x,y)$$
$$-\sum_{s_y=1}^{S_y} c_{s_y}\left\{\sum_{q=0}^{1}\sigma^1_{qs_y}K^1_{qs_y}\left(x-\frac{L_x}{2},y\right)+\sigma^3_{qs_y}K^1_{qs_y}\left(x+\frac{L_x}{2},y\right)\right\}$$
$$-\sum_{s_x=1}^{S_x} c_{s_x}\left\{\sum_{q=0}^{1}\sigma^2_{qs_x}K^2_{qs_x}\left(x,y-\frac{L_y}{2}\right)+\sigma^4_{qs_x}K^2_{qs_x}\left(x,y+\frac{L_y}{2}\right)\right\}$$
$$=G_p\left(x-x_0,y-y_0\right), \qquad [8.35]$$

where the integral kernels are $K^1_{0s_y}(x-x',y)=\int_{-L_y/2}^{L_y/2}T_{s_y-1}(2y'/L_y)G_p(x-x',y-y')dy'$, $K^1_{1s_y}(x-x',y)=\partial K^1_{0s_y}(x-x',y)/\partial x'$, $K^2_{0s_x}(x,y-y')=\int_{-L_x/2}^{L_x/2}T_{s_x-1}(2x'/L_x)G_p(x-x',y-y')dx'$,

$K^2_{1s_y}(x,y-y') = \partial K^2_{0s_y}(x,y-y')/\partial y'$ and $J_{m_x m_y}(x,y) = \int_{-L_x/2}^{L_x/2}\int_{-L_y/2}^{L_y/2} T_{m_x-1}(2x'/L_x)T_{m_y-1}(2y'/L_y)G_p(x-x',y-y')dx'dy'$.

The boundary conditions lead to equations similar to [8.35], except for the displacement, which is zero, and the location of the considered point (at the edge and not in the plate). The coefficients of the various approximations can be used to write equations [8.33] and [8.35] and the eight boundary conditions of the corresponding collocation points. Equation [8.33] is written for $M_x \times M_y$ points $Q_{ij}^{M_x M_y}$ defined for the direct product of the two Chebyshev supports $E_T^{M_x}$ and $E_T^{M_y}$ as $Q_{ij}^{M_x M_y} = \left(L_x/2x_i^{M_x}, L_y/2y_j^{M_y}\right)$ for $i \in [1, N_x]$ and $j \in [1, N_y]$. Similarly, [8.35] is written for $N_x \times N_y$ points $Q_{ij}^{N_x N_y}$ defined for the support $E_T^{N_x}$ and $E_T^{N_y}$ as $Q_{ij}^{N_x N_y} = \left(L_x/2x_i^{N_x}, L_y/2y_j^{N_y}\right)$ for $i \in [1, N_x]$ and $j \in [1, N_y]$. Finally, the eight equations along the edge are written at the corresponding collocation points (which all belong to the Chebyshev supports). For example, along Γ_1, we have S_y points $Q_j^{S_y}$ defined for the support $E_T^{S_y}$ so that $Q_j^{S_y} = \left(L_y/2y_j^{S_y}\right)$ for $j \in [1, S_y]$. Finally, a system of simultaneous equations with $M_x \times M_y + N_x \times N_y + 4(S_x + S_y)$ unknowns is obtained. It can be solved using a Gauss method without particular difficulty.

The calculation of the resonances of the coupled plate is performed by searching for the complex frequencies, which cancel the determinant of the system. This calculation can be carried out using Newton's iterative method (the difficulty lies in initializing the algorithm but this is another subject). Figure 8.16 presents a mapping in the lower left quadrant of the Argand plane of the determinant of the linear system obtained after discretization by a series of polynomials of degree 13 for all unknowns (if we consider that a minimum of π points per wavelength is required, this method allows us to estimate up to the mode (4,4) of a plate with a thickness of 7.4 cm). In this map, the resonances are located at the center of the zones, where the contour lines wrap around the corresponding zero. The resonances near the real axis correspond to the spectrum of the structure (with added mass and damping effects). What we can notice immediately is that the spectrum has a very significant richness.

When we plot the modal deformation of the two modes located in Figure 8.16 by small thick circles for frequencies with an imaginary part of approximately -1,000 Hz and a real part of approximately 4,000 and 5,000 Hz, the mode obtained for the frequency $f = 4,827 - \imath 1,025$ Hz represented in Figure 8.17(a) and the mode obtained for the frequency $f = 3,717 - \imath 1,053$ Hz represented in Figure 8.17(b), we identify each deformation as a mode (1-3). This numerical result shows that a mode (or an entity identified as such) of a coupled structure can no longer be considered as the usual resonance frequency–resonance mode pair. This is related to what was mentioned at the beginning of the section, namely that the two spectral families (the structure spectrum and the diffraction spectrum) blend, making the interpretation of the identified modes very difficult, since we cannot clearly separate the origin (structure or fluid) of the mode. Several modes identified as the modes $(1 - 1)$, $(2 - 1)$, $(2 - 2)$ and $(1 - 3)$, identified as squares, stars, hexagons and circles in the figure, respectively, are contained in this map, showing that this duplication of modes is not limited to a single mode.

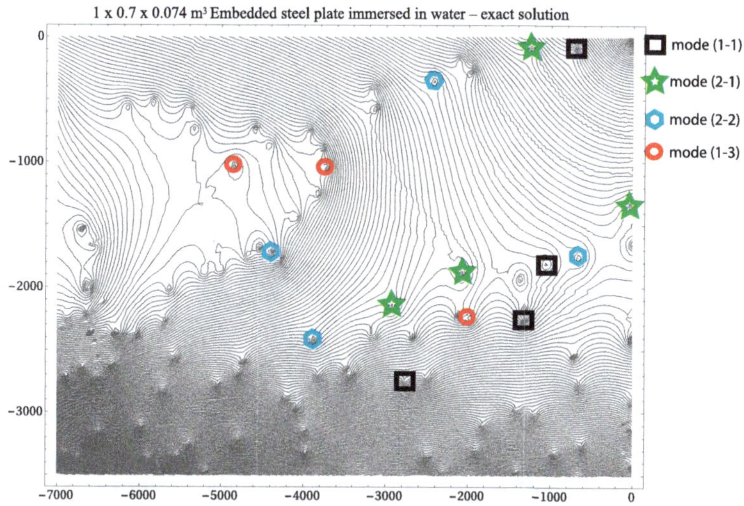

Figure 8.16. *Map of the determinant of the linear system obtained after discretization by a series of polynomials of degree 13 of the integral equations at the edge of a clamped rectangular steel plate ($1 \times 0.7 \times 0.074$ m^3) immersed in water. For a color version of the figure, see www.iste.co.uk/anselmet/acoustics.zip*

This phenomenon is strictly comparable to what occurs when an elastic structure of revolution (cylinder or sphere) is immersed in a fluid. We then observe that when the resonances coincide with a structure mode and a wavelength, which correspond to the perimeter of the body (and its multiples), we obtain a multiplication of resonances whose frequential repartition has a high regularity and follows a relatively regular trajectory (Regge trajectory) [UBE 77, DER 79, GES 87]. Naturally, in the case of a baffled plate, we cannot really refer to creeping waves (which correspond to waves which "turn" around the surface of a body that presents a uniform curvature) and thus to maximum coupling, especially since the thin plates are dispersive media for which the wave propagation velocity depends on the frequency, and only in a few cases we get a perfect coincidence of waves.

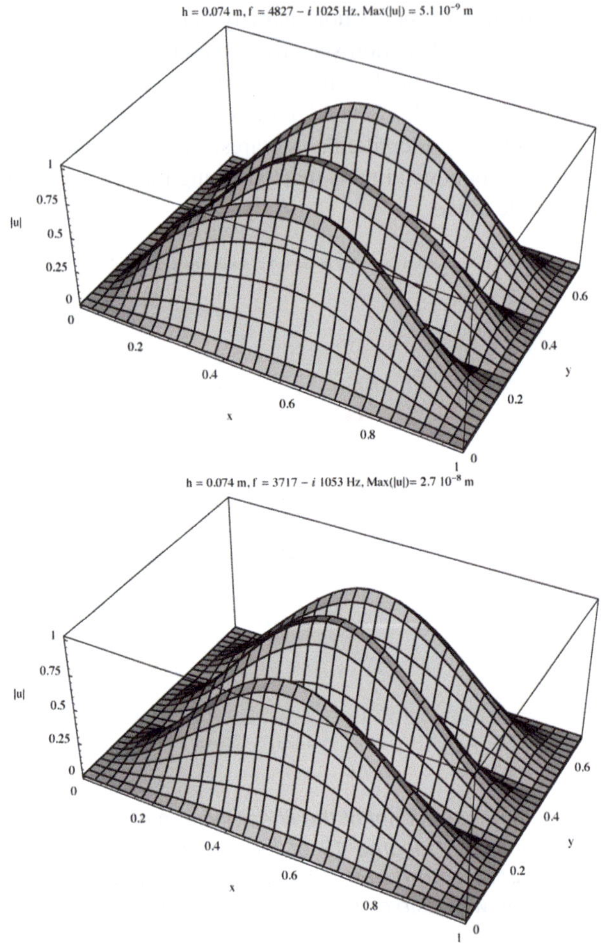

Figure 8.17. *Clamped steel plate ($1 \times 0.7 \times 0.074$ m^3) in water.*
a) resonance mode shape $f = 4,827 - \imath 1,025$ Hz.
b) resonance mode shape $f = 3,717 - \imath 1,053$ Hz

8.6.1.2. *Calculation of resonances using higher order perturbation*

In order to better understand this phenomenon, we can focus on solving the higher order approximation [8.26] which consists of finding for each mode $(m - n)$ the resonance pulsations $\hat{\omega}_{mn}$, solutions to the equation:

$$\hat{\omega}^2_{mn}(1 - \epsilon\beta^{mn}_{\hat{\omega}_{mn}}) - \left(\tilde{\omega}^{(0)}_{mn}\right)^2 = 0, \qquad [8.36]$$

where $\epsilon = 2\rho_0/(\rho_p h)$. We recall that $\omega_{mn}^{(0)}$ is the eigen pulsation of the mode $(m-n)$ *in vacuo* (which is purely real for an elastic plate). Let us establish the mode resonance function $(m-n)$ whose complex zeros, for each mode $(m-n)$, are the sought resonance pulsations:

$$F^{mn}(z) = z^2\left(1 - \epsilon\beta^{mn}(z)\right)\left(\omega_{mn}^{(0)}\right)^2, \qquad [8.37]$$

where β_ω^{mn} was written as $\beta^{mn}(z)$, a function of the variable z. This equation, which appears to be very simple, is in practice extremely difficult to solve. In fact, $\beta^{mn}(z)$ is calculated from the integral:

$$\beta_\omega^{mnpq}(\omega) = L_x^2 Ly^2 \int_0^1\int_0^1\int_0^1\int_0^1 U_{mn}^{(0)}(x,y) \times \frac{e^{\imath\frac{\omega}{c_0}\sqrt{(x-x')^2+(y-y')^2}}}{2\pi\sqrt{(x-x')^2+(y-y')^2}} U_{pq}^{(0)}(x',y')dxdx'dydy', \qquad [8.38]$$

where we highlight the fact that the modes of the plate are double and thus involve two pairs of double indices $(m-n)$ and $(p-q)$. Naturally, to solve equation [8.36], we only deal with the case $m = p$ and $n = q$, which slightly simplifies the calculations, and we note that in this case $\beta^{mnmn}(\omega) = \beta^{mn}(\omega)$. The evaluation of this quadruple integral, which can only be performed digitally, can be considerably simplified if the modes are given in the form of separate variables $U_{mn}^{(0)}(x,x) = X_m(x)Y_n(y)$. A simple linear transformation given by $X = x - x'$ and $Y = y - y'$ [MAN 63] leads to:

$$\beta^{mnpq}(\omega) = 4L_x^2 Ly^2 \int_0^1\int_0^1 A_{mp}(X)B_{nq}(Y)\frac{e^{\imath\frac{\omega}{c_0}\sqrt{X^2+Y^2}}}{2\pi\sqrt{X^2+Y^2}}dXdY,$$

$$A_{mp}(X) = \int_0^{1-X} X_m(X+x')X_p(x')dx', B_{nq}(Y) = \int_0^{1-Y} Y_n(Y+y')Y_q(y')dy'.$$

In the case of rectangular plates, whose modes are given by the formulas presented in section 7.2.2, using almost any formal calculation software, we can easily obtain the relatively simple analytic results for $A_{mp}(X)$ and $B_{nq}(Y)$, which allow us to efficiently calculate $\beta^{mnpq}(\omega)$. For example, for a supported plate, for $X_m(x) = \sin(m\pi x)$, we obtain:

$$A_{mp}(X) = \frac{-2p\sin(m\pi X) + (m+p)\sin((mX+m-p)\pi) + (p-m)\sin((mX+m+p)\pi)}{2(m-p)(m+p)\pi},$$

and

$$A_{mm}(X) = \frac{\sin(m\pi X) - \sin(m\pi(X+2)) + 2m\pi\cos(m\pi X)}{4m\pi},$$

with similar results for the function $B_{nq}(Y)$. The case of the clamped plate, which we are focusing on here, is considerably more complex. We always assume that the modes are given by approximate relations [7.19], [7.20] and [7.21]. This only gives us results for $A_{mm}(X)$. For $m = 1$:

$$\begin{aligned}A_{11}(X) = 0.982654\,[&(0.394309 - 0.5x)\cos(\gamma_1\pi x)\\&+(0.114517 - 0.00882547x)\cosh(\gamma_1\pi x)\\&+0.209549\sin(\gamma_1\pi x) - 0.101976\sinh(\gamma_1\pi x)],\end{aligned}$$

where $\gamma_1 \approx 1.505618$.

For an odd m:

$$\begin{aligned}A_{mm}(X) = \frac{1}{2(\pi+2m\pi)}&\left[-(1+2m)\pi(x-1)\cos\left(\gamma_m\pi x\right) - 2\sin\left(\gamma_m\pi(x-1)\right)\right.\\&+\frac{\sinh\left(\frac{\gamma_m\pi}{2}\right)}{\sin\left(\frac{\gamma_m\pi}{2}\right)}\left\{\frac{\sin\left(\frac{\gamma_m\pi}{2}\right))}{\sinh\left(\frac{\gamma_m\pi}{2}\right)}\left(-(1+2m)\pi(x-1)\cosh\left(\gamma_m\pi x\right)\right.\right.\\&-2\sinh\left(\gamma_m\pi(x-1)\right)) + (2-2\imath)\left[\sin\left(\frac{1+\imath}{2}\gamma_m\pi(1-(1-\imath)x)\right)\right.\\&-\sin\left(\frac{1+\imath}{2}\gamma_m\pi(-1+(1+\imath)x)\right) + \sinh\left(\frac{1+\imath}{2}\gamma_m\pi(1-(1-\imath)x)\right)\\&\left.\left.\left.-\sinh\left(\frac{1+\imath}{2}\gamma_m\pi(-1+(1+\imath)x)\right)\right]\right\}\right],\end{aligned}$$

where $\gamma_m = m + 1/2$ and, for an even m:

$$\begin{aligned}A_{mm}(X) = \frac{1}{2(\pi+2m\pi)}&\left[-(1+2m)\pi(x-1)\cos\left(\gamma_m\pi x\right)\right.\\&+2\sin\left(\gamma_m\pi(x-1)\right)\end{aligned}$$

$$+\frac{\sinh\left(\frac{\gamma_m\pi}{2}\right)}{\sin\left(\frac{\gamma_m\pi}{2}\right)}\left\{\frac{\sin\left(\frac{\gamma_m\pi}{2}\right))}{\sinh\left(\frac{\gamma_m\pi}{2}\right)}((1+2m)\pi(x-1)\cosh\left(\gamma_m\pi x\right)\right.$$

$$-2\sinh\left(\gamma_m\pi(x-1)\right))+(2-2\imath)\left[-\sin\left(\frac{1+\imath}{2}\gamma_m\pi(1-(1-\imath)x)\right)\right.$$

$$+\sin\left(\frac{1+\imath}{2}\gamma_m\pi(-1+(1+\imath)x)\right)+\sinh\left(\frac{1+\imath}{2}\gamma_m\pi(1-(1-\imath)x)\right)$$

$$\left.\left.-\sinh\left(\frac{1+\imath}{2}\gamma_m\pi(-1+(1+\imath)x)\right)\right]\right\}\Bigg].$$

In order to compare these results with those from the exact equation solutions, we present a resonance map in the lower left quadrant of the Argand plane. To do this, we construct a function which should allow us to locate the resonances of any mode. It is given by the product of the eigenvalue equations of the considered modes, whose general expression can be given by [8.37] taken at $z = 2\pi f$. For example, when we focus on the resonance frequencies of the modes $(1-1)$ to $(6-5)$, we can construct the function $\mathcal{F}_{11}^{65}(f) = \prod_{(m,n)=(1-1)}^{(m,n)=(6-5)}\left(f^2\left(1-\epsilon\beta^{mn}(f)\right)-f_{0mn}^{(2)}\right)^2$ for which it is trivial to see that the frequencies where it cancels (the resonances of the coupled clamped plate) correspond to frequencies for which all resonance functions $F^{mn}(2\pi f)$, given by [8.37], for $(m-n)=(1-1),...,(5-6)$, cancel. We present in Figure 8.18 a map in the lower left quadrant of the Argand plane. Similarly, as for Figure 8.16, the resonances identifiable on this map are located in the center of the zones, where the contour lines wrap around the corresponding zero. In this figure, several resonances have been marked by squares for the mode $(2-1)$, by hexagons for the mode $(2-2)$ and circles for the mode $(1-3)$. Overall, the resonances near the real axis (the smallest imaginary part) correspond to the resonances observed in Figure 8.16. And while the higher order modes present a greater dispersion, the mode separation remains well described.

8.6.1.3. *Examples of resonances for a plate immersed in a heavy fluid*

Several resonance frequencies calculated by finding the zeros of the resonance function [8.37] have been estimated numerically using a Newton-type method [PRE 92]. They are compared with those obtained by calculating the zeros of the linear system (here, once again using a Newton-type method), which have been presented in the map in Figure 8.16. The results are given in Table 8.1 for a clamped plate ($1\times 0.7\times h$ m^3) made of steel ($E = 200$ GPa, $\nu = 0.3$, $\rho_p = 7{,}800$ kg/m^3) in water ($\rho_0 = 1{,}000$ kg/m^3, $c_0 = 1{,}500$ m/s) for four different thicknesses. The results obtained *in vacuo* and at the order 0 of the expansion do not coincide exactly (approximately 0.5% precision) with the exact solution due to the approximation of modes of the plate *in vacuo* by a product of separate variables. These results show that Warburton's approximation of the resonance equation provides us with excellent estimates of the plate resonances, even for relatively high values of the ϵ parameter.

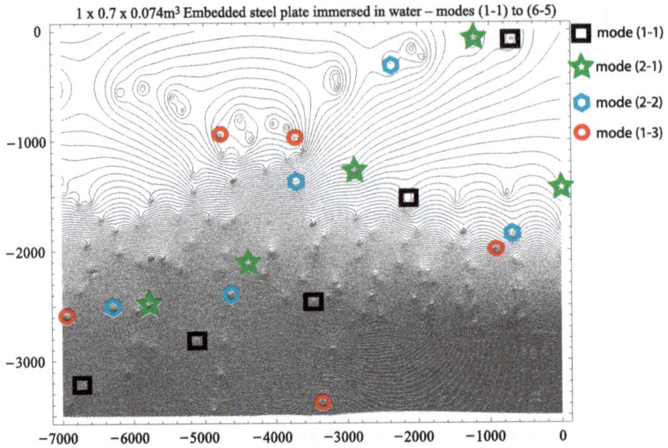

Figure 8.18. *Map of the function $\mathcal{F}_{11}^{65}(f)$ of a clamped rectangular steel plate ($1 \times 0.7 \times 0.074$ m^3) immersed in water for the modes $(1-1)$ to $(6-5)$. For a color version of the figure, see www.iste.co.uk/anselmet/acoustics.zip*

		Exact		Order of expansion	
		in vacuo	fluid loaded	0	1
h=5mm	Mode 1 1	68.86	$17.2 - \imath\, 0.11$	69.06	$17.3 - \imath\, 0.11$
$\epsilon \approx 51.3$ m^{-1}	Mode 1 2	166.05	$61.87 - \imath\, 0.0017$	166.57	$62.06 - \imath\, 0.0017$
	Mode 1 3	313.32	$135.01 - \imath\, 0.82$	314.1	$125.9 - \imath\, 2.78$
h=1cm	Mode 1 1	137.7	$47.1 - \imath\, 0.78$	138.12	$47.6 - \imath\, 0.77$
$\epsilon \approx 25.6$ m^{-1}	Mode 1 2	332.1	$163.5 - \imath\, 0.08$	333.1	$164.0 - \imath\, 0.07$
	Mode 1 3	626.6	$347.6 - \imath\, 4.9$	628.2	$332.9 - \imath\, 16.3$
h=3cm	Mode 1 1	413.1	$220.2 - \imath\, 13.9$	414.3	$222. - \imath\, 13.6$
$\epsilon \approx 8.5$ m^{-1}	Mode 1 2	996.3	$677.0 - \imath\, 12.87$	999.4	$679.9 - \imath\, 11.9$
	Mode 1 3	1,879.9	$1{,}648.1 - \imath\, 8.52$	1,884.7	$1{,}453.7 - \imath\, 60.8$
h=10cm	Mode 1 1	1,377.1	$1{,}089.2 - \imath\, 199.1$	1,381.2	$1{,}093.6 - \imath\, 195.4$
$\epsilon \approx 2.7$ m^{-1}	Mode 1 2	3,321	$3{,}122.4 - \imath\, 559.1$	3,331.5	$3{,}137.5 - \imath\, 570.6$
	Mode 1 3	6,266.4	$6{,}273.4 - \imath\, 368.7$	6,282.2	$6{,}279.5 - \imath\, 368.7$

Table 8.1. *Comparison of exact resonances and resonances approximated using Warburton's approximation for a clamped steel plate ($1 \times 0.7 \times h$ m^3) in water (in Hz)*

We present in Figure 8.19 the evolution of the resonance frequencies of a steel plate (1 m $\times$ 0.7 m) in water for the mode (1 3) when the thickness of the plate varies. This figure presents a comparison of the exact values ($\hat{f}_{13}^{e}$) and those obtained using Warburton's approximation ($\hat{f}_{13}^{1}$). Apart from the fact that the frequencies derived from exact calculations and by perturbation are very close, in each figure, we can note the presence of two disjoint curves. The real part increases almost linearly from the origin[8] and then appears to separate into two disjoint curves (denoted by "sup" and "inf") at a thickness of approximately 7 cm for this example. Similarly, the imaginary part (that characterizes damping) clearly shows two curves that intersect at approximately the same thickness of $h_m \approx 7$ cm. In practice, for this thickness, the mode (1 3) clearly exists for two different frequencies as seen in Figures 8.17, which show two modes that are calculated for the thickness $h = 7.4$ cm. Even if the very high values that take imaginary parts rapidly reach the numerical limits of number representation, making it difficult to demonstrate, all modes exhibit a comparable behavior.

These curves show that one thickness plays a specific role: it corresponds to the point where the imaginary parts intersect and where the real parts are at their closest. As we have already discussed, the plates are dispersive media for which the bending wave celerity depends on frequency as well as on the thickness of the plate. We have already defined a critical pulsation given by equation [8.8], which showed that for the geometrical and mechanical characteristics, given at the coincidence frequency, the coupling is at its maximum. Similarly, if the frequency is fixed, there is a thickness for which the bending wavelengths in the infinite plate and the compression wavelengths in the fluid are equal and ensure maximum coupling. When the plate is of finite dimensions, the spectrum is fixed by the geometry of the problem and for each mode we can find a thickness h_c which corresponds to maximum coupling. For a clamped plate, whose eigenfrequencies f^0_{mn} are given by equation [7.22], we aim to find h_c such that $\omega_c = 2\pi f^{(0)}_{mn}$, that is:

$$h_c = \frac{c_0}{\pi}\sqrt{\frac{12(1-\nu^2)\rho_p}{R}}\left(\left(\frac{G(m)}{\ell_x}\right)^4 + \left(\frac{G(n)}{\ell_y}\right)^4 + 2\frac{H(m)}{\ell_x^2}\frac{H(n)}{\ell_y^2}\right)^{-\frac{1}{4}}. \quad [8.39]$$

For the considered plate and the mode (1 3), we obtain a thickness of $h_c \approx 6$ cm. This result can be further refined by an iterative calculation, assuming that the resonance frequency of the coupled mode significantly differs from that of the mode *in vacuo* and that, after having calculated h_c satisfying $\omega_c = \omega^0_{mn}$, we aim to find a finer estimate of h^1_c such that:

$$h^1_c = h_c \Re\left(\sqrt{1-\epsilon\beta^{mn}_{\omega_c}}\right), \quad [8.40]$$

where we have calculated $\epsilon = 2\rho_0/(\rho_p h_c)$ and ω_c by taking $h = h_c$ given by [8.39]. For the mode (1 3), we always obtain a thickness value of maximum coupling $h^1_c \approx$ 7 cm which is very close to the observed value. In order to fix these ideas, we present in Table 8.2 the results calculated for the first nine modes of our clamped steel plate of 1 m $\times$ 0.7 m $\times$ h_c m in water.

8 We recall that the eigenfrequencies of a plate *in vacuo* are directly proportional to the thickness.

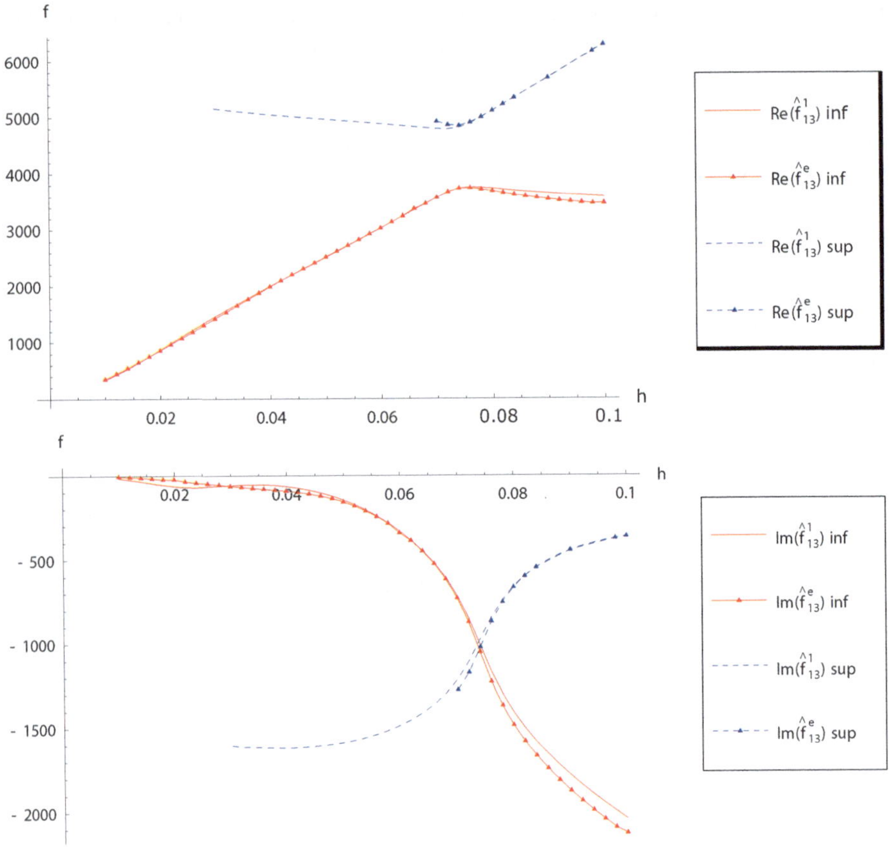

Figure 8.19. *Clamped steel plate in water* ($1 \times 0.7 \times h$ *m*^{3}). *The evolution of real and imaginary parts of resonances of the mode* $(1-3)$ *when the thickness varies. A comparison of exact values* ($\hat{f}^e_{13}$) *and higher order approximations* ($\hat{f}^1_{13}$). *Top: real part. Bottom: imaginary part. For a color version of the figure, see www.iste.co.uk/anselmet/acoustics.zip*

modes (m-n)	(1-1)	(1-2)	(1-3)	(2-1)	(2-2)	(2-3)	(3-1)	(3-2)	(3-1)
h_c (cm)	13	8.4	6.1	10.2	7.5	5.8	8.1	6.5	5.3
h_c^1 (cm)	13.9	9.2	6.9	11.3	8.4	6.5	9.1	7.4	6.1

Table 8.2. *Thicknesses of maximum coupling* h_c *given by [8.39] and* h_c^1 *given by [8.40] for the first modes of a clamped steel plate* ($1 \times 0.7 \times h_c$ *m*^{3}) *in water*

8.6.2. *Location of resonances of a coupled plate*

As we observed in the previous examples, when the fluid density is comparable to the structure density, the nature of the spectrum of the structure changes considerably. We can show [MAT 09] that, for a supported plate[9], $\beta^{mn}(z)$ and the resonance function $F^{mn}(z)$ are, for $\epsilon \neq 0$, entire functions of exponential type (namely entire functions of the order 1 and normal σ type) [LEV 80, LEV 96] which, thus, have an infinite number of roots. This theory is actually very simple, it is based on Picard's great theorem, which indicates that, if an entire function $f(z)$ has an essential singularity at a point z_w, then, for any open set which contains z_w, $f(z)$ takes all the possible values, with the exception of one, and an infinite number of times (the archetypal function is the function $\exp(1/z)$ which has an essential singularity at the origin and takes all the values, except 0). In addition, an entire function is either a polynomial or has an essential singularity at infinity. Thus, if the function $F^{mn}(z)$ is an entire function, if it is not a polynomial and if 0 is not the exception of Picard's theorem, then it has an infinite number of roots.

Two cases are considered, without and with coupling. In the first case, without coupling and thus for $\epsilon = 0$, the function $F^{mn}(z)$ is a polynomial of degree 2 and has two zeros. In the second case, with coupling, we can show that $\beta^{mn}(z)$ is an entire exponential-type function (at the order 1 and of $\sigma = \max(a, b)$ type, where $a = L_x$ and $b = L_y$)[10], thus, for $\epsilon \neq 0$, the entire function $F^{mn}(z)$ is not a polynomial. As there exists at least one complex root, 0 is not the exception of Picard's theorem, and $F^{mn}(z)$ has an infinite number of roots.

We can obtain an even finer result. In fact, we can show that the entire exponential-type functions $F^{mn}(z)$, for $\epsilon \neq 0$, and $\beta^{mn}(z)$ belong to the class C Cartwright functions. This result is fundamental, because it indicates, according to the Cartwright–Levinson theorem [LEV 96, theorem 1 p. 127], that, if we denote:

– $n_+(r, \alpha)$ the number of zeros of $F^{mn}(z)$ in the sector $\{z : |z| \leq r, |\arg(z)| \leq \alpha\}$;

– $n_-(r, \alpha)$ the number of zeros in the sector $\{z : |z| \leq r, |\pi - \arg(z)| \leq \alpha\}$,

then, the density of the set of zeros $n_\pm(r, \alpha)/r$ has the limit $\lim_{r\to\infty} n_\pm(r, \alpha)/r = c_0\sigma/\pi$. This shows that, for each mode, in the presence of coupling[11], we obtain not only an infinite number of resonance frequencies but also that these frequencies are

9 For a clamped or free plate, even if the demonstration cannot be carried out rigorously because of the calculation complexity, intuitively, the result is the same, since the two systems only differ in a part of the boundary conditions and since this phenomenon is caused by fluid-structure coupling.

10 Which corresponds to $\max_{|z|=r} |F^{mn}(z)| \approx C\exp(\sigma r)$, where C is a constant.

11 In the absence of coupling, we have only two roots $z = \pm\omega_{mn}^{(0)}$.

close to the real axis and that they are spaced apart from each other by a distance of[12] $\Delta\omega \approx 2c_0\pi/\max(a,b)$.

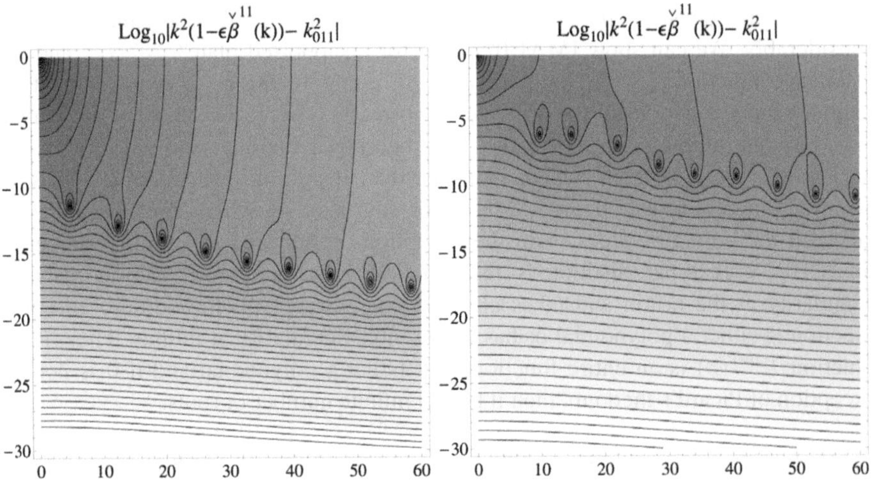

Figure 8.20. *Iso-amplitudes of the function $k^2(1-\epsilon\beta^{mn}(k)) - k^2_{0mn}$ for the mode (1-1). a) plate in contact with air ($\epsilon \approx 0.17$ m^{-1}) b) plate in contact with water ($\epsilon \approx 128.2$ m^{-1})*

Figures 8.20 and 8.21 represent the iso-amplitudes of the resonance functions plotted against $k = \omega/c_0$ for a supported steel plate of $a = 1$ m $b = 0.7$ m and $h = 2$ mm in contact with air and water for the mode (1 1), Figure 8.20, and for the mode (1 3), Figure 8.21. We can easily observe that in air (for a weak coupling) the spectrum has two distinct families, the one with the modes of structures that are slightly offset from the real axis (whose zero is very close to the origin for the mode (1 1) and slightly offset from the origin for the mode (1 3)), and the one with the diffraction modes (for $-20 < \Im(k) < -10$). However, when the coupling significantly increases, the diffraction mode family approaches the real axis, the structure mode family keeps shifting increasingly until the two families mix, as clearly shown in Figure 8.21.

When ϵ approaches zero (which corresponds to the case addressed by Sanchez [SAN 89]), we can identify in Figure 8.22 the spectral family which corresponds to the diffraction spectrum of the structure whose resonances (for

12 We must pay attention to the fact that since the roots are in complex pairs with anti-Hermitian symmetry, each time the distance from the origin increases by $c_0 2\pi/\max(a,b)$, we get two extra roots at ω, which correspond to the limit σ/π of the density of the set of zeros fixed by the Cartwright–Levinson theorem.

$-30 < \Im(k) < -20$) are identical for all the modes (we must also note that the resonances are separated by a distance $\Delta k \approx 6$ in agreement with the Cartwright–Levinson theorem).

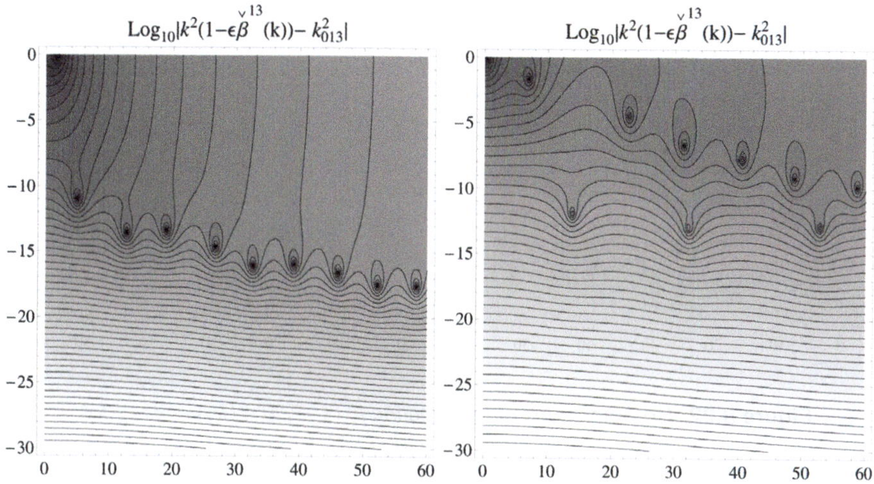

Figure 8.21. *Iso-amplitudes of the function $k^2(1 - \epsilon\beta^{mn}(k)) - k^2_{0mn}$ for the mode (1 3). a) plate in contact with air ($\epsilon \approx 0.17$ m^{-1}) b) plate in contact with water ($\epsilon \approx 128.2$ m^{-1})*

Thus, the diffraction spectrum plays an important role here. This spectrum, which consists of strongly damped complex resonances, exists for all rigid bodies immersed in a fluid. The observed resonances correspond to the frequencies whose wavelengths correspond to the characteristic dimensions of the diffracting object, such as the perimeter for a ball and the sum of the sides for a cuboid. The theoretical elements developed here show that coupling as a phenomenon leads to the fact that each resonance of the structure becomes coupled with all the diffraction resonances. This coupling, which is especially important when the fluid has a density comparable to the density of the structure, differs for all diffraction resonances. In practice, the coupling transforms the two distinct resonance families, the spectrum of the vibrating structure and the diffraction spectrum, into a much denser family, since each component of the vibration spectrum is associated with a new spectral family.

In the most common case of coupling with a low-density fluid such as air, these spectral families consist only of one term (and its anti-Hermitian symmetric) with a small imaginary part and the coupling effect can be described by a slight modification of the apparent stiffness and the mass. For the coupling of a metal plate, with a thickness in centimeters, in contact with water, this multiplication of resonances becomes significant and the usual modal analysis proves to be insufficient.

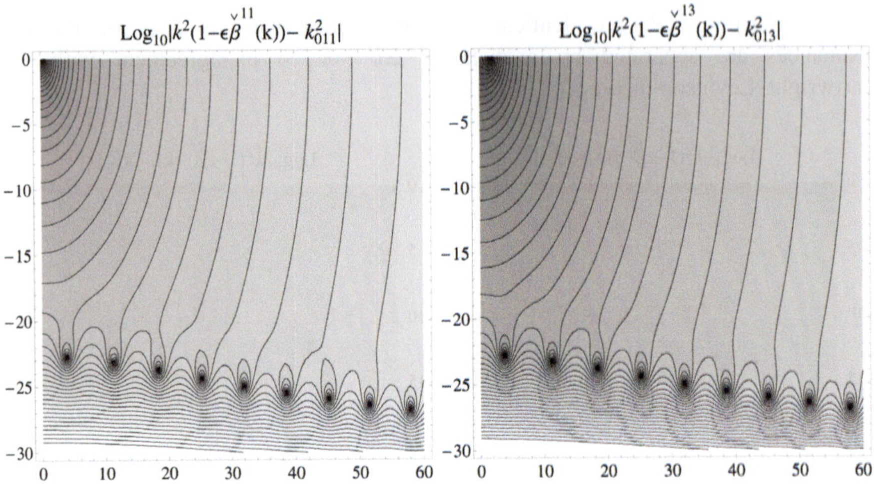

Figure 8.22. *Iso-amplitudes of the function $k^2(1 - \epsilon\beta^{mn}(k)) - k_{0mn}^2$ for the case $\epsilon = 10^{-5}$ m^{-1}. a) mode (1 1). b) mode (1 3)*

8.7. Vibrations of a thin plate in a turbulent flow

We consider a flat rectangular plate in contact with a fluid, in a turbulent flow parallel to the x axis of Mach M_0, with a non-negligible density on the top side, while its lower side is in contact with a fluid that is light enough so that its influence can be neglected. The plate is composed of a purely elastic material with a Young's modulus of $E = 200$ GPa, a Poisson's ratio of $\nu = 0.3$ and a density of $\rho_p = 7,800$ kg/m^3 . The plate with a thickness of $h = 0.5$ cm occupies the domain Σ which is defined in Cartesian coordinates (O, x, y, z) by $-L_x/2 < x < +L_x/2, -L_y/2 < y < +L_y/2, z = 0$ with $L_x = 0.7$ m and $L_y = 1$ m. $\partial\Sigma$ is the boundary of Σ with an outer normal. Σ' is the complement to Σ of the plane $z = 0$. $D = Eh^3/(12(1-\nu^2))$ is the bending stiffness of the plate. The plate is clamped at all four edges along $\partial\Sigma$ and prolonged by an infinite rigid flat screen, which occupies Σ'.

To simplify, we assume here that the vibration of the plate is low enough for it to be decoupled from the flow. At each point, the plate is excited by a pressure fluctuation which is regarded as a random process. We let $f(M, t)$ be a realization of this process and $F(M, \omega)$ be its temporal Fourier transform. We call $\Gamma(M, M')$ the Green kernel of the plate.

The Fourier transform of a realization of a displacement can be given by:

$$u(M, \omega) = \int F(Q, \omega)\Gamma(M, Q, \omega)dQ.$$

The interspectrum of the displacement is given by:

$$u(M,\omega)u^*(M',\omega) = \int\int \Gamma(M,Q,\omega)F(Q,\omega)F^*(Q',\omega)\Gamma^*(M',Q',\omega)dQdQ'.$$

The average (in the Fourier domain) of the interspectrum of the displacement is the interspectral density of the displacement $S_u(M,M',\omega) = \overline{u(M,\omega)u^*(M',\omega)}$. Since the only forcing term $F(Q,\omega)$ is random, we obtain:

$$S_u(M,M',\omega) = \int\int \Gamma(M,Q,\omega)\overline{F(Q,\omega)F^*(Q',\omega)}\Gamma^*(M',Q',\omega)dQdQ'. \quad [8.41]$$

We call $\Phi_p(Q,Q',\omega) = \overline{F(Q,\omega)F^*(Q',\omega)}$ the interspectral density of turbulent wall pressure fluctuations. Thus, we must estimate the interspectral density of the excitation $\Phi_p(Q,Q',\omega)$ and the Green kernel of the plate $\Gamma(M,Q,\omega)$.

8.7.1. *Interspectral density: simple models*

One of the main difficulties for modeling this spectral density is the fact that, in a given frequency band, a number of vortices contribute to the sought value. The small structures, located near the wall, move slowly, whereas the large structures, further away from the wall, move faster. If the ratio $\omega = U_c/\ell$ (convection velocity/ characteristic size) is identical for each structure, they all contribute to the turbulent wall pressure field. Generally, we write that:

$$\Phi_p(Q,Q',\omega) = \Phi(Q,\omega)C(Q,Q',\omega)e^{\imath\phi(Q,Q',\omega)},$$

where $\Phi(Q,\omega)$ is the point pressure spectrum. Generally, we consider a model of established turbulence and thus $\Phi(Q,\omega) = \Phi(\omega)$ does not depend on the point considered. $C(Q,Q',\omega)$ is a coherence function that reflects the similarity of the pressure between the points Q and Q'. $\phi(Q,Q',\omega)$ describes the phase relation between the points. To better understand these ideas, we consider a flow over a plane (O,x,y) along the x direction. The phase, which characterizes the wave propagation, depends primarily on the separation $r_x = x - x'$. According to the Corcos hypothesis, it is linked to the convection velocity U_c through a convective wave number $k_c = \omega/U_c(\omega)$, $\phi(Q,Q',\omega) = \omega r_x/U_c(\omega)$. $C(Q,r_x,r_y,\omega)$ is linked to two integral length scales L_x and L_y. We obtain $L_x(Q,\omega) = \int_0^\infty C(Q,r_x,r_y=0,\omega)dr_x$ and a similar relation for the transverse direction. The correlation in the flow direction is obviously much greater than in the transverse direction. For a more comprehensive presentation of these notions, the readers are invited to refer to sections 9.6 and 10.2.

In the Corcos model, we assume that the coherences in x and y are decoupled (which is obviously false but allows us to separate the variables) and have an exponential dependence. Always with $r_x = x - x'$ and by defining in a similar way $r_y = y - y'$, we obtain:

$$\begin{aligned}\Phi_P(Q, Q', \omega) &= \Phi(\omega) \mathrm{e}^{-|r_x|/L_x} \mathrm{e}^{-|r_y|/L_y} \mathrm{e}^{\imath k_c r_x} \\ &= \Phi(\omega) \mathrm{e}^{-|r_x| \alpha_x k_c} \mathrm{e}^{-|r_y| \alpha_y k_c} \mathrm{e}^{\imath r_x k_c},\end{aligned} \quad [8.42]$$

where $\Phi(\omega) = \Phi_P(0, 0, \omega)$ corresponds to the point spectrum. It can be measured by a microphone placed on the surface at a given point of the wall, however, we must always bare in mind that the microphone has a surface which can be greater than the influence surface of the convected vortex which sonifies it. This results in spatial filtering, which induces a reduction in the high-frequency content of the wall fluctuation spectrum. In practice, we must perform measurements with microphones for which pressure is measured through a fine needle. The parameters α_x, α_y, U_c and $\Phi(\omega)$ are identified from measurements. We refer the readers to the work by Blake [BLA 86] which describes in detail the measurement techniques. In many cases, we can choose $\alpha_x = 0.1$, $\alpha_y = 7\alpha_x$ and $U_c = 0.7 U_0$.

Sometimes, we take $\Phi(\omega) = 1$ (white noise), if this approximation is valid at low frequencies, it is completely false at high frequencies and this is all the more so when the Mach number is low. An alternative is to use for $\Phi(\omega) = \Phi_P(0, 0, \omega)$ the point spectrum given by the Durant–Robert model:

$$\log_{10} \Phi_e(f) = -5.1 - 0.9 \log_{10} f_e - 0.34 \left(\log_{10} f_e\right)^2 - 0.04 \left(\log_{10} f_e\right)^3, \quad [8.43]$$

where $\Phi_e(f) = \Phi(\omega/2\pi) U_0 / q_0^2 R$ and $f_e = f R / U_0$ represent the spectral density and the frequency, normalized with the flow external scales. The spectral density is normalized by the flow dynamics $q_0 = 1/2 \rho_0 U_0^2$.

We present in Figure 8.23 the amplitude of the spectrum $\Phi(\omega/2\pi)$ as a function of the frequency for a turbulent boundary layer in water at 20 m/s and a boundary layer thickness of $\delta = 7.5$ cm.

A very convenient way to represent these models is to define them in the domain of the spatial wave number. Using the spatial Fourier transform, with $k_\omega = \omega / U_c$, for the Corcos model, we obtain:

$$\hat{\Phi}_P(k_x, k_y, \omega) = \frac{\Phi(\omega)}{k_c^2} \frac{4 \alpha_x \alpha_y}{\left(\alpha_y^2 + \left(\frac{k_y}{k_c}\right)^2\right)\left(\alpha_x^2 + \left(\frac{k_x}{k_c} - 1\right)^2\right)}.$$

This model suffers from many flaws: it describes the spatial correlations poorly, the "coherence spot" is in the form of a diamond, elongated in the direction of the flow (instead of a more common form, such as an ellipse). At very low spatial frequencies, we cannot obtain a spectrum which behaves as k^2 so that it, therefore, does not satisfy the Kraichnan–Phillips theorem; the Corcos model overestimates the spectrum at low wave numbers. The latter very poorly (almost not at all) describes the acoustic phenomena (eigennoise of the turbulence) at low wave numbers (near ω/c_0). However, it describes the convective peak (around ω/U_c) very well and it is easy to implement because of the separation of variables.

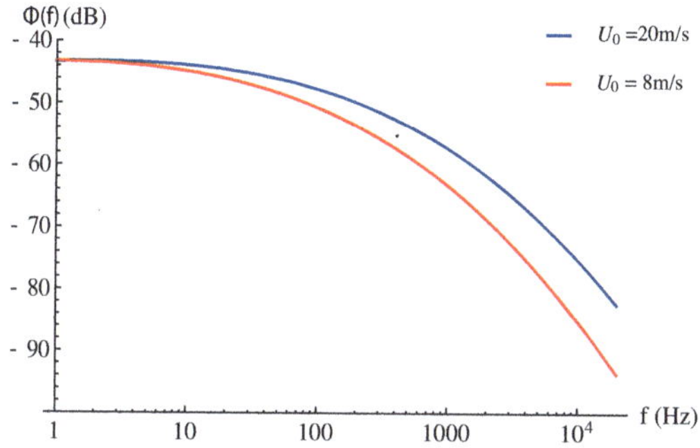

Figure 8.23. *Point spectrum of the Durant–Robert model. For a color version of the figure, see www.iste.co.uk/anselmet/acoustics.zip*

A way to improve the behavior at low spatial frequencies, while still not satisfying the Kraichnan–Phillips theorem, is to substitute the Corcos model (which allows separation of space variables) with a so-called "elliptical" model [MIL 11], the Mellen model $\Phi_P(Q,Q',\omega) = \Phi(\omega)\exp(\imath\omega r_x/U_c(\omega) - \sqrt{(\alpha_\omega r_x)^2 + (\beta_\omega r_y)^2})$ where $\alpha_\omega = \alpha_x\omega/U_c$ and $\beta_\omega = \alpha_y\omega/U_c$. In the wave number domain, for the Mellen model, we have:

$$\hat{\Phi}_P(k_x,k_y,\omega) = \frac{\Phi(\omega)}{k_c^2}\frac{2\pi\left(\alpha_x\alpha_y\right)^2 k_c^3}{\left(\left(\alpha_x\alpha_y k_c\right)^2 + \left(\alpha_x k_y\right)^2 + \alpha_y^2\left(k_x - k_c\right)^2\right)^{3/2}}.$$

For $k_x = 0$ and $k_y = 0$, for the Corcos model, we obtain $\hat{\Phi}_P(k_x,k_y,\omega)k_c^2\Phi(\omega) = 4/\left(\alpha_x\alpha_y + \alpha_y/\alpha_x\right) \approx 0.6$. For the Mellen model, we obtain $\hat{\Phi}_P(k_x,k_y,\omega)k_c^2\Phi(\omega) = 2\pi/(1 - \alpha_y^2/\alpha_x^2\sqrt{\alpha_y^2(1+\alpha_x^2)}) \approx 0.2$, this is a level reduced by about 10 dB. Other elliptical models exist in the literature, such as the Smol'yakov–Tkachenko model, and to further study these models, the readers are encouraged to refer to section 10.2.

8.7.2. *Green's representation of a coupled plate*

Green's kernel of the plate $\Gamma(Q, Q'\omega) = \Gamma(Q - Q', \omega)$ and the acoustic pressure $p(M)$ satisfy the following system of differential equations:

$$D\left(\Delta_Q^2 - \frac{\rho_p h\omega^2}{D}\right)\Gamma(Q,\omega) = \delta(Q) - P(Q), Q \in \Sigma,$$

$$\text{where } P(Q) = \lim_{M\in\Omega^+\to Q\in\Sigma} p(M)$$

$$\Delta p(M) + k_f^2 p(M) + 2\imath k_f M_0 \frac{\partial p}{\partial x} - M_0^2 \frac{\partial^2 p(M)}{\partial x^2} = 0,$$

$$\frac{\partial p(Q)}{\partial \vec{n}_Q} = \begin{cases} \rho_0\left(-\imath\omega + c_0 M_0 \frac{\partial}{\partial x}\right)^2 \Gamma(Q,\omega), Q \in \Sigma \\ 0, Q \in \Sigma' \end{cases}$$

To these equations, we must add the boundary conditions and the Sommerfeld conditions. $k_p = \sqrt[4]{\frac{\rho_p h\omega^2}{D}}$ is the wave number of the infinite plate *in vacuo* and $k_f = \frac{\omega}{c_0}$ is the acoustic wave number.

We use Green's representation of pressure with Green's kernel of the convected Helmholtz equation given by [4.17] $G_e(M, M') = -e^{\imath k_f \tilde{R}}/(4\pi\sqrt{1-M^2}R^*) - e^{\imath k_f \tilde{R}_i}/(4\pi\sqrt{1-M^2}R_i^*)$, where $\tilde{R}_i$ and R_i^* are the images of $\tilde{R}$ and R^* with respect to the horizontal plane (we use $-z$ in the place of z). We obtain a system of integro-differential equations, which governs the displacement of the plate and the surface pressure:

$$D\left(\Delta^2 - \frac{\rho_p h\omega^2}{D}\right)\Gamma(Q,\omega) = \delta(Q) - P(Q),$$

$$\begin{aligned} P(Q) &= \frac{\rho_0\omega^2}{\beta}\int_\Sigma G_e(Q,Q')\Gamma(Q',\omega)dQ' \\ &+ \frac{2\imath\rho_0\omega c_0 M_0}{\beta}\int_\Sigma G_e(Q,Q')\frac{\partial\Gamma(Q',\omega)}{\partial x'}dQ' \\ &- \frac{\rho_0 c_0^2 M_0^2}{\beta}\int_\Sigma G_e(Q,Q')\frac{\partial^2\Gamma(Q',\omega)}{\partial x'^2}dQ'. \end{aligned}$$

We must note that, due to the flow, the plate can efficiently radiate at very low frequency ($\omega \to 0$). These equations are difficult to solve. A common approximation, if we are at a low Mach number, involves neglecting the last two integrals in the surface pressure (their influence is mainly felt near the maximum response of the plate and strongly decreases with frequency). We have thus reduced our task to solving the problem of a plate coupled to a fluid (only the definition of Green's kernel of the Neumann problem for the Helmholtz equation changes).

To simplify this problem, we will assume that there is no aeroacoustic coupling between the plate and flow, and that the turbulent excitation acts on the plate as a random process outside of the system. The response of the plate to turbulent excitation is given by the interspectral power density of the displacement (equation [8.41]):

$$S_u(M;M',\omega) = \int_\Sigma \int_\Sigma \Gamma(Q;M,\omega)\Phi_P(Q;Q',\omega)\Gamma^\star(Q';M',\omega)dQdQ'.$$

The spectral power density is described by a Corcos model given by [8.42] where $\Phi(\omega) = \Phi_P(0,0,\omega)$ is always the point spectrum given by the Durant–Robert model [8.43]. As previously, we can solve these integral equations using the Chebyshev spectral approximation. We can easily show that the numerical evaluation of the spectral power density of the displacement can be reduced to:

$$S_u(M;M,\omega) = \Phi(\omega)\frac{L_x^2}{4}\frac{L_y^2}{4}\sum_{n_x,n_y}^{N_x,N_y}\sum_{n'_x,n'_y}^{N_x,N_y} u_{n_x n_y}(M)\, u^\star_{n'_x n'_y}(M)\, \mathcal{I}_{n_x n_y n'_x n'_y}(\omega),$$

where $\mathcal{I}_{n_x n_y n'_x n'_y}(\omega)$ is a quadruple integral. The variable separation of the displacement and the Corcos model transform it into a double integral product $\mathcal{I}_{n_x n_y n'_x n'_y}(\omega) = \mathcal{I}^x_{n_x n'_x}(\omega) \times \mathcal{I}^y_{n_y n'_y}(\omega)$ given by:

$$\mathcal{I}^x_{n_x n'_x}(\omega) = \int_{-1}^{1}\int_{-1}^{1} T_{n_x-1}(X)T_{n'_x-1}(X')\mathrm{e}^{-\alpha_x\frac{\omega|X-X'|}{U_c}}\mathrm{e}^{\imath\frac{\omega(X-X')}{U_c}}dXdX'$$

$$\mathcal{I}^y_{n_y n'_y}(\omega) = \int_{-1}^{1}\int_{-1}^{1} T_{n_y-1}(Y)T_{n'_y-1}(Y')\mathrm{e}^{-\alpha_y\frac{\omega|Y-Y'|}{U_c}}dYdY',$$

where $\mathcal{I}^x_{n_x n'_x}$ and $\mathcal{I}^y_{n_y n'_y}$ can be easily calculated using a Chebyshev–Gauss quadrature [ABR 65]:

$$\int_{-1}^{1}\frac{f(yx)}{\sqrt{1-x^2}}dx \approx \sum_{i=1}^{N} w_i f(x_i^N), w_i = \frac{\pi}{N}, x_i^N = \cos\left(\frac{2i-1}{2N}\pi\right).$$

We obtain:

$$\mathcal{I}^y_{n_y n'_y}(\omega) = \frac{\pi^2}{NN'} \sum_{i=1}^{N} \sum_{j=1}^{N'} T_{n_y-1}(y_i^N) \sqrt{1 - (y_i^N)^2} T_{n_x y-1}(x_j^{N'}) \times \sqrt{1 - (x_j^{N'})^2} \mathrm{e}^{-\alpha_x \frac{\omega |x_i^N - x_j^{N'}|}{U_c}}$$

Due to the analytic expression of the Chebyshev polynomials $T_n(y) = \cos(n \arccos(y))$, this last relation becomes:

$$\mathcal{I}^y_{n_y n'_y}(\omega) = \frac{\pi^2}{NN'} \sum_{i=1}^{N} \sum_{j=1}^{N'} \cos \frac{(n_y - 1)(2i-1)}{N}\pi \sin \frac{2i-1}{N}\pi \times \cos \frac{(n'_x - 1)(2i-1)}{N'}\pi \sin \frac{2j-1}{N'}\pi \,\mathrm{e}^{-\alpha_y \frac{\omega \left|\cos\left(\frac{2i-1}{N}\pi\right) - \cos\left(\frac{2j-1}{N'}\pi\right)\right|}{U_c}}. \quad [8.44]$$

A similar equation can be obtained for $\mathcal{I}^x_{n_x n'_x}$. The only difficulty lies in choosing the order of the Chebyshev–Gauss quadrature N and N' for $\mathcal{I}^y_{n_y n'_y}(\omega)$ and the corresponding values for $\mathcal{I}^x_{n_x n'_x}$. We recall that a Gauss quadrature of the order n is exact for a polynomial of degree $2n - 1$. Of course, these orders depend on the degree of the polynomial that will be integrated. For example, N depends on n_y. In this particular case of integration, due to the presence of the square root and exponentials, it is necessary to take a sufficiently high value. Choosing $N = 3n_y$ leads to excellent results (an error of less than 0.01%).

Figure 8.24 shows an example of the spectral density of the vibration velocity of a plate (we recall that at each frequency ω, the vibration velocity $v(\omega)$ can be calculated from the displacement by $v(\omega) = -\imath\omega u(\omega))$ excited by a turbulent boundary layer when the fluid is water with a density of $\rho_0 = 1,000$ kg/m^3 and an acoustic wave celerity of $c_0 = 1,500$ m/s with a flow rate of 10 m/s. The spectral power density of the vibration velocity $S_v = -\omega^2 S_u$ can be calculated for two configurations. The black curve is the response of the plate coupled to the water, whereas the red discontinuous curve represents the response of the plate *in vacuo* excited by a force which represents the same turbulent field. Both curves were calculated at the point with coordinates $x = 0.9L_x, y = 0.9L_y$, which allow us to highlight the fact that all the vibration modes of the plate are excited by the turbulent flow.

The response of the plate is at its maximum near the eigenmodes of the plate as well as when I_{mnij} is at its maximum. We can show that this happens when the excitation wave number $k_e = \omega/U_c$ is near a eigen wave number of the plate. For a given flow rate, there is always a frequency ω_h for which this phenomenon, called the hydrodynamic coincidence, appears. This hydrodynamic coincidence corresponds to the fact that the flow rate is equal to the bending wave propagation velocity in the plate (which is a dispersive medium). Conversely, we can calculate, for a given flow rate, the hydrodynamic coincidence frequency. We get $U_c(\omega_h) = c_p(\omega_h)$. This is:

$$k_p = \sqrt{\omega\sqrt{\rho_p h/D}} = \omega/c_p$$

$$c_p = \frac{\omega}{\sqrt{\omega\sqrt{\rho_p h/D}}} = \sqrt{\omega}/\sqrt[4]{\rho_p h/D}$$

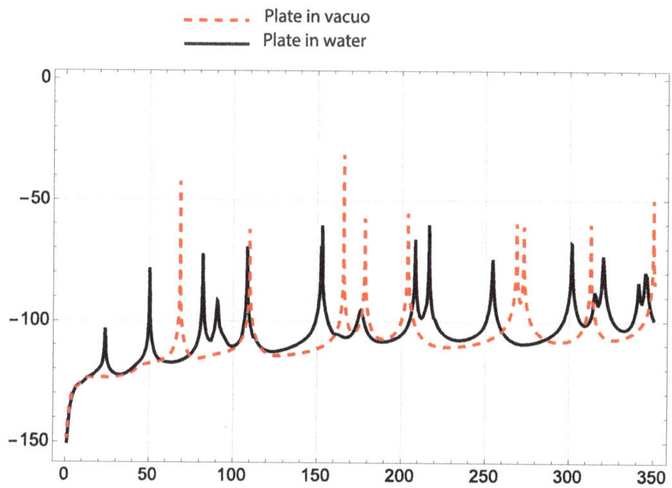

Figure 8.24. *Spectral density of the vibration velocity of a plate excited by a turbulent boundary layer (continuous curve: plate charged by the fluid; discontinuous curve: plate in vacuo). For a color version of the figure, see www.iste.co.uk/anselmet/acoustics.zip*

This is $\omega_h = U_c^2\sqrt{\rho_p h/D}$. This hydrodynamic coincidence is strictly analogous to the coincidence frequency $\omega_c = c_0^2\sqrt{\rho_p h/D}$, it is naturally lower for a subsonic flow. For the plate considered here, the critical frequency in water is approximately 45 kHz, 2,330 Hz in air and the hydrodynamic coincidence frequency here is at 2 Hz and thus the frequency is too low to be visible in the response of the plate. As clearly shown by Juvé *et al.* [HAB 99], when the convection velocity is significantly lower than the free wave velocity in the plate (or at a frequency lower than the hydrodynamic coincidence frequency), the calculated vibroacoustic response strongly depends on the model chosen and, especially, on its description of low wave numbers near the acoustic peak. In the case of hydrodynamic coincidence, the vibroacoustic response practically does not depend on the chosen excitation model, because in this case the response of the plate is governed by the energy near the convective peak that all the models describe quite similarly, as shown in section 10.2.

8.8. Aeroelastic coupling and sloshing

8.8.1. *Sloshing*

Any structure with a non-circular section is likely to be exposed to the sloshing effect (electric cables loaded with ice, bridge decks in the wind, cooling towers, bridge piers in a current, turbine blades and wings). This effect is caused by a variation of the lift and drag forces of non-circular sections during the movements of this section, which produces a self-sustained oscillation. We can find very detailed developments in the work by Blevins [BLE 90a], which inspired the following work.

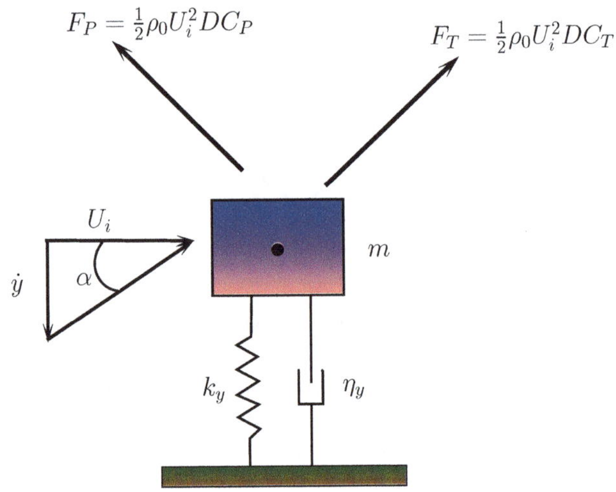

Figure 8.25. *Geometry of sloshing*

To simplify the analysis, we examine a steady state for which the forces induced by the fluid motion are solely due to the instantaneous velocity. In particular, we assume that the movement of the structure does not affect the flow. We focus on a mechanical 1-D model consisting of a mass (m) – spring (k_y) – damper (c_y) system subject to constant uniform flow with a velocity of U of a fluid with a mass ρ_0. c_y is the proportionality constant which we can write in the form of $c_y = 2m\zeta_y\omega_y$, where

$\omega_y^2 = k_y/m$ is the eigenfrequency of the spring-mass system and ζ_y is the damping factor with $\zeta_y = \eta_y/2$, if $\eta_y \ll 1$ is the loss factor. Here, we have a fixed $\eta_y = 0.05$.

The forces acting on this system are the lift forces per unit length $F_P = \frac{1}{2}\rho_0 U_i^2 D C_P$ and drag forces per unit length $F_T = \frac{1}{2}\rho_0 U_i^2 D C_T$, where D represents a characteristic length of the body. The stability of this model is studied by developing a quasi-static model of the aerodynamic forces and trying to find the response near equilibrium. During a displacement, the angle of incidence of the fluid is $\alpha = \arctan(\dot{y}/U)$, where $\dot{y}$ is the vertical velocity. We count the positive displacement toward the bottom in accordance with the conventions used in aeronautics. $U_i^2 = U^2 + \dot{y}^2$ is the resulting incident velocity. The vertical force F_y acting on the body is obviously the result of the drag and lift forces $F_y = -F_P \cos\alpha - F_T \sin\alpha = 1/2\rho_0 U^2 D C_y$, where the vertical force coefficient $C_y = -U_i^2/U^2(C_P \cos\alpha + C_T \sin\alpha)$ depends on the shape of the body, the angle of incidence and the Reynolds number.

For low angles of incidence, α, U_i and C_y can be developed in a Taylor series:

$$\begin{aligned}
\alpha &= \frac{\dot{y}}{U} + \mathcal{O}(\alpha^2) \\
U_i &= U + \mathcal{O}(\alpha^2) \\
C_y(\alpha) &= C_y(\alpha)|_{\alpha=0} + \left[\frac{\partial C_y(\alpha)}{\partial \alpha}\right]_{\alpha=0} \alpha + \mathcal{O}(\alpha^2) \\
&= -C_P(\alpha)|_{\alpha=0} - \left[\frac{\partial C_P(\alpha)}{\partial \alpha} + C_T\right]_{\alpha=0} \alpha + \mathcal{O}(\alpha^2)
\end{aligned}$$

because $\partial(C_P \cos\alpha + C_T \sin\alpha)/\partial\alpha = \partial C_P/\partial\alpha \cos\alpha + C_P \sin\alpha + \partial C_T/\partial\alpha \sin\alpha + C_T \cos\alpha$.

The movement equation of the mass-spring-damper system subjected to the vertical force is given by:

$$\begin{aligned}
m\ddot{y} + 2m\zeta_y\omega_m\dot{y} + k_y y = F_y &= \frac{1}{2}\rho_0 U^2 D C_y \\
&= -\frac{1}{2}\rho_0 U^2 D C_P(\alpha)|_{\alpha=0} \\
&\quad + \frac{1}{2}\rho_0 U^2 D \left[\frac{\partial C_y(\alpha)}{\partial \alpha}\right]_{\alpha=0} \frac{\dot{y}}{U} + \mathcal{O}(\alpha^2).
\end{aligned}$$

In this equation, we eliminate the terms that are of an order higher than one in α and we obtain:

$$m\ddot{y} + 2m\omega_m \left(\zeta_y - \frac{\rho_0 U D}{4m\omega_y} \left[\frac{\partial C_y(\alpha)}{\partial \alpha} \right]_{\alpha=0} \right) \dot{y} + k_y y = -\frac{1}{2}\rho_0 U^2 D C_P(\alpha)|_{\alpha=0}.$$

The term $\zeta_T = \zeta_y - \rho_0 U^2 D/(4m\omega_y)[\partial C_y(\alpha)/\partial\alpha]_{\alpha=0}$ characterizes the damping of the movement and, under certain conditions, can lead to an unstable movement. It is trivial to show that the resulting movement, which is the sum of a constant term and a damped oscillation, can be written as:

$$y = \frac{1}{2k_y}\rho_0 U^2 D C_P(\alpha)|_{\alpha=0} + A_y \mathrm{e}^{-\zeta_T \omega_y t} \sin\left[\omega_y \sqrt{1 - \zeta_T^2} t + \phi_y \right].$$

It is obvious that, if ζ_T is negative, the amplitude of the movement increases over time. Conversely, if ζ_T is positive, the movement decreases over time and the movement is stable. If $\zeta_T = 0$, the movement will be stable for:

$$\frac{\partial C_y(\alpha)}{\partial \alpha} < 0 \text{ or } \frac{\partial C_P(\alpha)}{\partial \alpha} + C_T > 0.$$

Any non-circular section has a domain of incidence angle α for which $\frac{\partial C_y(\alpha)}{\partial \alpha} > 0$. The instability appears for $\zeta_T = 0$, or for a critical velocity U_c such that:

$$U_c = \frac{4m(\omega_y \zeta_y)}{\rho_0 D} \frac{1}{\frac{\partial C_y(\alpha)}{\partial \alpha}}.$$

For example, for a square section, we obtain $\partial C_y(\alpha)/\partial\alpha \approx 3$ (unstable) for $Re = 10^5$, for any wing profile [PAR 99], we obtain $\partial C_y(\alpha)/\partial\alpha \approx -6$ (stable) for $Re > 10^3$. For a rectangular section with a height D and a length $2/3D$, we get $\partial C_y(\alpha)/\partial\alpha \leq 0$ (partially stable) and for a rectangular section with a height $2/3D$ and a length D, we obtain $\partial C_y(\alpha)/\partial\alpha > 0$ (unstable).

8.8.2. *Convective instability*

Another sloshing effect reveals itself when an elastic structure is in contact with a flow. This effect is not linked to the variation of aerodynamic efforts or to excitation by a turbulent boundary layer, however, it is related to an instability that arises in the structure due to the flow. This instability results from an exchange of energy between the flow and the elastic structure. This effect is responsible for garden hoses lashing, flags snapping in the wind or wings galloping. Most of the phenomena (the exchange of energy between the flow and the vibration of the structure) can be described for a simple geometry in the presence of potential flow (incompressible). This problem was treated analytically by de Langres [DEL 02], and what follows was largely inspired by this article.

We will try to demonstrate the appearance of wall waves that are "convectively" unstable (the flow transports the waves downstream and locally the response of the system eventually decreases) and absolutely unstable (the waves invade the whole area and the amplitude of the response of the structure always grows).

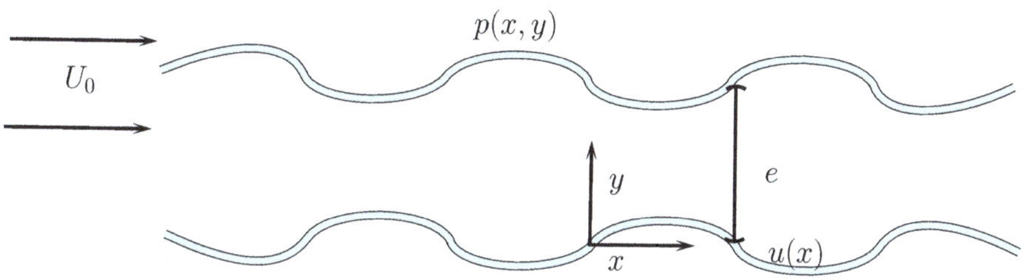

Figure 8.26. *Geometry of aeroelastic coupling*

To get a better understanding of these ideas, we examine two elementary (but not trivial) examples, one of an infinite plane, in which we introduce a membrane tension, in a flow, and one of a channel with elastic walls containing a flowing fluid. With D as the bending stiffness and T as the membrane tension, the equations of the model are:

$$D\frac{\partial^4 u(x,t)}{\partial x^4} - T\frac{\partial^2 u(x,t)}{\partial x^2} + m\frac{\partial^2 u(x,t)}{\partial t^2} = p(x, y = 0, t)$$

$$\{+p(x, y = e, t) \text{ for the channel}\}$$

$$\Delta\phi(x, y, t) = 0$$

$$\frac{\partial\phi(x, y = 0, t)}{\partial y} = \left(\frac{\partial}{\partial t} + U_0\frac{\partial}{\partial x}\right)u(x, t)$$

and $$\frac{\partial\phi(x, y = e, t)}{\partial y} = \pm\left(\frac{\partial}{\partial t} + U_0\frac{\partial}{\partial x}\right)u(x, t) \text{ for the channel.}$$

The sign $\pm$ must be selected depending on the type of wall movement that we are considering (sinusoidal or varicose wave). To these equations, we add the conservation of energy for $\phi(x, y, t)$. The potential $\phi(x, y, t)$ is linked to the pressure in the fluid by the relation $p(x, y, t) = -\rho_0\left(\partial\phi(x, y, t)/\partial t + U_0\partial\phi(x, y, t)/\partial x\right)$.

Initially, we consider the plate as being in contact only with the flow. We focus on waves which propagate in the structure in the form of

$u(x,t) = \exp(\imath kx)\exp(-\imath\omega t)$, with $k \in I\!R$. The continuity condition of the displacements normal to the wall, therefore, gives us for the potential:

$$\frac{\partial}{\partial y}\phi(x,y=0,t) = \left(\frac{\partial}{\partial t} + U_0\frac{\partial}{\partial x}\right)u(x,t) = (-\imath\omega + U_0\imath k)\mathrm{e}^{\imath kx}\mathrm{e}^{-\imath\omega t}.$$

Thus, that is $\phi(x,y,t) = (-\imath\omega + U_0\imath k)/\gamma\exp(\imath kx + \gamma y)\exp(-\imath\omega t)$. The equation of the potential, $\Delta\phi(x,y,t) = 0$, gives us $(-k^2 + \gamma^2)\mathrm{e}^{\imath kx+\gamma y}$ $\exp(-\imath\omega t) = 0$. This is $\gamma = |k|$. From there, we can deduce that:

$$p(x,y=0,t) = -\rho_0\left(\frac{\partial\phi(x,y,t)}{\partial t} + U_0\frac{\partial\phi(x,y,t)}{\partial x}\right) = \rho_0\frac{(\omega - U_0k)^2}{|k|}\mathrm{e}^{\imath kx}\mathrm{e}^{-\imath\omega t}.$$

The dispersion relation of the coupled strained plate is thus $\mathcal{D}(\omega,k) = Dk^4 + Tk^2 - m\omega^2 - \rho_0(\omega - U_0k)^2/|k| = 0$. We can write this dispersion relation in the more general form $K(k) - \omega^2M(k) - C(k)(\omega - U_0k)^2 = 0$, where $K(k)$ reflects the effect of mechanical elasticity, $M(k)$ reflects the effect of mechanical mass and $C(k)$ reflects the effects of mass and stiffness introduced by the fluid. We solve this dispersion relation with respect to ω and, very simply, we obtain:

$$\omega^{\pm} = \frac{U_0kC(k) \pm \sqrt{K(k)(M(k)+C(k)) - U_0^2k^2M(k)C(k)}}{M(k)+C(k)}.$$

The condition for the system to be stable is that the imaginary part of ω must be negative. This means $K(k)(M(k)+C(k)) - U_0^2k^2M(k)C(k) < 0$. For a real wave number k, the flow rate from which we have an instability is, therefore, such that:

$$U_c = \min_{k\in I\!R}\left(\frac{K(k)(M(k)+C(k))}{k^2M(k)C(k)}\right)^{1/2}.$$

Below this flow rate, the system remains stable. To characterize the instability, we can study the momentum response $G(x,t)$ of the system. If $\lim_{t\to\infty}|G(0,t)| \to 0$, we are talking of convective instability and if $\lim_{t\to\infty}|G(0,t)| \to \infty$, we are talking of absolute instability. In practice, we can study the nature of the instability without examining the behavior of the momentum response. For this, we focus on the frequency ω_0 and the wave number k_0, implicitly defined by:

$$\mathcal{D}(\omega_0,k_0) = 0 \text{ and} \frac{\partial\mathcal{D}}{\partial k}(\omega_0,k_0) = 0.$$

The absolute or convective nature of the instability depends on the sign of the imaginary part $\Im(\omega_0)$. It is absolute if $\Im(\omega_0) > 0$, and convective if $\Im(\omega_0) < 0$. We aim to find out when the instability changes its nature. The velocity from which a convective instability becomes absolutely unstable is called the transition velocity U_T. We can show (this is difficult) that it will appear as the triple root of the dispersion relation:

$$\mathcal{D}(\omega_T, k_T, U_T) = 0, \frac{\partial \mathcal{D}}{\partial k}(\omega_T, k_T, U_T) \text{ and } \frac{\partial^2 \mathcal{D}}{\partial k^2}(\omega_T, k_T, U_T) = 0,$$

always for a real k_T.

In the case of a plate with no tension, we have the dispersion relation $\tilde{k}^4 - \tilde{\omega}^2 - (\tilde{\omega} - \tilde{U}\tilde{k})^2/|\tilde{k}| = 0$, with $\tilde{k} = km/\rho_0$, $\tilde{\omega} = \omega m^{5/2}/\rho_0^2 D^{1/2}$ and $\tilde{U} = Um^{3/2}/(\rho_0 D^{1/2})$. The convective instability occurs for:

$$\tilde{U}_C = \min_{\tilde{k} \in I\!R} \left(\frac{\tilde{k}^4 \left(1 + \frac{1}{|\tilde{k}|}\right)}{k^2 \left(\frac{1}{|\tilde{k}|}\right)} \right)^{1/2} = \min_{\tilde{k} \in I\!R} \left(\tilde{k}^2 \left(|\tilde{k}| + 1 \right) \right)^{1/2},$$

that is $\tilde{U}_C = 0$: a plate coupled to a flow always presents a convective instability due to its dispersive nature (regardless of the frequency, there is always a wave which propagates at a non-zero phase velocity).

We can show that the transition to absolute instability occurs for a flow rate of:

$$\tilde{U}_T = \frac{2^{1/2} 5^{5/4}}{3^{3/4}} \left(2 - \frac{\sqrt{15}}{2} \right)^{3/2} \approx 0.074.$$

In physical units, for a steel plate with a thickness of 1 cm in water, this corresponds to $U_0 \approx 0.074 \times \rho_0 \sqrt{D}/m^{3/2} \approx 14 m/s$, this is approximately 27 knots, which is a value commonly reached by ship. In general, the shells are thicker than 1 cm, which shifts the transition toward higher velocities.

For a strained membrane without stiffness, we obtain the dispersion relation $\tilde{k}^2 - \tilde{\omega}^2 - (\tilde{\omega} - \tilde{U}\tilde{k})^2/|\tilde{k}| = 0$, with $\tilde{k} = km/\rho_0$, $\tilde{\omega} = \omega m^{3/2}/\rho_0 T^{1/2}$ and $\tilde{U} = Um^{1/2}/T^{1/2}$. The convective instability appears for:

$$\tilde{U}_C = \min_{\tilde{k} \in I\!R} \left(\frac{\tilde{k}^2 \left(1 + \frac{1}{|\tilde{k}|}\right)}{k^2 \left(\frac{1}{|\tilde{k}|}\right)} \right)^{1/2} = \min_{\tilde{k} \in I\!R} \left(|\tilde{k}| + 1 \right)^{1/2},$$

that is $\tilde{U}_C = 1$: a membrane coupled to a flow presents convective instability only when the flow rate exceeds the wave propagation velocity in the membrane ($c_t^2 = T/m$).

We can show that the transition to absolute instability occurs for a flow rate of:

$$\tilde{U}_T = \left(6\sqrt{3} - 9\right)^{1/2} \approx 1.18.$$

The absolute instability threshold is very close to the wave propagation velocity in the membrane. This explains why the sails of a boat, when they are not taut enough (or when the flags are only slightly strained), flap in the wind and why even a low sail or flag tension stops the oscillations of the cloth and flapping.

8.8.3. *Kelvin–Helmholtz instability*

The previous results can be used to interpret the Kelvin–Helmholtz instability, which appears at the interface of two fluids separated by an infinitely thin surface which has a surface tension that corresponds to a membrane without a mass in contact with two fluids. The dispersion relation of the separation surface, whose geometry is shown in Figure 8.27, regarded as a membrane with a mass m, is simply the dispersion relation of a strained membrane without stiffness, in which we have taken into account the fluid at rest. Thus, we have:

$$Tk^2 - \left(m + \rho_1 \frac{1}{|k|}\right)\omega^2 - \rho_1 \frac{(\omega - U_0 k)^2}{|k|} = 0.$$

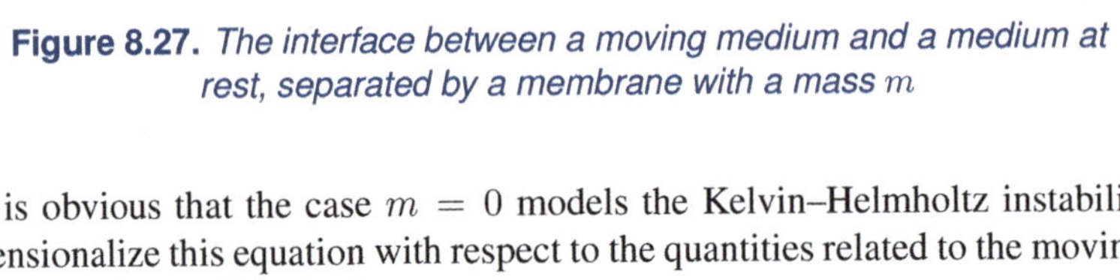

Figure 8.27. *The interface between a moving medium and a medium at rest, separated by a membrane with a mass m*

It is obvious that the case $m = 0$ models the Kelvin–Helmholtz instability. We adimensionalize this equation with respect to the quantities related to the moving fluid with the density ρ_0. We obtain:

$$\hat{k}^2 - \left(1 + \frac{\chi}{|\hat{k}|}\right)\hat{\omega}^2 - \frac{(\hat{\omega} - \hat{U}\hat{k})^2}{|\hat{k}|} = 0,$$

where $\chi = \rho_1/\rho_0$. The critical velocity is:

$$\hat{U}_C = \min_{\hat{k} \in \mathbb{R}} \left(\frac{\hat{k}^2 + \hat{k}(1+\chi)}{\hat{k} + \chi}\right)^{1/2}.$$

$\hat{U}_C$ varies significantly from $\hat{U}_C = 1$ for $\chi = 0$, which corresponds to the flow over a flexible membrane, to $\hat{U}_C = 0$ as soon as the area at rest is taken into account (i.e. $\chi > 0$). We can show that when $\chi > 1/3$, the instability is always absolute. In practice, this instability can explain the formation of waves at the interface of two fluids in relative movement with respect to each other, such as the waves at the surface of the sea (partly) or the oscillations of clouds observed at the interface of different atmospheric layers.

9

Basic Theoretical Aeroacoustics Models

9.1. Preamble

In this chapter and the two that follow, we will present and discuss the main theoretical developments which constitute the basis of aeroacoustics. The goal is not to be exhaustive, but to provide the ingredients necessary to the understanding of the most important phenomena, and with the objective of either implementing experimental processes adapted to the analysis of specific practical problems, or using in a reasoned, and thus in a reasonable, manner the numerical modeling tools that have almost become common usage because of their implementation in commercial computational codes.

Throughout this, we are, therefore, going to present the derivation of Lighthill's equation, as well as subsequent developments that make it possible to take into account, for example, the influence of walls or flow unsteadiness, as well as the nonlinear coupling between turbulent velocity fields and sound waves. In this chapter, we will then examine the various analytical forms that have been proposed to model the source terms in idealized turbulence (i.e. homogeneous and isotropic). We will see that the various approximations which have been developed over the years can result in radiated noise levels that greatly differ from one another.

It should be noted that, although this is a relatively new specialization, aeroacoustics has already notably been the subject of a number of books ([GOL 76] and [HOW 03], in particular), of five general review articles published in the *Annual Review of Fluid Mechanics* [FFO 77, GOL 84, TAM 95a, WAN 06, WEL 97] and some others more specific (such as on active noise control), as well as several review articles published in the journal *Progress in Aerospace Sciences*.

Finally, the few illustrations that follow allow the visualization of sound wave fields that can be generated by typical flows such as mixing layers and jets. Therefore, subjects related to these acoustic waves, their origin and some of their main properties, are going to be studied here.

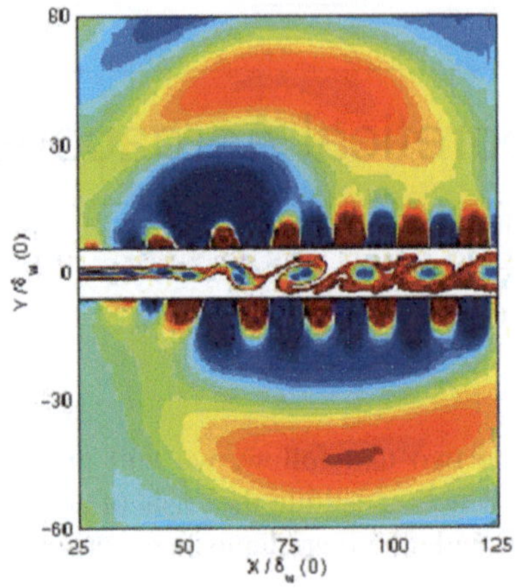

Figure 9.1. *Noise generated in a mixing layer by the pairing of vortices. For a color version of the figure, see www.iste.co.uk/anselmet/acoustics.zip*

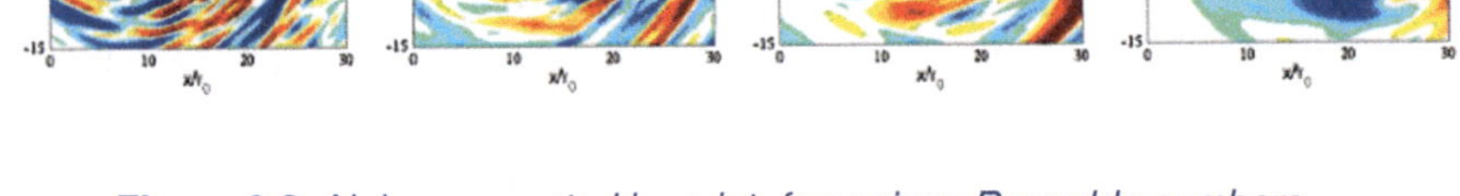

Figure 9.2. *Noise generated by a jet, for various Reynolds numbers. For a color version of the figure, see www.iste.co.uk/anselmet/acoustics.zip*

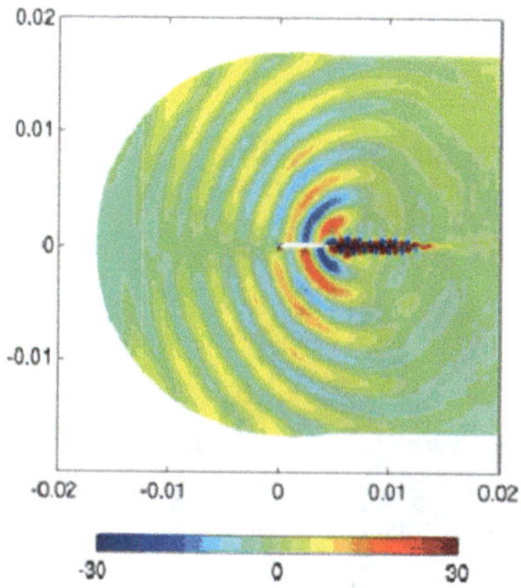

Figure 9.3. *Noise generated by an aircraft wing and its wake. These three numerical results have been obtained by C. Bailly's group, École Centrale de Lyon. For a color version of the figure, see www.iste.co.uk/anselmet/acoustics.zip*

9.2. Lighthill's equation and some of the generalizations that have followed

The analogy proposed by Lighthill [LIG 52] is the first simple formulation of the coupling between the generation of noise by turbulent fluid flow and its propagation. This wave equation can be easily solved in full form if Green's function associated with the problem under study is known, which is the case in free space. In order to take into account in the propagation operator the effects of the mean flow on the radiated sound field, the Lighthill analogy has been redrafted notably by Phillips [PHI 60], and then by Lilley [LIL 73]. These analytical approaches have made it possible to bring forward the dimensional laws of the evolution with the velocity U of the noise production. Thus, the acoustic intensity of a jet varies according to a law in U^8 law in subsonic regime, and a U^3 law in supersonic regime. The use of a statistical description of turbulence to model the acoustic source terms [PRO 52, LIG 54], associated with the explicit accounting of the mean shear, also allowed for the definition of the directivity laws of acoustic radiation of jets [RIB 64, RIB 69, RIB 77]. It can be finally noted that there are numerous extensions of the analogy. We will cite Powell's reformulation of the source terms into vorticity terms [POW 64], recovered by Howe [HOW 75], and the developments of [CUR 55, FFO 69, COR 63, CHA 80, FAR 88] which include the presence of immobile or moving solid walls. These are the main different approaches, to be associated with the different analytical formulations used to model the fourth-order correlations of the turbulent velocity field which are involved in the quadrupole source term, and we will now present them in a synthetic manner. In the text that follows, we will confine ourselves (in general) to subsonic flows and without thermal phenomena.

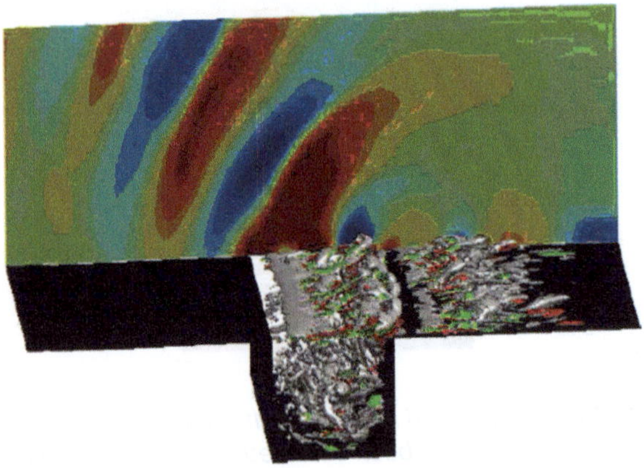

Figure 9.4. *Noise generated by the interaction of a flow with a surfacing cavity. Numerical works from X. Gloerfelt, École des Arts et Métiers, Paris. For a color version of the figure, see www.iste.co.uk/anselmet/acoustics.zip*

The Lighthill equation [LIG 52] is obtained from the equation of conservation of mass:

$$\frac{\partial \rho}{\partial t} + \frac{\partial(\rho u_i)}{\partial x_i} = 0$$

and from the Navier–Stokes equation:

$$\frac{\partial(\rho u_i)}{\partial t} + \frac{\partial(\rho u_i u_j)}{\partial x_j} = -\frac{\partial p}{\partial x_i} + \frac{\partial \tau_{ij}}{\partial x_j},$$

where, for a Newtonian fluid, the viscous stress tensor τ_{ij} is written as:

$$\tau_{ij} = \mu \left(\frac{\partial u_i}{\partial x_j} + \frac{\partial u_j}{\partial x_i} \right) - \frac{2}{3}\mu \frac{\partial u_k}{\partial x_k} \delta_{ij}.$$

The difference between the time derivative of the first equation and the divergence of the second equation gives:

$$\frac{\partial}{\partial t}\left(\frac{\partial \rho}{\partial t} + \frac{\partial(\rho u_i)}{\partial x_i} \right) - \frac{\partial}{\partial x_i}\left(\frac{\partial(\rho u_i)}{\partial t} + \frac{\partial(\rho u_i u_j)}{\partial x_j} \right) = \frac{\partial}{\partial x_i}\left(\frac{\partial p}{\partial x_i} - \frac{\partial \tau_{ij}}{\partial x_j} \right),$$

where

$$\frac{\partial^2 \rho}{\partial t^2} - \frac{\partial^2(\rho u_i u_j)}{\partial x_i \partial x_j} = \frac{\partial^2}{\partial x_i \partial x_j}\left(p\delta_{ij} - \tau_{ij}\right),$$

and, by introducing the reference velocity c_0 and T_{ij} the Lighthill tensor:

$$\frac{\partial^2 \rho}{\partial t^2} - c_0^2 \frac{\partial^2 \rho}{\partial x_i \partial x_i} = \frac{\partial^2 T_{ij}}{\partial x_i \partial x_j},$$

with $c_0^2 = p_0/\rho_0$, the speed of sound in the medium at rest where it propagates, and $T_{ij} = \rho u_i u_j + \left(p - \rho c_0^2\right)\delta_{ij} - \tau_{ij}$. Naturally, as proposed in Chapter 4, this equation can be rewritten in terms of fluctuations (or variations) of ρ with respect to its value ρ_0 in the medium at rest and not subject to the effect of sources, $\rho' = \rho - \rho_0$. This amounts to more clearly highlighting that equation [9.1] is a wave propagation equation corresponding to the variations in ρ' (that some also call ρ_a for *acoustic fluctuations of* ρ):

$$\frac{\partial^2 \rho'}{\partial t^2} - c_0^2 \frac{\partial^2 \rho'}{\partial x_i^2} = \frac{\partial^2 T_{ij}}{\partial x_i \partial x_j}, \qquad [9.1]$$

with then $T_{ij} = \rho u_i u_j + \left((p - p_0) - c_0^2(\rho - \rho_0)\right)\delta_{ij} - \tau_{ij}$.

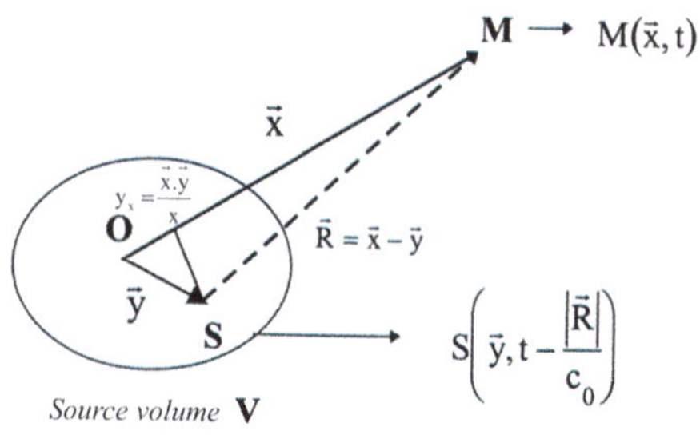

Figure 9.5. *Notation used for the Lighthill expansion*

There is of course an equivalent notation to [9.1], but it makes use of the pressure fluctuation p' as main variable instead of ρ': $\frac{1}{c_0^2}\frac{\partial^2 p'}{\partial t^2} - \frac{\partial^2 p'}{\partial x_i^2} = \frac{\partial^2 T_{ij}}{\partial x_i \partial x_j}$.

Lighthill's original idea was thus to isolate in the left-side member of equation [9.1] the free-space acoustic propagation operator and to consider the right one as a source term, which is here of quadrupolar nature (see, for example, the presentation of the different types of sources and their properties which is detailed in Chapter 5).

Recall that Green's function associated with a free space is given by[1], $G(\vec{x},t/\vec{y},\tau) = 1/(4\pi R)\delta(t-\tau-R/c_0)$, which characterizes the impulse response of the d'Alembert equation $1/c_0^2\partial^2 G/\partial t^2 - \Delta G = \delta(\vec{x}-\vec{y})\delta(t-\tau)$. We then get the analytical expression of the far-field sound pressure radiated by source S (or, more accurately, by the distribution throughout the whole volume V of the infinitesimal sources S) :

$$p'(\vec{x},t) = \frac{1}{4\pi}\int\limits_{V(\vec{y})}\int\limits_{\tau} S(\vec{y},\tau)\delta(t-\tau-\frac{R}{c_0})\frac{d\vec{y}}{R}d\tau.$$

Let here:

$$p'(\vec{x},t) = \frac{1}{4\pi c_0^2}\int\limits_{V(\vec{y})} \frac{\partial^2 T_{ij}}{\partial y_i \partial y_j}(\vec{y},t-\frac{R}{c_0})\frac{d\vec{y}}{R},$$

where $R = \left|\vec{R}\right| = |\vec{x}-\vec{y}|$ is the distance between the different points of the distributed source S and the observer M (see Figure 9.5). From a far-field perspective and with a low Mach number, it can be shown that the effect of retarded times which appear in the previous equation can be overlooked on the one hand, and that the spatial derivatives can be replaced by temporal derivatives ([PRO 52, LIL 93], notably) on the other hand. As a matter of fact, if $x \gg y$ (where similarly, x is the shorthand notation for $|\vec{x}|$), it can be written that $R = |\vec{x}-\vec{y}| \approx x - (\vec{x}.\vec{y})/x + \mathcal{O}(y^2/x)$ and that:

$$\frac{\partial F(t-R/c_0)}{\partial y_i} = -\frac{R_i}{Rc_0}\frac{\partial F(t-R/c_0)}{\partial t} \approx -\frac{1}{c_0}\frac{x_i}{x}\frac{\partial F(t-R/c_0)}{\partial t},$$

(this is what is commonly called the far-field approximation), such that:

$$p'(\vec{x},t) \approx \frac{1}{4\pi c_0^2 x}\frac{x_i x_j}{x^2}\int\limits_{V(\vec{y})} \frac{\partial^2 T_{ij}}{\partial t^2}(\vec{y},t-\frac{R}{c_0})d\vec{y}. \qquad [9.2]$$

It may be noted that the approximation $t - R/c_0 = t - |\vec{x}-\vec{y}|/c_0 \approx t - x/c_0 + \vec{x}.\vec{y}/xc_0 + ...$ also implies that the retarded time $(\vec{x}.\vec{y})/xc_0 \approx L/c_0$ be compared with a characteristic turbulence time, which is generally taken equal to L/u_{rms}. Therein u_{rms} represents the mean square root value of the velocity fluctuations and L is a length that gives the order of magnitude of the fluid volume ($V \approx L^3$) which corresponds to the acoustic sources (to quantify the influence of

1 For more details on Green's functions and their properties, the readers can consult Chapters 3 and 4.

retarded times between the origin point and the other points of the source S, as it can be seen in Figure 9.5). The ratio of these two times then equals $(L/c_0)/(L/u_{rms}) = M_t$, the turbulent Mach number. Thus, the previous expansions remain valid as long as $M_t \ll 1$, which thus defines another limit to the compact source approximation. In practice, this remains valid as long as the common Mach number of the mean flow, $M = U/c_0$, is less than about 2.

In far-field, the sound intensity $I(\vec{x})$ perceived at an observation point, or the acoustic energy radiated by a flow, can then be written from the autocorrelation function of fluctuations of the sound pressure $I(\vec{x},\tau)$, which is given by $I(\vec{x},\tau) = \overline{p'(\vec{x},t)p'(\vec{x},t-\tau)}/(\rho_0 c_0)$, that is:

$$I(\vec{x}) = \frac{\overline{p'^2(\vec{x},t)}}{\rho_0 c_0}.$$

The relationship that defines $I(\vec{x},\tau)$ can naturally be rewritten in the spectral space, in order to obtain the expression of the spectrum, $p_s(\vec{x},\omega)$, of the fluctuations of the radiated noise:

$$p_s(\vec{x},\omega) = \frac{1}{2\pi}\int_{-\infty}^{+\infty} I(\vec{x},\tau)e^{i\omega\tau}d\tau,$$

a very important quantity, which we will later consider in detail. In addition, an order of magnitude of the total power radiated by the fluid volume can then be obtained from the integral of this quantity over the sphere of radius x (if we consider first the radiation of a single vortex, of characteristic size l, typically equal to the integral scale of the turbulence, which will be accurately defined in section 9.3):

$$4\pi x^2 I \approx 4\pi x^2 \frac{\overline{p'^2}}{\rho_0 c_0} \approx 4\pi x^2 \frac{c_0^3}{\rho_0}\overline{\rho'^2} \approx \rho_0 l^2 \frac{U^5}{c_0^5}U^3,$$

since

$$\rho'(\vec{x},t) \approx \frac{1}{4\pi c_0^4 x}\frac{x_i x_j}{x^2}\int_{V(\vec{y})} \frac{\partial^2 T_{ij}}{\partial t^2}(\vec{y}, t-\frac{R}{c_0})d\vec{y} \approx \frac{1}{4\pi}\frac{1}{c_0^4 x}\frac{\rho_0 U^2}{(l/U)^2}l^3.$$

It should be recalled that, in the context of this book, the compressibility effects present in high-speed flows ($M \approx 1$ or $M > 1$) and thermal effects are not discussed or considered, so that only the terms related to velocity are retained in the expression of T_{ij}. If we consider that the total volume V contains a number V/l^3 of independent vortices, then the acoustic power generated by this volume of turbulent fluid is equal to $V\rho_0 U^5/c_0^5 \times U^3/l$. Thus, the acoustic power radiated by the volume V varies in subsonic flow such as U[8], a very exciting result that has from the beginning contributed to the success of Lighthill's theory. In addition, this calculation allows the flow acoustic effectiveness η to be inferred, which is the ratio between the radiated acoustic energy and the mechanical energy injected into the flow, that is for its part, proportional to $V\rho_0 U^3/l$: $\eta = P_{\text{acoustic}}/P_{\text{mecanical}} \approx U^5/c_0^5 = M^5$. It therefore appears that phenomena related to fluid mechanics constitute a very inefficient noise generator since, if for example, the Mach number is of the order of 0.4 at most (a common value in applications related to land vehicles such as trains and automobiles), then η remains smaller than 0.01 (i.e. 1%). This has two consequences that can be immediately deducted, and that will be considered in detail later in this chapter:

1) The analytical expansions that attempt to calculate $I(\vec{x},\tau)$ in a more precise manner or the acoustic spectrum associated with $p_s(\vec{x},\omega)$ must take into account in a very fine manner the different mechanisms (and we will see that these mechanisms are complex ...) related to turbulence which are responsible for the generation of noise. The reason is that they are not commensurate with the usual mechanisms that generate the turbulent kinetic energy.

2) The numerical tools that are developed to simulate the acoustic radiation must also be sophisticated, because they must be able to correctly produce and propagate the pressure field acoustic disturbances whose amplitude is very low compared with the turbulence kinetic energy (and naturally also the kinetic energy of the mean motion of the fluid).

On the other hand, since it characterizes the solution of the previously obtained far-field Lighthill equation (equation [9.2]), the autocorrelation function $I(\vec{x},\tau)$ can be expressed directly with respect to the source term:

$$I(\vec{x},\tau) \approx \frac{x_i x_j x_k x_l}{16\pi^2 c_0^5 \rho_0 x^6} \iint_{V(\vec{y})} \overline{\frac{\partial^2 T_{ij}}{\partial t^2}(\vec{y}',t')\frac{\partial^2 T_{kl}}{\partial t^2}(\vec{y}'',t''+\tau)d\vec{y}'d\vec{y}''}, \qquad [9.3]$$

where t' and t'' are the travel times of the waves originated from the localized source points in $\vec{y}'$ and $\vec{y}''$ to reach the observation point located in $\vec{x}$. The approximation $p'(\vec{x},t) = \rho'(\vec{x},t)c_0^2$ which binds p' and ρ' in far-field has been used of course. In addition, regarding observation times, we can write $t' = t - |\vec{x} - \vec{y}'|/c_0$ and $t'' = t - |\vec{x} - \vec{y}''|/c_0$.

In the particular case of steady mean flows (case of turbulence grid or turbulent jets studied here notably), Goldstein [GOL 76] shows that the hypothesis of stationary turbulence allows the correlation of the second derivatives of the source term (see below the detailed computation) to be transformed into a fourth-order time derivative of a fourth-order correlation of the velocity field which is in fact a spatio-temporal correlation (relating to two points and two times). The last equation can then be written as follows:

$$I(\vec{x},\tau) \approx \frac{x_i x_j x_k x_l}{16\pi^2 c_0^5 \rho_0 x^6} \iint\limits_{V(\vec{y})} \frac{\partial^4}{\partial\tau^4} \overline{T_{ij}(\vec{y}',t')T_{kl}(\vec{y}'',t''+\tau)} d\vec{y}' d\vec{y}''.$$

In order to better visualize the exact nature of this fourth-order correlation function, a first change of variable is achieved, $t^* = t - |\vec{x} - \vec{y}'|/c_0$, which makes it possible to write:

$$\begin{aligned}
&\overline{T_{ij}(\vec{y}',t')T_{kl}(\vec{y}'',t''+\tau)} \\
&= \lim_{T\to\infty} \frac{1}{2T} \int_{-T}^{T} T_{ij}\left(t - \frac{\left|\vec{x}-\vec{y}'\right|}{c_0}\right) T_{kl}\left(t + \tau - \frac{\left|\vec{x}-\vec{y}''\right|}{c_0}\right) dt, \\
&= \lim_{T\to\infty} \frac{1}{2T} \int_{-T}^{T} T_{ij}(t^*) T_{kl}\left(t^* + \tau - \frac{\left|\vec{x}-\vec{y}''\right| - \left|\vec{x}-\vec{y}'\right|}{c_0}\right) dt^*, \\
&= \overline{T_{ij}(\vec{y}',t) T_{kl}\left(\vec{y}'', t + \tau - \frac{\left|\vec{x}-\vec{y}''\right| - \left|\vec{x}-\vec{y}'\right|}{c_0}\right)}.
\end{aligned}$$

It then yields for the autocorrelation function of the acoustic pressure:

$$\begin{aligned}
I(\vec{x},\tau) \approx{}& \frac{x_i x_j x_k x_l}{16\pi^2 c_0^5 \rho_0 x^6} \\
&\times \iint\limits_{V(\vec{y})} \frac{\partial^4}{\partial\tau^4} \overline{T_{ij}(\vec{y}',t) T_{kl}(\vec{y}'', t + \tau - \frac{\left|\vec{x}-\vec{y}''\right| - \left|\vec{x}-\vec{y}'\right|}{c_0})} d\vec{y}' d\vec{y}''.
\end{aligned} \qquad [9.4]$$

If the source separation vector is now defined such that $\vec{\zeta} = \vec{y''} - \vec{y'}$, in far-field, this verifies $\left|\vec{\zeta}\right| << |\vec{x}|$ and since $\left|\vec{x} - \vec{y'}\right|$ varies as $\left|\vec{x} - \vec{y'}\right| \approx x - \vec{x}.\vec{y'}/x$, it can be written, in the case of large observation distances x, $(\left|\vec{x} - \vec{y'}\right| - \left|\vec{x} - \vec{y''}\right|)/c_0 \approx \vec{x}.\vec{\zeta}/xc_0$, such that the expression of the function $I(\vec{x},\tau)$ becomes:

$$I(\vec{x},\tau) \approx \frac{x_i x_j x_k x_l}{16\pi^2 c_0^5 \rho_0 x^6} \iint\limits_{V(\vec{y})} \frac{\partial^4}{\partial\tau^4} \overline{T_{ij}(\vec{y'},t) T_{kl}(\vec{y'} + \vec{\zeta}, t + \tau + \frac{\vec{x}.\vec{\zeta}}{xc_0})} d\vec{y'} d\vec{\zeta},$$

that can also be rewritten in the compact form:

$$I(\vec{x},\tau) \approx \frac{x_i x_j x_k x_l}{16\pi^2 c_0^5 \rho_0 x^6} \iint_{V(\vec{y})} \frac{\partial^4}{\partial \tau^4} R_{ijkl}(\vec{y'},\vec{\zeta},\tau + \frac{\vec{x}.\vec{\zeta}}{xc_0}) d\vec{y'} d\vec{\zeta}, \qquad [9.5]$$

where the correlation function $R_{ijkl}(\vec{y'},\vec{\zeta},\tau + (\vec{x}.\vec{\zeta})/xc_0)$ is, therefore, a fourth-order velocity correlation function, which implies two different spatial points and two different times. It is, therefore, a particularly complex statistical object, which is not frequently studied because it does not play a significant role in classical studies of turbulence, and this all the more since it can be seen that its fourth-order time derivative is comprised in equation [9.5]. It is nonetheless by means of this complex statistical tool that noise generation is carried out within a turbulent flow. As we will see in the next few sections, these correlations are often modeled as being the product of a spatial correlation and a temporal correlation (the latter being independent of the scale of the vortices that are being considered). It was only very recently, on the basis of the latest theoretical developments on turbulence – which are still widely controversial – that more sophisticated models have been proposed (see section 9.6).

It is, therefore, as we had already pictured by considering the orders of magnitude of the phenomena involved in aeroacoustics, from the perspective of the analytical modeling of turbulence, a problem especially difficult for which there is presently no fully satisfactory solution, even though significant progress has been made for about 40 years. This is what we will discuss in the next few sections of this chapter. Nevertheless, it seems opportune to point out, in the few lines that conclude this present section, that the Lighthill formulation needs to be completed in order to take into account other commonly encountered factors, which are related to the characteristics of fluid flows in real situations, and in particular in transport-related industries.

Thus, to complete the Lighthill equation, which was established in free space, Curle [CUR 55] and Ffowcs Williams and Hawkings [FFO 69] have recovered the same reasoning considering the presence of solid walls and that of mass flow variations. If Q represents the mass flow introduced by unit volume, the continuity equation becomes:

$$\frac{\partial \rho}{\partial t} + \frac{\partial(\rho u_i)}{\partial x_i} = Q,$$

while the Navier–Stokes equation, denoting by f_i the ith component of the external force per unit volume, becomes:

$$\frac{\partial(\rho u_i)}{\partial t} + \frac{\partial(\rho u_i u_j)}{\partial x_j} = f_i - \frac{\partial p}{\partial x_i} + \frac{\partial \tau_{ij}}{\partial x_j},$$

so that the equation governing the propagation of sound waves is then written as:

$$\frac{\partial^2 \rho}{\partial t^2} - c_0^2 \frac{\partial^2 \rho}{\partial x_i \partial x_i} = \frac{\partial Q}{\partial t} - \frac{\partial f_i}{\partial x_i} + \frac{\partial^2 T_{ij}}{\partial x_i \partial x_j}.$$

It therefore appears that, in the most general case, it is important to consider not only the quadrupolar contribution of Reynolds stresses (Lighthill tensor), but also the monopolar source related to mass flow variations as well as the dipole source related to the forces exerted by the walls on the flow. Figure 9.6(a) shows the placement within the flow of a turbojet of the different noise sources. In addition, as we will see in Chapter 11, recent works based on direct homogeneous and isotropic turbulence numerical simulations [WHI 00], nonetheless confined to a finite cross-sectional region in order for sound waves to propagate only in the third direction of space, in plane-waves, show that turbulence then corresponds to a dipolar source and not to a quadrupolar source.

It seems, therefore, very important to stress these points right now, to keep in mind that, for applications in the field of land transportation, for example, an identification of the nature of the sources seems essential before seeking to apply a more or less complicated analytical model to represent the fourth-order correlations that appear in equation [9.5]. An experimental way to characterize the sources in a global manner (that is without trying to locate them physically) appears to be based on the variation of the intensity radiated with regard to the velocity since, as shown in Figure 9.6(b), monopolar, dipolar and quadrupolar-type sources radiate aeroacoustics powers which vary, respectively, as the 4th, 6th and 8th-powers of the general velocity of the flow. Thus, if the objective being sought is to reduce aeroacoustic noises, the approach appears rather complex since it must account for the different sources of noise, and of their relative importance depending on the velocity of the flow (or that of the vehicle), as it is also shown in Figure 9.7.

EXERCISE.– Calculate in detail the fourth-order correlation term.

SOLUTION.– *Let W_{ijkl} be the correlation tensor of source terms appearing in relation [9.3]:*

$$W_{ijkl} = \overline{\frac{\partial^2 T_{ij}}{\partial t^2}(y', t') \frac{\partial^2 T_{kl}}{\partial t^2}(y'', t'' + \tau)}.$$

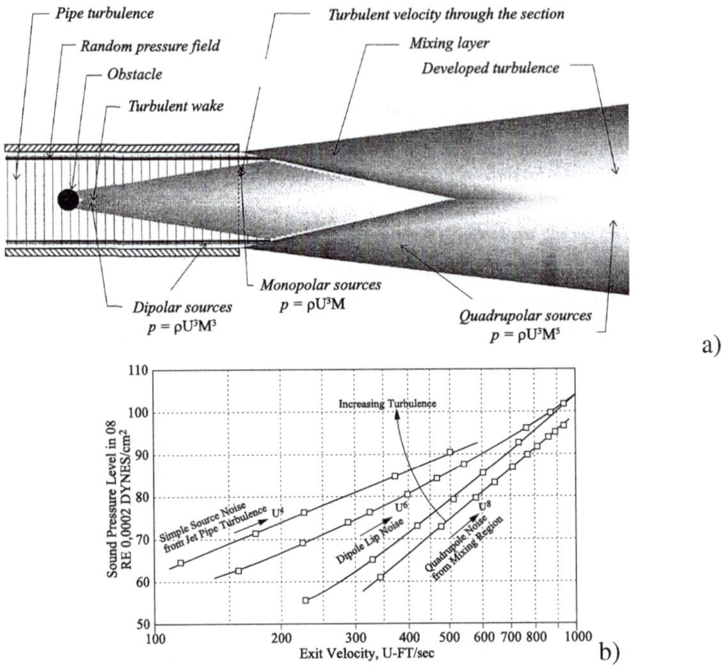

Figure 9.6. *Schematic illustration of the noise produced by turbulent jets with its different sources; a). According to [LAZ 97]. Evolution with the flow velocity of the different aeroacoustics mechanisms for a turbulent jet; b). According to [FFO 65a]*

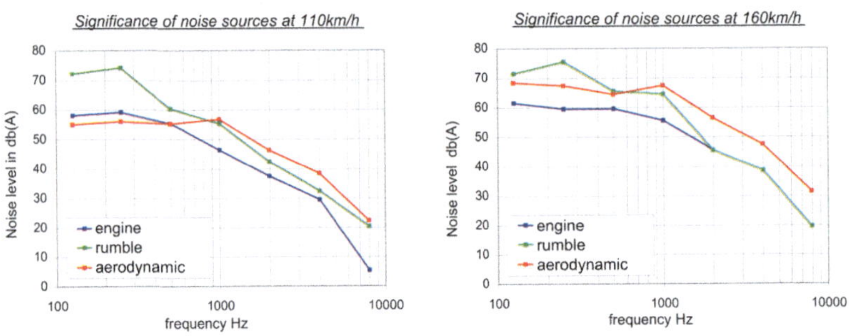

Figure 9.7. *Evolution of different types of noise with the velocity for an automobile vehicle. For a color version of the figure, see www.iste.co.uk/anselmet/acoustics.zip*

By definition of the average operator, W_{ijkl} is more accurately written in the form:

$$W_{ijkl} = \lim_{T\to\infty} \frac{1}{2T} \int_{-T}^{+T} \frac{\partial^2 T_{ij}}{\partial t^2}(y',t') \frac{\partial^2 T_{kl}}{\partial t^2}(y'',t''+\tau)dt. \qquad [9.6]$$

If the observed mechanism is assumed to be stationary, then the statistical moments are independent of the computing time. Thus, the temporal derivation can be achieved as well as with respect to time t as with respect to retarded time τ, such that $\partial T_{kl}(y'',t''+\tau)/\partial t = \partial T_{kl}(y'',t''+\tau)/\partial \tau$. Thus, equation [9.6] is written as:

$$W_{ijkl} = \lim_{T\to\infty} \frac{1}{2T} \int_{-T}^{+T} \frac{\partial^2 T'_{ij}}{\partial t^2} \frac{\partial^2 T''_{kl}}{\partial \tau^2} dt = \lim_{T\to\infty} \frac{1}{2T} \frac{\partial^2}{\partial \tau^2} \int_{-T}^{+T} \frac{\partial^2 T'_{ij}}{\partial t^2} T''_{kl} dt.$$

A first integration by parts leads to write:

$$\int_{-T}^{+T} \frac{\partial^2 T'_{ij}}{\partial t^2} T''_{kl} dt = \left[\frac{\partial T'_{ij}}{\partial t} T''_{kl} \right]_{-T}^{T} - \int_{-T}^{+T} \frac{\partial T'_{ij}}{\partial t} \frac{\partial T''_{kl}}{\partial t} dt,$$

such that, the integration with respect to the retarded time t of the first left member being zero, the previous equation is simply written:

$$W_{ijkl} = -\lim_{T\to\infty} \frac{1}{2T} \frac{\partial^2}{\partial \tau^2} \int_{-T}^{+T} \frac{\partial T'_{ij}}{\partial t} \frac{\partial T''_{kl}}{\partial t} dt.$$

By reiterating this and under the assumption of a stationary turbulence, we finally get for the source terms:

$$W_{ijkl} = \overline{\frac{\partial^2 T_{ij}}{\partial t^2}(y',t') \frac{\partial^2 T_{kl}}{\partial t^2}(y'',t''+\tau)} = \frac{\partial^4}{\partial \tau^4} \overline{T_{ij}(y',t') T_{kl}(y'',t''+\tau)}.$$

Finally, to conclude this general introductory section, note that there exist more elaborate versions of the Lighthill equation. These versions notably take into account the effect of nonlinear coupling between the acoustic waves that propagate and the actual propagation speed of these waves (which takes place in the left side member of equation [9.1]) which is equal to c_0^2 only as a first approximation. Thus, for example, Lilley [LIL 73] in 1973 has suggested the following equation to consider this effect:

$$\frac{D}{Dt}\left(\frac{D^2\Pi}{Dt^2} - \frac{\partial}{\partial x_j}\left(c_0^2 \frac{\partial \Pi}{\partial x_j}\right)\right) + 2\frac{\partial u_k}{\partial x_j} \frac{\partial}{\partial x_k}\left(c_0^2 \frac{\partial \Pi}{\partial x_j}\right) = -2\frac{\partial u_j}{\partial x_k} \frac{\partial u_i}{\partial x_j} \frac{\partial u_k}{\partial x_i},$$

where $\Pi = (1/\gamma)\ln p$ and $D/Dt = \partial/\partial t + u_k\partial/\partial x_k$. However, taking into account the approximations that are made in traditional analytical models (that we will present in the next few sections), which in common applications seem to be largely dominating over this nonlinear propagation effect, we will no longer discuss the problem here. The interested readers may consult, in particular, articles [RIB 77, GOL 84, COL 97, GOL 03] or [COL 04].

Another type of complex situation that we will discuss further in Chapter 10 corresponds to the case where the medium in which acoustic waves propagate contains density inhomogeneities ρ_0, which is particularly the case of the atmosphere and marine environments, where the density has significant gradients according to the vertical direction due to thermal effects or salinity. In these cases, as we will see, obtaining a realistic propagation equation and taking into account the effects of soil and terrain are themselves problems which are still the focus of specific research.

9.3. Reminder of some notions on turbulence which will be useful here

In the Lighthill equation, a number of velocity products such as u_iu_j appear as the essential source terms of the absence of thermal phenomena, of the presence of walls or other complexities. These velocity products become very significant when the flow that generates the noise is turbulent, as illustrated by equations [9.3] and [9.5]. It is, therefore, important to recall here briefly the few basics about turbulence which will be useful later. Generally, a flow becomes turbulent when the Reynolds number which is associated to it, $Re = U_eL/\nu$, where U_e is a characteristic velocity (for example, bulk velocity in the case of a flow in a pipe), L, a characteristic length scale (the diameter of the pipe) and ν, the kinematic viscosity of the fluid, is greater than a certain critical value. In the case of, for example, a pipe whose walls can be regarded as smooth, the critical Reynolds number, beyond which the flow becomes turbulent, is 2,300. This has been highlighted in the 1890s by O. Reynolds at the University of Manchester. When the flow is turbulent, at any point, any variable such as each of the components of velocity u_i or the pressure p must then be considered as a random variable, with fluctuations $u_i^{'}$ with regard to the mean value U_i: $u_i = U_i + u_i^{'}$. It should be noted that the mean is being considered within the statistical sense, that is with respect to many realizations of the same flow, but with random initial and boundary conditions because they are impossible to control (for example, by an experimenter) in every detail.

It is then shown that, from the instantaneous Navier–Stokes equation:

$$\frac{\partial u_i}{\partial t} + u_j\frac{\partial u_i}{\partial x_j} = -\frac{1}{\rho}\frac{\partial p}{\partial x_i} + \nu\frac{\partial^2 u_i}{\partial x_j \partial x_j},$$

it yields for a stationary mean flow and without any density variation, after decomposition of each variable into mean and fluctuations, and then taking the mean of the equation:

$$U_j \frac{\partial U_i}{\partial x_j} = -\frac{1}{\rho}\frac{\partial P}{\partial x_i} + \nu \frac{\partial^2 U_i}{\partial x_j \partial x_j} - \frac{\partial \overline{u_i' u_j'}}{\partial x_j}.$$

Additional unknowns, the quantities $\overline{u_i' u_j'}$ called the Reynolds stresses R_{ij}, therefore, appear in this balance equation for first-order statistical quantities that are the mean velocities; from a statistical point of view, these quantities are second-order moments. Thus, turbulence appears as a mathematically open problem since, by including the averaged continuity equation, $\partial U_i / \partial x_i = 0$, 4 + 6 = 10 unknowns are then generally obtained, for only four equations. Hence, the compulsory necessity to use models to obtain closure of the problem in order to numerically simulate turbulent flows. Chassaing's [CHA 00], Schiestel's [SCH 06] or Pope's [POP 00] books present this matter in a detailed manner.

Nevertheless, turbulence cannot be considered as being only associated with purely statistical fluctuations without internal correlation (spatial or temporal) since large-scale structures are also present. This aspect also contributes to make the turbulence problem particularly complex. Turbulence must then also be characterized by means of characteristic length scales, as well as in a much finer manner it is necessary to analyze the way, on average, the velocity field at a point $\vec{x}$ is more or less strongly correlated with the velocity at the neighboring points $\vec{y}$, such that $\vec{y} = \vec{x} + \vec{r}$. Without going into the details of the mathematical expansions, it can be shown that, assuming homogeneous and isotropic turbulence, then the correlations in two points of any component of the velocity can be expressed from two correlation functions only, denoted by f and g, which just depend on the modulus r of $\vec{r}$. Function $f(r)$ is called the longitudinal correlation and function $g(r)$ is the transverse correlation (see Figure 9.8): in the case of f, the displacement $\vec{r}$ is parallel to the velocity component being considered whereas, for g, $\vec{r}$ is in the plane perpendicular to the velocity component being considered. In homogeneous and isotropic turbulence, it can also be shown that f and g are related by the relation $g(r,t) = f(r,t) + (r/2)\partial f(r,t)/\partial r$. The time dependence in the notation $f(r,t)$ reflects the fact that, in general, the function f depends on the point $\vec{x}$ which is being considered, or rather, in general, on its position according to the general axis of the flow (which then corresponds to grid turbulence decreasing downstream) which thus corresponds to different instants if we consider ourselves in a mobile reference with the mean flow velocity. It is in this framework that all homogeneous and isotropic turbulence theories were developed in the 1940s–1950s. Nowadays, decreasing turbulence is the one obtained from a direct numerical simulation of the Navier–Stokes equations for a numerical box, if no forcing is imposed at the large-scale level.

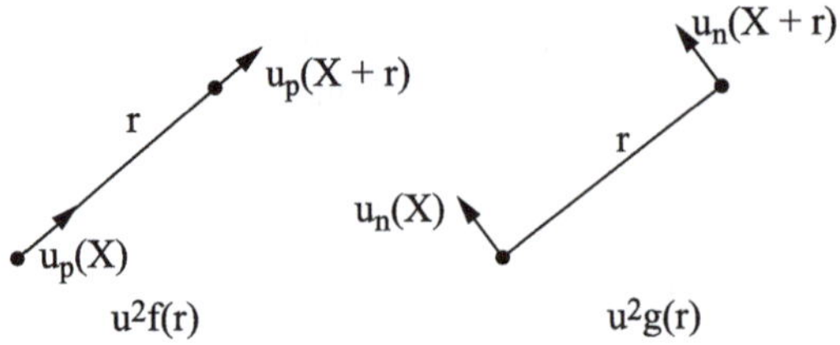

(a) Schematic diagrams for the longitudinal f (on the left) and transversal g (on the right) correlations. The notation u_p refers to the parallel velocity whereas u_n refers to the normal velocity.

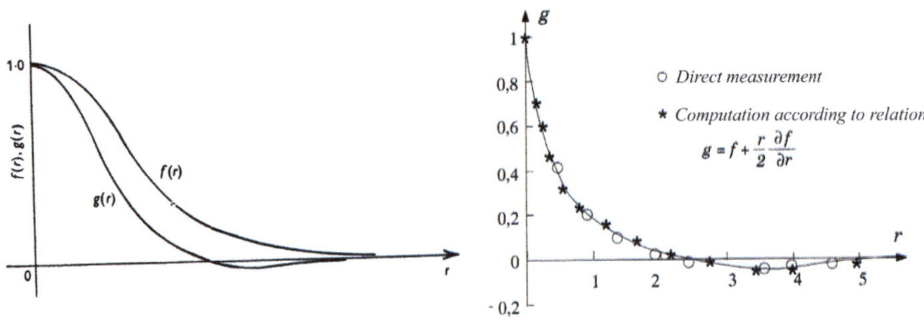

(b) Typical experimental results for the correlation functions f and g (normalized by the variances of the corresponding velocity fluctuations), as obtained in grid turbulence.

Figure 9.8. *A few results which illustrate the basic properties of the two point correlations in homogeneous and isotropic turbulence. According to [CHA 00]*

From functions f and g, the integral scales of turbulence are then defined, respectively, longitudinal L_f and transversal L_g, from integration with respect to all the values of r: $L_f(t) = \int_0^\infty f(r,t).dr$ and $L_g(t) = \int_0^\infty g(r,t).dr$. The isotropic relationship between f and g requires that $L_f = 2L_g$. These two scales are naturally related by multiplicative coefficients close to 1 to the integral scale l that is most often defined from the kinetic energy of turbulence k^2, and for the dissipation rate of this kinetic energy of turbulence, ϵ: $l = (C_\mu)^{3/4}k^{3/2}/\varepsilon$, where C_μ designates the constant of the model $k-\epsilon$, whose value is 0.09. Most often, the analytical modeling of the correlation f is the Gaussian distribution: $f(r) = \exp(-\pi r^2/(4L_f^2)) \approx \exp(-r^2/L_f^2)$. However, there are other formulations, as we will see later. Generally speaking, in homogeneous and isotropic turbulence, it can be written that the correlation function $Q_{i,j}(\vec{r},t) = \overline{u_i^{'}(\vec{x}+\vec{r},t).u_j^{'}(\vec{x},t)}$ can be reduced to $Q_{i,j}(\vec{r},t) = q^1(r,t)r_ir_j + q^2(r,t)\delta_{ij}$, with $q^1(r,t) = \overline{(u_1^{'})^2}(f(r,t) - g(r,t))/r^2$ and

2 The quantities k and ϵ are essential for the characterization of a turbulent flow [CHA 00, SCH 06, POP 00]. These are defined as $k = 1/2\overline{u_i^{'}u_i^{'}}$ and $\epsilon = \nu\overline{(\partial u_i^{'}/\partial x_j)(\partial u_i^{'}/\partial x_j)}$.

$q^2(r,t) = \overline{(u_1')^2} g(r,t)$, and $g(r,t) = f(r,t) + r/2\partial f(r,t)/\partial r$ (taking into account the fact that, also, $\overline{(u_1')^2} = \overline{(u_2')^2} = \overline{(u_3')^2}$, the direction 1 being then, by convention, the direction of the mean flow velocity, associated with the x-axis). The last relation derives from the equation of continuity (which defines the incompressibility of the flow).

Furthermore, Figure 9.9 illustrates what double correlations in space and time are, usually called spatio-temporal correlations when they are of the kind $u'(\vec{x},t)u'(\vec{x}+\vec{R},t+\tau)$ and that the separation $\vec{R}$ is carried by the longitudinal axis of the flow. The notion of convection velocity of the large vortices of the flow is then easily brought forward. This figure also illustrates the fact that, once the convection velocity of large vortices is taken into account, the space-time correlation can be written, as a first approximation, as the product of a space correlation and a time correlation.

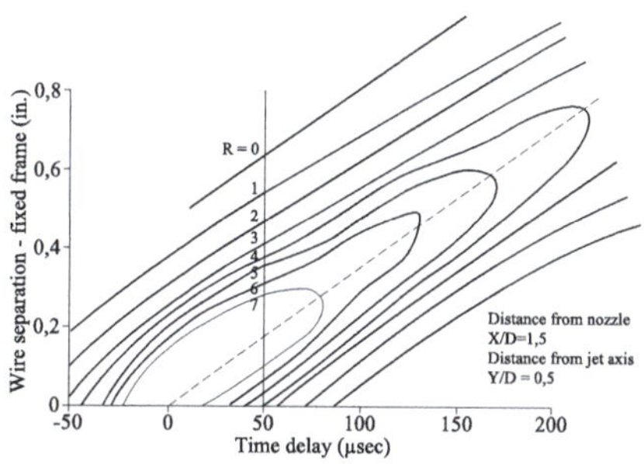

(a) Isovalue contours for the raw data of $u'(\vec{x},t)u'(\vec{x}+\vec{R},t+\tau)$.

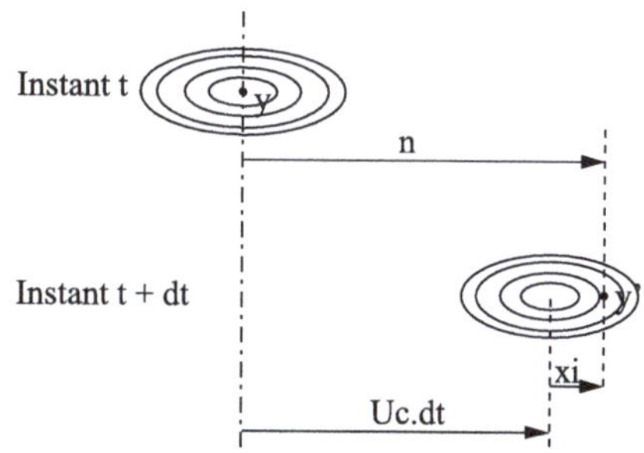

(b) Diagram explaining the concept of frozen turbulence convection by the large vortices of the flow and the associated convection velocity (U_c).

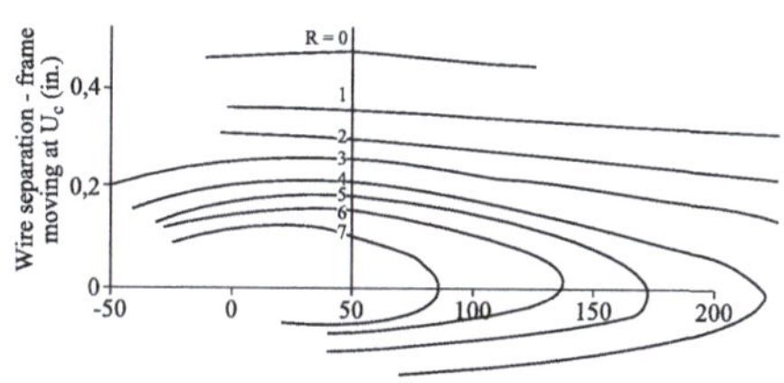

(c) Correlation isovalue contours in the system moving at the velocity U_c.

Figure 9.9. *Typical example of double space- and time-correlations obtained in a jet. According to [DAV 63]*

The homogeneous and isotropic turbulence which we have already mentioned, but without really defining it, is an idealized situation for which the statistical properties of turbulence are invariant by translation and rotation in space. In theory, a real homogeneous and isotropic turbulence excludes situations where there are mean velocity gradients (also known as shears), which occurs only for grid turbulence flows, and is thus quite far from the real situations (even if they mainly remain academic compared to real applications, particularly in the transportation field) that we will address here such as jets and boundary layers. Unfortunately, it is generally only in this very limited context that advanced analytical developments can be achieved. In the case of homogeneous turbulence, which is a less restrictive consideration than isotropy, but still simple enough to enable analytical developments, it suffices that shears (if they exist) be constant within the flow. In reality, the real turbulence in jets and boundary layers, for example, is rather closer to being axisymmetric (see, for example, [ANT 91]), that is to say such that there exists a particular direction (which is in general the longitudinal direction x along which the flow evolves) and the other two directions that present similar properties between them, as illustrated in Figure 9.10. This figure presents the distribution of invariants II_b and III_b of the anisotropy tensor b_{ij} (=$R_{ij} - 2/3k\delta_{ij}$) associated with the Reynolds tensor R_{ij} for a boundary layer (more or less the same thing is obtained for a jet). These invariants are defined by $II_b = -1/8(\lambda_1^2 + \lambda_2^2 + \lambda_3^2)$ and $III_b = 1/24(\lambda_1^3 + \lambda_2^3 + \lambda_3^3)$, where λ_1, λ_2 and λ_3 are the eigenvalues of b_{ij}. The two curved lateral branches correspond to axisymmetric turbulence, while isotropic turbulence corresponds to the origin point O(0,0) of this diagram. For the axisymmetric branch with $III_b > 0$, the turbulence is preferentially organized *in cigars* whereas for that with $III_b < 0$, the turbulence is preferably organized *in disks*; this is all the more pronounced that $|III_b|$ is significant. On the other hand, it is also important to note here that this diagram has very strong implications for turbulence models (see, for example, [CHA 00] or [POP 00]) since it implies that the different terms of R_{ij} must be such that III_b and II_b always lie within the area bounded by the three branches of the Lumley diagram.

In the case of axisymmetric turbulence, theoretical developments comparable to those conducted for homogeneous and isotropic turbulence can also be conducted but they become significantly more complicated, therefore only a very small number of problems have been tackled within this framework. It can be notably mentioned the fact that, in general, $\overline{u_3'^2} = \overline{u_2'^2}$ and $\overline{u_1'^2} \approx 1.5$ to $2\overline{u_2'^2}$ (*cigars*, $III_b > 0$: boundary layers and jets) or $\overline{u_1'^2} \approx 0.5$ to $0.9\overline{u_2'^2}$ (*disks*, $III_b < 0$: mixing layers), and that it is then necessary to have two independent functions available to characterize the correlations functions (these two functions include any of the two variances, $\overline{u_1'^2}$ or $\overline{u_2'^2}$, as well as the two distinct independent length scales – since the relation $L_f = 2L_g$ is no longer available). In addition, for practical applications, and especially in the transportation field, there are also other complications within flows that come across, as in the wakes of obstacles or in separated parts of the flow over a backward-facing step, and these more or less strong fluid recirculations certainly cannot be taken into account in the analytical approaches. All of this should be kept in mind when considering the different approaches to modeling or the experimental validations that have been developed following Lighthill's work.

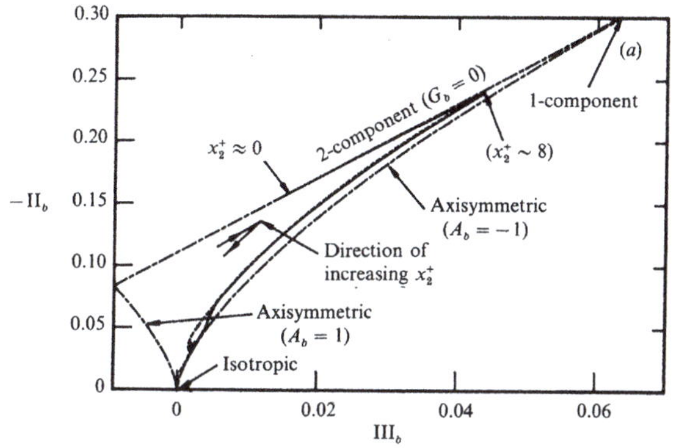

Figure 9.10. *Map of the invariants of the anisotropy tensor b_{ij} (also called the Lumley diagram) for a turbulent boundary layer. The two branches corresponding to axisymmetric turbulence have, respectively, as equations $III_b = 2(-II_b/3)^{3/2}$ and $III_b = -2(-II_b/3)^{3/2}$. According to [ANT 91]*

9.4. The Proudman model for homogeneous and isotropic turbulence

Very shortly after the historic publication of the equation corresponding to the Lighthill analogy, Proudman has published (the same year, in 1952, [PRO 52]), a statistical analysis that allows, in decaying homogeneous and isotropic turbulence (grid turbulence), to predict the aeroacoustics power radiated in this situation. It is as well a historical contribution since the analytical works that followed it have only contributed to refine Proudman's results in less academic situations. Note that in the summary that we are presenting here, the influence of the decay of the turbulence downstream the grid does not appear. Proudman starts from a slightly different form of relation [9.3], by writing the acoustic intensity $I(\vec{x}, t)$ radiated by a volume $V$of turbulence as:

$$I(\vec{x}, t) = \frac{\rho_0}{16\pi^2 c_0^5}\frac{1}{x^2}\iint_V \overline{\left[\frac{\partial^2}{\partial t^2}(u_x^2 - \overline{u_x^2})\right]_{t-\xi/c_0}\left[\frac{\partial^2}{\partial t^2}(u_x'^2 - \overline{u_x^2})\right]_{t-\xi'/c_0}} d\vec{y}d\vec{y}',$$

where the first brace is relative to quantities evaluated at point y and time $t - \xi/c_0$, and the second at point y' and time $t - \xi'/c_0$, with $\xi = |\vec{x} - \vec{y}|$, $\xi' = \left|\vec{x} - \vec{y}'\right|$ and $\tau = (\xi - \xi')/c_0$. The double integral can be rewritten [PRO 52, LIL 93] defining the amount C (which corresponds to the quantity denoted by U in Lilley's theoretical developments [LIL 93]) such as:

$$C = \overline{\left[\frac{\partial^2}{\partial t^2}(u_x^2 - \overline{u_x^2})\right]\left[\frac{\partial^2}{\partial t^2}(u_x'^2 - \overline{u_x^2})\right]},$$

where the two quantities are evaluated at the same time, with the approximation that the structures which most contribute to the generation of noise (whose typical size is denoted by l) are such that the correlation at two different instants remains in fact close to 1 for all delay instants τ such that τ is smaller than l/c_0, and we then get (with $\vec{R} = \vec{y'} - \vec{y}$):

$$I(\vec{x}, t) = \frac{\rho_0}{16\pi^2 c_0^5} \frac{1}{x^2} \int \left[\int C d\vec{R}\right]_{t-\xi/c_0} d\vec{y}.$$

All the modeling is then carried forward to the optimal evaluation of the correlation C. Proudman [PRO 52, LIL 93] shows that, under the conventional hypotheses of homogeneous and isotropic, and joint-normal, turbulence (we will return to the joint-normal assumption in section 10.2, where the works carried out, in a manner apparently independent of the problem posed by the Lighthill relation, by Hill [HIL 95, HIL 97], will be presented), C can be expressed with respect to only the longitudinal spatial correlation $f(r)$ and its derivatives, so that the acoustic power radiated by volume V becomes:

$$p_s = \int I(\vec{x}) d\vec{x} = \alpha \frac{\rho_0 u_{rms}^8}{c_0^5 L_f}, \qquad [9.7]$$

where u_{rms} is the standard deviation of the velocity fluctuations, L_f represents the integral scale of turbulence (which is defined such that $f(r) = \exp(-\pi r^2/4L_f^2)$) and the constant α (since then known as Proudman's) is defined (here) by:

$$\alpha = \frac{8}{5} \int_0^\infty f^2 G^2 z^4 dz, \qquad [9.8]$$

with $z = r/L_f$ and $G = f''' + (4/z)f'' - (4/z^2)f'$, the $'$ here referring to the derivatives of f with respect to z. Note that Proudman has used the Millionshtchikov hypothesis for joint-Gaussian variables, $\overline{(p'q')_A(p''q'')_B} = \overline{p'q'} \cdot \overline{p''q''} + \overline{p'p''} \cdot \overline{q'q''} + \overline{p'q''} \cdot \overline{p''q'}$ (where the notations p' and p'' indicate the quantities, respectively, evaluated in A and B). In this context, he obtained that the constant $\alpha = 13.5$. We will see later (section 9.5) that with the same general physical approximations, but with a finer consideration of the effect of the retarded time, Lilley [LIL 93] obtains an expression of p_s formally similar to [9.7], but with a value of α quite markedly different from that obtained by Proudman. Similarly, if it is used in [9.7] another analytical form for the correlation f such as the one known as Heisenberg's (see additional note that follows), a different value is obtained for α (37.5 for Heisenberg). On the other hand, Proudman's reasoning correctly results in an 8th-power function variation of the velocity for the acoustic power radiated by quadratic sources, such as the experimental observations were already showing then (see also Figure 9.6(b)).

Additional note

Generally speaking, following the theoretical and experimental studies of turbulence that have been developed over the years, various more or less advanced analytical formulations of the correlation $f(r)$ have emerged, based on these different results. These different models have allowed for different forms of spatial correlation to be obtained, as well as and above all, for different forms of the spectrum of kinetic energy of turbulence $E(k)$ to be obtained. The variable k refers to the modulus of the wave number being considered (the integral over all values of k being equal to the kinetic energy of turbulence). As we will see in the examples that follow (section 9.6), the model of $E(k)$ is the one often indicated then, knowing that the spatial correlation $f(r)$ is therefrom deduced by the conventional relation $\overline{(u_1^{'})^2} f(r) = (2/r^2) \int_0^\infty (E(k)/k^2)(\sin(kr)kr - \cos(kr))dk$.

If the most conventional law is that which corresponds to the Kolmogorov spectrum (Figure 9.11), the correlation known as Heisenberg has also often been referred to. It is calculated from the analytical spectrum model $E(k)$ which has been suggested by Heisenberg : $E(k) = Ak^{-7}(Bk^{-4}+1)^{-4/3}$, where the coefficients A and B vary with the Reynolds number (see Figures 9.12(a) and 9.12(b)). If, when considering a flow with an infinite Reynolds number, the formula proposed by Kolmogorov in 1941 is correctly restored, $E(k) \approx \varepsilon^{2/3}k^{-5/3}$ (where ϵ is the dissipation rate of the kinetic energy of turbulence), when the constants A and B are those for large but finite values of the Reynolds number, a non-negligible deviation from the law in $k^{-5/3}$ can be seen, when the wave numbers decrease, that is to say for large scales of about the same size as the integral scale l. This is the choice that has been made here for obtaining $\alpha = 37.5$. Furthermore, the Kolmogorov length scale η is also defined in a conventional manner in turbulence as $\eta = (\frac{\nu^3}{\varepsilon})^{1/4}$. This scale is generally regarded as characteristic of the smallest vortices within a turbulent flow: it is at the level of these vortices of size η that the turbulent energy is dissipated into heat due to the effect of the viscosity forces. Figure 9.11 shows that when spectra $E(k)$ are made dimensionless using the scale η, their form becomes universal (that is independent of both the type of flow being considered and the turbulence Reynolds number R_λ defined as $R_\lambda = (u_{rms}\lambda)/\nu$, where λ, the Taylor microscale, is given by the relation $1/\lambda^2 = -(1/2)(\partial^2 f(r)/\partial r^2)_{r=0} = \overline{(\partial u'/\partial x)^2}/\overline{u'^2}$ (where we also use the usual notation such that u is identified to u_1, v and w then, respectively, identified to u_2 and u_3)) over a wide range of scales and very close to the law proposed by Kolmogorov: these scales constitute what is known as the inertial region. For small scales, $E(k)$ follows a law different from the Kolmogorov law because the effect of the viscosity forces becomes then very important: this is known as the dissipative region. For very large scales, due to the generation mechanisms of turbulence, spectra $E(k)$ also deviate from the Kolmogorov law. Note that Figure 9.11 uses logarithmic scales, whereas Figures 9.12(a) and 9.12(b) use linear scales (the analytical adjustments for g use the isotropic relationship previously given, which relates f and g).

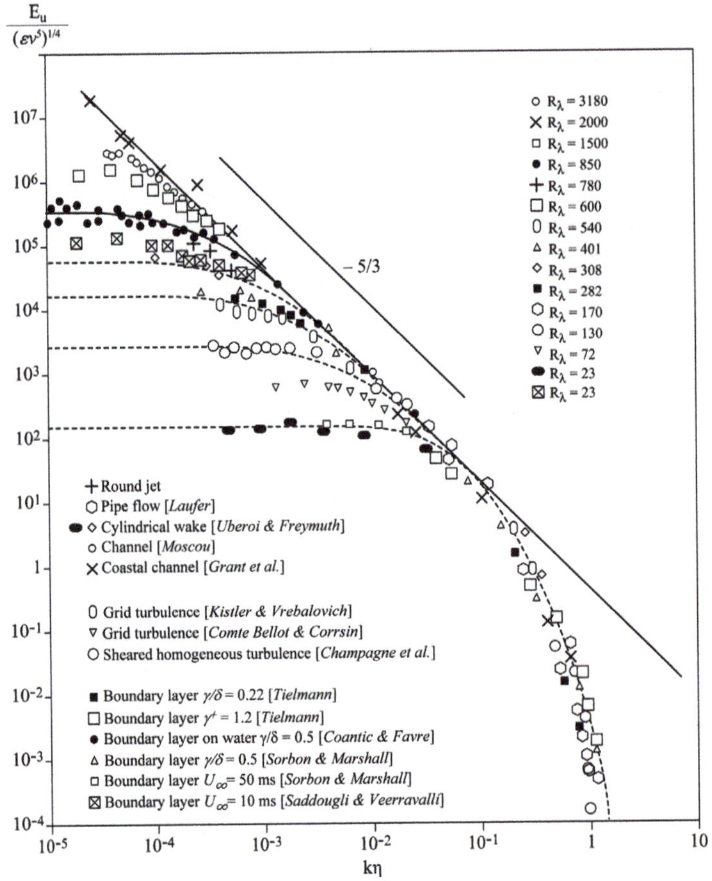

Figure 9.11. *Energy spectra $E(k)$ non-dimensionalized by the Kolmogorov scale η, obtained for flows of different types and Reynolds numbers R_λ ranging from about 25 to about 3,000. According to [CHA 00]*

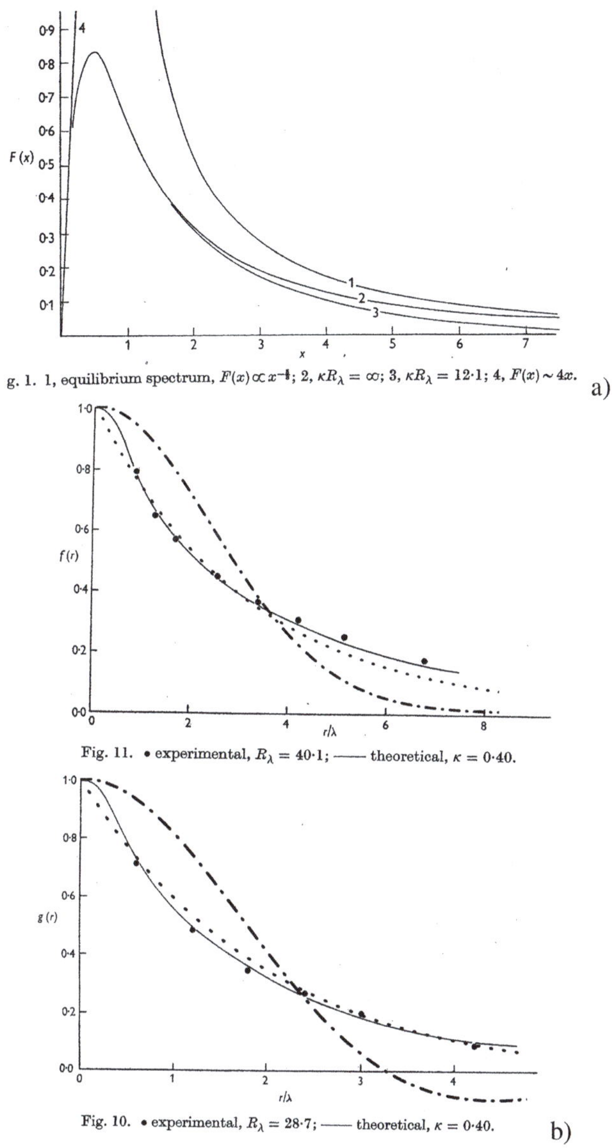

Figure 9.12. *Energy spectra $E(k)$ modeled according to the Heisenberg theory (the variable x represents the wave number k non-dimensionalized in a specific manner); a). Comparison of the correlation curves f and g derived from the Heisenberg model (——), from the law in $\exp(-z)$ (······) and from the one in $\exp(-\pi z^2/4)$ (-·-·-), with experimental data obtained in grid turbulence (•) with a small Reynolds number; b). According to [PRO 51]*

9.5. The Lilley model for homogeneous and isotropic turbulence

Lilley, in 1993, recovered Proudman's analytical developments in a methodical manner, but reconsidering some approximations. From his analysis, it seems paramount to us to bring forward the following points:

1) A finer manipulation of the fourth-order correlations allows him to reveal explicitly in the formula giving p_s the contribution of the flatness coefficient (which he names T_1) of the longitudinal velocity fluctuations.

2) The value of the frequency Ω which quantifies the time correlation decay law is a crucial parameter of his approach, and more specifically, the value of the Strouhal number for turbulent fluctuations, S_T, which he defines with $S_T = \Omega L/u_{rms}$.

3) The acoustic power radiated by a volume of homogeneous and isotropic turbulence is indeed given by relation [9.7], but the value of Proudman's constant α is in a matter of fact given by:

$$\alpha = 1.80(T_1 - 1)(S_T)^4, \qquad [9.9]$$

such that a small deviation from the actual value of S_T with regard to its approximate value of 1 may cause a very large variation in the value of α (for example, for $T_1 = 3$ and $S_T = 1$, $\alpha = 3.6$ while, for $T_1 = 3$ and $S_T = 1.25$, $\alpha = 8.8$).

Generally speaking, there is no universal criterion to evaluate *a priori* the characteristic frequency Ω. According to Khavaran *et al.* [KHA 94], when computing the aeroacoustic noise on the basis of a $k - \epsilon$-based turbulence model (in a situation not very far from being homogeneous and isotropic), it might be considered that Ω is given by $2\epsilon/k$, whereas in the case of a jet, it will probably be preferable to take Ω equal to the mean shear.

Another result physically very important obtained by Lilley is that, in the term in the form of an integral of the relation that defines α (Proudman had obtained, see p. 382, $\alpha = 8/5 \int_0^\infty f^2G^2z^4dz$), which he expressed as $\int_0^\infty z^2(3f^2 + 4fg + 8g^2)dz$ (where f is the longitudinal correlation function and g is the transversal correlation function, with for homogeneous and isotropic turbulence: $g = f + (r/2)f'$), most of the scales have a contribution to this integral globally equal to zero. As a matter of fact, if the relation $\int_0^\infty (3z^2f^2 + 2z^3f(df/dz))dz = \int_0^\infty d(z^3f^2)/dzdz = 0$ is used, we get that $\int_0^\infty z^2(3f^2 + 4fg + 8g^2)dz = 2\int_0^\infty z^4(df/dz)^2dz$, such that then $\alpha = 32/15(T_1 - 1)S_T^4 \int_0^\infty z^4(df/dz)^2dz$, which leads to the value of α specified above, if for $f(r)$ an expression such as $f(r) = \exp(-\pi r^2/4L^2)$ is considered. The most interesting aspect of this calculation lies in clearly highlighting the fact that the larger scales (Figure 9.13) are those that especially contribute to the power of the radiated noise: it is thus not important to have a very fine representation of the form of f at small scales, but on the contrary it is essential to have a representation that ensures the continuity of the derivatives of f at small scales. That is why, in most studies, as many authors before Lilley had devised, choosing a correlation in the form of $f(r) = \exp(-\pi r^2/4L^2)$ is much more satisfactory for aeroacoustics calculations than choosing $f(r) = \exp(-r/L)$ though the latter is a much better representative of large Reynolds number turbulence. This result also explains why (homogeneous and isotropic) turbulence is a relatively ineffective noise generator. In addition, Lilley's analytical study clearly shows why, as other authors had pointed out, the dominant contribution to the acoustic fluctuations spectrum originates from vortices slightly larger sized than the turbulence most energetic vortices (the maximum being obtained, here, for $r = 1.38L$).

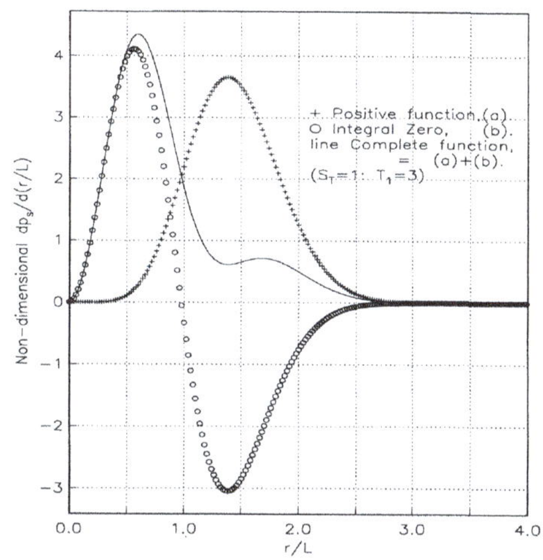

Figure 9.13. *Function that acts as integrand for the computation of α in the calculations developed by Lilley [LIL 93]*

9.6. The recent models and a few experimental validations

The most recent analytical models have strived to better represent the time correlation. In effect, as we will see with the typical example in Figure 9.14, the main flaw of analytical formulations based on space and time correlation products to approximate source terms involved in relation [9.2] is that they poorly reproduce the low-frequency and high-frequency behaviors of noise spectra, while the value of the frequency corresponding to the maximum of the noise spectrum is in general (more or less) overvalued.

To the best of our knowledge, Zhou and Rubinstein [ZHO 96] have been the first to suggest that the fact of decomposing the double space and time correlations into products of a scale correlation and a time correlation (totally independent of the scale that is being considered) is a pretty poor approximation of the turbulent reality and that, on the contrary, it is necessary to take into account the fact that the time correlation depends strongly on the scale being considered. These authors then identified two kinds of possible refinements to this approximation, which they have identified with the hypotheses (more or less conventional in turbulence theories) of, respectively, *sweeping* and *straining*. The term *sweeping* reflects the fact that, for a scale r, the vortices are simply washed away (or swept away) by the largest energetic scales of turbulence – we are in this case faced with rather Eulerian considerations of turbulence, the characteristic velocity being then u_{rms} so that $(1/\tau)_r \approx u_{rms}/r$-, while the term *straining* reflects the fact that, for the scale r, vortices are distorted by local shear at this scale – this is then a rather Lagrangian perspective of turbulence, fluid particles at this scale being thus during the course of their displacement directly influenced by the local shear at this scale, so that $1/\tau_r \approx |(u(x+r) - u(x))^3|^{1/3}/r$, that is since $|(u(x+r) - u(x))^3| \approx \epsilon.r$, $(1/\tau)_r \approx \epsilon^{1/3}.r^{-2/3}$ (see, for example, Frisch's book [FRI 95] for more details).

The analytical developments of [ZHO 96] show that the *sweeping* approximation corresponds to Proudman's theoretical approach [PRO 52], resulting in particular in a prediction of $\alpha = 12.0$, while the *straining* approximation corresponds to Lilley's theoretical approach [LIL 93], resulting in particular in a prediction of $\alpha = 3.6$. These works have then been subsequently completed by those of [RUB 00], which showed that in the *sweeping* approximation, the high-frequency behavior of the noise spectrum evolves in $\omega^{-4/3}$, while in the *straining* approximation it evolves in $\omega^{-7/2}$. These authors have also explained, by mentioning the principles of triad interaction widely used in spectral approaches to turbulence modeling (see, for example, [FRI 95]), why the turbulence is very effective as a noise generating mechanism. This is because, in order to generate noise, two interacting modes k_1 and k_2 (respectively, associated with velocity fluctuations $u_i(k_1, \omega_1)$ and $u_k(k_2, \omega_2)$ in the source region, where ω_1 and ω_2 are the reciprocals of the associated decorrelation times) must satisfy the constraint that $k_1 + k_2 = \omega\vec{n}/c_0$ and $\omega_1 + \omega_2 = \omega$ where $\vec{n}$ represents the unit vector of the propagation direction so that the interaction of these two modes generates an acoustic mode (of frequency ω) $\vec{U}\exp(\imath\omega\vec{n}.\vec{x}/c_0 - \imath\omega t)$ in the far-field (with the vector $\vec{U}$ parallel to the vector $\vec{n}$). In general, even if the interaction between the modes $u_i(k_1, \omega_1)$ and $u_k(k_2, \omega_2)$ certainly generates a sound wave of wave number $\omega\vec{n}/c_0$ and frequency ω, this interaction does not generate an incompressible turbulence mode $\vec{U}(\omega\vec{n}/c_0, \omega)$ because a turbulent mode with such a wave number $\omega\vec{n}/c_0$- frequency ω combination of a sound wave does not exist in the turbulent velocity field (ω, which is a true frequency, is fully defined from the inverses of the decorrelation time ω_1 and ω_2 as mentioned above).

The results on the basic properties of turbulence [PRA 93, SMA 01] show that the hypothesis of decorrelation driven by the effect of *sweeping* (this theory was originally proposed by Kraichnan, see, for example, [CHE 89]) is in fairly good agreement with the experimental results. We will not further indulge into the details of this subject that would drive us otherwise to considerations too far remote from our present topic, but it seemed important to emphasize here that these ideas are still quite controversial within the specialists community of turbulence; therefore, it is not at all surprising that the modeling of the fourth-order velocity correlations, double in space and time, be still quite approximate for applications in aeroacoustics.

More recently, inspired by the above considerations, Woodruff *et al.* [WOO 01] have implemented an aeroacoustics noise computation for jets. They have tested several formulations for time correlations, all of the type $\exp(-(t/\tau_0)^2/\sigma^2(k))$, that is involving a factor $\sigma(k)$ that takes into account the considerations defined at the beginning of this section, in order to account for the fact that the time correlation is specific to each scale (and therefore to the wave number) which is considered. Figure 9.14 shows the results obtained for the sound intensity radiated by a volume V of turbulence, when different analytic forms are used for the correlation at two instants and for the turbulence spectrum. This is due to the fact that when working in Fourier space, it is naturally the turbulence spectrum that appears in the expression of $I(\vec{x},\Omega)$ instead of the correlation in two spatial points (in the notations of these authors, Ω is actually the common frequency ω). It is clear that it is the choice of the time correlation that more clearly influences the results, and in particular the position (here in terms of $\Omega\tau_0$) of the peak. Figure 9.15 is, meanwhile, typical of the results obtained by [WOO 01] for the noise spectrum radiated by the jet, correctly illustrating the effect of directivity in a jet flow. This figure shows a slight improvement in the prediction of the position of the peak of the noise spectrum when the formulation where $\sigma(k)$ is proportional (as for the *straining decorrelation* approximation) to $k^{-2/3}$ is used. Nevertheless, these authors are overall fairly disappointed by their results, and they think that, to improve such predictions, one should more widely find inspiration in the works of Tam [TAM 99] who used a specific formula for two-point and two-time correlations that allows, notably, that the specific properties of turbulence in highly sheared regions of a jet be taken into account in an *ad-hoc* manner (by comparison and adjustment on experimental data).

The results presented below, from [KER 04], more accurately illustrate all this matter, directly comparing spatio-temporal correlation measurements with different analytical formulations. Since part of the results obtained concerns a heated supersonic jet, these works also help to illustrate the manner that the effects related to the Mach number ($M \approx 1$) of the flow and to thermal transfers can more or less modify the results with respect to the academic situation of a subsonic flow without heat transfer which has been addressed so far. All these measurements were performed by laser-Doppler anemometry. Figure 9.16(a) first presents the overall appearance of the spatio-temporal correlations, where from are derived the convection velocity values of the large-scale vortices in the flow. This convection velocity is shown in Figure 9.16(b) with respect to the distance to the axis of the jet ($r*$ here represents the normalized distance relatively to the position – at the section x/D being considered – of the axis of the mixing layer that is developed from the edges of the nozzle – as shown in Figure 9.16(b).

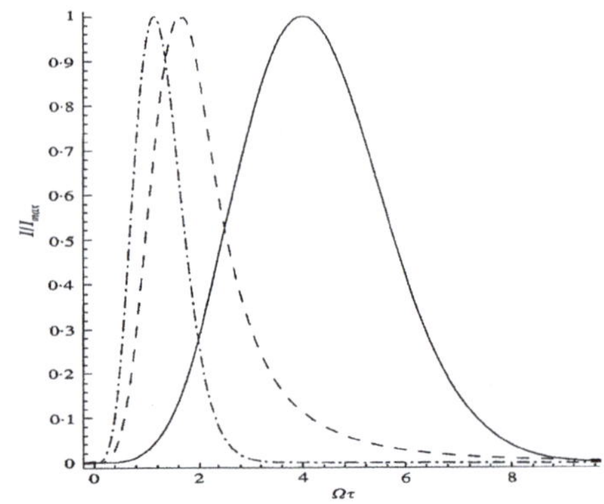

Figure 9.14. *Intensities of aeroacoustic noise radiated by a turbulence volume, obtained by [WOO 01] using different analytical formulations: —— Both spatial and temporal Gaussian correlations, and not correlated ($\sigma(k) = 1$) ; - - - - - - Spatial correlation associated with the Kolmogorov spectrum for spatial correlation, and $\sigma(k) \approx k^{-2/3}$; -· - · -· - Spatial Gaussian correlation and $\sigma(k) \approx k^{-2/3}$*

A relatively similar decrease is observed for the convection velocity (symbols) and the mean velocity (solid line), and this regardless of the section of the jet under study. On the internal side of the mixing layer ($r* < 0$), the convection velocity does not exceed 90% of the exit velocity at the nozzle whereas, on the external side ($r* > 0$), it is locally higher than the mean velocity. For the prediction of the noise radiated by the flow, a single unique value of the convection velocity is usually adopted. The choice of this value is still often discussed. If the admitted value is that where the turbulent field is characterized by the most energetic large-scale structures, then the value to be adopted is that obtained on the axis of the mixing layer. In doing so, the convection velocity of the large-scale structures estimated in this case is of the order of 0.6 times the exit velocity of the jet. This value is actually the one typically found in the literature in perfectly relaxed subsonic or supersonic regime. On the other hand, if the value to be adopted is that where the convection velocity is equal to the local mean velocity, then these results show that this convection velocity is approximatively equal to 0.65 times the exit velocity of the jet, thus a slightly higher value. Regardless of which of the two values is adopted, these results indicate that the convection velocity at a point of the flow is, therefore, biased toward the local mean velocity. As a result, this suggests that the turbulence small scales are moving at convection velocities close to that of the mean flow. Large scales move in contrast with their own velocity. To better clarify this distribution of velocities relatively to the scales of the flow, a frequency study of the convection velocity would, therefore, be necessary. This point is rather critical since, as we will see in Chapter 10, the concept of local convection velocity is introduced in order to take into account its convective nature in the characterization of the turbulent field, or in other words that of the aeroacoustic sources. The directivity of the acoustic far-field is then governed essentially by convective amplifications of the radiated field, more significant still in the case of high-speed jets.

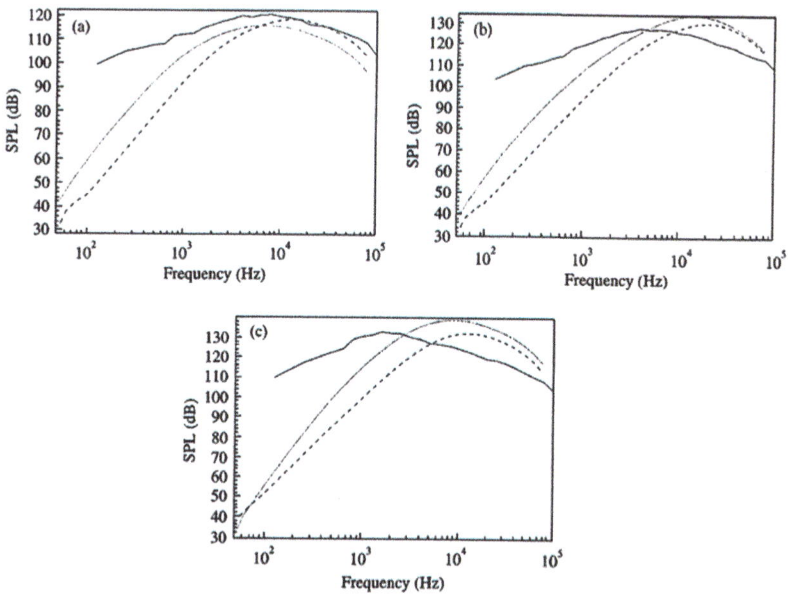

Figure 9.15. *Noise spectra obtained in a jet by [WOO 01] for different directivities) a) $\theta = 93°$; b) $\theta = 123°$; c) $\theta = 139°$. —— Experimental data; – - - - Gaussian correlations both in space and in time, and not correlated ($\sigma(k) = 1$) ; · · · · · · Spatial correlation associated with the Kolmogorov spectrum for the spatial correlation, and temporal Gaussian correlation with $\sigma(k) \approx k^{-2/3}$*

Returning now to the modeling of the correlations involved in the analytical developments considered in this chapter (equation [9.5]), the comparison of the experimental data can be achieved with different spatial correlation formulae, such as, in particular, that proposed by Ribner [RIB 69]:

$$R_{ijkl}(\vec{x}, \vec{\xi}, \tau) = \exp\left[-\pi\left(\frac{\xi_1^2}{L_{11}^2} + \frac{\xi_2^2}{L_{22}^2} + \frac{\xi_3^2}{L_{33}^2}\right) - \omega_\xi^2 \tau^2\right],$$

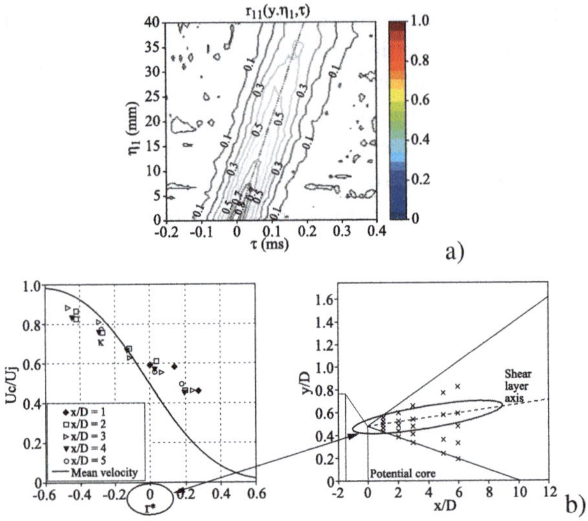

Figure 9.16. *Iso-contours of the spatio-temporal correlation of the velocity at a given point in the mixing layer that is developed just after the jet nozzle (without thermal effect); a). Transversal distribution of the convection velocity of large-scale vortices. (U_j is the exit velocity of the jet and D the nozzle diameter); b). According to [KER 04]. For a color version of the figure, see www.iste.co.uk/anselmet/acoustics.zip*

where ξ_i represent the components of the space-separation vector $\vec{\xi}$ and L_{ii} represent the integral lengths of the three components u_i of the velocity field according to the direction i, or a formula proposed by Kerherve [KER 04], on the basis of second-order correlations modeled as $\overline{u'_i(\vec{x})u'_j(\vec{x}+\vec{\xi})} = \exp(-\pi\xi^2/L^2)$, with L then obtained by $L = \sqrt{(L_{11}^2 L_{12}^2 L_{13}^2(\xi_1^2+\xi_2^2+\xi_3^2))/(L_{12}^2 L_{13}^2 \xi_1^2 + L_{11}^2 L_{13}^2 \xi_2^2 + L_{11}^2 L_{12}^2 \xi_3^2)}$. By comparing this turbulent correlation model in the isotropic and anisotropic cases, it appears that this last formulation allows the structure modification of the spatial decay to be taken into account. This modification is reflected in the anisotropic case by a reduction in the spatial extent on which the turbulent field remains correlated more pronounced for the longitudinal component of the Reynolds stress correlation tensor. Always in the anisotropic case, this model allows to take into account the vortex containing nature of the turbulent field, which is reflected by the presence of damped oscillations in the spatial decays. It should be noted that the model proposed by Ribner does, however, not account for this vortex character of the turbulent field. Indeed, it is then observed (Figure 9.17(a)) that the refined model provides results that agree much more with the measurements than the conventional Ribner model. With regard to the temporal second-order correlation of the component i of the velocity, three models have been examined by this author:

1) the model proposed by Ribner [RIB 69], $\overline{u_i'(\vec{x},t)u_i'(\vec{x},t+\tau)} = \exp(-\omega_{ti}^2\tau^2)$;

2) the model proposed by Chu [CHU 66], $\overline{u_i'(\vec{x},t)u_i'(\vec{x},t+\tau)} = \exp(-\omega_{ti}\tau)$;

3) the model proposed by Bailly *et al.* [BAI 97], $\overline{u_i'(\vec{x},t)u_i'(\vec{x},t+\tau)} = 1/\cosh(\frac{2}{5}\omega_{ti}\tau)$.

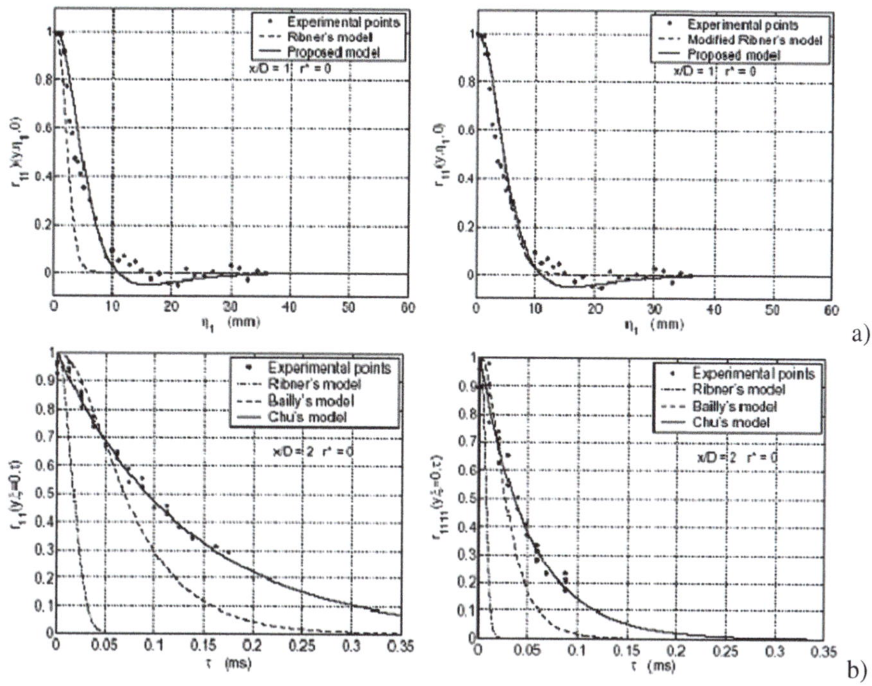

Figure 9.17. *2nd-order space-correlations of the velocity in a jet (without thermal effect); (a). 2nd-order time-correlations of velocity in a jet (without thermal effect); (b). According to [KER 04]*

In all three cases, the characteristic frequency ω_{ti} is then considered as equal to $2\pi/\tau_{\xi i}$, where $\tau_{\xi i}$ designates the decay time of the velocity component i in the mobile coordinate system – at the convection speed U_c – (ξ, τ). Figures 9.17(a) and 9.17(b) illustrate the results obtained in a jet without thermal effect for, respectively, the space- and time-correlation models.

Based on these correlation models, it is then possible to calculate the noise radiated by the jet (applying the methods known as hybrid which are described in Chapter 11), since the source terms corresponding to the Lighthill tensor can then be calculated at each point in the calculation domain comprised within the volume where the jet is developed, and then be used to calculate the propagation of sound waves in the infinite medium surrounding the jet. In the case that we will consider here, this calculation was conducted for the jet with the heating effects found at a high Mach number ($M \approx 2$), which makes it possible to correctly differentiate aerodynamic effects on the one hand and thermal (or entropic) effects on the other hand, and to compare the results of these numerical previsions with measurements in the jet noise. In this case, three types of sources are effectively present, purely aerodynamic, purely entropic and mixed, since we get:

$$\frac{\partial^4}{\partial\tau^4}R_{ijkl}(y,\xi,\tau) = \frac{\partial^4}{\partial\tau^4}\left[\overline{A'_{ij}A''_{kl}}\right] + \frac{\partial^2}{\partial\tau^2}\left[\overline{S'_{ij}S''_{kl}}\right] + \frac{\partial^3}{\partial\tau^3}\left[\overline{A'_{ij}S''_{kl}}\right] + \frac{\partial^3}{\partial\tau^3}\left[\overline{S'_{ij}A''_{kl}}\right], \quad [9.10]$$

where $S_{ij}(y,t) = \partial(p - \rho c_0^2)\delta_{ij}/\partial t$ refers to the fluctuations of entropy. Meanwhile, it should be noted that, given their expression (respectively, $\partial^4/\partial\tau^4$, $\partial^2/\partial\tau^2$ and $\partial^3/\partial\tau^3$), these three contributions do not evolve the same way with respect to the velocity (or to the Mach number). Purely entropic sources have a fourth-power evolution and mixed a sixth-power one, while aerodynamic sources have, as we have already seen, an 8-th power behavior. This is what can be seen, notably in Figure 9.18. Lilley's article (1996) can be referred to for additional observations. The entropy fluctuations are then linked to the turbulent fluxes of temperature, through the approximation $S \approx \rho_0/\overline{T}\overline{u'_i\partial T'/\partial x_i}$ (where T' represents temperature fluctuations and $\overline{T}$ represents the mean temperature), such that it is then necessary to model turbulent temperature fluxes ($\equiv \overline{u'_1T'}$), and then to weight them by means of a space-time correlation function considered as being the same as that obtained (and developed) for the velocity field, in order to model the purely entropic sources. The same is true for the sources known as mixed, which are based on the product of the variances of velocity fluctuations and temperature fluxes, weighted by the same correlation function. In order to simplify the computation of turbulent temperature fluxes, quite a complex specific problem (which will, therefore, not be addressed or commented here ...), these were simply linked to the kinetic energy of turbulence k and to the gradients of mean longitudinal velocity and temperature, according to the conventional approach used in turbulence (see specific works on turbulence modeling, [CHA 00, SCH 06, POP 00]): $\overline{u'_2T'} = \overline{u'_1u'_2}(\partial\overline{T}/\partial x_2/\partial\overline{U_1}/\partial x_2) \approx (k/3)(\partial\overline{T}/\partial x_2/\partial\overline{U_1}/\partial x_2)$. The following figures illustrate the results obtained. Generally speaking (Figures 9.18, 9.19 and 9.20), it appears that the principal contribution to noise is that of aerodynamic origin. In addition, the frequency ω_ξ is calculated by $\omega_\xi = C_\omega 2\pi\frac{\varepsilon}{k}$, the optimal value of the constant C_ω being adjusted by comparison with the experimental data (see Figure 9.19(a)). The optimal value of the analytical formulation that gives the integral scale is also adjusted by comparison with experimental data (see Figures 9.19(b) and 9.20)).

These different results show that, if the position of the frequency peak of the acoustic radiation is satisfactorily predicted by this type of modeling, the peak levels (and the noise levels in general) are rather largely overestimated. This may be related to the effect of the convection velocity of the structures responsible for the noise generation, since their convection velocity is included as a fifth-power factor (as we will see in section 10.1 which describes the analytical modeling of the radiated noise for a real case such as jets, that is to say with, notably, the presence of a mixing layer in which the velocity gradients are high and generate non-negligible refraction and directivity effects for the radiated noise).

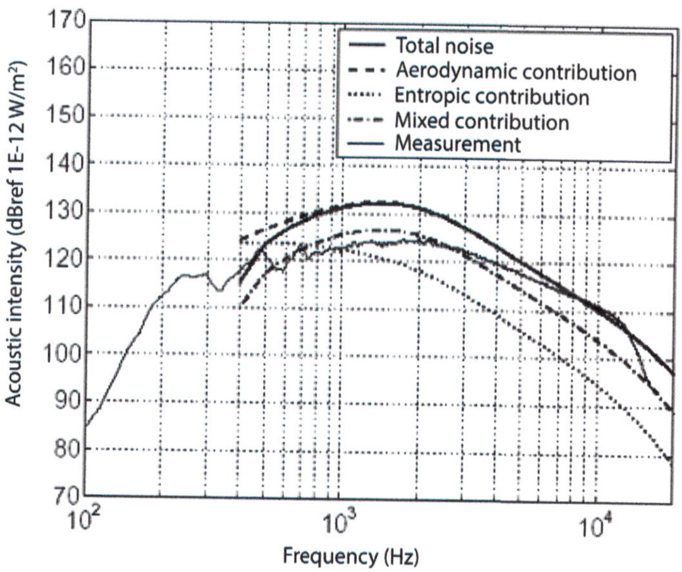

Figure 9.18. *Acoustic intensity spectra at* 40° *relatively to the axis of the jet (with thermal effects, results for the nominal model). According to [KER 04]*

Here we will conclude this section, in order to not overburden it unnecessarily, but it is clear that there is still need for fundamental works on the basic properties of turbulence in order to develop realistic analytical models for aeroacoustics applications. The question is to know up to what level of sophistication it is possible to go from an analytical point of view, especially in the light of the results that direct numerical simulations of turbulent flows are now providing (see Chapter 11), which reveal all the complexity of the phenomena involved.

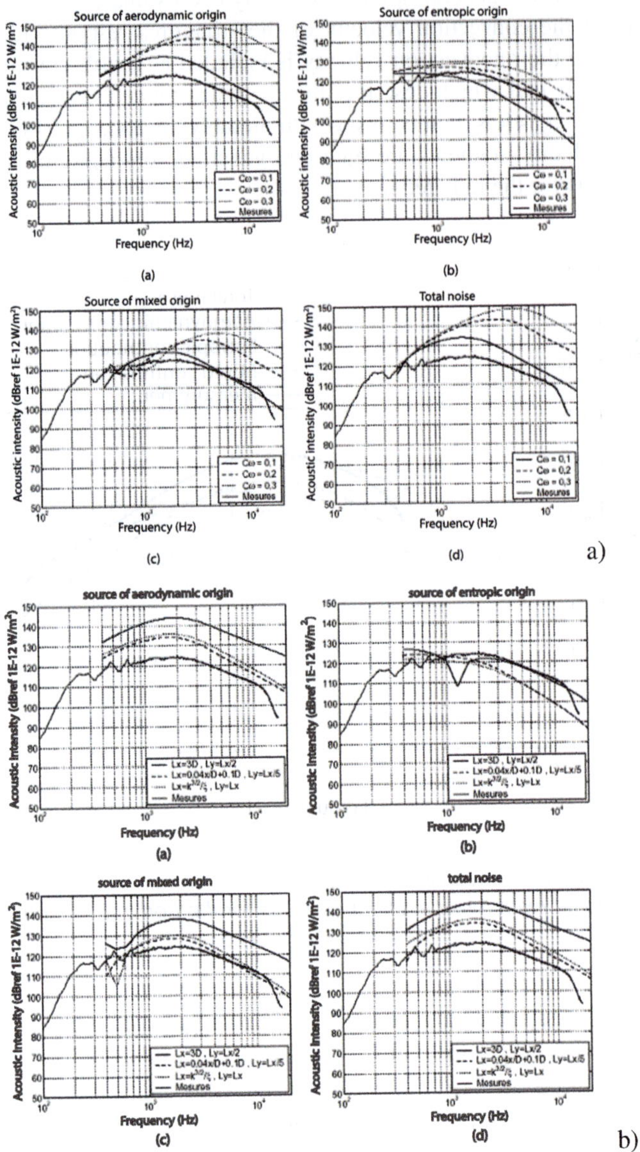

Figure 9.19. *Acoustic intensity spectra at* $40°$ *relatively to the axis of the jet (with thermal effects, study of the influence of the coefficient value* C_w*); a). Acoustic intensity spectra at* $40°$ *relatively to the axis of the jet (with thermal effects, study of the influence of the relation giving the integral scale* L_x*); b). According to [KER 04]*

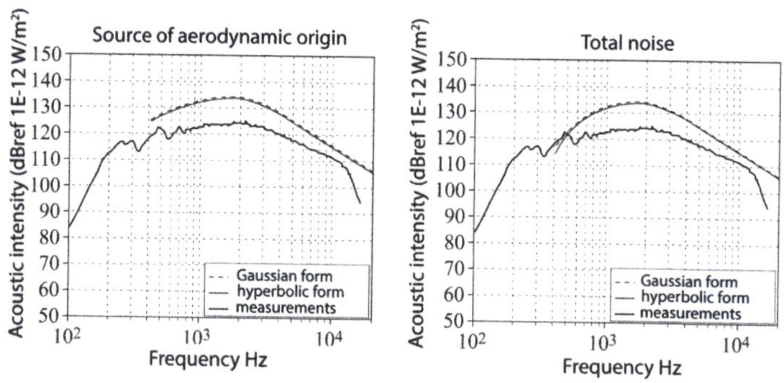

Figure 9.20. *Acoustic intensity spectra at* 40° *relatively to the axis of the jet (with thermal effects, study of the influence of the analytical formulation of the spatial correlation). According to [KER 04]*

9.7. The Powell–Howe equation for vorticity-generated sound

Although it is somewhat distant from the problems and the models described here, it seemed important to also mention the approach developed by Powell [POW 64] and Howe ([HOW 75] in particular, then largely recovered in the book, [HOW 03]), with the aim of linking radiated acoustic fields and flow dynamics. As a matter of fact, the presence of coherent vortex structures in flows playing a significant role in acoustic radiations has led Powell to characterize noise emissions in terms of the velocity curl. He introduces, therefore, this curl in the acoustic source terms from a reformulation of the Navier–Stokes equations. This alternative to the Lighthill analogy is known under the generic term of *vortex sound*. This model is mainly used for the study of aeroacoustics phenomena for simple flows, such as the noise generated by the vortex appearing on the trailing edge of an aircraft wing or a vortex interacting with an aircraft wing. Additional examples may include noise generated by the flow in pipes perforated with one or more holes such as a flute for which the source term is easily identifiable because it is related to a single (or a very small number of) vortex structure(s) whose vorticity is quite easily modeled in an analytical manner.

Thus, we will consider a local vorticity source contained within a turbulent fluid volume. In order to simplify the beginning of the analytical development, this will be, for example, a single vortex as shown in Figure 9.21. This vortex (of size l and with an associated turbulent velocity u) corresponds to a local (or distributed) vorticity source denoted by $\vec{\omega}(\vec{y})$, to which a velocity field given by the Biot and Savart relation is associated:

$$\vec{u}(\vec{x},t) = \overrightarrow{\text{curl}} \int \frac{\vec{\omega}(\vec{y},t)}{4\pi\,|\vec{x}-\vec{y}|} d^3\vec{y}.$$

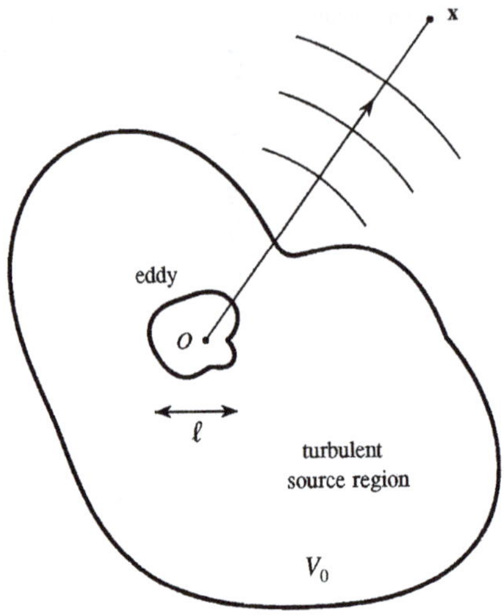

Figure 9.21. *Diagram illustrating the configuration taken into account by the Powell–Howe approach [HOW 03]*

The basis of the analytical development of the approach proposed by Powell and Howe is thus to express the components u_i and u_j of the velocity vector involved in the Lighthill tensor, $T_{ij} = \rho u_i u_j + \left((p - p_0) - c_0^2(\rho - \rho_0)\right)\delta_{ij} - \tau_{ij}$, by their expression taken from the above relation. To this end, the total velocity vector (denoted by $\vec{v}$) in the flow must also be decomposed into its rotational part denoted by $\vec{u}$ (which is, therefore, generated by the vorticity sources and has an incompressible nature, $\text{div}(\vec{u}) = 0$) and its irrotational part (which comes from the effects of compressibility in the flow and therefore does not satisfy $\text{div}(\vec{v}) = 0$, but which derives from a potential function φ such that its contribution to the velocity field is given by $\overrightarrow{\text{grad}}(\varphi)$), and we then get the general expression of the total velocity field, $\vec{v} = \vec{u} + \overrightarrow{\text{grad}}(\varphi)$. Since $\text{div}(\vec{u}) = 0$, the continuity equation is then expressed as:

$$\Delta\varphi + \frac{1}{\rho}\frac{D\varphi}{Dt} = 0,$$

where the notation Δ represents the Laplacian (similarly, the $\overrightarrow{\text{grad}}$ operator will also be subsequently denoted, for convenience, by ∇). It is then obtained that:

$$\frac{\partial^2(u_i u_j)}{\partial x_i \partial x_j} = \text{div}(\vec{\omega} \wedge \vec{u}) + \Delta(\frac{1}{2}u^2),$$

which allows the radiated acoustic pressure to be expressed (using also the fact that $p(x,t) = c_0^2(\rho - \rho_0))$ by $p(x,t) = p_1(x,t) + p_2(x,t)$, with:

$$p_1(\vec{x},t) = \frac{-\rho_0 x_i}{4\pi c_0 |\vec{x}|^2} \frac{\partial}{\partial t} \int (\vec{\omega} \wedge \vec{u})_i(\vec{y}, t - \frac{|\vec{x}|}{c_0} + \frac{\vec{x}.\vec{y}}{c_0 |\vec{x}|}) d^3\vec{y}, \text{ and}$$

$$p_2(\vec{x},t) = \frac{\rho_0}{4\pi c_0^2 |\vec{x}|} \frac{\partial^2}{\partial t^2} \int \frac{1}{2} u^2(\vec{y}, t - \frac{|\vec{x}|}{c_0} + \frac{\vec{x}.\vec{y}}{c_0 |\vec{x}|}) d^3\vec{y}.$$

When overlooking the contribution of retarded times $\vec{x}.\vec{y}/c_0 |\vec{x}|$ between the various points located inside the vortex, it can then be shown that $\int \vec{\omega} \wedge \vec{u} d^3\vec{y} = 0$, because $\vec{u}$ evolves as $1/y^3$ when $|y| \to \infty$. By limited expansion of the quantity present in the relation that gives p_1, we get:

$$(\vec{\omega} \wedge \vec{u})(\vec{y}, t - \frac{|\vec{x}|}{c_0} + \frac{\vec{x}.\vec{y}}{c_0 |\vec{x}|}) = (\vec{\omega} \wedge \vec{u})(\vec{y}, t - \frac{|\vec{x}|}{c_0})$$
$$+ \frac{\vec{x}.\vec{y}}{c_0 |\vec{x}|} \frac{\partial}{\partial t} \left\{ (\vec{\omega} \wedge \vec{u})(\vec{y}, t - \frac{|\vec{x}|}{c_0}) \right\} + \ldots,$$

such that, when $|x| \to \infty$,

$$p_1(\vec{x},t) \approx \frac{-\rho_0 x_i x_j}{4\pi c_0^2 |\vec{x}|^3} \frac{\partial^2}{\partial t^2} \int y_i (\vec{\omega} \wedge \vec{u})_j(\vec{y}, t - \frac{|\vec{x}|}{c_0}) d^3\vec{y} \approx \frac{\ell}{|\vec{x}|} \rho_0 u^2 M^2.$$

The order of magnitude of p_2 is then given by the fact that the Navier–Stokes equation is written in the present context as:

$$\frac{\partial \vec{u}}{\partial t} + \nabla \left(\int \frac{dp}{\rho} + \frac{1}{2} v^2 + \frac{\partial \varphi}{\partial t} \right) = -\vec{\omega} \wedge \vec{u} - \vec{\omega} \wedge \nabla\varphi - \nu \overrightarrow{\text{curl}} \vec{\omega},$$

such that (by achieving the scalar product with $\vec{u}$):

$$\frac{\partial}{\partial t} \left(\frac{1}{2} u^2 \right) + \text{div} \left\{ \vec{u} \left(\int \frac{dp}{\rho} + \frac{1}{2} v^2 + \frac{\partial \varphi}{\partial t} \right) \right\} = -\vec{u}.\vec{\omega} \wedge \nabla\varphi - \nu \vec{u}.\overrightarrow{\text{curl}} \vec{\omega}$$
$$= -\vec{u}.\vec{\omega} \wedge \nabla\varphi + \nu(\text{div}(\vec{u} \wedge \vec{\omega}) - \omega^2).$$

Then, integrating over the whole space and noting that the contribution of the terms corresponding to the divergence of $\vec{u} \left(\int dp/\rho + 1/2v^2 + \partial\varphi/\partial t \right)$ cancels out at least as fast as $1/y^3$ when $|y| \to \infty$, where also $\vec{\omega} = 0$,

$$\frac{\partial}{\partial t} \int \frac{1}{2} u^2(\vec{y}, t) d^3\vec{y} = -\int (\vec{u}.\vec{\omega} \wedge \nabla\varphi + \nu\omega^2)(\vec{y}, t) d^3\vec{y} \approx \ell^2 u^3 M^2 + \frac{\ell^2 u^3}{Re},$$

where the Reynolds number, calculated here as $Re = ul/\nu$, is at least of the order of 10^4 since we are in a situation corresponding to a turbulent flow regime. Then, still overlooking the retarded times, we get that:

$$p_2(\vec{x}, t) \approx \frac{\ell}{|\vec{x}|}\rho_0 u^2 M^4 + \frac{\ell}{|\vec{x}|}\frac{\rho_0 u^2 M^2}{Re},$$

so that $p_2 \ll p_1$ for the turbulent flows considered here, such that $Re \gg 1$ and $M \ll 1$. Therefore, for subsonic flows, the contribution $\text{div}(\rho_0 \vec{\omega} \wedge \vec{u})$ is the essential noise source, corresponding to the Lighthill quadripole, $\partial^2(\rho_0 u_i u_j)/\partial x_i \partial x_j$.

Once these preliminary relationships established, the expansion itself is carried out fairly simply. To this end, it is necessary to consider the variable B, the total enthalpy, defined by:

$$B = \int \frac{dp}{\rho} + \frac{1}{2}v^2,$$

which is in this context the reference acoustic variable instead of p (or $c_0^2(\rho - \rho_0)$) which was used in the Lighthill approach. In fact, this quantity appears naturally in the formulation known as Crocco's equation for the momentum conservation equation of the fluid:

$$\frac{\partial \vec{v}}{\partial t} + \vec{\omega} \wedge \vec{v} + \nabla B = -\nu \left(\overrightarrow{\text{curl}}\vec{\omega} - \frac{4}{3}\nabla(\text{div}\vec{v}) \right),$$

of which only the following simplified version will be of interest, where the terms related to the viscous friction are ignored:

$$\frac{\partial \vec{v}}{\partial t} + \vec{\omega} \wedge \vec{v} + \nabla B = -\nu \overrightarrow{\text{curl}}\vec{\omega}.$$

In the irrotational regions, Crocco's equation is reduced to $\partial \vec{v}/\partial t = -\nabla B$, such that $B = -\partial \varphi/\partial t$ in these regions where $\vec{\omega} = 0$. Thus, B is constant in the stationary and irrotational flow regions and therefore represents the sound waves when one is located at a great distance from the acoustic sources. If the flow is at rest in the far-field, the acoustic pressure is then given by $p = \rho_0 B = -\rho_0 \partial\varphi/\partial t$. To calculate the pressure in terms of B anywhere else in the flow, it is important to use the fact that $\int dp/\rho = B - v^2/2$. By differentiating over time and by using Crocco's equation, it then yields:

$$\frac{1}{\rho}\frac{\partial p}{\partial t} = \frac{\partial B}{\partial t} - \vec{v}.\frac{\partial \vec{v}}{\partial t} = \frac{\partial B}{\partial t} - \vec{v}.(-\nabla B - \vec{\omega} \wedge \vec{v} - \nu \overrightarrow{\text{curl}}\vec{\omega}) = \frac{DB}{Dt} + \nu \vec{v}.\overrightarrow{\text{curl}}\vec{\omega}.$$

The small viscous contribution can be overlooked in flow situations of high Reynolds number flows, and we then get that $(1/\rho)(\partial p/\partial t) = DB/Dt$. It is then relatively easy to obtain the equivalent of the Lighthill relation. To this end, Crocco's relation must be multiplied by the density ρ and the divergence be taken:

$$\text{div}\left(\rho\frac{\partial\vec{v}}{\partial t}\right) + \nabla.(\rho\nabla B) = -\text{div}(\rho\vec{\omega}\wedge\vec{v}).$$

Using the formulation of the continuity equation, such as $\text{div}(\vec{v}) = -(1/\rho).(D\rho/Dt)$, then writing successively that:

$$\begin{aligned}\text{div}(\rho\frac{\partial\vec{v}}{\partial t}) &= \nabla\rho.\frac{\partial\vec{v}}{\partial t} + \rho\frac{\partial}{\partial t}\text{div}\vec{v} = \nabla\rho.\frac{\partial\vec{v}}{\partial t} - \rho\frac{\partial}{\partial t}\left(\frac{1}{\rho}\frac{D\rho}{Dt}\right)\\ &= \nabla\rho.\frac{\partial\vec{v}}{\partial t} - \rho\frac{\partial}{\partial t}\left(\frac{1}{\rho}\frac{\partial\rho}{\partial t}\right) - \frac{\partial\vec{v}}{\partial t}.\nabla\rho - \rho\vec{v}.\nabla(\frac{1}{\rho}\frac{\partial\rho}{\partial t})\\ &= -\rho\frac{D}{Dt}\left(\frac{1}{\rho}\frac{\partial\rho}{\partial t}\right) = -\rho\frac{D}{Dt}\left(\frac{1}{\rho c^2}\frac{\partial p}{\partial t}\right) = -\rho\frac{D}{Dt}\left(\frac{1}{c^2}\frac{DB}{Dt}\right),\end{aligned}$$

the equation called *vortex sound* is finally obtained (by dividing by ρ) for a homentrope flow (i.e. whose entropy is the same at any point) :

$$\left(\frac{D}{Dt}\left(\frac{1}{c^2}\frac{D}{Dt}\right) - \frac{1}{\rho}\nabla.(\rho\nabla)\right)B = \frac{1}{\rho}\text{div}(\rho\vec{\omega}\wedge\vec{v}). \qquad [9.11]$$

From this equation, it can be derived that the source expressed in terms of vorticity that appears in the right member cancels out in the irrotational regions of the flow. If $\vec{\omega} = 0$ everywhere and there is no solid boundary in motion, the total enthalpy B is constant and there is no acoustic wave that propagates in the fluid. If no acoustic wave can cross through the boundaries of the volume, then in the case of a homentrope fluid, noise generation is only possible if a vorticity source in motion is present (the member on the right side of the equation giving thus the analytical formulation of the sound source).

When observing a flow with a low Mach number, so that $c/c_0 \approx 1 + \mathcal{O}(M)$ and $\rho/\rho_0 \approx 1 + \mathcal{O}(M)$, equation [9.11] then becomes:

$$\left(\frac{1}{c_0^2}\frac{\partial^2}{\partial t^2} - \Delta\right)B = \text{div}(\vec{\omega}\wedge\vec{v}),$$

such that, in far-field, the acoustic pressure is given by $p(x,t) \approx \rho_0 B(x,t)$.

Furthermore, considering an irrotational fluid defined by its velocity potential $\varphi_0(x)$, so that $U = \nabla\varphi_0$, and considering an elementary disturbance of the velocity potential such as $\varphi = \varphi_0(x) + \varphi'(x,t)$, then the general equation describing the evolution of φ is:

$$\frac{1}{c^2}\frac{\partial^2\varphi}{\partial t^2} + \frac{1}{c^2}\frac{D}{Dt}\left(\frac{1}{2}(\nabla\varphi)^2\right) + \frac{1}{c^2}\frac{\partial}{\partial t}\left(\frac{1}{2}(\nabla\varphi)^2\right) - \Delta\varphi = 0.$$

The linearized version of this equation that describes the propagation of small amplitude sound waves can then be obtained. Nevertheless, when $\omega = 0$, it is easier to consider the linearized equation for $B = -\frac{\partial\varphi'}{\partial t} = -\dot{\varphi}$, which is written as:

$$\left(\frac{D}{Dt}\left(\frac{1}{c^2}\frac{D}{Dt}\right) - \frac{1}{\rho}\nabla.(\rho\nabla)\right)\dot{\varphi} = 0,$$

where from is derived the linearized equation for φ', which is then the acoustic variable:

$$\left\{\left(\frac{\partial}{\partial t} + \vec{U}.\nabla\right)\left[\frac{1}{c^2}\left(\frac{\partial}{\partial t} + \vec{U}.\nabla\right)\right] - \frac{1}{\rho}\nabla.(\rho\nabla)\right\}\varphi' = 0.$$

The quantities c and ρ as well as U which are included in the propagation operator being the values relative to the undisturbed reference medium.

Although this approach seems quite distinct from the general concerns addressed in this chapter, as mentioned at the beginning of this section, it allows a number of practical problems of interest to be analytically addressed. In these problems, a precise physical origin generates vorticity (eventually overlapping fluctuations of turbulent origin): it is the case for flows around specific obstacles such as spheres, cylinders and airplane wings, be they in infinite medium or placed near a wall. Thus, ignoring the then secondary turbulence effects, it is possible to obtain a simple and quick formulation of the generation of acoustic waves, a bit like the theory of potential flows (to which this approach can be also related) allows, in fluid mechanics, the calculation of the streamlines around these objects, or even that the lift or drag be also calculated thereof. Naturally, even if all details cannot be taken into consideration, the orders of magnitude and the dominant trends can. Howe's book [HOW 03] and the review article [HOW 08] present several examples of problems that are solved by this approach. We will also present in section 10.4.3 an example using this approach for the calculation of noise associated with the interaction between a blade (of a rotating propeller or a helicopter rotor) and an aircraft wing and the associated vortex shedding. This phenomenon is often called *blade-vortex interaction*.

10

A Few Situations Closer to Reality

In this chapter, we will focus on more realistic situations, corresponding to jet flows on the one hand and boundary layer flows on the other hand, as well as on noise radiated by flames. We will consider which additional complexities these three classes of situations must take into account. Finally, we will examine two other situations in which the deviations from the simplifying hypotheses of Chapter 9 are even more significant, such that other theoretical approaches adapted from those described in Chapter 9 had to be developed: it concerns the noise generated by rotating blades or vanes, on the one hand, and the noise radiated outdoors, for example by roads, motorways, railways or airports, which propagates in the atmosphere over large distances, on the other hand.

10.1. The Ribner model for jets

The analytical developments that were conducted after Lighthill's [LIG 52] pioneering work have mainly tried to take into account the effect of the shear of the mean velocity field, which is particularly significant for jets (which were the main thrust for the application of all these theories since it was then necessary to reduce to an acceptable level the noise generated by aircraft engines). Consequently, Lighthill [LIG 54] and Lilley [LIL 73], but especially Ribner [RIB 64, RIB 69], have developed analytical models in which the velocity correlations that appear in the Lighthill tensor are explicitly decomposed with the objective to separate each instantaneous velocity u_i into the sum of the associated mean velocity U_i and the velocity fluctuations u'_i.

Thus, the tensor of the source terms' correlations consists of a combination of the correlations of the sources having a quadrupole origin. In the case of an axisymmetric jet, Ribner shows that, among all these possible correlations, only those defined for the indices i, j, k and l equal in pairs contribute to the radiated acoustic field. The remaining correlation terms are thus representative of the mechanisms of noise with highly characteristic directivity and related to the nature of the quadrupole source. A simple formulation for the directionality of each of these contributions can be obtained from equation [9.5] by writing the acoustic power $P(y,\theta)$ radiated in the observation direction θ by a fluid volume at point y in the following manner, where $\xi = \zeta - U_c\tau$ represents the separation in the mobile reference frame at velocity U_c:

$$P(y,\theta) \propto \cos^4\theta I_{1111} + \cos^2\theta\sin^2\theta(2I_{1212} + 2I_{1313} + I_{1122} + I_{1133}) + \sin^4\theta(\frac{3}{8}I_{2222} + \frac{3}{8}I_{3333} + \frac{1}{2}I_{2323} + \frac{1}{4}I_{2233}), \qquad [10.1]$$

with $I_{ijkl} = \int_V \partial^4/\partial\tau^4 R_{ijkl}(y,\xi,\tau)d\xi$. Figure 10.1 illustrates the directivity of the different quadrupoles in this expression where the angle θ localizes the position of the observation point with respect to the main axis of the jet.

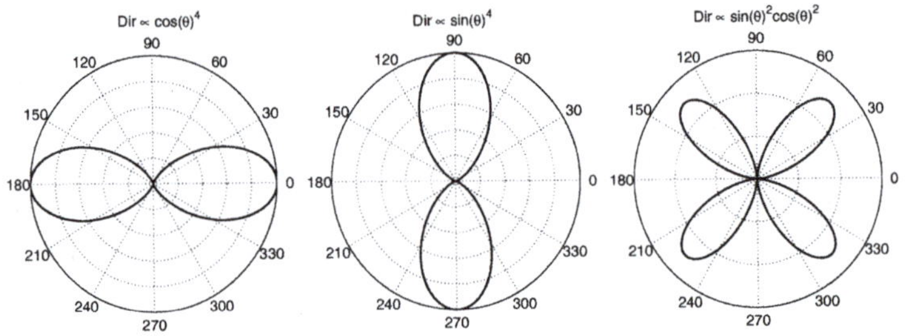

Figure 10.1. *Directivity of a longitudinal quadrupole along the axis of the jet (left), radial (center) and mixed (right) (According to [KER 04])*

Lighthill had already suggested the existence of two noise components, one contributing to the high frequencies generated by the fine turbulence and the other contributing to the low frequencies and having as origin the interaction of the turbulent field with the sheared mean flow. In addition, Lilley proposed the terminology of shear noise and self-noise for these two components whose generation mechanisms and directivity intrinsically appear different. Ribner was the first to give a mathematical description of these two noise components. The latter used the Reynolds decomposition of the instantaneous velocity u in its local mean component U and turbulent component u_t (previously u', this notation becoming confusing hereafter as symbols $'$ and $''$ will be used to locate the points that are considered in the correlations at two points in space). If the coordinate $i = 1$ refers to the main flow direction, the mean velocity component U_1 (denoted by U further in the text) is one order of magnitude higher than the other components of the mean velocity. The component i of velocity can then be written as $u_i = U_i\delta_{1i} + u_{ti}$, where $\delta_{1i} = 1$ if $i = 1$ and 0 otherwise. Therefore, the turbulent correlation tensor can be written in its reduced form in the following manner:

$$\begin{aligned} R^*_{ijkl}(y,\xi,\tau) &= \overline{u'_i u'_j u''_k u''_l} \\ &= \overline{(U\delta_{1i} + u_{ti})'(U\delta_{1j} + u_{tj})'(U\delta_{1k} + u_{tk})''(U\delta_{1l} + u_{tl})''}, \end{aligned}$$

where the symbols $'$ and $''$ indicate that the quantity is, respectively, estimated in $(y;t)$ and $(\xi; t+\tau)$. That is, in a more expanded form:

$$\begin{aligned} R^*_{ijkl}(y,\xi,\tau) = {} & \overline{u'_{ti}u'_{tj}u''_{tk}u''_{tl}} + U'(\delta_i\overline{u'_{tj}u''_{tk}u''_{tl}} + \delta_j\overline{u'_{ti}u''_{tk}u''_{tl}}) \\ & + U''(\delta_k\overline{u'_{ti}u'_{tj}u''_{tl}} + \delta_l\overline{u'_{ti}u'_{tj}u''_{tk}}) + U'^2\delta_{ij}\overline{u''_{tk}u''_{tl}} \\ & + U''^2\delta_{kl}\overline{u'_{ti}u'_{tj}} + U'U''(\delta_{ik}\overline{u'_{tj}u''_{tl}} + \delta_{jl}\overline{u'_{ti}u''_{tk}} + \delta_{jk}\overline{u'_{ti}u''_{tl}} \\ & + \delta_{il}\overline{u'_{tj}u''_{tk}}) + U'^2U''^2\delta_{ijkl}. \end{aligned}$$

The terms in U'^2 and U''^2 as well as in $U'^2U''^2$ are constant with the retarded time separation and, as a result, they do not bring any contribution to acoustic radiation (factors eliminated by the derivation in $\partial^4/\partial\tau^4$). It is thus possible to retain only the useful terms in this last expression, that is:

$$\begin{aligned} R^*_{ijkl}(y,\xi,\tau) = {} & \overline{u'_{ti}u'_{tj}u''_{tk}u''_{tl}} + U'(\delta_i\overline{u'_{tj}u''_{tk}u''_{tl}} + \delta_j\overline{u'_{ti}u''_{tk}u''_{tl}}) \\ & + U''(\delta_k\overline{u'_{ti}u'_{tj}u''_{tl}} + \delta_l\overline{u'_{ti}u'_{tj}u''_{tk}}) + U'U''(\delta_{ik}\overline{u'_{tj}u''_{tl}} \\ & + \delta_{jl}\overline{u'_{ti}u''_{tk}} + \delta_{jk}\overline{u'_{ti}u''_{tl}} + \delta_{il}\overline{u'_{tj}u''_{tk}}). \end{aligned}$$

The first fourth-order correlation term is representative of the self-noise associated with the interactions of turbulence with itself. It is usually proposed as a formulation for the force of this noise source for a given fluid volume:

$$Q_t = \rho_0 \frac{\partial^2(u_{ti}u_{tj})}{\partial x_i \partial x_j}.$$

The second-order correlation term is representative of the shear noise resulting from the source term:

$$Q_t = 2\rho_0 \frac{\partial u_{t2}}{\partial x_1}\frac{\partial U}{\partial x_2},$$

due to the interaction of the turbulent field with the mean flow. The amplitude of this component depends entirely on the velocity gradient of the mean flow. When the isotropic turbulence hypothesis is omitted, third-order correlations appear. If the turbulent field is assumed to be isotropic, the integration of these correlations over a source volume is zero. Although this last hypothesis is erroneous in the case of jet mixing layers in particular, the information concerning the interpretation of the noise component associated with these third-order correlations is scarce. Their contribution has so far not been established yet. By retaining only the correlation terms contributing to the field radiated in the case of an isotropic turbulence, the list presented in Table 10.1 is obtained.

		Shear noise	Self-noise	Weight.
$\overline{u_1'u_1'u_1''u_1''}$	$\int =$	$4U'U''\overline{u_{t1}'u_{t1}''}$	$+ \quad \overline{u_{t1}'^2u_{t1}''^2}$	1
$\overline{u_1'u_2'u_1''u_2''}$	$\int =$	$U'U''\overline{u_{t2}'u_{t2}''}$	$+ \quad \overline{u_{t1}'u_{t2}'u_{t1}''u_{t2}''}$	4
$\overline{u_1'u_3'u_1''u_3''}$	$\int =$	$U'U''\overline{u_{t3}'u_{t3}''}$	$+ \quad \overline{u_{t1}'u_{t3}'u_{t1}''u_{t3}''}$	4
$\overline{u_1'u_1'u_2''u_2''}$	$\int =$		$\overline{u_{t1}'^2u_{t2}''^2}$	2
$\overline{u_1'u_1'u_3''u_3''}$	$\int =$		$\overline{u_{t1}'^2u_{t3}''^2}$	2
$\overline{u_2'u_3'u_2''u_3''}$	$\int =$		$\overline{u_{t2}'u_{t3}'u_{t2}''u_{t3}''}$	4
$\overline{u_2'u_2'u_3''u_3''}$	$\int =$		$\overline{u_{t2}'^2u_{t3}''^2}$	2
$\overline{u_2'u_2'u_2''u_2''}$	$\int =$		$\overline{u_{t2}'^2u_{t2}''^2}$	1
$\overline{u_3'u_3'u_3''u_3''}$	$\int =$		$\overline{u_{t3}'^2u_{t3}''^2}$	1

Table 10.1. *Identification of the various terms involved in the Ribner approach [RIB 69], depending on the values of i, j, k and l*

The notation $\int =$ indicates that the contribution to the acoustic field of the left term is identical to that of the right term. The coefficients included in the right column represent the number of possible permutations of the indices $ijkl$ in R^*_{ijkl} which give the same contribution. Using formulation equation [10.1] for the acoustic power per volume unit, $P(y, \theta)$, a first estimation of the self-noise directivity and that of shear noise can be obtained. The summation of the different correlation terms mentioned above allows us in effect to write (by referring to Table 10.2) the directivity of the two noise components:

$$\Xi(\theta)_{self} \propto A,$$

$$\Xi(\theta)_{shear} \propto \frac{B}{2}(\cos^4\theta + \cos^2\theta),$$

where A and B are two proportionality constants. In the case of homogeneous and isotropic turbulence in an axisymmetric jet[1], the self-noise is, therefore,

1 It should be emphasized here that the concepts of *axisymmetric jet* and *axisymmetric turbulence* are different. An axisymmetric circular jet is such that the azimuthal mean velocity is zero and the derivatives of all averaged quantities with respect to the azimuthal angle are

omni-directional whereas the directivity of the shear noise is identical to that of the field radiated by a dipole oriented along the axis of the jet as shown in Figure 10.2. For axisymmetric turbulence, the directivity of the self-noise becomes that of a dipole oriented following the radial direction of the jet, whereas the shear noise is always identical to the dipole aligned along the axis of the jet.

$ijkl$	X_{ijkl}	C_{ijkl} identical	Weight	R_{ijkl}
1111	$\cos^4\theta$	C_{1111}	1	$\overline{u_{t1}'^2 u_{t1}''^2} + 4U'U''\overline{u_{t1}' u_{t1}''}$
1212	$(1/2)\cos^2\theta\sin^2\theta$	$C_{1212}, C_{1221}, C_{2121}, C_{2112}$	4	$\overline{u_{t1}' u_{t2}' u_{t1}'' u_{t2}''} + U'U''\overline{u_{t2}' u_{t2}''}$
1313	$(1/2)\cos^2\theta\sin^2\theta$	$C_{1313}, C_{1331}, C_{3131}, C_{3113}$	4	$\overline{u_{t1}' u_{t3}' u_{t1}'' u_{t3}''} + U'U''\overline{u_{t3}' u_{t3}''}$
1122	$(1/2)\cos^2\theta\sin^2\theta$	C_{1122}, C_{2211}	2	$\overline{u_{t1}'^2 u_{t2}''^2}$
1133	$(1/2)\cos^2\theta\sin^2\theta$	C_{1133}, C_{3311}	2	$\overline{u_{t1}'^2 u_{t3}''^2}$
2323	$(1/8)\sin^4\theta$	$C_{2323}, C_{2332}, C_{3232}, C_{3223}$	4	$\overline{u_{t2}' u_{t3}' u_{t2}'' u_{t3}''}$
2233	$(1/8)\sin^4\theta$	C_{2233}, C_{3322}	2	$\overline{u_{t2}'^2 u_{t3}''^2}$
2222	$(3/8)\sin^4\theta$	C_{2222}	1	$\overline{u_{t2}'^2 u_{t2}''^2}$
3333	$(3/8)\sin^4\theta$	C_{3333}	1	$\overline{u_{t3}'^2 u_{t3}''^2}$

Table 10.2. *Summary of the various contributions that play a role in the Ribner approach [RIB 69], according to the values of i, j, k and l*

The weighting term (weight) corresponds to the number of permutations, and X_{ijkl} represents the contribution $x_i x_j x_k x_l / x^4$ (given the axisymmetric property of jets). It is remarkable to note that the overall result for the self-noise term due to the turbulence only is the consequence of a *miraculous balance* such that $[\cos^4\theta + 2\cos^2\theta\sin^2\theta(7/16 + 7/16 + 1/16 + 1/16) + \sin^4\theta(12/32 + 12/32 + 7/32 + 1/32)] = 1$, thus eliminating any directional effect, which is in agreement with the isotropy hypothesis for the turbulent velocity fluctuations.

Note that the estimation of self-noise and shear noise directivity so far does not take into account the effects due to the convection of the sources and the refraction of acoustic waves in the presence of the mean flow. The directivity of jet noise results from the combination of these two effects. Without taking into account the convection or the refraction effects, the resulting radiation field can be considered as the basic acoustic field, represented in Figure 10.3.

This decomposition of the mixing noise into self-noise and shear noise has since then been challenged by some authors who suggest that these two components cannot correspond to separate noise generation mechanisms, but are merely mathematical

---

zero. On the contrary, axisymmetric turbulence refers to, as explained in section 9.3, turbulence situations for which two velocity variances are almost equal (and either larger or smaller than the third one). Axisymmetric turbulence is, therefore, to be compared with isotropic turbulence for which the three velocity variances are equal.

constructions. They thus prefer to talk about noise components relative to fine turbulence and large-scale turbulence. Therefore, although questionable, the terms shear noise and self-noise are retained here to refer to the second-order and fourth-order correlations, respectively, but at the same time accounting for uncertainties regarding the origin of these fictitious sources.

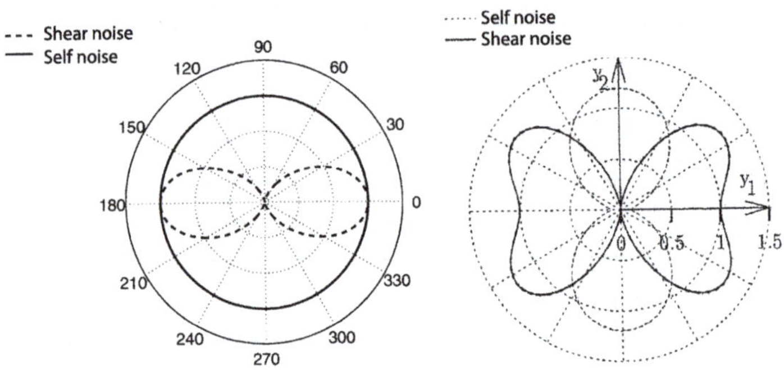

Figure 10.2. *Directivity of the components of self-noise and shear noise for isotropic a) and axisymmetric turbulence b). According to [KER 04]*

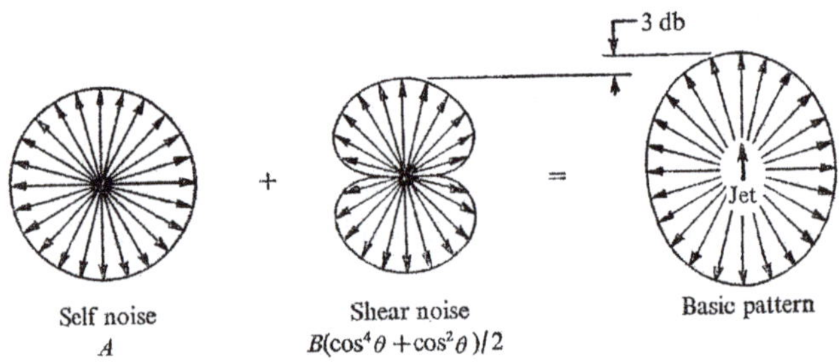

Figure 10.3. *Decomposition and directivity of the acoustic field radiated by a jet (on the right) into its two noise components: self-noise (left) and shear noise (center), for an isotropic turbulence noise, as presented by [RIB 69]*

In fact, the effects of convection of noise sources as well as the refraction effects modify the directivity of the basic acoustic field that we have just seen. Under the effect of convection, the quadrupole sources tend to align themselves according to the direction of the main flow. The acoustic intensity of jet noise is directly proportional to the convection factor established by Ffowcs Williams and Maidanik [FFO 65b] (see note presented on page 411), which corrects the Doppler convection factor originally proposed by Lighthill by taking into account the lifetime of the turbulent structures in the form βM_c. By comparing the experimental results obtained for turbojets, Ribner showed that a value of 0.55 for the coefficient β allows an acceptable directivity to be found. In parallel, the presence of velocity gradients induces refraction effects of the acoustic waves radiated by these sources all the more important that the wave frequency is high. This refraction effect by the mean flow is directly contained in the source term of the Lighthill equation. Assuming the contributions to the radiated field of sources of entropic and viscous origins negligible compared to those of convective origin, the source term can actually be written:

$$T_{ij} = \frac{\partial^2 \rho u_i u_j}{\partial x_i \partial x_j} = \left[\frac{\partial^2 \rho u_i u_j}{\partial x_i \partial x_j}\right]_{U=0} - 2U\frac{\partial^2 \rho}{\partial x_i \partial t} - U^2\frac{\partial^2 \rho}{\partial x_1^2}.$$

The principal direction of the flow is assumed to be carried here by the direction x_1. The index $U = 0$ indicates that the term is evaluated in the coordinate system moving with the mean flow. This relationship shows how the mean flow affects density gradients. By inserting this expression into the Lighthill equation, and assuming the isentropic energy transformations ($\partial p' \approx c_0^2 \partial \rho'$) and with $D/Dt = \partial/\partial t + U\partial/\partial x_1$, this results in writing:

$$\frac{1}{c^2}\frac{D^2 p}{Dt^2} - \Delta p = \left[\frac{\partial^2 \rho u_i u_j}{\partial x_i \partial x_j}\right]_{U=0}.$$

This equation is nothing more than a wave equation established in the mobile system at velocity U. For a sheared flow with $U = U(x_2)$, this equation then predicts the refraction of sound waves. Note here that the incompressibility hypothesis generally used for low-speed flows removes these refraction effects by elimination of the convection terms. These refraction effects are more or less significant depending on the frequency of the radiated acoustic wave. It has been shown in particular that, by propagating through the mixing layer, acoustic waves of wavelength lower than or of the same order of magnitude as the thickness of this layer will be strongly affected by the density gradients (identical effect is found for light waves in the atmosphere in the presence of temperature gradient). It is, therefore, clear that the wave outbound refraction of the jet will be larger if the acoustic wavelength is small. Concerning now the two components of the mixing noise, the self-noise being a mechanism of noise generation rather oriented toward high frequencies, the effects of refraction will be in overall more significant for this noise component than for shear noise. However, the refraction effects on the self-noise and shear noise are nonetheless hardly quantifiable because of the complexity of the quadrupole nature of these two mechanisms in real jet flow in particular. It has been shown that the spatial organization of the quadrupoles is also an important factor, the effects of refraction being particularly more significant for longitudinal quadrupoles aligned with the direction of the flow.

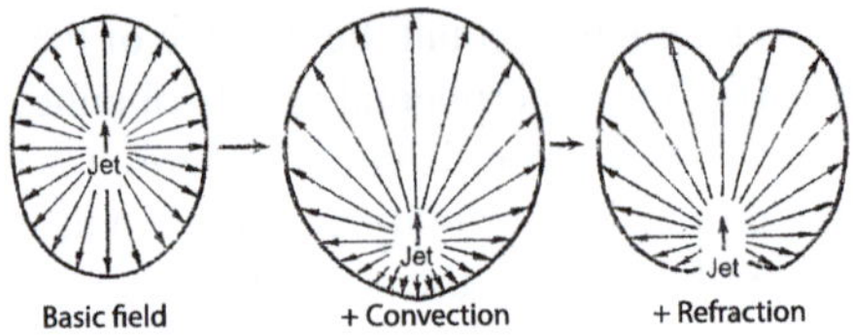

Figure 10.4. *Convection (center) and refraction (right) effects on the global acoustic field (left) radiated (according to [RIB 69])*

The combined effects of convection and refraction on the overall directivity of the jet are illustrated in Figure 10.4 as proposed by Ribner for an axisymmetric jet and a homogeneous and isotropic turbulence. From the nearly-ellipsoidal form of the intrinsic radiated field without taking into account the convection nor the refraction, the directivity of the jet noise after correcting these two effects presents a symmetrical nature and a region called silence region in the observation directions near the axis of the jet. This particular region is a consequence of the curvature of the acoustic rays radiated by the sources in the outbound part of the jet. Taking into account the effects of both convection and refraction, Ribner has shown that the directionality of the jet can then be given by:

$$\Xi_{jet}(\theta) \propto \frac{1}{C_D^5}(A + B\frac{\cos^4\theta + \cos^2\theta}{2}).$$

This synthetic formula assumes in particular that in any cross-section of the jet, the longitudinal mean velocity follows a Gaussian distribution with respect to the relative position to the axis. By expressing the coordinates x_i of the position of the source in spherical coordinates, we thus get, after integration (with respect to the angle ϕ considering the axisymmetry of the jet) of basic relation [9.5] transformed according to this approach, the expression of the radiated intensity, $I(x,\theta) = I_{self}(x,\theta) + I_{shear}(x,\theta)$, where

$$I_{self}(x,\theta) = A = \frac{3\sqrt{2}}{4\pi^2 x^2}\frac{\rho_0}{c_0^5}\frac{1}{C_D^5}L^3 u_{rms}^4 \Omega^4$$

represents the contribution of the turbulence interacting with itself, and

$$\begin{aligned}I_{shear}(x,\theta) &= B(cos^4\theta + cos^2\theta)/2\\ &= \frac{3}{8\pi^2 x^2}\frac{\rho_0}{c_0^5}\frac{\cos^4\theta + \cos^2\theta}{2C_D^5}L^3 u_{rms}^2 U^2 \frac{\sigma}{(1+\sigma)^{3/2}}\Omega^4\end{aligned}$$

represents the contribution due to the mean velocity shear in interaction with the turbulence. The coefficient σ appears in the Gaussian law characterizing the radial distribution of the mean velocity, such as:

$$U'(y + r_2/2)U''(y - r_2/2) \equiv (U(y))^2 \exp(-\frac{\sigma \pi r_2^2}{L^2}),$$

while the factor C_D defined by:

$$C_D(M_c, \theta) = \left[(1 - M_c \cos\theta)^2 + (\beta M_c)^2\right]^{1/2}$$

accounts for the convection velocity of the structures responsible for the generation of noise and the angle θ is formed between the axis of the jet and the line joining the observer and the region where the sources are located (x is actually the distance between the observer and the region where the sources are). It should be noted that, in this approach, a weaker approximation (and largely reused since then) than that used by Proudman makes it possible to separate the space separation variables and the time separation variables that occur in the integrand of relation [9.5], since the approximation $\langle u_i u'_k \rangle = R_{ik}(r, \tau) = R_{ik}(r)g(\tau)$, with $g(\tau) = \exp(-\Omega^2\tau^2)$ is used, where Ω represents a characteristic frequency of turbulence. Conversely, Proudman considered that g(τ) remains equal to 1 for the sizes l of vortices that most contribute to the generation of aeroacoustic noise (comparing, as explained in section 9.4, τ and l/c_0). It is also important to note that it is found that directional effects do not exist in the term I_{self}, which indeed corresponds to what had been considered in Chapter 9 for homogeneous and isotropic turbulence.

The comparisons carried out by Ribner with experimental data obtained by Chu [CHU 66] in a subsonic jet corroborate the empirical formulation given above for the directivity of the jet, as well as those carried out with the measurements for turbojets, even if the minimum obtained at angles close to 0° cannot be reproduced (Figure 10.5).

Additional note:

Without going into the details of the calculation which are rather tedious and require working in spectral space starting from the Fourier transform of relation [9.5], it can be shown that the convection effect of the sources (assuming that this convection can be quantified by a single-velocity U_c which applies regardless of the size of the source and hence of the frequency of the radiated noise) is felt for a fixed observer by:

– a Doppler effect such as the frequency perceived toward θ emanating from a source point moving at velocity U_{ci} in the direction i is $f(1 - M_c \cos\theta)$, where f is the frequency emitted by the source and M_c designates the Mach number associated with the convection velocity U_{ci}. Thus, the true intrinsic frequencies of turbulence are hidden unless one of the two points is moved relatively to a coordinate system that follows the movement of turbulent structures. A solution to take into account the intrinsic frequencies consists of introducing the concept of the mobile reference frame initially proposed by Lighthill (which is to consider the spatial separation $\xi = \zeta - U_c\tau$ that we have used from the beginning of this section 10.1);

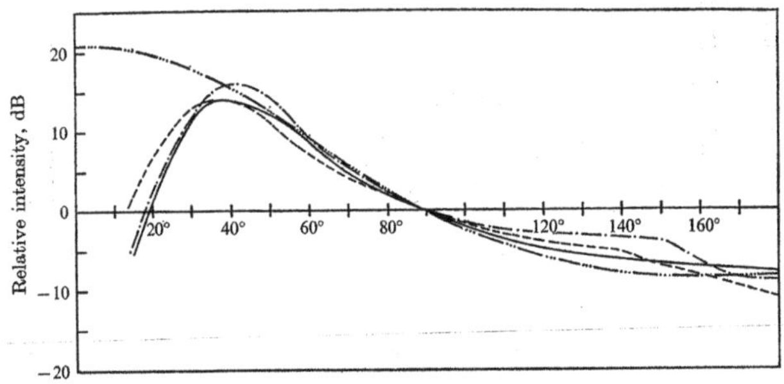

Figure 10.5. *Comparison of the formula obtained by Ribner for the noise radiated by a jet,* $\Xi_{jet}(\theta) \propto 1/C_D^5(A + B(\cos^4\theta + \cos^2\theta)/2)$*, with* $A = B$ *and* $\beta^2 = 0.3$*, with measurements for different types of turbojets by Pietrasanta:* —— - ——, $M_c = 0.76$ *(engine type J33-A-20);* ———, $M_c = 0.78$ *(engine type J34-WE-34) ;* - - - -, $M_c = 0.84$ *(engine type J48 P-8);* —— - - - ——, $M_c = 0.82$ *(according to [RIB 69])*

– an amplification effect linked to the convection of the sources: by introducing this change of variable in the Fourier transform of relation [9.5] that gives the expression of the radiated acoustic intensity, then returning to physical space after a few computation stages that we will not indicate here, we then show this amplification effect linked to the convection of sources, as highlighted by [FFO 65b], such that the sound pressure correlation function is then written:

$$I(\vec{x},\tau) \approx \frac{x_i x_j x_k x_l}{16\pi^2 c_0^5 \rho_0 x^6} \iint_{V(\vec{y})} \frac{1}{\Theta^5}\frac{\partial^4}{\partial t^4} R_{ijkl}(\vec{y'},\vec{\xi},t + \frac{1}{\Theta}\frac{\vec{x}.\vec{\xi}}{xc_0})\Bigg|_{t=\tau/\Theta} d\vec{y'}d\vec{\xi},$$

where Θ is the amplification factor related to the sources convection, $\Theta = 1 - M_c\cos\theta$. The factor in Θ^{-5} takes into account the amplification effect itself presented in section 10.3 (but only in subsonic regime), but there are important additional mechanisms of sound generation. One of them, which is called Mach noise wave, is a simple generalization of what we have just exposed and we will, therefore, present it below in its main lines. The second is most frequently referred to by the term *screech noise*. This noise corresponds to intense harmonic radiation observed upstream of the flow, which shows a large number of frequency peaks, and is the result of a feedback loop. As a matter of fact, the interaction of turbulent structures with the quasi-periodic network of shock cells can produce an acoustic wave that propagates upstream of the jet. The loop closes at the level of the lip of the nozzle where the mixing layer is then excited by the acoustic wave. Various studies have shown the complexity of the mechanisms related to the *screech* noise, resulting in large part from the high complexity of shock cells that may appear in supersonic jets (see the works of Délery [DEL 08] or Sellier [SEL 00] in particular). We, therefore, refer the interested readers to the following those studies ([RAM 94, RAM 97a, RAM 97b, RAM 98, TAM 87, TAM 88, TAM 89, TAM 95a, TAM 95b, BER 06] particularly), because it seems irrelevant to further talk about it here. Figure 10.6 illustrates, in spectral space and for an angle of 30°, the typical distribution of these three noises with different origins. Generally speaking, the noise generated by supersonic flows originates mainly in the large scales (or large vortices) of the flow, be they linked to turbulence or waves resulting from instability mechanisms. Thus, the *turbulent mixing noise* appears as having a relatively low overall contribution. Instead, the *broadband shock noise*, which results from the interaction of the large scales (or large vortices) with shock waves, has a very significant contribution, as well as the noise corresponding to the *screech noise*. Finally, it should be noted that the relative weight of these three sources of acoustic radiation strongly depends on the observation angle. Thus, toward downstream, the *turbulent mixing noise* is still dominant while in the upstream direction the *broadband shock noise* instead is. In the case of circular jets, the *screech noise* radiates primarily in the upstream direction. For more details, the readers may consult the works of Tam [TAM 91, TAM 95a] or Milllet [MIL 03].

When studying the flow of a supersonic jet with a large Mach number, the theory developed by Goldstein and Howes [GOL 73a] consists of a first generalization of the Ribner theory. It tends to take into account the fact that the propagation equation being considered must be more general than the one deriving from the Lighthill theory. Namely because the fluid medium in which the sound waves propagate presents variations of density, and therefore also of speed of sound c_0, since the flow is then supersonic, even if the analysis is restricted here to situations without shock waves. The propagation equation that is used is that proposed by Lilley in 1973, which we saw in the previous chapter of this work, taking into account the fact that the dominant mean shear is dU/dx_2:

$$\frac{D}{Dt}\left(\frac{D^2\Pi}{Dt^2} - c_0^2\frac{\partial^2\Pi}{\partial x_j^2}\right) - 2\gamma\frac{dU}{dx_2}\frac{\partial}{\partial x_1}\left(\frac{Du_2'}{Dt}\right) = \gamma\frac{D}{Dt}(\frac{\partial u_j'}{\partial x_k}\frac{\partial u_k'}{\partial x_j}),$$

where $\Pi = (1/\gamma)\ln p$, γ is the ratio of the specific heats and $D/Dt = \partial/\partial t + u_k\partial/\partial x_k$. We can then derive the propagation equation for the variable $\Gamma = D\Pi/Dt$:

$$\frac{D^2\Gamma}{Dt^2} - c_0^2\frac{\partial^2\Gamma}{\partial x_j^2} = 4\gamma\frac{\partial}{\partial x_1}(\frac{dU}{dx_2}\frac{Du_2'}{Dt}) + \gamma\frac{\partial^2}{\partial x_k\partial x_j}(\frac{Du_k'u_j'}{Dt}),$$

which constitutes a generalization of the Lighthill equation. Goldstein and Howes [GOL 73a] have thereof proposed an approximate solution valid in the low frequencies domain, which, in infinite field where the medium is at rest, so that $\Gamma \approx (\gamma/\rho_0 c_0^2)\partial p'/\partial t$, makes it possible to obtain the analytical expression of the acoustic spectral density per unit volume of the source, S_a (where Ξ_θ represents the angular directivity of shear noise):

$$\Xi_\theta = (\cos^4\theta + \cos^2\theta)/2,$$

and C_D is the amplification factor due to convection previously defined):

$$S_a(x\,|y\,,\omega) = \frac{\omega^4 C_D^2}{32\pi^3 \rho_0 c_0^5 x^2}\Bigg[\iint R_{1111} e^{iC_D\omega\tau} d\tau d\xi - D_\theta \frac{16}{3}\left(\frac{dU}{dx_2}\right)\iint \xi_2^2 R_{11} e^{iC_D\omega\tau} d\tau d\xi\Bigg].$$

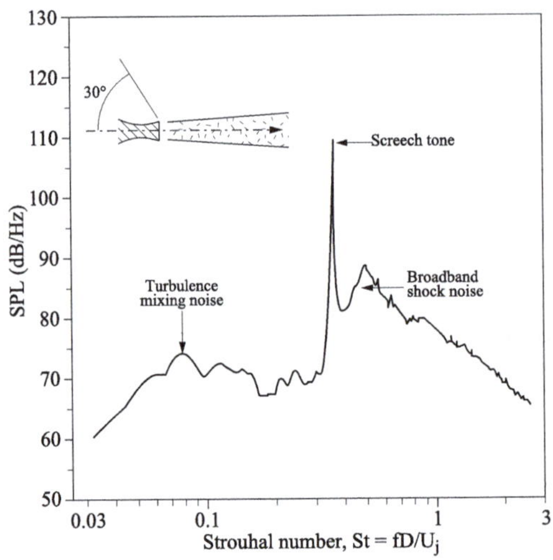

Figure 10.6. *Typical example of noise levels, respectively, associated with turbulence noise, the interaction noise with shock waves and the screech noise (according to [TAM 95a])*

We find in this expression, as in the results obtained by Ribner, the contributions of the self-noise (associated with the fourth-order correlation R_{1111}) and shear noise (associated with the second-order correlation R_{11} and the mean shear dU/dx_2). Assuming a Gaussian correlation in time, it is then obtained that:

$$S_a(x,\omega) = \frac{1}{8\pi^3\sqrt{\pi}\rho_0 c_0^5 x^2}\int_V \frac{\rho\overline{u'^2}L^3}{\Omega}\omega^4 \times \left[\frac{\overline{u'^2}C_D^2}{16}\exp(-\frac{C_D^2\omega^2}{8\Omega^2}) - D_\theta\frac{L^2C_D^2}{3\pi}(\frac{dU}{dx_2})^2\exp(-\frac{C_D^2\omega^2}{4\Omega^2})\right]dy.$$

The radiated acoustic intensity $I(x)$ is then derived from the time correlation (and defining $\tau = 0$) obtained by inverse Fourier transform:

$$I(x) = \frac{1}{\pi^3\rho_0 c_0^5 x^2}\int_V \frac{\rho^2\overline{u'^2}L^3\Omega^4}{C_D^3}\left[\frac{3\overline{u'^2}}{2\sqrt{2}} + \frac{L^2}{\pi}(\frac{dU}{dx_2})^2 D_\theta\right]dy.$$

Compared with the results obtained by Ribner, the essential difference lies in the fact that the amplification factor appears here to the third power, while it appears to the fifth power in Ribner's. This greatly improves the numerical predictions on the effect of wave directivity, as we will see later. In addition, the relation between the shear noise and self-noise is then equal to $1/2\sqrt{2\pi}$ when starting from the Lighthill equation, while the result we have just established gives $2\sqrt{2/3\pi}$.

A further phenomenon is to be taken into account when the convective Mach number M_c is supersonic ($M_c = U_c/c_0$, where U_c is as previously the convection velocity of the sources). In effect, in this case, the sources are advected at a velocity equal to or greater than the speed of sound, and the experiments then show that the effectiveness of acoustic radiation is even greater. It is this specific effect that characterizes the noise of Mach waves. This phenomenon has been addressed by [FFO 65b] (see also [BAI 96]). The expression of the resulting noise intensity I_M is:

$$I_M(x) = \frac{c_0^2 p_0^2}{32\pi^2\rho_0 x^2}\int_{V*}\frac{\cos^2\hat{\theta}\sin^2\hat{\theta}}{\hat{C}_D^5}\left[\frac{1}{c_0^2}\frac{dU}{dx_2}\right]^2 M_c^3\Omega dy,$$

and the associated spectrum:

$$S_{aM}(x,\omega) = \frac{p_0^2}{32\pi^2\rho_0}\int_{V*}\frac{\cos^2\hat{\theta}\sin^2\hat{\theta}}{\hat{C}_D^5}\omega^2 G(\omega)\Omega\delta_1^2 M_c dy.$$

The angle $\hat{\theta}$ designates the relative angle between the observer's position (x) and the position of the source, $\hat{C}_D$ is the normalized convection factor $C_D/\beta M_c$, while δ_1

represents the thickness of the mixing layer. The integral relates only to the fraction of volume $V*$ where M_c is greater than 1. The function $G(\omega)$ corresponds to the Fourier transform of the pressure autocorrelation $g(\tau)$ such that $\overline{p'p''}(\eta,\tau) = \overline{p'p''}(\eta,0)g(\tau)$:

$$G(\omega) = \left[1 - \frac{5}{8}(1 - \frac{\omega^2}{2\omega_M^2})\right] \frac{1}{2\sqrt{\pi}\omega_M} \exp(-\frac{\omega^2}{4\omega_M^2}),$$

where ω_M is the frequency defined by $\omega_M^2 = 2U_c^2/9\delta_1^2$.

This formula, therefore, does only apply to the regions of the jet where $M_c > 1$, so that the implementation in a computational code of such a contribution must first detect these regions.

The results of Figure 10.7 correctly illustrate the trends obtained, namely for a flow with a Mach number of 2. Thus, the Ffowcs Williams and Maidanik model is the most efficient for the angles ($20° < \theta < 60°$) corresponding to the dominant Mach wave radiation. On the contrary, for angles greater than $60°$, for which the shear noise is the dominant one, the Goldstein and Howes model is the most efficient, and in particular in relation to the Ribner model. In general, for the very small angles, the attenuation experimentally observed is due to refraction effects, which are not taken into account by the models discussed here. Nevertheless, it would seem that, for this case, the effects of Mach waves dominate for the very small angles relatively to shear noise. Thus, for high-Mach-number jets, even in the absence of shock waves, it is important to use specific approaches other than the Ribner model to take into account the effects of compressibility, which may have different origins depending on the angle that is chosen to observe the acoustic radiation. In the presence of shock waves, the problem is even more complex, as we have already mentioned, and we therefore refer the interested readers to the bibliographic references that we have previously indicated.

10.2. Problems and models specific to the interactions with walls (boundary layers)

This section concerns the specific effects present when there is a wall along which the flow develops. We will begin with the presentation of results, due to Hill and his collaborators, on the analytical modeling of the terms involved in the evaluation of the second-order structure function of the pressure fluctuations. As far as we know, in the light of the bibliographic references which are cited by these authors, their works are not directly inspired by the Lighthill equation. On the other hand, naturally, since they concern the second-order statistics in two space points of the pressure fluctuations, the results of these authors have a direct family link with the subject we are discussing here. Presenting them, therefore, seems to be unavoidable, even if it is the dynamic pressure that is considered here and not the acoustic pressure only. However, when considering the vibrations induced by a turbulent boundary layer flow on a wall, these two types of pressure fluctuations play a significant role and it is often very complicated to differentiate them. Thus, as we will see later, if the acoustic pressure corresponds to relatively small energy levels compared to the dynamic pressure, it is nevertheless very often contributing the most to the vibration induced on the walls of a transport vehicle (for example, car, train and plane) – and therefore to the discomfort of the passengers – because the acoustic pressure corresponds to waves that repeat in a regular manner. These excitations in the form of waves are, therefore, likely to cause wall resonances much more harmful than the dynamic pressure that corresponds on its part to disordered random fluctuations. Then, in the major part of the present section, we will present the specific models developed in this case, basing ourselves on a certain number of results presented in the first part of this section.

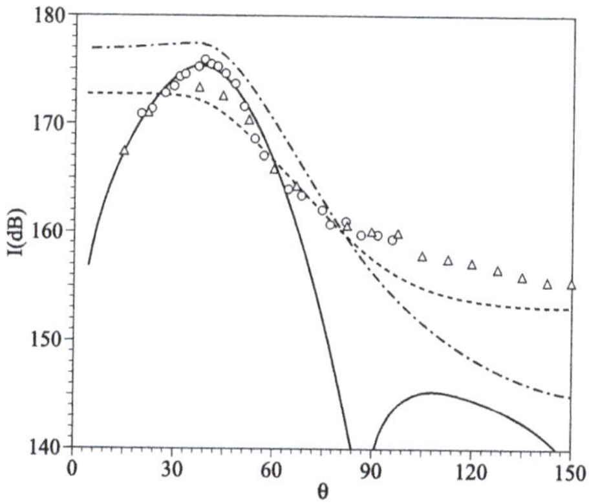

Figure 10.7. *Angular distribution of the noise radiated by a jet at Mach 2. The symbols represent the experimental points. The lines, the numerical results : — -—, Ribner's Model ;- - - - -, Goldstein and Howes's Model;——, Ffowcs Williams and Maidanik's Model (according to [BAI 96])*

Without going into the historical and theoretical details, it seems worth mentioning that early works on the correlations of pressure in two space points date from 1949, in an article published by Obukhov [OBU 49], and that they were afterward completed by Batchelor [BAT 51] and by Obukhov and Yaglom [OBU 51]. By the way, it is also interesting to note that these two communities, that of acoustics and that of turbulence experts, did appear to just recognize Batchelor's article as the only common heritage and that, apart from this article, they widely ignored each other. This was the case until turbulence experts such as Lele, Moin or Sarkar [COL 97, MIT 99, SAR 93, WAN 06], due to developments in direct numerical simulations, used their data to study the fine mechanisms of aeroacoustics noise generation (see Chapter 11), while validating or invalidating the analytical modeling that we have examined in the previous sections (9.4, 9.5 and 9.6). The explanation that we can propose thereof is that acoustics experts seek to model the effect of turbulence only for its overall contribution to the sound waves that are propagated far away from the sources (that is outside the turbulence volume), whereas turbulence experts seek, on their part, to theoretically model the properties of pressure fluctuations inside the turbulence volume itself. Nonetheless, given the refinements now developed to model the turbulence and source term of the Lighthill equation, it looks surprising that these two communities still appear to be claiming historically different contributions. Moreover, as explained in detail by Hill and Wilczak [HIL 95], the approach developed by [OBU 49], [BAT 51] and [OBU 51] consists of considering the Poisson equation for pressure p :

$$\frac{1}{\rho}\frac{\partial^2 p}{\partial x_i \partial x_i} = -\frac{\partial u_i}{\partial x_j}\frac{\partial u_j}{\partial x_i}, \qquad [10.2]$$

from which is obtained, using the homogeneity hypothesis, a relation for the structure function $D_p(\vec{r}) = \overline{(p(\vec{x}) \ - \ p(\vec{x}+\vec{r}))^2}$:

$$\frac{\partial^4 D_P}{\partial r_i \partial r_i \partial r_k \partial r_k} = -2Q,$$

where $Q(\vec{r}) = \overline{[\partial^2(u_i u_j)\partial x_i \partial x_j]_{\vec{x}}\,[\partial^2(u_k u_l)/\partial x_k \partial x_l]_{\vec{x}+\vec{r}}}$. If isotropy is now used (see [MON 75], for example, for the details that this implies), it is then possible to integrate the previous relation and obtain for $D_P(r)$ (where only the modulus r of the separation vector $\vec{r}$ appears):

$$D_P(r) = \frac{1}{3r}\int_0^r \left(y^4 - 3ry^3 + 3r^2y^2\right) Q(y)dy + \frac{r^2}{3}\int_r^\infty yQ(y)dy. \qquad [10.3]$$

Any analytical modeling then lies in the behavior of the quantity Q, which involves the fourth-order time derivatives of velocity fluctuations in two points. Hill and Wilczak [HIL 95], followed by Hill [HIL 96] and Hill and Boratav [HIL 97], then use, to simplify the form of relation [10.3], a number of arguments quite similar to those used by Lilley [LIL 93] notably, such as the joint-Gaussian hypothesis in particular, to express (after having written a few pages of calculations ...) the numerous fourth-order correlations that appear in [9.5] using only the second-order correlations, f and g. The problem is obviously difficult, and the degree of validity of the various approximations proposed by these authors seems to depend largely on the Reynolds number. A similar tendency is obtained by Antonia *et al.* [ANT 99], as well as by Pearson [PEA 99]: these authors show, for example, that when the separation r tends to infinity, D_P tends toward the variance $\overline{p'^2}$ of pressure fluctuations, whereas D_u (the second-order structure function for the velocity longitudinal fluctuations) tends toward the variance $\overline{u'^2}$; if $\overline{p'^2}$ and $\overline{u'^2}$ are as expected on the basis of dimensional arguments, such as $\overline{p'^2} \approx (\overline{u'^2})^2$, the proportionality constant varies, on its part, from approximately 0.4 to 1.5 when the turbulence Reynolds number R_λ varies from about 40 to about 400. Thus, in our opinion, there should be enough material to refine the analytical modeling that we have discussed in the previous chapters of this book, relying on the results briefly mentioned in this present section; this seems to be an approach that has never been undertaken until now. It should not be forgotten that the second-order structure function and the spectrum of any quantity are Fourier transforms of each other, naturally.

However, the most important problem that is found in all the areas relating to transport (whether on land, air, sea or underwater) is actually that of the interactions with the walls. This issue could be the subject of a specific chapter, but we will discuss here only points that seem to us essential to understand the approach and the approximations used. To analytically study the pressure field on a wall in the presence of a boundary layer flow, in general the initial step (see [GIO 99, BUL 96, AST 91, KIM 89] in particular) would be to use the Poisson equation [10.2]. It is then shown, as in Ribner's analysis, that there exist two separate contributions, one linked to the interaction of the fluctuating turbulent velocity field with itself, and the other related to the interaction of the turbulent velocity field with the mean velocity field (and more specifically with the shear $\partial\overline{U}/\partial y$, where y represents the direction normal to the wall). Thus, the Poisson equation for pressure fluctuations is decomposed into:

$$\frac{1}{\rho}\frac{\partial^2 p'}{\partial x_i \partial x_i} = -2\frac{\partial U_i}{\partial x_j}\frac{\partial u'_j}{\partial x_i} - \frac{\partial^2 (u'_i u'_j - \overline{u'_i u'_j})}{\partial x_i \partial x_j}. \qquad [10.4]$$

The first contribution corresponds to the so-called fast pressure fluctuations $p'^{(1)}$, while the second contribution corresponds to pressure fluctuations known as slow $p'^{(2)}$ (it should meanwhile be noted that this distinction is also involved, in a traditional manner, in the single point modeling of turbulence and that the contributions related to $p'^{(1)}$ and $p'^{(2)}$ play each a very important and specific role therein – see, for example, [CHA 00, COU 89, POP 00] or [SCH 06]). Even if the topic has been controversial for some time, according to the results of Kim [KIM 89] obtained through direct numerical simulations of a plane channel flow, it seems that these two contributions are roughly of the same order of magnitude in terms of contributions to the variance of the wall pressure fluctuations. For these two contributions, it also appears that the most intense sources are situated at a distance of $y^+ = yU_\tau/\nu = 20$ from the wall (where U_τ is the friction velocity at the wall, $U_\tau = (\tau_p/\rho)^{1/2}$, τ_p being the wall shear stress, U_τ is also connected to the friction coefficient C_f by $U_\tau = U_e\sqrt{C_f/2}$). In both cases, they are directly related to the various coherent structures that exist in the neighborhood of the wall of a boundary layer. These are the structures which are also directly related to the production mechanisms of turbulent kinetic energy. These sources are advected at a velocity close to the mean velocity at this position.

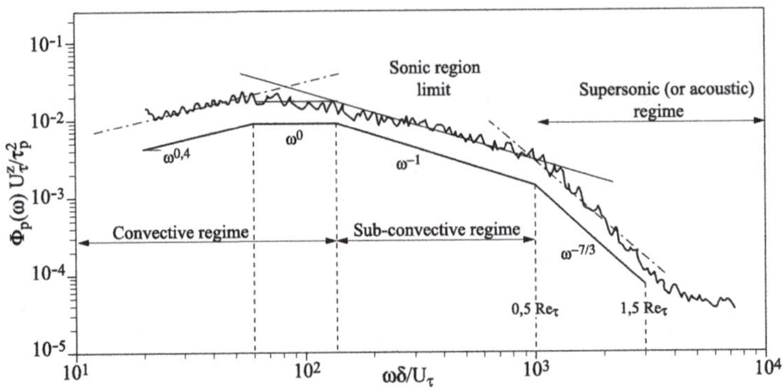

Figure 10.8. *Spectrum of wall pressure fluctuations in a turbulent boundary layer, non-dimensionalized with mixed variables ($Re_\tau = U_\tau \delta/\nu$) and in logarithmic scales (at a constant wave number, such that $k\delta^* \approx 1.5$) (according to [AST 91])*

From a spectral perspective, different types of sources can be found in the spectrum of wall pressure fluctuations (presented for a wave number k such that $k\delta^* \approx 1.5$, with δ^* the displacement thickness of the boundary layer, which is about 8 times smaller than the thickness of the boundary layer, δ). Therefore, the lower frequencies region of Figure 10.8 shows a behavior in $\omega^{+0.4}$ which reaches the frequency corresponding to the maximum of the spectrum, where some kind of plateau can be found, which is rooted in the most energetic structures (in the sense of the turbulent energy) of the outer region of the boundary layer. Next, a decay regime can be observed in ω^{-1} associated with the contribution of the sources situated in the logarithmic region of the mean velocity profile, followed by a decay regime in $\omega^{-7/3}$ which originates from sources located in the region that we have discussed previously ($y^+ \approx 20$, that is to say in what is commonly called the buffer layer region[2]). These observations are in particularly good agreement with the fact that the spectra obtained for different Reynolds numbers flows are, overall, correctly grouped when using the mixed variables δ and U_τ in order to make them dimensionless, whereas only low frequencies are correctly grouped when using δ and U_e (the external velocity), and only high frequencies are when using ν/U_τ and U_τ.

2 The buffer layer region is located between the logarithmic region – where turbulence effects largely dominate viscosity effects – and the viscous sublayer where the mean velocity profile is linear because viscosity effects then largely dominate turbulence effects; see [COU 89, POP 00] for more details on the specific properties of turbulent boundary layers.

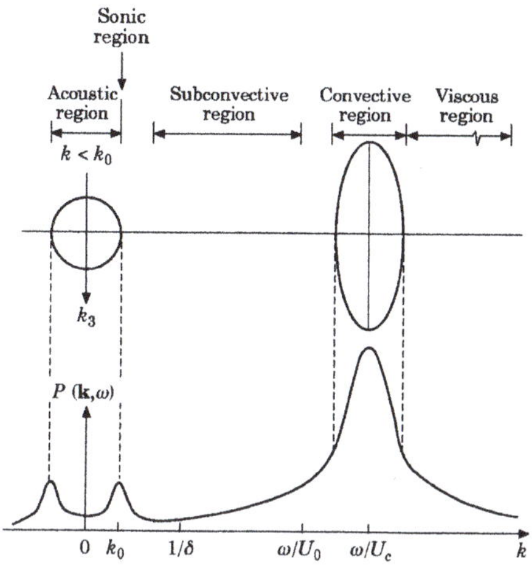

Figure 10.9. *Schematic spectrum of wall pressure fluctuations in a turbulent boundary layer, in linear scales (at constant frequency, such that $\omega\delta/U_e < 1$) (according to [BUL 96])*

On the other hand, when the distribution of wall pressure fluctuation wave numbers is represented (at least schematically, as shown in Figure 10.9), we notice that there are wave numbers much smaller than those corresponding to the regimes visualized in Figure 10.8. This corresponds to an acoustic regime in addition to the subconvective and convective regimes that could be observed in Figure 10.8. Generally, it can be considered that this acoustic regime begins at $k_0 = \omega/c_0$ (for frequencies such as $\omega\delta/U_e < 1$). It can thus be observed that the pressure fluctuations spectrum in terms of wave number–frequency is particularly complex, contrary to what we had encountered so far (even though until then the main objective consisted only of modeling the correlations that occur as sources of acoustic pressure fluctuations). The difficulty comes from the fact that, due to the presence of the wall, the so-called dynamic pressure (i.e. non-acoustic) and the acoustic pressure are intimately intertwined, at least when carrying out (for example with a microphone inserted into a very small cavity dug into the wall) the measurement of the wall pressure fluctuations. This is all the more true as, in boundary layers, there are sources of different sizes and energies that are localized at different positions ranging from the wall up to the border of the boundary layer. The wave numbers that are considered are the contributions from the components in the horizontal plane $x - z$, that is k_x and k_z. Taking these difficulties into account, added to the fact that there are, as we have already seen at the beginning of this section, pressure fluctuations referred to as fast pressure fluctuations and others as slow, it is virtually impossible to develop a theoretical approach based on what has been presented in Chapter 9, even if authors such as Kraichnan [KRA 56] have dedicated themselves to this difficult task, starting from the Poisson equation described above.

Thus, we will only present here the two analytical formulations which generally give the best results, that is to say those proposed by Corcos [COR 64] and Chase [CHA 80] . The approach followed by Corcos extends the one that we have outlined in Chapter 9. It means that this author proposes an analytical modeling of the spectrum $\Phi_p(k,\omega)$ in wave number–frequency based on a product of space and time correlations (the time correlation being actually modeled by the frequency spectrum, $\Phi(\omega)$):

$$\Phi_p(r_x, r_z, \omega) = \Phi(\omega)\exp(-\alpha_x\left|\omega r_x\right|/U_c).\exp(-\alpha_z\left|\omega r_z\right|/U_c).\exp(-i\omega r_x/U_c),$$

hence the expression of $\Phi_p(k,\omega)$ can be derived:

$$\Phi_p(\vec{k},\omega) = \frac{4\alpha_x\alpha_z\omega^2\Phi(\omega)}{U_c^2\left[\alpha_x^2\omega^2/U_c^2 + (k_x - \omega/U_c)^2\right]\left[\alpha_z^2\omega^2/U_c^2 + k_z^2\right]},$$

where the constants α_x and α_z are adjusted from experimental data, such as Farabee and Casarella's [FAR 91]. The most common values are 0.1 and 0.77, respectively, or also 0.32 and 0.7. In accordance with the experimental data available when it was proposed, this model assumes that the correlation decays according to the longitudinal $\vec{x}$ and transversal $\vec{z}$ directions are decorrelated, and the phase difference between two pressure signals is only proportional to r_x. Thus, the structures are convected in an overall manner at a constant mean velocity U_c. It should be noted that this formulation does not involve any of the characteristic properties of the boundary layer, such as its thickness δ or the friction velocity U_τ. According to him, Chase started from the Poisson equation for pressure [10.4], thus following the same path as Kraichnan [KRA 56], showing explicitly the contribution related to the interaction of the fluctuating turbulent velocity field with itself in this equation, and that related to the interaction of the turbulent velocity field with the mean velocity field (and more specifically with the shear $\partial\overline{U}/\partial y$). After a few pages of calculations, he obtained the following expression for $\Phi_p(k,\omega)$:

$$\frac{\Phi_p(k,\omega)}{\rho_0^2U_\tau^3\delta^3} = \left[(k_+\delta)^2 + 1/b^2\right]^{-5/2}\left[\frac{C_M(k_x\delta)^2k^2}{\left|k^2 - k_0^2\right| + \varepsilon^2k_0^2}\right.$$
$$\left. + C_T(k\delta)^2\frac{(k_+\delta)^2 + 1/b^2}{(k\delta)^2 + 1/b^2}(c_1 + \frac{c_2\left|k^2 - k_0^2\right|}{k^2} + \frac{c_3^2}{\left|k^2 - k_0^2\right| + \varepsilon^2k_0^2})\right],$$

where k_+ is defined by $k_+^2 = \left[(\omega - U_ck_x)^2/(hU_\tau)^2 + k^2\right]$, $k_0 = \omega/c_0$ and $c_1 = 1 - c_2 - c_3$. The constants b, c_2, c_3, C_M, C_T and h are also determined by fitting experimental data. The most common values are, respectively, 0.75, 1/6, 1/6, 0.1553, 0.0047 and 3. The terms associated with C_M represent the contributions originating from the interactions between shear and turbulence (known as fast pressure fluctuations), while those associated with C_T result from the interactions of turbulence with itself (known as slow pressure fluctuations). The value of U_c, the convection velocity of the most energetic vortices, is generally established as equal to 0.75 or $0.8.U_e$.

Note that other models have also been proposed more recently, with the aim of improving the behaviors associated with these two historical models (see in particular [GRA 97] and [ARG 06] for more details). The models proposed by the following authors can in particular be cited:

– Ffowcs Williams modifies the Corcos model taking into account an acoustic term obtained with the Lighthill analogy. For low Mach-number flows, Hwang and Geib propose the following simplified formula:

$$\Phi_p^{FW}(k_x, k_z, \omega) = \Phi_p^{Corcos}(k_x, k_z, \omega)\frac{k_x}{k_c}.$$

In the neighborhood of $k_c = \omega/U_c$, the Ffowcs Williams model coincides with Corcos's, but it leads to a more realistic decrease level for small wave numbers. However, it is sometimes pointed out that with this expression for the wave number spectrum, the Parseval relation is not respected.

– Efimtsov follows the same path as Corcos, but defines correlation lengths depending on the thickness of the boundary layer. The empirical constants of the model are determined through measurements on an airplane.

– Cockburn and Robertson propose an exponential decay inter-spectrum model which takes into account the thickness of the boundary layer, but in a different form from Efimtsov's. Again, measurements carried out on the nose of an aircraft contributed to defining the constants.

– Witting proposes a model based on the physics of the boundary layer: turbulent gusts (*bursts* and *sweeps*) are modeled as convected dipoles. The parameters of this model are fitted on published measurements. This model provides results very close to the Corcos model.

– Smol'yakov and Tkachenko fit an exponential curve to the results of their measurements of the space pressure correlation. They do not assume a purely longitudinal and purely transverse separation function expression, but take a combined form in $\exp(-\sqrt{r_x^2/L_x^2 + r_z^2/L_z^2})$. This elliptical form has also been deemed more representative of reality, in particular by comparison to the results from large-eddy simulations (LES).

The behaviors of the two historical models, as well as of a few other slightly less well known that we have just briefly mentioned, are compared in the graphs in Figure 10.10 (where wave numbers are rendered dimensionless such that the convective peak is located in 1). In the first instance, it is clear that a number of models (including Corcos's) are not at all consistent with the behavior in k^2 of the very small wave number spectrum whereas other models (such as Chase's) represent very well this evolution that is observed in the experimental data (and which was also obtained by Kraichnan by theoretical means). Second, the dependence in δ (that the Corcos model does not take into account) is felt differently depending on the model being considered (by comparing the results presented on the two graphs, which are obtained for two different values of $\omega\delta/u_{\tau}$): these differences mainly relate to the way how the convective peak is more or less enlarged, to the existence or not of the very low wave number region where we should see a white noise behavior (for $\omega/c_0 < k \ll \omega/U_c$), as well as to the possibility of accounting for the subsonic region ($k < \omega/c_0$).

These spectral wave number distributions are quantities that are particularly difficult to measure. Thus, different measurement methods of wave number spectra can be used (see [ARG 06, GIO 99] or [BUL 96] for more details). The first method consists of a direct measurement of the wave number spectral density, using discrete spatial Fourier transform on the spectral pressure densities. The difficulty in implementing this technique explains the scarcity and the recent character of the publications on this type of measurements. In effect, it requires measurement points very close to each other, in order to be able to measure small wavelengths (according to the Shannon sampling theorem). Thus, at 1 kHz, for a convection velocity in the order of 35 m/s characteristic of car velocities, the wavelength is of the order of a few centimeters; at 10 kHz, it is equal to a few millimeters. It is, therefore, necessary to have a very large number of measurement points. In addition, to avoid aliasing, the measured signal must be filtered in terms of wave numbers, in order to attenuate the large wave numbers (small wavelengths) which cannot be sufficiently well captured without error, due to the spacing between the sensors.

The second technique, which is the most commonly used, consists of using pressure sensor antennas. This network of sensors can be one-dimensional or two-dimensional, but also corresponds to a more sophisticated geometric arrangement. The principle of these antennas is to obtain, by linear combination of signals, a wave number spectrum for discrete values of the wave number depending on the spacing between the sensors. It should be noted that wave number filtering techniques using a network of sensors are effective only if we have an *a priori* knowledge of the wave number spectrum to be measured. This is not necessarily the case, for example in studies where flows are three-dimensional and separated. Within the context of studies on boundary layers, many authors have focused on wave number spectra of pressure fluctuations. Based on the Corcos hypothesis of decoupling between longitudinal and transverse fluctuations, they have ensured to measure only one-dimensional spectra in these two directions. But this independence hypothesis has been rejected by Chase, whose modeling takes account of the angular dependence of the wave number vector and the discussion is still far from over. In the case of flow studies more complex than the single boundary layer case, for which it may be difficult to define a longitudinal and/or a transverse direction, the focus should absolutely fall on wave number spectra in two dimensions, and therefore we should measure the pressures on a surface and not on one (or two orthogonal) line(s). To this end, some authors have developed a method to obtain the wave number spectrum in all directions, by using a number of pressure sensors positioned on a rotating platform. If, in the past, the maximum wave number was limited by the size of the sensors, the current possibilities in the field of remote sensors allow today for the use of this type of device to be also considered.

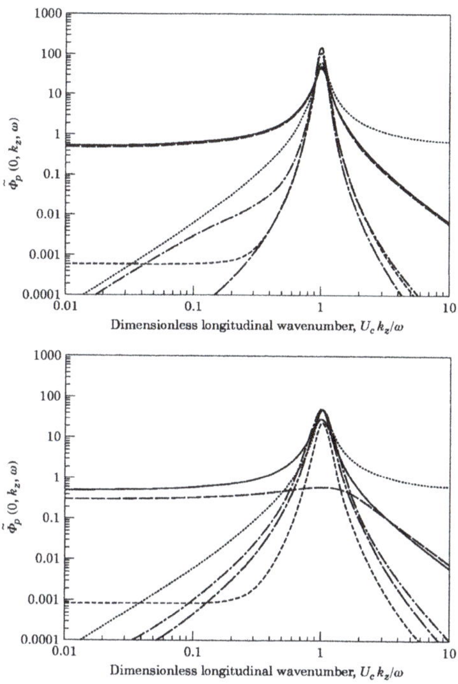

Figure 10.10. *Schematic comparisons of the wave number spectra of the wall pressure obtained by semi-empirical formulations: ——, Corcos; — —, Efimtsov; - - - - -, Smol'yakov and Tkachenko; ········, Ffowcs Williams; — · — · — Chase I; – · – · – Chase II (for the frequencies $\omega\delta/U_\tau = 248$ at the top and $\omega\delta/U_\tau = 24.8$ at the bottom) (according to [GRA 97])*

Whatever the geometry envisaged for the antenna, it is important to have a high spatial resolution, a low background noise and a wide range of measurements. However, the pressure sensors traditionally used in fluid mechanics have an excessive background noise and acoustic sensors (1" or $1/2$" condenser sensors) have an excessive sensitivity. If 1/4" and 1/8" condenser sensors might be suitable in terms of sensitivity, the former are often too large and the latter still too expensive to be mounted on an antenna. The main problem often comes then from the size of the sensors. Indeed, when they are not sufficiently pointwise, they cause a spatial averaging of the pressure over their surface. Thus, the small wavelengths are no longer resolved and spectra obtained in high frequencies are lower than the actual level. A possible solution to reduce the size of the sensitive surface of the sensors can be achieved using a pinhole assembly. It consists of drilling a very small hole on the surface and then of connecting this hole to a conventional microphone through a cavity of very small dimensions to repel the resonances toward the high frequencies.

All these constraints, to which must be added the need for the additional use of very sophisticated signal processing, explain the difficulty and thus the relative scarcity of such measurements. In particular, if we consider that an antenna network of sensors may require the simultaneous use of 30–60 sensors (or more) that on the other hand should be regularly re-calibrated.

All this illustrates the complexity of this type of flows and the actual need to obtain new information, either experimental or originating from numerical simulations of turbulence, to improve the domains of validity of the semi-empirical formulations that we have presented. This problem is especially very important for applications which consider the wall itself, that is for which the objective is to calculate the elastic behavior of plates excited by the pressure fluctuations generated by a turbulent boundary layer flow. To this end, the Corcos and Chase models are essentially used, the Corcos formulation having the advantage of its simplicity and the low number of parameters it contains. In particular, in the general case, there is often no specific information neither about the value of δ nor about that of U_T and it is then preferable to use this model which is more robust because less (or not at all) sensitive to quantities that would be poorly evaluated from empirical criteria. Since the Chase formula involves the product $U_T\delta$ to the cube, we can picture the errors that can be obtained using values of U_T or δ improperly evaluated (or poorly estimated).

10.3. Flame-generated noise

It is now commonly accepted that the noise generated by a flame is originated from the temporal fluctuations in the gas volume flux produced by the flame. For example, when considering the flames of torches, which correspond to what is known as the premixed flames type. That is to say that the combustion reaction occurs within a medium where reactive gases are already premixed, their combustion triggers only because, at a given point, the temperature exceeds the activation temperature of the reaction. On the contrary, for the other major type of combustion, in which flames are referred to as diffusion flames, the reactive gases are brought by separate ducts and the mixture of these gases must first be achieved before they can burn. Interested readers will find more details in the book published by Borghi and Champion [BOR 00] ; the combustion regime is that of wrinkled flames and the flow volume is proportional to the flame surface. A temporal variation of this amount, therefore, results in sound emission. Multiple causes can be at the origin of this variation of the flame surface: flow turbulence, hydrodynamic instabilities, lift-off instabilities, interactions between flame fronts, etc. Thus, although the purely acoustic aspect of the noise emission by combustion is relatively well understood, the underlying phenomena are complex and most of them not fully understood to date.

The first observations date back to the 19th Century where it was noted, on the one hand, that a jet of hydrogen burning in a tube, open at both ends, could produce a sound with a well-defined frequency, and, on the other hand, that a similar phenomenon occurred by placing in the tube an electrically heated grid instead of the flame. This demonstrated that these phenomena are the result of a thermoacoustic instability which is amplified if heat and acoustic pressure oscillate in phase in the tube that then works as a resonator. It is worth noting that, more recently, with the advent of space propulsion, the excitement of acoustic vibrations by combustion emerged in a more critical manner in rocket engines, as the resonances between combustion and acoustics could generate large mechanical vibrations to destroy the combustion chamber. This specific application falls outside the scope of this book, and we will, therefore, not develop this aspect of the phenomena. Nevertheless, it is clear that the noise generated by the flames and the acoustic coupling that may occur constitute an area that is still the subject of sustained academic research.[3]

The monopolar character of combustion noise has been clearly highlighted by Thomas and Williams [THO 66] through the study of an unsteady spherical flame. The experimental device of these authors consists of a small soap bubble filled with a hydrocarbon-air mixture. A spark produced in the center of the bubble then initiates the combustion and a spherical flame propagates radially outward. The process occurs at constant pressure, and the volume of the soap bubble increases due to the thermal expansion of the burnt gases. The surface of the bubble moves by compressing the layer of air that surrounds it, similarly to an inflatable balloon that is inflated, and this compression is then propagated into the ambient air in the form of a pressure wave. The size of the soap bubble being small compared to the acoustic wavelength, the wave is spherical and divergent, such that the emission by an acoustically compact volume source is really monopolar. The sound pressure radiated in the far-field by this monopole is then given, as we have already seen, by the formula $p'(r,t) = \rho_0/(4\pi r) d\dot{V}/dt$, where $\dot{V}$ represents the volume of the source and r is the distance to the source. For this experiment, the calculation of the rate of volume production is simple: the volumes of burnt gas are $(4/3)\pi R^3$, where R is the radius of the spherical flame. Before burning, this gas occupied a volume

3 Thermoacoustics will not be further discussed hereafter. Chapter 7 [SWI 07] of the book edited by Rossing [ROS 07] is devoted to thermoacoustic phenomena.

$(4/3)\pi R^3/E$, where $E = \rho_f/\rho_b$ is the ratio of the densities of fresh and burnt gases (typically, $E = 6$ to 10). Therefore, the increase in volume due to the combustion is thus $\Delta V = (4/3)\pi R^3(1 - 1/E)$, which gives a rate of volume production, $\dot{V} = d/dt(\Delta V) = 4\pi R^2(dR/dt)(E-1)/E$.

By carrying forward this term into $p'(r,t) = \rho_0/(4\pi r)d\dot{V}/dt$, it is then obtained

$$p' = \frac{\rho_0}{r}\frac{E-1}{E}\left[2R(\frac{dR}{dt})^2 + R^2\frac{d^2R}{dt^2}\right]. \qquad [10.5]$$

By measuring the temporal evolution of the radius of the flame from a film, Thomas and Williams have calculated the sound pressure from this expression. As shown in Figure 10.11, the calculated pressure is very close to the experimentally measured value.

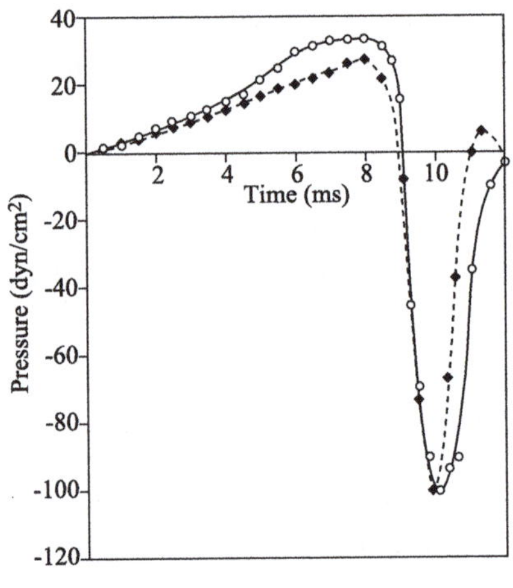

Figure 10.11. *Acoustic pressure issued by a spherical ethylene-air flame. + + +, measured pressure; o o o, pressure calculated from the relation (10.5) (according to [THO 66])*

In the case of turbulent flames, which is the most usual situation for practical applications, the structure and geometry are much more complex than for the spherical flame. Approaches and theoretical developments are consequently more complex, which resulted, historically, in many studies that cannot possibly be recalled in detail here. For example, interested readers may consult Truffaut's thesis [TRU 98]. At the present time, the basic theoretical developments (see in particular [TRU 98] or [BUI 08]) consider a combination of the equations of conservation:

– for the mass, $\frac{D\rho}{Dt} + \rho\mathrm{div}\vec{v} = 0$;

– for the momentum, while overlooking the viscous effects, $\rho\frac{D\vec{v}}{Dt} = -\overrightarrow{\mathrm{grad}}p$;

– for the energy, $\rho C_p\frac{DT}{Dt} = \dot{Q} + \frac{Dp}{Dt} + \mathrm{div}\vec{J_q}$, with $\dot{Q} = \sum_i h_i\varpi_i$;

– for chemical species, $\rho\frac{DY_i}{Dt} = \varpi_i + \mathrm{div}\vec{J_i}$.

where Y_i and W_i are, respectively, the mass fraction and molar mass of the component indicated by the index i, h_i and ϖ_i are its formation enthalpy by unit volume and its reaction rate, and by virtue of the properties of a perfect gas mixture, $p = \rho\Re/W$, with $W = (\sum_i Y_i/W_i)^{-1}$ and $\Re/W = C_p - C_v$. $\vec{J_q}$ and $\vec{J_i}$ are, respectively, the heat flux and mass flux of the species i.

Considering that the combustion takes place in a space region that does not exchange either heat or mass with the ambient medium, it can be considered that diffusive fluxes are not directly involved in the expansion of the fluid and can, therefore, be overlooked. By rewriting the conservation of energy in a different form that uses the general relation:

$$dT = dp(\frac{\partial T}{\partial p})_{\rho,Yi} + d\rho(\frac{\partial T}{\partial \rho})_{p,Yi} + \sum_i dY_i(\frac{\partial T}{\partial Y_i})_{p,\rho,Yj\neq i},$$

and by rewriting the partial derivatives from the state equation, which gives:

$$\frac{DT}{Dt} = \frac{T}{p}\frac{Dp}{Dt} - \frac{T}{\rho}\frac{D\rho}{Dt} - T\sum_i \frac{W}{W_i}\frac{DY_i}{Dt},$$

it is thus obtained:

$$\frac{1}{\gamma\rho}\frac{Dp}{Dt} - \frac{1}{\rho}\frac{D\rho}{Dt} = \frac{\dot{Q}}{\rho C_p T} + \frac{1}{n}\sum_i \frac{\varpi_i}{W_i},$$

where $n = \rho/W$ is the number of moles per unit volume and $\gamma = C_p/C_v$ is the ratio of the specific heats. The last term in the above equation involves the molar production rate at a constant density, which can be denoted by $\dot{n}_\rho$. Thus, if it is assumed that the combustion occurs in a region in which the Mach-number M is very small compared to unity, so that pressure variations of acoustic and thermodynamic origins can be ignored compared to the relative changes in unity order density, we obtain:

$$-\frac{1}{\rho}\frac{D\rho}{Dt} = \frac{\dot{Q}}{\rho C_p T} + \frac{\dot{n}_\rho}{n}.$$

This shows that the volume production by combustion is due to thermal and molar expansions.

Considering here too, the conventional approximations of the combustion volume with a very small size when compared to the wavelength, and the medium characterized by its density ρ_0 and a speed of sound c_0 constant, the equation of wave propagation in this medium can finally be derived:

$$\frac{1}{c_0^2}\frac{\partial^2 p'}{\partial t^2} - \Delta p' = \rho_0 \frac{\partial}{\partial t}\left(\frac{\dot{Q}}{\rho C_p T} + \frac{\dot{n}_\rho}{n}\right).$$

This allows the verification, through a more accurate and broader approach than that envisaged with the experiment on the spherical flame, that the combustion really constitutes a source of the monopolar type for sound waves. Using the formalism of Green functions in free space, we then get the expression of the sound pressure at a large distance (r) from the source:

$$p'(r,t) = \frac{\rho_0}{4\pi r}\left[\frac{d}{dt}\int_{V_c}\left(\frac{\dot{Q}}{\rho C_p T} + \frac{\dot{n}_\rho}{n}\right)d\tau\right]_{t-r/c_0},$$

where the integral covers the entire volume of the combustion zone. The radiated acoustic pressure can then be calculated for any type of subsonic combustion if the source terms can be modeled. In the case of premixed flames, and ignoring the diffusive flux terms as indicated above, it thus yields:

$$p'(r,t) = \frac{\rho_0}{4\pi r}\left[\frac{d}{dt}\int_{\upsilon}\frac{1}{\upsilon}\frac{D\upsilon}{Dt}d\tau\right]_{t-r/c_0},$$

let

$$p'(r,t) = \frac{\rho_0}{4\pi r}\left[\frac{d}{dt}\int_{\upsilon}\left(\frac{1}{\upsilon}\frac{\partial \upsilon}{\partial t} + \rho\vec{v}.\overrightarrow{\text{grad}\upsilon}\right)d\tau\right]_{t-r/c_0},$$

where $\upsilon = 1/\rho$ represents the mass volume. If this equation is rewritten in the coordinate system linked to the flame (by denoting $\vec{D}$ the local velocity of the flame), we get:

$$p'(r,t) = \frac{\rho_0}{4\pi r}\left[\frac{d}{dt}\int_{\upsilon}\left(\frac{1}{\upsilon}\frac{\partial \upsilon}{\partial \hat{t}} + \rho(\vec{v}-\vec{D}).\widehat{\overrightarrow{\text{grad}\upsilon}}\right)d\tau\right]_{t-r/c_0},$$

where the partial derivative (denoted with ^) with respect to time and the gradient operators are then evaluated in the mobile system. The first source term reflects the contribution of unsteady local effects, such as fluctuations in the burning rate due to the flame expansion or to the formation of pockets of combustion; the second represents the contribution of the convective effects such as the creation or destruction of flame surface at constant velocity flame. The change in the number of moles is included in the variation of v with the molar mass.

In the common regime of premixed wrinkled flames (see [BOR 00]), the integration volume involved in the equations above is given as the product of the total flame surface, S, by the laminar flame thickness d_L, and all the time scale characteristics of the turbulent flow, t_i, are large compared to the transit time in the laminar flame, t_L. Thus, the term of unsteady fluid expansion, $\frac{1}{v}\frac{\partial v}{\partial \hat{t}} = \mathcal{O}(t_i^{-1})$, is negligible compared to the convective term, $\rho(\vec{v} - \vec{D})\widehat{\overrightarrow{\text{grad}}}v = \mathcal{O}(t_L^{-1})$. In addition, in this regime, the flame thickness d_L is much smaller than the characteristic sizes of vortices, l_i (it should be noted here that the index i refers to the i-th generation of vortices among all those that are created, according to Kolmogorov's image of a vortex cascade, by successive fractioning of the large vortices – with a size close to the integral turbulence scale – down to the smallest vortices – with a size neighboring the Kolmogorov scale η). Thus, inside the flame, the gradients in the direction parallel to the front, $\widehat{\overrightarrow{\text{grad}}}_{||} = \mathcal{O}(l_i^{-1})$, are negligible compared with those in the normal direction, $\partial/\partial n = \mathcal{O}(d_L^{-1})$. The relation above can then be reduced to:

$$p'(r,t) = \frac{\rho_0}{4\pi r}[\frac{d}{dt}(S\int_0^{d_L} \rho(\vec{v} - \vec{D}).\vec{n}\frac{\partial v}{\partial n})\text{dn}]_{t-r/c_0},$$

where $\vec{n}$ represents the unit vector in the direction normal to the flame front. The integration is performed through the thickness of the flame front and dn is a length element in the direction normal to the front. The conservation of mass gives $\rho(\vec{v} - \vec{D})\vec{n} = \rho_0 U_L$. Thus, after integration, the acoustic pressure finally becomes:

$$p'(r,t) = \frac{\rho_0}{4\pi r}(\frac{T_b W_{init}}{T_{init} W_b} - 1)U_L \dot{S}(t - r/c_0),$$

such that the mean acoustic intensity radiated by the flame, $I = \overline{p'^2}/\rho_0 c_0$, is equal to

$$I = \frac{\rho_0}{16\pi^2 r^2 c_0}(\frac{T_b W_f}{T_0 W_b} - 1)^2 U_L^2 \overline{\dot{S}^2},$$

where the bar denotes a time average and the indices f and b, respectively, refer to the fresh gas and burnt gas. The acoustic power , $P = 4\pi r^2 I$, is then equal to

$$P = \frac{\rho_0}{4\pi c_0}(\frac{T_b W_0}{T_0 W_b} - 1)^2 U_L^2 \overline{\dot{S}^2}.$$

Similar results have also been obtained by other authors such as Clavin and Siggia [CLA 91], based on less rigorous reasoning.

It should finally be noted that if a monopolar-based radiation should in theory be isotropic because the source may be assumed as a source point, there exist, however, significant directional effects of various origins (sources flow convection, sound diffraction and acoustic shadow effects on the burner, etc.). The maximum intensity, as shown in Figure 10.12, is usually obtained for an angular position comprised between 80 and 50 degrees relative to the axis of the burner, whereas a minimum exists on the axis, more pronounced in the opposite direction to the flow. And thus, in reality, the radiation diagram is not different from that of a dipole, for which the intensity is zero on the axis.

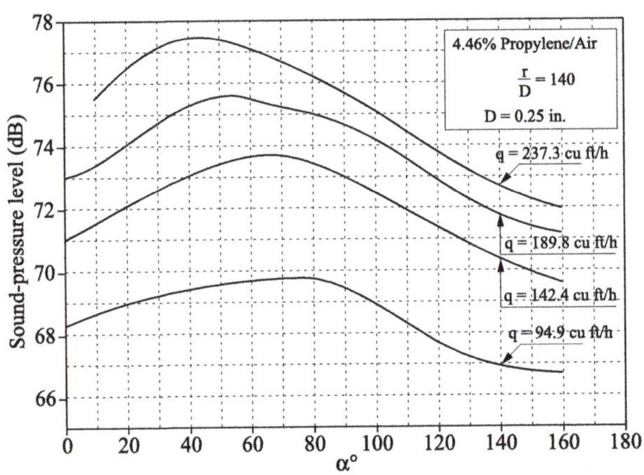

Figure 10.12. *Directivity of the combustion noise (according to [SMI 63])*

10.4. Noise generated by blades or propellers in motion: the Kirchhoff method

When there are walls in rotation, such as helicopter wings and engine aircraft propellers, an approach which generalizes Lighthill's theoretical developments [LIG 52] should be used taking into account the additional contributions from the movement of the walls. We have briefly mentioned these theoretical developments before, but it is absolutely necessary to expose them in detail now. It concerns essentially Curle's works [CUR 55] and, above all, Ffowcs Williams and Hawkings's [FFO 69], as well as their reformulations in terms of Khirchhoff integrals [FAR 88, LYR 94, FRE 96, BRE 98, FAR 01]. Readers interested in the details of the establishment of this formulation can refer to section 4.8.4.

10.4.1. *Noise generated by a solid body in motion, in the temporal domain*

For the formulation in terms of the Kirchhoff integral, we must consider the acoustic intensity that is radiated through a fictitious surface which encompasses the source's. Thus, the complete calculation of the source's acoustic radiation, based on nonlinear propagation equations, is carried out only within this volume. Outside of this volume, the linear propagation equations are the only ones to be solved, using the formulation based on the Green function in free space. In the situation being considered here, the surface will be naturally the physical surface corresponding to the body in motion (the wing or blade), but it is also possible to use this approach in situations in which there is no solid body, but just a virtual surface that surrounds the region of flow where the sources are. In the latter case, the benefits, as well as the difficulties of the approach, have been properly brought forward (see [FRE 96] in particular, in the case of the noise radiated by a jet). Thus, if the surface is called S, it will be considered that it is represented by the function $f(x,t)$ such that $f(x,t)$ cancels out at any point of S, f is negative inside the surface and f is positive outside the surface. Then, denoting by ϕ the function (the pressure fluctuation p' here, but, since the approach is much more general, it is preferable to keep ϕ in what follows) which is transported by the acoustic propagation equation, which is, therefore, linear in any point outside S, we get:

$$\frac{1}{c_0^2}\frac{\partial^2\phi}{\partial t^2} - \Delta\phi = 0.$$

This function ϕ should nonetheless be extended to any points of the space, which leads us to define a generalized function $\tilde{\phi}$ such that:

$$\tilde{\phi}(x,t) = \begin{cases} \phi(x,t) & \text{for } f > 0 \text{ outside of } S \\ 0 & \text{for } f < 0 \text{ inside of } S \end{cases}.$$

The generalized function $\tilde{\phi}$ is discontinuous and non-differentiable (in the common sense) over the surface S (that is for $f = 0$). Consequently, generalized derivatives have to be defined[4] that will be denoted with a bar such as:

$$\frac{\bar{\partial}\tilde{\phi}}{\partial t} = \frac{\partial\tilde{\phi}}{\partial t} + \tilde{\phi}\frac{\partial f}{\partial t}\delta(f) \text{ and } \frac{\overline{\partial^2}\tilde{\phi}}{\partial t^2} = \frac{\partial^2\tilde{\phi}}{\partial t^2},$$
$$\overline{\overrightarrow{\text{grad}}}\tilde{\phi} = \overrightarrow{\text{grad}}\tilde{\phi} + \phi\vec{n}\delta(f) \text{ and } \bar{\Delta}\tilde{\phi} = \Delta\tilde{\phi} + \phi_n\delta(f) + \overline{\overrightarrow{\text{grad}}}.[\phi\vec{n}\delta(f)],$$

where the notation $\delta(f)$ represents the Kronecker symbol function associated with f and $\vec{n}$ is the unit normal vector external to the surface S, the function f can be (arbitrarily) chosen such that $\overrightarrow{\text{grad}}f = \vec{n}$ (that is such that $\left|\overrightarrow{\text{grad}}f\right| = 1$) on the surface. The derivative of ϕ with respect to the normal to the surface is represented by the notation ϕ_n. With these relations, while the generalized function $\tilde{\phi}$ satisfies the propagation equation, we get:

$$\begin{aligned}\frac{1}{c_0^2}\frac{\overline{\partial^2}\tilde{\phi}}{\partial t^2} - \bar{\Delta}\tilde{\phi} &= \frac{1}{c_0^2}\frac{\partial^2\tilde{\phi}}{\partial t^2} - \Delta\tilde{\phi} - \phi_n\delta(f) - \overline{\overrightarrow{\text{grad}}}.[\phi\vec{n}\delta(f)]\\ &= -\phi_n\delta(f) - \overline{\overrightarrow{\text{grad}}}.[\phi\vec{n}\delta(f)] \neq 0.\end{aligned}$$

This equation is valid in the whole space. As we have seen before and in the other chapters, it is then common to use the Green function in unlimited domain, equal to $\delta(g)/(4\pi r)$, where g is given by $g = \tau - t + r/c_0$ and $r = |x - y|$ (the time and space variables for the observer and the source being (x, t) and (y, τ), respectively). Thus, from the previous equation, it follows that:

$$4\pi\tilde{\phi}(x,t) = -\iint \frac{\phi_n}{r}\delta(f)\delta(g)dyd\tau - \overline{\overrightarrow{\text{grad}}}.\iint \frac{\phi\vec{n}}{r}\delta(f)\delta(g)dyd\tau.$$

The integration signs involve the whole space and the time interval $]-\infty, t]$. By using the fact that the volume element is given by $dy = dfdS$ where dS represents an element of the surface S defined by $f = 0$, and noting that the presence of $\delta(f)$ restricts the integrals to $f = 0$, it then yields (see [FAR 88] for more details about the intermediate steps) that:

$$4\pi\tilde{\phi}(x,t) = -\int_{f=0} \frac{[\phi_n]}{r}dS - \overline{\overrightarrow{\text{grad}}}.\int_{f=0} \frac{[\phi]\,\vec{n}}{r}dS,$$

4 In this section, we have preferred the use of this notation rather than that defined in sections 3.2.3 and 3.2.4 by formulas [3.28], [3.19], [3.20] and their applications, because even though these notations are formally identical, those used here are more common in the specialized literature.

where the notation in brackets indicates an assessment of the variable at the retarded time, that is $[\phi] = \phi(y, t - r/c_0)$. By including the gradient under the integral operator, and keeping the terms in $1/r$ and $1/r^2$, we obtain:

$$4\pi\tilde{\phi}(x,t) = \begin{cases} 4\pi\phi(x,t) & f > 0 \\ 0 & f < 0 \end{cases}$$

$$= -\int\limits_{f=0} \frac{\left[\frac{1}{c_0}\phi_\tau \cos\theta - \phi_n\right]}{r} dS - \int\limits_{f=0} \frac{[\phi]\cos\theta}{r^2} dS,$$

with the notation $\phi_\tau = \partial\phi(y,\tau)/\partial\tau$, the angle θ designating the angle between the external normal n and the direction of the acoustic radiation, $r = x - y$. This relation represents what is usually called the Kirchhoff formula for a fixed surface S. It is a full representation of the function ϕ for the points external to S, which is expressed in terms of quantities evaluated on the surface S.

We will now consider cases in which the surface S is mobile. With regard to the previous calculations, the normal displacement velocity of the interface defined by $f = 0$, v_n, must then be included so that:

$$\frac{\bar{\partial}\tilde{\phi}}{\partial t} = \frac{\partial\tilde{\phi}}{\partial t} + \tilde{\phi}\frac{\partial f}{\partial t}\delta(f) = \frac{\partial\tilde{\phi}}{\partial t} - \phi v_n \delta(f) \text{ and}$$

$$\frac{\overline{\partial^2}\tilde{\phi}}{\partial t^2} = \frac{\partial^2\tilde{\phi}}{\partial t^2} - \left.\frac{\partial\phi}{\partial t}\right|_x v_n\delta(f) - \frac{\bar{\partial}}{\partial t}\left[\phi v_n \delta(f)\right].$$

It thus yields:

$$\frac{1}{c_0^2}\frac{\overline{\partial^2}\tilde{\phi}}{\partial t^2} - \bar{\Delta}\tilde{\phi} = -\left(\phi_n + \frac{1}{c_0}M_n \left.\frac{\partial\phi}{\partial t}\right|_x\right)\delta(f)$$

$$-\frac{1}{c_0}\frac{\bar{\partial}}{\partial t}\left[M_n\phi\delta(f)\right] - \overline{\overrightarrow{\text{grad}}}.[\phi\vec{n}\delta(f)],$$

in which M_n is the Mach number associated with the normal displacement velocity of the surface S, $M_n = v_n/c_0$. By introducing the Green function here too in infinite field, we then obtain:

$$4\pi\tilde{\phi}(x,t) = -\iint \frac{1}{r}\left(\phi_n + \frac{1}{c_0}M_n \left.\frac{\partial\phi}{\partial\tau}\right|_{y,\tau}\right)\delta(f)\delta(g)dyd\tau$$

$$-\frac{1}{c_0}\frac{\bar{\partial}}{\partial t}\iint \frac{1}{r}M_n\phi\delta(f)\delta(g)dyd\tau - \overline{\overrightarrow{\text{grad}}}.\iint \frac{\phi\vec{n}}{r}\delta(f)\delta(g)dyd\tau,$$

that we can rewrite, as previously, as

$$
\begin{aligned}
4\pi\tilde{\phi}(x,t) = &-\iint \frac{1}{r}\left(\phi_n + \frac{1}{c_0}M_n \left.\frac{\partial\phi}{\partial\tau}\right|_{y,\tau}\right)\delta(f)\delta(g)dyd\tau \\
&+\frac{1}{c_0}\frac{\bar{\partial}}{\partial t}\iint \frac{1}{r}(\cos\theta - M_n)\phi\delta(f)\delta(g)dyd\tau \\
&+\iint \frac{\phi}{r^2}\cos\theta\delta(f)\delta(g)dyd\tau.
\end{aligned}
$$

This last expression involves two integrals, denoted by I_1 and I_2, which are defined schematically by:

$$
I_1 = \iint Q_1\delta(f)\delta(g)dyd\tau \text{ and } I_2 = \frac{1}{c_0}\frac{\bar{\partial}}{\partial t}\iint Q_2\delta(f)\delta(g)dyd\tau.
$$

The computation of these two integrals is somewhat tricky and it does not seem worthwhile to present it in detail here. It is exposed in the article [FAR 88]. Thus, in subsonic situation, that is when $M_n < 1$, and assuming that the surface S is rigid, it yields that:

$$
4\pi\tilde{\phi}(x,t) = \int_S \left[\frac{E_1}{r(1-M_r)}\right]_{\tau *} dS + \int_S \left[\frac{\phi E_2}{r^2(1-M_r)}\right]_{\tau *} dS,
$$

with $M_r = -1/c_0\partial/\partial\tau\,|x-y|$. M_t is the Mach number based on the tangential component of the velocity of the surface. The quantities E_1 and E_2 are then defined as:

$$
\begin{aligned}
E_1 = &(M_n^2 - 1)\phi_n + M_n M_t \overrightarrow{\text{grad}}\phi \frac{M_n}{c_0}\dot{\phi} + \frac{1}{c_0(1-M_r)^2}\left[\dot{M}_r(\cos\theta - M_n)\phi\right] \\
&+\frac{1}{c_0(1-M_r)}\left[(\dot{n}_r - \dot{M}_n - \dot{n}_M)\phi + (\cos\theta - M_n)\dot{\phi} + (\cos\theta - M_n)\phi\dot{\sigma}\right] \\
E_2 = &\cos\theta + \frac{1}{1-M_r}\left[2M_r\cos\theta - M_n M_r - M_n\right] \\
&+\frac{1}{(1-M_r)^2}\left[(\cos\theta - M_n)(M_r^2 - M^2)\right].
\end{aligned}
$$

The quantities with a point over them represent the partial derivatives with respect to time related to the source (i.e. τ).

When the same approach using a generalized function is used to rewrite the analytical development presented by Ffowcs Williams [FFO 69], the generalized variables must be considered:

$$\tilde{\rho} = \begin{cases} \rho & f > 0 \\ \rho_0 & f < 0 \end{cases}, \tilde{\rho}\tilde{u}_i = \begin{cases} \rho u_i & f > 0 \\ 0 & f < 0 \end{cases}, \tilde{P}_{ij} = \begin{cases} P_{ij} & f > 0 \\ 0 & f < 0 \end{cases},$$

in which the function f characterizes the solid in motion, that is to say that $f = 0$ represents the mobile surface in an external stationary fluid flow. In this context, the notation P_{ij} is simply the quantity $p'\delta_{ij}$ and the generalized continuity equation is written as:

$$\frac{\bar{\partial}\tilde{\rho}}{\partial t} + \frac{\bar{\partial}\tilde{\rho}\tilde{u}_i}{\partial x_i} = (\rho'\frac{\partial f}{\partial t} + \rho u_i \frac{\partial f}{\partial x_i})\delta(f),$$

with density fluctuation $\rho' = \rho - \rho_0$. The generalized momentum equation, on its part, is written as:

$$\frac{\bar{\partial}\tilde{\rho}\tilde{u}_i}{\partial t} + \frac{\bar{\partial}\tilde{\rho}\tilde{u}_i\tilde{u}_j}{\partial x_j} + \frac{\bar{\partial}\tilde{P}_{ij}}{\partial x_j} = (\rho u_i \frac{\partial f}{\partial t} + (\rho u_i u_j + P_{ij})\frac{\partial f}{\partial x_j})\delta(f).$$

As in the calculations performed by Lighthill, taking the sum of the derivative with respect to time of the equation of continuity and the divergence of the momentum equation, we then obtain:

$$\begin{aligned} \frac{1}{c_0^2}\frac{\partial^2}{\partial t^2}p'(x,t) - \Delta p'(x,t) &= \frac{\partial^2}{\partial x_i \partial x_j}\left[T_{ij}H(f)\right] \\ &\quad -\frac{\partial}{\partial x_i}\left[(P_{ij}n_j + \rho u_i(u_n - v_n))\delta(f)\right] \\ &\quad +\frac{\partial}{\partial t}\left[(\rho_0 v_n + \rho(u_n - v_n))\delta(f)\right], \end{aligned}$$

where u_n is the velocity of the fluid according to the normal to the surface ($f = 0$) and v_n is the normal component of the displacement velocity of the surface. The notation $H(f)$ represents the Heaviside function defined in section 3.2.3, equation [3.21] ($H(f) = 0$ for $f < 0$ and $H(f) = 1$ for $f > 0$). Meanwhile, it should be noted that the surface $f = 0$ in general corresponds exactly to the surface of the solid body in motion but that the theoretical development presented above does not require it. Thus, u_n and v_n are not necessarily equal.

It is then possible to compare the developments obtained according to the Kirchhoff integral approach (by letting $\phi = p'$) and according to the approach presented in [FFO 69]. This detailed comparison is presented in the article [BRE 98]. The term source for the Kirchhoff integral formulation is then expressed as:

$$Q_{Kir} = -(\frac{\partial p'}{\partial t}\frac{M_n}{c_0} + \frac{\partial p'}{\partial n})\delta(f) - \frac{\partial}{\partial t}\left[p'\frac{M_n}{c_0}\delta(f)\right] - \frac{\partial}{\partial x_i}(p' n_i \delta(f)),$$

such that the difference with regard to the term obtained following the approach presented in [FFO 69] is simply equal to:

$$\frac{\bar{\partial}^2 \rho u_i u_j}{\partial x_i \partial x_j} H(f).$$

Its amplitude depends explicitly on the choice of the Kirchhoff surface over which the integral is calculated. If this integral is computed on a surface located in the regions of linear propagation (i.e. containing no source), the two approaches lead to the same results. On the other hand, if the integral is calculated on a surface that is located in a region where there are sources, then the results could be significantly different. This is not surprising in light of the comments we made earlier in this section progressively along with the analytical developments, but it has the merit of having been clearly established by Brentner and Farassat [BRE 98]. The other interest of the comparison lies in the fact that the formulation obtained by following the approach of Ffowcs Williams [FFO 69] allows for the proper definition of the various known contributions for the different acoustic sources. In this context, the latter are usually referred to as, namely, the noise due to the thickness of the airfoil (or *thickness noise*, which is of the monopolar type), the noise due to the load on the airfoil (or *loading noise*, which is of the dipolar type) and the shear noise (which is a quadrupole type). The thickness noise is due to the mass of the fluid displaced by the blades. The loading noise is produced by the forces exerted on the fluid by the blades. These forces can be decomposed in a mean part and in fluctuations. The mean loading noise usually dominates for an isolated propeller. The effect of fluctuating loads can become dominant for large incidence or in the case of interactions of the blades with an obstacle (wing, fuselage, landing gear, etc.), or even in a counter-rotating doublet (interaction between the two rows of blades). Thickness and loading noise directivity diagrams are different in the subsonic regime: the thickness noise is maximum in the plane of the propeller and the mean loading noise radiates more downstream of the propeller. Note that the influence of fluctuating loads can also occur around the axis of an isolated propeller, region where the mean loading noise and thickness noise do not radiate. In subsonic regime, shear noise is negligible and is of the same order of magnitude as the transonic thickness noise. Thus, as the relative Mach at the end of the blade is always less than 0.9 for propellers of light aircrafts, the shear noise is negligible.

Figures 10.13, 10.14 and 10.15 illustrate the results obtained by both approaches, and depending on the position chosen for the surface which is used in the calculation of the Kirchhoff integral. When the surface of integration is correctly chosen, containing no source, the results obtained by the two approaches are the same (Figure 10.13), while, when the surface of integration is chosen too close to the blade, that is in a region in which there are sources, the Kirchhoff method gives very different results (Figure 10.14).

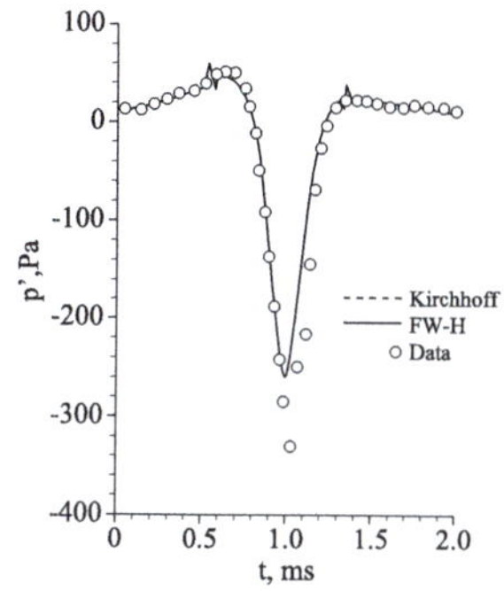

Figure 10.13. *Acoustic pressures measured (data) and obtained by numerical simulations, for an observer situated in the far-field plane of the blade, either by the Kirchhoff method, or by the Ffowcs Williams and Hawkings method (FW-H) (according to [BRE 98])*

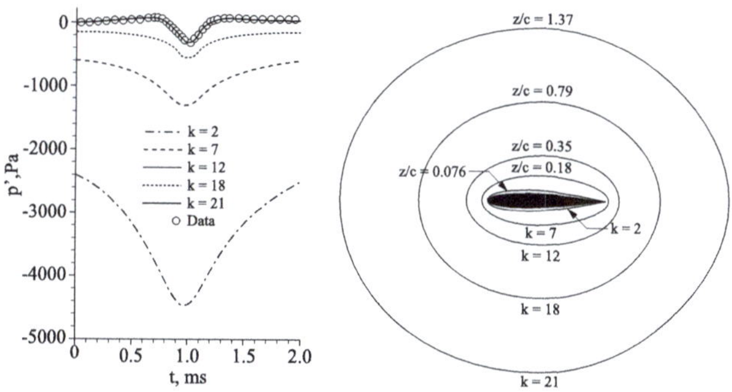

a) b)

Figure 10.14. *Acoustic pressures obtained by numerical simulations, for an observer located within the far-field plane of the blade, by the Kirchhoff method a), and using the different positions of the integration surface reported on the diagram b) (according to [BRE 98])*

Finally, Figure 10.15 illustrates, on the one hand, the relative importance of the different types of sources obtained for this transonic flow regime (Mach number equal to 0.9) and, on the other hand, the fact that, here also in far-field, certain simplifying approximations are quite justified.

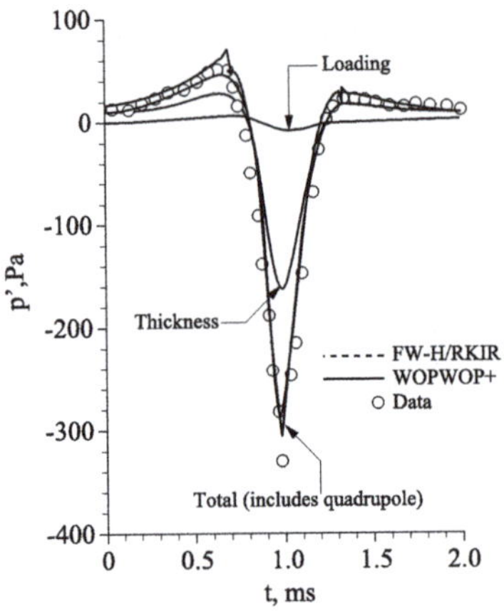

Figure 10.15. *Acoustic pressures obtained by numerical simulations, according to the origin of the sources, for an observer located in the far-field plane of the blade by the Ffowcs Williams and Hawkings method (FW - H). The FW-H/RKIR code uses the full theoretical development, while the WOPWOP code overlooks the directional and the retarded time effects in the Lighthill tensor (according to [BRE 98])*

10.4.2. *Noise generated by a set of rotating blades and fixed cascading blades, in the frequency domain*

To complete this section, it is first necessary to indicate that the theoretical developments that we have just presented, as well as their implementation in sophisticated computational codes, still correspond to studies conducted in research or research and development contexts instead located upstream of practical applications. They consider equations and reasonings conducted in the time domain. On the contrary, when we are directly focusing on the study of complete groups of turbines or multi-stage – or cascading – compressors, this kind of approach cannot yet be implemented and we have, therefore, still to make use of less complex approaches and tools. The theoretical formulations that are then used are based on equations written in the frequency domain (see [KOR 05], [LEW 94] or [LEW 07], in particular). This corresponds to the fact that these situations are most often associated with sound radiation where characteristic frequencies dominate. This is what, historically, motivated, as early as the 1930s, theoretical developments carried out in the frequency domain [GUT 48], [ERN 36], [DEM 37], [LOW 65]. Subsequently, it is only later, from the 1970s, that the theoretical approaches that were outlined above, in the time domain, were developed.

Some simple geometric considerations first contribute to understanding how the interaction of rotating blades and fixed vanes connected to the stator produces sound radiation at specific frequencies. These constitute, at first glance, the principal acoustic signature for rotor-stator systems. Next, more refined methods allow the respective contributions of quantities such as loading noise and thickness noise to be predicted. These are the different considerations that we present below, in order of increasing complexity.

Thus, when considering a B-blade rotor, rotating at an angular velocity $\Omega = 2\pi N$, where N is the rotation velocity usually given in rpm, in interaction with the V blades of a stator, the moving rotor blades generate a noise that, for a fixed observer, will essentially consist of the motion frequency of the blades equal to BN, and its harmonics, $f_a = nBN$, $n \geq 1$ integer. For an observer associated with the rotor, the interaction with the stator is manifested V times per round, that is with the frequency VN. This means that the sound source due to the load of the moving blades has harmonic frequencies of VN, $f_s = jVN$ (j being integer). The value $j = 0$ corresponds to the mean load, which produces the compression or the useful thrust. The transition from frequencies f_s at the source to acoustic frequencies f_a can be interpreted as a Doppler effect associated with the rotational motion. Thus, if we consider a rotating mode in $\exp(\imath\omega t - m\theta)$, with angular phase velocity $\Omega_\phi = \omega/m$, an acoustic wave characterized by $\omega_a = 2\pi f_a$ and m in the fixed coordinate system, and that has a pulse $\omega_s = 2\pi f_s$ in the rotating coordinate system and the same spatial structure and therefore the same m, this mode will be characterized by $m = NB - jV$. This result simply derives from the law of combination of velocity, which is written as Absolute velocity = Relative velocity + Frame velocity. That is here, $\Omega_a = \Omega_s + \Omega$ and thus $f_a/m = f_s/m + N$. Which, with laws defining f_a and f_s, correctly yields the announced result. This result is fundamental, because it explains the structure in dominant lines of the noise generated in these situations. Thus, on the one hand, angular modes m generated at a given acoustic frequency nBN are such that $m = nB - jV$, in which n is fixed and where $j = \cdots -2, -1, 0, 1, 2, \cdots$ is the rank of the load harmonic. In addition, on the other hand, a harmonic load at frequency $f_s = jVN$ contributes to each acoustic line $f_a = nBN$ by means of the angular mode $m = nB - jV$.

If now, as most often is the case, the rotor-stator assembly is inserted into a pipe that acts as cowling, the propagation in this duct fairing the machine causes the elimination of evanescent waves and produces the elimination of certain modes[5]. This is true after a certain distance, which is very short in practice because the decay of such a mode is very high (several tens of dB/m) as soon as the frequency is significantly lower than its cutoff frequency. The waves that propagate and therefore create the perceived noise are those whose cutoff frequency is less than the emitted frequency nBN. As there is no criterion upon the radial mode μ, the lowest cutoff

5 Waveguide propagation is presented in section 5.9.

frequency of an angular mode m, the one determining its propagation, corresponds to $\mu = 1$. Thus, relying on the propagation condition which is usually written in the form $R|\Omega_\phi| > c_0$ (where R is the radius of the pipe), we get

$$R\frac{2\pi nBN}{|m|} > c_0, \text{ that is } nBM_{rot} > |m|,$$

in which $M_{rot} = 2\pi RN/c_0 = \Omega R/c_0$ is the peripheral rotation Mach number. It is concluded that if the rotor is isolated, that is to say that it undergoes no interaction, then only the mean load $j = 0$ exists, and therefore $m = nB$. As a result, the propagation condition becomes $M_{rot} > 1$, so that an isolated rotor sound can propagate if the rotor is supersonic at the end of the vane. Therefore, a perfectly isolated subsonic rotor does not make any noise. However, this is hardly achieved in practice, namely because of flow distortions and the cranks holding the shaft. Nevertheless, it also results that the mean load is actually the dominant sound source for transonic rotors, as it is far greater than load fluctuations generated by the various interactions. Generally speaking, it can then be written that $NBM_{rot} > |nB - jV|$, that is $-nBM_{rot} < nB - jV < nBM_{rot}$ and $nB(1 - M_{rot}) < jV < nB(1 + M_{rot})$.

These inequalities, therefore, provide the loading harmonic domain generating propagative modes at frequency nBN. This domain is centered on this frequency and its width is proportional to the peripheral rotation Mach number M_{rot}. It can be verified that the harmonic of rank $j = 0$ is included in this range only if $M_{rot} > 1$.

These few results having been established, we will now go back to the source terms, and see how, in the frequency domain, which is obviously the best suited in this case, the acoustic energy radiated at the different frequencies can be calculated, with both loading and thickness as origins. To this end, we will first consider a non-streamlined rotor. Based on what we have seen at the beginning of this section about noise generated by blades or propellers in motion, we can write the sound pressure at a point identified by its position vector x and at time t as:

$$p(x,t) = \int_A \int_\tau \rho_0 V_n \frac{DG}{Dt} d\tau dA + \int_A \int_\tau f_i \frac{\partial G}{\partial y_i} d\tau dA, \qquad [10.6]$$

where the first term of the right member of equation [10.6] represents the thickness noise and the second term represents the loading noise. As we have already seen, the Green function G ($G(y,\tau;x,t) = 1/(4\pi S_0)\delta(t-\tau-\sigma_0/c_0)$, where S_0 and σ_0 are defined just below) describes the response in x at t of a pulse emitted in y at τ, and the f_i are the components of the forces by element surface applied to the fluid. Using a Fourier transform, we then obtain the spectral components of the sound pressure:

$$P(x,\omega) = \int_A \int_\tau \rho_0 V_n \frac{DG_\omega}{Dt} d\tau dA + \int_A \int_\tau f_i \frac{\partial G_\omega}{\partial y_i} d\tau dA, \qquad [10.7]$$

in which the expression of the Fourier transform G_ω of G is given by:

$$G_\omega = \frac{1}{4\pi T S_0} \exp\left(-i\omega\left(\tau + \frac{\sigma_0}{c_0}\right)\right) = \frac{1}{4\pi T S_0} \exp\left(-iK\omega\left(c_0\tau + \sigma_0\right)\right),$$

where σ_0 is the normalized distance defined by:

$$\begin{aligned} \sigma_0 &= \frac{M(x_1 - x_{10}) + S_0}{1 - M^2} \\ &= \frac{M(x_1 - x_{10}) + \sqrt{(x_1 - x_{10})^2 + (1 - M^2)((x_2 - x_{20})^2 + (x_3 - x_{30})^2)}}{1 - M^2} \end{aligned}$$

and K is the wave number $K = \omega/c_0$. Indices i in x_i represent the i-th component of the position vector of the observer x, and index 0 refers to the position of the source (such that x_{i0} is identified with y_i). The direction x_1 corresponds to the axis that carries the velocity (equal to $U = M.c_0$).

The material derivative of G_ω is then given by:

$$\frac{DG_\omega}{Dt} = \alpha G_\omega, \text{ with } \alpha = -i\frac{c_0 K}{1 - M^2}\left[1 + M\frac{x_1 - x_{10}}{S_0}(1 - i\frac{1 - M^2}{KS_0})\right]$$

and the spatial derivatives by:

$$\frac{\partial G_\omega}{\partial y_i} = E_i G_\omega, \text{ with } \begin{cases} E_1 = i\dfrac{K}{1 - M^2}\left[M + \dfrac{x_1 - x_{10}}{S_0}(1 - i\dfrac{1 - M^2}{KS_0})\right] \\ E_2 = iK\dfrac{x_2 - x_{20}}{S_0}(1 - i\dfrac{1 - M^2}{KS_0}) \\ E_3 = iK\dfrac{x_3 - x_{30}}{S_0}(1 - i\dfrac{1 - M^2}{KS_0}) \end{cases}$$

Assuming the sound sources in uniform rotation at angular velocity $\Omega = 2\pi N$ around the axis Ox_1, and if the rotor has B equidistant blades, and further considering in relations [10.6] and [10.7] the surface A as relative to a single one, then the emitted acoustic frequencies are limited to the harmonics of the nB-rank of the rotation since

$\omega = nB\Omega$. A source located at y is then identified by its coordinates (x_1, r, ψ). By expanding the force f_i into Fourier series,

$$f_i = \sum_{s=-\infty}^{+\infty} F_i^{(s)} e^{is\psi} \quad and \quad F_i^{(s)} = \frac{1}{2\pi} \int_0^{2\pi} f_i e^{-is\psi} d\psi,$$

$F_i^{(s)}$ then designates the loading harmonic of rank s, depending on the position on the surface A of a blade but not on ψ. If a point on A is identified at $\tau = 0$ by its coordinates (x_1, r, ψ_c), the values of ψ_c, comprised between $\psi_1(r)$ and $\psi_2(r)$, describe the angular extent of the transverse projection of the chord line of the airfoil at a wingspan r: in practice, all these angles are similar (and around zero, for example), such that $\psi = \Omega\tau + \psi_c = 2\pi N\tau + \psi_c$, and the quantities independent of ψ also are of τ.

The last step of the expansion consists of calculating the far-field approximation, that is for observation distances considered large compared to the dimensions of the source and compared to the wavelength $\lambda = 2\pi/K$. We then get $x_{10}/S_0, x_{20}/S_0, x_{30}/S_0 \ll 1$ and $KS_0 \gg 1$. The E_i and α are then independent of x_{20} and x_{30}, that is ψ and τ, so that:

$$P(x,\omega) = \int_A \rho_0 V_n \alpha \left[\int_{\tau=0}^{T} G_\omega d\tau \right] dA + \sum_{s=-\infty}^{+\infty} \int_A F_i^{(s)} E_i \left[\int_{\tau=0}^{T} G_\omega e^{is\psi} d\tau \right] dA.$$

The period T associated with the rotation blades is defined by $T = 1/(BN)$. The integrals over time are calculated by developing S_0 and τ_0 at the first order, zero-order being:

$$S = \sqrt{x_1^2 + (1-M^2)(x_2^2 + x_3^2)} \quad and \quad \sigma = \frac{Mx_1 + S}{1 - M^2}.$$

It yields, finally therefore:

$$P(x, nB\Omega) = P_e(x, nB\Omega) + P_c(x, nB\Omega),$$

with the thickness P_e and the loading P_c noises, respectively, given by:

$$P_e = -i\rho_0 \frac{Bf}{2(1-M^2)} \left[1 + M\frac{x_1}{s}\right] e^{-iK\sigma} e^{-inB(\varphi - \pi/2)}$$
$$\times \int_A V_n e^{i(KA_1 + nB\psi_c)} \mathrm{J}_{nB} \left(Kr\frac{d}{s}\right) dA$$

and

$$P_c = i\frac{Bf}{2cS(1-M^2)}e^{-iK\sigma}\sum_{s=-\infty}^{+\infty} e^{-i(nB-s)(\varphi-\pi/2)}$$
$$\times \int_A \left\{\left[M+\frac{x_1}{s}\right]F_1^s + (1-M^2)\frac{x_2}{s}F_2^s\right.$$
$$\left. +(1-M^2)\frac{x_3}{s}F_3^s\right\} e^{i(KA_1+nB\psi_c)}\mathrm{J}_{nB-s}\left(Kr\frac{d}{s}\right) dA,$$

where $A_1 = x_{10}/(1-M^2)(M+x_1/s)$ expresses the fact that the blades are not plane (they are not contained in the rotation plane $x_{10} = 0$). The coordinates of the observer are referenced in cylindrical coordinates by x_1, d, φ. And it is reminded that $K = 2\pi f/c_0(2\pi nBN)/c_0$, such that $Krd/s = nBM_r d/s$, in which $M_r = \Omega r/c_0$ is the Mach number at radius r. In these expressions, the Bessel functions of the first kind also appear, denoted by $\mathrm{J}_n(z)$. The Fourier transform $P(nB\Omega, x)$ is calculated on positive and negative frequencies so that the acoustic spectrum, expressed only on the positive frequencies, is given by:

$$2\left|P(x, nB\Omega)\right|^2 = 2\left|P_e(x, nB\Omega) + P_c(x, nB\Omega)\right|^2, \qquad (n > 0).$$

The two previous relations are becoming simpler in some cases. If overlooking the flow velocity ($M \approx 0$ and $1 - M^2 \approx 1$), we then have $s = \sigma = D, Krd/S = nBM_r \sin\theta$ and $A_1 = x_{10}\cos\theta$. It follows that the component $P\exp(\imath\omega t)$ of the sound pressure at pulse $\omega = nB\Omega$ contains as a factor the term $(1/D)\exp(\imath\omega(t - D/c_0))$, characteristic of a spherical wave. And the term between brackets of P_c is simply written as $F^{(s)}x/D$. If $F^{(s)}$ is decomposed in terms of axial thrust $F_T^{(s)}$ and drag $F_D^{(s)}$, the expression of the loading noise P_c is then identical to Lowson's [LOW 65]. If, moreover, the rotor is isolated, only the mean loading remains (s =0) and P_c is reduced to the result that Gutin had established for a propeller since 1936 [GUT 48].

If now considering a rotor that is cowled by a duct assumed as cylindrical, infinitely long and with perfectly rigid walls, the previous expansions using the Green function should be revisited, making use of the Green function (in harmonic regime, without flow, the calculation of this Green function is presented in section 5.9.3) which is then appropriate,

$$G(y,\tau;x,t) = -\frac{i}{2}\sum_{m=-\infty}^{+\infty}\sum_{\mu=1}^{+\infty}\frac{\mathrm{J}_m(k_T r)\mathrm{J}_m(k_T d)}{\Gamma}\frac{e^{im(\psi-\varphi)}}{\Delta}\delta(t-\tau-\frac{k}{K}\frac{x_1-x_{10}}{c_0}),$$

in which K, k_T and k are, respectively, the total, transverse and longitudinal wave numbers, $\Delta = (K^2 - (1 - M^2)k_T^2)^{1/2}$ and $\Gamma = \pi(R^2 - \frac{m^2}{k_T^2})\mathrm{J}_m^2(k_T R)$.

By following the same procedure as previously, with the exception that there is no far-field approximation, the following source terms (and neglecting the radial force) are then obtained:

– source associated with thickness, $E_{m\mu} = \int\limits_A V_n \mathrm{J}_m(k_T r) e^{i(nB\psi_c + kx_{10})} dA$;

– sources due to the thrust, $T_{m\mu}^{(s)} = \int\limits_A F_T^{(s)} \mathrm{J}_m(k_T r) e^{i(nB\psi_c + kx_{10})} dA$;

– sources due to the drag, $D_{m\mu}^{(s)} = \int\limits_A \frac{1}{r} F_D^{(s)} \mathrm{J}_m(k_T r) e^{i(nB\psi_c + kx_{10})} dA$.

We then get the expressions of P_e and P_c:

$$P_e = -\frac{1}{2}\rho_0 a B \sum_{\mu=1}^{+\infty} \frac{K - Mk}{\Gamma\Delta} E_{nB\mu} \mathrm{J}_{nB}(k_T d) e^{-i(nB\varphi + kx)}, \qquad [10.8]$$

$$P_c = \frac{1}{2} B \sum_{\mu=1}^{+\infty} \sum_{s=-\infty}^{+\infty} \frac{1}{\Gamma\Delta} (k T_{m\mu}^{(s)} + m D_{m\mu}^{(s)}) \mathrm{J}_m(k_T d) e^{-i(m\varphi + kx)}, \qquad [10.9]$$

where $m = nB - s$.

It particularly stands out that the component $P \exp(\imath\omega t)$ of the total sound pressure at the pulsation $\omega = nB\Omega$ at the reception point referenced by its coordinates (d, φ, x) is the sum of terms $C\mathrm{J}_m(k_T d) \exp(\imath\omega t - m\varphi - kx)$, which represents helicoidal waves (or spinning waves) in the duct (with $m = nB$ for the thickness noise). The acoustic field is thus decomposed on the eigenfunctions of the waves equation in cylindrical geometry; m is the angular wave number and μ is the radial mode. These results are valid inside an infinite conduct. Radiation in the free field can be obtained by slicing it into $x = L$ and assuming that reflections on the aperture are low enough in order for the solution given by [10.8] and [10.9] to remain valid. The simplest calculation is then obtained by considering that the incident acoustic field is equivalent to a set of monopoles in the output section, showing phase relations between them. This is a generalization of the classical problem of the radiation of a baffled plane piston. An elementary wave of the type $C\mathrm{J}_m(k_T d) \exp(\imath\omega t - m\varphi - kx)$ gives finally in the far-field, at a point of coordinates (D, θ, ϕ) referenced now with respect to the center of the output section and no longer of the rotor:

$$p(D,\theta,\phi,t) = i^{m+1}\frac{CkR^2}{2D}e^{\imath\omega(t-D/c_0)}e^{-\imath(m\phi+kL)}J_m(k_TR)$$
$$\cdot\left[\frac{2KR\sin\theta}{(k_TR)^2-(KR\sin\theta)^2}\cdot J_m^{'}(KR\sin\theta)\right],$$

in which $J_m^{'}(z)$ is the derivative of $J_m(z)$. The angle θ is limited to $[0,2\pi]$ due to the hypothesis of embedding in a perfectly rigid plane. The bracketed term determines the directivity[6].

Another important point which should also be considered is the cutoff (or extinction) property of the duct. Cutoff phenomena play an essential role in the noise emission of a cowled rotor. They are derived from the dispersion relation between the wave numbers K, k_T and k previously introduced, $(K-Mk)^2 = k^2+k_T^2$. It is a second-degree equation in k, whose discriminant Δ^2 must be positive so that k be real and for the wave to travel. If the duct is perfectly rigid, the cancelation condition of the wall vibration velocity gives $J_m^{'}(k_TR) = 0$, that is $k_T = \chi_{m\mu}/R$, where $\chi_{m\mu}$ is the abscissa of the μth zero of the derivative $J_m^{'}(z)$. This shows that k_T, and thus k, Δ and Γ depend on m and μ, although these indices have been omitted in the above to simplify the writing. The condition $\Delta^2 > 0$ thus implies: $K \geq \sqrt{1-M^2}k_T$, or $|f| \geq f_c$, with $f_c = \sqrt{1-M^2}c_0\chi_{m\mu}/(2\pi R)$, f_c being the cutoff frequency of the acoustic mode (m,μ).

Finally, we can consider what comes out from the previous formulations when studying thickness and mean loading noises on the one hand, and interaction noise, on the other hand. Thickness and mean loading noises correspond both to the fact that $s = 0$, that is also $m = nB$, the harmonic rank of the rotation velocity being considered. Since $f = nBN$ and $\chi_{m\mu} > \chi_{m1} > |m|$, but χ_{m1} is not much larger than $|m|$, a necessary (but not strictly sufficient) condition for $\Delta^2 > 0$ is thus that $|n|\,BN > \sqrt{1-M^2}(c_0\,|m|)/(2\pi R)$, that is, since $m = nB$, $2\pi RN/c_0 > \sqrt{1-M^2}$, or $M_R > \sqrt{1-M^2}$, or $M_{rel} > 1$ (with $M_{rel} = \sqrt{M_R^2+M^2}$, the relative Mach number at the extremity of the blade). Thus, the studied terms radiate in free-field for transonic rotors only (assuming that the fairing is long enough). They then constitute the primary sound source. Consequently, the tonal noise of a subsonic rotor is only due to the interactions between the rotor and stator or flow distortion (loading rank harmonics, s, non-zero). Since $m = nB - s$, the previous inequality is then written $|n|\,BM_R/\sqrt{1-M^2} > |nB-s|$, where $\left[1-M_R/\sqrt{1-M^2}\right]|n|\,B < |s| < \left[1+M_R/\sqrt{1-M^2}\right]|n|\,B$. This relation, therefore, defines the troublesome loading harmonics range, which is centered on the rank nB component : $0 < s_{\text{min}} < |s| < s_{\text{max}}$.

6 An example of such a calculation is presented in section 4.8.4.

Figure 10.16 shows the results obtained with such approaches, for the noise generated by a three-stage turbine, including three rotors and three stators; their numbers of blades are given in the legend of the figure. Here, the principal topic of interest is the noise corresponding to a single frequency peak, the one that is the most energetic. This frequency peak has origins in the different contributions which correspond either to the relation $m_1 = B_1 - jV_1$ (that is origin at the level of the first rotor), with $j = 2$ and $m_1 = 47$, $j = 3$ and $m_1 = 9$, $j = 4$ and $m_1 = -29$, $j = 5$ and $m_1 = -67$, or to $m_2 = B_1 - jV_2$ (that is origin at the level of second rotor), with $j = 1$ and $m_2 = 6$. The general agreement between the numerical results and experimental data is good, given the fact that the numerical model does not take into account certain interactions, and in particular the effects related to the turbulence which tends to diffuse acoustic waves and nonlinear interactions between modes. The latter contribute to reducing the noise generated during the propagation between the various stages of the turbine.

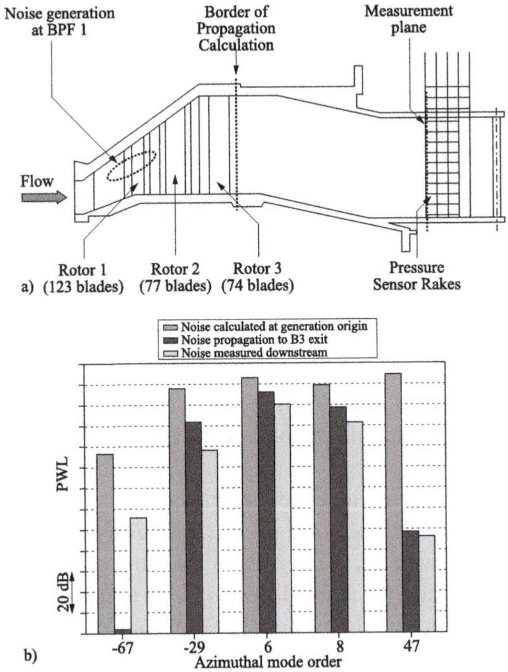

Figure 10.16. *Typical results obtained for the noise associated with a system comprising three rotors (of, respectively, 123, 77 and 74 blades) associated with three stators (38, 117 and 126 blades) (a), for different observation sections and by comparison with experimental data (b). The main concern here is the most energetic frequency peak, which corresponds to the 123rd harmonic of the rotation velocity (according to [KOR 05])*

10.4.3. *Noise generated by blade–vortex interaction, using the vortex sound generation method*

To complete this section 10.4, which acknowledges the very large complexity that corresponds to this kind of configuration, we will indicate that, on the one hand, it is important to take into account the broadband noise associated with the different interactions between the rotating airfoil and the flow more or less disturbed in its neighborhood because of the presence of other blades of the same rotor or the stators, as well as in combination with the presence of a potential cowl. This problem is generally addressed from the approaches presented at the beginning of this section (Ffowcs Williams and Hawkings' equation or the Kirchhoff method presented in section 10.4.1), but adapted to the complexity of the particular situation (for example, helicopter blades or rotor blades and cascade stators) which is often much more complex than the idealized situation that we had seen at the beginning of this chapter. It is then necessary to reconsider the expansion, adapting it to the particular configuration (in terms of the Green function notably). Review journal articles have been written by Brentner and Farassat [BRE 03] or Lewy [LEW 07], notably.

On the other hand, an important contribution also comes from what is commonly referred to as blade–vortex interaction noise. This noise is the dominant noise for a helicopter in descending phase, where it becomes very penalizing for people on the ground. When this noise is generated, it is the one that dominates all other sources and it should, therefore, be known how to precisely predict it in order to try to reduce it as much as possible. It results from the interaction between the vortices produced by the previous blade on the blade considered, which is just behind. One of the most secure ways to fight it consists of giving a shape to the blades such that the wake of the previous blade lies outside the path of the blade that follows it, generally just below it. The difficulty at the level of the numerical predictions lies especially in the complexity of the flow generated by the rotation of a helicopter blade. It also lies in the fact that the blades suffer a significant deformation during their motion. This fact is such that the accurate prediction of the fluctuating pressures on the surface of each blade is very tricky and generally requires the interlocking of several computational submodules, each specialized in a specific field (computation of the motion of rotor blades, computation of their wakes, determination of the vortices present in these wakes, computation of the pressure on the blades during their rotation, taking into account the interactions with these vortices, and then, finally, computation of the noise radiated by the rotor blades, see [YU 00], [BRE 03] or [LEW 07], in particular). Our aim here is not to go into the details of all these sequences, but rather to illustrate the mechanism of sound production by the phenomenon of blade–vortex interaction on a simple case, by implementing the Howe–Powell approach presented in Chapter 9.7. To this end, we will need some of the results that have been presented at the beginning of section 10.4.1, which will be applied in the context of the Howe–Powell approach.

The situation being considered is simply that of a single airfoil inside a uniform velocity field. In this case, it is well known that, even assuming the flow without viscosity and therefore without any effects due to turbulence, a vortex appears on the trailing edge of the airfoil. The appearance of this vortex is often interpreted in terms of the Kutta condition on the trailing edge, reflecting the absence of discontinuities in the velocity and pressure fields whether following the contour on the side of the extrados or following it on the side of the intrados (see [CRI 85] or [PAR 99]).

Generally speaking, by considering the equation for the total enthalpy B:

$$\left(\frac{1}{c_0^2}\frac{\partial^2}{\partial t^2} - \Delta\right) B = \mathrm{div}(\vec{\omega} \wedge \vec{v}),$$

such that, in far-field, acoustic pressure is given by $p(x,t) \approx \rho_0 B(x,t)$, we then get the formulation which gives the value of $p(x,t)$:

$$\begin{aligned}\rho_0 B(\vec{x},t) = &- \int\int_V (\rho\vec{\omega} \wedge \vec{v})(\vec{y},t).\frac{\partial G}{\partial \vec{y}}(\vec{x},\vec{y},t,\tau) d^3\vec{y} d\tau \\ &+ \int\int_S (\rho B)(\vec{y},\tau)\frac{\partial G}{\partial \vec{y}}(\vec{x},\vec{y},t,\tau).d\vec{S}d\tau \\ &+ \int\int_S G(\vec{x},\vec{y},t,\tau)(\rho\frac{\partial \vec{v}}{\partial \tau})(\vec{y},\tau).d\vec{S}d\tau \\ &+\mu \int\int_S \vec{\omega}(\vec{y},\tau) \wedge \frac{\partial G}{\partial \vec{y}}(\vec{x},\vec{y},t,\tau).d\vec{S}d\tau, \end{aligned} \quad [10.10]$$

written in the form of the Kirchhoff integral according to the principles presented in section 10.4.1 (see also [HOW 08]). The notation $d\vec{S}$ refers to a surface element whose normal is directed to the ambient volume V. The surface integral of $\rho_0 B$ in the right-hand side of the equation above can be eliminated according to the principle that normal differentials $\partial G/\partial x_n$ and $\partial G/\partial y_n$ cancel out, for $\vec{x}$ and $\vec{y}$ located on S. By rearranging the remaining terms, we then obtain:

$$\begin{aligned}\rho_0 B(\vec{x},t) = &\int\int_S G(x,y,t,\tau)(\rho\frac{\partial \vec{v}}{\partial \tau})(\vec{y},\tau).d\vec{S}d\tau \\ &- \int\int_V (\rho\vec{\omega} \wedge \vec{v})(\vec{y},\tau).\frac{\partial G}{\partial \vec{y}}(\vec{x},\vec{y},t,\tau) d^3\vec{y} d\tau \\ &+\mu \int\int_S \vec{\omega}(\vec{y},\tau) \wedge \frac{\partial G}{\partial \vec{y}}(\vec{x},\vec{y},t,\tau).d\vec{S}d\tau. \end{aligned} \quad [10.11]$$

The surface integral involving the normal component of $\partial\vec{v}/\partial\tau$ characterizes the influence of the surfaces in motion and is equivalent to a distribution of monopole sources on the surface S. The volume integral that follows represents the noise radiated by sources due to vortices which are located in the fluid, taking into account the surface S whose presence determines the expression of the Green function G. The last integral is associated with the contribution of the friction forces on the surface S. The Green function, considering the origin of the reference O inside the surface S, the source points y close to S and the observer situated in x in the far-field (see Figure 10.17), is given by

$$G(x, y, t, \tau) = 1/(4\pi x)\delta(t - \tau - x/c_0) + (x_i Y_i)(4\pi c_0 x^2)\delta'(t - \tau - x/c_0),$$
$$\text{with } x = |\vec{x}| \rightarrow \infty. \qquad [10.12]$$

The notation δ' indicates the derivative with respect to time t of the Dirac function δ and the notation Y corresponds to the point y considered at time τ. The first term in the expression above does not involve the source points y, and it therefore characterizes the monopolar component of the sources. On the contrary, the second term represents the dipole sources. The question about knowing which one of these two contributions is dominant depends on the properties of the sources that are in the expression $p(x, t)$ above.

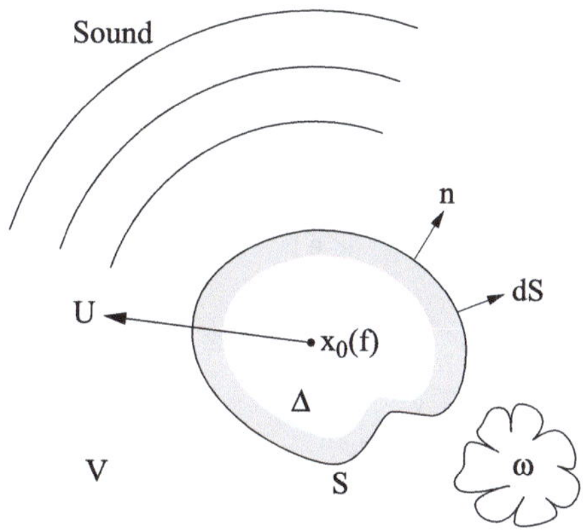

Figure 10.17. *Geometric configuration for the computation of the sound resulting from a blade–vortex interaction (according to [HOW 08])*

Including this expression of G [10.12] in relation [10.11] that gives $p(x,t)$ then yields:

$$p(\vec{x},t) \approx \frac{\rho_0}{4\pi x}\left[\frac{\partial}{\partial t}\int_S \vec{V}.d\vec{S}\right] + \frac{\rho_0 x_i}{4\pi c_0 x^2}\frac{\partial}{\partial t}\left[\frac{\partial}{\partial t}\int_S Y_i\vec{v}.d\vec{S} - \int_S \frac{DY_i}{Dt}\vec{v}.d\vec{S}\right]$$

$$-\frac{\rho_0 x_i}{4\pi c_0 x^2}\frac{\partial}{\partial t}\left[\int \vec{\omega}\wedge\vec{v}.\vec{\nabla}Y_i d^3\vec{y} - \nu\int_S \vec{\omega}\wedge\vec{\nabla}Y_i.d\vec{S}(\vec{y})\right], x \to \infty. \quad [10.13]$$

In expression [10.13], the quantities in brackets are evaluated at the retarded time $t - x/c_0$. The first term represents the omnidirectional emission of a monopole produced by the volumetric pulsations of the medium S. The other terms correspond to dipoles and they are *a priori* smaller in a factor $\mathcal{O}(M)$. Nevertheless, they become the dominant terms when the volume of the solid which suffers deformation is either constant or slightly changing. The terms which contain $v.ds$ originate from the translations and rotations, as well from the deformations, of dS. The volume integral of the quantity $\vec{\omega}\wedge\vec{v}.\vec{\nabla}Y_i$ corresponds to the contribution of a quantity that characterizes in magnitude and amplitude the rate at which the vortex of intensity ω cuts the streamlines of the flow. This term also exists for a compact solid stationary medium. The last term corresponds, again, to the contribution of the frictional forces.

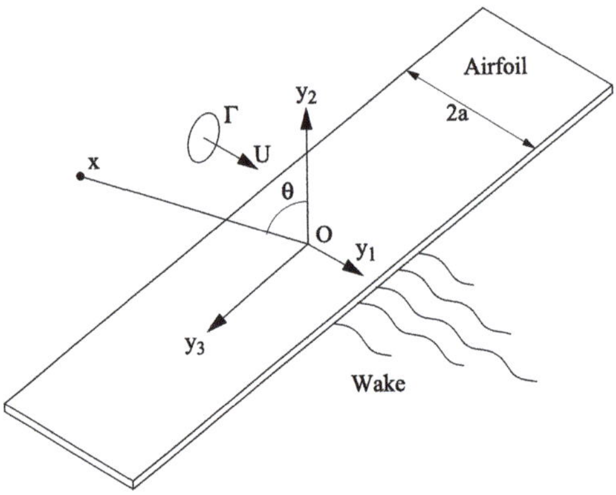

Figure 10.18. *Geometric configuration for the computation of the sound generated by an airfoil in interaction with a vortex. According to [HOW 08]*

If now, returning to the case being considered here, namely of an aircraft wing airfoil (of chord $2a$) and the vortex of intensity Γ it generates downstream (Figure

10.18) in its wake, the surface force responsible for the noise emission is reduced to $F_i = -\rho_0 \int\limits_V (\vec{\omega} \wedge \vec{v})(\vec{y}, t).\vec{\nabla} Y_i d^3 y_i$.

Then, $\text{div}(\vec{\omega} \wedge \vec{v}) \equiv \dfrac{\partial(\vec{\omega} \wedge \vec{v})_i}{\partial x_i}$, $(\vec{\omega} \wedge \vec{v})_i = -\pi R^2 \Gamma U \dfrac{\partial}{\partial x_i}(\delta(x_1 - Ut)\delta(x_2 - h)\delta(x_3))$, where U refers to the forward velocity of the airfoil, R, the radius of the vortex generated downstream, and h, the position (according to the vertical axis of index 2) of its center ($R \ll h \ll a$). Such that the radiated acoustic pressure is then:

$$
\begin{aligned}
p(x,t) &\approx -\frac{\rho_0 \cos\theta}{4\pi c_0 x} \frac{\partial}{\partial t} \left[\int (\vec{\omega} \wedge \vec{v})_i \frac{\partial Y_2}{\partial y_i} d^3 \vec{y} \right] \\
&\approx -\frac{\rho_0 \sqrt{2a} \cos\theta}{4\pi c_0 x} \frac{\partial}{\partial t} \left[\int (\vec{\omega} \wedge \vec{v})_i \frac{\partial \Re(\sqrt{y_1 + i y_2 + a})}{\partial y_i} d^3 \vec{y} \right] \\
&= \frac{3}{16\sqrt{2}} \left(\frac{R}{a}\right)^2 \frac{\rho_0 U^2 \Gamma \cos\theta}{c_0 x} \Re \left\{ \frac{1}{\left(\frac{U\,[t]}{a} + 1 + \frac{ih}{a}\right)^{5/2}} \right\}, x \to \infty,
\end{aligned}
$$

in which θ is the angle between the observation direction and the normal to the wing. The acoustic radiation thus corresponds to a dipole produced by the interaction between the solid that represents the wing and the vortex generated in its wake. The axis of this dipole is normal to the wing and its amplitude is proportional to the unsteady lift produced by the motion of the vortex. The pressure signature thus obtained is shown in Figure 10.19 for the case in which $h/a = 0,25$. The dotted line corresponds to the pressure generated by the vortex near the wing when the Kutta condition is not imposed (there is no vortex shedding), it is equal and opposite to the pressure radiated by the vorticity of the wake. Considering a wing simplifies the analytical expansions because the three-dimensional effects are negligible. On the contrary, the situation for a blade is more complicated because of its transverse elongation which is reduced and because of the additional effects related to the attaching mechanism on the hub, which does not make it possible to address the problem using the approach presented here, and requires resorting to numerical simulation computation. In this case, the vortex of intensity Γ is primarily that generated by the blade located at the forefront, even if, in reality, it originates from the very complex mechanisms related to the wakes of all the blades that interact with one another, considering also that the blades are deformed and cannot be approached by planar surfaces.

10.5. Noise generated and propagating in the outer atmosphere: accounting for the thermal stratification and for likely obstacles

The last complex situation that we will consider here concerns the propagation of sound waves, more commonly known as noise, in the outer atmosphere, over distances that may be very large, so that the properties of thermal stratification or non-uniform distribution of the wind velocity in the concerned part of the atmospheric boundary layer can no longer be neglected. Thus, in the following sections, we will first examine the main consequences resulting there from with respect to the value of the speed of sound c_0, then we will make a brief description of the main analytical and modeling approaches that have been developed to address this case. Indeed, the lower part of the atmosphere, or troposphere, is the seat of many complex physical phenomena that interact and substantially affect the noise levels at medium and large distance. The propagation of road noise is thus often determined by geometrical (source height and size of likely acoustic screens), surface (soil and heterogeneities) and topographic (terrain and obstacles) conditions.

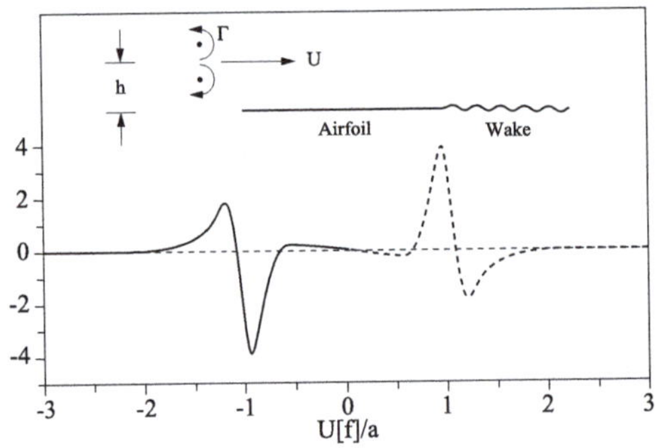

Figure 10.19. *Acoustic pressure signature resulting from the wing–vortex interaction (for $h/a = 0.25$). ——- pressure distribution $p(x,t)/\left[(\frac{R}{a})^2\frac{\rho_0 U^2 \Gamma \cos\theta}{c_0 x}\right]$; - - - - pressure in the absence of vortex shedding (according to [HOW 08])*

In addition, in this case, the dimensions of the sound source are in general very small compared to that of the considered wavelength λ, so that each wavefront has the same amplitude and phase in all points on its surface (spherical or cylindrical). Thus, at large distances from the source, it is natural to consider wave fronts as almost plane, such that the sound wave tends then toward a plane wave. The propagation of these waves is then often identified with that of rays, as it is commonly the case in optics, so that specific tools have been developed for this kind of situations such as those briefly mentioned in section 4.7 . When the source is placed exactly on the surface of a flat and perfectly reflective ground area, sound propagates in a hemisphere rather than in a sphere, but the decay laws remain the same and can be extended to cases of sources exactly localized at intersection points of infinitely rigid planes.

The principal practical applications considered here are those associated with the noise generated by large infrastructures such as roads, highways, airports and railway lines, but this may also refer to public works sites or quarries.

10.5.1. *Characteristic properties of the atmospheric boundary layer and impacts on sound propagation*

As we have indicated earlier, various properties of the lower layer of the atmosphere, as well as of the ground, affect the propagation of sound waves. This is what we are going to consider in this section, in which we will focus only on the main results and on the most general principles. For more information on the statistical properties of atmospheric turbulence or on the propagation of waves in the turbulent atmosphere, we can refer to the works of Panofsky and Dutton [PAN 84], Delmas *et al.* [DEL 05] or Tatarski [TAT 71] and Monin and Yaglom [MON 75]. In the case of syntheses specific to sound propagation and its practical modeling in the atmosphere or in submarine environment, Attenborough's [ATT 07] and Kuperman and Roux's [KUP 07] articles, respectively, both integrated in the book edited by Rossing [ROS 07], are also recommended. For reasons of space, we will examine here mainly the atmospheric propagation, even though submarine propagation has also been the subject of numerous studies and the richness of the phenomena involved and the concerned applications is very large.

It is usual to consider that the acoustic propagation velocity c at a given point $\vec{x}$, according to temperature $T(\vec{x})$ and wind velocity $V(\vec{x})$ at this point, is given by

$$c(\vec{x}) = \sqrt{\gamma RT(\vec{x})} + V(\vec{x})\cos\theta,$$

where θ designates (in a horizontal plane) the angle between the propagation direction of the wave (that is of the ray which approximates it) and the direction of the wind, while γ is the ratio of the specific heats, and R is the constant of perfect gases. It is then derived there from that, according to the variations of temperature and the wind velocity in the region of the atmospheric boundary layer being considered,

$$\frac{\partial c}{\partial z} = \frac{c}{2T}\frac{\partial T}{\partial z} + \vec{u}.\frac{\partial \vec{V}}{\partial z},$$

in which $\vec{u}$ refers to the unit vector in the propagation direction of the sound.

There are then several situations, as illustrated in Figures 10.20(a)–(c) for the influence of the temperature gradient and that of the wind on the propagation of acoustic waves (or rays). It can be observed that, following the Snell–Descartes refraction law, $n_1.\sin\theta_1 = n_2.\sin\theta_2$, sound waves are deflected to the areas of smaller celerity (with n_1 and n_2 which represent, respectively, the refraction indices in media 1 and 2, and θ_1 and θ_2 the angles of the rays with respect to the normal of the interface between these two media). The appearance of shadow regions can be further seen in two situations, regions that sound rays cannot directly reach or regions from which no ray directly originates.

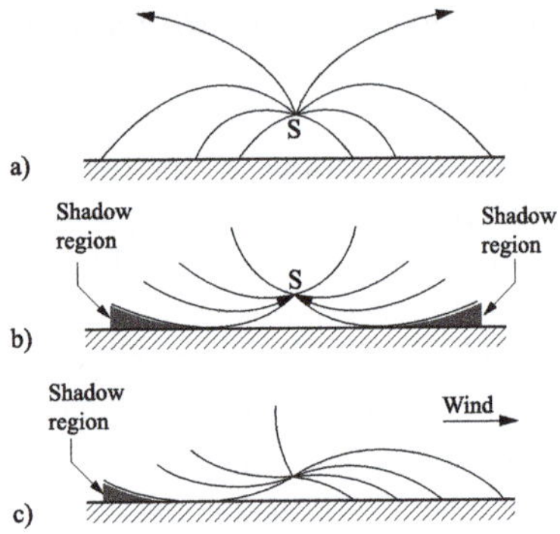

Figure 10.20. *Profiles of sound rays in the case $\partial T/\partial z > 0$ – a). Profiles of sound rays in the case $\partial T/\partial z < 0$ – b). Profiles of sound rays in the case without thermal gradient, in the presence of wind – c) (according to [GAU 99])*

In addition, the turbulence within the atmospheric boundary layer as well directly influences the propagation of sound waves. Generally, this is characterized by means of the refraction index of the medium (n) which, as for light rays, is then defined by $n = c_0/c$, where c_0 is the celerity in the first medium – or reference medium, which corresponds to the ideal situation in which there is no turbulence – and c is the propagation celerity in the new medium (that is when turbulence is present). Thus, we can then write $n(\vec{x}) = \overline{n(\vec{x})} + \mu(\vec{x})$, with $\overline{n(\vec{x})} = c_0/c(\vec{x}) \approx 1$ and $\mu(\vec{x}) << 1$.

Furthermore, if assuming that $T(\vec{x}) = \overline{T(\vec{x})} + T'(\vec{x})$ and $V(\vec{x}) = \overline{V(\vec{x})} + V'(\vec{x})$, where T' and u' designate the temperature T and the velocity V fluctuations, the fluctuations are then given by $\mu(\vec{x}) = -\dfrac{T'(\vec{x})}{2T_0} - \dfrac{V'(\vec{x})}{c_0}$. In addition, if we write that, by convention, the acoustic pressures with and without turbulence are, respectively, $p_0 = A_0 \exp(iS_0)$ and $p = A\exp(iS)$, with S the phase and A the amplitude, the parameters of intensity $\Gamma(r)$ and diffraction $\Lambda(r)$, after traveling a path of length r along the ray, can then be defined, whose expressions are:

$$\Gamma^2(r) = k_0^2 \overline{(\int_0^r \mu(r')dr')^2}$$

$$\Lambda(r) = N \int_0^r (\frac{R_F(r')}{L})^2 dr'.$$

N is a normalization factor and k_0 is the wave number of the angular frequency f under consideration (with $k_0 = \omega/c_0$ and $\omega = 2\pi f$). The quantity R_F is usually called the Fresnel radius. Its value characterizes the extent of the area for which the route of the diffused ray differs from the direct one by about half a wavelength. The optimal size of the vortices to generate fluctuations in the case of spherical wavefronts is approximately that of a Fresnel zone, that is $(\lambda r)^{1/2}$, where λ is the considered wavelength and r is the distance between the source and receiver.

In the case of a correlated Gaussian function, we then get $\Gamma^2(r) = 2k_0^2\overline{\mu^2}rl_\mu = \sqrt{\pi}\overline{\mu^2}k_0^2 rL$ and $\Lambda = \dfrac{r}{k_0 L^2}$, in which L is the integral scale of turbulence. The integral scale l_μ, characteristic of the index fluctuations, is defined from the autocorrelation function of the fluctuations of index $\psi_n(r)$ by $l_\mu = \dfrac{1}{\overline{\mu^2}} \int_0^\infty \psi_n(r')dr'$. For a Gaussian correlation function, it yields that $l_\mu = L\sqrt{\pi}/2 \approx 0.9L$.

Under these conditions, there are different propagation regimes that are usually presented in the form of a diagram (Λ, Γ). This diagram (see Figure 10.21) presents three space regions that are delimited by the straight lines $\Lambda = 1, \Gamma = 1$ and $\Lambda\Gamma = 1$:

– Region [1] is the area of geometric acoustics ($\Lambda \ll 1$ and $\Lambda\Gamma \ll 1$). In this region, the diffraction by heterogeneities of turbulence is negligible and the notion of acoustic ray is preserved despite the presence of turbulence (near-field). This region is characterized by a phase variance $\overline{S^2}$ much larger than that of log-amplitude $\overline{\chi^2} = \overline{(\ln A - \ln A_0)^2}$, whose expressions in this region are expressed, respectively, as: $\overline{S^2} = \Gamma^2 = \sqrt{\pi}\overline{\mu^2}k_0^2 rL$ and $\overline{\chi^2} = \Lambda^2\Gamma^2/6 = 8/3\sqrt{\pi}\overline{\mu^2}r^3/L^3$.

– Region [2] represents the diffraction area without saturation ($\Lambda \gg 1$ and $\Gamma \ll 1$). In this region, it is usual to apply the Rytov approximation which assumes that the amount of energy diffused by the heterogeneities of turbulence is low compared with the quantity of energy carried by the incident wave. The phase and the log-amplitude fluctuations are equal up to a distance rs known as the saturation distance. They are then written as $\overline{S^2} = \overline{\chi^2} = \sqrt{\pi}\overline{\mu^2}k_0^2 rL$.

– Region [3] is the saturated zone ($\Gamma \gg 1$ and $\Gamma\Lambda \gg 1$). The behavior of $\overline{\chi^2}$ defined above in the diffraction region is valid only if the intensity of turbulence is small from the point of view of its effect on the pressure field. For large values of Γ and $\Lambda\Gamma$, the amplitude fluctuations saturate with the propagation distance. This corresponds to the convergence of $\overline{\chi^2}$ toward a horizontal asymptote and therefore a sound field dominated by phase fluctuations is obtained. For sound path lengths that are larger still, the laws of optics predict that amplitude fluctuations reach a maximum level and then drop slightly as turbulence intensifies; this phenomenon is called supersaturation.

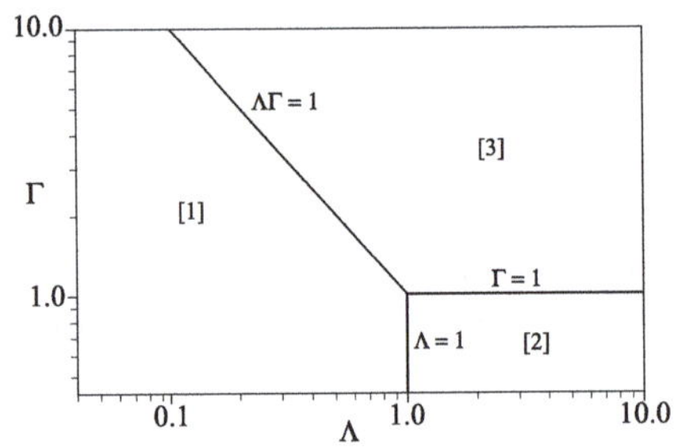

Figure 10.21. *Diagram (Λ, Γ) for a spherical wave and a Gaussian correlation function. $f = 4{,}000\ Hz$, $r = [20 - 400]\ m$, $\overline{\mu^2} = 10^{-6}$, $L = 1.1\ m$ (according to [GAU 99])*

The log-amplitude saturation phenomenon of the sound wave has sometimes driven some authors to take into account only the phase fluctuations for long-range propagation (typically beyond a hundred meters for frequencies above a hundred Hertz). Thus, the amplitude fluctuations can be neglected and we can consider a uniformly distributed phase on $[-\pi, \pi]$.

These results are valid only within the line of sight region, in which a large dependence of the acoustic pressure field on frequency can be observed. In the acoustic shadow region, the dependence of the pressure field on frequency is much lower. Some other mechanisms are involved in the description of the acoustic propagation in this region, including the diffusion of the sound energy by turbulent heterogeneities that present a stochastic spatial distribution. A formulation can be given therefrom, by analogy with the laws of optics. Some values of the length scales of the diffusing structures are effectively more relevant than others in the process of diffusion of sound energy toward the low-intensity region (gray area). The experimental results regarding this particular point are often in agreement with the values provided by the optical laws of x-ray diffraction by crystal structures, or Bragg reflection condition (see Figure 10.22). If λ is the considered wavelength and θ is the diffusion angle, this condition then gives the spatial periodicity D of the most diffusing structures. It is written as $\lambda = 2D\sin(\theta/2)$. For a frequency of 500 Hz and an angle of $10°$, the Bragg condition predicts a structure size $D = 4$ m. However, for source-receiver geometries that are commonly considered, those playing the most important role are those comprised between 2 and 5 m. The existence of a privileged diffusion region has also sometimes been highlighted, located downstream of the limit ray originating from the source and of that traced in the opposite direction from the receiver under the same conditions of refraction. Thus, it is customary to consider that this distribution volume is roughly located just above the limit ray, halfway between source and receiver. In addition, some have also supported the thesis of a noise level in overall reinforced in the neighborhood of the limit ray (appearance of caustics) with regard to that existing in the region of the line of sight.

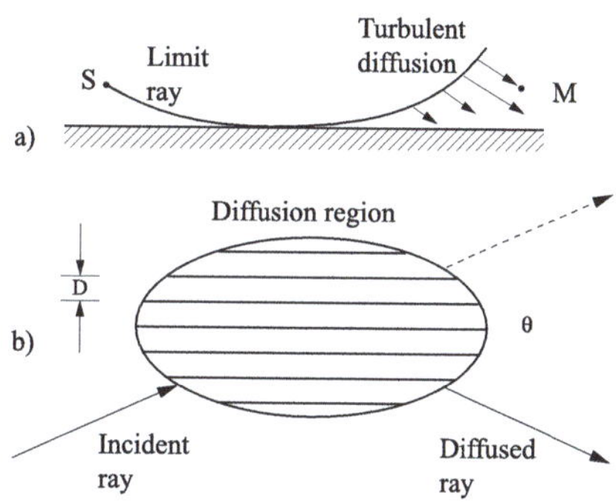

Figure 10.22. *Sonorization of the shadow region by turbulent diffusion (a) and diagram of the Bragg reflection condition (b) (according to [GAU 99])*

In order to synthesize the situations (such as highlighted by Wiener and Keast [WIE 59] notably), we can refer to the diagram representation in Figure 10.23. Thus, for small propagation distances (corresponding to region 1, known as line of sight), each point can be reached by a ray emerging from the source; the effects of the velocity gradient and turbulence on the mean levels are negligible, and the relative levels (to the propagation in a homogeneous atmosphere) of sound pressure are zero. In the presence of a strongly negative sound speed gradient, a brutal transition can be observed between a sonorized region and a lower sound energy region; it is region 2, or the transition area, in which the acoustic rays begin to be diverted and to no longer reach the receiver directly. In region 3, the receiver can be found largely below the limit ray and is deeply situated in the shadow area. Because of the diffusion of the sound energy by the atmospheric turbulence, the sound level presents a minimum and it depends quite weakly on the distance, as well as the height and frequency.

We have just seen that the quantity $\overline{\mu^2}$, the variance of the refractive index, plays a significant role in quantifying the influence of turbulence on the propagation of sound waves. It has, therefore, appeared worthwhile to conclude this section with the calculation of this quantity. To this end, as previously stated, it is considered that the propagation direction and the wind direction are in the same horizontal plane (independence of the quantities in the azimuthal angle) and the angle θ is defined as that between the mean direction of the wind and the direction of the sound propagation. By including the respective fluctuations V' and T' such that $V = V_0 + V'$ and $T = T_0 + T'$ (the index 0 always indicates a reference value), and according to the dependence of the sound speed on temperature and on the wind velocity, the effective sound speed c is then written $c = c_0 + c' = c_0(1 + T/T_0)^{1/2} + v_0 \cos\theta(1 + V'/V_0)$.

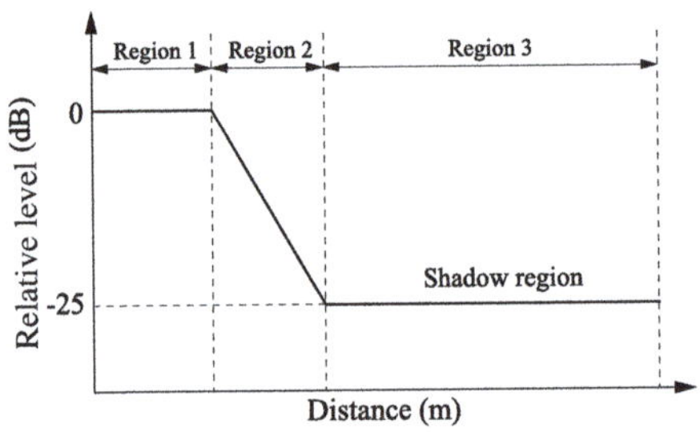

Figure 10.23. *Illustration of the various zones of influence of the atmospheric turbulence depending on the distance to the source (according to [GAU 99])*

Since $\overline{V'} = \overline{T'} = 0$ and the variances can be written as, respectively, $\sigma_V^2 = \overline{V'^2}$ and $\sigma_T^2 = \overline{T'^2}$, we can define the cross-correlation term R between the wind and the temperature fluctuations such as $R = \overline{V'.T'}/\sigma_V \sigma_T$. By rewriting that $n = c/c_0 \approx (c_0 + c')/c_0 = 1 + \mu$, we then get $\overline{\mu^2} = ((\sigma_V \cos\theta)/c_0)^2 + (\sigma_V \sigma_T R \cos\theta)/(c_0 T_0) + (1/2\sigma_T/T_0)^2$. Some authors have suggested generalizing this expression to include the dependency to the elevation angle, the angle between the horizontal plane and the propagation direction if it does not belong to this plane. In practice, as there is in general a preferential interest for the horizontal acoustic propagation (even if the refraction phenomenon appears to challenge this hypothesis), the elevation angle will be considered as null. We then make use of the common relations that give the typical values of the velocity and temperature variances, $\sigma_V^2 \approx \sigma_u^2 + \sigma_v^2 \approx (5.4u_*^2 + 0.2w_*^2) + (4.6u_*^2 + 0.2w_*^2)$, in which u and v refer to the longitudinal and lateral components of the wind, u_* and w_* are the conventional friction velocity and a velocity characterizing the buoyancy effect that may exist if the thermal gradients are significant (or the wind velocities low), $w_* = (Q\beta z)^{1/3}$ (with Q, the surface heat flux and $\beta = g/T_0$). There are also comparable relations for the quantities that involve temperature variations, $\sigma_T^2 \approx 4T_*^2$ and $\overline{T'u'} \approx 2.5u_*T_*$, where $T_* = -Q/u_*$.

Thus, finally, the quantity $\overline{\mu^2}$ can be written as the sum of six different contributions A_i, $\overline{\mu^2} = \sum_{i=1}^{6} A_i$, where A_i are defined in Table 10.3.

Nevertheless, it is quite usual to consider that the thermal contributions are negligible, on the one hand, and that the variances of the longitudinal and lateral velocity components are equal, on the other hand. We then get $\overline{\mu^2} \approx 5.4u_*^2/c_0^2 + 0.2w_*^2/c_0^2 \approx \sigma_u^2/c_0^2$. Although neglecting the specific contributions of the temperature-related terms may seem fairly coarse *a priori*, this is quite justified. Indeed, even on a sunny day, when the air heated near the ground is convectively unstable and rises in altitude, the air motions arising therefrom (called *thermals*) are characterized by both an increase in the velocity contribution (characterized by w_*/c_0) and in the thermal contribution (T_*/c_0). Therefore, the variance of the wind can quite naturally dominate the variance of the refraction index during a sunny day, and this is achieved through the contribution of the terms in w_*/c_0, with the turbulence, originally convective, dominating the turbulence due to the shear in the atmospheric boundary layer. As a matter of fact, it is commonly found that the temperature variance term compared to the velocity variance term is approximately an order of magnitude smaller at the height of 2 m even for extremely large values of the surface flux ($Q = 0.5$ K.m/s) as it can happen at mid-day during summer.

Index	A_i	Interpretation
1	T_*^2/T_0^2	Temperature variance
2	$5.4(u_*^2/c_0^2)\cos^2\theta$	Longitudinal velocity variance, small-scale
3	$4.6(u_*^2/c_0^2)\sin^2\theta$	Lateral velocity variance, small-scale
4	$2.5(u_*T_*/c_0T_0)\cos\theta$	Wind-temperature covariance
5	$0.2(w_*^2/c_0^2)\cos^2\theta$	Longitudinal velocity variance, large-scale
6	$0.2(w_*^2/c_0^2)\sin^2\theta$	Lateral velocity variance, large-scale

Table 10.3. *Different contributions A_i participating to the value of $\overline{\mu^2}$*

Slightly surprising at first glance, the situations in which thermal fluctuations have a more significant effect on the acoustic diffusion than dynamic fluctuations (wind) are more frequently found during the night. Thus, within a very stable nocturnal boundary layer, as can be found during clear and calm nights, turbulent mixing disappears and highly positive temperature gradients develop. In the presence of such temperature contrasts, even very small air motions generate temperature variances for a given point in space.

Regarding characteristic length scales which are involved in the analytical expression of the autocorrelation function for fluctuations of indices $\psi_n(r)$, that is written generally as $\psi_n(\xi) = \overline{\mu^2} \exp(-\xi^2/L^2)$, it emerges that it is necessary to relate L (which is an effective length scale) to the correlation functions of the longitudinal and lateral velocity components as well as to the temperature. Neglecting the cross-terms between the velocity fluctuations and the temperature fluctuations, and further assuming that each quantity individually taken has an autocorrelation function that is Gaussian, then it is common ([DAI 78]) to write:

$$\psi_n(\xi) = (\frac{\sigma_V \cos\theta}{c_0})^2 \exp(-\xi^2/L_V^2) + (\frac{\sigma_T}{2T_0})^2 \exp(-\xi^2/L_T^2),$$

in which L_V and L_T represent, respectively, the dynamic and thermal integral scales. From this relation, it can naturally be derived that, if $L_V = L_T$, then the effective value L is also equal to L_V. On the contrary, if L_V and L_T are not equal, then L will rather take the value of L_V or L_T according to the relative importance of the corresponding variance terms. In general (e.g. [WIL 94]), the integral scale of the horizontal component of the wind is greater than that of the temperature.

On the other hand, it also often appears that turbulence is markedly anisotropic. Very close to the ground and when conditions are almost neutral, the instabilities due to shear predominate and the vortices have an elongation in the longitudinal direction; conversely, but to a lesser extent, the elongation may be vertical in the case of a turbulence of mainly thermal origin. In order to model these effects, it is usual (e.g. [WIL 94]) to use the matrix notation to express the autocorrelation function $\psi_n(x, y, z)$:

$$\psi_n(\xi) = \overline{\mu^2} \exp(-\overline{\overline{X}} L^{-2} X),$$

in which the double bar indicates the transposition and X is the matrix:

$$X = \begin{bmatrix} x \\ y \\ z \end{bmatrix}, \text{ with } L^{-2} = \begin{bmatrix} 1/L_x^2 & 0 & 0 \\ 0 & 1/L_y^2 & 0 \\ 0 & 0 & 1/L_z^2 \end{bmatrix}.$$

When considering the case of horizontal sound propagation and when expressing, as we have already seen, by θ the angle that the propagation direction makes with the mean wind in the reference frame of axes (x,y,z), these matrix relations lead to the definition of the parallel ($L_{//}$, in the horizontal propagation plane) and perpendicular ($L_\perp$, normal to the propagation plane) effective length scales. Their expressions are, respectively:

$$L_{//} = (\frac{\cos^2\theta}{L_x^2} + \frac{sin^2\theta}{L_y^2})^{-1/2},$$

and

$$L_{\perp} = (\frac{\sin^2\theta}{L_x^2} + \frac{\cos^2\theta}{L_y^2} + \frac{1}{L_z^2})^{-1/2}.$$

Compared to what we previously had for the intensity parameter Γ^2 that characterizes the fluctuations of the phase signal and for the diffraction parameter Λ, we then get:

$$\Gamma^2 = \sqrt{\pi}\overline{\mu^2}k_0^2 r L_{//} \ \ and \ \ \Lambda = 4r/(k_0 L_{\perp}^2).$$

Wilson and Thomas [WIL 94] could then model the dependence of the effective length scales $L_{//}$ and $L_{\perp}$ on the angle θ. Values are calculated for an inversion height equal to 1,000 m and $u_* = 0.5$ m/s, and this for two different heights, $z = 2$ m and $z = 10$ m, and two types of meteorological conditions: the practically neutral conditions ($T_* = -0.01K$) and well-developed thermal convection ($T_* = -0.3K$). From these results, it appears that:

– $L_{//}$ depends strongly on meteorological conditions, but weakly on height. The reason thereof is that $L_{//}$ is essentially linked to the large atmospheric vortices, whose horizontal scale depends on climatic conditions;

– $L_{\perp}$ strongly depends on the height but not on meteorological conditions. This is explained by the fact that $L_{\perp}$ is mainly affected by smaller vortices, generated by shear flow instabilities near the groundfloor;

– $L_{//}$ presents a peak (at least a tendency to emerge) for $\theta = 0°$ or $180°$; whereas the same phenomenon is observed for $L_{\perp}$ at the value of $\theta = 90°$.

This agrees very well with other results which have shown that the effective length scale for acoustic phase fluctuations is 50–150 m during convective conditions, at about a height of 2 m above the ground, and that the scale decreases in the direction perpendicular to the wind.

Thus, the model proposed by Wilson and Thomas [WIL 94] predicts that the effective horizontal length scale depends mainly on the inversion height, while the vertical scale depends mainly on the altitude z. This results from the fact that the large phase fluctuations are governed by the horizontal and large-scale turbulence structures, while the high amplitude fluctuations are determined by the vertical and small-scale turbulence structures. This could explain why other studies have experimentally shown that length scales of the order of the height (approximately 1 m) should be used, whereas in most meteorological situations we generally resort to substantially larger scales. Similarly, the variance of the refraction index experimentally estimated by acousticians is usually located in the range $[10^{-6} - 10^{-5}]$, whereas the usual experimental values of the turbulence intensity obtained from the velocity variances measured in the atmospheric boundary layer are at least an order of magnitude larger. Despite these underestimations of the scale and the turbulence intensity, a reason for which acousticians have often obtained a good agreement between the predictions of their models and the experimental data could be that if acoustic diffusion originates from a small range of vortex sizes, a good agreement with the experimental results can nevertheless be obtained by just choosing a length-scale characteristic of this spatial domain and a variance representing only the energy contained in this domain. This idea is reinforced by the fact that most of the acoustic studies have considered a turbulent length scale of the order of the acoustic wavelength, and that, in wide-angle diffusion problems, the bulk of the diffusion of sound comes from atmospheric vortices whose size is comparable to the wavelength [WIL 96].

In practice, since it is simpler to measure temporal statistics of fluctuating micro-meteorological fields than spatial statistics, Taylor's hypothesis of frozen turbulence is often adopted and homogeneous and isotropic turbulence considered. The value of the integral turbulence scale is then given by [DAI 78] $L = \overline{V}\tau_0$, in which τ_0 is the characteristic time scale associated with the temporal measurement of the correlation function ψ_n by adopting a Gaussian form for it, $\exp(-\tau^2/\tau_0^2)$.

After this presentation of the main phenomena, we are now going to look in more detail into how these different pieces of phenomenological information can guide the development of physical models that will be used as the basis for the numerical computations of the propagation of sound waves in the atmosphere.

10.5.2. *Models of sound wave propagation in the atmosphere*

We have just examined in a global manner how the propagation of acoustic rays in the atmosphere can be altered by the presence of temperature gradients or refraction index fluctuations due to turbulence. Embleton's article [EMB 96] presents a synthetic review of the richness of phenomena that can occur and of the situations that may come across. We are now going to present some of the principal models that have been developed to help calculate this in a quantitative manner.

10.5.2.1. *Ray tracing methods*

Since we are talking about acoustic rays from the beginning of this section, it has appeared logical to present first the method known as the ray tracing method. It is an analytical method, which is directly inspired by Rudnick's expansions for the computation of the pressure field without turbulence, in the presence of a finite impedance ground, overlooking ground waves and surface waves. The configuration under study is shown in Figure 10.24, in which S is the source and R is the receiver, r_d represents the direct path, r_r represents the reflected path, and l_m represents the maximal transverse separation between the direct wave and reflected wave.

When considering a sound wave of spherical nature, emitted from a source point S, the sound pressure above the soil with impedance is written as the superposition of a spherical wave originating from S propagating in free space and a spherical wave originating from the image of S with respect to the ground propagating also in free space but corrected by a factor of $Q = R_p + (1 - R_p)F$, in which F is a complex function that reflects the interaction of the wavefront with the soil with impedance. The sound pressure at a point $\vec{x}$ of the propagation medium is then given by equation [PIE 77]:

$$p(\vec{x}) = \frac{A_d}{r_d}\exp(ik_d r_d) + R_p \frac{A_r}{r_r}\exp(ik_r r_r) + (1 - R_p)F\frac{A_r}{r_r}\exp(ik_r r_r),$$

where A_i is the considered wave amplitude ($i = d$ or r–direct or reflected), r_i is the length of the acoustic ray, k_i is the associated wave vector and R_p is the plane-wave reflection coefficient. This reflection coefficient is defined according to the reduced impedance of the material constituting the soil (ratio of the surface impedance Z of the latter and the specific impedance of air Z_0) by the following relation (where reflection is assumed to be specular, and thus that $\phi_r = \phi_i$, ϕ_i being the angle between the surface and the incident ray):

$$R_p(\phi_i) = \frac{\xi(\phi_i)\sin\phi_i - 1}{\xi(\phi_i)\sin\phi_i + 1},$$

with

$$\xi(\phi_i) = \frac{Z}{Z_0 \sin\phi_t} = \frac{Z}{Z_0\sqrt{1 - (\frac{k_0}{k})^2\cos^2\phi_i}},$$

in which the indices i, r and t, respectively, refer to the incident, reflected and transmitted (in the material) rays.

By making a number of probabilistic hypotheses about the acoustic propagation in turbulent environments, we then get [DAI 78] a statistically random field, whose variance is written:

$$\overline{p^2} = \frac{2}{r_d r_r}\left\{\frac{\sigma_A^2}{2}(\frac{r_r}{r_d} + |Q|^2\frac{r_d}{r_r}) + \frac{r_r}{2r_d}(1 - |Q|\frac{r_d}{r_r})^2 + |Q| \right.$$
$$\left. + |Q|\,(1 + \sigma_A^2\rho_A)\cos(\phi + \gamma)\exp(-\sigma_s^2(1 - \rho_s))\right\},$$

where $Q = |Q|\exp(i\gamma)$ *and* $\phi = k(r_r - r_d)$. The amplitude of the field is $A = 1 + a$ in which a is the amplitude fluctuation; its phase is written $S = S_0 + S'$, with S_0 the turbulenceless phase and S' the phase fluctuations.

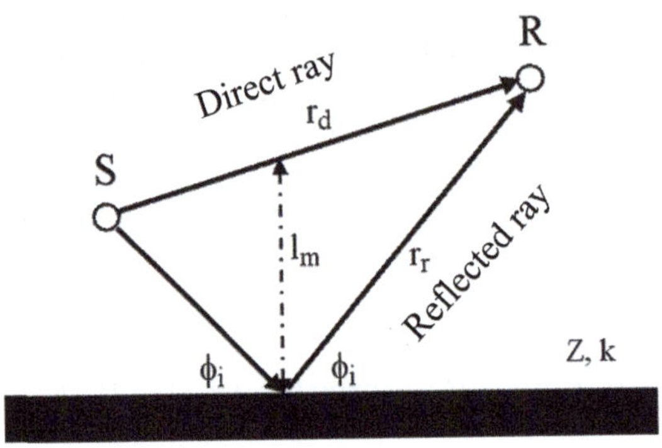

Figure 10.24. *Outline diagram of the ray tracing method*

The quantities σ_A^2 and σ_s^2 are the amplitude and field phase variances, ρ_A and ρ_s are the amplitude and phase covariances between the direct wave and reflected wave. The calculation of σ_A^2 and σ_s^2 is achieved in the small fluctuation Rytov approximation. For a homogeneous and isotropic turbulence with Gaussian correlation of the index fluctuations, there is an analytical expression of the phase variance as well as an expression of the log-amplitude variance. With regard to the calculation of ρ_A and ρ_s, the approach is much more empirical and must be adapted to the configuration of the problem (these quantities depend notably on the maximum distance l_m between the direct path and the path of the reflected wave). This is one of the weaknesses of this method, although some more recent studies (see [OST 98] in particular) now allow a better modeling of the turbulence using this method.

Generally speaking, the results provided by this method are relatively correct, despite the significant number of approximations and hypotheses. It makes it possible to account for impedance discontinuity, as well as for refraction and turbulence phenomena. In addition, it also allows the case of the propagation over a surface of extended, and possibly multilayer, reaction to be addressed. Finally, although the introduction of a screen in the method is possible, the effect of the presence of this screen on the refraction profiles cannot be taken into account by this model which assumes a homogeneous profile. It is for these reasons, notably, that more complex but also more specific stochastic methods have been developed. From these, there are a few that we are now going to present.

10.5.2.2. *The Fast Field Program method*

Originally, this method was developed to predict the propagation of acoustic waves in underwater media. The method then spread more widely, and, in particular, it can be applied to the computation of the propagation of sound waves in external media. This method amounts to solving the wave propagation equation in two dimensions (in cylindrical symmetry) which is written for a point source (by replacing, according to common usage, the partial derivative with respect to time by $k^2 p$):

$$k^2 p + \frac{\partial^2 p}{\partial r^2} + \frac{1}{r}\frac{\partial p}{\partial r} + \frac{\partial^2 p}{\partial z^2} = -\frac{2}{r}\delta(r)\delta(z - z_s),$$

where z_s represents the height from the ground at which the source is positioned and r is the longitudinal distance (i.e. according to the axis parallel to the ground). The Fast Field Program consists of computing the Hankel transform (which was defined in [7.16]) of the previous relation, which is written:

$$\frac{\partial^2 P}{\partial z^2} + (k^2(z) - K^2)P = -2\delta(z - z_s),$$

in which $P(z, K)$ is the order-zero Hankel transform of the pressure $p(r, z)$, which, once again, is given by:

$$P(z, K) = \int_0^\infty p(r, z)\mathrm{J}_0(Kr)rdr.$$

This new propagation equation now depends only on variable z and the problem is reduced to a one-dimensional problem. The value of $P(z_r, K)$ at the point where is located the receiver is obtained by the Thomson–Haskell Matrix (THM) method from the value $P(z_r, K)$, calculated stepwise from one stratum to another [WES 91]. Finally, the sound pressure $p(r, z)$ is obtained by means of the inverse Hankel transform:

$$p(r, z) = \int_0^\infty P(z, K)\mathrm{J}_0(Kr)KdK.$$

Over the years, this approach has been largely improved, notably with the introduction of accounting for the atmospheric turbulence [LES 95].

10.5.2.3. *Parabolic approximation-based methods*

In numerous situations, unfortunately, the methods previously presented prove to be excessively approximate and it then becomes important to really solve an equation of propagation. However, the most often used is a parabolic approximation, which helps to approximate the pressure sound p by transforming the starting elliptic equation, written in a compact manner under the form $k^2 p + \Delta p = 0$, into a parabolic equation that is better adapted to numerical computations. This results in particular from the fact that the parabolic equation is then of the first order and not of the second order, as is the case with the real propagation equation which is elliptical, according to the main propagation direction (x or r). The simplification results from the fact that it is certain that the propagation takes place toward the front following the direction x (or r) corresponding to the ray. As a result, the solution is greatly simplified numerically, because the solution is calculated at each new point $x + dx$ (or $r + dr$) from that in point x (or r), using a step method as is usually done for an evolution over time. Instead, for the basic equation, which is elliptical in nature, there is a need to solve the equation simultaneously in all points x (or r), which is naturally much more complex, and even sometimes impossibl [7]. This approximation is made possible by the fact that the sound speed c depends primarily on the position in z and much less on the position in x. First, this approximation has been proposed in the context of submarine acoustic propagation, which presents many points in common with the propagation in the atmosphere, by [TAP 77]. We would also like to quickly mention that a very large number of works have been dedicated, and for a very long time, to submarine sound propagation (see in particular Kuperman and Roux' review article [KUP 07] in [ROS 07]). These works on submarine sound propagation essentially had military applications as motivation at the outset. Thus, as shown in Figure 10.25, the presence of the acoustic channel resulting from the stratification of the deep waters of the ocean allows following at a very long distance, from regions near the equator, the motion and the activity of submarines navigating in the polar regions. Indeed, according to the Snell–Descartes law, acoustic rays are deflected toward the medium in which the celerity of waves is the smallest. That is why the waves emitted at the level of the surface of the ocean in polar regions are then audible (and thus easily detectable) at depths of around 1,300 m and more at the level of the equator.

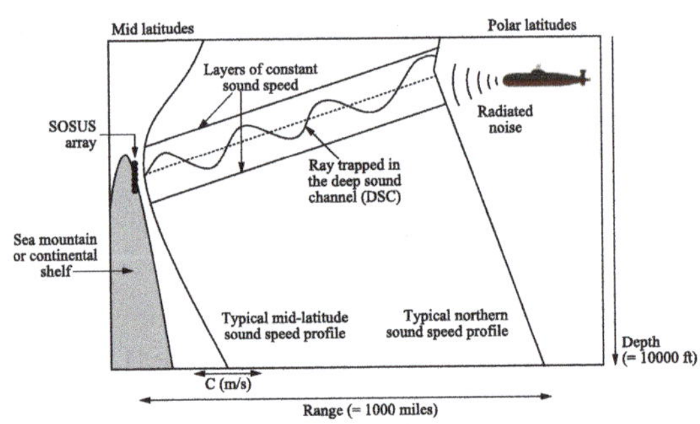

Figure 10.25. *Illustration of the use of the phenomenon of the acoustic channel corresponding to the waveguide mechanisms resulting from the stratification of deep waters in the ocean, for military purposes and for the very long-distance monitoring of submarines motion (according to [KUP 07])*

7 There are several ways to approach the starting equation in order to transform it into a parabolic one. For more details, the readers can refer, notably, to Tappert's article [TAP 77].

Several methods have been proposed to solve this new equation but we will expose only one thereof, which seems to be the most efficient at the present time, and that is commonly referred to as the *split-step Padé* method. As we have just seen, it is thus usual, in this current context, to write the equation of propagation of sound waves p under the hypothesis of azimuthal symmetry, which leads us to write this equation in cylindrical coordinates:

$$k^2 p + \frac{\partial^2 p}{\partial r^2} + \frac{1}{r}\frac{\partial p}{\partial r} + \frac{\partial^2}{\partial z^2} = 0.$$

Sound pressure is then separated (*split*) into two components: a propagative cylindrical wave, represented by the asymptotic expression of the Bessel function (or Hankel in 2D) of the first kind and order-zero in far-field ($k_0 r \gg 1$, with $k = \pi f / c(r, z) = k_0 n(r, z)$), and an envelope function $u(r, z)$ which varies very slightly with the distance:

$$p(r, z) = u(r, z) \mathrm{H}_0^{(1)}(k_0 r) \approx \frac{1}{\sqrt{r}} u(r, z) \exp(ik_0 r),$$

in which the carrier wave number is chosen close to the dominant horizontal wave number in the spectral decomposition of $u(r, z)$ (see [GAU 02, LIH 06] and [OST 97], in particular). A good approximation for this carrier wave number is then $k_0 = \omega / \overline{c}$, where $\overline{c}$ is the mean speed of sound. Assuming then that $n(r, z)$ weakly depends on the distance and that the retro-propagated acoustic energy is negligible, then the evolution of $u(r, z)$ is governed by the unidirectional wide-angle parabolic equation:

$$\frac{\partial u(r, z)}{\partial r} = ik_0(Q - 1)u(r, z),$$

in which the differential operator Q is defined by the relation $Q^2 = (1/k_0^2)\partial^2/\partial z^2 + n^2 = 1 + \xi + \eta$, defining $\xi = (1/k_0^2)\partial^2/\partial z^2$ and $\eta = n^2 - 1$.

The difficulty in solving the parabolic equation which has just been written lies in the computation of the pseudo-differential operator Q. There are various developments for the operator Q, each of them leading to a number of different expressions of the parabolic equation, with, for each, its advantages and disadvantages. As already stated above, we will focus here only on the approximation which appears as the most efficient at the present time, and commonly referred to as the *split-step Padé* method. In this method, the first step is to integrate the parabolic equation and then carry out the *Padé* approximation. The existing field is thus considered at an arbitrary distance r_0, assuming that Q varies very little on the interval $[r_0, r_0 + \Delta r]$, such that the approximated solution of the parabolic equation at step $r_0 + \Delta r$ is given by $u(r_0 + \Delta r, z) = u(r_0, z). \exp(ik_0 \Delta r(Q - 1))$.

By defining $\Im = Q^2 - 1$, the original advantage of this method is not to approach the operator Q, but the exponential operator $\exp(\sigma(Q-1))$ directly, which results in a computation procedure simpler than Cranck–Nicholson's. The latter is obtained by other methods that focus on the expansion of (Q-1) in the exponential, as outlined notably by Ostashev [OST 97], as well as on a wider angle aperture. This is achieved by means of a second-order Padé expansion:

$$\exp(\sigma(Q-1)) = \exp(\sigma(\sqrt{1+\Im}-1)) \approx \frac{1+p_1\Im+p_2\Im^2}{1+q_1\Im+q_2\Im^2}. \qquad [10.14]$$

When the modulus of this rational fraction is fixed at unity, the stability of the scheme is ensured. If the expansion is limited to the second-order, the computation of the coefficients p_1, p_2, q_1 and q_2 does not require the use of a sophisticated technique such as those that have to be used in other approximation methods. As a matter of fact, it suffices to carry out a fourth-order Taylor series expansion on $\exp(\sigma(\sqrt{1+\Im}-1))$, and then to identify with the rational fraction that appears in relation [10.14]. We can thus derive the expressions of the coefficients p_1, p_2, q_1 and q_2:

$$p_1 = \frac{3+\sigma}{4}; p_2 = \frac{\sigma^2+6\sigma+3}{48}; q_1 = \frac{3-\sigma}{4}; q_2 = \frac{\sigma^2-6\sigma+3}{48}.$$

Thus, the marching algorithm is expressed according to the coefficients p_1, p_2, q_1 and q_2, of $\eta = n^2 - 1$ and of the operator $\xi = (1/k_0^2)\partial^2/\partial z^2$:

$$(1+q_1(\eta+\xi)+q_2(\eta+\xi)^2)u(r_0+\Delta r, z) = (1+p_1(\eta+\xi)+p_2(\eta+\xi)^2)u(r_0, z),$$

in which $(\eta+\xi)^2$ is developed of course in $(\eta+\xi)^2 = \eta^2 + 2\eta\xi + \xi^2$.

The presence of a soil with impedance is described by the boundary condition for sound pressure: $p(r,0) = Z_s V_n(r,0)$ where Z_s is the surface impedance and V_n is the velocity normal to this surface that demarcates the soil. The Euler equation for a monochromatic wave is written on the edge as $\partial p/\partial n = \imath\omega\rho V_n$. Thus, when the sound pressure is expressed as the product of a carrier wave $\exp(\imath k_0 r)$ and an envelope function $u(r,z)$ slowly varying according to r, $p(r,z) = u(r,z)\exp(\imath k_0 r)$, and if defining $\beta = \rho c_0/Z_s = 1/Z_n$ (reduced admittance of the soil), the boundary condition on the edge is written:

$$\frac{\partial u}{\partial z} + ik_0\beta u = 0.$$

This is a Fourier-type boundary condition; it involves the field u as well as its partial derivative with respect to z. The value of the admittance β is assumed to be identical at each step of the computation, and thus independent of the angle of incidence. This amounts to making the hypothesis of a localized-type reaction for the soil under consideration. Finally, the problem of the propagation of sound in the atmosphere above a finite impedance ground is described by the system of equations:

$$\begin{cases} (1+q_1(\eta+\xi)+q_2(\eta+\xi)^2)u(r_0+\Delta r,z) = (1+p_1(\eta+\xi) \\ \qquad +p_2(\eta+\xi)^2)u(r_0,z), \\ Q^2 = 1+\eta+\xi, \text{ with } \eta = n^2-1, \text{ and } \xi = \dfrac{1}{k_0^2}\dfrac{\partial^2}{\partial z^2}, \\ \dfrac{\partial u}{\partial z} + ik_0\beta u = 0. \end{cases}$$

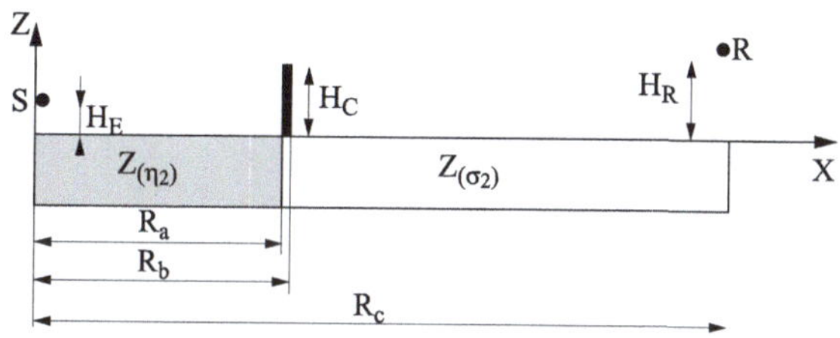

Figure 10.26. *General configuration for the study of the effect of a screen on atmospheric sound propagation by the* split-step *Padé method. According to [GAU 02]*

By outlining the atmospheric turbulence in a overall manner (as it was presented in section 10.5.1), using a spectral distribution (or a spatial correlation law) of the turbulent energy – and therefore also of the refraction index – associated with an integral scale L and a variance $\overline{\mu^2}$, it is possible to simulate the effect produced on the propagation of sound waves by different complexities (atmospheric turbulence only, presence of screens represented in general by a surface impedance discontinuity, etc.). The mean velocity and mean temperature distributions in the atmospheric boundary layer are represented overall through their influence on the speed of sound, which is given by the law $c(z) = c_0 + a\ln(z/z_0)$, in which z_0 and a are commonly called the roughness parameter and the refraction parameter. Figures 10.26, 10.27 and 10.28 illustrate the results obtained in the case of a screen placed at different distances from the source and for different heights (Figure 10.26). First, Figure 10.27 gives the acoustic intensity levels for different distances to the source and different altitudes, with or without atmospheric turbulence. It can thus be first observed that an impedance discontinuity actually allows the presence of a screen to be represented in a satisfactory manner. This is mainly due to the associated diffracting effects which are comparable, at least in low frequencies where these diffracting effects dominate[8]. At a higher frequency, where diffusion effects in the

8 The problem of screens has been very widely studied in the past, and it seems necessary to mention, by the way, that other approaches may be used to model the influence of screens on

shadow region just downstream the screen dominate, the analogy is nonetheless worse. The comparison of the *split-step Padé* method with other models is illustrated in Figure 10.28, showing a very good agreement between these different methods all based on the parabolic equation approximation for the two situations considered here, with and without atmospheric turbulence.

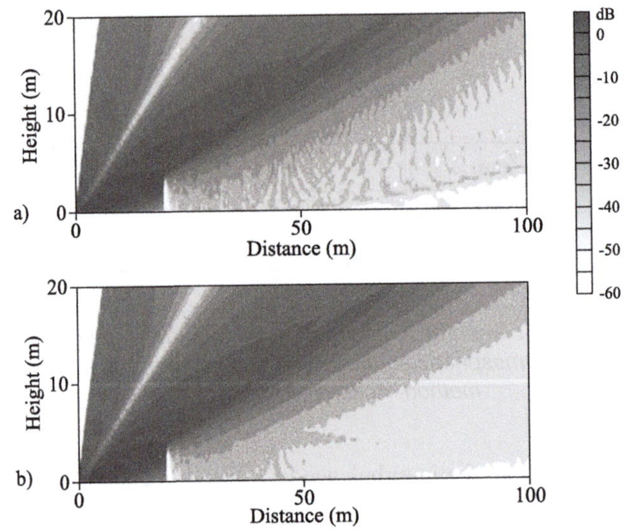

Figure 10.27. *Modeling of the effects induced by the presence of a screen on atmospheric sound propagation by the* split-step Padé *method.* $f = 5\ kHz$, $H_s = 0.03\ m$, $\sigma_1 = \sigma_2 = 3\ 10^5\ kPa\ s\ m^{-2}$, *and (a) without atmospheric turbulence or (b) with atmospheric turbulence and* $a = -2\ m/s$, $R_b = 20\ m$, $H_b = 4\ m$. *Turbulence is modeled by a Von Karman-type spectrum with* $\overline{\mu^2} = 8\ 10^{-6}$, $L = 1.1\ m$ *and the internal scale contributing in the Von Karman-type spectrum* $l_0 = 0.05\ m$ *(according to [GAU 02])*

sound propagation. Attenborough's article [ATT 07] presents a detailed synthesis thereof. This article also discusses other phenomena such as surface waves, which cannot be mentioned here for lack of space.

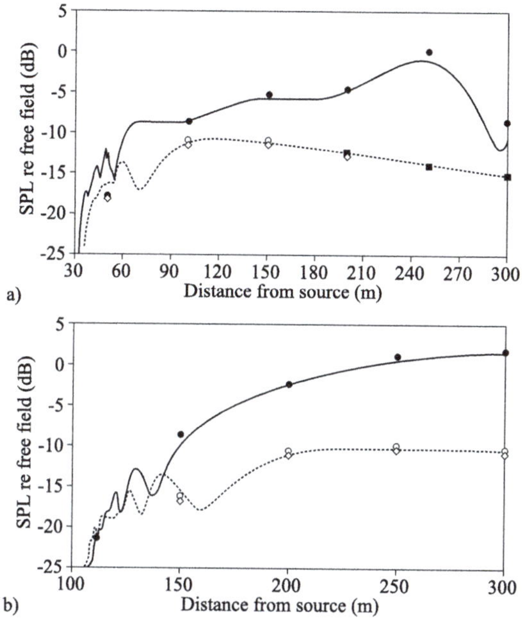

Figure 10.28. *Comparison of the predicted noise levels (thin line: $a = 0$ m/s and thick line: $a = 2\ m/s$) by the* split-step Padé *method with the predictions of other models also based on the parabolic equation approach $f = 1\ kHz$, $H_s = H_r = 2\ m$, $\sigma_1 = \sigma_2 = 3\ 10^2\ kPa\ s\ m^{-2}$, and (a) $R_b = 30$ m, $H_b = 4\ m$ or (b) $R_b = 100\ m$, $H_b = 6\ m$ (according to [GAU 02])*

11

Implementation and Usage of Numerical Simulations

In this chapter, we will discuss the numerical implementation of the models presented in the previous chapters, insisting, in particular, on the most recent numerical results that, from direct numerical simulations or from large eddy simulations of turbulence, allow either the validation of analytical models or the identification and localization of the sources and, thus, a better distinguishing of the mechanisms which exist within the flows and that really generate aeroacoustic noise. This presentation does not pretend to be exhaustive, given the complexity of the phenomena and the situations that we come across, which can include a much wider range of subjects (such as noise generated by risers in the oil industry, thermoacoustic phenomena, etc.) than that which will be covered here. It pretends to be only an introduction that lays the foundations and opens the way to theoretical developments or toward the thoughtful usage of numerical tools. These tools now start to be widely distributed and commonly used in industrial and university environments (Fluent, Star CCM+, computational codes developed by research organisms such as the ONERA and the CEA in France and more).

Historically, after the Lighthill analogy, it is the emphasis given to the presence of coherent structures in turbulent flows that gave new life to aeroacoustics. In 1971, Crow and Champagne [CRO 71] already suggest that the acoustic radiation from a jet can be modified by acting on the ordered structures of the flow, while Bishop *et al.* [BIS 71] identify macroscales as the main sources of acoustic radiation in supersonic jets. Thus, in 1974, when Brown and Roshko [BRO 74] observe coherent structures in mixing layers, even at high Reynolds numbers, has arisen the question of identifying whether these structures constitute the main noise sources of the jets. From this point on, an attempt has been made to distinguish the acoustic contribution of the large structures, with an almost deterministic behavior, from that of small almost-random turbulent structures. As a matter of fact, it is notable that, in addition to the defects or deficiencies in modeling that we have presented in Chapter 9, the Lighthill equation does not explicitly contain any contribution related to the presence of these organized large-scale structures. Nevertheless, this is one of the principal dilemmas of any approach to turbulence modeling, whether it is addressing analytical modeling (as for the source term of the Lighthill equation) or statistical modeling with the purpose of calculating, using a $k - \varepsilon$ model or a second-order model, the characteristics of a turbulent flow: how to realistically account for the presence of these coherent structures? The influence of the Reynolds number should be highlighted here. Indeed, even if it seems that coherent structures always come across, the development of small turbulence scales is limited by viscous effects at low Reynolds numbers. Studies undertaken to bring forward the relative contributions of large and small turbulent structures have thus often consisted of varying the Reynolds number of the jets in order to compare the characteristics of their respective radiations.

Fortunately, for some 15 years, we can make use of particularly effective new numerical tools that allow overcoming most of the weaknesses of one-point statistical models, which consist of direct numerical simulations (DNS) or large eddy simulations (LES). Naturally, these tools have been widely used since then in aeroacoustics and they have notably helped in advancing our understanding of the mechanisms that generate acoustic radiation, as we are going to see it now, even if only a few typical examples can be mentioned here. The readers interested in more details are invited to consult the specialized literature [WAG 07, TAM 95a, WAN 06, WEL 97].

11.1. Hybrid methods

These numerical predictions of the acoustic field are characterized by a two-step approach. In the first step, the aerodynamic flow field is calculated and then the sound field is obtained in the second stage, using an acoustic analogy and the previously computed aerodynamic field data to evaluate the sound sources. In this approach, taking account of the acoustic quantities during the velocity field calculation is no longer a priority. It is then no longer necessary to solve the compressible equations and solving a simplified system can be considered, assuming a constant density within the flow, for example, for low Mach number flows. The computation of the flow velocity field can be performed either by DNS or LES, or by solving the averaged equations coupled with a turbulence model. In the first two cases, the computed velocity fields are instantaneous, whereas in the latter case, the turbulent field is described only by statistical quantities. Historically, it is through these latter approaches that first aeroacoustics computations have been performed. Even based on numerous approximations (whose most important aspects have been covered in Chapter 9), these techniques, in general called hybrid methods, are still being extensively used because they are easy to employ in complex geometries such as those which an engineer must face. They are now also implemented in commonly used numerical tools in industrial environments.

For these velocity field simulations and considering the averaged equations in a stationary context, with a $k - \epsilon$ stationary-type turbulence model, the knowledge of the flow is reduced to statistically averaged quantities, which are used to estimate the value of the acoustic source terms. The statistical modeling of acoustic sources no longer allows the evaluation of an instantaneous acoustic quantity. Only the second-order statistical acoustic quantities, such as the acoustic intensity, become accessible through mathematical treatment. Therefore, Ribner [RIB 64] proposed a complete approach to express the acoustic intensity radiated per unit source volume as a function of the statistically averaged quantities of the turbulent flow assumed as known, by adopting a description based on homogeneous and isotropic turbulence. Its model allows accessing the overall radiated acoustic intensity and its spectrum, by modeling a number of local quantities (turbulence characteristic frequency, turbulence integral length scale, large structures convection velocity, etc.) from the statistically averaged quantities provided by $k - \epsilon$-type turbulence computational codes. Later, other models, such as Goldstein and Rosenbaum's [GOL 73b], based on the axisymmetric description of turbulence, which involves the usage of two integral turbulence scales, have contributed to refine the method. Such models, as we have seen in Chapter 9, are now able to satisfactorily reproduce the main characteristics of the noise emission of academic turbulent flows (jets, mixing layers, etc.). In particular, the comparison of a cold jet and a heated jet (Figure 11.1) shows the increase in noise levels at low frequencies and the offset of the acoustic intensity spectrum toward lower frequencies when the mean exit temperature is increased.

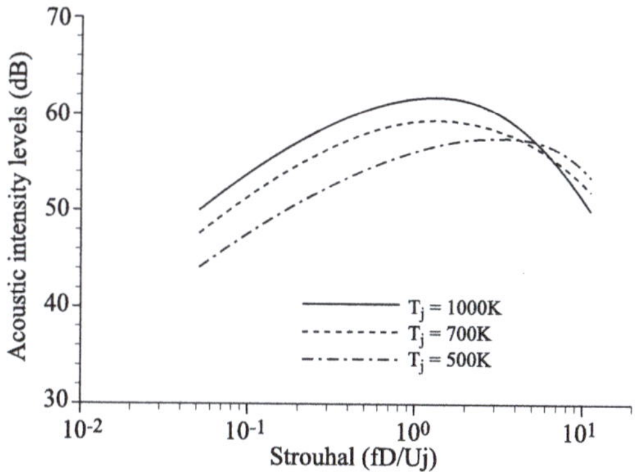

Figure 11.1. *Acoustic intensity spectra radiated at 90° with respect to the axis, for jets with different exit temperatures (according to [FOR 99])*

However, this type of model requires the adjustment of a number of parameters and a global calibration factor, obtained by comparison with experimental results. Such a degree of empiricism obviously severely limits the opportunities for adaptation of such models to even slight modifications in the flow pattern. This constitutes the main handicap of this type of approache (for example, it is now well known – [CHA 00] and [POP 00], for example – that the characteristics of flows with recirculation regions, such as flows on backward facing steps and flows on walls with a curvature or in the presence of pressure gradients, are poorly reproduced by $k - \epsilon$-type models). We have discussed in this regard, in section 9.6, the difficulties that exist in deriving, even for simple situations, general (if not universal) analytical formulations which allow the modeling of the source terms of the Lighthill equation from approximated formulae for the temporal and spatial correlations.

To conclude, it should be noted, to further establish a link with DNS/LES-based approaches that the stochastic noise generation and radiation (SNGR) model constitutes an alternative to access instantaneous quantities using the statistical data provided by a solution of the averaged equations. In this model, the synthetic turbulent field ([BAI 94] and [LAM 06], in particular) is rebuilt from statistical distributions and characteristic turbulence properties deduced from the Kolmogorov theory. The radiated acoustic field is then obtained from the linearized Euler equations and the instantaneous source terms evaluated from the synthetic turbulent field. In the case of a free jet, the acoustic results show a satisfactory qualitative agreement with the experimental measurements. Note that this method has also been applied to the calculation of sound radiation of confined flows and that the numerical estimations present a good agreement with the experimental results.

11.2. Direct numerical simulations/large eddy simulations

The first domain in which the significance of DNS/LES is particularly evident is the generation of acoustic radiation. Thus, if the experimental studies such as those we have mentioned at the beginning of this chapter have made it possible to acknowledge that, as early as the 1970s, the coherent structures are largely responsible for the generation of noise, only the DNS/LES have contributed to analyze this in a very fine manner (resulting from the implementation of specific numerical techniques – see, for example, [TAM 95b], [WEL 97] or [COL 04], required in particular to correctly transport into far-field pressure fluctuations of very low level compared with the velocity fluctuations present in near-field). These techniques are also required to allow the sound waves to escape the computation domain (without being reflected on its boundaries). This has been achieved notably by Colonius *et al.* [COL 97] for a mixing layer, by Bastin *et al.* [BAS 97] for a jet plane, by Mitchell *et al.* [MIT 99] for an axisymmetric jet or by Hulshoff *et al.* [HUL 01] for flows in highly variable section pipes (which as a matter of fact provides an overview of the upstream part of a jet nozzle). All these works have particularly helped to show that the pairing mechanism of the primary vortices (particularly stable and clearly identifiable in these cases) largely contributes to the noise radiated by this type of flow. Nevertheless, these works have also shown that this mechanism alone cannot explain the far-field radiated noise.

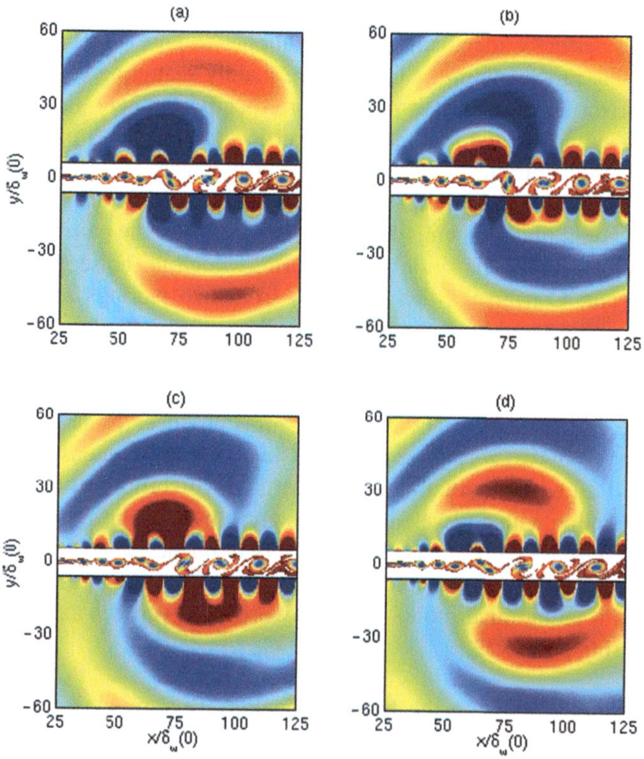

Figure 11.2. *Four representations of the pairing region in a mixing layer excited at f_0 and $f_0/2$ (where f_0 is its fundamental frequency, and $\delta_\omega(0)$ its original vorticity thickness). The vorticity field is represented in the center and the dilatation field (div$\vec{u}$) elsewhere (the dilatation field is directly related to the sound waves that propagate); the levels are ranging from -3 to +3 s^{-1}. The time interval between two successive figures is equal to $T_p/4$. The cycle (a)-(b)-(c)-(d)-(a) therefore describes a pairing complete period T_p (according to [BOG 00])*

The example we have chosen to illustrate this point was obtained by Bogey [BOG 00]. This author has simulated, with increasing degree of complexity, the pairing of two isolated vortices, then a two-dimensional mixing layer and finally an axisymmetric jet. For each of these situations, he has then solved (it is thus referred to as a hybrid method) the equation for the propagation of acoustic waves until the far-field, from the velocity field that the DNS (1st test case) (in near-field) and the LES (two other test cases) have provided him. The first test case has mainly been employed in the validation of the transition from near-field computation to far-field computation. The case of the mixing layer and in particular when it is excited in order to freeze the axial position where the pairing of vortices is situated (which are all perfectly identical) really highlights (Figure 11.2) the fact that the sound waves that propagate in the far-field are mainly scattering off from the region where the first vortex pairing takes place. More specifically, here, the duration of the acoustic emission associated with a pair of vortices is equal to a pairing period T_p. Thus, the wavefronts produced by successive pairing match perfectly together. It can be observed in particular that there is no discontinuity in the acoustic field, because a discontinuity would imply an emission period smaller than the pairing period, that is to say the existence of a transition period during which no noise generation would any longer occur. Conversely, the radiations from two pairings may not interfere because interference would require two simultaneous sources of radiating noise, with a scattering off duration greater than the pairing period. The absence of interference is also related to the displacement of the acoustic sources that only occurs over a very small distance during pairing. The two wave fronts produced by pairing furthermore appear to come from two very close positions. In addition, the acoustic wavelength is directly associated with the rotation velocity of the two vortices. In order to produce an acoustic radiation with a wavelength that is given by the pairing frequency fp, the rotation velocity must correspond to half of this frequency due to the symmetry of the quadrupole source (indeed, the author analytically calculates the sound field produced by a rotating quadripole). In other words, the vortices achieve a half spin during the pairing period T_p.

In this example, applying the Lighthill analogy (Figure 11.3) gives very good results but only when using, for the computation of the source term, the total velocity fields and not only the velocity fluctuations. Otherwise, when considering only the fluctuations of the velocity field to calculate the source terms of the Lighthill equation, as reported by Ribner in the case of the jet (see section 10.1, Figures 10.3 and 10.4), the convection and the refraction effects by the mean velocity field (and its shear) are not taken into account (in particular, the wavefronts predicted with the fluctuating velocities only are circular whereas they should not be). Furthermore, this work actually shows that the formulation using the time derivatives, and that using the space derivatives give quite similar results in the very far-field while in the less distant field, only the formulation in terms of the spatial derivatives is correct (in comparison with the results of the direct computation), as Lilley had correctly pointed out [LIL 93] (see Chapter 9, in particular).

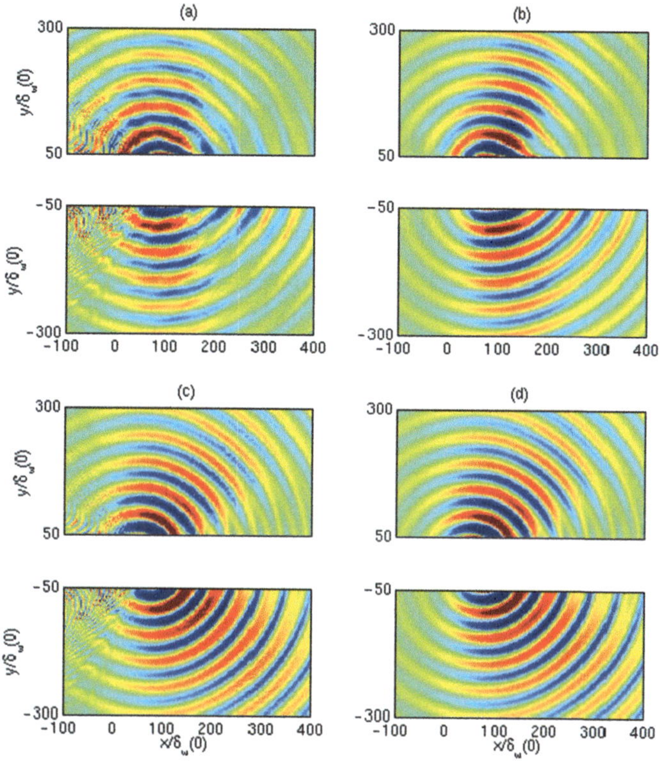

Figure 11.3. *Representation of the pressure field in the excited mixing layer calculated at the same instant by the Lighthil analogy: (a) and (b) source terms based on the Lightill tensor involving the total velocities (mean + fluctuating) and (c) and (d) source terms only involving fluctuating velocities. For (a) and (c), the formulation with spatial derivatives has been used while for (b) and (d), the formulation with the temporal derivatives has been used (according to [BOG 00])*

By simulating afterward, using a three-dimensional code, an axisymmetric turbulent jet which can develop either in a natural manner or by being excited at its preferential frequency f_0 (such that the associated Strouhal number is $St = 0.34$) and its harmonic $f_0/2$, Bogey [BOG 00] shows that the pairing of purely axisymmetric vortices does not constitute the predominant mechanism of sound generation in jets. Even though, when they occur at a frequency comprised within the privileged frequency range of the acoustic radiation, here at $St = 0.34$, the sound field located downstream is similar in terms of level to the field obtained for a natural jet. However, regarding the directionality, although the sound field produced by the pairing of the excited annular vortices is typical of the axisymmetric quadrupole sources, with an extinction angle close to $80°$, this is not found in the case of the jet presenting a natural aerodynamic development. In the case of angles smaller than $60°$, the two directivity curves are similar and show the same maximum around $40°$. This agreement can be explained by the refraction effects that are identical in both cases (the levels calculated for the jet excited at f_0 and $f_0/2$ are here only 2 dB lower than the levels of the randomly excited jet). On the contrary, for angles greater than $60°$, the two directivities are markedly different, with levels that are of the order of 6 dB stronger in the case of the jet developing naturally.

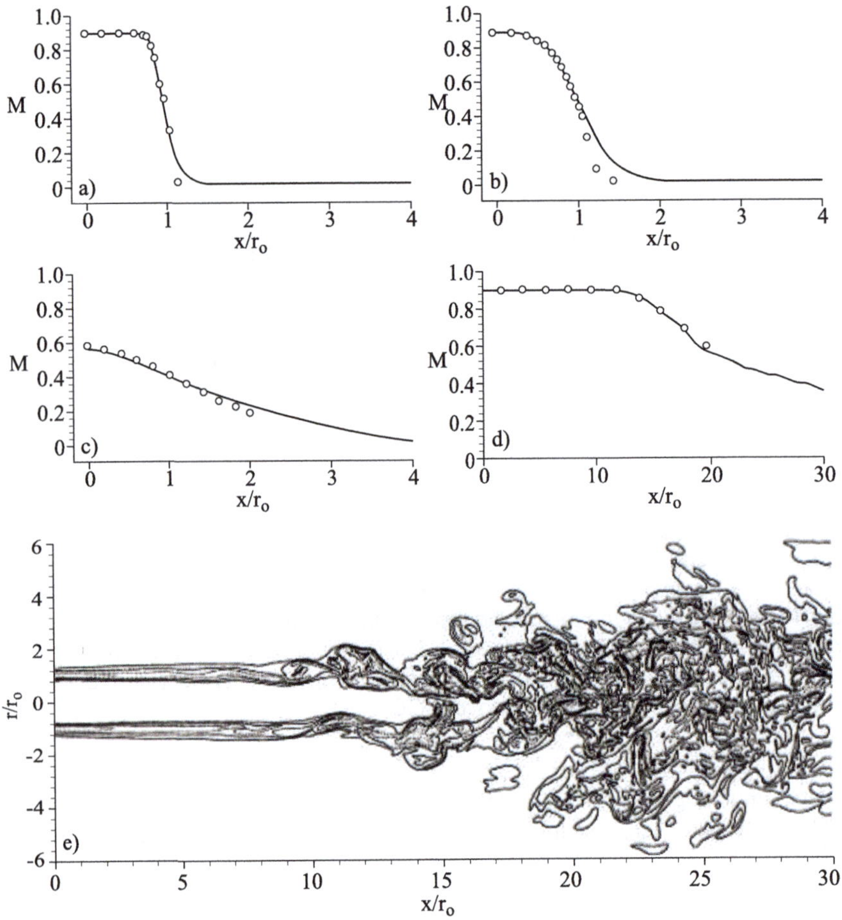

Figure 11.4. *Distribution of the mean velocities a) and iso-vorticity contours b) in the DNS of the jet at Mach 0.9 ($Re = 3600$) which has been studied by Freund [FRE 01]*

This seems to indicate that sound radiation of jets is indeed produced by the pairing of vortical structures, at the end of the potential cone, when velocity fluctuations are characterized by Strouhal numbers 0.2 < St < 1.5. Extending this study with the study of the noise of annular vortex pairing occurring for higher Strouhal numbers, in order to bring forward the evolution law of the sound level relatively to the frequency associated with vortices, would probably be very interesting.

DNS are also a very powerful tool to validate the analytical models discussed in Chapter 9, the experimental data in this field appearing to be practically non-existent. Zhou *et al.* [ZHO 94] report a few two-point correlation measurements, in addition to the works that we have already mentioned in Chapter 9, which mainly concern the noise radiated by jets of different characteristics, wall pressure fluctuations and the analysis of the hypothesis of *sweeping decorrelation*. This was done, in particular, by Sarkar and Hussain [SAR 93], Witkowska *et al.* [WIT 95] and Seror *et al.* [SER 01], for homogeneous and isotropic turbulence in a cubic box. These works have especially focused on validating the relations proposed by Proudman and the Lilley, showing, for example, that the Proudman constant α has a value comprised between 2 and 3, which corresponds to the predictions of Lilley's formula. They have also shown that the acoustic spectrum reaches its maximum for frequencies significantly greater than that corresponding to the maximum of the turbulent energy spectrum, the vortices that make the major contribution to noise generation being those of sizes comprised between the size of the most energetic scales and the size of the scales that most contribute to the vorticity. However, it is quite difficult to give a precise value (that would not depend on the Reynolds number nor on the forcing conditions at the boundaries of the computational domain) to the Strouhal number S_T which is included in the formula proposed by Lilley (see p. 386). Nevertheless, it seems that the formulation of the correlations suggested by Proudman, which leads to a spectrum in $\omega^{-7/3}$ in the regime of the frequencies higher than that corresponding to the maximum of the spectrum, is in much better agreement with the data originated from DNS than the formulation proposed by Lilley (which leads to a spectrum in $\omega^4(1+2\omega^2)^{-3}$).

To complement what has just been said, we present hereafter a few typical results obtained by Freund [FRE 01] from DNS of a jet flow initially at Mach 0.9; for a nozzle Reynolds number equal to 3,600. These results (Figures 11.4, 11.5, 11.6 and 11.7) illustrate the fact that overall, with a relatively high Reynolds number, it happens that it is in the turbulent events present in the peripheral region of the jet (that where shears, and therefore the production of turbulence kinetic energy, are the strongest) that noise sources are mainly located even if there is no perfect correlation between the highest levels of the turbulent energy and the most intense sources. These sources are advected (Figure 11.8) at a velocity corresponding to about Mach 0.5, irrespective of the longitudinal position x being considered.

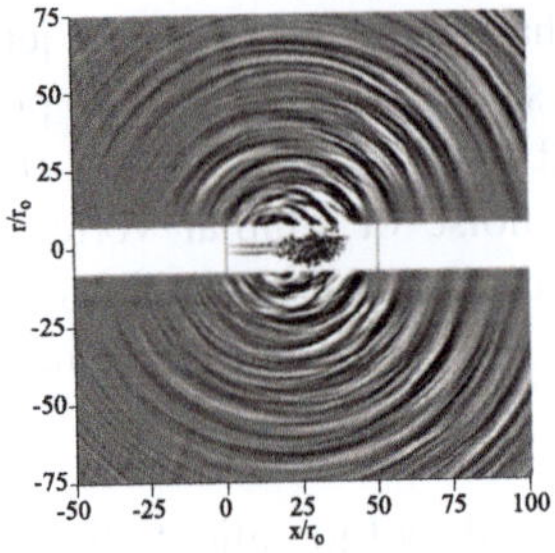

Figure 11.5. *Representation of the noise radiated in far-field (grayscale between black, div$\vec{u} < -0.0005c_0/r_0$ and white, div$\vec{u} > 0.0005c_0/r_0$) (according to [FRE 01])*

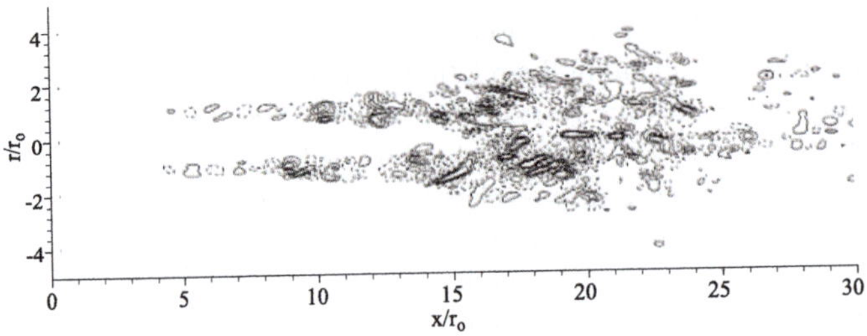

Figure 11.6. *Localization of the noise sources S resulting from the Lighthill tensor (according to [FRE 01]). The sources are such that:*

$$\frac{\partial^2 p}{\partial t^2} - c_0^2 \frac{\partial^2 p}{\partial x_i \partial x_i} = S, \text{ with } S = c_0^2 \left[\frac{\partial^2 \rho u_i u_j}{\partial x_i \partial x_j} - \frac{\partial^2 \tau_{ij}}{\partial x_i \partial x_j} \right] + \frac{\partial^2}{\partial t^2}(p - \rho c_0^2),$$

$$\text{and } \widehat{S}_n = \frac{\partial^2 \widehat{p}_n}{\partial r^2} + \frac{1}{r}\frac{\partial \widehat{p}_n}{\partial r} + \left[\omega^2 - k^2 - \frac{n^2}{r^2} \right] \widehat{p}_n, \text{ where}$$

$$\widehat{p}_n(k, r, \omega) = \sum_{j=0}^{N_\theta - 1} \int_{-\infty}^{\infty} \int_{-\infty}^{\infty} p(x, r, \theta_j, t) w(t) e^{i\omega t} e^{ikx} e^{in\theta_j} dt dx$$

In addition to these results, Freund [FRE 03] has performed a very detailed analysis of the source terms for the same jet flow. Figures 11.9 and 11.10 illustrate this, respectively, for the directivity effect of the various source terms involved in the Lighthill equation and for the frequency distribution of these terms. It appears, on the one hand, that the results are consistent with Ribner's theoretical developments for the directivity effect (Figure 11.9), with in particular, a pronounced extinction of the *shear noise* term (the one that involves mean velocities) for an angle close to 90° and a generally very small contribution of the terms of entropic origin except for angles close to 0° (i.e. along the longitudinal axis). These global trends are also found in spectral distributions (Figure 11.10), which illustrate the important directional effect that exists for the *shear noise*. This effect affects both the global levels and the position of the frequency peak associated with each angle. The term that involves only the velocity fluctuations (*self-noise*) is instead not particularly directionally pronounced, both on the overall levels and on the position of the frequency peak associated with each angle.

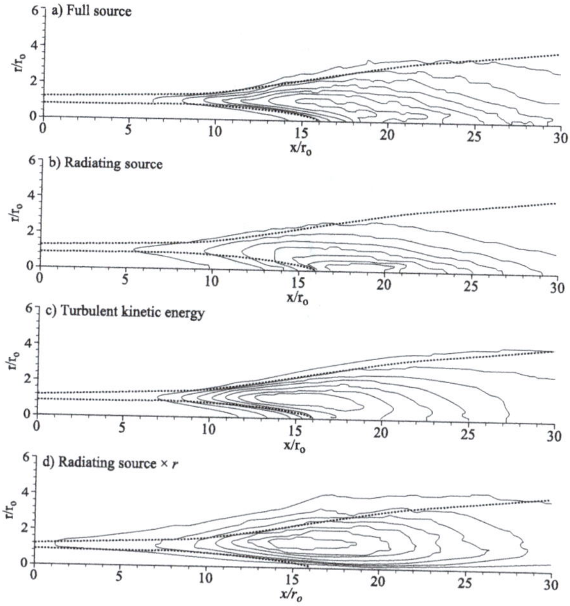

Figure 11.7. *Iso-levels of the RMS values of the noise sources and comparison with the turbulence kinetic energy field. The radiating sources are such that $\omega > kc_0$ (according to [FRE 01])*

Other works that also appear as very original and interesting are Whitmire and Sarkar's [WHI 00]. These authors have focused on a rather special situation, since it concerns a homogeneous and isotropic turbulence volume – in cross-section, according to the axes x and z, of $2\pi \times 2\pi$ and with a length along the y axis equal to 12.8 – which is inserted in a parallelepiped domain with a cross-section equal to $2\pi \times 2\pi$ such that the acoustic waves can propagate only along the large length of this domain (the complete computational domain has a total length along the y-axis of 68.2). These authors show analytically that, in this situation, the Lightill source terms correspond to dipole sources (taking into account the specific form that then takes the associated Green function, since the sound waves that propagate are then plane waves, and they are calculating waves which are averaged over each section perpendicular to the propagation axis (y)):

$$G(y, \xi) = \frac{1}{2ik} \exp(ik\,|y - \xi|).$$

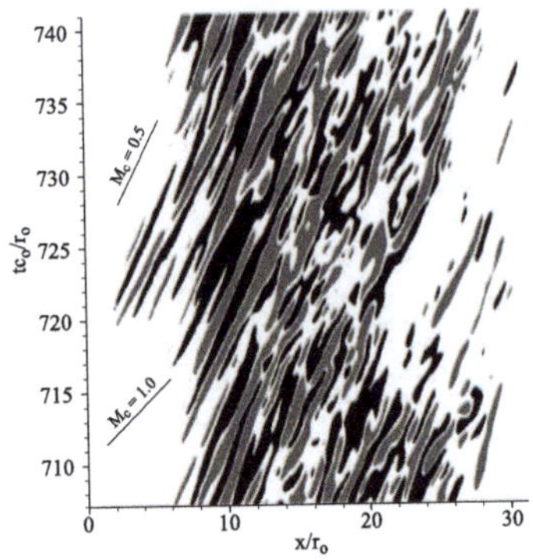

Figure 11.8. *Spatio-temporal diagram characterizing the advection of the sources (according to [FRE 01])*

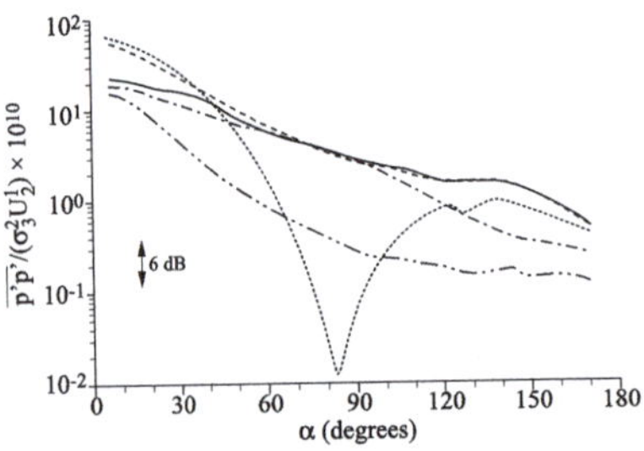

Figure 11.9. *Directivity diagram of the noise scattered off by a jet, according to the nature of the sources, at 120 diameters from the nozzle. ———, full source term, ········, source term involving the mean velocities, — · — · —, source term only involving velocity fluctuations, — · · — · · —, source term of entropy origin alone ($(p' - c_0^2\rho')\delta_{ij}$), — — —, full source term minus the source term of entropy origin (according to [FRE 03])*

This expression is here given relatively to the wave number $k(= \omega/c_0)$ since it corresponds to the Lightill equation written in Fourier space:

$$k^2 \widehat{\langle \rho \rangle'} + \frac{d^2 \widehat{\langle \rho \rangle'}}{dy^2} = -\frac{\widehat{\langle T \rangle'}}{c_0^2},$$

where $\widehat{\langle \rho \rangle'}$ represents the Fourier transform of the acoustic fluctuation of ρ averaged in a plane and $\widehat{\langle T \rangle'} = \rho_0 \langle (\partial^2 (\widehat{u'_i u'_j})) / \partial x_i \partial x_j \rangle)$. In particular, the acoustic power then radiated should vary as the 6th power of the velocity (as opposed to the 8th power characteristic of quadrupole sources). The principal question investigated by these authors is: what does the Lighthill analogy become in this case? Is it still valid or not? This study has seemed to be original, but above all, fairly closely related to situations found in the transportation field since the presence of the side walls in vehicles should guide the propagation of sound waves.

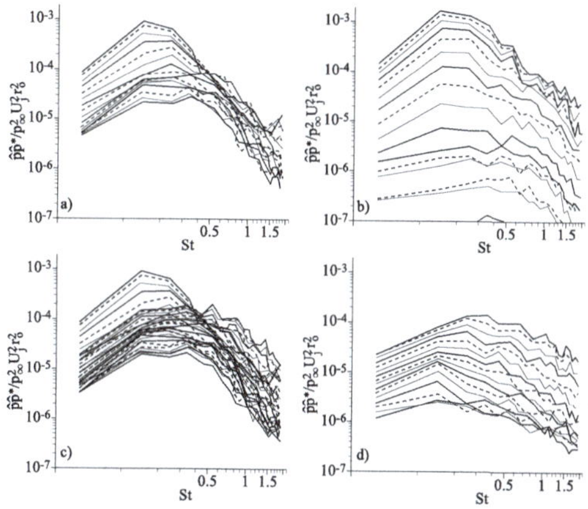

Figure 11.10. *Spectral distribution of noise scattered off by a jet, according to the nature of the sources, at 120 nozzle diameters for the full source term (a), the source term involving the mean velocities (b), the source term only involving the velocity fluctuations (c) and the source term of entropy origin alone (d). The curves correspond to angles ranging from $18°$ (upper curves) up to $102°$ (bottom curves) in steps of $6°$ (according to [FRE 03])*

Figure 11.11 (which represents only one half of the computational domain along the z-axis) illustrates the properties related to sound propagation in such a situation. In particular, in the absence of organized large-scale coherent structures and because of the guidance by the boundaries of the computational domain (on which symmetry conditions are being imposed), the sound waves present shapes quite significantly different from those obtained in the excited mixing layer. However, in this situation, the Lighthill analogy also gives satisfactory results, even if the source terms related to the turbulent velocity field are then of the dipole type and not quadrupole.

11.3. Conclusion

The results presented in Chapters 9, 10, 11 clearly illustrate the high complexity of the phenomena involved in aeroacoustics. They show in particular that, on the one hand, the detail of the physical mechanisms responsible for the generation of aeroacoustic noise is not yet well understood – even if coherent large-scale structures play a very important role therein – and that, on the other hand, there is to date no satisfactory analytical formulation – which seems quite logical since concepts such as that of *sweeping decorrelation* are implied when addressing fourth-order velocity correlations with space and time separations. These concepts have not yet found a definitive solution even from the point of view of the basic studies in homogeneous and isotropic turbulence. This reflects the fact that the noise radiated by turbulent flows just originates equally both from the statistical nature and the organized nature (in large-scale vortices) of turbulence. This is thus cruelly highlighting our still significant weaknesses in correctly modeling the different aspects of turbulence.

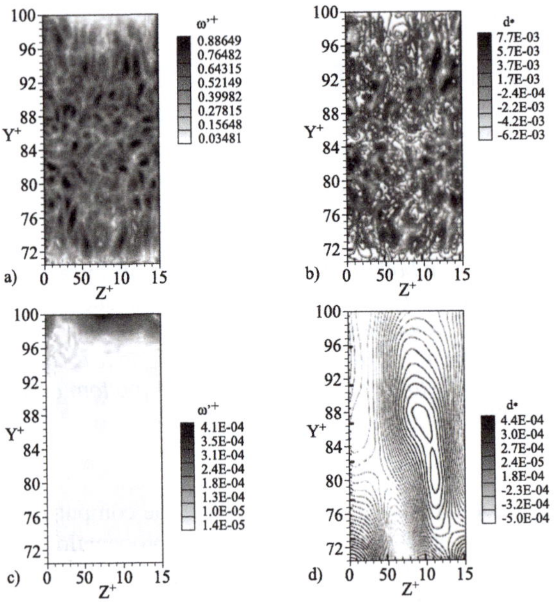

Figure 11.11. *Vorticity and dilatation fields in the turbulence volume (a) and (b), respectively, and in the far-field (c) and (d). The different quantities are made dimensionless by means of the turbulence integral scale and the speed of sound (according to [WHI 00])*

In less academic situations involving turbulence and especially in the case of applications in the transportation field, it is clear that the presence of more or less properly identifiable coherent structures and that of significant shear on the edge of wakes originating from, for example, the mirrors of a car may make the analytical formulations presented in section 9 unrealistic. In addition, the concept of far-field (and, as a result, the validity of the approximations resulting therefrom) appears as unrealistic and this may also become an important cause of errors in the estimations of overall noise levels – without even mentioning the predictions of the directivity of the radiated acoustic waves. Finally, considering the fluid-structure coupling that appears nowadays absolutely unavoidable in many cases, whether in the automotive field (phenomena related to aeroelasticity, in which the vibration-type motion must instead be considered) or in the aeronautical field (where in contrast the large deformations of airplane wings or helicopter blades should be taken into account), all of this will give increasingly more importance to numerical simulation methods and it is, therefore, in this area that most progress and advances are expected in the coming years.

As a result, this should nevertheless encourage the utmost caution when using commercial numerical modeling codes (such as Fluent and Star CCM+, or those developed by research organisms such as the ONERA and the CEA in France) that allow, easily and without requiring any special knowledge, to generate numerous and detailed results but whose representativity in relation to the physical reality is far from being guaranteed.

Bibliography

Books and documents on acoustics, vibrations and aeroacoustics

[ABR 65] ABRAMOWITZ M., STEGUN I.A., *Handbook of Mathematical Functions*, Dover, New York, 1965.

[ACH 82] ACHENBACH J.D., GAUTESEN A.K., MCMAKEN H., *Ray Methods for Waves in Elastic Solids: with Applications to Scattering by Cracks*, Pitman Advanced Pub. Program, 1982.

[APP 02] APPEL W., *Mathématiques pour la physique*, H & K Éditions, Paris, 2002.

[ARV 00] ARVESON W., *A Short Course on Spectral Theory*, Springer, New York, 2000.

[AST 10] ASTM INTERNATIONAL, Standard test method for measuring vibration-damping properties of material, available at www.astm.org.Standards/E756.htm, 2010.

[BEN 91] BENDER C.M., ORSZAG S., *Advanced Mathematical Methods for Scientists and Engineers I*, McGraw-Hill, New York, 1991.

[BER 91] BERNADI C., MADAY Y., *Approximations spectrales de problèmes aux limites elliptiques*, Springer, Paris, 1991.

[BER 93] BERANEK L.L., *Acoustics Measurements, Revised Edition*, Acoustical Society of America, Woodbury, 1993.

[BLA 86] BLAKE W.K., *Mechanics of Flow-Induced Sound, Vibration, Vol. 1: General Concepts, Elementary Sources. Vol. 2: Complex Flow-Structure Interactions*, Academic Press, NewYork, 1986.

[BLE 90a] BLEVINS R.D., *Flow-Induced Vibration*, Krieger Publishing Company, Florida, 1990.

[BOU 01] BOUTET DE MONVEL L., Cours de Maîtrise: Introductions aux dérivées partielles, Institut de Mathématiques de Jussieu, Université Pierre et Marie Curie, 2001.

[BRE 90] BREKHOVSKIKH L.M., GODIN O.A., *Acoustics of Layered Media I : Plane, Quasi-Plane Wave, II : Point Sources, Bounded Beams)*, Springer, Berlin, 1990.

[BRI 46] BRILLOUIN L., *Wave Propagation in Periodic Structures*, McGraw-Hill, New York, 1946.

[BRU 83] BRUNEAU M., *Introduction aux Théories de l'Acoustique*, Université du Maine Editeur, 1983.

[CHA 13] CHAIGNE A., KERGOMARD J., *Acoustique des instruments de musique*, seconde édition revue et augmentée, Belin, Paris, 2013.

[CHL 09] CHLADNI E.F.F., *Traité d'Acoustique*, Chez Coucier, Paris, 1809.

[CHR 82] CHRISTENSEN R.M., *Theory of Viscoelasticity, an Introduction*, Academic Press, New York, 1982.

[CRE 05] CREMER L., HECKL M., PETERSSON B.A.T., *Structure-Borne Sound,* 3rd ed., Springer, Berlin, 2005.

[CRI 92] CRIGHTON D.G., DOWLING A.P., FFOWCS WILLIAMS J.E. *et al.*, *Modern Method in Analytical Acoustics*, Springer, London, 1992.

[CRO 98] CROCKER M.J., *Handbook of Acoustics*, Wiley Interscience, New York, 1998.

[COU 53] COURANT R., HILBERT D., *Methods of Mathematical Physics, Volumes 1 & 2*, John Wiley & Sons, New York, 1953.

[DRI 73] DRION CH., FERNET E., *Traité de physique élémentaire*, 4th ed., Masson, Paris, 1873.

[EGO 92] EGOROV Y.V., SHUBIN M.A. (eds.), *Foundations of the Classical Theory of Partial Differential Equations*, (Encyclopaedia of Mathematical Sciences, 30), Springer, Berlin, 1992.

[FAH 07] FAHY F., GARDONIO P., *Sound and Structural Vibration,* 2nd ed., Academic Press, Oxford, 2007.

[FAS 07] FASTL H., ZWICKER E., *Psycho-acoustics, Fact and Models,* 3rd ed., Springer, Berlin, 2007.

[FEY 79] FEYNMAN R.P. , LEIGHTON R.B., SANDS M., *Le cours de Physique de Feynman, T. 1 (Mécanique) et 2 (Electromagnétisme)*, InterEditions, Paris, 1979.

[FIL 94] FILIPPI P.J.T., BERGASSOLI A., GUYADER J.-L. *et al.*, *Acoustique Générale*, Les Editions de Physique, Les Ulis, 1994.

[FIL 99] FILIPPI P.J.T., BERGASSOLI A., HABAULT D. *et al.*, *Acoustics: Basic Physics Theory & Methods*, Academic Press, San-Diego, 1999.

[FIL 08] FILIPPI P.J.T., *Vibrations et vibro-acoustique des structures minces*, Hermes Lavoisier, Paris, 2008.

[FLE 91] FLETCHER N.H., ROSSING T.D., *The Physics of Musical Instruments*, Springer, New York, 1991.

[FLU 73] FLÜGGE W., *Stresses in Shells*, Springer, Berlin, 1973.

[GER 93] GERADIN M., RIXEN D., *Théorie des vibrations*, Masson, Paris, 1993.

[GES 87] GESPA N., *La diffusion acoustique par des cibles élastiques de forme géométrique simple, Théories et expériences*, Publication du CEDOCAR, 1987.

[GOL 76] GOLDSTEIN M.E., *Aeroacoustics*, McGraw-Hill, New York, 1976.

[GUY 02] GUYADER J.-L., *Vibrations des milieux continus*, Hermes Lavoisier, Paris, 2002.

[HAB 99] HABAULT D. (ed.), *Fluid-Structure Interaction in Acoustics*, CISM Course, Lectures, Springer, New York, 1999.

[HAR 93] HARDIN J.C., HUSSAINI M.Y. (eds), *Computational Aeroacoustics*, Springer, New York, 1993.

[HEL 68] HELMHOLTZ H., *Théorie physiologique de la musique*, Masson, Paris, 1868.

[HOW 03] HOWE M.S., *Theory of Vortex Sound*, Cambridge Texts in Applied Mathematics, Cambridge University Press, 2003.

[HUB 94] HUBBARD H.H. (ed.), *Aeroacoustics of Flight Vehicles, Theory, Practice. Volume 1: Noise Sources; Volume 2: Noise Control*, Acoustical Society of America, Woodbury, 1994.

[JUN 86] JUNGER M.C., FEIT D., *Sound, Structure, their Vibration,* 2nd ed., MIT Press, Cambridge, 1986.

[KRE 93] KREYSZIG E., *Advanced Engineering Mathematics,* 7th ed., John Wiley & Sons, New York, 1993.

[LAN 89a] LANDAU L.D., LIFSHITZ E.M., *Physique Théorique, tome 6. Mécanique des Fluides*, Mir, Moscow, 1989.

[LAN 89b] LANDAU L.D., LIFSHITZ E.M., *Physique Théorique, tome 7. Théorie de l'Elasticité*, Mir, Moscow, 1989.

[LAN 02] LANCASTER P., *Lambda-matrices, Vibrating Systems,* 2nd ed., Dover, New York, 2002.

[LEI 93a] LEISSA A.W., *Vibration of Plates*, Acoustical Society of America, (originally issued by NASA, 1969), Woodbury, 1993.

[LEI 93b] LEISSA A.W., *Vibration of Shells*, Acoustical Society of America, (originally issued by NASA, 1973), Woodbury, 1993.

[LES 88] LESUEUR C., *Rayonnement acoustique des structures*, Éditions Eyrolles, Paris, 1988.

[LEV 80] LEVIN B.Y., *Distribution of Zeros of Entire Functions, Revised Edition*, American Mathematical Society, Providence, 1980.

[LEV 96] LEVIN B.Y., *Lectures on Entire Functions*, American Mathematical Society, Providence, 1996.

[LEW 01] LEWY S., *Acoustique industrielle et aéroacoustique*, Hermes, Paris, 2001.

[LOV 44] LOVE A.E.H., *A Treatise on the Mathematical Theory of Elasticity*, Dover, New York, 1944.

[MOR 48] MORSE P.M., *Vibration and Sound*, McGraw-Hill, New York, 1948.

[MOR 53] MORSE P.M., FESHBACH H., *Method of Theoretical Physics vol. I & vol. II.*, McGraw-Hill, New York, 1953.

[MOR 68] MORSE P.M., INGARD K.U., *Theoretical Acoustics*, McGraw-Hill, New York, 1968.

[NAS 85] NASHIF A.D., JONES D.I.G., HENDERSON J.P., *Vibration Damping*, John Wiley & Sons, New York, 1985.

[NAY 73] NAYFEH A.H., *Perturbation Methods*, Wiley-Interscience, New York, 1973.

[NAY 93] NAYFEH A.H., *Introduction to Perturbation Techniques*, Wiley Classics Library, New York, 1993.

[PAR 84] PARTON V., PERLINE P., *Méthodes de la Théorie Mathématique de l'Élasticité, tome 1 & 2*, Mir, Moscow, 1984.

[PER 60] PERSOZ B., *Introduction à l'étude de la rhéologie*, Dunod, Paris, 1960.

[PIE 81] PIERCE A.D., *Acoustics. An Introduction to its Physical Principle and Applications*, McGraw-Hill, New York, 1981.

[PRE 92] PRESS W.H., TEUKOLSKY S.A., VETTERLING W.T. *et al.*, *Numerical Recipes. The Art of Scientific Computing,* 2nd ed., Cambridge University Press, Cambridge, 1992.

[ROD 78] RODDIER F., *Distributions et transformation de Fourier*, McGraw-Hill, Paris, 1978.

[ROS 07] ROSSING T.D. (ed.), *Springer Handbook of Acoustics*, Springer, New York, 2007.

[SAK 65] SAKS S.,ZYGMUND A., *Analytic Functions*, Paristwowe Wydawnictwo Nankowe, 1965.

[SAL 09] SALENÇON J., *Viscoélasticité pour le calcul des structures*, Editions de l'Ecole Polytehnique, Palaiseau, 2009.

[SAN 89] SANCHEZ-HUBERT J., SANCHEZ-PALENCIA E., *Vibration, Coupling of Continuous Media - Asymptotic Methods*, Springer, Berlin, 1989.

[SCH 64] SCHWARTZ L., *Méthodes Mathématiques pour les Sciences Physiques*, Hermann, Paris, 1964.

[STR 45] STRUTT J.W., RAYLEIGH L., *The Theory of Sound*, Dover, New York, 1945.

[TIM 40] TIMOSHENKO S., *Theory of Plates, Shells*, McGraw-Hill, New York, 1940.

[WAG 07] WAGNER C., HUTTL T., SAGAUT P., (eds), *Large-eddy Simulation for Acoustics*, Cambridge University Press, Cambridge, 2007.

[WAT 44] WATSON G.N., *A Treatise on the Theory of Bessel Functions,* 2nd ed., Cambridge Mathematical Library, Cambridge, 1944.

[ZEN 48] ZENER C., *Elasticity and Anelasticity of Metals*, University of Chicago Press, 1948.

Books on turbulence and problems linked to its modeling

[BOR 00] BORGHI R., CHAMPION M., *Modélisation et théorie des flammes*, Editions Technip, Paris, 2000.

[CHA 00] CHASSAING P., *Turbulence en mécanique des fluides*, Editions Cépaduès, Toulouse, 2000.

[COU 89] COUSTEIX J., *Turbulence et couche limite*, Editions Cépaduès, Toulouse, 1989.

[DEL 05] DELMAS R., MÉGIE G., PEUCH V.H., *Physique et chimie de l'atmosphère*, Belin, Paris, 2005.

[DEL 08] DÉLERY J., *Traité d'aérodynamique compressible, Ouvrage en 3 volumes*, Hermes Lavoisier, Paris, 2008.

[FRI 95] FRISCH U., *Turbulence: the Legacy of A.N. Kolmogorov*, Cambridge University Press, Cambridge, 1995.

[MON 75] MONIN A.S., YAGLOM A.M., *Statistical Fluid Mechanics*, MIT Press, Cambridge, 1975.

[PAN 84] PANOFSKY H.A., DUTTON J.A., *Atmospheric Turbulence*, Wiley Interscience, New York,1984.

[PAR 99] PARASHIVOIU I., *Aérodynamique subsonique*, Presses de l'Ecole Polytechnique de Montréal, 1999.

[POP 00] POPE S.B., *Turbulent Flows*, Cambridge University Press, Cambridge, 2000.

[SCH 06] SCHIESTEL R., *Méthodes de modélisation et de simulation des écoulements turbulents*, Hermes Lavoisier, Paris, 2006.

[SEL 00] SELLIER A., *Introduction aux écoulements compressibles et aux fluides hétérogènes*, Editions de l'Ecole Polytechnique, Palaiseau, 2000.

[TAT 71] TATARSKI V.I., *The Effects of the Turbulence Atmosphere on Wave Propagation*, Jerusalem IPST Keter Press, 1971.

Recent theses in French (non-exhaustive list)

[ARG 06] ARGUILLAT B., Etude expérimentale et numérique de champs de pression pariétale dans l'espace des nombres d'onde, avec application aux vitrages automobiles, Thesis, Ecole Centrale de Lyon, 2006.

[AST 91] ASTOLFI A., Etude du champ convectif des fluctuations de pression pariétale et de ses relations avec le champ des fluctuations de vitesse sous une couche limite turbulente, Thesis, University of Aix-Marseille II, 1991.

[BAI 94] BAILLY C., Modélisation du rayonnement acoustique des écoulements turbulents libres subsoniques et supersoniques, Thesis, Ecole Centrale de Paris, 1994.

[BER 06] BERLAND J., Modélisation des erreurs numériques dans une simulation des grandes échelles et étude du screech dans un jet rectangulaire supersonique, Thesis, Ecole Centrale de Lyon, 2006.

[BOG 00] BOGEY C., Calcul direct du bruit aérodynamique et validation de modèles acoustiques hybrides, Thesis, Ecole Centrale de Lyon, 2000.

[DUR 99] DURANT C., Etude expérimentale de l'excitation et de la réponse vibroacoustique d'une conduite sollicitée par un écoulement interne, Thesis, Ecole Centrale de Lyon, 1999.

[FOR 00] FORTUNÉ V., Etude par simulation numérique directe du rayonnement acoustique de couches de mélange isothermes et anisothermes, Thesis, University of Poitiers, 2000.

[GAU 99] GAUVREAU B., Influence des conditions micrométéorologiques sur l'efficacité des écrans acoustiques, Thesis, University of du Maine, 1999.

[GIO 99] GIOVANNELLI G., Etude expérimentale et modélisation des fluctuations de pression pariétale aux petits nombres d'onde de l'écoulement turbulent sur de longs cylindres, Thesis, University of Aix-Marseille II, 1999.

[GLO 01] GLOERFELT X., Bruit rayonné par un écoulement affleurant une cavité: Simulation aéroacoustique directe et application de méthodes intégrales, Thesis, Ecole Centrale de Lyon, 2001.

[KER 04] KERHERVÉ F., Caractérisation statistique des sources aéroacoustiques en jets supersoniques par vélocimétrie laser à effet Doppler : application à la prédiction du bruit de jet, Thesis, University of Poitiers, 2004.

[LAM 06] LAMAISON G., Génération d'un champ turbulent stochastique à partir d'un modèle spectral avec application à l'aéroacoustique, Thesis, Ecole Centrale de Lyon, 2006.

[LAZ 97] LAZURE H., Relation entre la structure aérodynamique d'écoulements turbulents incompressibles et le bruit émis, Thesis, University of Poitiers, 1997.

[MIL 03] MILLET C., Rayonnement des ondes d'instabilités dans les jets supersoniques, Thesis, Ecole Nationale Supérieure de l'Aéronautique et de l'Espace, 2003.

[NOG 99] NOGER C., Contribution à l'étude des phénomènes aéroacoustiques se développant dans la "baignoire" et autour des pantographes du TGV, Thesis, University of Poitiers, 1999.

[PER 04] PÉROT F., Calcul du rayonnement acoustique d'écoulements turbulents basé sur des analogies acoustiques couplées aux simulations aérodynamiques instationnaires, Thesis, Ecole Centrale de Lyon, 2004.

[TRU 98] TRUFFAUT J.M., Etude expérimentale de l'origine du bruit émis pas les flammes de chalumeaux, Thesis, University of Aix-Marseille I, 1998.

[ZOG 06] ZOGHAIB L., Optimisation vibroacoustique de structures amorties par des traitements en élastomère, Thesis, University of Aix-Marseille II, 2006.

Articles and theses in English

[ADH 00] ADHIKARI S., Damping models for structural vibration, PhD Thesis, University of Cambridge, 2000.

[ANT 91] ANTONIA R.A.,KIM J., BROWNE L.W.B., "Some characteristics of small-scale turbulence in a turbulent duct flow", *J. Fluid Mech.*, vol. 233, pp. 369–388, 1991.

[ANT 99] ANTONIA R.A., BISSET D.K., ORLANDI P. *et al.*, "Reynolds number dependence of the second-order turbulent pressure structure function", *Physics of Fluids*, vol. 11, pp. 241–243, 1999..

[ATT 07] ATTENBOROUGH K., "Sound propagation in the atmosphere", *Springer Handbook of Acoustics*, ROSSING T.D. (ed.), Springer, New York, pp .113–147, 2007.

[BAI 96] BAILLY C., CANDEL S., LAFON P., "Prediction of supersonic jet noise from a statistical acoustic model, a compressible turbulence closure", *J. Sound and Vibration*, vol. 194, pp. 219–242, 1996.

[BAI 97] BAILLY C., LAFON P., CANDEL S., "Subsonic, supersonic jet noise predictions from statistical source models", *AIAA Journal*, vol. 35, pp. 1688–1699, 1997.

[BAS 97] BASTIN F., LAFON P., CANDEL S., "Computation of jet mixing noise due to coherent structures: the plane jet case", *J. Fluid Mech.*, vol. 335, pp. 261–304, 1997.

[BAT 51] BATCHELOR G.K., "Pressure fluctuations in isotropic turbulence", *Proc. Camb. Phil. Soc.*, vol. 47, pp. 359–374, 1951.

[BER 73] BERT C.W., "Material damping: an introductory review of mathematical models, measures, experimental techniques", *J. Sound, Vibration*, vol. 29, no. 2, pp. 129–153, 1973.

[BER 01] BERTHELOT Y.H., "Surface acoustic impedance, causality", *J. Acoust. Soc. Am.*, vol. 109, no. 4, pp. 1736–1739, 2001.

[BIS 71] BISHOP K.A., FFOWCS WILLIAMS J.E., SMITH W., "On the noise sources of the unsuppressed high-speed jet", *J. Fluid Mech.*, vol. 50, pp. 21–31, 1971.

[BLE 90b] BLEVINS R.D., "Damping of structures", in [BLE 90a], 1990.

[BRE 98] BRENTNER K.S., FARASSAT F., "An analytical comparison of the acoustic analogy and Kirchhoff formulation for moving surfaces", *AIAA Journal*, vol. 36, pp. 1379–1386, 1998.

[BRE 03] BRENTNER K.S., FARASSAT F., "Modeling aerodynamically generated sound of helicopter rotors", *Prog. Aerospace Sciences*, vol. 39, pp. 83–120, 2003.

[BRO 74] BROWN G.L., ROSHKO A., "On density effects, large structure in turbulent mixing layers", *J. Fluid Mech.*, vol. 64, pp. 75–816, 1974.

[BRU 94] BRUNEAU M., "Acoustique des cavités: modèles et applications", *Journal de Physique IV*, vol. 04, no. C5, pp. C5-675-C5-684, 1994.

[BUI 08] BUI T.P., SCHRODER W., MEINKE M., "Numerical analysis of the acoustic field of reacting flows via acoustic perturbation equations", *Computers and Fluids*, vol. 37, pp. 1157–1169, 2008.

[BUL 96] BULL M.K., "Wall-pressure fluctuations beneath turbulent boundary layers: some reflections on forty years of research", *J. Sound, Vibration*, vol. 190, pp. 299–315, 1996.

[CAR 82] CARGILL A.M., "The radiation of high frequency sound out of jet pipe", *J. of Sound, Vibration*, vol. 83, no. 3, pp. 313–337, 1982.

[CHA 80] CHASE D.M., "Modeling the wave-vector-frequency spectrum of turbulent boundary layer", *J. Sound, Vibration*, vol. 70, pp. 29–67, 1980.

[CHA 82] CHASE D.M., NOISEUX C.F., "Turbulent wall pressure low wavenumbers: relation to nonlinear sources in planar, cylindrical flow", *J. Acoust. Soc. Amer.*, vol. 72, pp. 975–982, 1982.

[CHA 01] CHAIGNE A., LAMBOURG CH., "Time-domain simulation of damped impacted plates. I. Theory, experiments", *J. Acoust. Soc. Amer.*, vol. 109, no. 4, pp. 1422–1432, 2001.

[CHE 89] CHEN S., KRAICHNAN R.H., "Sweeping decorrelation in isotropic turbulence", *Physics of Fluids*, vol. A1, pp. 2019–2024, 1989.

[CHU 66] CHU W.T., Turbulence measurements relevant to jet noise, UTIAS Report, 119, Institute for Aerospace Studies, University of Toronto, 1966.

[CLA 91] CLAVIN P., SIGGIA E.D., "Turbulent premixed flames, sound generation", *Combustion Science and Technology*, vol. 78, pp. 147–155, 1991.

[COL 97] COLONIUS T., LELE S.K., MOIN P., "Sound generation in a mixing layer", *J. Fluid Mech.*, vol. 330, pp. 375–409, 1997.

[COL 04] COLONIUS T., LELE S.K., "Computational aeroacoustics: progress on nonlinear problems of sound generation", *Prog. Aerospace Sciences*, vol. 40, pp. 345–416, 2004.

[CON 09] CONARD N.J., MALINA M., MÜÜNZEL S.C., "New flutes document the earliest musical tradition in southwestern Germany", *Nature*, vol. 460, pp. 737–740, 2009.

[COR 63] CORCOS G.M., "Resolution of pressure in turbulence", *J. Acoust. Soc. Amer.*, vol. 35, pp. 192–199, 1963.

[COR 64] CORCOS G.M., "The structure of the turbulent pressure field in boundary-layer flows", *J. Fluid Mech.*, vol. 18, pp. 353–378, 1964.

[CRA 70] CRANDALL S.W., "The role of damping in vibration theory", *J. Sound, Vibration*, vol. 11, no. 1, pp. 3–18, 1970.

[CRI 75] CRIGHTON D.G., "Basic principles of aerodynamic noise generation", *Prog. Aerospace Sciences*, vol. 16, pp. 31–96, 1975.

[CRI 85] CRIGHTON D.G., "The Kutta condition in unsteady flow", *Ann. Rev. Fluid Mech.*, vol. 17, pp. 411–445, 1985.

[CRO 71] CROW S.C., CHAMPAGNE F.H., "Orderly structure in jet turbulence", *J. Fluid Mech.*, vol. 48, pp. 547–591, 1971.

[CUR 55] CURLE N., "The influence of solid boundaries on aerodynamic sound", *Proc. Roy. Soc. Lond. A*, vol. 231, pp. 505–514, 1955.

[DAI 78] DAIGLE G.A., PIERCY J.E., EMBLETON T.F.W., "Effects of atmospheric turbulence on the interference of sound waves near a hard boundary", *J. Acoust. Soc. Am.*, vol. 64, pp. 622–630, 1978.

[DAV 63] DAVIES P.O.A.L., FISHER M.J., BARRATT M.J., "The characteristics of the turbulence in the mixing region of a round jet", *J. Fluid Mech.*, vol. 15, pp. 337–367, 1963.

[DEL 70] DELANY M.E., BAZLEY E.N., "Acoustical properties of fibrous absorbent materials", *Applied Acoustics*, vol. 3, pp. 105–116, 1970.

[DEL 02] DE LANGRES E., "Absolutely unstable waves in inviscid hydroelastic systems", *J. Sound, Vibration*, vol. 256, no. 2, pp. 299–317, 2002.

[DEM 37] DEMING A.F., "Noise from propellers with symmetrical sections at zero blade angle, I and II", *NACA TN 605–679*, pp. 1–31 and pp. 1–20, 1937 and 1938.

[DER 79] DEREM A., "Relation entre la formation des ondes de surface et l'apparition des résonances", *Revue du Cethedec*, vol. 58, pp. 43–79, 1979.

[DUR 00] DURANT CH., ROBERT G., FILIPPI P.J.T. *et al.* "Vibroacoustic response of a thin cylindrical shell excited by a turbulent internal flow: comparison between numerical prediction and experimentation", *J. Sound, Vibration*, vol. 229, pp. 1115–1155, 2000.

[EMB 96] EMBLETON T.F.W., "Tutorial on sound propagation outdoors", *J. Acoust. Soc. Am.*, vol. 100, pp. 31–48, 1996.

[ERN 36] ERNSTHAUSEN W., "The source of propeller noise", *Luftfahrtforschung*, vol. 13, pp. 433–440, 1936.

[FAR 88] FARASSAT F., MYERS M.K., "Extension of Khirchhoff formula to radiation from moving surfaces", *J. Sound, Vibration*, vol. 123, pp. 451–460, 1988.

[FAR 91] FARABEE T.M., CASARELLA M.J., "Spectral features of wall pressure fluctuations beneath turbulent boundary layers", *Physics of Fluids*, vol. A3, pp. 2410–2420, 1991.

[FAR 96a] FARASSAT F., The Kirchhoff formulas for moving surface in aeroacoustics – the subsonic, supersonic cases, NASA technical memorandum 110285, 1996.

[FAR 96b] FARASSAT F., MYERS M.K., "A study of wave propagation in a duct, mode radiation", *2nd AIAA/CEAS Aeroacoustics Conference AIAA paper 96-1677*, 1996.

[FAR 01] FARASSAT F., "Acoustic radiation from rotating blades - the Kirchhoff method in aeroacoustics", *J. Sound, Vibration*, vol. 239, pp. 785–800, 2001.

[FFO 65a] FFOWCS WILLIAMS J.E., GORDON C.G., "Noise of highly turbulent jets at low exhaust speeds", *AIAA J.*, vol. 3, pp. 791–793, 1965.

[FFO 65b] FFOWCS WILLIAMS J.E., MAIDANIK G., "The Mach wave field radiated by supersonic turbulent shear flows", *J. Fluid Mech.*, vol. 21, pp. 641–657, 1965.

[FFO 69] FFOWCS WILLIAMS J.E., HAWKINGS D.L., "Sound generation by turbulence, surfaces in arbitrary motion", *Phil. Trans. Roy. Soc. Lond. A*, vol. 264, pp. 321–342, 1969.

[FFO 77] FFOWCS WILLIAMS J.E., "Aeroacoustics", *Ann. Rev. Fluid Mech.*, vol. 9, pp. 447–468, 1977.

[FFO 96] FFOWCS WILLIAMS J.E., "Aeroacoustics", *J. Sound, Vibration*, vol. 190, pp. 387–398, 1996.

[FOR 99] FORTUNÉ V., GERVAIS Y., "Numerical investigation of the noise radiated from hot subsonic turbulent jets", *AIAA J.*, vol. 37, pp. 1055–1061, 1999.

[FRE 96] FREUND J.B., LELE S.K., MOIN P., "Calculation of the radiated sound field using an open Khirchhoff surface", *AIAA J.*, vol. 34, pp. 909–916, 1996.

[FRE 01] FREUND J.B., "Noise sources in a low-Reynolds-number turbulent jet at Mach 0.9", *J. Fluid Mech.*, vol. 438, pp. 277–305, 2001.

[FRE 03] FREUND J.B., "Noise-sources turbulence statistics, the noise from a Mach 0.9 jet", *Physics of Fluids*, vol. 15, pp. 1788–1799, 2003.

[GAU 02] GAUVREAU B., BÉRENGIER M., BLANC-BENON P. *et al.*, "Traffic noise prediction with the parabolic equation method: validation of a split-step-Padé approach in complex environments", *J. Acoust. Soc. Am.*, vol. 112, pp. 2680–2687, 2002.

[GOL 73a] GOLDSTEIN M.E., HOWES W.L., "New aspects of subsonic aerodynamic noise theory", *NASA TN D-7158*, 1973.

[GOL 73b] GOLDSTEIN M.E., ROSENBAUM B., "Effect of anisotropic turbulence on aerodynamic noise", *J. Acoust. Soc. Amer.*, vol. 54, pp. 630–645, 1973.

[GOL 84] GOLDSTEIN M.E., "Aeroacoustics of turbulent shear flows", *Ann. Rev. Fluid Mech.*, vol. 16, pp. 263–285, 1984.

[GOL 03] GOLDSTEIN M.E., "A generalized acoustic analogy", *J. Fluid Mech.*, vol. 488, pp. 315–333, 2003.

[GRA 97] GRAHAM W.R., "A comparison of models for the wavenumber-frequency spectrum of turbulent boundary layer pressures", *J. Sound, Vibration*, vol. 206, pp. 541–565, 1997.

[GUT 48] GUTIN L., "On the sound field of a rotating propeller", *NACA TM1195*, vol. 12, pp. 76–83, 1948.

[HAB 98] HABAULT D., FILIPPI P.J.T., "Light fluid approximation for sound radiation, diffraction by thin elastic plates", *J. Sound and Vibration.*, vol. 213, no. 2, pp. 333–374, 1998.

[HIL 95] HILL R.J., WILCZAK J.M., "Pressure structure functions, spectra for locally isotropic turbulence", *J. Fluid Mech.*, vol. 296, pp. 247–269, 1995.

[HIL 96] HILL R.J., "Pressure-velocity-velocity statistics in isotropic turbulence", *Physics of Fluids*, vol. 8, pp. 3085–3093, 1996.

[HIL 97] HILL R.J., BORATAV O.N., "Pressure statistics for locally isotropic turbulence", *Phys. Rev. E*, vol. 56, pp. R2363–R2366, 1997.

[HOC 99] HOCTER S.T., "Exact, approximate directivity patterns of the sound radiated from a cylindrical duct", *Journ. Sound, Vibration*, vol. 227, no. 2, pp. 397–407, 1999.

[HOC 00] HOCTERN S.T. "Sound radiated from a cylindrical duct with Keller's geometrical theory", *Journ. Sound, Vibration*, vol. 231, no. 5, pp. 1243–1256, 2000.

[HOW 75] HOWE M.S., "Contributions of the theory of aerodynamic sound, with application to excess jet noise, the theory of the flute", *J. Fluid Mech.*, vol. 71, pp. 625–673, 1975.

[HOW 08] HOWE M.S., "Rayleigh lecture 2007: flow-surface interaction noise", *Journ. Sound, Vibration*, vol. 314, pp. 113–146, 2008.

[HUL 01] HULSHOFF S.J., HIRSCHBERG A., HOFMANS C.C., "Sound production of vortex-nozzle interactions", *J. Fluid Mech.*, vol. 439, pp. 335–352, 2001.

[KHA 94] KHAVARAN A., KREJSA E.A., KIM C.M., "Computation of supersonic jet mixing noise for an axisymmetric convergent-divergent nozzle", *AIAA J. Aircraft*, vol. 31, pp. 603–609, 1994.

[KIM 89] KIM J., "On the structure of pressure fluctuations in simulated turbulent channel flow", *J. Fluid Mech.*, vol. 205, pp. 421–451, 1989.

[KOR 05] KORTE D., HÜTTL T., KENNEPOHL F. *et al.*, "Numerical simulation of multistage turbine sound generation, propagation", *Aerospace Science, Technology*, vol. 9, pp. 125–134, 2005.

[KRA 56] KRAICHNAN R.H., "Pressure fluctuations in turbulent flow over a flat plate", *J. Acoust. Soc. Amer.*, vol. 28, pp. 378–390, 1956.

[KRI 84] KRIEGSMANN G.A., NORRIS A., REISS E.L., "Acoustic scattering by baffled membranes", *J. Acoust. Soc. Amer.*, vol. 75, no. 3, pp. 685–694, 1984.

[KUN 99] KUNG S.W., SINGH R., "Development of approximate methods for the analysis of patch damping design concept", *Journ. Sound, Vibration*, vol. 219, no. 5, pp. 785–812, 1999.

[KUP 07] KUPERMAN W.A., ROUX P., "Underwater acoustics", *Springer Handbook of Acoustics*, ROSSING T.D. (ed.), Springer, New York, pp. 149–204, 2007.

[LAM 35] LAMSON H.W., "A simple and precise standard of musical pitch", *J. Acoust. Soc. Amer.*, vol. 7, pp. 51–55, 1935.

[LES 95] L'ESPÉRANCE A., GABILLET Y., DAIGLE G.A., "Outdoor sound propagation in the presence of atmospheric turbulence: experiments and theoretical analysis with the Fast Field Program", *J. Acoust. Soc. Am.*, vol. 98, pp. 570–579, 1995.

[LEW 94] LEWY S., "Prévision du bruit de raie émis par un rotor: application à l'aéronautique", *J. Phys. IV*, vol. 4, pp. C5.55-C5-65, 1994.

[LEW 07] LEWY S.,. "Prediction of turbofan rotor or stator broadband noise radiation", *Acta Acustica United with Acustica*, vol. 93, pp. 275–283, 2007.

[LIF 99] LIFSHITZ R., ROUKES M.L., "Thermoelastic damping in micro- and nanomechanical systems", *Physical review* B, vol. 6, no. 8, pp. 5600–5609, 1999.

[LIG 52] LIGHTHILL M.J., "On sound generated aerodynamically. Part I. General theory", *Proc. R. Soc. Lond. A*, vol. 211, pp. 564–587, 1952.

[LIG 54] LIGHTHILL M.J., "On sound generated aerodynamically. Part II. Turbulence as a source of sound", *Proc. R. Soc. Lond. A*, vol. 222, pp. 1–32, 1954.

[LIH 06] LIHOREAU B., GAUVREAU B., BÉRENGIER M. *et al.*, "Outdoor sound propagation modeling in realistic environments: application of coupled parabolic, atmospheric models", *J. Acoust. Soc. Am.*, vol. 120, pp. 110–119, 2006.

[LIL 73] LILLEY G.M., "On the noise from jets", *AGARD CP-131*, pp. 13.1–13.12, 1973.

[LIL 93] LILLEY G.M., The radiated noise from isotropic turbulence revisited, ICASE Report 93-75, 1993.

[LIL 96] LILLEY G.M., "The radiated noise from isotropic turbulence with applications to the theory of jet noise", *J. Sound, Vibration*, vol. 190, pp. 463–476, 1996.

[LOW 65] LOWSON M.V., "The sound field for singularities in motion", *Proc. R. Soc. Lond. A*, vol. 286, pp. 559–572, 1965.

[LYR 94] LYRINTZIS A.S., "Review: the use of Khirchhoff's method in computational aeroacoustics", *J. Fluids Engin.*, vol. 116, pp. 665–676, 1994.

[MAN 63] MANGIAROTTY R.A., "Acoustic radiation damping of vibrating structure", *J. Acoust. Soc. Am.*, vol. 35, pp. 369–377, 1963.

[MAR 92] MARCELIN J.L., TROMPETTE PH., SMATI A., "Optimal constrained layer damping with partial coverage", *Finite Elements in Analysis, Design*, vol. 12, pp. 273–280, 1992.

[MAR 11] MARIANI R., BELLIZZI S., COCHELIN B. *et al.*, "Toward an adjustable non linear low frequency acoustic absorber", *J. Sound, Vibration*, vol. 330, no. 1, pp. 5245–5258, 2011.

[MAT 96] MATTEI P.-O., "A two-dimensional Tchebycheff collocation method for the study of the vibration of a baffled fluid-loaded rectangular plate", *Journal of Sound, Vibration*, vol.196, no. 4, pp. 407–427, 1996.

[MAT 06] MATTEI P.-O., "Directivity of a simplified aeroengine", *Acta Acustica United with Acustica*, vol. 92, no. 2, pp. 225–236, 2006.

[MAT 07] MATTEI P.-O., "High order perturbation expansion for the spectral analysis of fluid-loaded vibrating structure", *Acta Acustica United with Acustica*, vol. 93, no. 4, pp. 303–313, 2007.

[MAT 09] MATTEI P.-O., "Multiple resonances in fluid-loaded vibrating structures", *Acta Acustica United with Acustica*, vol. 95, no. 2, pp. 314–324, 2009.

[MIK 90] MIKI Y., "Acoustical properties of porous materials - modifications of Delany-Bazley models", *J. Acoust. Soc. Jpn. (E)*, vol. 11, no. 1, pp. 19–24, 1990.

[MIL 11] MILLER T.S., GALLMAN J.M., MOELLE M.J., "Review of turbulent boundary layer models for acoustic analysis", *49th AIAA Aerosapce Science Meeting*, Orlando, Fl, AIAA-2011-1083: pp.1–20, 2011.

[MIT 99] MITCHELL B.E., LELE S.K., MOIN P., "Direct computation of the sound generated by vortex pairing in an axisymmetric jet", *J. Fluid Mech.*, vol. 383, pp. 113–142, 1999.

[OBU 49] OBUKHOV A.M., "Pressure fluctuations in turbulent flows", *Dokl. Akad. SSSR*, vol. 66, pp.17–20, 1949.

[OBU 51] OBUKHOV A.M., YAGLOM A.M., "The microstructure of turbulent flow", *Prikl. Mat. Mekh.*, vol. 15, pp. 3–26, 1951.

[OST 97] OSTASHEV V.E., JUVÉ D., BLANC-BENON P., "Derivation of a wide-angle parabolic equation for sound waves in inhomogeneous moving media", *Acta Acustica United with Acustica*, vol. 83, pp. 455–460, 1997.

[OST 98] OSTASHEV V.E., GOEDECKE G.H., "Interference of direct and ground reflected waves in a turbulent atmosphere", *Proc. 8th Intern. Symp. on long range sound propagation, Penn State College*, pp. 313–325, 1998.

[PEA 99] PEARSON B.R., Experiments on small-scale turbulence, PhD Thesis, Newcastle University, Australia, 1999.

[PHI 60] PHILLIPS O.M., "On the generation of sound by supersonic turbulent shear layers", *J. Fluid Mech.*, vol. 9, pp. 1–28, 1960.

[PIE 77] PIERCY J.E., EMBLETON T.F.W., SUTHERLAND L.C., "Review of noise propagation in the atmosphere", *J. Acoust. Soc. Am.*, vol. 61, pp. 1403–1418, 1977.

[POW 64] POWELL A., "Theory of vortex sound", *J. Acoust. Soc. Amer.*, vol. 36, pp.177–195, 1964.

[PRA 93] PRASKOVSKY A.A.,GLEDZER E.B., KARYAKIN M.Y. *et al.*, "The sweeping decorrelation hypothesis and energy-inertial scale interaction in high Reynolds number flows", *J. Fluid Mech.*, vol. 248, pp. 493–511, 1993.

[PRO 51] PROUDMAN I., "A comparison of Heisenberg's spectrum of turbulence with experiment", *Proc. R. Soc. Lond. A*, vol. 47, pp.158–176, 1951.

[PRO 52] PROUDMAN I., "The generation of noise by isotropic turbulence", *Proc. R. Soc. Lond. A*, vol. 214, pp.119–132, 1952.

[RAM 94] RAMAN G., RICE E.J., "Instability modes excited by natural screech tones in a supersonic rectangular jet", *Physics of Fluids*, vol. 6, pp. 3999–4008, 1994.

[RAM 97a] RAMAN G., "Screech tones from rectangular jets with spanwise oblique shock-cell structures", *J. Fluid Mech.*, vol. 330, pp.141–168, 1997.

[RAM 97b] RAMAN G., "Cessation of screech in underexpanded jets", *J. Fluid Mech.*, vol. 336, pp. 69–90, 1997.

[RAM 98] RAMAN G., "Advances in understanding supersonic jet screech: review and perspective", *Prog. Aerospace Sci.*, vol. 34, pp. 45–106, 1998.

[RAO 01] RAO M.D., "Recent applications of viscoelastic damping for noise control in automobiles, commercial airplanes", *Proceedings of the India-USA Symposium on Emerging Trends in Vibration, Noise Engineering, Journal of Sound and Vibration*, vol. 262, no. 3, pp. 457474, 2003.

[RIB 64] RIBNER H.S., "The generation of sound by turbulent jets", *Advances in Applied Mechanics*, Academic Press, New York, vol. 8, pp. 103–182, 1964.

[RIB 69] RIBNER H.S., "Quadrupole correlations governing the pattern of jet noise", *J. Fluid Mech.*, vol. 38, pp.1–24, 1969.

[RIB 77] RIBNER H.S., "On the role of the shear term in jet noise", *J. Sound, Vibration*, vol. 52, pp.121–132,1977.

[ROS 92] ROSSING T.D., RUSSEL D.A., BROWN D.E., "On the acoustics of tuning fork", *American Journal of Physics*, vol. 60, pp. 620–626, 1992.

[RUB 00] RUBINSTEIN R., ZHOU Y., "The frequency spectrum of sound radiated by isotropic turbulence", *Physics Letters A*, vol. 267, pp. 379–383, 2000.

[RUS 99] RUSSEL D.A., "Acoustic monopoles, dipoles, quadrupoles: an experiment revisited", *American Journal of Physics*, vol. 67, no. 8, pp. 660–664, 1999.

[RUS 00] RUSSEL D.A., "On the sound field radiated by a tuning fork", *American Journal of Physics*, vol. 68, no. 12, pp. 1139–1145, 2000.

[SAR 93] SARKAR S., HUSSAINI M.Y., Computation of the sound generated by isotropic turbulence, ICASE Report 93–74, 1993.

[SCH 97] SCHINDEL D.W., HUTCHINS D.A., SMITH S.T., "A study of material at high temperature using minimized resonant tuning forks and noncontact capacitance transducers", *J. Acoust. Soc. Amer.*, vol. 102, no. 3, pp.1296–1309, 1997.

[SER 01] SEROR C., SAGAUT P., BAILLY C. *et al.*, "On the radiated noise computed by large-eddy simulation", *Physics of Fluids*, vol.13, pp. 476–487, 2001.

[SMA 01] SMALLEY R.J., ANTONIA R.A., "Velocity-velocity difference correlations and the sweeping decorrelation hypothesis", *Proc. AFMC 14 Conference*, Adelaide, pp. 789–792, 2001.

[SMI 63] SMITH T.J.B., KILHAM J.K., "Noise generation by open turbulent flames", *J. Acoust. Soc. Amer.*, vol. 35, pp. 715–724, 1963.

[SOC] Société Méditerranéenne des Applications du Caoutchouc (SMAC), http://www.smac.fr/.

[SWI 07] SWIFT G., "Thermoacoustics", in ROSSING T.D. (ed.), *Springer Handbook of Acoustics*, Springer, New York, pp. 239–255, 2007.

[TAM 87] TAM C.K.W., "Stochastic model theory of broadband shock associated noise from supersonic jets", *Journ. Sound, Vibration*, vol. 116, pp. 265–302, 1987.

[TAM 88] TAM C.K.W., "The shock-cell structures, screech tone frequencies of rectangular and non-axisymmetric supersonic jets", *Journ. Sound, Vibration*, vol. 121, pp.135–147, 1988.

[TAM 89] TAM C.K.W., HU F.Q., "On the three families of instability waves of high-speed jets", *J. Fluid Mech.*, vol. 201, pp. 447–483, 1989.

[TAM 91] TAM C.K.W., "In aeroacoustics of flight vehicles", in HUBBARD H.H. (ed.), *Jet Noise Generated by Large-scale Coherent Motion*, vol. 1, pp. 311–390, 1991.

[TAM 95a] TAM C.K.W., "Supersonic jet noise", *Ann. Rev. Fluid Mech.*, vol. 27, pp.17–43, 1995.

[TAM 95b] TAM C.K.W., "Computational aeroacoustics: issues, methods", *AIAA J.*, vol. 33, pp. 1788–1796, 1995.

[TAM 99] TAM C.K.W.,AURIAULT L., "Jet mixing noise from fine-scale turbulence", *AIAA J.*, vol. 37, pp.145–153, 1999.

[TAP 77] TAPPERT F.D., "The parabolic approximation method", KELLER J.B., PAPADAKIS J.S. (eds), *Wave Propagation and Underwater Acoustics*, Springer, pp. 224–287, 1977.

[THO 66] THOMAS A., WILLIAMS G.T., "Flame noise: sound emission from spark-ignited bubbles of combustible gas", *Proceed. Royal Soc. London A*, vol. 294, pp. 449–466, 1966.

[TRU 98] TRUFFAUT J.M., SEARBY G., BOYER L., "Sound emission by non-isomolar combustion at low Mach numbers", *Combustion Theory, Modelling*, vol. 2, pp. 423–428, 1998.

[UBE 77] ÜBERALL H., DRAGONETTE L.R., FLAX L., "Relation between creeping waves, normal modes of vibration of a curved body", *J. Acoust. Soc. Amer.*, vol. 61, pp. 711–715, 1977.

[VIV 74] VIVOLI J., FILIPPI P., "Eigenfrequencies of thin plates and layer potentials", *J. Acoust. Soc. Amer.*, vol. 55, no. 3, pp. 562–567, 1974.

[WAN 06] WANG M., FREUND J.B., LELE S.K., "Computational prediction of flow generated sound", *Ann. Rev. Fluid Mech.*, vol. 38, pp. 483–512, 2006.

[WEL 97] WELLS V.L., RENAUT R.A., "Computing aerodynamically generated noise", *Ann. Rev. Fluid Mech.*, vol. 29, pp.161–199, 1997.

[WES 91] WEST M., SACK R.A., WALKDEN F., "The Fast Field Program (FFP). A second tutorial: application to long range sound propagation in the atmosphere", *Appl. Acoust.*, vol. 33, pp. 199–228, 1991.

[WHI 00] WHITMIRE J., SARKAR S., "Validation of acoustic-analogy predictions for sound radiated by turbulence", *Physics of Fluids*, vol.12, pp. 381–391, 2000.

[WIE 59] WIENER F.M., KEAST D.N., "Experimental study of the propagation of sound over ground", *J. Acoust. Soc. Am.*, vol. 31, pp. 724–733, 1959.

[WIL 94] WILSON D.K., THOMSON D.W., "Acoustic propagation through anisotropic, surface-layer turbulence", *J. Acoust. Soc. Am.*, vol. 96, pp. 1080–1095, 1994.

[WIL 96] WILSON D.K., "A brief tutorial on atmospheric boundary layer turbulence for acousticians", *Proc. 7th Intern. Symp. On Long Range Sound Propagation*, Lyon, France, pp. 111–121, 1996.

[WIT 95] WITKOWSKA A., BRASSEUR J.C., JUVÉ D., "Numerical study of noise from isotropic turbulence", *J. Comput. Acoust.*, vol. 5, pp. 317–336, 1995.

[WOD 98] WODTKE H.-W., LAMANCUSA J.S., "Sound power minimization of circular plates through damping layer placement", *J. Sound and Vibration*, vol. 215, no. 5, pp.1145–1163, 1998.

[WOJ 04] WOJTOWICKI J.-L., JAOUEN L., PANNETON R., "New approach for the measurement of damping properties of materials using the Oberst beam", *Review of Scientific Instruments*, vol. 75, no. 8, pp. 2569–2574, 2004.

[WOO 98] WOODHOUSE J., "Linear damping models for structural vibration", *J. Sound, Vibration*, vol. 215, no. 3, pp. 547–569, 1998.

[WOO 01] WOODRUFF S.L., SEINER J.M., HUSSAINI M.Y. *et al.*, "Implementation of new turbulence spectra in the Lighthill analogy source terms", *J. Sound, Vibration*, vol. 242, pp.197–214, 2001.

[YU 00] YU Y.H., . "Rotor blade-vortex interaction noise", *Prog. Aerospace Sci.*, vol. 36, pp. 97–115, 2000.

[ZHO 94] ZHOU Y., PRASKOVSKY A., ONCLEY S., On the Lighthill relationship and sound generation from isotropic turbulence, ICASE Report 94–92, 1994.

[ZHO 96] ZHOU Y., RUBINSTEIN R., "Sweeping and straining effects in sound generation by high Reynolds number isotropic turbulence", *Physics of Fluids*, vol. 8, pp. 647–649, 1996.

[ZOG 04] ZOGHAIB L., MATTEI P.-O., "A component study of the vibroacoustic damping of a non-baffled free plate partially covered with elastomer patches", *Actes du 7ème CFA/DAGA'04*, Strasbourg, pp. 309–310, 2004.

[ZOG 14] ZOGHAIB L., MATTEI P.-O., "Modeling and optimization of local constraint elastomer treatments for vibration, noise reduction", *J. Sound, Vibration*, vol. 333, pp. 7109–7124, 2014.

[ZOG 15] ZOGHAIB L., MATTEI P.-O., "Damping analysis of a free aluminum plate", *J. Vib., Control*, vol. 21, pp. 2083–2098, 2015.

Index

D

E

F

G

N, O, P

Q, R

S

T

V

W, Y, Z

Other titles from

iSTE

in

Waves

2014

JARRY Pierre, BENEAT Jacques N.
RF and Microwave Electromagnetism

LAHEURTE Jean-Marc
UHF RFID Technologies for Identification and Traceability

SAVAUX Vincent, LOUËT Yves
MMSE-based Algorithm for Joint Signal Detection, Channel and Noise Variance Estimation for OFDM Systems

TING Michael
Molecular Imaging in Nano MRI

VALIÈRE Jean-Christophe
Acoustic Particle Velocity Measurements using Laser: Principles, Signal Processing and Applications

VANBÉSIEN Olivier, CENTENO Emmanuel
Dispersion Engineering for Integrated Nanophotonics

2013

BENMAMMAR Badr, AMRAOUI Asma
Radio Resource Allocation and Dynamic Spectrum Access

BOURLIER Christophe, PINEL Nicolas, KUBICKÉ Gildas
Method of Moments for 2D Scattering Problems: Basic Concepts and Applications

GOURE Jean-Pierre
Optics in Instruments: Applications in Biology and Medicine

LAZAROV Andon, KOSTADINOV Todor Pavlov
Bistatic SAR/GISAR/FISAR Theory Algorithms and Program Implementation

LHEURETTE Eric
Metamaterials and Wave Control

PINEL Nicolas, BOURLIER Christophe
Electromagnetic Wave Scattering from Random Rough Surfaces: Asymptotic models

SHINOHARA Naoki
Wireless Power Transfer via Radiowaves

TERRE Michel, PISCHELLA Mylène, VIVIER Emmanuelle
Wireless Telecommunication Systems

2012

LI Jun-chang, PICART Pascal
Digital Holography

2011

BECHERRAWY Tamer
Mechanical and Electromagnetic Vibrations and Waves

GOURE Jean-Pierre
Optics in Instruments

LE CHEVALIER François, LESSELIER Dominique, STARAJ Robert
Non-standard Antennas

2010

BEGAUD Xavier
Ultra Wide Band Antennas

MARAGE Jean-Paul, MORI Yvon
Sonar and Underwater Acoustics

2009

BOUDRIOUA Azzedine
Photonic Waveguides

BRUNEAU Michel, POTEL Catherine
Materials and Acoustics Handbook

DE FORNEL Frederique, FAVENNEC Pierre-Noël
Measurements using Optic and RF Waves

2008

FILIPPI Paul J.T.
Vibrations and Acoustic Radiation of Thin Structures

2006

BOUCHET Olivier *et al.*
Free-Space Optics

BRUNEAU Michel, SCELO Thomas
Fundamentals of Acoustics

GUILLAUME Philippe
Music and Acoustics

GUYADER Jean-Louis
Vibration in Continuous Media

Printed and bound by CPI Group (UK) Ltd, Croydon, CR0 4YY
13/09/2026
14961025-0004